How to Evaluate, Simplify, and Solve ————

The most common directions in algebra are *evaluate, simplify,* and *solve.*

EVALUATE: To find the value (usually numerical) of an expression.
To replace variables and constants with numbers in a formula or expression.
To replace the function variable with numbers or another expression.

SIMPLIFY: To change an expression, making it less complicated.

The meaning of *simplify* expands as you progress through the book.

Chapter 1   Do the operations shown, such as $+$, $-$, $\times$, and $\div$.
Do the operations shown, taking into account the order of operations.
Eliminate common factors in a fraction, leaving the fraction in lowest terms.
Use the distributive property to eliminate parentheses from an expression.
Do the operations shown using number properties, such as the associative properties of $+$ and $\times$ or the commutative properties of $+$ and $\times$.
Add (or subtract) like terms.
Carry out any of the above steps on the expressions on either side of an equation before solving the equation.

Chapter 2   Do the operations needed, after substituting numbers into an expression or function.

Chapter 3   Add (or subtract) like terms in a polynomial.
Arrange terms in descending order of exponents on whichever variable is first alphabetically.
Do the operations shown with power expressions, leaving no zero or negative exponents.
Do the operations shown with power expressions, using the properties of exponents.

Chapter 5   Eliminate common units (feet, inches, etc.) from the numerator and denominator of a fraction.
Eliminate common factors in a ratio, leaving the ratio in lowest terms.
Eliminate common units from a unit analysis and do the operations.
Factor rational expressions and eliminate common factors.
Apply the properties of fractions to adding, subtracting, multiplying, and dividing rational expressions.

Chapter 6   Do the operations shown with power expressions, using the properties of exponents.
Do the operations shown in and with square root expressions.
Apply properties of square roots to variable expressions.
Rationalize the denominator of an expression.

Chapter 7   Use the properties of exponents on the definition of a logarithm.

SOLVE: To isolate the specified variable on just one side of an equation or inequality.
To find the common ordered pair in tables for the left and right sides of an equation.
To find the intersection of the graphs of the left and right sides of an equation.
To draw on a number line the solution set to a linear inequality in one variable.
To isolate the specified variable on just one side of a formula.
To find the values of both variables in a system of two equations.
To find the values of three variables in a system of three equations.
To draw on rectangular coordinate axes the solution set to a linear inequality in two variables.

2e

Intermediate Algebra
A Just-in-Time
Approach

Alice Kaseberg
Lane Community College

Brooks/Cole
Thomson Learning™

Australia • Canada • Mexico
Singapore • Spain • United Kingdom • United States

Publisher: *Robert W. Pirtle*
Marketing Manager: *Leah Thomson*
Advertising: *Samantha Cabaluna*
Marketing Assistant: *Debra Johnston*
Editorial Assistant: *Erin Wickersham*
Production Coordinator: *Keith Faivre*
Production Service: *Lifland et al., Bookmakers*
Manuscript Editor: *Sally Lifland*
Interior Design: *Vernon T. Boes*

Cover Design: *Christine Garrigan*
Cover Illustration: *Harry Briggs*
Art Editor: *Lisa Torri*
Interior Illustration: *Scientific Illustrators, Cyndie C. H. Wooley*
Print Buyer: *Vena Dyer*
Typesetting: *The Beacon Group*
Cover Printing: *Phoenix Color Corporation*
Printing and Binding: *Transcontinental*

For more information, contact:
BROOKS/COLE
511 Forest Lodge Road
Pacific Grove, CA 93950 USA
www.brookscole.com

For permission to use material from this work, contact us by
Web: www.thomsonrights.com
fax: 1-800-730-2215
phone: 1-800-730-2214

Credits: Page 312, Figure 1, Lick Observatory Photograph; Figure 2, NOAA/National Climatic Data Center.

Printed in Canada

10 9 8 7 6 5 4 3 2 1

Library of Congress Cataloging-in-Publication Data

Kaseberg, Alice.
    Intermediate algebra : a just-in-time approach/Alice Kaseberg. -- 2nd ed.
        p. cm.
    Includes indexes.
    ISBN 0-534-35748-2 (hc. : alk. paper)
    1. Algebra.   I. Title.

QA154.2.K35   1999
512.9--dc21   99-048029

THIS BOOK IS PRINTED ON ACID-FREE RECYCLED PAPER

Contents

1 Expressions and Equations   1

Fundamental skills are reviewed and then associated with problem solving, writing skills, and the three ways of thinking about algebra: numeric, symbolic, and graphic. Linear equations and inequalities are solved with tables, symbols, and graphs. The important role of inverses in equation solving is emphasized.

2 Equations, Functions, and Linear Functions   73

Strategies are presented for modeling with linear equations. After an introduction to functions, special linear functions are discussed. Study of numerical sequences provides a basis for distinguishing linear data from the nonlinear data introduced in later chapters.

Quadratic Functions: Applications and Solutions to Equations   150

A review of the polynomial operations and the square root function leads to solving quadratic equations three ways: with tables and graphs, by factoring and taking square roots, and with the quadratic formula. Sequences are used to identify quadratic functions. Work with quadratic inequalities builds on the work with inequalities in Chapter 1.

Quadratic Functions: Special Topics   240

Previous work with polynomials is extended to cubic and higher order expressions. Tools used to take a detailed look at quadratic equations include quadratic modeling, predicting graphs from equations, operating with complex numbers, and finding minimum and maximum values.

8 Systems of Equations 549

The chapter presents four different ways to solve systems of two linear equations, as well as three different solution outcomes. Work is then extended to solving systems of three equations with matrices, solving systems of nonlinear equations (including the geometric curves called conic sections), and solving systems of inequalities.

Preface

Intermediate Algebra: A Just-in-Time Approach is a nontraditional approach to algebra, based on

- the premise that concept development and understanding of mathematical thinking are facilitated by problem solving and discovery,
- agreement with the reforms advocated by organizations such as the National Council of Teachers of Mathematics (NCTM) and the American Mathematical Association of Two Year Colleges (AMATYC),
- the availability of technology and its considered use, and
- the mastery of certain basic skills.

The material is personalized by a 30-year career in mathematics, including experience at the community college, high school, and junior high levels; a curiosity about what mathematics is good for (which led me to degrees in business administration, mathematics, and engineering science); and, most importantly, an appreciation for problem solving, hands-on ways to present mathematics, and the beauty and wonder of mathematics inspired by the work of George Polya, W. W. Sawyer, and M. C. Escher.

I wrote the text because I want students to appreciate applications in mathematics, to understand rather than memorize skills, and to be prepared for indepth function work in college algebra. I also want both the veteran and the novice instructor to think about new connections, new methods, and new ways of learning.

The backgrounds of developmental algebra students are diverse. There are those who have never had algebra, those who have been out of school for a while and need to relearn forgotten math skills, those who are fresh out of high school but need to review, and those who failed algebra in high school.

Intermediate Algebra: A Just-in-Time Approach is intended for students who have passed an Introductory Algebra course. The text is sufficiently different in its approach to provide new learning and thinking experiences for students with some algebra background. These students will be especially surprised to find an approach that is not a rehash of their high school algebra. The text is also sufficiently complete in its presentation and practice of basic skills to accommodate the student with only an Introductory Algebra background.

Because students' levels of optimism and energy are highest at the beginning of the term, Chapter 1 introduces problem solving and connections among

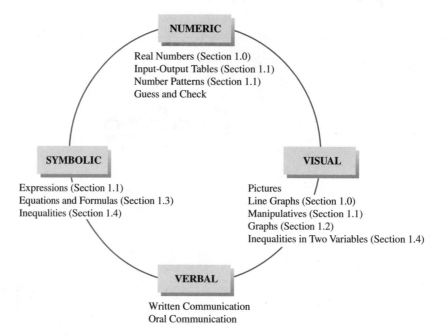

Figure A

numeric, visual, verbal, and symbolic information (see Figure A). From the beginning of this text, students are asked to make connections among tabular, graphical, verbal, and symbolic information; to integrate ideas from algebra and geometry (and, through projects, ideas from trigonometry, probability, and statistics); and to apply mathematics to real-world settings. There is more reading than in many math books, because the text includes the thinking that leads to a concept, the connections between the concept and other mathematics, and the applications for the mathematics. Where possible, material is presented through a discovery approach: exploration, question, summary, example. This approach takes longer than the traditional mode of statement and example.

The second edition includes considerably more skill building than the first edition did, as well as more explicit connections among objectives, examples, exercises, and tests. The text has been completely revised to provide a more accessible reading level and a friendlier appearance.

Pedagogy

Objectives

The learning outcomes for each section are listed at the beginning of the section. They serve as a summary for both students and instructors and coordinate with the titles on the examples.

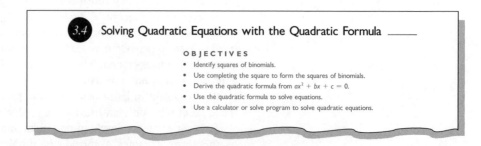

Projects

Projects are intended for group work or for individual effort. They may be more complicated problems related to the topic at hand, activity-based problems using manipulatives, or real-world applications that require research outside class. Projects suited to in-class group work are marked with an asterisk in the Index of Projects. The Index of Projects is repeated in the *Instructor's Resource Manual.*

---

31. Draw sketches of quadratic functions that explain why quadratic equations can have only 0, 1, or 2 possible solutions.

32. Explain the relationship between the vertex of a parabolic graph and the set of outputs, or range, for the function.

33. The factors of a quadratic expression are $(ax + b) \cdot (cx + d)$. How might the factors be used to find the x-intercepts of $y = (ax + b)(cx + d)$?

34. Describe the relationship between the value of a in $y = ax^2 + bx + c$ and whether the parabola opens up or opens down.

35. If a graph of a quadratic equation has the y-axis as an axis of symmetry and has $x = 4$ as one x-intercept, what will be another x-intercept?

36. If a graph of $f(x)$ has the y-axis as a line of symmetry, what may be said about $f(a)$ and $f(-a)$?

Projects

37. *Positions on a Parabola.* Match each phrase with one of these three positions on a parabola: x-intercept, y-intercept, or vertex. Make a sketch to clarify your answer. Answers may vary depending on your sketch.

a. The highest profit in a parabolic business profit curve

b. The place where a ball on a parabolic path falls to the ground when thrown

c. The start-up costs in a business, where the total cost graph is a parabola

d. The time it takes for a porpoise to re-enter the water after a parabolic leap

e. The down payment in a purchase plan where total cost is a parabola

f. The highest point on a jet of water

g. The number of sales required to change from loss to profit on a parabolic business income curve

h. The maximum height in the parabolic path of a ball

i. The initial height of a golf tee, where the path of the ball is a parabola

j. The lowest point in a parabolic cable on a suspension bridge

k. The initial cost in a parabolic cost curve

l. The point where a diver enters the water, when the curve shows his parabolic path

38. *Estimating Reaction Time.* Turn your hand so that you can catch a ruler between your thumb and fingers. Have someone hold a ruler just above finger level and drop it between your outstretched thumb and fingers (see the figure). Catch the ruler. Use the point on the ruler at which your fingers catch it as an estimate of distance d in inches. Use $d = \frac{1}{2}gt^2$, with $g = 32.2$ ft/sec^2, to estimate your reaction time t. (*Hint:* How many inches in a foot?)

Warm-ups

The Warm-up at the beginning of each section is designed to serve as a class opener, reviewing important concepts and linking prior and upcoming topics. Warm-ups tend to be skill-oriented; they generally connect to the algebra needed to solve text examples. The answers to the Warm-up appear in the Answer Box at the end of the section.

WARM-UP

Solve $f(x) = 0$. Without graphing $f(x)$, find the vertex from the x-intercepts.

1. $f(x) = x^2 - 3x - 10$

2. $f(x) = 2x^2 - 3x - 5$

3. $f(x) = x^2 - 2$

4. $f(x) = -16.1x^2 + 400x - 1493$

Small-Group Work

Some sections contain introductory questions or activities. These are intended to be done in class in small groups.

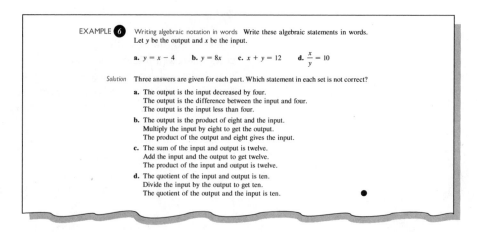

In Section 1.1, for example, multiple solutions are given to Example 6; students may work together to decide which statement is not correct. Example 8 has a similar format. The examples demonstrate how each student may contribute to the class and, in turn, learn from others. It is important to emphasize that students improve their own understanding by helping others.

Problem Solving

George Polya's four-step approach to problem solving—understanding the problem, making a plan, carrying out the plan, and checking the solution—is introduced in Section 1.1 and revisited where appropriate. The text then focuses on planning strategies. Section 1.1 introduces the strategies of *looking for a number pattern, making a table of inputs and outputs,* and *using manipulatives. Finding number patterns* continues in Sections 2.4, 3.2, 4.1, and 7.0. *Making a graph* first appears in Section 1.2. *Working backwards* is the fundamental idea used in solving equations and formulas in Section 1.3. *Choosing a test number or ordered pair and checking it* is used in Section 1.4 in drawing a line graph and identifying half-planes for two-variable inequalities and in Section 8.6 in solving systems of inequalities. *Guessing and checking,* which is a natural extension of choosing a test number for inequalities, is essential in writing equations in Section 2.0 and in building and solving systems of equations in Section 8.0. *Making a systematic list* is an essential component of factoring in Section 3.0.

Polya's Problem-Solving Steps

The problem-solving steps introduced in this section are based on those proposed by George Polya in his book *How to Solve It,* first published in 1945 by Princeton University Press. Polya earned his Ph.D. in mathematics in 1912, taught for 26 years at the Swiss Federal Institute of Technology, and left Europe in 1940 because "Hitler was too close." He retired in 1953, at age 65, after teaching at Stanford. Polya was a prolific writer and an active educator and speaker well into his 90s. *How to Solve It* has been published in 15 languages and is still in print.

Polya suggested that the elements of problem solving can be summarized by understanding the problem, making a plan, carrying out the plan, and then checking the solution.

- To *understand the problem,* we must consider the assumptions. In the Warm-up, we are not told that we are counting. We assume that the number below each design somehow describes the design and think about what feature of the design is being described.

- Our *plan* includes finding a pattern, either from our past experience or by looking for changes in the designs or numbers. How does each design differ from the previous design? How does each number differ from the previous number?

- As we *carry out the plan,* we look to see that we are satisfying the assumptions.

- Finally, we *check the solution.* We return to the original problem. Does the numerical answer match the answer found from the designs? Are there other processes that confirm the result?

Explorations

Some examples are intended to be used in class for individual or group exploration. The solutions to these exploratory examples are included in the Answer Box at the end of the section.

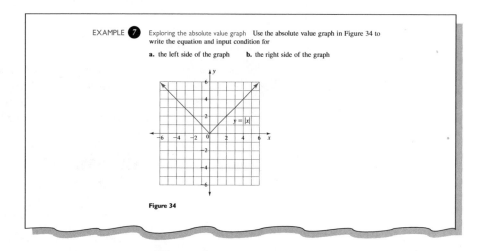

EXAMPLE 7 Exploring the absolute value graph   Use the absolute value graph in Figure 34 to write the equation and input condition for

a. the left side of the graph   b. the right side of the graph

Figure 34

Applications

To encourage creative thinking and depth in understanding, the text often poses a variety of questions about a single application setting. In addition, several applications, such as the cost of transcripts, are repeated throughout the text so that students may observe the continuity and connections among topics.

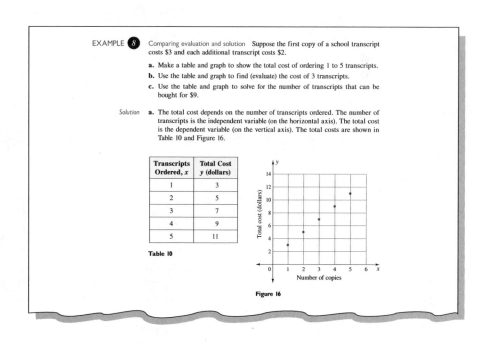

EXAMPLE 8 Comparing evaluation and solution   Suppose the first copy of a school transcript costs $3 and each additional transcript costs $2.

a. Make a table and graph to show the total cost of ordering 1 to 5 transcripts.

b. Use the table and graph to find (evaluate) the cost of 3 transcripts.

c. Use the table and graph to solve for the number of transcripts that can be bought for $9.

Solution a. The total cost depends on the number of transcripts ordered. The number of transcripts is the independent variable (on the horizontal axis). The total cost is the dependent variable (on the vertical axis). The total costs are shown in Table 10 and Figure 16.

Transcripts Ordered, x	Total Cost y (dollars)
1	3
2	5
3	7
4	9
5	11

Table 10

Figure 16

Think about it

"Think about it" questions are included within the reading material to encourage students to relate examples to prior material, to extend examples, and to practice verbalization skills. Answers to the questions are provided in the Answer Box.

> Linear Function   A **linear function** can be written $f(x) = mx + b$. The constants m and b may be any real number. The variable, x, has 1 as its exponent.
>
> *Think about it 2:* Why would a vertical line not be a function?

Answer Boxes

Answers to the Warm-up and Explorations, as well as the "Think about it" questions, are placed in the Answer Box at the end of the section (just before the exercises). By providing answers as feedback, the Answer Box permits the text to be used in class or as a laboratory manual for group work or independent study.

> **ANSWER BOX**
>
> **Warm-up 1.** 0.75 **2.** 0 **3.** undefined **4.** $-\frac{1}{2}$ **5.** $\frac{2}{3}$ **6.** 0.1 **Think about it 1:** No. If a graph has two intersections with the vertical axis, it has two outputs for one input—that is, two y-values for $x = 0$. In that case, it cannot be a function. **Think about it 2:** A vertical line would fail the vertical-line test. For each input x there is not exactly one output y.

Examples

Each example begins with a title, which states the purpose of the example. Usually these titles relate back to the objectives for the section.

> EXAMPLE **3**   Completing the square   Complete the square of these expressions and then check with a table.
>
> **a.** $x^2 + 3x$, with decimals
>
> **b.** $x^2 + \dfrac{bx}{a}$, with fractions
>
> Solution   **a.** The coefficient on the x term is 3, so to complete the square we add the square of 1.5: $1.5^2 = 2.25$. The binomial square is
>
> $$x^2 + 3x + 2.25 = (x + 1.5)^2$$
>
> To check, we draw the square shown in Figure 44.

Sequences

Number patterns are used to identify and distinguish linear, quadratic, and exponential functions.

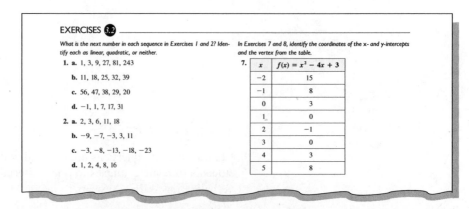

> EXERCISES **3.2**
>
> What is the next number in each sequence in Exercises 1 and 2? Identify each as linear, quadratic, or neither.
>
> **1. a.** 1, 3, 9, 27, 81, 243
>
> **b.** 11, 18, 25, 32, 39
>
> **c.** 56, 47, 38, 29, 20
>
> **d.** −1, 1, 7, 17, 31
>
> **2. a.** 2, 3, 6, 11, 18
>
> **b.** −9, −7, −3, 3, 11
>
> **c.** −3, −8, −13, −18, −23
>
> **d.** 1, 2, 4, 8, 16
>
> In Exercises 7 and 8, identify the coordinates of the x- and y-intercepts and the vertex from the table.
>
> **7.**
>
> | x | $f(x) = x^2 - 4x + 3$ |
> | --- | --- |
> | −2 | 15 |
> | −1 | 8 |
> | 0 | 3 |
> | 1 | 0 |
> | 2 | −1 |
> | 3 | 0 |
> | 4 | 3 |
> | 5 | 8 |

Input-Output Relationships and Functions

The text uses number patterns to introduce the concept of input-output relationships in Section 1.1 and then uses input-output relationships to lead to functions in Section 2.1.

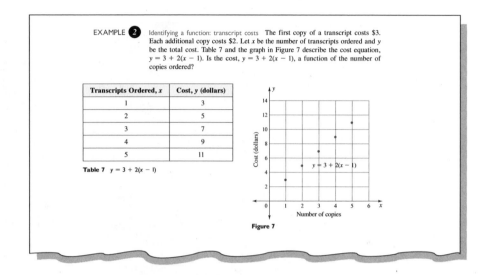

EXAMPLE **2**   Identifying a function: transcript costs   The first copy of a transcript costs $3. Each additional copy costs $2. Let x be the number of transcripts ordered and y be the total cost. Table 7 and the graph in Figure 7 describe the cost equation, $y = 3 + 2(x - 1)$. Is the cost, $y = 3 + 2(x - 1)$, a function of the number of copies ordered?

Transcripts Ordered, x	Cost, y (dollars)
1	3
2	5
3	7
4	9
5	11

Table 7 $y = 3 + 2(x - 1)$

Figure 7

Tables and Graphs

Extensive use is made of data in tabular form. Tables encourage organization of information and promote observation of patterns. They also prepare students for spreadsheet technology. Where appropriate, a graph is related to the table, to underscore the connections among algebra, geometry, statistics, and the real world. Numbers and their corresponding equations and graphs are employed to emphasize the fact that algebra is the transition language between arithmetic and analysis.

Aids are provided to help students read tables and graphs. For example, where students might think a parabolic graph showed a path, the horizontal axis is numbered with digital clocks.

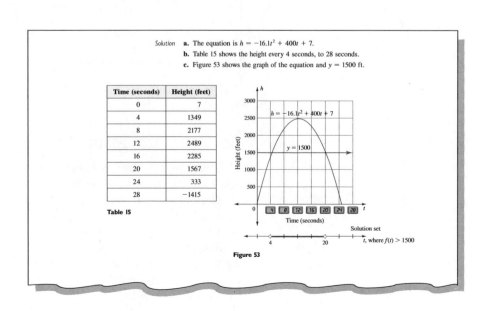

Solution   **a.** The equation is $h = -16.1t^2 + 400t + 7$.

   b. Table 15 shows the height every 4 seconds, to 28 seconds.

   c. Figure 53 shows the graph of the equation and $y = 1500$ ft.

Time (seconds)	Height (feet)
0	7
4	1349
8	2177
12	2489
16	2285
20	1567
24	333
28	-1415

Table 15

Figure 53

Hands-On Materials

The text supports use of an assortment of hands-on materials. Section 1.1 suggests algebra tiles for adding like terms; Section 7.0 uses paper folding; and Section 7.1 includes an exploration with coin tossing. Many of the projects in the Exercises involve hands-on materials.

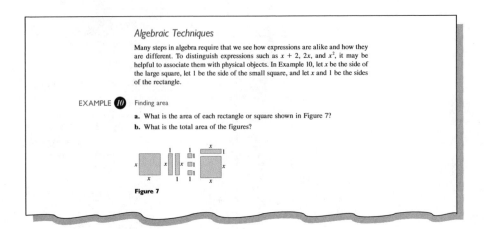

Algebraic Techniques

Many steps in algebra require that we see how expressions are alike and how they are different. To distinguish expressions such as $x + 2$, $2x$, and x^2, it may be helpful to associate them with physical objects. In Example 10, let x be the side of the large square, let 1 be the side of the small square, and let x and 1 be the sides of the rectangle.

EXAMPLE **10**   Finding area

a. What is the area of each rectangle or square shown in Figure 7?

b. What is the total area of the figures?

Figure 7

Exercises

Tables and graphs in the exercises give students practice in skills such as solving equations. They also help students to learn graphing technology.

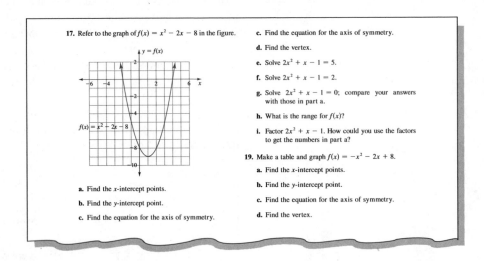

17. Refer to the graph of $f(x) = x^2 - 2x - 8$ in the figure.

$y = f(x)$

$f(x) = x^2 - 2x - 8$

a. Find the x-intercept points.

b. Find the y-intercept point.

c. Find the equation for the axis of symmetry.

c. Find the equation for the axis of symmetry.

d. Find the vertex.

e. Solve $2x^2 + x - 1 = 5$.

f. Solve $2x^2 + x - 1 = 2$.

g. Solve $2x^2 + x - 1 = 0$; compare your answers with those in part a.

h. What is the range for $f(x)$?

i. Factor $2x^2 + x - 1$. How could you use the factors to get the numbers in part a?

19. Make a table and graph $f(x) = -x^2 - 2x + 8$.

a. Find the x-intercept points.

b. Find the y-intercept point.

c. Find the equation for the axis of symmetry.

d. Find the vertex.

Calculator Techniques

The graphing calculator is essential. Graphs and tables are provided in many examples so that students who are just learning the technology are not handicapped by a lack of understanding of the calculator. Graphs in the text are not reproductions of those made by graphing calculators because the calculator resolution is not sufficient to provide clear details and smooth curves.

Suggestions for use of graphing calculators in general are included throughout the text in Graphing Calculator technique boxes, identified by the graphing calculator icon. For a quick reference to techniques, look at the Calculator Objectives at the back of the book.

> Graphing Calculator Technique:
> Finding the Intersection of Graphs
>
> Read in your calculator manual about the option for finding the intersection of graphs. You will need to enter the equations into ⟨Y=⟩ and graph the equations with a window that will show the intersections of the graphs.
>   After selecting the INTERSECTION option, select the graphs, usually by moving the cursor arrow up or down and pressing ⟨ENTER⟩. If there is more than one point of intersection, you need to indicate one point, usually in response to GUESS. Trace to your chosen point and press ⟨ENTER⟩. Repeat for other intersections of the same or different graphs.

Reading Aids

A large capital letter extending down two lines (called a drop cap) signals a transition for the reader. The drop cap has been proven to aid readability. Here it leads the reader into an explanatory introduction to the next example.

> A t times we want to describe regions in the coordinate plane with inequalities. After graphing the equation corresponding to the inequality, we can use test points to locate regions to be shaded, as we did in Section 1.4.

Mid-Chapter Test

To keep students engaged and build their confidence, a Mid-Chapter Test is included in each chapter. This test gives students the opportunity to check their progress. All answers appear in the back of the book, in the Selected Answers section.

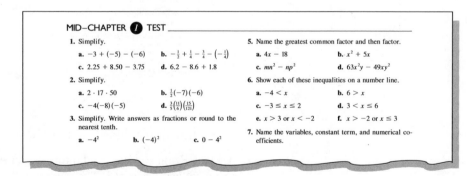

> **MID–CHAPTER ❶ TEST**
>
> **1.** Simplify.
>   **a.** $-3 + (-5) - (-6)$ **b.** $-\frac{1}{2} + \frac{1}{4} - \frac{3}{4} - \left(-\frac{1}{4}\right)$
>   **c.** $2.25 + 8.50 - 3.75$ **d.** $6.2 - 8.6 + 1.8$
>
> **2.** Simplify.
>   **a.** $2 \cdot 17 \cdot 50$ **b.** $\frac{1}{2}(-7)(-6)$
>   **c.** $-4(-8)(-5)$ **d.** $\frac{3}{5}\left(\frac{11}{6}\right)\left(\frac{15}{121}\right)$
>
> **3.** Simplify. Write answers as fractions or round to the nearest tenth.
>   **a.** -4^2 **b.** $(-4)^2$ **c.** $0 - 4^2$
>
> **5.** Name the greatest common factor and then factor.
>   **a.** $4x - 18$ **b.** $x^2 + 5x$
>   **c.** $mn^2 - np^2$ **d.** $63x^2y - 49xy^2$
>
> **6.** Show each of these inequalities on a number line.
>   **a.** $-4 < x$ **b.** $6 > x$
>   **c.** $-3 \le x \le 2$ **d.** $3 < x \le 6$
>   **e.** $x > 3$ or $x < -2$ **f.** $x > -2$ or $x \le 3$
>
> **7.** Name the variables, constant term, and numerical coefficients.

Chapter Summary, Chapter Review Exercises, and Chapter Test

Every chapter ends with a Chapter Summary, Chapter Review Exercises, and a Chapter Test. The student is provided with answers to the odd-numbered Review Exercises and answers to all of the Chapter Test questions.

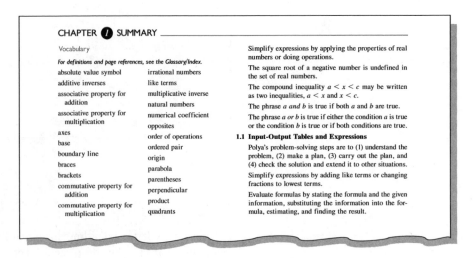

CHAPTER **1** SUMMARY

Vocabulary

For definitions and page references, see the Glossary/Index.

absolute value symbol	irrational numbers
additive inverses	like terms
associative property for addition	multiplicative inverse
associative property for multiplication	natural numbers
	numerical coefficient
axes	opposites
base	order of operations
boundary line	ordered pair
braces	origin
brackets	parabola
commutative property for addition	parentheses
	perpendicular
commutative property for multiplication	product
	quadrants

Simplify expressions by applying the properties of real numbers or doing operations.

The square root of a negative number is undefined in the set of real numbers.

The compound inequality $a < x < c$ may be written as two inequalities, $a < x$ and $x < c$.

The phrase *a and b* is true if both *a* and *b* are true.

The phrase *a or b* is true if either the condition *a* is true or the condition *b* is true or if both conditions are true.

1.1 Input-Output Tables and Expressions

Polya's problem-solving steps are to (1) understand the problem, (2) make a plan, (3) carry out the plan, and (4) check the solution and extend it to other situations.

Simplify expressions by adding like terms or changing fractions to lowest terms.

Evaluate formulas by stating the formula and the given information, substituting the information into the formula, estimating, and finding the result.

Cumulative Review

To help students maintain skills, a set of Cumulative Review Exercises is placed at the end of each even-numbered chapter. A Final Exam Review is included after the last chapter.

CUMULATIVE REVIEW OF CHAPTERS 1 AND 2

1.

Input x	Input y	Output xy	Output x + y	Output x − y
−2	4			
−3	7			
2		−6		
−3		6		
−1			−7	
	−2		−7	
	−2		1	
2			−7	

In Exercises 2 and 3, match each expression with one of the given words.

2. Choose from

   whole numbers, quotient, difference, natural numbers, rational numbers

b. Two numbers or expressions, *a* and *b*, that are multiplied to obtain the product *ab*

c. Two numbers, *n* and $1/n$, that multiply to 1

d. Removing a common factor from two or more terms

e. Collections of objects or numbers

Simplify the expressions in Exercises 4 to 9.

4. $3[4 - 2(5 - 8) - 6]$

5. $a(b + c) - b(a + c) + c(a - b)$

6. πr^2 for $r = 7.5$ feet

7. $\frac{1}{4}\pi d^2$ for $d = 15$ feet

8. $\dfrac{ac - bc}{c}$

9. $\dfrac{16 + 21x}{6}$

Final Exam Review

A Final Exam Review, containing exercises requiring both short and long answers, follows Chapter 8. The review is divided by chapter and may be used as a source of additional exercises or cumulative review material.

Glossary/Index

For the convenience of students and instructors, essential vocabulary is defined and referenced by page in the combined Glossary/Index. Vocabulary is considered essential if it is listed in the Chapter Summary. Nonessential terms are referenced only by page in the Glossary/Index.

Ancillaries to Accompany the Text

For Instructors

ANNOTATED INSTRUCTOR'S EDITION (AIE)   For the instructor's convenience, the *Annotated Instructor's Edition* includes the answer to each exercise adjacent to that exercise; answers too long to fit in the available space are included in the Additional Answers section at the very end of the AIE. Annotations in the margin offer planning hints and teaching strategies, to supplement the more detailed planning and teaching information provided in the *Instructor's Resource Manual*.

INSTRUCTOR'S RESOURCE MANUAL (IRM)   The *Instructor's Resource Manual* starts out with suggestions on planning and teaching a developmental algebra course. This extensive discussion is followed by section-by-section comments and lesson plans, plus key worksheets and overhead transparency masters. Included are fresh ideas for the experienced instructor, as well as extensive guidelines for successful mathematics instruction for the novice instructor. The IRM is designed to both provide inservice support and function as a daily reference.

The variety of instructional methods mentioned in the AIE and the IRM give you the permission and tools to be the teacher you want to be: lecturer, instructor/leader, coach, resource person, and/or facilitator of group work.

ASSESSMENT MATERIALS   Test items, sample tests, and a project for each chapter are included in the assessment materials.

For Students

STUDENT'S SOLUTION MANUAL   Complete worked-out solutions for all odd-numbered exercises are provided in the separate *Student's Solution Manual*. For Mid-Chapter Tests and Chapter Tests, complete worked-out solutions are given for all problems.

VIDEOTAPES   A set of videotapes presents examples from each section as well as new examples. A videotape icon in the margin points out the examples covered in tapes.

Acknowledgments

I would like to thank the following reviewers and class-testers for their helpful comments and significant contributions, both to this edition and to the first edition:

Rick Armstrong
Florissant Valley Community College

Joann Bossenbroek
Columbus State Community College

Bradd Clark
University of Southwestern Louisiana

Robert Davidson
East Tennessee State University

Marsha Davis
Eastern Connecticut State University

John Dersch
Grand Rapids Community College

Margaret Donaldson
Eastern Tennessee State University

Megan Florence
Albuquerque Technical-Vocational Institute

Judy Kasabian
El Camino College

Linda Knauer
*Pellissippi State Technical
Community College*

Jaclyn LeFebure
Illinois Central College

Lalo Mata
Hartnell College

Marveen McCready
Chemeketa Community College

Greg Perkins
Hartnell College

Pat Stone
Tomball College

Katherine Struve
Columbus State Community College

Deborah White
College of the Redwoods

Robert Wynegar
University of Tennessee, Chattanooga

My deepest gratitude goes to my husband, Rob Bowie, and to Cosmo—the regular occupants of the empty chair in my home office. Thanks go to our parents for their patience and understanding over this past seven years. A special thank-you is due to Bob Pirtle and everyone at Brooks/Cole Publishing Company and to Sally Lifland and her crew at Lifland et al., Bookmakers.

I appreciate the tremendous work done by everyone using the first edition. Fixing what was left undone has been a matter of pride. The second edition serves as my thank-you to those who provided feedback and encouragement.

Alice Kaseberg

To the Student

"Just in time" is an industrial engineering term that describes a modern inventory management scheme in a manufacturing plant. Materials used in the manufacturing process are scheduled for purchase and delivery at the precise moment at which they are needed. The old inventory method of accumulating huge stockpiles of all manufacturing materials was costly, and during the recessions of recent decades many manufacturers either changed to the just-in-time inventory method or went out of business.

Intermediate Algebra: A Just-in-Time Approach presents material for an intermediate algebra course. "Just in time" describes this algebra in two ways. First, it refers to the book's novel approach to the study of algebra, in which you work on real-world problems, learning algebraic principles and procedures just in time as you need them. Second, it describes the book's new curriculum, which allows you to concentrate on what you need to know in this age of modern technology, just in time for the twenty-first century.

Intermediate Algebra: A Just-in-Time Approach will help you to understand algebraic concepts, not memorize skills. Tables and graphs are used from the beginning to give numerical and visual meaning to algebra. Because of the new graphing and algebraic technology, the focus in algebra now is on when to use it and what it means. Estimating skills are more important than ever because you need to know whether your results are reasonable and whether they are meaningful. The text permits you to use calculator technology regularly.

Beginning the New Term

Problem Solving

Section 1.1 introduces problem-solving steps in a mathematical context. Right now, think about these four basic problem-solving steps in the context of your planning for the next few months:

1. *Understand the problem.* Your problem is that you don't have enough time to do everything you would like to do.

2. *Make a plan.* One time management strategy is to make a list of everything you have to do and when you have to do it. Make a chart showing each waking hour for the next seven days. For each hour, write in what you plan to do with that time. (There are many problem-solving strategies for planning. For a more complete listing, see Problem Solving in the Preface.)

3. *Carry out the plan.* Follow your plan (your schedule) for a week. Write notes on it to indicate when you varied from the plan and why.

4. *Check and refine.* After one week, review the plan. Did you get everything done that you needed to do? What do you need to change in your plan? Are your school load and work load reasonable? Redo the schedule to accommodate needed changes.

Time Management

Make sure your course load is sensible. Exceptionally few people can productively manage 60 hours per week of classes, study, and work. Check how sensible your plan for this term is:

Multiply your number of credits hours by 3, and then add your number of hours of employment.

For most people, 40 to 45 hours is a reasonable commitment to school and employment.

Make or buy a calendar for the term, with spaces large enough for noting assignments, tests, appointments, and errands. Keep the calendar with you at all times.

Sticking to Your Plan

Successful students plan their time carefully. Because studying is most productively done in the daytime and in hour-long segments, plan to study between classes. Unless you have a pressing need to leave campus, stay an extra hour and study again after your last class. Not only will a schedule help you be more efficient; it will also remind you of your priorities and prevent you from avoiding tasks that need to be done.

Are You in the Right Course?

Each mathematics course has one or more prerequisite courses. Having passed a placement test does not ensure that you are prepared to succeed. If you have studied the background material recently, then usually with time, effort, confidence, and patience you will be able to learn the new material. If you have had a semester or a quarter or a summer break since your last math course, it is necessary to review. If the review provided in the text is not sufficient for you to recall, say, operations with fractions, you should immediately seek advice from your instructor or outside help. Use your prior book as a reference. If you took the prerequisite course more than a year ago, you may want to retake it before going on.

Beginning the Course

Here are a number of different issues for you to think about as you begin this course.

Getting a Good Start

To succeed, you need to attend class, read the book before class, and do the homework in a timely manner. Plan your study time. Some students set up their schedules to have the hour after math class free, to review notes and start the assignment.

Success also depends on being prepared with the proper equipment: an appropriate calculator, a six-inch ruler also marked in centimeters, and graph paper. Do all your graphs on graph paper.

Keep in mind that your first homework paper is a "grade application," just as a cover letter and resume are part of a job application. First impressions count. Neatness and completeness make a lasting impression on the instructor. So does having homework ready to turn in as you walk into class.

Taking Notes

Observe the five R's of note-taking (the Cornell system):

1. *Record.* Write down the ideas and concepts in a lecture. (Reading the text ahead of time will help identify these items.) Don't recopy notes.

2. *Reduce.* Summarize notes immediately after class (or as soon as possible); highlight important items.

3. *Recite.* Say out loud in your own words the main ideas and concepts.

4. *Reflect.* Think about how the material fits in with what you already know.

5. *Review.* Once a week, go over the ideas from each class so far in the term.

After the First Class

As you review your course syllabus, write test dates and other deadlines on your calendar. If you are working or taking classes at two schools, make sure there are no schedule conflicts with final exams. Talk with your instructor this week to resolve any scheduling problems.

Homework

Do the homework as one of the steps in your learning—not just as a requirement of the course. Work on assignments as soon as possible, right after class or early in the day or weekend. This gives you the option of going back later and spending more time on a difficult exercise. During long study periods, build in breaks to keep yourself fresh: Work for an hour, do another subject for an hour, and then come back.

Make the homework meaningful. Write notes to yourself on homework papers. Highlight exercises that were difficult and that you want to review again later. Highlight formulas or key steps. Re-read the objectives. Summarize the definitions and solution methods in your own words to be sure you understand. Describe how the current section fits in with prior sections.

Using the Answer Box and Answer Section Effectively

Practice working quickly. Do not work with the answers in front of you. Wait to check your answers until you have finished several exercises or half or more of the homework assignment. Let your own reasoning tell you whether something is correct.

Preparing for the Next Day's Class

Each of the following steps will get you progressively more prepared for your next class.

- Skim first. Read objectives. List unfamiliar words and identify new skills or concepts.
- Scan the section and look for definitions of vocabulary.
- Write vocabulary words and definitions on note cards.
- Outline the section, including summaries of skills and applications. (For your convenience in outlining, the objectives, headings, and example titles are in color in the text.)
- Read through the steps in several examples.
- Try the homework ahead of time.

What to Expect from This Course

Learning Styles

Because you and your classmates have different cultural backgrounds, with a wide variety of past and present life experiences, no single example or presentation will appeal to all of you. Consider how you best learn directions to a friend's house—in words over the phone (verbally), from a map (visually), or from having been there with someone else (kinesthetically). As a student of algebra, you may prefer words (a verbal approach), drawings, pictures, and graphs (a visual approach), getting up and moving around (a kinesthetic approach), or a combination of these approaches. To be successful in mathematics, you need to know your learning style and focus on those ways of learning information that best fit your style. It is advantageous, in the long run, to begin to learn in the other styles also. To help you achieve success with algebra, *Intermediate Algebra: A Just-in-Time Approach* presents concepts in as many ways as space permits.

Independent Thinking

Although you will find the examples helpful, this text is designed to encourage your independent thinking. Look for patterns and relationships; discover concepts for yourself; seek out applications that are meaningful to you. Try to work through examples on your own first, before you look at the solution. The more involvement you have with the material, the more useful it will be and the longer you will remember it.

Groups

You are encouraged to work with others throughout this course. One of the most important benefits of working on mathematics with other people is that you clarify your own understanding when you explain an idea to someone else. This is especially true for the kinesthetic learner.

Alternative Approaches

Those of you who have had algebra before may find many familiar concepts in the text. Some concepts will appear just the way you learned them the first time; others will be presented quite differently. You are being asked to learn alternative approaches, not to discard your former skills.

Alternative approaches are important for several reasons. Your old way may not work in all situations. The new way may help introduce a later concept; it may be more efficient or give more useful results. Acknowledgment of alternative approaches validates your own discoveries. New and often better ways to do mathematics are being discovered all the time.

Now, get on with the course. Come back to the following suggestions if you run into difficulty at a later time.

Preventing Big Problems

Stuck on the Homework?

Suppose you took notes in class, read the section, and still are stumped by an exercise. If you understand the directions but can't get the problem to work, try it again on a clean sheet of paper. If you don't know how to do an exercise, summarize the relevant information and drawings and go on to another exercise. Be sure to read the exercise aloud before you give up. Sometimes we hear things that we miss when reading.

Sometimes we get too close to a problem and overlook the obvious. A fresh point of view may help. Come back later. If necessary, call another student. If you are off-campus, call your instructor during office hours to get a hint or suggested strategy. Many instructors also welcome e-mail questions.

Obtain help from your teacher or from the resource center as you need it. Don't wait until just before a test.

Falling Behind?

If you find yourself falling behind, let your instructor know that you are trying to catch up. Set up a plan that allows two to three days for each missed assignment. Most important, do current assignments first, even if you have to skip a few problems because you missed material. Work immediately after the class session. By doing the current assignment first, you will stay with the class. If you gradually complete missed work, within a reasonable amount of time you will be completely caught up. Do not skip class because you are behind or confused.

Forgetting Material?

Many students select one or two exercises from each section and write them on $3'' \times 5''$ cards, with complete solutions on the back. These "flash cards" may then be shuffled and practiced at any time for review. Cards provide an excellent way to study for tests and the final exam. Include vocabulary words in your card set.

Strategies for Taking and Learning from Tests

Prepare Yourself Academically

1. Attend class, and do the homework completely and regularly. If there are exercises on the homework that you do not know how to do, get help—from a classmate, the teacher, or another appropriate source.

2. Work under time pressure on a regular basis. Set yourself a limited amount of time to do portions of the homework. Use a time limit when doing review exercises or the practice tests at the middle and end of each chapter. Working in one- or two-hour blocks of time is usually more productive than spending all afternoon and evening on math one day a week.

Prepare Yourself Physically

3. Get a good night's sleep. Being rested helps you think clearly, even if you know less material.

Prepare Yourself Mentally

Psych yourself up! This is especially important if you have your test later in the day.

4. If you have a test at 8:00 A.M., use the last few minutes before bed to get everything ready for the next day. Make your lunch, set your books or pack on a chair by the door, set out your umbrella or appropriate weather gear, and make sure you have change for the bus or train or that your car's tires, battery, and gasoline level are okay.

5. Plan 10 or 15 minutes of quiet time before the test. Try to arrive early, if possible.

6. Mentally picture yourself taking the test.

    a. Imagine writing your name on the test.

    b. Imagine reading through the test completely to see where the instructor put various types of questions.

    c. Imagine writing notes, formulas, or reminders to yourself on the test.

    d. Imagine working your favorite type of problem first.

Take the Test Right

7. Arrive early. Be ready—pencil sharpened and homework papers ready to turn in.

8. Concentrate on doing the steps that you imagined in item 6 above.

9. Work quickly and carefully through those problems you know how to solve. Don't spend over two minutes on one problem until you have tried every problem.

10. Be confident that, having prepared for the test, you can succeed.

Learn from the Test

11. After you turned in the test, did you remember information that would have helped you on the test? Would reading through the test more thoroughly at the start have given you time to recall information you needed?

12. Before you forget, look up and write down anything that you needed to know for the test but did not know.

13. When you get the test back, look at each item you missed. Which ones did you know how to do, and which ones did you not know how to do? Figure out what caused you to miss the ones you knew how to do. Carefully re-work on paper the ones you did not know how to do, getting help as needed.

14. Write down what you will do differently in preparing for the next test.

Expressions and Equations

Wind (miles per hour)

Current temperature (°Fahrenheit)	5	10	15	20	25	30
35	33	21	16	12	7	5
30	27	16	11	3	0	−2
25	21	9	1	−4	−7	−11
20	16	2	−6	−9	−15	−18
15	12	−2	−11	−17	−22	−26
10	7	−9	−18	−24	−29	−33
5	1	−15	−25	−32	−37	−41
0	−6	−22	−33	−40	−45	−49
−5	−11	−27	−40	−46	−52	−56

Table I   Wind-chill apparent temperature chart

How do you feel on a cold day when the wind is blowing? On a cold day when the wind is calm? Table I shows the temperature you feel (apparent temperature) as a result of the cooling provided by the wind. On a hot day, this cooling effect can be created with a fan. The temperatures are shown in degrees Fahrenheit. Observe how the wind drops many temperatures below the freezing point for water, 32°F. Which causes a greater drop in apparent temperature, a 30-mile-per-hour wind at 25°F or a 10-mile-per-hour wind at −5°F?

This chapter introduces algebra in four ways: symbolically (with the traditional algebraic notation), numerically (with input-output tables, number patterns, and other numerical methods), visually (with graphs and models), and verbally (in words with phrases and sentences). We review operations with and properties of real numbers. We work with sets of numbers, inequalities, and intervals. We evaluate and simplify expressions and solve formulas, as well as equations in one variable.

1.0   Review of Real Numbers

OBJECTIVES

* Identify sets of numbers.
* Add, subtract, multiply, and divide integers and real numbers.
* Identify properties of real numbers.
* Use the order of operations including square root, exponents, and absolute value.
* Match inequalities and their number-line graphs.

> ### WARM-UP
> 1. Locate the numbers $\{-3, 6, -2, -4, -6\}$ on a number line.
> 2. Which number is farther from zero, -4 or 6? Show why on your number line from Exercise 1.
> 3. If the distance from 0 to 1 is 1 unit, how far is it from -4 to 6?

THIS SECTION REVIEWS the names of the various sets of real numbers, operations with positive and negative real numbers, properties of real numbers, order of operations, and line graphs. The descriptions that follow may seem boring, but the vocabulary of number sets and their properties make it possible to give definitions, describe patterns, write equations, and find solutions—in other words, to do algebra!

Sets of Numbers

A **set** is *a collection of objects or numbers*. The **real numbers** are *the set of numbers that can be located with a point on the number line*. The numbers shown in Figure 1, $-3\frac{1}{2}$, $-\sqrt{4}$, -0.5, $\sqrt{2}$, and π, are all real numbers.

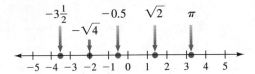

Figure 1   Real number line

Many of the sets of numbers contained in the real numbers can be described by a listing of the numbers placed in braces { }. The **natural numbers** are *the set containing the numbers for counting*, $\{1, 2, 3, \ldots\}$. When we *add zero to the set of natural numbers*, we have the set of **whole numbers,** $\{0, 1, 2, 3, \ldots\}$. The positive and negative numbers in Table 1 are integers. The **integers** are *the whole numbers and their opposites*, $\{\ldots, -3, -2, -1, 0, 1, 2, 3, \ldots\}$.

By placing each integer over 1, $\{\ldots, -\frac{3}{1}, -\frac{2}{1}, -\frac{1}{1}, \frac{0}{1}, \frac{1}{1}, \frac{2}{1}, \frac{3}{1}, \ldots\}$, we show that integers can be written as rational numbers. **Rational numbers** *can be written $\frac{a}{b}$, where a and b are integers and b is not zero*.

The **square root** of x, $\sqrt{x}$, is *the real number that, when multiplied by itself, gives the radicand, x*. For example, $\sqrt{3} \cdot \sqrt{3} = 3$. The $\sqrt{}$ symbol is called the square root sign or radical. Some square roots, such as $\sqrt{4}$ and $\sqrt{9}$, are rational;

others are irrational. The **irrational numbers** are *the set of numbers that cannot be written as a rational number.* Irrational numbers include pi (π), $\sqrt{2}$, $\sqrt{3}$, and $\sqrt{5}$.

Here is a summary of the real numbers.

Set of Real Numbers

Sets of rational numbers (which can be written $\frac{a}{b}$ with a and b integers, b not zero) include

- natural numbers $\{1, 2, 3, \ldots\}$
- whole numbers $\{0, 1, 2, 3, \ldots\}$
- integers $\{\ldots, -3, -2, -1, 0, 1, 2, 3, \ldots\}$
- square roots with exact decimal values $\{\sqrt{1}, \sqrt{4}, \sqrt{9}, \sqrt{16}, \ldots\}$

Sets of irrational numbers (which cannot be written $\frac{a}{b}$ with a and b integers, b not zero) include

- square roots without exact decimal values $\{\sqrt{2}, \sqrt{3}, \sqrt{5}, \ldots\}$
- pi (π), equal to $3.14159265358\ldots$

Student Note: Try the problem posed in the example before reading the solution. You might want to cover up the solution with paper or a bookmark. The answers to selected examples are in the Answer Box just before the exercise set.

The square root of a negative number, such as $\sqrt{-4}$, is not a real number and is **undefined** (*has no value or meaning*) when we work within the set of real numbers. Note that $-\sqrt{4}$ is defined; $-\sqrt{4} = -2$ (more on square roots of negative numbers later).

EXAMPLE 1   Identifying sets of numbers   Write the names of all the sets of numbers to which each number belongs.

a. 5   **b.** $\frac{1}{4}$ **c.** $\sqrt{49}$ **d.** $\sqrt{11}$ **e.** 1.5

Solution   **a.** 5 belongs to the sets of real, rational, integer, whole, and natural numbers.
b. $\frac{1}{4}$ belongs to the sets of real and rational numbers.
c. $\sqrt{49}$ equals 7 and belongs to the sets of real, rational, integer, whole, and natural numbers.
d. $\sqrt{11}$ does not have an exact square root and is irrational.
e. 1.5 equals $\frac{3}{2}$ and belongs to the sets of real and rational numbers. ●

Operations with Positive and Negative Real Numbers

As mentioned earlier, the positive and negative numbers in the wind-chill chart in Table 1 are integers. The exploration in Example 2 is designed to start you thinking about operations with integers.

EXAMPLE 2   Exploring wind chill: reading a table and finding temperature change

a. What is the wind-chill apparent temperature at 25°F with a wind of 30 miles per hour (mph)?
b. By how much did the wind in part a change the temperature?
c. What is the wind-chill apparent temperature at −5°F with a wind of 10 mph?
d. By how much did the wind in part c change the temperature?

Solution   *Hint:* To find the change in temperature, subtract the starting temperature from the ending temperature. See the Answer Box. ●

Parts b and d in Example 2 required subtraction of integers. We can model this subtraction of integers by finding the starting temperature on a number line (thermometer) and counting the number of degrees to the ending temperature. If the temperature falls, the change is negative; if the temperature rises, the change is positive.

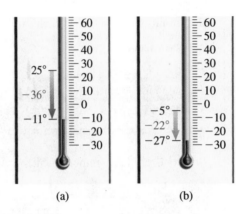

(a)   (b)

Figure 2

Here is a summary of the operations with positive and negative real numbers.

Summary of Operations with Positive or Negative Real Numbers

- Addition of Positive and Negative Real Numbers
  If the signs are alike, add the numbers and place the common sign on the answer:

  $$-6 + (-8) = -14 \qquad 21 + 17 = 38$$

  If the signs are different, subtract the number portion and place the sign of the number farthest from zero on the answer:

  $$15 + (-10) = +(15 - 10) = 5 \qquad 4 + (-5) = -(5 - 4) = -1$$

- Subtraction
  Two numbers are **opposites**, or **additive inverses**, *if they add to zero.* Opposites are the same distance from zero on the number line and have opposite signs. Subtraction is the inverse operation to addition. To subtract two numbers, change subtraction of a number to addition of the opposite number:

  $$6 - 5 = 6 + (-5) = 1 \qquad 6 - (-4) = 6 + (+4) = 10$$
  $$-11 - 25 = -11 + (-25) = -36$$
  $$-27 - (-5) = -27 + (+5) = -27 + 5 = -22$$

- Multiplication of Positive and Negative Real Numbers
  If two numbers have like signs, their product is positive:

  $$(-2)(-3) = +6 \qquad 3(5) = 15$$

  If two numbers have unlike signs, their product is negative:

  $$(-4)(+3) = -12 \qquad (+6)(-3) = -18$$

  The product of zero and any number is zero:

  $$0(6) = 6(0) = 0$$

> • Division
>
> Two numbers are **reciprocals** *if their product is* 1. A reciprocal is also known as a **multiplicative inverse.** Division is the inverse operation to multiplication. To divide two numbers, change division by a number to multiplication by the reciprocal:
>
> $$-5 \div \tfrac{1}{3} = -5 \cdot \tfrac{3}{1} = -15 \qquad 6 \div \tfrac{2}{3} = 6 \cdot \tfrac{3}{2} = \tfrac{18}{2} = 9$$
>
> The inverse of an operation performed after a given operation will nullify the original operation:
>
> $$6 + 4 - 4 = 6 \qquad 5 \cdot 8 \div 8 = 5$$

Mathematicians use **simplify** as *a shorthand way to write directions.* Simplify can mean "to perform the indicated operations," as in Example 3.

EXAMPLE 3   Practicing operations with real numbers   Simplify by performing these operations.

a. $4 + (-3)$ **b.** $4(-3)$ **c.** $4 \div \dfrac{-4}{5}$ **d.** $2 \cdot (-4)$

e. $(-6) + (-9)$ **f.** $(-6) \div \dfrac{-3}{4}$ **g.** $-6 - (-3)$ **h.** $-6 - 3$

i. $-3 + 2 - 2$ **j.** $-6 \div 2 \cdot 2$

Solution   **a.** $4 + (-3) = 1$ **b.** $4(-3) = -12$

c. $4 \div \dfrac{-4}{5} = 4 \cdot \dfrac{-5}{4} = \dfrac{4 \cdot (-5)}{4} = -5$

d. $2 \cdot (-4) = -8$

e. $(-6) + (-9) = -15$

f. $-6 \div \dfrac{-3}{4} = -6 \cdot \dfrac{-4}{3} = \dfrac{-6 \cdot (-4)}{3} = \dfrac{24}{3} = 8$

g. $-6 - (-3) = -6 + (+3) = -3$

h. $-6 - 3 = -6 + (-3) = -9$

i. $-3 + 2 - 2 = -3$

j. $-6 \div 2 \cdot 2 = -6$

Properties of Real Numbers

ASSOCIATIVE AND COMMUTATIVE PROPERTIES   Real numbers have properties for addition and multiplication that permit us to rearrange terms or to change the order in which we add or multiply. The commutative property of addition and the commutative property of multiplication allow us to rearrange the terms in an expression.

EXAMPLE 4   Arranging numbers to make mental calculation easier   Find shortcuts for doing these problems.

a. $3 + 6 + 4 + 2 + 7 + 8$ **b.** $5 \cdot 19 \cdot 2$

Solution   **a.** $3 + 6 + 4 + 2 + 7 + 8 = 3 + 7 + 6 + 4 + 2 + 8$
$$= 10 + 10 + 10 = 30$$

b. $5 \cdot 19 \cdot 2 = 19 \cdot 5 \cdot 2 = 19 \cdot 10 = 190$

In part a of Example 4, the numbers that add to 10 were added first. The associative property for addition permits us to add the numbers in a convenient order. In part b of Example 4, we multiplied the 5 and 2 first and then the 19 and 10. The associative property of multiplication permits multiplying numbers in a convenient order.

The following properties are true for all real numbers, a, b, and c.

Associative Property for Addition	$a + (b + c) = (a + b) + c$ $2 + (3 + 4) = (2 + 3) + 4$

Associative Property for Multiplication	$a \cdot (b \cdot c) = (a \cdot b) \cdot c$ $3 \cdot (4 \cdot 5) = (3 \cdot 4) \cdot 5$

Commutative Property for Addition	$a + b = b + a$ $6 + 8 = 8 + 6$

Commutative Property for Multiplication	$a \cdot b = b \cdot a$ $6 \cdot 8 = 8 \cdot 6$

EXAMPLE **5**

Using the properties of real numbers   Do these problems mentally by rearranging and regrouping the numbers.

a. $\$4.25 + \$3.98 + \$3.75$ b. $(-2) + (-5) + (-8) + 5$
c. $-4.5 + 2.8 - 5.5 + 3.2$ d. $\frac{1}{2} + \frac{1}{4} + \frac{3}{4} + \frac{1}{4}$
e. $1\frac{2}{3} - 2\frac{1}{2} + \frac{1}{3}$ f. $(-2)(7)(-5)$

Solution   a. $\$4.25 + \$3.98 + \$3.75 = \$4.25 + \$3.75 + \$3.98$
$$= \$8.00 + \$3.98 = \$11.98$$

b. $(-2) + (-5) + (-8) + 5 = (-2) + (-8) + (-5) + 5$
$$= -10 + 0 = -10$$

c. $-4.5 + 2.8 - 5.5 + 3.2 = -4.5 - 5.5 + 2.8 + 3.2$
$$= -10 + 6 = -4$$

d. $\frac{1}{2} + \frac{1}{4} + \frac{3}{4} + \frac{1}{4} = \frac{1}{2} + 1 + \frac{1}{4} = 1\frac{3}{4}$

e. $1\frac{2}{3} - 2\frac{1}{2} + \frac{1}{3} = 1\frac{2}{3} + \frac{1}{3} - 2\frac{1}{2} = 2 - 2\frac{1}{2} = -\frac{1}{2}$

f. $(-2)(7)(-5) = (-2)(-5)(7) = 10 \cdot 7 = 70$

BASES AND EXPONENTS   The **exponent** is *the small raised number or expression to the right* of a number or expression called the base. The **base** is *the number to which the exponent is applied.* In 4^2, the 4 is the base and the 2 is the exponent. In this case, the exponent means to multiply the base by itself: $4 \cdot 4 = 16$.

Together the base and exponent are called the **power of a number.** The expression 4^2 is called the *second power of four.* The expression 4^3 is called the *third* *power of four.* In 4^3, the 4 is the base and the 3 is the exponent. The exponent means to write the base three times and then multiply. In general,

---

When the exponent is an integer, n, the base, b, is written n times:

$$b^n = \underbrace{b \cdot b \cdot b \cdot \cdots \cdot b}_{n \text{ factors}}$$

(See page 8 for the definition of factors.) In letters, $x \cdot x = x^2$ and $x \cdot x^2 = x^3$.

EXAMPLE **6**   Identifying bases and applying exponents   Name the base in each expression and then write the expression without exponents.

a. $(-2)^3$ b. $(-3)^2$ c. -3^2 d. $0 - 3^2$ e. $2x^2$

Solution   a. The base is -2:

$$(-2)^3 = (-2)(-2)(-2) = -8$$

b. The base is -3:

$$(-3)^2 = (-3)(-3) = +9$$

c. The base is 3:

$$-3^2 = -3 \cdot 3 = -9$$

d. The base is 3:

$$0 - 3^2 = 0 - 9 = -9$$

e. The base is x:

$$2x^2 = 2xx$$

●

Think about it 1: How is $(2x)^2$ the same as or different from $2x^2$?

DISTRIBUTIVE PROPERTY   Another property of real numbers, the distributive property, justifies multiplying out parentheses and factoring.

Distributive Property of Multiplication over Addition

$a(b + c) = ab + ac$

$3(4 + 5) = 3 \cdot 4 + 3 \cdot 5$

Because a is multiplying both b and c, we say that this property distributes a over the sum, or addition, of b and c.

EXAMPLE **7**   Applying the distributive property   Multiply in these problems.

a. Change to a sum: $x(x - 2)$.

b. Find the area of the rectangle, where length and width are in feet:

$x + 4$

2 ☐

c. Using the table form for multiplication is like finding a rectangular area. Multiply the numbers on the left and top to obtain the inner "area" (in the white area).

Multiply	x^2	$+3x$	$+2$
x			

Solution **a.** $x(x - 2) = x^2 - 2x = x^2 + (-2x)$

b.

$$\begin{array}{c|c} & x + 4 \\ \hline 2 & 2x + 8 \end{array}$$

Thus, 2 ft $\cdot$ (x ft + 4 ft) = $2x$ ft^2 + 8 ft^2.

c.

Multiply	x^2	$+3x$	$+2$
x	x^3	$+3x^2$	$+2x$

Thus, $x(x^2 + 3x + 2) = x^3 + 3x^2 + 2x$. ●

The following suggestions, illustrated in Figure 3, may help you remember the names of the properties. To *associate* means "to group together." The parentheses in the associative property group the numbers, just as business associates are grouped together. To *commute* means "to change positions." We may commute between home and work. To *distribute* means "to give out." We might, for instance, distribute cards to each player during a card game.

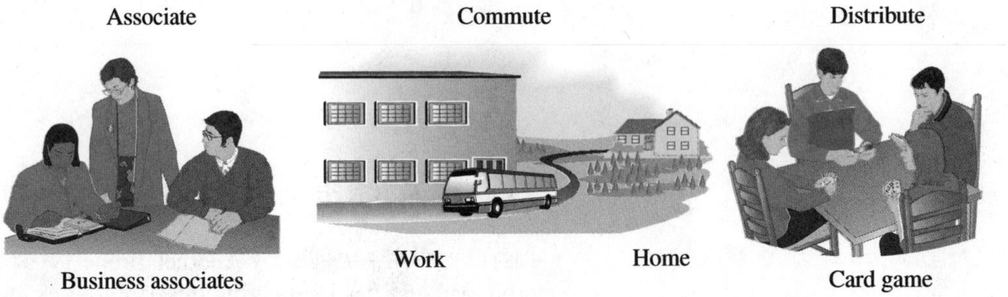

Associate   Commute   Distribute

Business associates   Work   Home   Card game

Figure 3

FACTORING **Factors** are *the numbers being multiplied.* In $a(b + c)$, there are two factors: a and $b + c$. In abc, there are three factors: a, b, and c.

EXAMPLE **8**   Identifying factors   How many factors are in each of these expressions?

a. $xy(z + 1)$ **b.** $a(b + c + d)$ **c.** $abcd$

Solution **a.** Three factors: x, y, and $z + 1$

b. Two factors: a and $b + c + d$

c. Four factors: a, b, c, and d ●

When we reverse the distributive property, $ab + ac = a(b + c)$, we are factoring. In this case, **factoring** is *changing the addition of ab and ac into the multiplication of a and (b + c).*

Distributive Property in Factored Form	$ab + ac = a(b + c)$ The letter a is called the greatest common factor.

Because factoring suggests a process, the word *factor*, like *simplify*, is a common direction in examples and exercises. Its meaning will vary somewhat with the topic being studied.

EXAMPLE   **Factoring**   Name the greatest common factor and then use it to factor.

a. Change to a product: $9a - 6$.

b. Find the dimensions (length and width):

$6x^2 + 12x$

c.

Factor			
	$2x^2$	$+6x$	$+2$

Solution   **a.** 3 is the greatest common factor:

$$9a - 6 = 3(3a - 2)$$

b. $6x$ is the greatest common factor:

$$x + 2$$

$6x$	$6x^2 + 12x$

Thus, $6x^2 + 12x = 6x(x + 2)$.

c. 2 is the greatest common factor:

Factor	x^2	$+3x$	$+1$
2	$2x^2$	$+6x$	$+2$

Thus, $2x^2 + 6x + 2 = 2(x^2 + 3x + 1)$.   ●

Think about it 2: In part b of Example 9, what are other possible dimensions? Why might the one shown in the solution be preferred?

Order of Operations

The **order of operations** is *an agreement about the order in which we work with numbers in algebra.*

Order of Operations	**1.** Calculate expressions within parentheses and other grouping symbols. **2.** Calculate exponents and square roots. **3.** Do the remaining multiplications and divisions in order of appearance, left to right. **4.** Do the remaining additions and subtractions in order of appearance, left to right.

EXAMPLE **10**   Applying the order of operations   Simplify the following expressions.

a. Perimeter of a photograph, in inches: $P = 2 \cdot 5 + 2 \cdot 7$

b. Outside surface area of a shoe box without lid, in square inches:
$S = 2 \cdot 4 \cdot 6 + 2 \cdot 4 \cdot 14 + 1 \cdot 6 \cdot 14$

c. Volume of a CD box, in cubic centimeters: $V = 1.0(14.2)(12.4)$

d. From solving an equation: $d = (-7)^2 - 4(4)(-2)$

e. Distance around a running track, in meters: $D = 2(100) + 2\pi(31.831)$

f. A puzzle: $A = 3 - \frac{8}{2} + 6 \cdot 3 + 3$

Solution

Student Note: The comments to the right tell you what to do to the given step to get to the next step.

a. $P = 2 \cdot 5 + 2 \cdot 7$ Multiply first.
$P = 10 + 14$ Add second.
$P = 24$ in.

b. $S = 2 \cdot 4 \cdot 6 + 2 \cdot 4 \cdot 14 + 1 \cdot 6 \cdot 14$ Multiply first.
$S = 48 + 112 + 84$ Add second.
$S = 244$ in^2

c. $V = 1.0(14.2)(12.4) = 176.08$ cm^3

d. $d = (-7)^2 - 4(4)(-2)$ Do the exponent first.
$d = 49 - 4(4)(-2)$ Do multiplication from left to right.
$d = 49 - (-32)$ Do subtraction.
$d = 81$

e. $D = 2(100) + 2\pi(31.831)$ Multiply first.
$D \approx 200 + 200$ Add second.
$D \approx 400$ m

f. $A = 3 - \frac{8}{2} + 6 \cdot 3 + 3$ Do division, then multiplication.
$A = 3 - 4 + 18 + 3$ Do subtraction.
$A = -1 + 18 + 3$ Do addition.
$A = 20$

In Example 10, we used a dot for multiplication of positive integers and parentheses for multiplication of numbers containing negative signs or decimals. There is no rule for this, but the choice makes the multiplications easier to read.

GROUPING SYMBOLS   **Grouping symbols,** such as parentheses, *indicate when an operation is to be done out of its normal order.* Grouping symbols are first in the order of operations. When several groupings are needed, we use brackets or braces. The preferred way to write multiple symbols is {[()]}, with **parentheses** (*rounded*) on the inside, **brackets** (*square-shaped*) used next, and **braces** (*like little wires*) used outermost. If the context is not confusing, double parentheses, (()), are acceptable.

Think about it 3: What is the value of $5 - \{2 - [3 - (4 + 5)]\}$?

If an absolute value, square root, or fraction contains operations, it is a grouping symbol. Note that we *place parentheses around numbers and operations within grouping symbols when using a calculator.*

The **absolute value symbol** is used in $|5 - 8|$. Find $5 - 8$ inside the absolute value, and then find the absolute value: $|-3| = 3$. On a calculator, write abs $(5 - 8)$.

The **square root symbol,** or radical sign, is used in $\sqrt{1 + 3}$. Add 1 and 3, and then take the square root: $\sqrt{4} = 2$. On a calculator, write $\sqrt{(1 + 3)}$.

The **horizontal fraction bar** is a grouping symbol in $\dfrac{8 - \sqrt{36}}{3 - 5}$. Find the value of the numerator and denominator separately, and then divide the two parts:

$$\frac{8 - \sqrt{36}}{3 - 5} = \frac{8 - 6}{-2} = -1$$

On a calculator, write $(8 - \sqrt{36})/(3 - 5)$.

Think about it 4: If the calculator places an opening parenthesis after the square root, you need to include a closing parenthesis: $(8 - \sqrt{(36)})/(3 - 5)$. Describe how 11 is obtained as a wrong answer from this entry: $(8 - \sqrt{(36)}/(3 - 5)$.

EXAMPLE **11**

Using grouping symbols   Simplify the following expressions.

a. Slope of a line: $m = \dfrac{-2 - 8}{1 - (-4)}$

b. Body mass index: $i = \dfrac{160(704.5)}{72^2}$

c. Distance formula: $d = \sqrt{(1 - (-4))^2 + (-2 - 8)^2}$

d. Mean absolute deviation: $d = \dfrac{|9.2 - 9.5| + |9.3 - 9.5| + |9.9 - 9.5|}{3}$

e. From solving an equation: $x = \dfrac{-(-7) + \sqrt{(-7)^2 - 4(4)(-2)}}{2(4)}$

f. A puzzle: $A = 5\{6 - 4[3 - 8(5 - 7) + 1] - 2\}$

Solution

a. $m = \dfrac{-2 - 8}{1 - (-4)}$ Find the numerator and denominator.

$m = \dfrac{-10}{5}$ Divide.

$m = -2$

b. $i = \dfrac{160(704.5)}{72^2}$ Find the numerator and denominator.

$i = \dfrac{112{,}720}{5184}$ Divide.

$i \approx 21.7$

c. $d = \sqrt{(1 - (-4))^2 + (-2 - 8)^2}$ Simplify inside the parentheses.

$d = \sqrt{5^2 + (-10)^2}$ Apply exponents.

$d = \sqrt{25 + 100}$ Simplify inside the radical.

$d = \sqrt{125} \approx 11.2$ Take the square root.

d. $d = \dfrac{|9.2 - 9.5| + |9.3 - 9.5| + |9.9 - 9.5|}{3}$ Do the absolute value.

$d = \dfrac{0.3 + 0.2 + 0.4}{3}$ Add the numerator.

$d = \dfrac{0.9}{3} = 0.3$ Do the division.

e. $x = \dfrac{-(-7) + \sqrt{(-7)^2 - 4(4)(-2)}}{2(4)}$ Simplify under the radical.

$x = \dfrac{-(-7) + \sqrt{49 - (-32)}}{2(4)}$ Simplify the numerator and denominator.

$x = \dfrac{7 + \sqrt{81}}{8}$ Take the square root.

$x = \dfrac{7 + 9}{8} = \dfrac{16}{8} = 2$ Divide and simplify.

f. $A = 5\{6 - 4[3 - 8(5 - 7) + 1] - 2\}$ Simplify inside the parentheses.
$A = 5\{6 - 4[3 - 8(-2) + 1] - 2\}$ Multiply by 8; simplify inside the brackets.
$A = 5\{6 - 4[20] - 2\}$ Multiply by 4; simplify inside the braces.
$A = 5\{-76\}$ Multiply by 5.
$A = -380$ ●

Line Graphs

We began this section by naming sets of numbers. We close the section by describing sets of numbers by their position on the number line.

INEQUALITIES   An **inequality** is *a statement that one quantity is greater than or less than another quantity.* The chart below gives a summary of the inequality symbols.

Inequality Symbol	Meaning
$<$	is less than (is to the left of on the number line)
$>$	is greater than (is to the right of on the number line)
$\leq$	is less than or equal to
$\geq$	is greater than or equal to

When the number line is vertical, $<$ implies *below* and $>$ implies *above.*

EXAMPLE **12**   **Using inequality symbols**   Describe how the numbers are related. Tell what inequality symbols make each of the statements true.

    a. $7 \square 6$ **b.** $-\frac{1}{2} \square -\frac{1}{4}$ **c.** $0.1 \square 0.05$ **d.** $-3 \square -3$

Solution   **a.** 7 is greater than 6; $>$ or $\geq$ **b.** $-\frac{1}{2}$ is less than $-\frac{1}{4}$; $<$ or $\leq$
         c. 0.1 is greater than 0.05; $>$ or $\geq$ **d.** -3 is equal to -3; $\geq$ or $\leq$ ●

We must be able to read and describe sets of numbers on a number line in order to effectively use a graphing calculator. Inequalities provide a convenient way to describe these sets of numbers. The equality sign in $\leq$ and $\geq$ means that we include the number in our set, which we indicate on a graph by using a solid dot. For the other inequality signs, $<$ and $>$, we use a small circle to indicate exclusion of the number.

EXAMPLE **13**   **Recognizing line graphs**   Match the inequality and the graph of the set of numbers it describes.

    a. $x > 2$ **b.** $x < 0$ **c.** $y > -2$
    d. $y < 0$ **e.** $x \geq -2$ **f.** $y \leq 2$

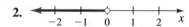
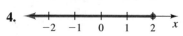
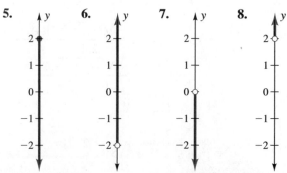

Solution   **a.** 3   **b.** 2   **c.** 6   **d.** 7   **e.** 1   **f.** 5   ●

LOGIC AND INEQUALITIES   The definitions of the inequalities $\leq$ and $\geq$ use the word *or*. Many definitions and other mathematical statements include the logic phrases *a or b* or *a and b*.

Logic: *a or b* versus *a and b*

> The phrase *a or b* is true if either the condition *a* is true or the condition *b* is true or if both conditions are true. The inequality $a \leq b$ is true if either *a* is less than *b* or *a* is equal to *b*, because the definition of the symbol $\leq$ contains the word *or*.
>
> The phrase *a and b* is true if both *a* and *b* are true. Because a number cannot be both greater than and equal to another number, the word *and* cannot be used in describing the inequality $\geq$.

And means that numbers must satisfy both conditions, whereas *or* means that either condition may be met. The inequality statements $x > 0$ and $x < 3$ are shown in the first two number lines in Figure 4. The *and* requires that we include only numbers in both sets (the numbers larger than zero and at the same time smaller than 3), as shown in the third number line of Figure 4. This inequality may be written as one statement, $0 < x < 3$, and is read *x is between 0 and 3*. The inequality signs indicate that the endpoints are not included in $0 < x < 3$.

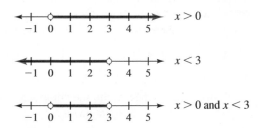

Figure 4

The statement $0 < x < 3$ is a compound inequality. A **compound inequality** *combines two inequality statements into one statement with two inequality signs in the same direction.* The combined inequality must be true even if the middle number or expression is left out. When we leave *x* out, $0 < 3$ is still true. The inequality signs in a compound inequality must always have the same direction. The statement $0 < x > -3$ is not acceptable.

Example 14 illustrates the use of *and* and *or* with inequalities.

EXAMPLE **14**

Recognizing line graphs   Match the inequality statement with its graph. The lines may represent either *x* or *y*. Then describe the inequalities in parts a to f in words.

a. $x > 3$ or $x < 1$ **b.** $-1 < y < 3$

c. $x < 3$ or $x > 5$ **d.** $1 \leq x \leq 5$

e. $y < 5$ or $y > 3$ **f.** $x \leq 3$ or $x \geq 5$

1.
```
←─┼──┼──┼──┼──┼──┼──┼──┼──┼──┼──┼→
 -3 -2 -1  0  1  2  3  4  5  6  7
```

2.

```
←———+———+———+———○———+———+———○———+———+———+———+———→
   -3  -2  -1   0   1   2   3   4   5   6   7
```

3.

```
←———+———+———○━━━━━━━━━━━○———+———+———+———+———→
   -3  -2  -1   0   1   2   3   4   5   6   7
```

4.

```
←———+———+———+———+———+———●———+———●———+———+———→
   -3  -2  -1   0   1   2   3   4   5   6   7
```

5.

```
←———+———+———○━━━━━━━━━━━━━━━━━━━○———+———+———→
   -3  -2  -1   0   1   2   3   4   5   6   7
```

6.

```
←———+———+———+———●━━━━━━━━━━━●———+———+———→
   -3  -2  -1   0   1   2   3   4   5   6   7
```

7.

```
←———+———+———+———●———+———●———+———+———+———→
   -3  -2  -1   0   1   2   3   4   5   6   7
```

8.

```
←———+———+———+———+———+———○———+———○———+———+———→
   -3  -2  -1   0   1   2   3   4   5   6   7
```

Solution   **a.** 2; the set of numbers less than 1 or greater than 3

b. 3; the set of numbers between -1 and 3

c. 8; the set of numbers less than 3 or greater than 5

d. 6; the set of numbers between 1 and 5, including 1 and 5

e. 1; the set of numbers less than 5 or greater than 3. The line graph in 1 includes all numbers because the set of numbers to the right of 3 is joined with the set of numbers to the left of 5, to cover the entire number line.

f. 4; the set of numbers less than or equal to 3 or greater than or equal to 5 ●

Think about it 5: Write the compound inequalities in parts b and d of Example 14 as two inequality statements.

ANSWER BOX

Warm-up: 1.

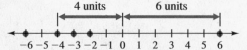

```
←———●———+———●———+———●———+———+———+———+———+———+———●———→
   -6  -5  -4  -3  -2  -1   0   1   2   3   4   5   6
```

2. 6 is farther from zero. **3.** 10 units **Example 2: a.** $-11°$F
b. Subtract the starting temperature from the ending temperature:
$-11° - 25° = -36°$. **c.** $-27°$F **d.** Subtract the starting temperature
from the ending temperature: $-27° - (-5°) = -27° + (+5°) =$
$-27 + 5 = -22°$. **Think about it 1:** $(2x)^2$ has a base of $2x$.
$(2x)^2 = 2x \cdot 2x = 4x^2$ and does not equal $2x^2$. **Think about it 2:**
$6(x^2 + 2x)$ and $x(6x + 12)$ are two possible pairs of dimensions. The
dimensions $6x(x + 2)$ have the greatest common factor, $6x$. **Think
about it 3:** $5 - \{2 - [3 - (4 + 5)]\} = 5 - \{2 - [3 - 9]\} =$
$5 - \{2 - [-6]\} = -3$. **Think about it 4:** $\sqrt{36}$ is divided by -2.
The result, -3, is subtracted from 8. **Think about it 5:** The compound
inequality $-1 < y < 3$ may be written as $-1 < y$ and $y < 3$. The
compound inequality $1 \le x \le 5$ may be written as $1 \le x$ and $x \le 5$.

EXERCISES 1.0

In Exercises 1 to 8, match each definition or related statement with a vocabulary word. Record any unfamiliar words and their definitions in your study notes.

1. Choose from
   whole numbers, natural numbers, rational numbers, opposites, reciprocals

   a. The set of numbers for counting, $\{1, 2, 3, \ldots\}$

   b. Two numbers that add to zero

   c. Two numbers that multiply to 1

   d. The set of natural numbers and zero, $\{0, 1, 2, 3, \ldots\}$

   e. The set of numbers that can be written $\dfrac{a}{b}$, a and b integers, b not equal to zero

2. Choose from
   multiplicative inverses, additive inverses, integers, real numbers, irrational numbers

   a. All the numbers that can be located with a point on the number line

   b. The whole numbers and their opposites, $\{\ldots, -3, -2, -1, 0, 1, 2, 3, \ldots\}$

   c. Real numbers that cannot be written as rational numbers

   d. Two numbers that multiply to 1

   e. Two numbers that add to zero

3. Choose from
   set, $\sqrt{-1}$, $\geq$, $<$, simplify

   a. is less than

   b. to perform the indicated operation

   c. is greater than or equal to

   d. a collection of objects or numbers

   e. is undefined in the real numbers

4. Choose from
   associative property for addition, commutative property for multiplication, distributive property of multiplication over addition, factors, factoring

   a. $c(d + e) = cd + ce$

   b. Two numbers being multiplied

   c. $b \cdot c = c \cdot b$

   d. Changing the addition of ab and ac into the multiplication of a and $(b + c)$

   e. $a + (b + c) = (a + b) + c$

5. Choose from
   associative property for multiplication, commutative property for addition, factored form of the distributive property, inequality, logic phrase "a or b"

   a. $a + b = b + a$

   b. A statement that one quantity is greater than or less than another quantity

   c. $b \cdot c + b \cdot e = b(c + e)$

   d. $a(bc) = (ab)c$

   e. Either a is true or b is true or both are true

6. Choose from
   unacceptable statement, dot on a number line, compound inequality, logic phrase "a and b," small circle on a number line

   a. Both a and b are true.

   b. The number at this position on the number line is excluded from the set.

   c. The number at this position on the number line is included in the set.

   d. It combines two inequality statements into one statement with two inequality signs in the same direction.

   e. $6 > 4 < 5$

7. Choose from
   positive, negative, zero

   a. The answer when two negative numbers are multiplied

   b. The answer when two negative numbers are added

   c. The answer when two positive numbers are added

   d. The answer when zero and any number are multiplied

   e. The answer when two positive numbers are multiplied

8. Choose from
   positive, negative, zero, one

   a. The answer when a negative and a positive number are multiplied

   b. The answer when zero and one are added

   c. The answer when two different negative numbers are divided

   d. The answer when a number and its reciprocal are multiplied

   e. The answer when a number and its opposite are added

In Exercises 9 to 12, which name does not describe the given number?

9. $\frac{1}{2}$: rational, integer, real

10. 1.5: natural, rational, real

11. $\sqrt{9}$: whole, real, irrational

12. $\sqrt{5}$: irrational, real, integer

In Exercises 13 and 14, assume that the order for subtraction and division of m and n is m − n and m ÷ n.

13. Add, subtract, multiply, and divide these pairs of numbers.

    a. −15 and −3   **b.** −6 and 2

    c. $\frac{1}{4}$ and $-\frac{1}{2}$ **d.** $-\frac{5}{12}$ and $-\frac{1}{3}$

    e. 2.5 and −0.25   **f.** $-\frac{3}{4}$ and $\frac{5}{6}$

14. Add, subtract, multiply, and divide these pairs of numbers.

    a. 4 and −12   **b.** $\frac{1}{2}$ and $\frac{1}{3}$

    c. $-\frac{2}{3}$ and $\frac{1}{6}$ **d.** 0.75 and −0.2

    e. −0.45 and −1.5   **f.** $-\frac{5}{8}$ and $-\frac{10}{9}$

15. Simplify.

    a. $-4 - (-3) + (-8)$

    b. $12 - (-4) + (-2)$

    c. $4.5 + (-3.2) - (-2.8)$

    d. $6.2 - (-1.4) + (-3.5)$

    e. $2\frac{1}{4} - 3\frac{1}{2} + 4\frac{1}{4}$

    f. $-3\frac{1}{2} + 2\frac{1}{4} - 5\frac{1}{4}$

16. Simplify.

    a. $8 - (-6) + (-2)$

    b. $15 + (-9) - (-8)$

    c. $-3.6 - (-4.8) + (4.3)$

    d. $-8.3 + (-3.8) - (-2.1)$

    e. $4\frac{1}{4} - 5\frac{1}{2} + 2\frac{1}{4}$

    f. $6\frac{1}{4} + 4\frac{1}{2} - 3\frac{1}{2}$

17. If possible, simplify or write without parentheses.

    a. $(-3)^2$ **b.** $(-2)^3$ **c.** -3^2

    d. $(3x)^2$ **e.** $3x^2$ **f.** $0 - 3x^2$

18. If possible, simplify or write without parentheses.

    a. -4^2 **b.** $(-4)^2$ **c.** $(-4x)^2$

    d. $-4x^2$ **e.** $(-2)^4$ **f.** $0 - 4x^2$

19. Simplify. Round to the nearest tenth.

    a. Area of Olympic basketball's 3-second restricted zone (a trapezoid), in square meters:
$A = \frac{1}{2}(5.8)(3.6 + 6.0)$

    b. Volume of a handball, in cubic inches:
$V = \frac{4}{3}\pi(1.875 \div 2)^3$

    c. From solving an equation:
$d = (-6)^2 - 4(8)(-3)$

    d. A puzzle: $x = 12 - 8 \div 4 + 5 \cdot 6 - 3^3$

20. Simplify. Round to the nearest tenth.

    a. At a range of 90 meters, the area of a circular archery target, in square centimeters:
$A = \pi(122 \div 2)^2$

    b. Surface area of a can of pumpkin, in square inches:
$S = 2\pi \cdot 2^2 + 2\pi(2)(4.5)$

    c. From solving an equation:
$d = (-5)^2 - 4(6)(-7)$

    d. A puzzle:
$x = 9 - 4(9 - 2) + 8 \div 4 - 2 \cdot 5$

21. Simplify.

    a. $x = \dfrac{3 - 12}{3^2 - 6}$

    b. $d = \sqrt{(6 - (-2))^2 + (3 - (-3))^2}$

    c. $m = \dfrac{3 - (-3)}{6 - (-2)}$

    d. $x = \dfrac{-(-2) + \sqrt{(-2)^2 - 4(3)(-5)}}{2(3)}$

    e. $d = \{|4.05 - 7.12| + |7.51 - 7.12| + |9.80 - 7.12|\} \div 3$

22. Simplify.

    a. $\dfrac{4 - 12}{2^3 - 6}$

    b. $d = \sqrt{(14 - 2)^2 + (2 - (-3))^2}$

    c. $m = \dfrac{2 - (-3)}{14 - 2}$

    d. $x = \dfrac{-(-2) - \sqrt{(-2)^2 - 4(3)(-5)}}{2(3)}$

    e. $d = \{|6.76 - 7.91| + |7.14 - 7.91| + |9.83 - 7.91|\} \div 3$

Exercises 23 to 28 contain visual models of the distributive property. Multiply or factor as indicated. Use the greatest common factor when factoring.

23. a.

Multiply	x	-2
$3x$		

b.

Factor	a	-1
$2a$	$2a^2$	$-2ab$

24. a.

Multiply	y	$+3$
$-2y$		

b.

Factor		
	$8x$	$-2x^2$

25. a. Find the area.

$2x + 4$

3	

b. Factor to find the dimensions.

$15y + 10$

26. a. Find the area.

$a + 3$

4	

b. Factor to find the dimensions.

$p + prt$

27. a. Find the area.

$x + 5$

x	

b. Factor to find the dimensions.

$ab + bc$

28. a. Find the area.

$y + 10$

y	

b. Factor to find the dimensions.

$x^2 + xy$

In Exercises 29 and 30, multiply to remove the parentheses.

29. a. $-2(3 - 4x)$ **b.** $-a(b - c)$

    c. $x(x^2 - 2x - 3)$

30. a. $-6(x - 4)$ **b.** $-x(x - 3)$

    c. $a(a^2 - 3a + 2)$

In Exercises 31 and 32, name the greatest common factor and then factor.

31. a. $6x - 54$ **b.** $15x - 225$ **c.** $6a^2 - 8a$

32. a. $5x - 75$ **b.** $11x - 121$ **c.** $10b^2 - 15b$

In Exercises 33 and 34, do the operations, naming each property (associative property for addition, associative property for multiplication, commutative property for addition, or commutative property for multiplication) as it is used.

33. a. $4 + (-2) + (-8) + 6$ **b.** $-5(7)(-2)(3)$

    c. $5 + (-2) + 15 + 12$ **d.** $\frac{1}{2}(19)(20)$

34. a. $-6 + 7 + 23 + 16$ **b.** $6(5)(-2)(-7)$

    c. $9(25)(4)\left(\frac{1}{2}\right)$

    d. $8 + (-7) + (-3) + 22$

Write the value of each side of the statements in Exercises 35 and 36. Then write $>$, $=$, or $<$ between the values.

35. a. $\frac{1}{2} + \frac{1}{2}$ □ $\frac{1}{2} \cdot \frac{1}{2}$ **b.** $\frac{1}{2} \div \frac{1}{2}$ □ $\frac{1}{2} + \frac{1}{2}$

    c. $\frac{1}{4} + \frac{1}{4}$ □ $\frac{1}{2} \cdot \frac{1}{2}$ **d.** $\frac{1}{2} - \frac{1}{4}$ □ $\frac{7}{8} - \frac{3}{8}$

    e. $0.3 \cdot 0.5$ □ $0.4 \cdot 0.4$

    f. $1.5 - 3.5$ □ $-2.5 - (0.5)$

36. a. $\frac{3}{4} \div \frac{1}{4}$ □ $\frac{2}{3} \div \frac{1}{3}$ **b.** $\frac{5}{8} - \frac{1}{4}$ □ $\frac{1}{4} \cdot \frac{1}{2}$

    c. $\frac{3}{4} + \frac{1}{4}$ □ $\frac{2}{3} + \frac{1}{3}$ **d.** $\frac{3}{8} + \frac{1}{8}$ □ $\frac{2}{3} - \frac{1}{3}$

    e. $0.8 + 0.7$ □ $2.5 - 0.8$

    f. $0.3 - 0.7$ □ $0.5 - 1.1$

In Exercises 37 to 50, complete the table for each inequality.

	Inequality	Number Line	Inequality in Words
37.	$x \leq 2$		x is less than or equal to 2
38.			
39.	$-1 < x < 5$		
40.			
41.	$x \geq -1$		
42.	$y < -2$		

	Inequality	Number Line	Inequality in Words
43.		$\leftarrow$ + + ○ + + + + + ○ + + $\rightarrow$ $-4\ -3\ -2\ -1\ \ 0\ \ 1\ \ 2\ \ 3\ \ 4\ \ 5\ \ 6$	
44.		$\leftarrow$ + + + + ● + ● + + + $\rightarrow$ $-4\ -3\ -2\ -1\ \ 0\ \ 1\ \ 2\ \ 3\ \ 4\ \ 5\ \ 6$	
45.		$\leftarrow$ + + + + + ○ + ○ + + $\rightarrow$ $-4\ -3\ -2\ -1\ \ 0\ \ 1\ \ 2\ \ 3\ \ 4\ \ 5\ \ 6$	
46.		$\leftarrow$ + + + ○ + + + + ○ + + $\rightarrow$ $-4\ -3\ -2\ -1\ \ 0\ \ 1\ \ 2\ \ 3\ \ 4\ \ 5\ \ 6$	
47.		$\leftarrow$ + + + ○ + + + ○ + + + $\rightarrow$ $-6\ -5\ -4\ -3\ -2\ -1\ \ 0\ \ 1\ \ 2\ \ 3\ \ 4$	
48.		$\leftarrow$ + + + + + + + + + + + $\rightarrow$ $-6\ -5\ -4\ -3\ -2\ -1\ \ 0\ \ 1\ \ 2\ \ 3\ \ 4$	
49.		$\leftarrow$ + ● + + + + ● + + + $\rightarrow$ $-6\ -5\ -4\ -3\ -2\ -1\ \ 0\ \ 1\ \ 2\ \ 3\ \ 4$	
50.		$\leftarrow$ + + + ● + + + + + ● + $\rightarrow$ $-5\ -4\ -3\ -2\ -1\ \ 0\ \ 1\ \ 2\ \ 3\ \ 4\ \ 5$	

51. The statement $4(x - 3) = 4 \cdot x - 4 \cdot 3 = 4x - 12$ contains a subtraction. Explain why we can apply the distributive property for multiplication over addition to an expression containing a subtraction.

52. The statement $\dfrac{x + 5}{2} = \dfrac{x}{2} + \dfrac{5}{2}$ contains a division. Explain why we can apply the distributive property for multiplication over addition to an expression containing a division.

1.1   Input-Output Tables and Expressions in Algebraic Notation ___

OBJECTIVES

- Build tables and find patterns from numerical data.
- Identify variables, constants, and numerical coefficients in algebraic expressions.
- Add like terms.
- Use the simplification property of fractions to find equivalent fractions.
- Evaluate expressions and formulas.

WARM-UP

How does each number shown below the design describe the design? Draw the next design in the pattern. What is the next number in each pattern?

1. Pattern is 1, 3, 5, 7, 9, ___ .

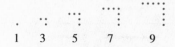

$\ \ \ 1\quad\ \ 3\quad\ \ \ 5\quad\ \ \ 7\quad\ \ \ 9$

2. Pattern is 4, 7, 10, 13, ___ .

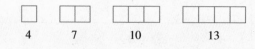

$\ \ \ \ 4\qquad\ 7\qquad\ \ 10\qquad\quad 13$

3. Pattern is 1, 3, 6, 10, ___.

1   3   6   10

4. Pattern is 3, 6, 9, 12, ___.

3   6   9   12

5. Pattern is 3, 5, 7, 9, ___.

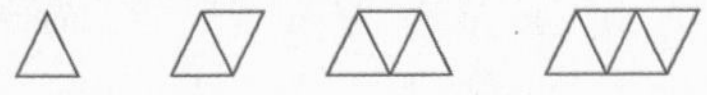

3   5   7   9

THE REVIEW IN Section 1.0 focused on sets of numbers, number names, and inequalities. After a discussion of Polya's problem-solving steps, this section progresses from a numerical emphasis (input-output tables) to three basic algebraic techniques: adding like terms, simplifying expressions, and evaluating expressions and formulas.

Polya's Problem-Solving Steps

The problem-solving steps introduced in this section are based on those proposed by George Polya in his book *How to Solve It,* first published in 1945 by Princeton University Press. Polya earned his Ph.D. in mathematics in 1912, taught for 26 years at the Swiss Federal Institute of Technology, and left Europe in 1940 because "Hitler was too close." He retired in 1953, at age 65, after teaching at Stanford. Polya was a prolific writer and an active educator and speaker well into his 90s. *How to Solve It* has been published in 15 languages and is still in print.

Polya suggested that the elements of problem solving can be summarized by understanding the problem, making a plan, carrying out the plan, and then checking the solution.

- To *understand the problem,* we must consider the assumptions. In the Warm-up, we are not told what we are counting. We assume that the number below each design somehow describes the design and think about what feature of the design is being described.

- Our *plan* includes finding a pattern, either from our past experience or by looking for changes in the designs or numbers. How does each design differ from the previous design? How does each number differ from the previous number?

- As we *carry out the plan,* we look to see that we are satisfying the assumptions.

- Finally, we *check the solution.* We return to the original problem. Does the numerical answer match the answer found from the designs? Are there other processes that confirm the result?

EXAMPLE **1**   Practicing problem solving   Apply the problem-solving steps to the first Warm-up pattern (repeated in Figure 4), and find the next number.

1   3   5   7   9

Figure 4

Solution   ***Understand:*** We assume that the numbers represent the numbers of dots. We assume that the pattern in the designs is from left to right.

Plan: We predict that the next design will be a corner shape with more dots. We look at the change in numbers of dots. We calculate the next number in the pattern by finding the difference between the numbers.

Carry out the plan: There are 2 dots added to each design as we move from left to right. The next design would have 11 dots. Each number increases by 2 as we move from left to right. Adding 2 to 9 gives 11. We may see the change by subtracting the numbers:

$$1, 3, 5, 7, 9, 11, \ldots$$
$$\lor \; \lor \; \lor \; \lor$$
$$2, 2, 2, 2, 2, \ldots$$

Check: The numbers are the *odd numbers*. The next odd number is 11, which matches our earlier work.   ●

Think about it 1: The 50th design in the pattern in Example 1 would contain how many dots?

EXAMPLE **2**   Practicing problem solving   Apply the problem-solving steps to the second Warm-up pattern (repeated in Figure 5), and find the next number.

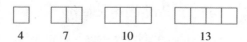

4   7   10   13

Figure 5

Solution   ***Understand:*** We assume that the numbers do not represent the numbers of squares, for that pattern would be 1, 2, 3, 4, not 4, 7, 10, 13. The designs contain squares, and a square has four sides. Perhaps the numbers have to do with the sides or line segments that make up the square.

Plan: We will count the number of line segments that make up each design. We will look for how the designs change and see whether that matches the change in the numbers.

Carry out the plan: The designs contain 4, 7, 10, and 13 line segments. Three segments are added each time as we move from left to right. The change between the numbers is also 3. The next design would have 16 line segments, and adding 3 to 13 gives 16 in the number pattern.

$$4, 7, 10, 13, 16, \ldots$$
$$\lor \; \lor \; \lor \; \lor$$
$$3, \; 3, \; 3, \; 3, \ldots$$

Check: We verify our work by adding 1 more square. The design with 5 squares will have 12 outer segments and 4 inner ones, for a total of 16 segments, which confirms our earlier work.   ●

Think about it 2: The 50th design in the pattern in Example 2 would contain how many line segments?

Input-Output Tables

Continuing the designs or lists of numbers is time consuming and thus is not a good way to find the nature of the 50th or 100th design in a pattern. A more useful approach is to write the numbers in an input-output table and focus on the relation between the design and its position in the set of designs.

An **input-output relationship** is *a rule, given in words, numbers, symbols, or algebraic equations, that associates sets of numbers.* The rule describes how to get from one set of numbers (the inputs) to the other set of numbers (the outputs).

In Examples 1 and 2, the inputs are the position, or order, in which the designs appear (1, 2, 3, 4, ...). The outputs are the numbers in the patterns. To build an input-output table, we list the inputs and outputs in columns.

EXAMPLE Building input-output tables   Write the number patterns in Examples 1 and 2 in an input-output table. Suggest ways of relating the input and outputs. Predict the output for an input of 100.

Solution   Table 2 is the input-output table for Example 1; Table 3 is the input-output table for Example 2.

Input: Position	Output: Number of Dots
1	1
2	3
3	5
4	7
5	9
6	11

Table 2

Input: Position	Output: Number of Line Segments
1	4
2	7
3	10
4	13
5	16

Table 3

In Table 2, the outputs are about twice the inputs. As we look down the table, we see that the output numbers change by 2 while the inputs change by 1. We might predict that the rule for the table would be the input multiplied by 2.

In Table 3, the outputs are about three times the inputs. As we look down the table, we see that the output numbers change by 3 while the inputs change by 1. We might predict that the rule for the table would be the input multiplied by 3.

See the Answer Box for the rules and for the outputs associated with an input of 100. ●

EXAMPLE Making a table and finding a rule   Make an input-output table for the third Warm-up pattern (repeated in Figure 6), suggest a rule, and predict the output for an input of 100.

```
 .    .:    .:.    .::.    .::.
      :.   .:.   .:::.   .:::::.
 1    3     6     10      15
```

Figure 6

Solution   To build the input-output table in Table 4, we use Figure 6 to match the number of dots (outputs) with a position (inputs).

Then we find the change in the outputs. In the table, we write the change between the output numbers, in a column to the right of the outputs. The outputs

Input: Position	Output: Number of Dots	Change in Output
1	1	
		〉 2
2	3	
		〉 3
3	6	
		〉 4
4	10	
		〉 5
5	15	

Table 4

are not a consistent amount apart, as they were in Examples 1 and 2. Thus, we need a different kind of rule. The output numbers get large quickly, so we might assume that the input is being multiplied, but not by the same number each time. Because multiplication is associated with the area of rectangles, the triangular arrangement of the dots suggests multiplication followed by division by 2. See the solution in the Answer Box.   ●

Vocabulary for Writing and Describing Algebraic Notation

We now review the words and symbols needed to describe operations and rules in algebraic notation.

Answers to Operations

> **A sum** is the answer to an addition problem.
>
> **A difference** is the answer to a subtraction problem.
>
> **A product** is the answer to a multiplication problem.
>
> **A quotient** is the answer to a division problem.

Unless otherwise indicated, assume that the order of the numbers for differences and quotients is that listed in the problem. The difference between a and b is $a - b$; the quotient of a and b is $a \div b$.

ADDITION AND SUBTRACTION   *Plus, added to, greater than, more than,* and *increased by* are words and phrases that describe addition. Subtraction words include *less, less than, fewer, minus,* and *decreased by.* If the word *than* appears in a subtraction setting, the order of the numbers is the reverse of the word order; for example, six less than the input, n, is $n - 6$.

EXAMPLE 5   Writing algebraic notation from words   Write these sentences in algebraic notation. Let x be the input number. Let y be the output number. Use the equal sign, =, to replace "is."

a. The output is the sum of the input and six.

b. The output is the quotient of the input and eight.

c. The difference between the input and output is four.

d. The output is five less than the product of the input and seven.

e. The input decreased by three gives the output.

Solution   See the Answer Box. (*Hint:* The five in part d follows the product of the input and seven.)   ●

EXAMPLE **6**   Writing algebraic notation in words   Write these algebraic statements in words. Let y be the output and x be the input.

a. $y = x - 4$ **b.** $y = 8x$ **c.** $x + y = 12$ **d.** $\dfrac{x}{y} = 10$

Solution   Three answers are given for each part. Which statement in each set is not correct?

a. The output is the input decreased by four.
The output is the difference between the input and four.
The output is the input less than four.

b. The output is the product of eight and the input.
Multiply the input by eight to get the output.
The product of the output and eight gives the input.

c. The sum of the input and output is twelve.
Add the input and the output to get twelve.
The product of the input and output is twelve.

d. The quotient of the input and output is ten.
Divide the input by the output to get ten.
The quotient of the output and the input is ten.

See the Answer Box.   ●

MULTIPLICATION AND DIVISION   *Double a number* and *twice a number* mean multiplication by 2, while *triple a number* means multiplication by 3. The words *at … each* suggest multiplication; for example, "5 candied apples at $3.59 each" means to multiply 5 times $3.59. The word *of* means multiplication in fraction, decimal, and percent notation; for example, $\frac{1}{4}$ of 8 means $\frac{1}{4}$ times 8. *Half of twelve* means $\frac{1}{2}$ times 12. *Divide 8 in half* means $8 \div 2 = 4$, whereas *divide 8 by a half* means $8 \div \frac{1}{2} = 8 \cdot 2 = 16$.

EXAMPLE **7**   Writing algebraic notation from words   Write these sentences in algebraic notation. Let x be the input and y be the output. Use the equal sign, $=$, to replace "is."

a. The sum of twice the input and six is the output.
b. The output is the difference between half of the input and four.
c. The output is the total cost of x cans of frozen juice at $1.49 each, less a $0.30 manufacturer's coupon.
d. The difference between seven and triple the input is the output.

Solution   **a.** $2x + 6 = y$ **b.** $y = \frac{1}{2}x - 4$
c. $y = 1.49x - 0.30$ **d.** $7 - 3x = y$ ●

EXAMPLE **8**   Writing algebraic notation in words   Write these algebraic statements in words. Let x be the input and y be the output.

a. $y = 3x + 4$ **b.** $\frac{1}{2}x - 5 = y$ **c.** $y = 4 - 2x$ **d.** $y = \dfrac{x - 3}{5}$

Solution   A set of three answers is given. Which statement in each set is not correct?

a. The output is three times the sum of the input and four. $3(x+4)$ ✗
The output is four added to three times the input. $y = 3x + 4$
The output is the sum of four and three times the input.
$y = 4 + 3x$

    b. Half the difference between the input and five is the output.
       Half the input less five is the output.
       The difference between half the input and five is the output.

    c. The output is four less than twice the input.
       The output is the difference between four and twice the input.
       The output is twice the input subtracted from four.

    d. The output is five divided by the difference between the input and three.
       The output is the quotient of the difference between the input and three, and five.
       The output is the quotient of three less than the input, and five.

See the Answer Box. ●

ALGEBRAIC NOTATION   The numbers and letters in algebraic notation have special names.

Selected Words in Algebraic Notation

> **A variable** is a letter or symbol that can represent any number from some set.
>
> **A constant** is a number, letter, or symbol whose value is fixed.
>
> **A numerical coefficient** is the sign and number multiplying the variable(s).

An **expression** is *any combination of signs, numbers, constants, and variables with operations such as addition, subtraction, multiplication, or division.*

EXAMPLE **9** Identifying constants, variables, and numerical coefficients   For each expression listed, name the constants, the variables, and the numerical coefficients.

    a. $2x + 6$ **b.** $1.49x - 0.30$ **c.** $7 - x$

Solution   **a.** 2 is a constant and the numerical coefficient of the variable x; 6 is a constant.

    b. 1.49 is a constant and the numerical coefficient of the variable x; -0.30 is a constant. The subtraction sign goes with the 0.30.

    c. 7 is a constant. The numerical coefficient of the variable x is -1. There is an assumed 1 in front of the x, and the subtraction sign goes with the variable x. ●

Algebraic Techniques

Many steps in algebra require that we see how expressions are alike and how they are different. To distinguish expressions such as $x + 2$, $2x$, and x^2, it may be helpful to associate them with physical objects. In Example 10, let x be the side of the large square, let 1 be the side of the small square, and let x and 1 be the sides of the rectangle.

EXAMPLE **10** Finding area

    a. What is the area of each rectangle or square shown in Figure 7?
    b. What is the total area of the figures?

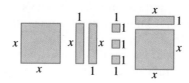

Figure 7

Solution   **a.** The large squares have area x^2 square units. The rectangles have area $x \cdot 1 = x$ square units. The small squares have area $1 \cdot 1 = 1$ square unit.

   b. There are two x^2 tiles, three x tiles, and three 1 tiles, so the total area is $2x^2 + 3x + 3$.   ●

In writing the total area of the tiles in Example 10, it is sensible to count the x^2 tiles together because they describe the same shape. The rectangles are a different shape, so they cannot be counted with either the x^2 tiles or the small square tiles.

ADDING LIKE TERMS   A **term** is *the product of a number (with or without a sign) and one or more variables (with or without exponents).* **Like terms** *have identical variable factors.* The large square tiles model one group of like terms. The rectangles model another set of like terms. The small square tiles model still a third type of like terms.

   When we count the like terms (like tiles) and write a single term to describe the total count, we are *adding like terms.*

EXAMPLE **11**   Adding like terms   Find the like terms and add them.

   a. $a + 2ab + 3ab - 4ba + 5b^2$ **b.** $x^2 + 2x - 3x^2 + 4x - 5$
   c. $2\pi r^2 + 2\pi rh$ **d.** $x^3 - 2x^2 + x - x^2 + 2x - 1$

Solution   **a.** $a + ab + 5b^2$. The terms containing ab and ba have the same variables as factors and are like terms.

   b. $-2x^2 + 6x - 5$

   c. Not like terms: 2 and π are constants, the variables are r^2 and rh and are not identical factors.

   d. $x^3 - 3x^2 + 3x - 1$

The -5 in part b and the -1 in part d *have no variables* and thus are called **constant terms.**   ●

SIMPLIFYING EXPRESSIONS   Recall that **simplify** is a short way of writing directions. Simplify has several meanings. In the last section, simplify meant to *apply the properties of real numbers or the order of operations to an expression.* Simplify also means *to add like terms* or *to change fractions to lowest terms* using the simplification property of fractions.

   When convenient, we write the reciprocal of a as $1/a$, $a \neq 0$. Because the product of a number and its reciprocal is 1, the expression

$$a \cdot \frac{1}{a} = \frac{a}{a} = 1$$

leads to the simplification property of fractions.

Simplification Property of Fractions

For all real numbers, $a \neq 0$, $c \neq 0$,

$$\frac{ab}{ac} = \frac{a}{a} \cdot \frac{b}{c} = 1 \cdot \frac{b}{c} = \frac{b}{c}$$

The simplification property says that if the numerator and denominator of any fraction contain common factors, these factors may be eliminated, giving a simpler fraction of the same value.

To simplify a fraction, we change the numerator and denominator to factors and eliminate common factors of the form a/a. This is called *changing to lowest terms.*

EXAMPLE

Changing to lowest terms   Identify the common factors in the numerator and denominator of each fraction, and simplify to lowest terms.

a. $\dfrac{16}{48}$ **b.** $\dfrac{450}{60}$ **c.** $\dfrac{25}{36}$ **d.** $\dfrac{abc}{cde}$ **e.** $\dfrac{xy + y^2}{y}$ **f.** $\dfrac{x + 3}{3}$

Solution   **a.** $\dfrac{16}{48} = \dfrac{\overline{(2 \cdot 2 \cdot 2 \cdot 2)}}{\overline{(2 \cdot 2 \cdot 2 \cdot 2)} \cdot 3} = \dfrac{1}{3}$ **b.** $\dfrac{450}{60} = \dfrac{②\cdot③\cdot 3 \cdot 5 \cdot ⑤}{②\cdot 2 \cdot ③\cdot ⑤} = \dfrac{15}{2}$

c. $\dfrac{25}{36} = \dfrac{5 \cdot 5}{6 \cdot 6} = \dfrac{5 \cdot 5}{2 \cdot 2 \cdot 3 \cdot 3}$

There are no common factors in 25 and 36, so $\frac{25}{36}$ cannot be simplified.

d. $\dfrac{a \cdot b \cdot ⓒ}{ⓒ \cdot d \cdot e} = \dfrac{ab}{de}$

e. $\dfrac{y(x + y)}{y} = \dfrac{x + y}{1} = x + y$

f. $\dfrac{x + 3}{3}$; the numerator and denominator have no common factors and so cannot be simplified.   ●

Parts c and f of Example 12 illustrate that *if there are no common factors in the numerator and denominator, the fraction cannot be simplified.*

EVALUATING EXPRESSIONS AND FORMULAS   Like *simplify*, *evaluate* is another short way of writing directions. To **evaluate**, *substitute numbers in place of the variables in expressions or formulas.*

EXAMPLE 13

Evaluating expressions   Evaluate these expressions.

a. $x^2 - 2x + 1$ for $x = 8$

b. $a^2 - 2ab + b^2$ for $a = 9$ and $b = 2$

c. $x^2 - y^2$ for $x = 9$ and $y = 2$

d. $\sqrt{a^2 + b^2}$ for $a = 3$ and $b = 4$

Solution   **a.** $x^2 - 2x + 1 = 64 - 16 + 1 = 49$

b. $a^2 - 2ab + b^2 = 81 - 36 + 4 = 49$

c. $x^2 - y^2 = 81 - 4 = 77$

d. $\sqrt{a^2 + b^2} = \sqrt{9 + 16} = \sqrt{25} = 5$ ●

When evaluating formulas, write the formula and then substitute the numbers before doing any calculation. Estimate the answer, especially if you are using a calculator. Your calculator work will be more accurate if you enter the entire expression rather than finding values of parts of it at a time. The following steps summarize these ideas.

To evaluate a formula:

1. State the formula and the information given.
2. Substitute the given information into the formula.
3. Estimate the result.
4. Find the result, using a calculator as needed. Note any special keystrokes.

EXAMPLE **14**

Evaluating formulas   Look in Appendix 1, as needed, for the formulas. Use a calculator, indicating any special keystrokes.

a. Find the volume of a golf ball (sphere) with a diameter of 1.68 inches.

b. Find the area of a trapezoid whose parallel sides are 5 feet and 8 feet and height is 4 feet.

c. Find S where $S = \dfrac{n}{2}[2a + (n - 1)d]$, $n = 50$, $a = 1$, and $d = 3$.

d. Find C where $C = \frac{5}{9}(F - 32)$ and $F = 212$.

e. Find z^2 where $z^2 = x^2 + y^2$, $x = 12$, and $y = 35$.

Solution   **a.** $V = \dfrac{4\pi}{3}\left(\dfrac{d}{2}\right)^3$, $d = 1.68$ in.

$$V = \dfrac{4\pi}{3}\left(\dfrac{1.68}{2}\right)^3$$

We estimate by rounding to the nearest whole number. For $\pi \approx 3$ and $(1.68/2) \approx 1$,

$$\dfrac{4(3)}{3}(1)^3 \approx 4$$

On a calculator,

4 $\boxed{\text{2nd}}$ $\pi/3$ * (1.68/2) $\boxed{\wedge}$ 3 =

Note the multiplication sign between the division by 3 and the parentheses around 1.68/2. Without the multiplication sign, some calculators recognize an "implied" multiplication first. These calculators multiply 3 times (1.68/2)^3 first and then do the division between 4π and 3.
   The result is

$$V \approx 2.48 \text{ in}^3$$

b. If a and b are the parallel sides and h is the height,

$$A = \tfrac{1}{2}h(a + b), \ a = 5 \text{ ft}, \ b = 8 \text{ ft}, \ h = 4 \text{ ft}$$
$$A = \tfrac{1}{2}(4 \text{ ft})(5 \text{ ft} + 8 \text{ ft})$$

We can do this problem mentally, omitting the estimate and calculator steps.

$$A = \tfrac{1}{2}(4)(5 + 8) \text{ ft}^2$$
$$A = 26 \text{ ft}^2$$

c. $S = \dfrac{n}{2}[2a + (n - 1)d]$, $n = 50$, $a = 1$, $d = 3$

$$S = \tfrac{50}{2}[2(1) + (50 - 1) \cdot 3]$$

We estimate

$$\tfrac{1}{2}(50)(50)(3) = \tfrac{1}{2}(7500) = 3750$$

On a calculator,

$$S = 50/2 * (2 + (49) * 3)$$

The result is

$$S = 3725$$

d. $C = \tfrac{5}{9}(F - 32),\ F = 212$
   $C = \tfrac{5}{9}(212 - 32)$

We estimate

$$\tfrac{1}{2}(180) = 90$$

On a calculator,

$$5/9 * (212 - 32)$$

The result is

$$C = 100$$

e. $z^2 = x^2 + y^2,\ x = 12,\ y = 35$
   $z^2 = 12^2 + 35^2$
   $z^2 = 144 + 1225 = 1369$

●

ANSWER BOX

Warm-up: 1. number of dots; 11 **2.** number of line segments; 16 **3.** number of dots in triangular array; 15 **4.** number of dots in rectangular array; 15 **5.** number of line segments; 11 **Think about it 1:** The 50th design would contain 99 dots. **Think about it 2:** The 50th design would contain 151 line segments. **Example 3:** The rule in Example 1 is "one less than twice the input." For an input of 100, the output is 199. The rule in Example 2 is "three times the input plus 1." For an input of 100, the output is 301. **Example 4:** The rule in Example 4 is "multiply the input by one more than the input, and then divide by two." The process for finding this rule will be described in Section 4.1, Example 3. For an input of 100, the output is $100(101)/2 = 5050$. **Example 5: a.** $y = x + 6$ **b.** $y = x/8$ **c.** $x - y = 4$ **d.** $y = 7x - 5$ **e.** $x - 3 = y$ **Example 6:** In each case, the last of the three statements is incorrect. **Example 8:** In each case, the first of the three statements is incorrect.

EXERCISES 1.1

In Exercises 1 to 4,

a. *Make an input-output table for the patterns.*

b. *Find the change in the outputs for each table.*

c. *Describe each table with a rule.*

d. *Predict the output for 50 as an input and for 100 as an input.*

1. The input is the set of natural numbers. The output is the number of dots. The output is 30 for the input 10.

Input: 1   2   3   4

2. The input is the set of natural numbers. The output is the number of line segments. The output is 21 for the input 10.

Input: 1   2   3   4

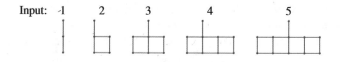

3. The input is the set of natural numbers. The output is the number of line segments. The output is 29 for the input 10.

Input: 1   2   3   4   5

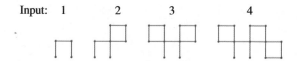

4. The input is the set of natural numbers. The output is the number of line segments. The output is 39 for the input 10.

Input: 1   2   3   4

In Exercises 5 to 12, write the sentences in algebraic notation.

5. Ten decreased by twice the input gives the output.

6. Twice the input increased by five gives the output.

7. The output is ten more than half the input.

8. The output is half the input decreased by six.

9. Two more than the quotient of the input and fifteen gives the output.

10. The output is sixteen less than the product of three and the input.

11. The output is fourteen fewer than the product of seven and the input.

12. Twelve added to the quotient of four and the input is the output.

In Exercises 13 to 20, write the algebraic statements in words. Let x be the input and y be the output.

13. $y = 2x - 5$

14. $y = \frac{1}{2}x + 7$

15. $x + y = 11$

16. $x - y = 40$

17. $2x - y = 5$

18. $3x + y = 10$

19. $\frac{x}{8} = y$

20. $y = \frac{x}{5} + 2$

In Exercises 21 to 26, write the applications in algebraic notation.

21. The output is the total cost of x cans of soup at \$1.29, less a \$0.45 manufacturer's coupon.

22. The output is the total cost of renting the Rug Doctor cleaning system for x hours at \$8.95 per hour, plus \$12 in cleaning solution.

23. You put \$1.25 into a parking meter. The output is the value of your time remaining on the meter after x minutes at \$0.05 per minute.

24. The output is the amount remaining on a \$20 phone card after you talk for x minutes at \$0.35 per minute.

25. The output is the total cost of taking x credit hours at \$125 per hour tuition and paying \$85 in fees.

26. The cell phone charge is \$9.50 per month and \$0.05 per minute. What is the total cost for x minutes of service each month?

In Exercises 27 to 29, match each vocabulary word with its definition. Record any unfamiliar words and their definitions in your study notes. Some words should be familiar from other courses.

27. Choose from
constant, even numbers, quotient, evaluate, input numbers

a. Substitute numbers into an expression or formula

b. Integers divisible by 2

c. The numbers in the first column of an input-output table

d. The answer to a division

e. A number, letter, or symbol whose value is fixed

28. Choose from
odd numbers, difference, lowest terms, numerical coefficient, like terms

a. The answer to a subtraction

b. Expressions with identical variable factors

c. Integers that are not divisible by 2

d. The sign and number by which a variable is multiplied

e. A fraction containing no common factors

29. Choose from
variable, expression, sum, product, square of x

a. The answer to a multiplication problem

b. A letter that can take on any number from a given set

c. Any combination of numbers and variables with operations

d. The number x multiplied by itself

e. The answer to an addition problem

30. *Simplify* does not include which of these directions?

    a. Apply the properties of real numbers.

    b. Apply the order of operations.

    c. Add like terms.

    d. Divide both sides of an equation by the same number.

    e. Change a fraction to lowest terms.

In Exercises 31 and 32, identify the variables, numerical coefficients, and constant term in each expression.

31. a. $2x + 3$ **b.** πr^2

    c. $4 - x$ **d.** $x^2 + x - 1$

32. a. $3x - 2$ **b.** $2\pi r$

    c. $9 - x$ **d.** $x^2 - x + 1$

33. Add like terms in the figure below.

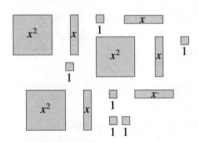

34. Add like terms in the figure below.

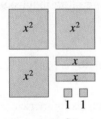

35. Add like terms in the figure below.

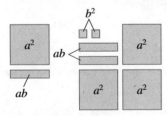

36. Add like terms in the figure below.

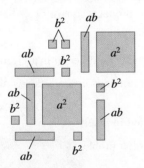

In Exercises 37 to 40, add like terms.

37. a. $3a - 2b + 4b - 2a$

    b. $6x^2 + 2x - 4x^2 - 3$

38. a. $4ab - 2ac + 3ba - 4ac$

    b. $6x^2 + 4x - 4x - x^2$

39. a. $6ac - 3ab + 2ca + 7bc$

    b. $x(4 + 2x) - x(x - 4)$

40. a. $7b - 4c + 3a - 6b$

    b. $x(3x - 1) - x(4 - 2x)$

In Exercises 41 to 46, name the greatest common factor and simplify the fraction by changing to lowest terms.

41. a. $\dfrac{15}{45}$ **b.** $\dfrac{48}{160}$ **c.** $\dfrac{5280}{3600}$

42. a. $\dfrac{84}{35}$ **b.** $\dfrac{52}{39}$ **c.** $\dfrac{1024}{256}$

43. a. $\dfrac{abe}{ben}$ **b.** $\dfrac{ab + bc}{b}$ **c.** $\dfrac{x - 2}{2}$

44. a. $\dfrac{cot}{ox}$ **b.** $\dfrac{ab + ac}{a}$ **c.** $\dfrac{x - 2}{x}$

45. a. $\dfrac{xy + xz}{xy}$ **b.** $\dfrac{x^2 + 2x}{2x}$ **c.** $\dfrac{xy^2 - 2x^2y}{2xy}$

46. a. $\dfrac{ab + bc}{ab}$ **b.** $\dfrac{y^2 + 3y}{3y}$ **c.** $\dfrac{2ab^2 - 2a^2b}{2ab}$

Evaluate the expressions in Exercises 47 to 52 for the given numbers.

47. a. $x^2 + 2x + 1$ for $x = 7$ **b.** $(x + 1)^2$ for $x = 7$

48. a. $x^2 + 9$ for $x = 5$ **b.** $(x + 3)^2$ for $x = 5$

49. a. $a^2 - b^2$ for $a = 9$, $b = 7$

    b. $(a - b)^2$ for $a = 9$, $b = 7$

50. a. $a^2 + 2ab + b^2$ for $a = 12$, $b = 3$

    b. $(a + b)^2$ for $a = 12$, $b = 3$

51. a. $9 - 3(x - 2)$ for $x = 8$

    b. $6(x - 2)$ for $x = 8$

52. a. $x^2 - x(4 - x)$ for $x = 5$

    b. $x(4 - x)$ for $x = 5$

Evaluate the formulas in Exercises 53 to 60 for the given numbers. Round to the nearest tenth.

53. Celsius temperature: $C = \frac{5}{9}(F - 32)$

    a. $F = 32$ **b.** $F = -40$ **c.** $F = 98.6$

54. Sum of a set of numbers: $S = \dfrac{n}{2}[2a + (n - 1)d]$

    a. $n = 100$, $a = 1$, $d = 2$

b. $n = 100$, $a = 2$, $d = 2$

c. $n = 100$, $a = 5$, $d = 5$

55. Last number in a set of numbers: $L = a + (n - 1)d$

    a. $n = 100$, $a = 1$, $d = 2$

    b. $n = 100$, $a = 2$, $d = 2$

    c. $n = 100$, $a = 5$, $d = 5$

56. Fahrenheit temperature: $F = \frac{9}{5}C + 32$

    a. $C = 100$ **b.** $C = 37$ **c.** $C = 20$

57. Volume of a sphere: $V = \frac{4}{3}\pi r^3$

    a. $r = 1.5$ cm   **b.** $r = 3$ cm   **c.** $r = 6$ cm

58. Surface area of a sphere: $S = 4\pi r^2$

    a. $r = 1.5$ cm   **b.** $r = 3$ cm   **c.** $r = 6$ cm

59. Distance between points (a, b) and (c, d):
$$D = \sqrt{(a - c)^2 + (b - d)^2}$$

    a. $(6, 3)$ and $(-2, -3)$

    b. $(14, 2)$ and $(2, -3)$

    c. $(-4, 3)$ and $(11, -5)$

60. Area of a triangle with sides a, b, and c:
$$S = \tfrac{1}{2}(a + b + c) \text{ and}$$
$$A = \sqrt{S(S - a)(S - b)(S - c)}$$

    a. $a = 3$, $b = 4$, $c = 5$; find S first.

    b. $a = 6$, $b = 8$, $c = 10$; find S first.

    c. $a = 5$, $b = 12$, $c = 13$; find S first.

Project

61. *Baking Time.*   A recipe on a box of gingerbread cake suggests baking the cake for 30 to 35 minutes in a 9-inch square pan or 35 to 40 minutes in an 8-inch square pan. What mathematical reasons are there for the fact that the larger pan takes a shorter baking time? Use reasoning to estimate how long cupcakes should be baked.

1.2   Coordinate Graphs and Equations

OBJECTIVES

* Identify horizontal and vertical axes, quadrants, the origin, and ordered pairs.
* Graph data and describe patterns in graphs.
* Build input-output tables from equations.
* Graph ordered pairs from input-output tables.
* Identify independent and dependent variables.
* Solve equations with tables and graphs.

WARM-UP

1. List the squares of the numbers from 1 to 15.

2. List the powers of 2 from 2^1 to 2^8.

3. What is the value of each of the following?

    a. $(-2)^2$ **b.** -2^2 **c.** $0 - 2^2$ **d.** $0 - (-2)^2$

THIS SECTION STARTS WITH a summary of the vocabulary and concepts needed to graph data on rectangular coordinate axes. Data from the wind-chill chart are graphed so that patterns can be observed. The section then introduces building tables and graphs from equations and solving equations from tables and graphs.

Rectangular Coordinate Graphs

VOCABULARY   The following paragraphs summarize the basic concepts and vo-cabulary needed for coordinate graphing.

In rectangular coordinate graphing, the **axes** are *two number lines placed at right angles so that they cross at zero.* The axes are **perpendicular** because *they cross at a 90° angle.* The axes lie on *a flat surface* called the **coordinate plane.**

Quadrants are *the four regions separated by the axes.* The quadrants, shown in Figure 8, are numbered counterclockwise because of applications in engineering and physics.

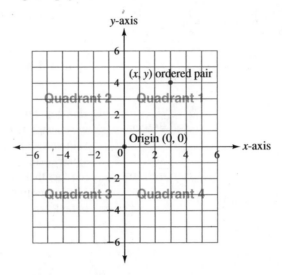

Figure 8

An **ordered pair** is *a pair of numbers (x, y) that uniquely describes each point in the coordinate plane.* The numbers in the ordered pair may be any real numbers (positive, negative, or zero). The **origin,** (0, 0), is at *the intersection of the two axes.*

LOCATING POINTS ON A GRAPH   In the ordered pair (x, y), the first number x, or the **x-coordinate,** *gives a position in the horizontal direction.* The second number y, or the **y-coordinate,** *gives a position in the vertical direction.*

The point labeled A in quadrant 1 of Figure 9 has the ordered pair (3, 2). Horizontally, it is 3 units to the right from the origin. Vertically, it is up 2 units from the origin.

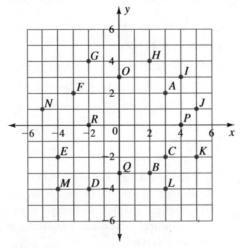

Figure 9

The point with the ordered pair (2, −3) is in quadrant 4. Horizontally, it is 2 units to the right from the origin. Vertically, it is down 3 units from the origin. The point (2, −3) is labeled B.

EXAMPLE **1** Finding and writing ordered pairs

a. Using Figure 9, locate the points named by these ordered pairs:

$$(2, 4), (4, 3), (-2, 4), (-2, -4), (3, -2), (-3, 2), (-4, -2)$$

b. Name the ordered pairs labeled *J* to *R* in Figure 9.

Solution **a.** (2, 4) is *H*; (4, 3) is *I*; (−2, 4) is *G*; (−2, −4) is *D*; (3, −2) is *C*; (−3, 2) is *F*; (−4, −2) is *E*.

b. *J* is (5, 1); *K* is (5, −2); *L* is (3, −4); *M* is (−4, −4); *N* is (−5, 1); *O* is (0, 3); *P* is (4, 0); *Q* is (0, −3); *R* is (−2, 0). ●

The ordered pair (*x*, *y*) is also called the Cartesian coordinates, after René Descartes (1596–1650). Descartes, a French philosopher and soldier, is credited with devising analytic geometry, a blending of algebra and geometry with graphing on the coordinate axes. The graphing calculator has given his contribution to mathematics even more significance to our study.

GRAPHING WIND-CHILL DATA   The next two examples return to the wind-chill apparent temperature chart, introduced on the chapter opener page and reproduced in Table 5.

Wind (miles per hour)

		5	10	15	20	25	30
	35	33	21	16	12	7	5
	30	27	16	11	3	0	−2
	25	21	9	1	−4	−7	−11
Current temperature (°Fahrenheit)	**20**	16	2	−6	−9	−15	−18
	15	12	−2	−11	−17	−22	−26
	10	7	−9	−18	−24	−29	−33
	5	1	−15	−25	−32	−37	−41
	0	−6	−22	−33	−40	−45	−49
	−5	−11	−27	−40	−46	−52	−56

Table 5 Wind-chill apparent temperature chart

EXAMPLE **2** Plotting data   Let *x* be the current temperature in degrees Fahrenheit and *y* be the wind-chill temperature in degrees Fahrenheit.

a. Suppose there is no wind; then the current temperature and the wind-chill temperature are the same. The far left column is both the input and the output for a zero wind. Write ordered pairs (*x*, *y*) for current temperatures of −5°, 5°, 15°, 25°, and 35°.

b. Graph the ordered pairs from part a. Number the axes from −40 to 40, counting by 10.

Solution **a.** (−5, −5), (5, 5), (15, 15), (25, 25), (35, 35)

b. The graph is shown in Figure 10. The ordered pairs, when plotted and connected from left to right, form a straight line.

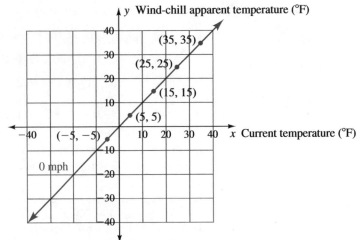

Figure 10

Think about it 1: What is the rule, or equation, for the line in Example 2?

EXAMPLE **3**

Plotting data   Refer to the wind-chill chart in Table 5. Let x be the current temperature and y be the wind-chill apparent temperature for 15 mph.

a. For 15 mph, one ordered pair is (35, 16). Write ordered pairs (x, y) for current temperatures of $-5°$, $5°$, $15°$, and $25°$.

b. Graph the ordered pairs on the same axes as in Figure 10.

c. What is the temperature change for a wind speed of 15 mph at 35°, at 15°, and at $-5°$? (*Hint:* To find the amount and direction of change, subtract the current temperature from the wind-chill apparent temperature.)

d. Complete this sentence with *less* or *greater*: For a given wind speed, the change in temperature is _____ for a colder current temperature than for a warmer current temperature.

Solution   **a.** $(-5, -40)$, $(5, -25)$, $(15, -11)$, $(25, 1)$

b. The graph is shown in Figure 11.

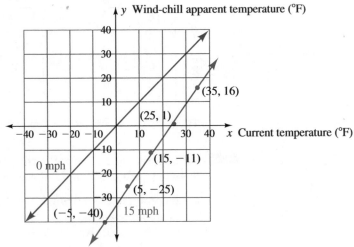

Figure II

c. At 35°, the temperature change is $16° - 35° = -19°$.
At 15°, the temperature change is $-11° - 15° = -26°$.
At $-5°$, the temperature change is $-40° - (-5°) = -35°$.

d. greater; the wind-chill apparent temperature drops more rapidly at colder temperatures.

Think about it 2: Assuming the lines are straight for other temperatures, estimate the temperature at which the lines for winds of 0 mph and 15 mph intersect. What does this intersection mean?

Repeat the graphing in Examples 2 and 3 by plotting ordered pairs on a graphing calculator. The Graphing Calculator Technique box includes steps and suggestions. Consult your calculator manual for details on keystrokes.

LABELING THE AXES   The axes on coordinate graphs may be labeled with any names or letters. If no names or letters are given in an application, use *x* and *y*. Because *x* and *y* are commonly placed on the axes, the *horizontal axis* is often called the **x-axis** and the *vertical axis* is often called the **y-axis**.

If available, always include units (such as $, pounds, months, or years). Write them in parentheses beside or below the name of the axis. In Examples 2 and 3, the axes were labeled with temperatures in degrees Fahrenheit (°F).

When writing numbers on the axes, choose an appropriate **scale**, or *spacing between the numbers*. One estimate for scale is

$$\frac{\text{Highest number} - \text{lowest number}}{10}$$

Round to the nearest easy counting number.

Graphing Calculator Technique:
Plotting Ordered Pairs with
the Statistics Function

> In general, calculators plot data entered in lists under STATISTICS. The procedure is as follows:
>
> Clear or shut off equations in $\boxed{Y =}$ and clear prior lists.
>
> For each (x, y), place x in one list and y in a second list.
>
> Set Xmin and Xmax to include the smallest and largest x.
>
> Set Xscl to 10% of (Xmax − Xmin) rounded to an easy counting number.
>
> Repeat for Ymin, Ymax, and Yscl.
>
> Turn on the statistical plot.
>
> Choose the type of graph.
>
> Choose the source or list for x and for y.
>
> Choose the type of mark for the graph.
>
> Press GRAPH to plot the data.
>
> *Note:* When you have finished, go back to the statistical plot and turn it off. Leaving a statistical plot on may create a "Dimension Mismatch" in the future and keep you from doing other graphing.

Building Input-Output Tables and Graphs from Equations

The rules and patterns for input-output tables (see Section 1.1, page 21) are most commonly written as equations. In the next few examples, we are given the equation. In later examples, we will build the equation.

An input-output table is a listing of some of the ordered pairs for an equation. A graph shows all points (and hence ordered pairs) for an equation in the region pictured by the graph.

Building a Table and Graph
from an Equation

> 1. To build an input-output table given the rule or equation, record selected numbers as inputs. Substitute the numbers for the input variable, evaluate, and record the results as the output.
>
> 2. To find ordered pairs from an input-output table, let the input be the first number in the ordered pair and the output be the second number in the ordered pair.
>
> 3. Graph by plotting the ordered pairs. If the points form a pattern, connect them and identify the resulting shape.

EXAMPLE Building tables and graphs from equations   For the two equations given, make a table with integer inputs, x, from -2 to 2, and graph the resulting input-output pairs. If the points form a recognizable pattern, connect them and identify the resulting shape.

   a. $y = 2x + 1$ **b.** $y = x^2$

Solution **a.** The table and graph are shown in Table 6 and Figure 12.

Input x	Output $y = 2x + 1$
-2	$2(-2) + 1 = -3$
-1	$2(-1) + 1 = -1$
0	$2(0) + 1 = 1$
1	$2(1) + 1 = 3$
2	$2(2) + 1 = 5$

Table 6

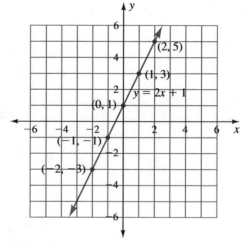

Figure 12

b. The table and graph are shown in Table 7 and Figure 13.

Input x	Output $y = x^2$
-2	$(-2)^2 = 4$
-1	$(-1)^2 = 1$
0	$(0)^2 = 0$
1	$(1)^2 = 1$
2	$(2)^2 = 4$

Table 7

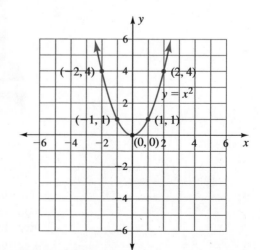

Figure 13

The points in the figure for part a appear to lie on a straight line. We connect the points left to right and extend the resulting line with arrows to indicate its infinite length. The points in the figure for part b appear to lie on a smooth curve called a **parabola**. We again connect the points left to right and extend the curve with arrows to indicate that it continues infinitely. ●

Think about it 3: Look at the graph for $y = 2x + 1$. Find an ordered pair for a new point on the graph and, with substitution, show that the ordered pair makes the equation true.

In Example 5, the origin is placed at a mathematically convenient location despite the fact that negative dimensions may not make sense. In Example 5, negative x's represent distances to the left of the center of the bridge.

EXAMPLE Building tables and graphs from equations: suspension bridge cables   A 50-foot suspension bridge across a canal has a cable anchored 20 feet above the road surface at each end of the bridge. Vertical cables connect the bridge to the suspended cable. (See Figure 14.) The vertical cables transform the suspended cable into a parabola. We assume that the center of the bridge is at the origin of a pair of axes.

The equation of the cable's height above the road is $y = \frac{4}{125}x^2$ for $-25 \le x \le 25$. (You will be able to find this equation yourself after you study Chapters 2 and 3.) Use the equation to make a table and graph for inputs from -20 to 20 in steps of 10.

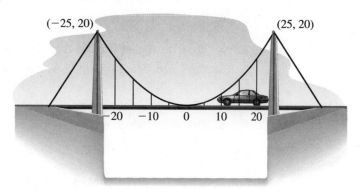

Figure 14

Solution   The table and graph are shown in Table 8 and Figure 15.

Distance from Bridge Center, x (feet)	Height, y (feet)
-20	12.8
-10	3.2
0	0.0
10	3.2
20	12.8

Table 8 $y = \frac{4}{125}x^2$

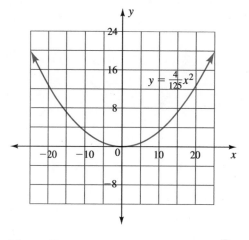

Figure 15   ●

Solving Equations from Tables and Graphs

INDEPENDENT AND DEPENDENT VARIABLES   When there is only one output for each input, we may describe the rule for the table in terms of independent and dependent variables. In Example 5, there is only one cable height at each position on the bridge. *The height of the bridge cable depends on the distance from the center of the bridge.* The phrase *depends on* leads us to say that the input, distance from the center of the bridge, is the independent variable. The output, height of the cable, is the dependent variable.

Independent and Dependent Variables

> The input to a rule is the independent variable. The output from a rule is the dependent variable. On a graph, the independent variable (input) is placed on the horizontal axis and the dependent variable (output) is placed on the vertical axis.

EXAMPLE   Naming independent and dependent variables   For each formula or sentence, name the independent and dependent variables.

a. A student's grade, G, in algebra depends on attendance, P.

b. The cost, c, of college tuition depends on the number, n, of credit hours taken.

c. Area of a circle of radius r: $A = \pi r^2$

d. Bowling handicap for an average below 200: $H = 0.80(200 - A)$

Solution   **a.** The independent variable is attendance, P; the dependent variable is grades, G.

b. The independent variable is number of credit hours, n; the dependent variable is cost, c.

c. The independent variable is radius, r; the dependent variable is area, A.

d. The independent variable is the average, A; the dependent variable is the handicap, H.   ●

Given the independent variable (input), we *evaluate* to find the dependent variable (output). Given the dependent variable (output), we *solve* to find the independent variable (input). Use Examples 7 and 8 to compare evaluating and solving equations.

EXAMPLE   Evaluating and solving equations: Fahrenheit and Celsius temperatures   Use Table 9 to do the following.

a. Evaluate $C = \frac{5}{9}(F - 32)$ for $F = 50°$ [that is, $C = \frac{5}{9}(50 - 32)$].

b. Evaluate $C = \frac{5}{9}(F - 32)$ for $F = 194°$ [that is, $C = \frac{5}{9}(194 - 32)$].

c. Solve $C = \frac{5}{9}(F - 32)$ for $C = 50°$ [that is, $50 = \frac{5}{9}(F - 32)$].

d. Solve $C = \frac{5}{9}(F - 32)$ for $C = 70°$. [that is, $70 = \frac{5}{9}(F - 32)$].

F	C
50	10
86	30
122	50
158	70
194	90

Table 9

Solution   To evaluate parts a and b, we find F in the table and look for the temperature, C.

a. $C = 10°$ **b.** $C = 90°$

To solve parts c and d, we find C in the table and look for the temperature, F.

c. $F = 122°$ **d.** $F = 158°$ ●

EXAMPLE **8**   *Comparing evaluation and solution*   Suppose the first copy of a school transcript costs \$3 and each additional transcript costs \$2.

a. Make a table and graph to show the total cost of ordering 1 to 5 transcripts.

b. Use the table and graph to find (evaluate) the cost of 3 transcripts.

c. Use the table and graph to solve for the number of transcripts that can be bought for \$9.

Solution   **a.** The total cost depends on the number of transcripts ordered. The number of transcripts is the independent variable (on the horizontal axis). The total cost is the dependent variable (on the vertical axis). The total costs are shown in Table 10 and Figure 16.

Transcripts Ordered, x	Total Cost, y (dollars)
1	3
2	5
3	7
4	9
5	11

Table 10

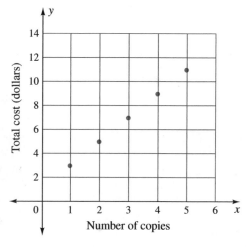

Figure 16

b. In Table 10, look in the input column to find 3 transcripts, and then look at the output. The input 3 matches with the output 7. Three transcripts will cost \$7. In Figure 16, find 3 on the horizontal axis, and then look for the point on the graph where $x = 3$. Again, for 3 transcripts, the total cost is \$7.

c. In Table 10, find \$9 in the output (total cost) column, and then look for the input. The \$9 cost matches with 4 transcripts. In Figure 16, find \$9 on the vertical axis, and then look for the point on the graph where $y = 9$.   ●

SOLUTION SET AND NUMBER OF SOLUTIONS   In Example 8, there was one solution to the number of transcripts that can be ordered for \$9. For many equations, there is more than one solution. For other equations, there is no solution. Solutions may be listed in a solution set.

Definition of Solution Set

> The **solution set** to an equation is the set of all numbers that, when substituted for the variable(s), make a true statement.

In Example 9, the equations have different numbers of solutions.

EXAMPLE

Solving equations with tables and graphs: suspension bridge cables, continued
Solve the following three equations to find the position, x, on the road surface for the given cable height.

a. The cable height is 8 feet: $\frac{4}{125}x^2 = 8$

b. The cable touches the surface: $\frac{4}{125}x^2 = 0$

c. The cable height is -4 feet (that is, 4 feet below the surface): $\frac{4}{125}x^2 = -4$

Solution   The table and graph, from Example 5, are reproduced here as Table 11 and Figure 17.

Distance from Bridge Center, x (feet)	Height, y (feet)
-20	12.8
-10	3.2
0	0.0
10	3.2
20	12.8

Table 11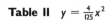

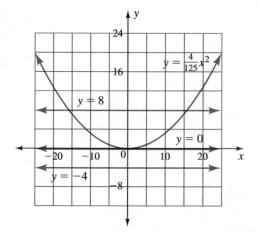

Figure 17

a. To solve $\frac{4}{125}x^2 = 8$, we look in the y column, the height of the cable, in Table 11 for the first pair of numbers between which 8 falls: 12.8 and 3.2. The input for $y = 8$ must be between $x = -20$ and $x = -10$. We might estimate $x = -15$ ft:

$$\frac{4}{125}(-15)^2 = 7.2$$

Finding that $x = -15$ is too low, we try $x = -16$:

$$\frac{4}{125}(-16)^2 = 8.192$$

Because $x = -16$ is close, we write $x \approx -16$.

In the graph in Figure 17, the line $y = 8$ intersects the curve $y = \frac{4}{125}x^2$ just to the left of $x = -15$. From the graph, $x \approx -16$.

There is a second position in the table where the output is 8 and a second point on the graph where $y = 8$. By reasoning similar to that above, a second solution is at $x \approx 16$. The set of approximate solutions is $\{-16, 16\}$.

b. From Table 11, the solution to $\frac{4}{125}x^2 = 0$ is $x = 0$. To solve $\frac{4}{125}x^2 = 0$ from the graph in Figure 17, we find $y = 0$ (the x-axis) and look for where $y = 0$ intersects the graph. The solution is at the origin, $(x, y) = (0, 0)$. The solution set is $\{0\}$.

c. The solution to $\frac{4}{125}x^2 = -4$ cannot be found from Table 11. There are no negative outputs from the equation $y = \frac{4}{125}x^2$. The line $y = -4$ does not intersect the graph of $y = \frac{4}{125}x^2$. There are no real-number solutions to $\frac{4}{125}x^2 = -4$. *The solution set is empty.* We write either an empty set $\{\ \}$ or the symbol $\varnothing$.   ●

Think about it 4: Do $\{0\}$ and $\{\ \}$ have the same meaning?

We could solve part a of Example 9 on a graphing calculator, either by graphing or by building a table.

To solve by creating a table, place the equation $y = \frac{4}{125}x^2$ in $\boxed{Y =}$. Go to TABLE SET-UP. Use -20 for a starting input, and let 10 be the change (Δ). Display the calculator table, and compare it with Table 11. Now go back to TABLE SET-UP and change to a starting input of -17 and a change of 0.5. Display the calculator table, and compare it with Table 12. The output is close to 8 at $x = -16$.

x	$f(x)$
-17	9.248
-16.5	8.712
-16	8.192
-15.5	7.688
-15	7.2

Table 12 $f(x) = \frac{4}{125}x^2$ with $\Delta x = 0.5$

To obtain a closer solution, reset the change to a smaller value.

To solve by graphing, place the equation $y = \frac{4}{125}x^2$ in $\boxed{Y =}$. Enter $y = 8$ as a second equation. Go to GRAPH SET-UP or WINDOW. Set the minimum for x at -20 and the maximum at 20, with a scale of 10. Set the minimum for y at -8 and the maximum at 24, with a scale of 8. Graph, and compare the result with Figure 17.* Trace to one intersection of $y = \frac{4}{125}x^2$ and $y = 8$. Zoom in to improve the estimate of x needed to obtain $\frac{4}{125}x^2 = 8$.

Graphing Calculator Technique:
Building a Table

Change the equation to $y =$ expression or expression $= 0$.

Press $\boxed{Y =}$ and enter an expression in Y_1.

Select TABLE SET-UP.

Enter the starting input for the table and enter the amount between inputs (Δx).

Press the TABLE key.

Graphing Calculator Technique:
Graphing an Equation

Change the equation to $y =$ expression or expression $= 0$.

Press $\boxed{Y =}$ and enter the expression after Y_1.

Press the GRAPH SET-UP or WINDOW key.

Set Xmin and Xmax to include the smallest and largest inputs.

Set Xscl to 10% of (Xmax $-$ Xmin) rounded to an easy counting number.

Set Ymin and Ymax to include the smallest and largest outputs.

Set Yscl to 10% of (Ymax $-$ Ymin) rounded to an easy counting number.

Press the GRAPH key.

Why bother? You might be asking "Why do all this table and graphing work when I know a way to solve the equation $\frac{4}{125}x^2 = 8$ by hand with algebraic notation?" The answer is twofold. First, equations in real applications are too messy to be solved by hand with algebraic notation. Second, many equations, such as $2^x = x^2$, cannot be solved at all with algebraic notation.

*Because there are two solutions, it is convenient to save the viewing window before zooming (look under ZOOM, MEMORY, STORE). To find the second point of intersection, return to the original viewing window with ZOOM, MEMORY, RECALL. Trace and zoom as needed.

EXAMPLE   Solving an equation with tables and graphs   Solve the equation $2^x = x^2$.

a. Complete Tables 13 and 14. For what values of x are the outputs the same? The first four rows of the first table are completed in case you have forgotten the meaning of negative and zero exponents.

x	$y = 2^x$
-3	$2^{-3} = \frac{1}{8} = 0.125$
-2	$2^{-2} = \frac{1}{4} = 0.25$
-1	$2^{-1} = \frac{1}{2} = 0.50$
0	$2^0 = 1$
1	
2	
3	
4	
5	

Table 13

x	$y = x^2$
-3	
-2	
-1	
0	
1	
2	
3	
4	
5	

Table 14

b. On the same axes, graph $y = 2^x$ and $y = x^2$ for integer inputs -3 to 5. What are the points of intersection?

c. Write the solution set to $2^x = x^2$.

Solution   a. The completed versions are shown in Tables 15 and 16.

x	$y = 2^x$
-3	$2^{-3} = \frac{1}{8} = 0.125$
-2	$2^{-2} = \frac{1}{4} = 0.25$
-1	$2^{-1} = \frac{1}{2} = 0.50$
0	$2^0 = 1$
1	$2^1 = 2$
2	$2^2 = 4$
3	$2^3 = 8$
4	$2^4 = 16$
5	$2^5 = 32$

Table 15

x	$y = x^2$
-3	9
-2	4
-1	1
0	0
1	1
2	4
3	9
4	16
5	25

Table 16

At $x = 2$, the outputs are both 4. At $x = 4$, the outputs are both 16. The tables show two solutions: $x = 2$ and $x = 4$. Is this all of the solutions?

b. Figure 18 shows the two points of intersection: (2, 4) and (4, 16). Like the tables, the graph gives the solutions $x = 2$ and $x = 4$. However, there is a third point of intersection with x between -1 and 0. This intersection is slightly closer to $x = -1$ than to $x = 0$. We might estimate the intersection at $\frac{3}{4}$ instead of $\frac{1}{2}$ and write $x \approx -0.75$.

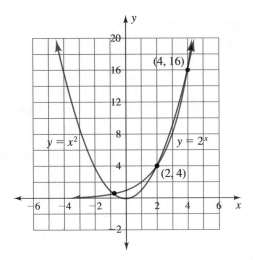

Figure 18

c. The set of approximate solutions is $\{-0.75, 2, 4\}$. ⬤

Think about it 5: Graph the equations $y = x^2$ and $y = 2^x$ on a graphing calculator. Trace and zoom to find the solution between -1 and 0. How might we have noticed this solution from the tables?

Summary: Evaluate and Solve

EVALUATING AN EQUATION When we evaluate an equation written in terms of x and y, we are given the input x and want to find the output y. To complete an input-output table, we take the input x and apply the rule to find y.

Given a table, we find the input in the first column and look for the matching output in the second column. Given a graph, we find the input x on the horizontal axis and look for the point on the graph that gives the output y. Finding y allows us to complete the ordered pair, $(x, \underline{\ \ })$. Evaluating means finding a value of the dependent variable given a value of the independent variable.

SOLVING AN EQUATION When we solve an equation written $y = \ldots$, we are given y, the output, and must find x, the input. Thus, we are completing $(\underline{\ \ }, y)$, or finding x when the rule is known. Solving means finding a value of the independent variable given a value of the dependent variable.

ANSWER BOX

Warm-up: 1. 1, 4, 9, 16, 25, 36, 49, 64, 81, 100, 121, 144, 169, 196, 225 **2.** 2, 4, 8, 16, 32, 64, 128, 256 **3. a.** 4 **b.** -4 **c.** -4 **d.** -4
Think about it 1: Because x and y are equal, the rule, or equation, is $y = x$. In Section 2.3, we will find equations for other graphs based on the wind-chill chart. **Think about it 2:** The two lines cross at about $80°$. Because $y = x$ on one line, the point of intersection is about $(80, 80)$. This means that a 15-mph wind will have little or no chilling effect at $80°$ Fahrenheit. Does this seem reasonable? Why or why not?
Think about it 3: One point on the line is $\left(-\frac{1}{2}, 0\right)$. Substitute $\left(-\frac{1}{2}, 0\right)$ into $y = 2x + 1$: $0 = 2\left(-\frac{1}{2}\right) + 1$, a true statement. **Think about it 4:** No; $\{0\}$ is the set containing the number 0, whereas $\{\ \}$ is a set with nothing in it. **Think about it 5:** $x \approx -0.767$. At $x = -1$, the output for x^2 is larger than that for 2^x. At $x = 0$, the output for x^2 is smaller than that for 2^x. Thus, a solution must lie between $x = -1$ and $x = 0$.

EXERCISES 1.2

1. Draw a set of axes and label these:

   a. vertical axis   **b.** quadrant 4

   c. origin   **d.** $(-1, 3)$

   e. $(-4, 0)$

2. Draw a set of axes and label these:

   a. quadrant 2   **b.** horizontal axis

   c. $(4, -1)$ **d.** $(0, -2)$

   e. origin

In Exercises 3 to 6, match each definition or related statement with a vocabulary word.

3. Choose from
origin, (x, y), x-axis, perpendicular, y-axis

   a. The vertical axis representing output

   b. The name of the point $(0, 0)$ where the axes cross

   c. An ordered pair

   d. The horizontal axis representing input

   e. Two lines forming a right angle

4. Choose from
evaluate, solve, independent variable, dependent variable, solution set

   a. All the numbers that make an equation true

   b. Given an input, find the output.

   c. Given an equation in which y depends on x and y, find x.

   d. What is y if y depends on x?

   e. What is x if y depends on x?

5. Choose from
scale, coordinate plane, { }, {0}, counterclockwise

   a. The set with zero in it

   b. The empty set

   c. The distance between numbers labeled on the axes

   d. The direction in which the quadrants are numbered

   e. The surface containing the horizontal and vertical axes

6. Choose from
Xmin, Yscl, Ymax, table change or Δx, $\boxed{Y =}$

   a. The distance between input numbers in a table

   b. The location for entering equations

   c. The distance between numbers on the vertical axis

   d. The smallest number on the horizontal axis

   e. The greatest number on the vertical axis

For Exercises 7 to 10, refer to the Celsius and Fahrenheit temperatures shown in the graph below.

In Exercises 7 and 8, locate the points by naming the closest letter.

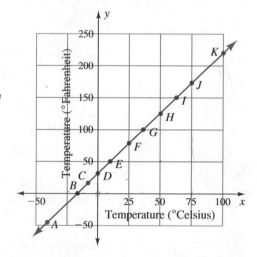

7. a. $(75, 167)$ **b.** $(10, 50)$

   c. Normal body temperature: $(37, 98.6)$

   d. Temperature at which water boils: $(100, 212)$

8. a. $(50, 122)$ **b.** $(-10, 14)$

   c. Temperature at which water freezes: $(0, 32)$

In Exercises 9 and 10, estimate the coordinates of the points from the graph. Your estimate should be within five units of the answer.

9. a. Point B

   b. The point at which the Celsius temperature equals the Fahrenheit temperature

10. a. Point F **b.** Point I

For Exercises 11 to 20, make an input-output table for each equation. Use as inputs the integers from -3 to 3.

11. $y = 2x + 3$ **12.** $y = 3x - 2$

13. $y = 3 - x$ **14.** $y = 4 - 2x$

15. $y = 3 - 2x$ **16.** $y = -2 - x$

17. $y = x^2 - x$ **18.** $y = 2x - x^2$

19. $y = 2 - x^2$ **20.** $y = x^2 + 2$

For Exercises 21 to 30, use graph paper and label the x-axis from -4 to 4 and the y-axis from -12 to 12.

21. Graph Exercise 11.   **22.** Graph Exercise 12.

23. Graph Exercise 13.   **24.** Graph Exercise 14.

25. Graph Exercise 15.   **26.** Graph Exercise 16.

27. Graph Exercise 17.　　**28.** Graph Exercise 18.

29. Graph Exercise 19.　　**30.** Graph Exercise 20.

31. a. Which of the graphs in Exercises 21, 23, 25, 27, and 29 are straight lines?

　　b. Which of the graphs in Exercises 21, 23, 25, 27, and 29 are parabolas?

　　c. Which equation, $y = ax + b$ or $y = ax^2 + bx + c$, makes a straight line?

32. a. Which of the graphs in Exercises 22, 24, 26, 28, and 30 are straight lines?

　　b. Which of the graphs in Exercises 22, 24, 26, 28, and 30 are parabolas?

　　c. Which equation, $y = ax + b$ or $y = ax^2 + bx + c$, makes a parabola?

The body mass index compares weight and height. An index in the range of 19 to 24 is recommended for good health. In Exercises 33 to 36, use the formula $I = \dfrac{W(704.5)}{H^2}$ to complete the tables for the given heights. The index, I, is in terms of weight in pounds, W, and height in inches, H. (Hint: Use the calculator table with the height substituted for H.) Round to the nearest tenth.

33. For height 62 inches:

Weight (lb)	Index
100	
110	
120	
130	
140	

34. For height 66 inches:

Weight (lb)	Index
110	
120	
130	
140	
150	

35. For height 70 inches:

Weight (lb)	Index
130	
140	
150	
160	
170	

36. For height 74 inches:

Weight (lb)	Index
150	
160	
170	
180	
190	

Use the tables in Exercises 33 to 36 to solve the equations in Exercises 37 to 40.

37. a. $23.8 = \dfrac{W(704.5)}{62^2}$　　**b.** $20.2 = \dfrac{W(704.5)}{62^2}$

38. a. $21.0 = \dfrac{W(704.5)}{66^2}$　　**b.** $24.3 = \dfrac{W(704.5)}{66^2}$

39. a. $23.0 = \dfrac{W(704.5)}{70^2}$　　**b.** $24.4 = \dfrac{W(704.5)}{70^2}$

40. a. $19.3 = \dfrac{W(704.5)}{74^2}$　　**b.** $23.2 = \dfrac{W(704.5)}{74^2}$

Use a table or graph from an earlier exercise to solve the equations in Exercises 41 to 48.

41. a. $2x + 3 = -3$　　**b.** $2x + 3 = 7$

42. a. $3x - 2 = -5$　　**b.** $3x - 2 = 7$

43. a. $3 - 2x = 5$　　**b.** $3 - 2x = -3$

44. a. $-2 - x = -3$　　**b.** $-2 - x = -5$

45. a. $x^2 - x = 2$　　**b.** $x^2 - x = 12$

　　c. $x^2 - x = -2$　　**d.** $x^2 - x = -0.25$

46. a. $2x - x^2 = -3$　　**b.** $2x - x^2 = 2$

　　c. $2x - x^2 = 1$　　**d.** $2x - x^2 = -15$

47. a. $2 - x^2 = 2$　　**b.** $2 - x^2 = -2$

　　c. $2 - x^2 = -14$　　**d.** $2 - x^2 = 4$

48. a. $x^2 + 2 = 0$　　**b.** $x^2 + 2 = 2$

　　c. $x^2 + 2 = 6$　　**d.** $x^2 + 2 = 11$

49. Explain how to use a table to evaluate the equation $y = 3x + 4$ at $x = 2$.

50. Explain how to use a graph to evaluate the equation $y = 3x + 4$ at $x = 2$.

51. Explain how to use a graph to solve the equation $3x + 4 = 7$.

52. Explain how to use a table to solve the equation $3x + 4 = 7$.

Name the independent and dependent variables in each formula or sentence in Exercises 53 to 60.

53. Area of a square of side s: $A = s^2$

54. Volume of a sphere of radius r: $V = \frac{4}{3}\pi r^3$

55. Area of an equilateral triangle of side s: $A = \dfrac{s^2\sqrt{3}}{4}$

56. Energy released from material of mass m (c is a constant): $E = mc^2$

57. The exercise heart rate, E, depends on the person's age, A.

58. The cost of riding the local transit system depends on the number of trips made.

59. The cost of shipping a package depends on the weight of the package.

60. The income tax rate depends on the adjusted gross income.

Complete each sentence in Exercises 61 to 66 with "finding the dependent variable" or "finding the independent variable."

61. Solving the equation $3x + 4 = -2$ is _____ .

62. Locating the output in a table when we are given the input is _____ .

63. Locating y, given the graph and x, is _____ .

64. Evaluating the equation $y = 3x + 4$ for $x = -2$ is

_____ .

65. Locating x, given the graph and y, is _____ .

66. Locating the input in a table when we are given the output is _____ .

Projects

67. *Graphing Calculator.* Use a table and graph to find the solution set for $3^x - 1 = x^3$.

68. *Wind-Chill Temperatures with Wind Speed Held Constant.*

  a. Copy the axes and labels from Figure 11 in Example 3.

  b. The temperature table is on page 33 and page 1. On the axes from part a, plot the temperatures, x, and

wind-chill apparent temperatures, y, for a wind speed of 20 mph. Extend the axes as needed.

  c. Plot the graph for 30 mph on the same axes.

  d. How are the graphs the same? How are they different?

69. *Wind-Chill Temperatures with Current Temperature Held Constant.* Return to the wind-chill apparent temperature chart on page 33 or page 1.

  a. Draw a new set of axes for the data. Let x be the wind speed and y be the wind-chill apparent temperature. Allow for any wind speed and temperature shown on the chart.

  b. For a constant current temperature of 35°, plot the apparent temperature for different wind speeds—that is, (5, 33), (10, 21), (15, 16), (20, 12), (25, 7), and (30, 5). Label this graph with 35°.

  c. Plot four other graphs on the same axes—one graph each for a current temperature of 25°, 15°, 5°, and −5°. Label each graph with its current temperature.

  d. List your observations about the graphs.

MID-CHAPTER **1** TEST _____

1. Simplify.

  a. $-3 + (-5) - (-6)$ **b.** $-\frac{1}{2} + \frac{1}{4} - \frac{3}{4} - \left(-\frac{1}{4}\right)$

  c. $2.25 + 8.50 - 3.75$ **d.** $6.2 - 8.6 + 1.8$

2. Simplify.

  a. $2 \cdot 17 \cdot 50$ **b.** $\frac{1}{2}(-7)(-6)$

  c. $-4(-8)(-5)$ **d.** $\frac{3}{5}\left(\frac{11}{6}\right)\left(\frac{15}{121}\right)$

3. Simplify. Write answers as fractions or round to the nearest tenth.

  a. -4^2 **b.** $(-4)^2$ **c.** $0 - 4^2$

  d. $8 - 6 \div 3 + 3 \cdot 4 + \frac{1}{2}(8 - 3)$ **e.** $\frac{3 - 6}{2 - (-7)}$

  f. $\frac{-(-2) - \sqrt{(-2)^2 - 4(5)(-3)}}{2(5)}$

  g. $\sqrt{(2 - (-7))^2 + (3 - 6)^2}$

  h. $\frac{|1.25 - 2.83| + |2.56 - 2.83| + |4.68 - 2.83|}{3}$

4. Multiply.

  a. $3(x + 2)$ **b.** $-3(x - 2)$

  c. $7 - 4(x + 5)$ **d.** $8 - 2(4 - x)$

5. Name the greatest common factor and then factor.

  a. $4x - 18$ **b.** $x^2 + 5x$

  c. $mn^2 - np^2$ **d.** $63x^2y - 49xy^2$

6. Show each of these inequalities on a number line.

  a. $-4 < x$ **b.** $6 > x$

  c. $-3 \leq x \leq 2$ **d.** $3 < x \leq 6$

  e. $x > 3$ or $x < -2$ **f.** $x > -2$ or $x \leq 3$

7. Name the variables, constant term, and numerical coefficients.

  a. $-4 - x$ **b.** $2x - x^2$

8. Remove parentheses and add like terms.

  a. $3(x^2 + 3x - 4) - x(x^2 + 3x - 4)$

  b. $a(a^2 - 2ab + b^2) - b(a^2 + 2ab - b^2)$

9. Simplify.

  a. $\frac{x^2yz}{xy^2z}$ **b.** $\frac{x^2yz}{(xy)^2z}$ **c.** $\frac{xy - yz}{xy}$

10. For each formula, name the independent and dependent variables. Then evaluate the formula for the given values of the variables.

a. Height of an equilateral triangle of side s: $h = \dfrac{s\sqrt{3}}{2}$, $s = 4$ yd

b. Area of a circle of side d: $A = \pi\left(\dfrac{d}{2}\right)^2$, $d = 5$ cm.

11. In which quadrants do ordered pairs have a negative first number?

12. Plot these ordered pairs and connect them in order:

$(-3, 2), (-2, -3), (3, -2), (2, 3), (-3, 2)$

Extension: What is the shape of the graph and why?

13. Make a table and graph for each equation.

a. $y = 4 - 3x$ **b.** $y = 3x - 2x^2$

14. Use the table and the graph below to solve the following equations. Circle the places where you find the answers.

Input x	Output $y = x^2 - 4x + 3$
−1	8
0	3
1	0
2	−1
3	0
4	3
5	8

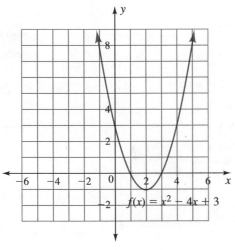

$f(x) = x^2 - 4x + 3$

a. $x^2 - 4x + 3 = 8$

b. $x^2 - 4x + 3 = 0$

c. $x^2 - 4x + 3 = -1$

15. Match each statement with one of the following inequalities:

$x > 0,\ x < 0,\ x \le 0,\ x \ge 0,\ y > 0,\ y < 0,\ y \le 0,$ $y \ge 0$

a. The outputs are positive.

b. The inputs are negative.

c. The outputs are negative.

d. The inputs are zero or positive.

e. The x-coordinate is in the first or fourth quadrant but not on the y-axis.

f. The y-coordinate is in the third or fourth quadrant or on the x-axis.

16. For the animal pen design shown, two square pens are added at once.

Number of Pairs of Pens	Total Number of Panels
1	7
2	
3	
4	
10	52
50	
100	

Top view

a. Complete the table.

b. Let x be the number of pairs of pens. Let y be the total number of panels needed to build the pens. Guess a rule for the pattern in the table.

In Exercises 17 to 24, match each of the word phrases with one of these algebraic expressions:

17. Four increased by double a number

18. Five times a number plus three

19. Five less than three times a number

20. Three more than half a number

21. Twice a number less four

22. Two less than four times a number

23. Five decreased by three times a number

24. One-half less four times a number

a. $5x + 3$

b. $4x - 2$

c. $\frac{1}{2} - 4x$

d. $3 + 2(x - 2)$

e. $5 - 3x$

f. $4 + 0.5(x - 2)$

g. $3x - 5$

h. $0.5x + 3$

i. $2x - 4$

j. $4 + 2x$

1.3   Solving Equations in One Variable and Formulas

OBJECTIVES

- Simplify expressions containing inverse operations.
- Solve equations using inverses.
- Solve formulas using inverses.
- Check the solution to a formula.
- Apply the addition property of solving equations.
- Apply the multiplication property of solving equations.

WARM-UP

In Exercises 1 to 5, match each expression with the equivalent expression from parts a to e:

a. $8 + \frac{1}{4}$ **b.** $-8 \cdot 4$ **c.** $8 \cdot (-4)$

d. $-8 + (-4)$ **e.** $-8 \cdot \left(-\frac{1}{4}\right)$

   1. $-8 - 4$ **2.** $-8 \div (-4)$ **3.** $-8 \div \frac{1}{4}$

   4. $8 - \left(-\frac{1}{4}\right)$ **5.** $8 \div \left(-\frac{1}{4}\right)$

Simplify or remove parentheses in Exercises 6 to 10.

   6. $9 - 4(x - 6)$ **7.** $5 - 2(7 - x)$

   8. $\frac{5}{9}(212 - 32)$ **9.** $\frac{9}{5}(100) + 32$

10. $\frac{n}{2}(2a + (n - 1)d)$, where $a = 5$, $d = 5$, and $n = 10$

THIS SECTION REVIEWS inverse operations. It shows how the order of operations and inverse operations may help us plan the steps to follow in solving an equation or formula. We apply Polya's problem-solving steps in solving and checking equations and formulas.

Inverse Operations and Inverses

Section 1.0 listed the rules for operations with real numbers. We can apply these rules to more problems if we change subtraction to addition of the opposite and change division to multiplication by the reciprocal. These changes are possible because subtraction is the inverse operation to addition and division is the inverse operation to multiplication.

Inverse Operations

An operation is undone by its inverse operation.

Subtraction is the inverse operation to addition:

$$a + b - b = a$$

Division is the inverse operation to multiplication:

$$a \cdot b \div b = a$$

EXAMPLE 1   Applying inverse operations   Simplify by first changing to the inverse operation.

a. $-8 - 4$ **b.** $-8 \div (-4)$ **c.** $-8 \div \frac{1}{4}$ **d.** $8 - \left(-\frac{1}{4}\right)$ **e.** $8 \div -\frac{1}{4}$

Solution   **a.** $-8 - 4 = -8 + (-4) = -12$

b. $-8 \div (-4) = -8 \cdot \left(-\frac{1}{4}\right) = +\frac{8}{4} = 2$

c. $-8 \div \frac{1}{4} = -8 \cdot 4 = -32$

d. $8 - \left(-\frac{1}{4}\right) = 8 + \frac{1}{4} = 8\frac{1}{4}$

e. $8 \div -\frac{1}{4} = 8 \cdot (-4) = -32$ ●

Because subtraction is the inverse operation to addition, *opposites* (numbers with opposite signs) are also called *additive inverses*.

Additive Inverses

> Additive inverses add to zero:
>
> $$a + -a = 0$$

A *reciprocal* is the number that multiplies x to give 1. The reciprocal of x is $\frac{1}{x}$. The reciprocal of $\frac{a}{b}$ is $\frac{b}{a}$. Because division is the inverse operation to multiplication, reciprocals are also called *multiplicative inverses*.

Multiplicative Inverses

> Multiplicative inverses multiply to 1:
>
> $$x \cdot \frac{1}{x} = 1 \qquad \frac{a}{b} \cdot \frac{b}{a} = 1$$

EXAMPLE Using inverses   Complete these statements.

a. $8 + \square = 0$ b. $4 \cdot \square = 1$ c. $\frac{5}{8} \cdot \square = 1$ d. $\frac{2}{3} + \square = 0$

e. $\frac{1}{3} \cdot \square = 1$ f. $1\frac{1}{2} \cdot \square = 1$

Solution   See the Answer Box.   ●

Solving Equations with the Inverse Order of Operations

Computers and calculators can now solve complicated equations using algebraic notation. You might ask, *What is left to learn?* The answer is that you need to

- know what it means to solve an equation.
- understand basic properties of equation solving.
- estimate solutions.
- plan and think step by step through a sequence of operations.
- solve common formulas.

In Section 1.2, we solved equations from tables and graphs. In Section 1.2, **solving an equation** meant *finding an input when given an output*. In this section, **solving an equation** means *using algebraic notation to isolate a variable on one side of the equation*. This process is based on the properties of equations.

PROPERTIES OF EQUATIONS   **Equivalent equations** *have the same set of solutions*. We obtain equivalent equations using the properties of equations.

Addition Property of Equations

> Adding the same number to both sides of an equation produces an equivalent equation:
>
> $$\text{If} \quad a = b, \quad \text{then} \quad a + c = b + c.$$

Because subtraction may be written as adding the opposite, this property also applies to subtraction:

$$\text{If} \quad a = b, \quad \text{then} \quad a - c = b - c.$$

Multiplication Property of Equations	Multiplying both sides of an equation by the same nonzero number produces an equivalent equation: $\quad$ If $\quad a = b, \quad$ then $\quad ac = bc \quad$ for $c \neq 0$.

Because division may be written as multiplying by the reciprocal, this property also applies to division:

$$\text{If} \quad a = b, \quad \text{then} \quad \frac{a}{c} = \frac{b}{c} \quad \text{for } c \neq 0.$$

EXAMPLE **3**   Solving equations with the properties of equations   Solve these equations using the addition and multiplication properties of equations.

a. $x - 3 = -4$

b. $\frac{1}{2}x = 8$

Solution   **a.** $\quad x - 3 = -4$ Add 3 to each side.

$\quad x - 3 + 3 = -4 + 3$ Note: $-3 + 3 = 0$

$\quad\quad\quad\quad x = -1$

b. $\quad \frac{1}{2}x = 8$ Multiply each side by 2.

$\quad 2\left(\frac{1}{2}x\right) = 2(8)$ Note: $2 \cdot \frac{1}{2} = 1$

$\quad\quad\quad\quad x = 16$

●

Think about it: Describe the role of inverses in solving the equations in Example 3.

MAKING A PLAN   In Example 3, only one step—one inverse operation—was required to solve each equation for x. When there is more than one operation in an equation, we need to know *which step to do first to solve the equation.*

Example 4 illustrates how inverse operations can apply to a sequence of several operations.

EXAMPLE **4**   Applying inverse operations to daily life   Here are three steps we might take in leaving home on a cold morning:

$\quad\quad$ Put on coat.$\quad\quad$ Fasten buttons.$\quad\quad$ Tie scarf.

What is the result when we reverse the order of the steps and do the inverse, or opposite, operation at each step?

Solution   The reverse order with inverse operations is

$\quad\quad$ Untie scarf.$\quad\quad$ Unbutton coat.$\quad\quad$ Take off coat.

These are the steps we take when we arrive at our destination.

●

By looking at the steps in the original activity (putting on, fastening, and tying), we can plan what to do to reverse the activity (untying, unfastening, taking off). This reversal of steps with the opposite operation is the basis for our plan in solving equations.

We will use the phrase **inverse order of operations** to describe *the reverse order of operations with a sequence of inverse operations.* In solving the equations in Examples 5, 6, and 7, we list the order of operations on *x* and then write a plan using the inverse order of operations.

EXAMPLE **5**   Making a plan   Solve $5x + 6 = 9$ for *x* with the inverse order of operations.

  a. Understand: Make an estimate.
  b. Plan: Identify the order of operations on *x* in the equation $5x + 6 = 9$. List the inverse order of opposite operations.
  c. Carry out the plan.
  d. Check.

Solution   **a.** ***Understand:*** Because $5(1) + 6 = 11$, $x \approx 1$ in $5x + 6 = 9$.

  b. ***Plan:*** The *x* is multiplied by 5, and then 6 is added. The inverse order of operations is to subtract 6 and then divide by 5.

  c. ***Carry out the plan:***

$$5x + 6 = 9 \qquad \text{Subtract 6 from both sides.}$$
$$5x + 6 - 6 = 9 - 6$$
$$5x = 3 \qquad \text{Divide by 5 on both sides.}$$
$$\frac{5x}{5} = \frac{3}{5}$$
$$x = 0.6$$

  d. ***Check:*** $5(0.6) + 6 \stackrel{?}{=} 9$ ✔

The symbol $\stackrel{?}{=}$ is used with each check.

In Example 5, we simplified $5x/5$ to *x*. We will use this type of simplification frequently in the following examples.

EXAMPLE **6**   Solving with a plan   Solve $9 - 4(x - 6) = 21$ for *x* with the inverse order of operations.

  a. Understand: Make an estimate.
  b. Plan: List the order of operations on *x*. List the inverse order of opposite operations.
  c. Carry out the plan.
  d. Check.

Solution   **a.** ***Understand:*** Because subtraction is the same as adding the opposite, the equation $9 - 4(x - 6) = 21$ can be written $9 + (-4)(x - 6) = 21$. The value of $(x - 6)$ needs to be negative in order to add an expression to 9 to get 21. Thus, *x* is smaller than 6.

  b. ***Plan:*** The order of operations on *x* is

  Subtract 6 from *x*.   Multiply the result by -4.   Add 9.

  The inverse order of opposite operations is

  Subtract 9.   Divide the result by -4.   Add 6.

c. *Carry out the plan:*

$$9 + (-4)(x - 6) = 21 \qquad \text{Subtract 9 from both sides.}$$
$$9 + (-4)(x - 6) - 9 = 21 - 9$$
$$-4(x - 6) = 12 \qquad \text{Divide by } -4 \text{ on both sides.}$$
$$\frac{-4(x - 6)}{-4} = \frac{12}{-4}$$
$$x - 6 = -3 \qquad \text{Add 6 on both sides.}$$
$$x - 6 + 6 = -3 + 6$$
$$x = 3$$

d. *Check:* $9 - 4(3 - 6) \overset{?}{=} 21$ ✔   ●

Inverse operations tell us that division by a/b is the same as multiplying by the reciprocal, b/a. We apply this fact in planning Example 7.

EXAMPLE **7**

Solving with a plan   Solve $\frac{5}{9}(x - 32) = 100$ for x with the inverse order of operations.

   a. Understand: Make an estimate.

   b. Plan: List the order of operations on x. List the inverse order of opposite operations.

   c. Carry out the plan.

   d. Check.

Solution   **a. *Understand:*** For us to multiply the number in parentheses by $\frac{5}{9}$ $\left(\text{approximately } \frac{1}{2}\right)$ and still have 100 as a result, x needs to be close to 200.

   b. *Plan:* The order of operations on x is

      Subtract 32.   Multiply by $\frac{5}{9}$.

   The inverse order of operations is

      Divide by $\frac{5}{9}$.   Add 32.

   Because multiplication by $\frac{9}{5}$ is equivalent to division by $\frac{5}{9}$, the inverse order of operations may be written as

      Multiply by $\frac{9}{5}$.   Add 32.

   c. *Carry out the plan:*

$$\frac{5}{9}(x - 32) = 100 \qquad \text{Multiply by } \frac{9}{5} \text{ on both sides.}$$
$$\frac{9}{5} \cdot \frac{5}{9}(x - 32) = 100 \cdot \frac{9}{5}$$
$$x - 32 = 180 \qquad \text{Add 32 on both sides.}$$
$$x - 32 + 32 = 180 + 32$$
$$x = 212$$

   d. *Check:* $\left(\frac{5}{9}\right)(212 - 32) \overset{?}{=} 100$ ✔   ●

You may know other methods for solving equations. Try several problems with the inverse order of operations method, and then you will be able to select the appropriate method when you need it. Knowing more than one method will increase your flexibility in solving equations and formulas.

Solving Formulas

The inverse order of operations works especially well in solving formulas. Solving formulas may be troublesome because the solution is an expression, often no simpler than the original formula.

In the following examples, we continue to perform operations on both sides.

EXAMPLE 8 Solving formulas with a plan   Solve $C = 2\pi r$ for r, using the inverse order of operations.

Solution ***Plan:*** The r is multiplied by 2π. To solve, we divide by 2π.

$$C = 2\pi r \qquad \text{Divide by } 2\pi.$$

$$\frac{C}{2\pi} = \frac{2\pi r}{2\pi}$$

$$\frac{C}{2\pi} = r$$

EXAMPLE 9 Solving with a plan   Solve $P = 2l + 2w$ for w.

Solution ***Plan:*** The w is multiplied by 2, and then $2l$ is added. To solve, we subtract $2l$ and divide by 2.

$$P = 2l + 2w \qquad \text{Subtract } 2l.$$

$$P - 2l = 2l + 2w - 2l$$

$$P - 2l = 2w \qquad \text{Divide by 2.}$$

$$\frac{P - 2l}{2} = \frac{2w}{2}$$

$$\frac{P - 2l}{2} = w$$

To solve the formula in Example 10, we have a choice. We may use the process for fractions shown in Example 7, or we may consider the factor $n/2$ as a multiplication by n and then a division by 2. Example 10 uses the latter procedure. Try the fraction form as an exercise.

EXAMPLE 10 Solving with a plan   Solve $S = \dfrac{n}{2}(2a + (n - 1)d)$ for d.

Solution ***Plan:*** The order of operations on d is

> Multiply by $(n - 1)$.   Add $2a$.   Multiply the result by n.   Divide by 2.

The inverse order of opposite operations is

> Multiply by 2.   Divide by n.   Subtract $2a$.   Divide by $(n - 1)$.

$$S = \frac{n}{2}(2a + (n - 1)d) \qquad \text{Multiply by 2 and simplify.}$$

$$2S = n(2a + (n - 1)d) \qquad \text{Divide by } n \text{ and simplify.}$$

$$\frac{2S}{n} = 2a + (n - 1)d \qquad \text{Subtract } 2a \text{ and simplify.}$$

$$\frac{2S}{n} - 2a = (n - 1)d \qquad \text{Divide by } n - 1.$$

$$\frac{\dfrac{2S}{n} - 2a}{n - 1} = d$$

Although the formula for d can be simplified, simplifying is not recommended at this time. Ways to simplify such expressions will be introduced in Chapter 5.

O ne important concept bears repeating: *To solve a formula means to isolate a particular variable on one side. The variable should not appear anywhere on the other side.* The thermodynamics formula $E = (T_h - T_c)/T_h$ contains the variable T_h twice. It would not be correct to solve for T_h and obtain

$$T_h = \frac{T_h - T_c}{E}$$

The variable T_h appears on both sides. Example 11 shows a correct solution of the formula for T_h. The solution is considerably more difficult than those in the other examples and is shown to illustrate the careful work needed in order to isolate one variable when it appears twice in a formula.

EXAMPLE 11   Solving when the variable appears twice   Solve $E = \dfrac{T_h - T_c}{T_h}$ for T_h.

Solution   **Plan:** Because T_h appears twice, we must rearrange the equation until T_h appears only once. We then apply the reverse order of operations.
Carry out the plan: We rearrange the equation until T_h appears only once.

$$E = \frac{T_h - T_c}{T_h}$$ Multiply both sides by T_h.

$$ET_h = T_h - T_c$$ Subtract T_h from both sides.

$$ET_h - T_h = -T_c$$ Factor T_h on the left side.

$$T_h(E - 1) = -T_c$$ Apply the reverse order of operations: divide by $(E - 1)$.

$$\frac{T_h(E - 1)}{E - 1} = \frac{-T_c}{E - 1}$$ Simplify.

$$T_h = \frac{-T_c}{E - 1}$$ Clear the negative sign: multiply by $\dfrac{-1}{-1}$.

$$T_h = \frac{-T_c(-1)}{(E - 1)(-1)}$$

$$T_h = \frac{T_c}{1 - E}$$

Example 11 is a good problem to use to test the symbolic manipulation feature on advanced calculators.

Checking Formula Solutions

Observe that our formula examples have no check statement. It is possible, however, to check work with formulas. The checking involves simplifications of expressions that may be less difficult than they appear.

Example 12 shows the steps in checking a solution, to remind us of the results when checking equations.

EXAMPLE 12   Checking solutions   Show that $x = 3$ is the solution to $9 - 4(x - 6) = 21$.

Solution
$$9 - 4(x - 6) = 21$$
$$9 - 4(3 - 6) = 21$$
$$9 - 4(-3) = 21$$
$$9 - (-12) = 21$$
$$21 = 21$$

The last line of the check is a statement saying a number is equal to itself. This is our confirmation that the answer $x = 3$ is a correct solution. We can generalize this substitution and result as follows.

Checking a Solution

> If a number or expression, x, is substituted into an equation or formula and the resulting equation simplifies to $a = a$, then x is a solution to the equation.

The next two examples check the answers in Examples 8 and 9. The check for Example 10 is left as an exercise.

EXAMPLE **13**   Checking a formula   Show that $r = \dfrac{C}{2\pi}$ is the solution to $C = 2\pi r$ solved for r.

Solution

$$C = 2\pi r \qquad\qquad \text{Substitute for } r.$$

$$C = 2\pi\left(\frac{C}{2\pi}\right) \qquad \text{Simplify.}$$

$$C = \frac{2\pi}{2\pi}\cdot C$$

$$C = C$$

The last line is $C = C$, so $r = C/2\pi$ is correct.   ●

EXAMPLE **14**   Checking a formula   Show that $w = \dfrac{P - 2l}{2}$ is the solution to $P = 2l + 2w$ solved for w.

Solution

$$P = 2l + 2w \qquad\qquad\qquad \text{Substitute for } w.$$

$$P = 2l + 2\left(\frac{P - 2l}{2}\right) \qquad \text{Simplify } \tfrac{2}{2}.$$

$$P = 2l + P - 2l \qquad\qquad \text{Add the } 2l \text{ terms.}$$

$$P = P$$

The last line is $P = P$, so our formula solution is correct.   ●

ANSWER BOX

Warm-up: 1. d **2.** e **3.** b **4.** a **5.** c **6.** $33 - 4x$ **7.** $-9 + 2x$ **8.** 100 **9.** 212 **10.** 275 **Example 2: a.** -8 **b.** $\tfrac{1}{4}$ **c.** $\tfrac{8}{5}$ **d.** $-\tfrac{2}{3}$ **e.** 3 **f.** $\tfrac{2}{3}$ **Think about it:** In each equation, we performed the opposite operation with the inverse number to solve for x.

EXERCISES **1.3**

1. Simplify by first changing to the inverse operation.

    a. $-4 - 6$ **b.** $-5 - (-3)$ **c.** $-\tfrac{1}{2} - \left(-\tfrac{3}{2}\right)$

    d. $100 \cdot \tfrac{1}{4}$ **e.** $\tfrac{5}{8} \div \tfrac{5}{8}$ **f.** $-6 \div \tfrac{1}{3}$

2. Simplify by first changing to the inverse operation.

    a. $5 - (-6)$ **b.** $-3 - (-3)$ **c.** $\tfrac{1}{4} - \tfrac{3}{4}$

    d. $4 \div \tfrac{1}{4}$ **e.** $-9 \div \tfrac{2}{3}$ **f.** $\tfrac{3}{4} \div \tfrac{3}{4}$

3. Complete these statements.

a. $-3 + \square = 0$ **b.** $5 \cdot \square = 1$ **c.** $-2 + \square = 1$

d. $8 \cdot \square = 0$ **e.** $\frac{7}{8} \cdot \square = 1$ **f.** $\frac{3}{8} + \square = 0$

4. Complete these statements.

a. $7 \cdot \square = 1$ **b.** $7 + \square = 0$ **c.** $7 + \square = 1$

d. $-\frac{2}{3} \cdot \square = 1$ **e.** $-12 + \square = 0$ **f.** $\frac{5}{8} + \square = 0$

5. Solve these equations.

a. $x - 8 = -4$ **b.** $x + 3 = -6$ **c.** $x - 6 = -8$

d. $\frac{1}{2}x = -6$ **e.** $\frac{3}{4}x = 21$ **f.** $\frac{3}{8}x = 12$

g. $-\frac{2}{3}x = 24$ **h.** $-\frac{1}{4}x = 8$ **i.** $\frac{7}{8}x = 35$

6. Solve these equations.

a. $x + 5 = -2$ **b.** $x - 12 = -3$ **c.** $x - 4 = -9$

d. $\frac{1}{4}x = 12$ **e.** $\frac{3}{4}x = 27$ **f.** $-\frac{1}{8}x = 40$

g. $\frac{2}{3}x = 18$ **h.** $-\frac{3}{8}x = 24$ **i.** $\frac{5}{8}x = 40$

In Exercises 7 to 12, which activities have a meaningful inverse order of operations? Describe the original activity and the inverse, if it exists. In which is the order not important?

7. Put on shirt, put on vest, put on jacket.

8. Shut off car, remove key, open door.

9. Take off lens cap, aim camera, take picture, advance film.

10. Fold clothes, put dishes away.

11. Dig hole, pile firewood, turn on sprinkler.

12. Put on sock, put on shoe, tie shoelace.

Write a plan and solve the equations in Exercises 13 to 16.

13. a. $2x - 4 = 8$ **b.** $2x - 4 = 6$

c. $2x - 4 = -6$ **d.** $2x - 4 = 0$

14. a. $\frac{1}{2}x + 1 = 5$ **b.** $\frac{1}{2}x + 1 = 3$

c. $\frac{1}{2}x + 1 = 0$ **d.** $\frac{1}{2}x + 1 = -4$

15. a. $\frac{5}{9}(x - 32) = -25$ **b.** $\frac{5}{9}(x - 32) = 10$

c. $\frac{5}{9}(x - 32) = 5$ **d.** $\frac{5}{9}(x - 32) = -30$

16. a. $0.80(200 - A) = 88$ **b.** $0.80(200 - A) = 56$

c. $0.80(200 - A) = 36$ **d.** $0.80(200 - A) = 20$

In Exercises 17 to 24, solve the equation for x.

17. $3x + 3 = 0$ **18.** $3x + 3 = -9$

19. $\frac{1}{3}x - 1 = 4$ **20.** $\frac{2}{3}x - 1 = 11$

21. $\frac{3}{4}x + 5 = 23$ **22.** $\frac{3}{4}x - 8 = 19$

23. $\frac{3}{8}x - 4 = 8$ **24.** $\frac{3}{8}x + 3 = 24$

In Exercises 25 to 38, simplify one side of the equation before solving.

25. $x - 2(4 - x) = -17$ **26.** $x - 2(4 - x) = 13$

27. $x - 3(4 + x) = 6$ **28.** $x - 3(4 + x) = -8$

29. $x + 3(5 - x) = -9$ **30.** $x + 3(5 - x) = 21$

31. $8 - 3(9 - x) = -14.5$ **32.** $7 - 2(8 - x) = -3.4$

33. $14 - 9(x + 2) = 45.5$ **34.** $17 - 7(x + 5) = 5.8$

35. $37.45 = x + 0.07x$ **36.** $50.40 = x + 0.05x$

37. $46.17 = x + 0.15x + 0.065x$

38. $63.70 = x + 0.15x + 0.075x$

In Exercises 39 to 58, solve the formula for the indicated variable.

39. Distance: $D = rt$ for t

40. Circumference: $C = \pi d$ for d

41. Linear equation: $y = mx + b$ for b

42. Electronics: $I = \dfrac{E}{R}$ for E

43. Volume of sphere: $V = \dfrac{4\pi r^3}{3}$ for r^3

44. Surface area of sphere: $A = 4\pi r^2$ for r^2

45. Thread on a screw: $pN = 1$ for p

46. Area of triangle: $A = \frac{1}{2}bh$ for b

47. Area of trapezoid: $A = \frac{1}{2}h(a + b)$ for b

48. Geometric sequence: $a_n = a_1 r^{n-1}$ for a_1

49. Arithmetic sequence: $a_n = a_1 + (n - 1)d$ for a_1

50. Sum of geometric sequence: $S = \dfrac{a_1}{r - 1}$ for a_1

51. Average: $A = \dfrac{a + b + c}{3}$ for a

52. Arithmetic sequence: $a_n = a_1 + (n - 1)d$ for d

53. Sum of arithmetic sequence: $S = \frac{1}{2}n(a_1 + a_n)$ for a_1

54. Skating party for 10 or more people: $C = 75 + 3.50(x - 10)$ for x

55. Thermodynamics: $E = \dfrac{T_h - T_c}{T_h}$ for T_c

56. Sound absorption: $T = \dfrac{0.16V}{S_e}$ for V

57. Sound velocity: $V = 344 + 0.6(T - 20)$ for T

58. Musical acoustics: $f = (2n - 1)\left(\dfrac{V}{4L}\right)$ for V

For Exercises 59 to 66, check the formula solution by substituting the second formula into the first formula.

59. $D = rt$ for r; $r = \dfrac{D}{t}$

60. $I = prt$ for r; $r = \dfrac{I}{pt}$

61. $A = \dfrac{1}{2}h(a + b)$ for h; $h = \dfrac{2A}{a + b}$

62. $S = \dfrac{n}{2}(2a + (n - 1)d)$ for d; $d = \dfrac{\dfrac{2S}{n} - 2a}{n - 1}$

63. $y = 3x - 3$ for x; $x = \frac{1}{3}(y + 3)$

64. $y = \frac{1}{3}x + 1$ for x; $x = 3(y - 1)$

65. $y = 2x + 4$ for x; $x = \frac{1}{2}(y - 4)$

66. $y = \frac{1}{2}x - 2$ for x; $x = 2(y + 2)$

67. Repeat Example 10, solving $S = \dfrac{n}{2}(2a + (n - 1)d)$ for d, but this time undo the multiplication by $n/2$ by multiplying both sides by the reciprocal, $2/n$.

Project

68. *Inverse Operations*

    a. List the reverse order with inverse operations for these steps. What do the original steps describe? What does the inverse describe?

> Turn on tape player.   Open tape box.
> Take out tape.   Close tape box.
> Insert tape into tape player.   Press play.

    b. Find and summarize a job-related order of operations and a meaningful inverse order of operations (example: delivery of petroleum-based fuel to a storage facility, including safety and environmental steps). Indicate any steps that must be taken that are not in the inverse.

1.4   Solving Inequalities in One and Two Variables

OBJECTIVES

- Solve inequalities in one variable with graphs and with algebraic notation.
- Find a number-line solution to one-variable inequalities.
- Write inequalities in two variables to describe quadrants and axes.
- Graph inequalities in two variables on coordinate axes.

WARM-UP

As a review of Section 1.0, draw a number-line graph for each of these inequalities.

1. $x > 3$ **2.** $x \le 5$

3. $-3 < x \le 2$ **4.** $x \le -1$ or $x > 1$

5. $x < -2$ or $x > -1$ **6.** $x > -2$ or $x \le 3$

Sketch a pair of axes containing the graphs of these equations.

7. $y = 3$ **8.** $y = -2$ **9.** $y = 0$

IN THIS SECTION, we solve inequalities in one variable with graphs and algebraic notation. We solve inequalities in two variables with graphs.

Solving Inequalities in One Variable by Graphing

The solution set to an inequality in one variable is another inequality. The graphical solution set can be drawn on a number line. The graph in Example 1 provides information about inequalities as well as equations.

EXAMPLE **1** Exploration: solving inequalities with a graph   Solve the equations in parts a to c from the graph in Figure 19.

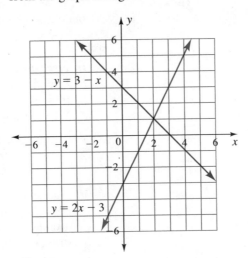

Figure 19

a. $2x - 3 = 0$ **b.** $3 - x = 0$ **c.** $2x - 3 = 3 - x$

Use the graph to find values of x for which the inequalities in parts d to g are true.

d. $2x - 3 > 0$ **e.** $3 - x < 0$

f. $3 - x > -2$ **g.** $2x - 3 < 3 - x$

Solution   **a.** The graph of $y = 2x - 3$ crosses the x-axis ($y = 0$) at $(1.5, 0)$. Thus, $x = 1.5$ solves $2x - 3 = 0$.

Check: $2(1.5) - 3 \overset{?}{=} 0$ ✔

b. The graph of $y = 3 - x$ crosses the x-axis at $(3, 0)$. Thus, $x = 3$ solves $3 - x = 0$.

Check: $3 - 3 \overset{?}{=} 0$ ✔

c. The graph of $y = 2x - 3$ crosses $y = 3 - x$ at $(2, 1)$. Thus, $x = 2$ solves $2x - 3 = 3 - x$.

Check: $2(2) - 3 \overset{?}{=} 3 - 2$ ✔

For the remaining solutions, see the Answer Box.   ●

There are two methods of solving an inequality with graphing. In both methods, we start by locating the point of intersection of the graphs of the left and right sides of the inequality. The first method then returns to a more symbolic approach, whereas the second method relies on the graph for information.

Solving an Inequality: Method 1

1. Graph the left and right sides of the equation separately.
2. Locate the point of intersection of the graphs. The x-coordinate is the starting point for the solution set.
3. Find the direction of the inequality by testing an x-coordinate on one *(left)* side of the intersection.

EXAMPLE **2** Solving inequalities with a graph   Solve these inequalities with Method 1.

a. $2x - 3 > 0$ **b.** $3 - x < 0$

Solution The graphs are shown in Example 1, Figure 19.

a. The graph of $y = 2x - 3$ intersects $y = 0$, the x-axis, at $(1.5, 0)$. A test point $x = 2$ is to the right of $x = 5$, and $2(2) - 3 > 0$ is true. Thus, $x > 1.5$ solves $2x - 3 > 0$; the solution is shown in Figure 20.

b. The graph of $y = 3 - x$ intersects $y = 0$, the x-axis, at $(3, 0)$. A test point $x = 4$ is to the right of $x = 3$, and $3 - 4 < 0$ is true. Thus, $x > 3$ solves $3 - x < 0$; the solution is shown in Figure 21.

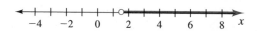

Figure 20

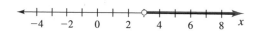

Figure 21   ●

The second method relies completely on the graph.

Solving an Inequality: Method 2

1. Graph the left and right sides of the equation separately.

2. Locate the point of intersection of the graphs. The x-coordinate is the starting point for the solution set.

3. Find the direction of the inequality from the positions of the graphs:

   a. For an inequality containing $<$ or $\leq$ to be true, the graph of the left side must be below the graph of the right side.

   b. For an inequality containing $>$ or $\geq$ to be true, the graph of the left side must be above the graph of the right side.

EXAMPLE **3** Solving inequalities with a graph   Solve these inequalities with Method 2.

a. $3 - x > -2$ **b.** $2x - 3 < 3 - x$

Solution **a.** The graph of $y = 3 - x$ intersects $y = -2$ at $(5, -2)$. The graph of $y = 3 - x$ is above $y = -2$ to the left of $x = 5$. Thus, $x < 5$ solves $3 - x > -2$. We check with a test point: $x = 4$ satisfies $x < 5$, and $3 - 4 > -2$ is true. Thus, the solution set is as shown in Figure 22.

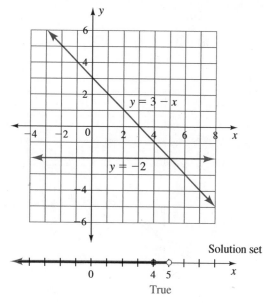

Figure 22

b. The graph of $y = 2x - 3$ intersects $y = 3 - x$ at $(2, 1)$. The graph of $y = 2x - 3$ is below the graph of $y = 3 - x$ to the left of $x = 2$. Thus, $x < 2$ solves $2x - 3 < 3 - x$. We check with a test point: $x = 1$ satisfies $x < 2$, and $2(1) - 3 < 3 - 1$ is true. Thus, the solution set is as shown in Figure 23.

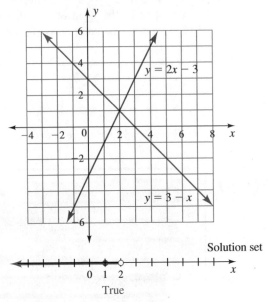

Figure 23

APPLICATION: WEDDING RECEPTION   The caterer for a wedding reception has three menu options: \$18 per person, \$21 per person, and \$25 per person. The total amount paid to the caterer will be the cost of a wedding cake (\$450) plus the cost of the menu choice. The total budget for food is \$3000. The number of people who could attend under each option is found by solving three inequalities:

$$18x + 450 \leq 3000$$
$$21x + 450 \leq 3000$$
$$25x + 450 \leq 3000$$

EXAMPLE **4**   Solving inequalities with a graph   Use Figure 24 to solve the inequalities graphically.

a. $18x + 450 \leq 3000$ **b.** $21x + 450 \leq 3000$ **c.** $25x + 450 \leq 3000$

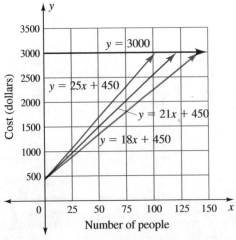

Figure 24

Solution

a. We trace along the line $y = 18x + 450$ to the intersection with $y = 3000$. The x-coordinate of the point of intersection is about 140. The graph of $y = 18x + 450$ is below $y = 3000$ to the left of the point of intersection, so the inequality $18x + 450 \leq 3000$ is true for $x \leq 140$.

b. We trace along the line $y = 21x + 450$ to the intersection with $y = 3000$. The x-coordinate of the point of intersection is about 120. The graph of $y = 21x + 450$ is below $y = 3000$ to the left of the point of intersection, so the inequality $21x + 450 \leq 3000$ is true for $x \leq 120$.

c. Similarly, with the point of intersection about 100, the inequality $25x + 450 \leq 3000$ is true for $x \leq 100$.

Graphing Calculator Solution

Repeating the process on a graphing calculator, we can find more accurate solutions: $x \leq 141$, $x \leq 121$, and $x \leq 102$, respectively. ●

APPLICATION: PHOTOCOPY CARD   A prepaid photocopy card costs $10. Each copy costs $0.10. The cost per copy is subtracted from the value of the card as copies are made. The value on the photocopy card after x copies have been made is $y = 10 - 0.10x$.

EXAMPLE **5**

Solving an inequality with a graph: photocopy card

a. Write an inequality that can be solved to find the number of copies that have been made if the value on the card is under $7.

b. Solve the inequality.

Solution

a. $10 - 0.10x < 7$

b. Figure 25 shows the graph of $y = 10 - 0.10x$ intersecting with $y = 7$ at $(30, 7)$. The value on the card, the graph of $y = 10 - 0.10x$, is below $y = 7$ to the right of the intersection—that is, for $x > 30$. Thus, for more than 30 copies, the card's value will be under $7.

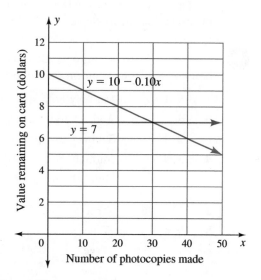

Figure 25 ●

Solving Linear Inequalities in One Variable with Algebraic Notation

For the most part, solving inequalities involves the same properties as solving equations. Steps in solving inequalities make equivalent inequalities. **Equivalent inequalities** *have the same solution set.*

Addition Property of Inequalities

> When a number is added to (or subtracted from) each side of an inequality, the result is an equivalent inequality. The direction of the inequality is not changed.
>   If $a < b$ and c is any real number,
>
> $$a + c < b + c$$

Multiplication Property of Inequalities I

> When an inequality is multiplied or divided by a positive number, the result is an equivalent inequality. The direction of the inequality sign is not changed.
>   If $a < b$ and $c > 0$,
>
> $$ac < bc$$

These properties hold for all other inequalities: $\leq$, $>$ and $\geq$.
We now return to our caterer situation.

EXAMPLE 6

Solving an inequality with algebraic notation   Solve $21x + 450 \leq 3000$.

Solution

$21x + 450 \leq 3000$	Subtract 450 from both sides.
$21x \leq 2550$	Divide both sides by 21.
$x \leq 121.4$	

The number of people who could attend at \$21 per person is $x \leq 121$. ●

The exceptions to solving inequalities are what cause difficulty. We explore the exceptions in Example 7.

EXAMPLE 7

Exploration: investigating operations on inequalities   Perform the indicated operation on each true statement. Is the resulting statement true or false?

a. $-5 < 8$ Multiply both sides by 2.
b. $-4 < 3$ Add -5 to both sides.
c. $-2 < 4$ Subtract 10 from both sides.
d. $-3 < 2$ Multiply both sides by -4.
e. $-6 < -2$ Divide both sides by -2.

Solution
a. $-10 < 16$ True
b. $-9 < -2$ True
c. $-12 < -6$ True
d. $12 < -8$ False
e. $3 < 1$ False

When we multiplied by a negative number in part d and divided by a negative number in part e, we obtained false statements. These statements would be true if we reversed the inequality signs.

$12 > -8$ True

$3 > 1$ True   ●

The difference between solving an equation and solving an inequality can be summarized as follows.

Multiplication Property of Inequalities II

When each side of an inequality is multiplied or divided by a negative number, reverse the direction of the inequality sign.
    If $a < b$ and $c < 0$, then

$$ac > bc$$

A change in the direction of an inequality sign is needed only for multiplication and division by a negative. For all other additions, subtractions, multiplications, and divisions, the inequality sign is unchanged.

In Example 8, we return to the photocopy card.

EXAMPLE 8

Solving an inequality with a negative coefficient on the variable term: photocopy card   The value on the prepaid photocopy card after x copies have been made is $y = 10 - 0.10x$. For how many copies will the value on the card be under $7?

Solution   The value on the card is to be less than $7, so we write $10 - 0.10x < 7$.

$10 - 0.10x < 7$	Subtract 10 from both sides.
$10 - 0.10x - 10 < 7 - 10$	Simplify.
$-0.10x < -3$	Divide both sides by -0.10, and reverse the inequality sign because of the division by a negative.
$\dfrac{-0.10x}{-0.10} > \dfrac{-3}{-0.10}$	Simplify.
$x > 30$	

A Strategy for Inequalities

Keeping a positive coefficient on the variable term avoids changing the inequality sign.

In Example 9, we add the term with a negative numerical coefficient to both sides to avoid division by a negative number.

EXAMPLE 9

Keeping a positive coefficient on the variable term   Solve $10 - 0.10x < 7$ by first adding $0.10x$ to each side.

Solution

$10 - 0.10x < 7$	Add $0.10x$ to each side.
$10 < 7 + 0.10x$	Subtract 7 from each side.
$3 < 0.10x$	Divide each side by 0.10.
$3 < 0.10x$	Simplify.
$30 < x$	

The result, $x > 30$, agrees with that in Examples 5 and 8.

Student Note: A common application of algebra is in rearranging equations to match an answer. Always first consider your answer to be correct, and then determine what might have been done by someone else to obtain the answer given.

Think about it: With algebraic notation, solve the alternative inequality for the photocopy machine, $7 > 10 - 0.10x$. Your answer should agree with that in Example 9.

Solving Linear Inequalities in Two Variables

Linear Inequality in Two Variables

> A linear inequality in two variables can be written $ax + by < c$, where a, b, and c are real numbers and a and b are not both zero.

The definition holds for all other inequalities: $\leq$, $>$, and $\geq$.

The solution set to a linear inequality in two variables is a half-plane. A **half-plane** is *the region on one side of a line*. The graph of every straight line creates two half-planes. *The line between the half-planes* is called the **boundary line**. The name *linear inequality* comes from the fact that the boundary is a line.

In Figure 26, the region to the right of the boundary line $3x + 2y = 6$ is the half-plane showing the solution to $3x + 2y > 6$. Together the boundary line and the shaded region are the solution set to $3x + 2y \geq 6$.

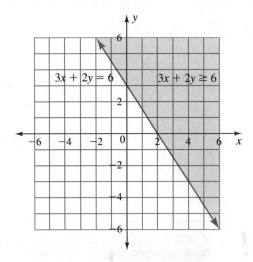

Figure 26

Solving a Linear Inequality in Two Variables

> 1. Graph the boundary line formed by replacing the inequality sign with an equal sign. Use a dashed line if the inequality is $<$ or $>$. Use a solid line if the inequality is $\leq$ or $\geq$.
>
> 2. Select a test point, not on the boundary line. Substitute the ordered pair for the test point into the inequality.
>
> 3. **a.** If the test point makes a true statement, shade the half-plane that contains the test point.
>
>   **b.** If the test point makes a false statement, shade the half-plane that does not contain the test point.

The inequality $ax + by < c$ is generally for noncalculator use. When we use a graphing calculator, we must solve the inequality for y.

Graphing Calculator Technique:
Graphing an Inequality in
Two Variables

Graphing calculators have a graphing option for inequalities, which starts with the equation of the boundary line and adds shading. Solve the boundary equation for *y*. Enter the equation in $\boxed{Y=}$. Check the manual for your calculator to find the location of the SHADE option. On some calculators, you simply move the cursor to the far left of Y_1 and press $\boxed{ENTER}$ until you see the correct shading option. Set an appropriate viewing window. Graph. A test point will help you identify the correct shading option.

EXAMPLE **10**   Shading the half-plane   Draw the indicated boundary line and shade the region described.

a. $y = 3, y \geq 3$ **b.** $y = x, y > x$ **c.** $x = -6, x \leq -6$

Solution

a.

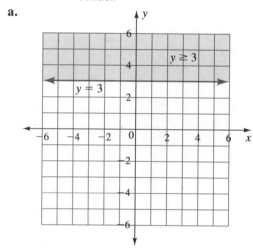

b.

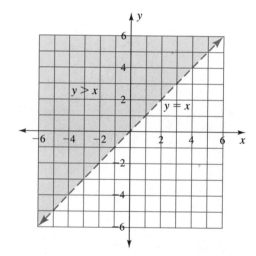

The line $y = x$ must be dashed to fit the inequality $y > x$.

c.

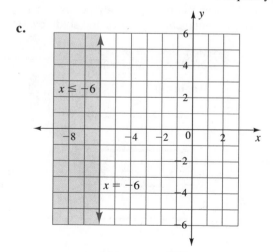

APPLICATION: MEETING A GOAL   When we have two different ways to meet a goal (such as a goal for dollars, calories, or enrollment), we can describe the possible outcomes with an inequality.

EXAMPLE  Writing and solving a two-variable inequality The university has at least 2400 freshmen who need to enroll in history courses. The school can offer large lectures of 300 students each or small research-based study courses of 40 students each.

a. Write an inequality that shows the numbers of each course the university can offer.

b. Solve the inequality with a graph.

c. Explain the meaning of the boundary line and half-plane in the problem setting.

Solution   **a.** Let x = number of large lectures. Let y = number of study courses. The total students served needs to be larger than 2400. The inequality is

$$300x + 40y > 2400$$

b. The solution is shown in Figure 27.

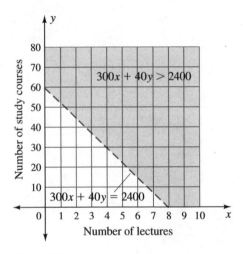

Figure 27

c. The boundary line shows that the university can serve 2400 students by offering 8 lectures or 60 study courses. Any point to the right shows a combination of lectures and study courses serving more than 2400 students. Only whole numbers of lectures and study courses would be meaningful.   ●

ANSWER BOX

Warm-up:

1.

2.

3.

4.

5.

6.

7–9.

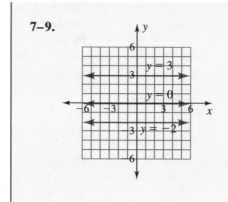

Think about it: $7 > 10 - 0.10x$, $0.10x + 7 > 10$, $0.10x > 3$, $\dfrac{0.10x}{0.10} > \dfrac{3}{0.10}$, $x > 30$

Example 1: d. $x > 1.5$
e. $x > 3$ **f.** $x < 5$ **g.** $x < 2$

EXERCISES 1.4

In Exercises 1 to 14, use the graphs to solve the inequalities, record the solution on a number line, and then solve each inequality with algebraic notation.

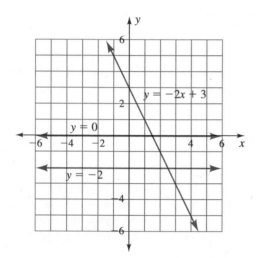

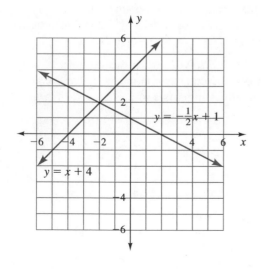

1. $-2 < -2x + 3$

2. $-2 > -2x + 3$

3. $-2x + 3 > 0$

4. $-2x + 3 < 0$

5. $x + 4 > -\frac{1}{2}x + 1$

6. $-\frac{1}{2}x + 1 > x + 4$

7. $0 < x + 4$

8. $x + 4 > 0$

9. $-\frac{1}{2}x + 1 \leq 0$

10. $-\frac{1}{2}x + 1 > 0$

11. $-2x + 3 \geq 3$

12. $-2x + 3 < -1$

13. $x + 4 \leq 4$

14. $x + 4 \geq -2$

In Exercises 15 to 24, graph the left and right sides of the inequalities and solve each inequality from the graph. Record the solution on a number line.

15. $1 - 2x < 3$ 16. $4 < x - 2$

17. $-2 < 1 - 2x$ 18. $x - 2 > -2$

19. $x < 1 - 2x$ 20. $x - 2 < -x$

21. $x \geq 1 - 2x$

22. $x - 2 \geq -x$

23. $x + 2 \leq 1 - 2x$

24. $2 - x \leq x - 2$

25. Graph each side of $x - 2 < x$. What do you observe about the solutions to this inequality?

26. Graph each side of $1 - 2x \le 4 - 2x$. What do you observe about the solutions to this inequality?

27. Use your graph in Exercise 25 to write an inequality that is never true.

28. Use your graph in Exercise 26 to write an inequality that is never true.

Solve the inequalities in Exercises 29 to 32 using any method.

29. $1 - 2x < x$ 30. $-2x + 3 > x$

31. $-\frac{1}{2}x + 1 < 4 - x$ 32. $4 - x < -2x + 3$

33. Explain why we need to reverse the inequality sign when we multiply an inequality on both sides by a negative. Illustrate your explanation with a numerical example as needed.

34. Explain how we can avoid multiplication or division by a negative when solving an inequality such as $3 - 2x > 5$.

In Exercises 35 to 40, draw the indicated line, and shade the half-plane described.

35. $x = 3, x \ge 3$ 36. $y = -2, y \ge -2$

37. $y = 1, y \ge 1$ 38. $x = -2, x \le -2$

39. $y = x, y \le x$ 40. $y = -x, y \ge -x$

In Exercises 41 to 48, graph the inequalities.

41. $x + y > 3$ 42. $x - y < 2$

43. $x - y \le 4$ 44. $x + y \ge -2$

45. $2x + 3y < 6$ 46. $2x - 3y > 6$

47. $3x - 2y \ge 12$ 48. $3x + 2y \le 12$

49. To meet its bid on an after-season bowl game, a school must earn $500,000 from ticket sales. Student tickets are $15, and regular tickets are $50. Write and graph an inequality showing the set of possible sales that would meet the goal. Explain the solution in terms of the problem setting.

50. A daily diet calls for 3500 mg of potassium. A serving of dried apricots contains 480 mg of potassium. A serving of dates contains 240 mg of potassium. Suppose these are the only sources of potassium in the diet. Write and graph an inequality to show the set of possible servings. Explain the solution in terms of the problem setting.

51. A maximum 2500-calorie diet in a hospital is from two sources: intravenous (I.V.) and a "total nutrition" liquid by mouth. If each I.V. solution contains 500 calories and the nutrition liquid contains 250 calories, write and graph an inequality that shows the possible sources of the calories.

52. A monthly budget permits spending a maximum of $75 on movies and books. Books cost an average of $15. Movies cost $5. Write and graph an inequality that shows the possible ways to spend up to $75. Explain the solution in terms of the problem setting.

53. Explain why some endpoints on line graphs are dots and others are small circles.

54. Explain why some boundary lines on graphs are dashed and others are solid.

55. Explain how to find which side of a boundary line to shade when graphing the solution set to an inequality.

56. Write an example in numbers that shows why we must change the direction of an inequality when multiplying or dividing by a negative number.

Project

57. ***Solving Compound Inequalities.*** The compound inequality $a < x < c$ may be written as two inequalities: $a < x$ and $x < c$. A compound inequality may be solved with algebraic notation by applying the properties of inequalities.

    Example: In $4 < x + 5 < 6$, subtract 5 from each part of the inequality to obtain $-1 < x < 1$. In $4 < -2x < 6$, divide both sides by -2 and reverse the inequality: $4/-2 > x > 6/-2$ or $-2 > x > -3$.

    The inequality is "solved" when the variable, x, is by itself between the other two numbers. Solve the compound inequalities in parts a to f.

    a. $10 < 5x < 20$ b. $3 < x - 2 < 7$

    c. $10 < -5x < 20$ d. $-4 < x + 6 < -1$

    e. $-3 < 2 - x < 4$ f. $-5 < 3 - 2x < -1$

    g. Explain how you would check your work.

    h. Explain how the graph of the inside expression, such as $y = x + 5$ in $4 < x + 5 < 6$, could be used to solve the inequality.

CHAPTER *1* SUMMARY _____

Vocabulary

For definitions and page references, see the Glossary/Index.

absolute value symbol	irrational numbers
additive inverses	like terms
associative property for addition	multiplicative inverse
associative property for multiplication	natural numbers
axes	numerical coefficient
base	opposites
boundary line	order of operations
braces	ordered pair
brackets	origin
commutative property for addition	parabola
commutative property for multiplication	parentheses
compound inequality	perpendicular
constant	power of a number
constant term	product
coordinate plane	quadrants
difference	quotient
distributive property of multiplication over addition	rational numbers
equivalent equations	real numbers
equivalent inequalities	reciprocal
evaluate	scale
exponent	set
expression	simplify
factor	solution set
factoring	solving an equation
fraction bar (horizontal)	square root
grouping symbols	square root symbol
half-plane	sum
inequality	term
input-output relationship	undefined
integers	variable
inverse order of operations	whole numbers
	x-axis
	x-coordinate
	y-axis
	y-coordinate

Concepts

See the vocabulary list for important definitions. In this chapter, algebra is presented visually, numerically, and symbolically (with algebraic notation).

1.0 Review of Real Numbers

Identify definitions and sets of numbers, properties of real numbers, and the order of operations.

Simplify expressions by applying the properties of real numbers or doing operations.

The square root of a negative number is undefined in the set of real numbers.

The compound inequality $a < x < c$ may be written as two inequalities, $a < x$ and $x < c$.

The phrase *a and b* is true if both *a* and *b* are true.

The phrase *a or b* is true if either the condition *a* is true or the condition *b* is true or if both conditions are true.

1.1 Input-Output Tables and Expressions

Polya's problem-solving steps are to (1) understand the problem, (2) make a plan, (3) carry out the plan, and (4) check the solution and extend it to other situations.

Simplify expressions by adding like terms or changing fractions to lowest terms.

Evaluate formulas by stating the formula and the given information, substituting the information into the formula, estimating, and finding the result.

1.2 Coordinate Graphs and Equations

The input-output relationship or rule for a table tells how to get from the input number to the output number across the table.

When graphing, place inputs (independent variables) on the horizontal axis and outputs (dependent variables) on the vertical axis.

To estimate the appropriate scale for a graph, find 10% of the difference between the highest number and lowest number to be graphed.

When we evaluate an expression, we are given an input and must find the output. When we solve an equation written $y = \ldots$, we are given an output and must find the input.

1.3 Solving Equations and Formulas

An operation is undone by its inverse operation.

Subtraction and addition are inverse operations; multiplication and division are inverse operations.

A number or expression is the solution to an equation (or formula) if its substitution into the equation (or formula) results in $a = a$.

1.4 Solving Inequalities

The solution set to inequalities in one variable may be shown on a number line.

A number may be added to or subtracted from both sides of an inequality without changing the direction of the inequality.

A positive number may be multiplied or divided on both sides of an inequality without changing the direction of the inequality. A negative number multiplied or divided on both sides of an inequality will change the direction of the inequality.

The solution set to inequalities in two variables is a half-plane.

CHAPTER ➊ REVIEW EXERCISES _____

For Exercises 1 to 12, use the vocabulary words from page 69.

1. List five properties of real numbers.

2. List five sets of numbers included in the real numbers.

3. List six grouping symbols.

4. List words related to rectangular coordinate graphing. Draw a picture and label the words on the picture.

5. List two words related to graphing inequalities in two variables.

6. List seven words describing numbers or letters or groupings of numbers and letters in algebraic notation.

7. List four words describing answers to operations.

8. List three phrases that have to do with solving equations.

9. List three words that describe inputs and outputs.

10. List two words that describe 2 and −2 as a pair of numbers.

11. List two words that describe $\frac{1}{2}$ and 2 as a pair of numbers.

12. List two words that describe the input and output variables.

For Exercises 13 to 16, complete the table.

	Input x	Input y	Output $x + y$	Output $x - y$
13. a.	−7	3		
b.	3	−7		
c.	7			4
14. a.	−3			−10
b.		7	10	
c.		−7		4
15. a.	3	4	12	
b.	−2	3		
c.	−7		21	
d.	−3	5		
16. a.	2		−6	
b.	4			1
c.		−2		−3
d.	3			2

Show how the real-number properties let us do the computations in Exercises 17 to 20 mentally. Tell what property you use.

17. $\left(\frac{2}{3} + 1\frac{1}{2}\right) + \frac{1}{2}$

18. $\frac{3}{8} + \frac{1}{3} + \frac{5}{8}$

19. $25 \cdot 13 \cdot 4$

20. $5(6 \cdot 7) = (5 \cdot 6) \cdot 7$

Simplify the expressions in Exercises 21 to 26.

21. Puzzle problem: $-2^2 + 5 - 3(4 - 5)$

22. Puzzle problem: $3\{6 - 2[8 - 3(9 - 12) + 3] - 1\}$

23. Surface area of a metal can:
$2\pi(1.5 \text{ in.})^2 + 2\pi(1.5 \text{ in.})(4 \text{ in.})$

24. Area of a trapezoid: $\frac{1}{2}(5 \text{ in.})(4.5 \text{ in.} + 7.6 \text{ in.})$

25. Eighth term of an arithmetic sequence:
$\$20.00 + (8 - 1)(-\$1.75)$

26. Sum of eight terms of an arithmetic sequence:
$\frac{1}{2}(8)(\$20 + \$7.75)$

In Exercises 27 and 28, write the algebraic notation in words.

27. a. $3 - 2x$ **b.** $3(x - 5)$

28. a. $4x + 5$ **b.** $5(x + 3)$

In Exercises 29 and 30, write the word phrases in algebraic notation.

29. Eight more than the quotient of a number and 15.

30. The product of six and the difference between 5 and a number.

In Exercises 31 and 32, eliminate parentheses as needed and add like terms.

31. $9 - 3(2x - 5y) - 4(5x - 3y)$

32. $a(b^2 + 3b - 1) - 3b(a^2 - 2a + 1) - ab(a - b)$

Evaluate the expressions in Exercises 33 and 34.

33. $a^2 - b^2$ for $a = -2$ and $b = -5$

34. $a^2 + 2ab + b^2$ for $a = -3$ and $b = -4$

Evaluate the formulas in Exercises 35 and 36.

35. $A = \pi r^2$ for $r = 3.5$ in.

36. $A = \frac{1}{2}bh$ for $b = 4.5$ cm and $h = 6.5$ cm

Build an input-output table and a graph for the equations in Exercises 37 to 40.

37. $y = -x$ **38.** $y = 4 - 2x$

39. $y = 30 - 3.5x$ **40.** $y = 150x + 300$

In Exercises 41 to 46, solve the equations as directed.

41. $2x + 2(1.5) = 11$ for x

42. $25 = \frac{1}{2}(2.5)(x)$ for x

43. $40 = \frac{5}{9}(F - 32)$ for F

44. $-40 = \frac{5}{9}(F - 32)$ for F

45. $20 + (n - 1)(-1.75) = 0.75$ for n

46. $60 + 3.75(n - 5) = 150$ for n

In Exercises 47 to 50, solve the formulas for the indicated letter.

47. $I = Prt$ for r

48. $A = \frac{1}{2}bh$ for b

49. $C = a + bY$ for b

50. $y = mx + b$ for b

In Exercises 51 and 52, find the greatest common factor for the terms shown and then write the expression as a product.

51. $6x^2 + 15x$ **52.** $12a^2b - 16ab^2$

Solve the inequalities in Exercises 53 to 56 from the graphs in the figure. Record your answer on a number line.

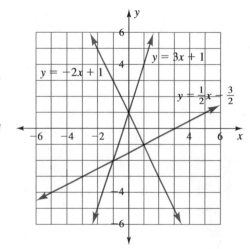

53. $3x + 1 > 4$ **54.** $-2x + 1 \le 3$

55. $\frac{1}{2}x - \frac{3}{2} \ge 0$ **56.** $\frac{1}{2}x - \frac{3}{2} < -1$

Solve the equations and inequalities in Exercises 57 to 59 from their graphs.

57. a. $3x + 1 = -2x + 1$ **b.** $3x + 1 > -2x + 1$

58. a. $\frac{1}{2}x - \frac{3}{2} = -2x + 1$ **b.** $\frac{1}{2}x - \frac{3}{2} \le -2x + 1$

59. a. $3x + 1 = \frac{1}{2}x - \frac{3}{2}$ **b.** $\frac{1}{2}x - \frac{3}{2} \ge 3x + 1$

For Exercises 60 to 62, solve the given exercise with algebraic notation.

60. Exercise 57   **61.** Exercise 58   **62.** Exercise 59

Solve the inequalities in Exercises 63 and 64 by graphing.

63. $y > x + 3$ **64.** $y \le 2x - 1$

65. Draw a picture or describe how these sets of numbers are related to each other: natural numbers, rational numbers, real numbers, integers, whole numbers.

66. Explain how solving an inequality with algebraic notation differs from solving an equation with algebraic notation.

CHAPTER ❶ TEST

1. Another name for additive inverse is _____.

2. A(n) _____ is used to locate a point on the coordinate graph.

3. A(n) _____ variable is the input, x, or set of numbers on the horizontal axis.

4. Draw a pair of rectangular coordinate axes. Show an example of and/or label the following: origin, quadrant 4, half-plane for $x < 0$, y-intercept.

5. Add, subtract, multiply, and divide (in the order given).

   a. -27 and -3 b. -1.5 and 0.5

   c. $1\frac{1}{2}$ and $-\frac{1}{4}$

6. Describe how a property of the real numbers lets us do these computations mentally.

   a. $(348 + 295) + 105$

   b. $230 + 689 + 70$

7. Simplify these expressions.

   a. $4\{2 - 3[4 + 5(2 - 5) + 5\}$

   b. -3^2

   c. $(-3)^2$

   d. $\frac{1}{2} \cdot 25(17 + 23)$

   e. $3x - (8 - 3x)$

8. Write in words: $4(x - 3)$.

9. Solve $A = \dfrac{a + b + c}{3}$ for b.

10. Solve $A = \dfrac{h}{2}(a + b)$ for b.

11. Find the greatest common factor in $15x^2y - 39xy^2$.

12. Use the graph in the figure to solve the following equations and inequalities. Show the solutions to the inequalities on a number line.

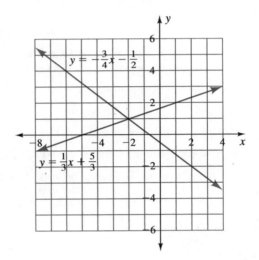

   a. $\frac{1}{3}x + \frac{5}{3} = 0$ b. $-\frac{3}{4}x - \frac{1}{2} = -2$

   c. $\frac{1}{3}x + \frac{5}{3} = -\frac{3}{4}x - \frac{1}{2}$ d. $-\frac{3}{4}x - \frac{1}{2} < -2$

   e. $\frac{1}{3}x + \frac{5}{3} < -\frac{3}{4}x - \frac{1}{2}$ f. $\frac{1}{3}x + \frac{5}{3} < 2$

13. Solve parts b and f in Exercise 12 using algebraic notation.

14. Show the solution set to $y \geq \frac{1}{3}x + \frac{5}{3}$ on a set of axes.

Equations, Functions, and Linear Functions

Relative humidity (%)

	0	10	20	30	40	50	60	70	80	90	100
125	111	123	141								
120	107	116	130	148							
115	103	111	120	135	151						
110	99	105	112	123	137	150					
105	95	100	105	113	123	135	149				
100	91	95	99	104	110	120	132	144			
95	87	90	93	96	101	107	114	124	136		
90	83	85	87	90	93	96	100	106	113	122	
85	78	80	82	84	86	88	90	93	97	102	108
80	73	75	77	78	79	81	82	85	86	88	91
75	69	70	72	73	74	75	76	77	78	79	80

Air temperature (°F)

Table 1   Source: *Heat Wave*, U.S. Department of Commerce, National Oceanic and Atmospheric Administration, NOAA/PA 85001.

More than 1250 people died in the United States during the heat wave of 1980. East of the Rocky Mountains, summer combines high temperatures with high humidities, causing even greater heat. The heat index chart in Table 1 shows the apparent temperature for various air temperatures (in °F) and relative humidities (%). The heat index (or apparent-temperature) values were found for shady, light-wind conditions. Exposure to full sunshine can increase the heat index values by up to 15°. Apparent temperatures above 105° are considered dangerous for lengthy exposure or physical activity.

This chapter covers special input-output relationships called functions and ways to write these relationships in function notation. The focus is on functions related to straight lines: linear functions. We look at number patterns, special straight lines (horizontal lines and the line $y = x$), and absolute value in terms of the function concept. The chapter opens with more practice in writing and solving linear equations.

2.0   Writing and Solving Linear Equations _____

OBJECTIVES

- Apply Polya's problem-solving steps.
- Build linear equations with guess and check.
- Write linear equations from word statements containing conditions.
- Graph linear equations containing conditions.
- Solve linear equations from graphs.

WARM-UP

Write each statement as a formula. Choose appropriate letters for variables.

1. Distance traveled is the product of rate and time.

2. Simple interest earned is the product of rate per year, amount invested, and number of years invested.

3. The circumference of a circle is the product of pi and the diameter.

4. The perimeter of a rectangle is the sum of twice the length and twice the width.

5. The area of a trapezoid is half the product of the height and the sum of the two parallel sides.

Write a sentence to translate each formula into words.

6. Total cost with sales tax: $C = x + 0.075x$

7. Area of a circle: $A = \pi r^2$

8. Volume of a box with rectangular sides: $V = l \cdot w \cdot h$

9. Surface area of a cylinder: $A = 2\pi r^2 + 2\pi rh$

10. Volume of a sphere: $V = \frac{4}{3}\pi r^3$

THIS SECTION INTRODUCES guess and check as a means of building expressions and equations from word-problem settings. Writing equations from sentences, which was introduced in Section 1.3, is now extended to include equations with conditions.

Linear Equations

Linear Equation in One Variable

> A **linear equation in one variable** can be written
>
> $$ax + b = 0$$
>
> where a and b are any real number and a is not zero.

We work with linear equations in one variable when we solve equations. These are examples of linear equations in one variable:

$$x = 4, \quad 2x - 3 = 0, \quad 2\pi r = 8$$

Linear Equation in Two Variables

A **linear equation in two variables** can be written

$$ax + by = c$$

where a, b, and c are any real number and a and b are not both zero.

We work with linear equations in two variables when we graph. These are examples of linear equations in two variables:

$$y = 2x - 3, \qquad C = 2\pi r$$

Substituting a number for one of the variables in an equation in two variables changes it to an equation in one variable. For example, if we let $y = 13$ in the two-variable equation $y = 2x - 3$, then we have $13 = 2x - 3$, which is an equation in one variable.

Building Linear Equations through Guess and Check

We begin building linear equations by identifying relationships within a problem setting using guess and check. When guess and check fails to give a solution, the work provides an estimate of the solution.

POLYA'S PROBLEM-SOLVING STEPS A good guess-and-check process follows Polya's four problem-solving steps: understand, plan, carry out the plan, and check.

- We move toward *understanding* by reading carefully.
- We *plan* by considering what might be reasonable inputs and preparing a table in which to record our guesses.
- We *carry out the plan* by working through the problem with the chosen input.
- We *check* by comparing the result with the conditions or requirements of the original problem.

In Examples 1 and 2, we use a table to organize our steps and record our guesses so that we learn from each guess. *Each row in the table represents a guess.*

EXAMPLE **1** Building equations with guess and check: producing chocolate bars   Janelle has 135 pounds of chocolate. Using a chocolate press (see Figure 1), she makes both 12-ounce bars and 1.5-ounce bars. She sells ten times as many small bars as large bars. Write and solve an equation to find how many of each she should make.

Solution ***Understand***: Janelle needs to make two types of candy bars, and the numbers of the two types are related by a factor of 10.

Plan: The input will be the number of large chocolate bars to be made. We set up a table with columns for the number of each type of candy bar and the number of ounces of chocolate used in making each type. The output will be the total ounces of chocolate. We change pounds to ounces in order to be consistent with the size of the chocolate bars:

$$135 \text{ lb} \cdot \frac{16 \text{ oz}}{1 \text{ lb}} = 2160 \text{ oz}$$

Carry out the plan: To get started, we guess an arbitrary number of large bars—say, 15. Because the number of small bars sold is ten times the number of large bars sold, we enter $10(15) = 150$ in Table 2 for the number of small bars. We calculate the total weight: 405 ounces.

Figure 1   Chocolate press

Our guess of 15 large bars produced too few total ounces, so we increase the guess to 50. Our third guess, 100, is too high. We modify the guess by looking at the amount of change that occurred between other guesses.

Number of Large Bars	Total Weight of Large Bars (ounces)	Number of Small Bars	Total Weight of Small Bars (ounces)	Total Weight of All Bars (ounces)
15	15(12) = 180	10(15) = 150	150(1.5) = 225	180 + 225 = 405 too low
50	50(12) = 600	10(50) = 500	500(1.5) = 750	600 + 750 = 1350 too low
100	100(12) = 1200	10(100) = 1000	1000(1.5) = 1500	1200 + 1500 = 2700 too high
75	75(12) = 900	10(75) = 750	750(1.5) = 1125	900 + 1125 = 2025 too low

Table 2

Leaving expressions in the column helps in identifying patterns and building equations. A good guessing strategy will get close to an answer in four or five guesses. We have not found an answer in four guesses, so we try a guess of x (see Table 3).

We added the weights of the small and large bars to get the total weight. We repeat this addition to build an equation, and then we solve the equation.

Number of Large Bars	Total Weight of Large Bars	Number of Small Bars	Total Weight of Small Bars	Total Weight of All Bars
x	$x(12)$	$10x$	$10x(1.5)$	2160

Table 3

$x(12) + 10x(1.5) = 2160$	Multiply 10(1.5).
$12x + 15x = 2160$	Add like terms.
$27x = 2160$	Divide both sides by 27.
$x = 80$	The number of large bars
$10x = 800$	The number of small bars

Check: $80(12) + 800(1.5) \overset{?}{=} 2160.$

Janelle should make 80 large chocolate bars and 800 small ones. ●

Think about it 1: What strategies might you use to continue moving toward a solution with guess and check in Table 2?

EXAMPLE ❷   Building equations with guess and check: appliance repair cost   J's Appliance Repair charges a $22 service call fee and $28 per hour for the repair. A repair required $85 in parts. How long was the repair job if the total bill was $205? Write an equation, using guess and check. Solve the equation.

Solution   The phrases "how long" and "$28 per hour" tell us that our variable should be the number of hours. In Table 4, we set up hours as our variable to guess.

Hours	Cost Charged at $28 per Hour	Service Call Fee (dollars)	Parts (dollars)	Total Cost (dollars)
2	$2(28) = 56$	22	85	$56 + 22 + 85 = 163$ too low
4	$4(28) = 112$	22	85	$112 + 22 + 85 = 219$ too high
x	$x(28)$	22	85	205

Table 4

We add the three costs—the charge at $28 per hour, the service call fee, and the parts—to obtain the total cost:

$28x + 22 + 85 = 205$	Add like terms.
$28x + 107 = 205$	Subtract 107 from each side.
$28x = 98$	Divide both sides by 28.
$x = 3.5$ hr	

●

Think about it 2: In Example 2, how do we know that a guess of 10 hours is too large?

GUESSING STRATEGIES   Examples 1 and 2 illustrate two important points about guessing strategies.

First, *any number can be used to get started.* Although making an estimate of the answer is a good idea, the first guess does not need to be close to the correct answer. If you have trouble estimating in an unfamiliar situation, any number—

such as the day of the month on which your birthday occurs—will get you started. The first guess will show whether the column headings are reasonable and whether the solution to the problem is larger or smaller than the guess.

Second, *doubling or halving may be useful in guessing possible solutions.* In Example 1, we doubled 50 large bars (too small) to get 100 large bars (too large). To find a middle number, we halved the distance between 50 and 100 and selected 75 as a next guess.

Think about it 3: In Example 1, a guess of 75 large bars was too low, and a guess of 100 large bars was too high. With halving, what are the next two guesses? In Example 2, a guess of 2 hours was too low, and a guess of double 2, or 4, was too high. With halving, what are the next two guesses?

Writing Linear Equations with Conditions

CONDITIONAL SETTINGS   Have you ever rented a videotape and had to pay a late fee when you returned it? If so, you have experienced a conditional setting.

To describe buying or renting when the unit cost changes, we use equations with conditions. A **condition** is *a limitation on the inputs, derived from the problem setting, that determines what equation is used to solve the problem.* Suppose rental of a videotape costs $3.00 for two days and a $4 late fee for each additional day. If x is the input, the video rental setting has the conditions that inputs are $1 \le x \le 2$ or $x > 2$.

EXAMPLE **3**   Finding conditions in problem settings   How are these examples the same? Describe the conditions on the inputs required by each setting.

a. Birthday party: $50 for up to 10 children, $4.50 for each additional child

b. Air compressor rental: $55 for up to 6 hours, $25 for each additional hour

c. Telephone call: $1.50 for the first minute, $0.10 for each additional minute

Solution   In each setting, there is a flat fee followed by a charge for additional children or units of time. If x is the input,

a. $1 \le x \le 10$ children; $x > 10$ children

b. $0 \le x \le 6$ hours; $x > 6$ hours

c. $0 < x \le 1$ minute; $x > 1$ minute   ●

EQUATIONS WITH CONDITIONS   Although creating a full guess-and-check table may not be appropriate in these problem settings, making a few entries in an input-output table can help us write equations.

EXAMPLE **4**   Writing equations for conditional settings   Table 5 shows the videotape rental cost for 1 to 5 days where renting a videotape costs $3.00 for two days and a $4 late fee for each additional day. Write two equations that describe the total cost of the rental.

Number of Days	Total Cost (dollars)
1	3
2	3
3	$3 + 4(1) = 7$
4	$3 + 4(2) = 11$
5	$3 + 4(3) = 15$

Table 5

Solution   The total cost is a flat fee for the first two days and then rises by \$4 for each additional day. Let x be the total number of rental days and y be the total cost.

$$y = 3, \qquad\qquad\quad 1 \le x \le 2$$
$$y = 3 + 4(x - 2), \quad x > 2$$

The inequality following the equation shows the days to which the equation applies.   ●

Think about it 4: Why is this equation wrong: $y = 3 + 4x$ for $x > 2$?

EXAMPLE 5

Writing equations for conditional settings   Write a set of equations for each of these settings.

a. Birthday party: \$50 for up to 10 children, \$4.50 for each additional child

b. Air compressor rental: \$50 for up to 6 hours, \$25 for each additional hour or part of an hour

c. Telephone call: \$1.50 for the first minute, \$0.10 for each additional minute or part of a minute

Solution   **a.** Let x = total number of children; let y = total cost in dollars.

$$y = 50, \qquad\qquad\qquad 1 \le x \le 10$$
$$y = 50 + 4.50(x - 10), \quad x > 10$$

b. Let x = total hours of rental. The phrase *part of an hour* means that any fraction of an hour is rounded up to the next highest hour. Let y = total cost in dollars.

$$y = 50, \qquad\qquad\quad 1 \le x \le 6$$
$$y = 50 + 25(x - 6), \quad x > 6$$

c. Let x = total minutes talked. The phrase *part of a minute* means that any fraction of a minute is rounded up to the next highest minute. Let y = total cost in dollars.

$$y = 1.50, \qquad\qquad\qquad x \le 1$$
$$y = 1.50 + 0.10(x - 1), \quad x > 1$$ ●

Graphs of Linear Equations with Conditions

In many settings, the points in a graph of equations with conditions are not connected with a straight line or smooth curve. Furthermore, the condition may cause a bend in the pattern formed by the points.

DOT GRAPHS   Because they relate to number of days and number of children, the video rental and party settings are meaningful only for positive integer inputs, and their graphs are dots.

EXAMPLE 6

Graphing equations for conditional settings   Graph the video rental and party cost equations.

Solution   The video rental equations are

$$y = 3, \qquad\qquad\quad 1 \le x \le 2$$
$$y = 3 + 4(x - 2), \quad x > 2$$

The graph is shown in Figure 2.

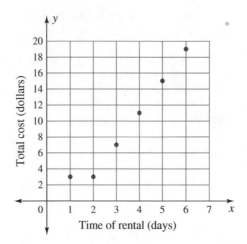

Figure 2

The party cost equations are

$$y = 50, \qquad\qquad 1 \le x \le 10$$
$$y = 50 + 4.50(x - 10), \quad x > 10$$

The graph is shown in Figure 3.

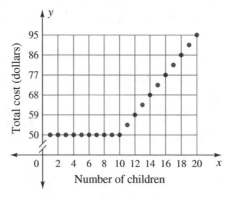

Figure 3

Think about it 5: What is the meaning of the double slash, //, on the vertical axis in the party cost graph in Figure 3? What is the spacing between the remaining numbers on the vertical axis, and why might that spacing have been chosen?

Dot Graphs

A **dot graph** is formed when inputs to an equation are limited to integers.

STEP GRAPHS   The inputs to the air compressor rental and telephone call settings are not limited to positive integers. The hours and minutes can be any positive numbers, and the graphs are step graphs—graphs formed by short, horizontal line segments.

EXAMPLE **7**   Graphing equations for conditional settings   Graph the air compressor and telephone call equations.

Solution   The air compressor rental equations are

$$y = 50, \qquad\qquad 0 \le x \le 6$$
$$y = 50 + 25(x - 6), \quad x > 6$$

Student Note: Remember that we round *x* before substituting, here and in the telephone call equations below.

The graph is shown in Figure 4.

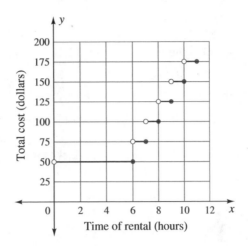

Figure 4

The telephone call equations are

$$y = 1.50, \qquad\qquad\qquad x \le 1$$
$$y = 1.50 + 0.10(x - 1), \quad x > 1$$

The graph is shown in Figure 5.

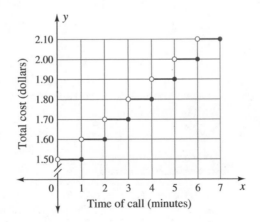

Figure 5

Think about it 6: Describe situations in which the small circle above zero in the graphs in Figures 4 and 5 might be dots—that is, situations in which there would be an output value when the input was zero.

Step Graphs

> A **step graph** is formed when fractional or decimal inputs are rounded to integers before they are substituted into the equation.

Think about it 7: In what way are all the graphs in Figures 2 through 5 the same?

Graphical Solutions of Equations Containing Conditions

Graphs give visual solutions to equations or to questions that may be difficult to describe with algebraic notation.

EXAMPLE 8

Solving equations from a graph Use Figures 2 and 3 to solve the equations in parts a and b. Check with algebraic notation.

a. $3 + 4(x - 2) = 19$

b. $50 + 4.50(x - 10) = 86$

c. If the party cost was $50, how many children attended?

Solution **a.** 19 on the vertical axis matches with 6 on the horizontal axis. The solution to $3 + 4(x - 2) = 19$ is $x = 6$.

$3 + 4(x - 2) = 19$	Multiply with the distributive property.
$3 + 4x - 8 = 19$	Add like terms.
$4x - 5 = 19$	Add 5 on each side.
$4x = 24$	Divide by 4 on each side.
$x = 6$	

b. 86 on the vertical axis matches with 18 on the horizontal axis. The solution to $50 + 4.50(x - 10) = 86$ is $x = 18$.

$50 + 4.50(x - 10) = 86$	Multiply with the distributive property.
$50 + 4.50x - 45 = 86$	Add like terms.
$5 + 4.50x = 86$	Subtract 5 from each side.
$4.50x = 81$	Divide by 4.50 on each side.
$x = 18$	

c. If the party cost $50, there could have been from 1 to 10 children attending. There is no way to work backwards from the total cost to a single number of children attending. ●

ANSWER BOX

Warm-up: 1. $D = r \cdot t$ **2.** $I = prt$ **3.** $C = \pi d$ **4.** $P = 2l + 2w$ **5.** $A = \frac{1}{2}h(a + b)$ **6.** Cost is sum of price and $7\frac{1}{2}\%$ of price. **7.** Area is product of pi and square of radius. **8.** Volume is product of length, width, and height. **9.** Surface area is product of 2π and square of radius added to product of 2π, radius, and height. **10.** Volume is product of $\frac{4}{3}$, pi, and cube of radius. **Think about it 1:** Once we know between what two numbers the answer is located, we can choose a number halfway as our next guess. **Think about it 2:** 10 hours at $28 per hour is $280, larger than the total bill of $205. **Think about it 3:** In Example 1, fractional chocolate bars don't make sense, so the guess halfway between 75 and 100 is 88, which is too high. The next guess, halfway between 75 and 88, is 82. In Example 2, the next guess is 3 (halfway between 2 and 4), which is too low. The following guess is 3.5 (halfway between 3 and 4). **Think about it 4:** For 3 days ($x = 3$), the rental charge would be $3 + 4(3) = \$15$. The \$4 is paid only for the number of days after the second day, $x - 2$, not all the days, x. **Think about it 5:** The double slash means that the spacing between 0 and 50 is not the same as the spacing between other numbers on the vertical axis. The other numbers are spaced 9 units apart because $2(4.50) = 9$. **Think about it 6:** There could be a charge for placing the rental order, whether or not you actually took the machine. There could be a connection charge for placing the phone call, even if no one answered the telephone. **Think about it 7:** Each starts with a horizontal section. The rest of the graph rises to the right, either in a line or as a series of steps.

EXERCISES 2.0

In Exercises 1 to 6, set up a guess-and-check table to solve each puzzle problem without an equation. In the last row, let your guess be x.

1. Three sides of a triangle add to 91. One side is three more than six times the smallest. The third side is three less than seven times the smallest. What is the measure of each side?

2. A \$1 million estate is divided three ways. The first part is \$50,000 more than the second part. The third part is \$30,000 less than the first part. How many dollars are in each part?

3. The total cost of a meal includes tip, sales tax, and the cost of the meal itself. DeMya spent \$33.88 altogether. Her tip was 15% of the price of the meal. The tip was 2.5 times the sales tax. What was the price of the meal?

4. Because of a delay in service, the store manager agrees to sell a new washing machine for \$350. This includes the delivery charge (\$24.95) and 7% sales tax; no tax is paid on the delivery charge. How much should he record as the purchase price?

5. Janelle makes $\frac{1}{2}$-ounce chocolate pieces for a business friend. She charges \$0.45 each to make the chocolates and to wrap them in foil. Special labels for the chocolates cost \$8.50 per 1000 labels. She can buy the labels only in sets of 1000 (and discards labels that are not used). How many chocolates can she make for the friend's budget of \$500?

6. A truck is rented for three days. The charge per day is \$37. The cost per mile is \$0.25. The rental agency gives a 5% discount because of a misaligned headlamp. If the payment after the discount is \$215.65, how many miles were driven?

In Exercises 7 to 16,

a. *Write equations with conditions on the inputs.*

b. *Explain whether a dot graph or a step graph is appropriate.*

c. *Draw a graph for the equations.*

7. A fax costs \$4 for up to two pages and \$0.50 for each page thereafter. What is the total cost, y, of x faxes?

8. A college transcript costs \$5 for the first copy and \$1 for each additional copy. If x is the total number of transcripts, what is the total cost, y?

9. Rental of a weed cutter costs \$20 for 3 hours plus \$5 for each additional hour or part of an hour. What is the total cost, y, of x hours of rental?

10. Rental of a pressure washer costs \$30 for the first 4 hours and \$10 per additional hour or part of an hour. If x is the total hours rented, what is the total cost, y?

11. A rental car costs \$65 for the first 100 miles and \$0.15 per mile thereafter. If x is the total miles driven, what is the total cost, y?

12. A taxi cab costs \$5 to hire and ride the first $\frac{1}{2}$ mile. Each additional $\frac{1}{10}$ mile costs \$0.75. If x is the total miles traveled, what is the total cost, y?

13. A party at an ice rink costs \$85 for up to 10 skaters and \$4.75 for each additional skater. What is the total cost, y, for x skaters?

14. A party at a roller skating rink costs \$40 for 10 skaters and \$4.50 for each additional skater. What is the total cost, y, for x skaters?

15. A cell phone costs \$19.95 each month for the first 100 minutes and \$0.25 for each additional minute or

portion of a minute. What is the total cost, y, for x minutes a month?

16. An e-mail account costs $9.95 each month for the first 10 hours and $2 for each additional hour or part of an hour. What is the total cost, y, for x hours a month?

For Exercises 17 to 26, write an equation and solve it both by graph and with algebraic notation.

17. In Exercise 7, how many pages can be sent by fax for a total cost of $8.50?

18. In Exercise 8, how many transcripts can be ordered for $11?

19. In Exercise 9, how many rental hours are obtained for $55?

20. In Exercise 10, how many rental hours are obtained for $80?

21. In Exercise 11, how many miles can be driven for $245?

22. In Exercise 12, how far can we travel by taxi for $12.50?

23. In Exercise 13, how many skaters can attend a party with a $100 budget?

24. In Exercise 14, how many skaters can attend a party with a $100 budget?

25. In Exercise 15, how many minutes can be bought on a $30 budget?

26. In Exercise 16, how many hours can be bought on a $30 budget?

Projects

27. *Postal Rates*

a. Complete this sentence, using the current postal rates for first-class postage: Mailing a letter costs _____ cents for the first ounce and _____ cents for each additional ounce or portion thereof up to _____ ounces.

b. Draw a graph of the total postage cost for 0 to 8 ounces.

28. *Guess-and-Check Practice.* Find a pair of numbers that fits each statement. As needed, guess a pair of numbers that fits the first condition; then see how closely your guess fits the second condition.

a. What two numbers have a sum of 6 and a product of 8?

b. What two numbers have a sum of 11 and a product of 30?

c. What two numbers have a sum of 12 and a difference of 6?

d. What two numbers have a difference of 5 and a sum of 7?

e. What two numbers have a product of 12 and a sum of 13?

f. What two numbers have a quotient of 5 and a sum of 30?

2.1   Functions

OBJECTIVES

- Identify a function.
- Describe a relationship with functions.
- Graph functions on coordinate axes.
- Apply the vertical-line test.
- Find the line of symmetry in tables and graphs.
- Evaluate functions written in function notation.
- Find the domain and range of a function.

WARM-UP

Complete the table for the equation given.

1.

Transcripts Ordered, x	Cost, y (dollars)
1	
2	
3	
4	
5	

Cost of transcripts, $y = 3 + 2(x - 1)$

2.

x	y
\	0
|	1
)	−1
	2
	−2

Parabola opening to the right, $x = y^2$

3.

Distance from Bridge Center, x (feet)	Height, y (feet)
−20	−12.8
−10	
0	
10	
20	

Height of suspension bridge cable, $y = \frac{4}{125}x^2$

THIS SECTION INTRODUCES functions and function notation. We examine tables and graphs of functions as well as methods of identifying functions and symmetry. We find the domain and range of functions.

Functions

For years, researchers have sought a cure for cancer, ways to encourage people to recycle, and methods to direct government spending to increase employment. A key element in this research is finding an association between action and results. In mathematics, we have a similar goal: identifying associations where, if we input x, only one outcome, y, can result.

Function

> A **function** is a relationship or association where for each value of input x there is exactly one value of output y.

Chapter 1 indicated that in some cases the output depends on the input, or y *depends on x*. Now we can use the term *function* to say that the output is a function of the input, or *y is a function of x*.

You probably have noticed that a hot day feels even hotter when the humidity is high. What may surprise you is that very low humidity will actually make a hot day seem cooler than the actual air temperature. We say that *for a given air temperature, the apparent temperature is a function of the relative humidity.*

The increase (or decrease) in temperature is measurable and is the basis for the heat index chart in Table 1 on the chapter opening page, where apparent temperature is related to relative humidity and air temperature. Example 1 shows data from the heat index chart and a graph for an air temperature of 85°F.

EXAMPLE Describing a relationship with a function: heat index   Table 6 and Figure 6 show the apparent temperature for several humidity levels and an air temperature of 85°F.

Relative Humidity (%)	Apparent Temperature (°F)
0	78
10	80
20	82
30	84
40	86
50	88
60	90
70	93
80	97
90	102
100	108

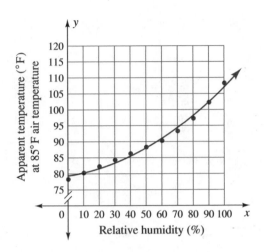

Figure 6

Table 6   Apparent temperatures for 85°F air temperature

a. Estimate the relative humidity when the apparent temperature is the same as the air temperature, 85°F.

b. For what levels of humidity is the apparent temperature cooler than the air temperature?

c. For what levels of humidity is the apparent temperature warmer than the air temperature?

d. For what levels of humidity is the apparent temperature warmer than 105°F? (This temperature greatly increases risk of sunstroke or heatstroke.)

e. For each input x, is there exactly one output y?

f. State the relationship in the table using the word *function*.

g. What other factors might change the apparent temperature?

Solution   a. approximately 35% (between 30% and 40% humidity)

b. approximately $x < 35\%$ c. approximately $x > 35\%$

d. approximately $x > 95\%$ e. Yes

f. The apparent temperature for an 85°F air temperature is a function of the relative humidity.

g. Wind, standing in the shade or sun, color and type of clothing   ●

IDENTIFYING FUNCTIONS   Mathematical functions are important because of the clear association between an input and its single output. In Examples 2 and 3, we look at the first two Warm-up tables to find out which one is a function and can be written *y is a function of x*. A table and graph are often helpful in finding whether *there is exactly one y for each x*.

EXAMPLE **2**   Identifying a function: transcript costs   The first copy of a transcript costs $3. Each additional copy costs $2. Let x be the number of transcripts ordered and y be the total cost in dollars. Table 7 and the graph in Figure 7 describe the cost equation, $y = 3 + 2(x - 1)$. Is the cost, $y = 3 + 2(x - 1)$, a function of the number of copies ordered?

Transcripts Ordered, x	Cost, y (dollars)
1	3
2	5
3	7
4	9
5	11

Table 7 $y = 3 + 2(x - 1)$

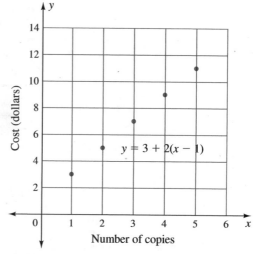

Figure 7

Solution   For each number of transcripts ordered there is exactly one cost. For each x there is only one y. This satisfies the requirement that for each input there is exactly one output. Thus, total cost is a function of the number of transcripts ordered. ●

Think about it 1: Why are the dots in the graph in Figure 7 not connected? Why do we write $2(x - 1)$ and not $2x$ in Example 2?

EXAMPLE **3**   Identifying a function   One parabola opening to the right is described by Table 8 and the graph in Figure 8. Is the parabola in this position a function?

Input, x	Output, y
0	0
1	1
1	−1
4	2
4	−2

Table 8 $x = y^2$

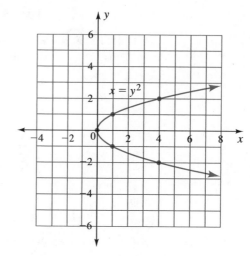

Figure 8

Solution   At $x = 1$, y may be either 1 or −1. At $x = 4$, y may be either 2 or −2. Only at $x = 0$ is there one output, $y = 0$, for one input. All other positive inputs of x have two outputs of y.

Part of the graph is above the *x*-axis, and another part is below the *x*-axis. For $x = 1$, there are two points that make the equation true: $(1, 1)$ and $(1, -1)$. Similarly, at $x = 4$ there are two points that make the equation true: $(4, 2)$ and $(4, -2)$. Because a function requires exactly one *y* for each *x*, this parabola is not a function.   ●

VERTICAL-LINE TEST   The function definition indicates that *a function must have exactly one output for each input*. If a table shows two outputs for any input, the rule for the table is not a function. If a graph shows two points with the same input, the rule for the graph is not a function.

We can identify a function from its graph with the vertical-line test.

Vertical-Line Test	A graph shows a function if every vertical line intersects the graph no more than once.

A parabola opening to the right, as shown in Figure 9, does not represent a function. At least one vertical line cuts the parabola in two places.

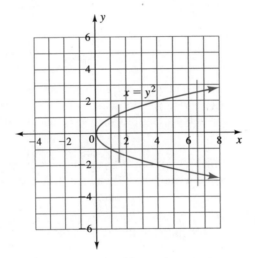

Figure 9

EXAMPLE **4**   Identifying functions from graphs   Use the vertical-line test to identify which of the following graphs are functions.

a.

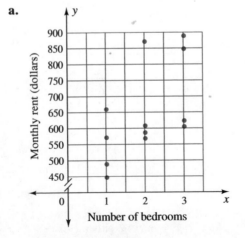

Apartment rent, 1999

b.

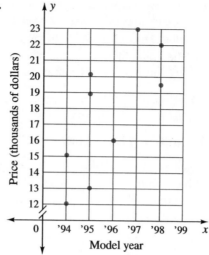

Cost of a used Jeep Grand Cherokee, 1999

c.

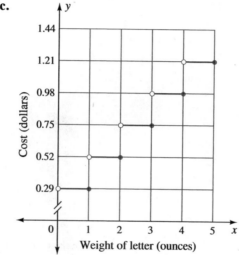

First-class postage, 1994

d.

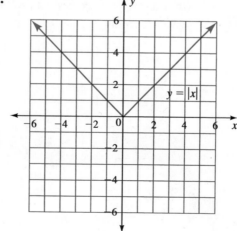

Solution **a. and b.** The graphs for apartment rent and the cost of a used Jeep Cherokee show more than one output (y) for at least one input (x). At least one vertical line passes through more than one point. Neither is a function.

c. The first-class postage cost is a function of weight in ounces. A vertical line appears to pass through two points, but one point is a dot and the other an open circle. Dots are included in the graph; open circles are excluded from the graph.

d. The graph of $y = |x|$, the absolute value of x, is intersected exactly once by any vertical line. The absolute value of x is a function. ●

Think about it 2: How would the graphs of more current data compare with the graphs in parts a, b, and c?

Visualizing Functions

There are two common ways to visualize a function. One is with a function machine, and the other is through mapping.

FUNCTION MACHINE A function machine is a mysterious box, containing a rule that takes an input and, after applying the rule, produces an output. Figure 10 illustrates a function machine for the rule $y = 2x$.

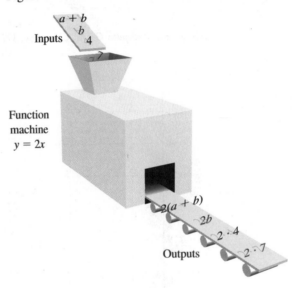

Figure 10 Function as a machine in a box

The first object into the box, 7, is the first out, $2 \cdot 7$. The second object into the box, 4, is the second out, $2 \cdot 4$. The third object, b, comes out as $2b$, and the fourth object, $a + b$, comes out as $2(a + b)$.

EXAMPLE **5**   Visualizing functions   What is the rule for each box?

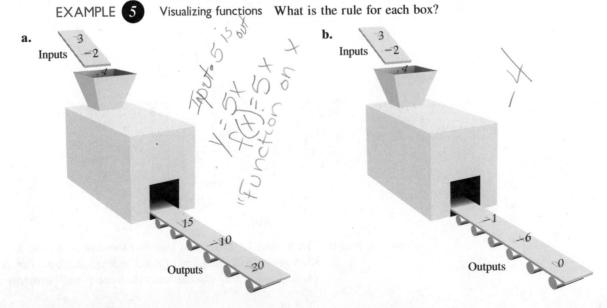

Solution   *Hints:* Part a is a multiplication; part b is a subtraction. See the Answer Box.

Student Note: Learn whichever visualization—function machine or mapping—makes more sense to you.

MAPPING   Mapping is a way of connecting inputs and outputs. The rule associates the input-output pairs. For a function, each input is connected to a single output. Figure 11 illustrates a mapping for the function $y = 2x$ and for the rule $x = y^2$ (not a function).

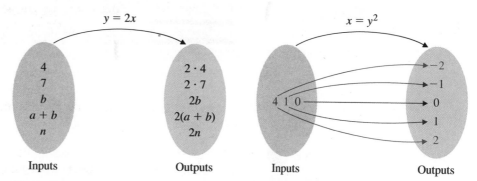

Figure 11   Rules as a mapping from input set to output set

EXAMPLE **6**   Visualizing functions   Use the rule to match each input with its output.

   a. $y = x^2$ **b.** $y = 2x - 1$

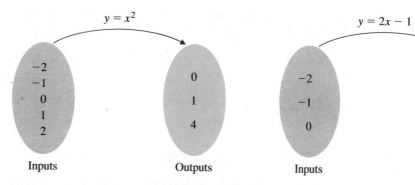

Solution   The matches are listed as ordered pairs.

   a. $(-2, 4)$, $(-1, 1)$, $(0, 0)$, $(1, 1)$, $(2, 4)$. Some of the outputs are used twice, but the rule, $y = x^2$, is a function because each input matches with exactly one output. The mapping is shown in Figure 12.

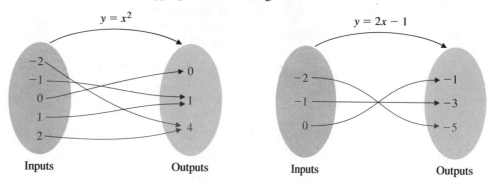

Figure 12   **Figure 13**

   b. $(-2, -5)$, $(-1, -3)$, $(0, -1)$. Each input matches with exactly one output. The mapping is shown in Figure 13.

Function Notation

We write functions in algebraic notation by indicating the name of the function (a letter, usually f) and the independent variable (the input, usually x). If y is a function of x, we write $y = f(x)$. The notation $f(x)$ started out being read *a function of x* and got shortened to *f of x*.

<table>
<tr><td>Function Notation</td><td>The notation $f(x)$ means *a function of x* and is read *f of x*.

The symbols $f(x)$ do not imply multiplication, and *of* in *f of x* does not mean multiplication.</td></tr>
</table>

The parentheses in $f(x)$ contain the variable or variables that control the behavior of the function.

EXAMPLE 7   Writing function notation   Write each of these sentences in symbols, using function notation. Select an appropriate letter for each variable.

a. The total cost of ordering transcripts is a function of the number ordered.

b. The apparent temperature for an air temperature of 85°F is a function of the humidity.

c. The payment required is a function of the credit card charge balance.

d. The amount earned at $6.75 per hour is a function of the number of hours worked.

Solution   a. Let c = total cost and n = number of transcripts ordered; $c = f(n)$.

b. Let t = apparent temperature and h = percent humidity; $t = f(h)$.

c. Let p = payment required and b = credit card charge balance; $p = f(b)$.

d. Let a = amount earned and n = number of hours worked; $a = f(n)$.   ●

Evaluating Functions

One of the most important reasons for studying functions is that the notation, $f(x)$, is a convenient way to indicate the evaluation of a function for a certain number or expression. *Function notation has the advantage of allowing us to simultaneously name the association and the input.*

Example 8 shows how function notation is used for evaluation.

EXAMPLE 8    Evaluating functions in words: setting an alarm   Suppose your electronic alarm clock permits you to set the time to get up for each day of the week. The alarm function might be written as follows:

$f(x)$ = alarm at 6:00 A.M. for x = Monday, Tuesday, Thursday, Friday

alarm at 7:00 A.M. for x = Wednesday

alarm at 8:00 A.M. for x = Saturday, Sunday

The notation f(Monday) refers to the alarm time for Monday. Find

a. f(Monday)

b. f(Saturday)

c. f(March)

d. f(Alex)

e. f(the day after Tuesday)

Solution **a.** f(Monday) = 6:00 A.M.

b. f(Saturday) = 8:00 A.M.

c. f(March) is undefined; the input set is restricted to the days of the week.

d. f(Alex) is also undefined.

e. f(the day after Tuesday) = f(Wednesday) = 7:00 A.M. ●

We call the rule in Example 8 a **conditional function**, because *the rule for the output depends on the type or value of the input*. The rule changes for different inputs.

EVALUATING FUNCTIONS GIVEN AS EXPRESSIONS   In Example 9, we return to the suspension bridge from Chapter 1 (see Figure 14) and find the height in feet of cable at several distances from the center of the bridge. When we write $f(x) = \frac{4}{125}x^2$ and ask for $f(-25)$, we want the x in $\frac{4}{125}x^2$ replaced by -25. Thus,

$$f(-25) = \frac{4}{125}(-25)^2 = \frac{4}{125}(625) = 20 \text{ ft}$$

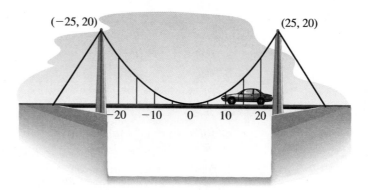

Figure 14

EXAMPLE **9**

Evaluating functions: tabulating suspension bridge cable heights   Find $f(-15)$, $f(-5)$, $f(5)$, $f(15)$, and $f(25)$, where the height in feet of the cable is given by the function $f(x) = \frac{4}{125}x^2$. Summarize these results, along with those from the Warm-up on page 85, in a table. Comment about any patterns in the table.

Solution

$f(-15) = \frac{4}{125}(-15)^2 = \frac{4}{125}(225) = 7.2 \text{ ft}$

$f(-5) = \frac{4}{125}(-5)^2 = \frac{4}{125}(25) = 0.8 \text{ ft}$

$f(5) = \frac{4}{125}(5)^2 = \frac{4}{125}(25) = 0.8 \text{ ft}$

$f(15) = \frac{4}{125}(15)^2 = \frac{4}{125}(225) = 7.2 \text{ ft}$

$f(25) = \frac{4}{125}(25)^2 = \frac{4}{125}(625) = 20 \text{ ft}$

Input, x	Output, $f(x)$
-25	20
-20	12.8
-15	7.2
-10	3.2
-5	0.8
0	0
5	0.8
10	3.2
15	7.2
20	12.8
25	20

Table 9 $f(x) = \frac{4}{125}x^2$

The height of the cable is the same for equal distances from the center of the bridge. This is apparent from the illustration in Figure 14 as well as from Table 9. The outputs match because the inputs are squared. Thus, the output for -15 is the same as the output for 15.   ●

The repeated outputs in Example 9 suggest how *function notation provides a convenient way to describe the equality of outputs*. The output is 7.2 for inputs $x = -15$ and $x = 15$. Using function notation, we state this equality as $f(-15) = f(15)$.

EVALUATING FUNCTIONS WITH VARIABLES AS INPUTS   As we saw in Example 8, the inputs to a function are not limited to numbers. In Example 10, the inputs are variables.

EXAMPLE **10**   Evaluating functions where inputs are variables   If $f(x) = x^2 - 2x - 3$, find the following. The answers to these evaluations will be expressions.

a. $f(a)$ **b.** $f(\pi)$ **c.** $f(\square)$
d. $f(-x)$ **e.** $f(a + b)$ **f.** $f(x + h)$

Solution   **a.** $f(a) = a^2 - 2a - 3$
b. $f(\pi) = \pi^2 - 2\pi - 3$
c. $f(\square) = \square^2 - 2\square - 3$
d. $f(-x) = (-x)^2 - 2(-x) - 3 = x^2 + 2x - 3$
e. $f(a + b) = (a + b)^2 - 2(a + b) - 3$
$\qquad\qquad = a^2 + 2ab + b^2 - 2a - 2b - 3$
f. $f(x + h) = (x + h)^2 - 2(x + h) - 3$
$\qquad\qquad = x^2 + 2xh + h^2 - 2x - 2h - 3$ ●

Caution: Three cautions are needed about Example 10.

• The notation $f(a + b)$ is function notation, with $a + b$ replacing the x in $f(x)$. The symbol $f(a + b)$ looks like part of the distributive property, $c(a + b) = ca + cb$, but it is not. One way to determine what a symbol means is by looking at the context. Compare the expression to others in the same problem. It is generally, but not always, safe to assume that the letter f refers to a function and is not a variable. Keep in mind that f is not reserved for use only as a function and, furthermore, that other letters may be used to name a function: $g(x)$, $P(x)$, $A(x)$, etc.

Student Note: For more on squaring binomials, see page 207.

• The solutions to parts e and f contain the squares of binomials, $(a + b)^2$ and $(x + h)^2$. Carefully review multiplying these:

$$(a + b)^2 = (a + b)(a + b) = a^2 + ab + ab + b^2 = a^2 + 2ab + b^2$$

Similarly,

$$(x + h)^2 = (x + h)(x + h) = x^2 + 2hx + h^2$$

• The distributive property of multiplication over addition is used in the solutions to parts e and f. Look for $2(a + b)$ in the solution to part e and for $2(x + h)$ in the solution to part f.

Domain and Range

For many functions, there are limitations on the inputs and/or outputs. In order to describe the restrictions on the sets of inputs or outputs of a function, we give the sets names: domain and range.

Domain and Range

> The **domain** is the set of all inputs to a function. The **range** is the set of all outputs.

EXAMPLE   Finding domain and range   What are the domain and range for each of these problem settings?

   a. The transcript costs in Example 2, page 87

   b. The alarm clock settings in Example 8, page 92

   c. The suspension bridge cable heights in Example 9, page 93

Solution **a.** The domain, or set of inputs, for the transcript costs is the set of positive integers. The range, or set of outputs, is the set of odd positive integers greater than 1.

   b. The domain for the alarm clock settings is the days of the week. The range is limited to 6, 7, or 8 A.M.

   c. The height function for the bridge cable has a domain between -25 and $+25$. The range is the set of numbers between 0 and 20. ●

The non-negative numbers are a common domain or range. The **non-negative numbers** are *the set of numbers greater than or equal to zero.*

EXAMPLE   Writing inequalities   Write each statement as an inequality. Assume that the variable x refers to inputs and the variable y refers to outputs. Identify each as a domain or range.

   a. x is positive.   **b.** y is positive.   **c.** y is non-negative.

   d. x is negative.   **e.** x is non-negative.

Solution **a.** $x > 0$, domain   **b.** $y > 0$, range   **c.** $y \geq 0$, range

   d. $x < 0$, domain   **e.** $x \geq 0$, domain ●

In Example 13, we use a table and graph to explore the domain and range of the squaring function, $f(x) = x^2$.

EXAMPLE **13** Finding the domain and range   What are the domain and range for $f(x) = x^2$? Use a graphing calculator or make a table and graph for selected integer inputs between -5 and 5.

Solution

Input x	Output $f(x) = x^2$
-5	25
-3	9
-1	1
0	0
1	1
3	9
5	25

Table 10

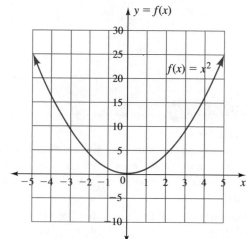

Figure 15

Both Table 10 and the graph in Figure 15 suggest that any real number may be used as an input, so the domain for $f(x) = x^2$ is the set of all real numbers. We use the symbol $\mathbb{R}$ to signify *the set of real numbers*. In symbols, we write

The domain of $f(x) = x^2$ is $\mathbb{R}$.

The outputs in both the table and the graph for $f(x) = x^2$ are either zero or positive. Thus, the range, or set of outputs, for $f(x) = x^2$ is the set of non-negative numbers. We may write the range as $f(x) \geq 0$.   ●

The words *domain* and *range* are not so strange if we consider what is being described. *Domain* represents the set of starting numbers, as in starting out from home. The word *domain* has the same root as domicile (legal word for home), domestic (pertaining to one's home or homeland), and dominion (territory under one ruler). *Range* represents the set of outcomes. The word *range* implies outside the home, as in "out on the range" or allowing an animal "to range."

Sometimes the domain or range is limited by the problem setting. We define the **relevant domain** as *the allowable or reasonable input set for a given problem setting*. Similarly, the **relevant range** is *the allowable or reasonable output set for a given problem setting*. The natural domain and range are the sets of numbers that are "legal" mathematically, although they may have no meaning in a problem setting.

EXAMPLE **14**   Finding the domain   What is the relevant domain for the area of a circle as a function of its radius?

Solution   We know that we can square any positive, zero, or negative number. However, when we use the area formula for a circle, $A = \pi r^2$, we exclude negative inputs for the radius r. There is no physical meaning to a negative radius. A zero radius generates a point with no area. Thus, the relevant domain for the area of a circle is the set of positive real numbers. We write the relevant domain as $r > 0$.   ●

ANSWER BOX

Warm-up: 1. Costs are \$3, \$5, \$7, \$9, and \$11. **2.** x is 0, 1, 1, 4, and 4. **3.** Distances are 12.8, 3.2, 0, 3.2, and 12.8.
Think about it 1: The dots are not connected because the inputs are only positive integers. An input such as 0 or $1\frac{1}{2}$ does not make sense in ordering transcripts. We write $2(x - 1)$ because we pay \$2 for each copy after the first one. **Think about it 2:**   The value of used Jeeps for the model years '94 to '98 would be lower, apartment rents would probably be higher, and postage would be higher.
Example 5: a. $y = 5x$ **b.** $y = x - 4$

EXERCISES **2.1** _____

In Exercises 1 to 6:

a. *Evaluate the equation for the given numbers. Record your answers in an input-output table. Assume x = input, y = output.*

b. *Is the equation a function? If so, write the function in function notation.*

1. The total cost of x compact disks at \$15 each with one \$4 discount coupon is $y = 15x - 4$, where y is in dollars. Evaluate for $x = 1, 2, 3, 4$.

2. The equation for the outer edge of the base of a cylinder of radius 3 feet is $x^2 + y^2 = 9$. Evaluate for $y = 3$, $-3, 0$.

3. $x = y^2 + 2$; evaluate for $y = -4, -2, 0, 2, 4$.

4. $3x^2 = y$; evaluate for $y = 27, 12, 3, 0$.

5. $y = 25 - x^2$ for x in the set $\{-5, -2, 0, 2, 5\}$

6. $x = 25 - y^2$ for x in the set {25, 16, 9, 0}

Which of Exercises 7 to 12 represent functions, $y = f(x)$?

7. $x =$ weight of a first-class letter
$y =$ postage cost of the letter in 1995, when rates were 32 cents for the first ounce and 23 cents for each additional ounce up to 11 ounces

8. $x =$ number of gallons of gasoline
$y =$ cost of the gasoline at $1.29 per gallon

9. $x =$ height of a child in third grade
$y =$ weight of the child

10. $x =$ shoe size
$y =$ cost of the shoe

11. $x =$ number of homework assignments completed by a student
$y =$ the student's score on a test

12. $x =$ state
$y =$ state capital of x

In Exercises 13 to 18, use the vertical-line test to find which graphs are functions.

13.

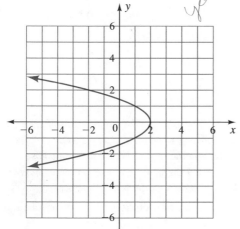

14.

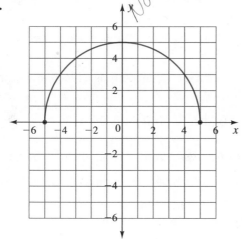

15.

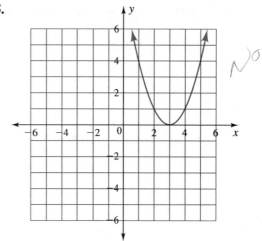

16.

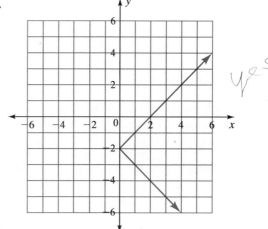

17.

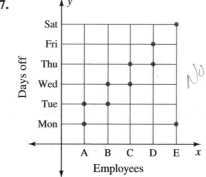

18.

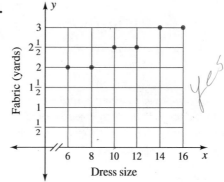

In Exercises 19 to 24, guess a rule that associates the input set with the output set.

19. {3, 4, 5, 6} with {5, 6, 7, 8}

20. {3, 4, 5, 6} with {1.5, 2, 2.5, 3}

21. {3, 4, 5, 6} with {1, $\frac{4}{3}$, $\frac{5}{3}$, 2}

22. {3, 4, 5, 6} with {0, 1, 2, 3}

23. {Texas, Ohio, Idaho, Florida} with {T, O, I, F}

24. {Texas, Ohio, Idaho, Florida} with {S, O, O, A}

Evaluate the function in Example 8 for the inputs in Exercises 25 and 26.

25. a. f(Thursday) **b.** f(September)

   c. f(the day after Sunday)

   d. f(the day before Thursday)

26. a. f(January) **b.** f(Wednesday)

   c. f(the day before Monday)

   d. f(the day after Friday)

In Exercises 27 to 32, evaluate the function for integers on the interval −2 to 4. Record the results in a table.

27. $f(x) = 5 + 2(x - 3)$ **28.** $f(x) = 25 + 4(x - 10)$

29. $f(x) = x^2 - 2x - 3$ **30.** $f(x) = x^2 + 4x - 5$

31. $f(x) = 8 - x^2$ **32.** $f(x) = 6 - x - x^2$

Evaluate the transcript cost function $f(x) = 3 + 2(x - 1)$ for the inputs in Exercises 33 and 34.

33. a. $f(1)$ **b.** $f(5)$

   c. $f(n)$ **d.** $f(n + m)$

34. a. $f(3)$ **b.** $f(4)$

   c. $f(m)$ **d.** $f(x - h)$

Evaluate the parabolic function $f(x) = x^2 + x - 2$ for the inputs in Exercises 35 and 36.

35. a. $f(3)$ **b.** $f(1)$

   c. $f(\square)$ **d.** $f(n)$

   e. $f(n - m)$

36. a. $f(-2)$ **b.** $f(2)$

   c. $f(\pi)$ **d.** $f(-n)$

   e. $f(x + h)$

A computer must be programmed to expect certain inputs—that is, relevant domains. For Exercises 37 to 40, give the number of digits expected for the relevant domains in programs seeking these data.

37. A telephone number plus area code

38. A telephone area code

39. A social security number

40. A postal zip code (There may be two answers.)

In Exercises 41 to 44, match the sets of inputs and outputs given with one of the graphs in Exercises 13 to 16.

41. $-5 \le x \le 5$ and $0 \le y \le 5$

42. $x \ge 0$ and y is any real number $\mathbb{R}$

43. x is any real number $\mathbb{R}$ and $y \ge 0$

44. $x \le 2$ and y is any real number $\mathbb{R}$

In Exercises 45 to 48, identify the domain and range. A table or graph may be helpful.

45. $f(x) = x^2 - 2$

46. $f(x) = x^2 + 2$

47. $f(x) = 6 - x^2$

48. $f(x) = 4 - x^2$

For each inequality in Exercises 49 to 56, assume x refers to inputs and y refers to outputs.
a. Is the set a domain or range?
b. Say whether each inequality includes positive numbers, negative numbers, negative numbers plus zero, or non-negative numbers.

49. $x \le 0$ **50.** $x < 0$

51. $y > 0$ **52.** $y \ge 0$

53. $x > 0$ **54.** $y < 0$

55. $y \le 0$ **56.** $x \ge 0$

Some relationships, especially formulas, are functions of more than one variable. If y is a function of both x and z, we write $y = f(x, z)$. For Exercises 57 to 62, give the related formula for each phrase and write it in function notation. You may need to look up the appropriate geometric formula in Appendix 1 at the back of the book.

57. a. Circumference of a circle as a function of radius

   b. Area of a circle as a function of radius

58. a. Perimeter of a rectangle as a function of length and width

   b. Area of a rectangle as a function of length and width

59. a. Volume of a rectangular box as a function of length, width, and height

   b. Volume of a cylinder as a function of radius and height

60. a. Area of a square as a function of length of side

   b. Perimeter of a square as a function of length of side

61. Simple interest earned in one year on money in a savings account as a function of the amount of money and the annual interest rate

62. a. Distance traveled at 55 miles per hour as a function of time

   b. Distance traveled in 3 hours as a function of rate (speed)

Solve each of these equations with algebraic notation, and check with the tables in your answers to Exercises 27 and 28. Extend your tables as needed.

63. a. $5 + 2(x - 3) = -3$ **b.** $5 + 2(x - 3) = 5$

   c. $5 + 2(x - 3) = -11$ **d.** $5 + 2(x - 3) = 13$

64. a. $25 + 4(x - 10) = -15$

   b. $25 + 4(x - 10) = -23$

   c. $25 + 4(x - 10) = -31$

   d. $25 + 4(x - 10) = 13$

Solve Exercises 65 to 68 with the tables in your answers to Exercises 29 to 32. It may be necessary to look for patterns and extend the table.

65. a. $x^2 - 2x - 3 = 0$ **b.** $x^2 - 2x - 3 = 21$

   c. $x^2 - 2x - 3 = 5$ **d.** $x^2 - 2x - 3 = 60$

66. a. $x^2 + 4x - 5 = -8$ **b.** $x^2 + 4x - 5 = 27$

   c. $x^2 + 4x - 5 = -5$ **d.** $x^2 + 4x - 5 = 0$

67. a. $8 - x^2 = 4$ **b.** $8 - x^2 = 7$

   c. $8 - x^2 = -8$ **d.** $8 - x^2 = -28$

68. a. $6 - x - x^2 = 6$ **b.** $6 - x - x^2 = 0$

   c. $6 - x - x^2 = -24$ **d.** $6 - x - x^2 = -6$

Projects

69. *Bus Stop.* What are the pros and cons of placing a bus stop before an intersection? After an intersection? Placement of a bus stop seems to be a function of what factors?

70. *Doors.* The direction in which a door opens is a function of its location. Describe the direction in which these doors open: exterior house door, movie theater door, bedroom door, bathroom door, closet door, cupboard door, door at the top of a staircase, door at the bottom of a staircase. State any reasons why the direction may be customary or mandatory.

2.2 Linear Functions

OBJECTIVES

- Find x- and y-intercepts of the graph of a linear function.
- Identify a linear function from a linear equation.
- Identify increasing and decreasing linear functions.
- Find the slope of a linear function from data points, a table, and a graph.
- Find the meaning of slope in an application setting.
- Find the slope of vertical and horizontal lines.
- Identify parallel and perpendicular lines.

WARM-UP

Simplify these expressions.

1. $\dfrac{-7.5 - (-11.25)}{10 - 5}$

2. $\dfrac{4 - 4}{0 - (-6)}$

3. $\dfrac{4 - (-6)}{-6 - (-6)}$

4. $\dfrac{3 - 6}{6 - 4}$

5. $\dfrac{3 - 1}{6 - 3}$

6. $\dfrac{70 - 50}{400 - 200}$

THIS SECTION INTRODUCES x-intercepts and y-intercepts, as well as linear functions and slope. We examine the slopes of horizontal, vertical, parallel, and perpendicular lines. We use *increasing* and *decreasing* to describe the direction of change shown by a line.

Intercepts

The graph for the photocopy card problem on page 61 of Section 1.4 is shown in Figure 16. The graph has been extended to include inputs to 100 copies. The graph crosses both horizontal and vertical axes. We now look at these crossing points more formally.

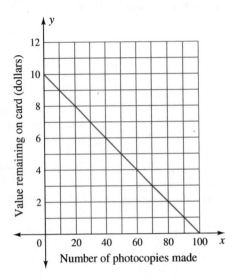

Figure 16

THE *x*-INTERCEPT   A graph crosses the *x*-axis at the *x*-intercept. In Figure 17, the point (*a*, 0) is an *x*-intercept point.

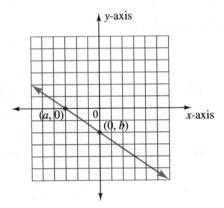

Figure 17

x-intercept

> The *x*-intercept point, (*a*, 0), is the intersection of a graph with the horizontal axis, or *x*-axis. The letter *a* denotes the *x*-intercept number, or ***x*-intercept.**

EXAMPLE **1**

Finding the *x*-intercept   What is the *x*-intercept point in the photocopy card graph in Figure 16? What does the intercept mean in the problem situation?

Solution   The graph crosses the horizontal axis at (100, 0). The *x*-intercept point is (100, 0). The intercept, 100, is the number of photocopies that can be made from a prepaid $10 card at $0.10 per copy.   ●

$\mathbf{T}$he x-intercept point has $y = 0$, so, in function notation, the x-intercept is where $f(x) = 0$. To find x-intercepts, set the function equal to zero and solve for x.

EXAMPLE 2 Finding x-intercept points Write each function as $f(x) = 0$ and solve for x.

a. $f(x) = 4x - 1$

b. $f(x) = 3 - \frac{3}{4}x$

Solution **a.**
$$f(x) = 4x - 1 \qquad \text{Set the function equal to zero: } f(x) = 0.$$
$$4x - 1 = 0 \qquad \text{Add 1 to each side.}$$
$$4x = 1 \qquad \text{Divide each side by 4.}$$
$$x = \tfrac{1}{4}$$

The x-intercept point is $(\frac{1}{4}, 0)$.

b.
$$f(x) = 3 - \tfrac{3}{4}x \qquad \text{Set } f(x) = 0.$$
$$3 - \tfrac{3}{4}x = 0 \qquad \text{Subtract 3 from each side.}$$
$$-\tfrac{3}{4}x = -3 \qquad \text{Multiply each side by } -\tfrac{4}{3}.$$
$$\left(-\tfrac{4}{3}\right)\left(-\tfrac{3}{4}x\right) = \left(-\tfrac{4}{3}\right)(-3) \qquad \text{Simplify fractions.}$$
$$x = 4$$

The x-intercept point is $(4, 0)$. ●

THE y-INTERCEPT A graph crosses the y-axis at the y-intercept. In Figure 17, the point $(0, b)$ is a y-intercept point.

---

y-intercept | The y-intercept point, $(0, b)$ is the intersection of a graph with the vertical axis, or $f(0)$. The letter b denotes the y-intercept number, or **y-intercept.**

EXAMPLE 3 Finding the y-intercept What is the y-intercept point in the photocopy card graph in Figure 16? What does the intercept mean in the problem situation?

Solution The graph crosses the vertical axis at $(0, 10)$. The y-intercept point is $(0, 10)$. The intercept, 10, is the value in dollars of the photocopy card when it is purchased. ●

$\mathbf{T}$he y-intercept point, $(0, b)$, has $x = 0$, so, in function notation, we find the y-intercept with $f(0)$.

EXAMPLE 4 Finding y-intercept points Let $x = 0$ in each function and find the y-intercept, $f(0)$. Then find the y-intercept point.

a. $f(x) = 4x - 1$

b. $f(x) = 3 - \frac{3}{4}x$

Solution **a.** $f(x) = 4x - 1$
$$f(0) = 4(0) - 1$$
$$f(0) = -1$$

The y-intercept point of $f(x) = 4x - 1$ is $(0, -1)$.

b. $f(x) = 3 - \frac{3}{4}x$
$$f(0) = 3 - \tfrac{3}{4}(0)$$
$$f(0) = 3$$

The y-intercept point of $f(x) = 3 - \frac{3}{4}x$ is $(0, 3)$. ●

Think about it 1: Is it possible for a function's graph to intersect the vertical axis in two places?

Linear Functions

A linear function is a function whose graph is a straight line. With the exception of the equation of a vertical line, $x = c$, all linear equations $ax + by = c$ may also be written as linear functions. We usually write $f(x) = mx + b$ for linear functions because this form excludes vertical lines.

Linear Function

> A **linear function** can be written $f(x) = mx + b$. The constants m and b may be any real number. The variable, x, has 1 as its exponent.

Think about it 2: Why would a vertical line not be a function?

To identify a linear function, use algebraic notation to obtain $f(x) = mx + b$.

EXAMPLE **5**   Identifying linear functions   Which of the following are linear functions?

a. $f(x) = 10 - 0.10x$ b. $f(x) = 3 + 2(x - 1)$ c. $f(x) = 4 - x^2$

Solution   a. In $f(x) = 10 - 0.10x$, $m = -0.10$ and $b = 10$. The function is linear.

b. $f(x) = 3 + 2(x - 1)$ Multiply with the distributive property.
   $f(x) = 3 + 2x - 2$ Add like terms.
   $f(x) = 2x + 1$ The function is linear, with $m = 2$ and $b = 1$.

c. $f(x) = 4 - x^2$ is not linear because x^2 does not have 1 as an exponent. ●

Increasing and Decreasing Functions

Many relationships are linear, are approximated by linear functions, or are linear in segments. In Example 6, we assume that the data can be connected with line segments so that we can look at the change from month to month. The prices in the example are for the first purchase of each month at a particular gas station. Similar data for 1995 and 1998 are included in the exercises.

EXAMPLE **6**   Exploring changes in graphs: gasoline prices   The prices given in Table 11 are
prices paid by the author for unleaded gasoline in 1985.

Month	Price Paid per Gallon (dollars)	Month	Price Paid per Gallon (dollars)
Jan.	1.16	July	1.26
Feb.	1.10	Aug.	1.15
March	1.10	Sept.	1.19
April	1.10	Oct.	1.19
May	1.19	Nov.	1.18
June	1.22	Dec.	1.18

Table II

a. Graph the data in Table 11.   **b.** When did prices increase?

c. When did prices decrease?   **d.** When did prices remain constant?

e. When did prices fall the fastest?

f. When did prices rise the fastest?

Solution   **a.** The prices are graphed in Figure 18.

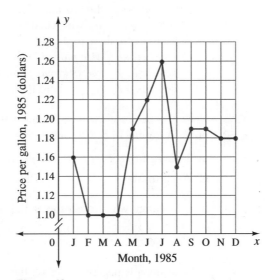

Figure 18

b. Prices increased from April to July and again from August to September.

c. Prices decreased from January to February, from July to August, and again from October to November.

d. The price was constant from February to April, from September to October, and from November to December.

e. From January to February, the price fell $0.06 per gallon. From July to August, the price fell $0.11 per gallon. It fell faster from July to August.

f. The price rose fastest from April to May, when it rose $0.09.   ●

Think about it 3: What assumptions are we making in Example 6 in connecting the data points with line segments? What limitations are there to our conclusions?

From April to July, the price of gasoline increased. The line segment between these months is *rising from left to right,* and so the price function is said to be an **increasing function.** The line segment from July to August is *falling from left to right,* and so the price function is a **decreasing function.** The graph is *horizontal* between February and April; and because the data did not indicate any change, the line segment represents a **constant function** (more on the constant function in Section 2.4).

Slope

The questions investigated in Example 6 focus on the direction and amount of change in the graph. We summarize these changes with the concept of slope. The slope of a linear function refers to the steepness of its graph.

Slope

Slope is a rate of change.

To obtain the rate of change, or slope, we divide the change in output by the corresponding change in input. In terms of variables, slope is the ratio of the change in y to the change in x:

$$\text{Slope} = \frac{\text{change in output}}{\text{change in input}} = \frac{\Delta y}{\Delta x} = \frac{y_2 - y_1}{x_2 - x_1}$$

The small triangle Δ is the Greek letter *delta* and means *the change in* the variable that follows it. The formula for slope reflects the fact that subtraction is the operation we associate with change.

The small numbers to the right of the variables are subscripts. **Subscripts** are *numbers of reduced size placed below and to the right of variables to distinguish a particular item from a group of similar items.* We generally assume that the point described by (x_1, y_1) is to the left of the one described by (x_2, y_2).

EXAMPLE **7**

Finding slope from data   Assume that the change in input between each data point in Figure 18 is 1 (month). What is the slope of each of these segments?

a. January to February   **b.** February to April

c. April to May   **d.** July to August

Solution   **a.** In a graph, we subtract the ending value from the beginning value:

$$\text{Slope} = \frac{1.10 - 1.16}{1} = -\frac{0.06}{1}, \text{ or } \$0.06 \text{ per month}$$

The slope is negative because the price dropped.

b. $\text{Slope} = \dfrac{1.10 - 1.10}{2} = \dfrac{0}{2} = 0$

The slope is zero because there is no change in price.

c. $\text{Slope} = \dfrac{1.19 - 1.10}{1} = \dfrac{0.09}{1}, \text{ or } \$0.09 \text{ per month}$

The slope is positive because the price rose.

d. $\text{Slope} = \dfrac{1.15 - 1.26}{1} = -\dfrac{0.11}{1}, \text{ or } -\$0.11 \text{ per month.}$

The slope is negative because the price dropped.   ●

In Example 8, we return to the photocopy card setting, naming the function $V(x)$ instead of $f(x)$.

EXAMPLE **8**   Finding slope from a table and graph: photocopy card   A prepaid photocopy card costs \$10. Each copy costs \$0.10. The cost per copy is deducted from the value of the card as copies are made. Table 12 and the graph in Figure 19 show the value remaining on the card after several copies have been made. Calculate the slope, first from the table and then from the graph.

Copies, x	Remaining Value, $V(x)$ (dollars)
0	10.00
10	9.00
20	8.00
30	7.00

Table 12   Photocopy card

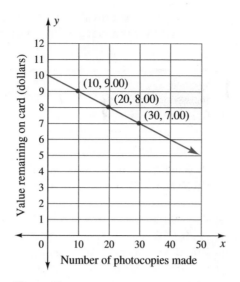

Figure 19

Solution   In Table 13, the change in y divided by the change in x is

$$m = \frac{\Delta y}{\Delta x} = \frac{-1.00}{10} = -0.10$$

Thus, the slope is $-\$0.10$ per copy. The slope is negative because it is the cost of each copy subtracted from the card's value.

Δx	Copies, x	Remaining Value, $V(x)$ (dollars)	Δy	$\dfrac{\Delta y}{\Delta x}$
10	0	10.00	−1.00	−0.10
10	10	9.00	−1.00	−0.10
10	20	8.00	−1.00	−0.10
	30	7.00		

Table 13   Photocopy card

In Figure 19, the line drops $\$1.00$ for each 10 copies made. Thus, the change in y over the change in x is $-1.00/10 = -0.10$, the same as the value from the table.   ●

In both the table and the graph in Example 8, the slope was the same, no matter what two points were chosen. This unchanging slope is an important property of straight lines.

Slope of Linear Functions

> If the slope between any two points of a function is constant, the function is linear.

You may find it helpful to think of slope from a graph as rise over run:

$$\frac{\text{Rise}}{\text{Run}} = \text{slope}$$

The change in y (rise) divided by the change in x (run) will give the same slope as before. Rise over run may also help identify the units on slope.

Units on Slope

> To find the units on slope, divide the units on the output axis by the units on the input axis.

In Example 8, the units on slope are cents per copy.
In Example 9, we name the function $P(x)$ instead of $f(x)$.

EXAMPLE Finding slope: chocolate bar profit Suppose deluxe chocolate bars for a fundraiser are sold at a profit of $0.75 each. Before these deluxe bars can be sold, an initial investment of $15 must be made for a display case. Calculate the slope of the line representing the total profit if x bars are sold. Use both Table 14 and the graph in Figure 20.

Bars Sold, x	Total Profit, $P(x)$ (dollars)
0	−15.00
5	−11.25
10	−7.50
15	−3.75
20	0

Table 14 Profit on sales

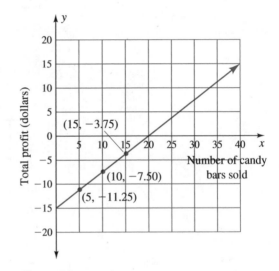

Figure 20

Solution

Δx	x	$P(x)$	Δy	$\Delta y/\Delta x$
5	0	−15.00	3.75	0.75
5	5	−11.25	3.75	0.75
5	10	−7.50	3.75	0.75
5	15	−3.75	3.75	0.75
	20	0		

Table 15 Profit on sales

In Table 15, the change in y divided by the change in x is

$$\frac{3.75}{5} = 0.75$$

The slope is the $0.75 profit per chocolate bar.

If we use the points $(5, -11.25)$ and $(10, -7.50)$ to find the slope from the graph, we have

$$\frac{y_2 - y_1}{x_2 - x_1} = \frac{-7.50 - (-11.25)}{10 - 5} = \frac{3.75}{5} = 0.75$$

As found from the table, the slope is the $0.75 profit per bar sold.   ●

In the photocopy card example, the graph was a decreasing function from left to right and the slope was negative. In the chocolate bar example, the graph increased from left to right and the slope was positive.

Slope and Linear Functions

An increasing linear function has a positive slope, and a decreasing linear function has a negative slope.

HORIZONTAL AND VERTICAL LINES   A portion of the gasoline price graph in Figure 18 was horizontal and had a zero slope. In Example 10, we find the slope of a horizontal line and a vertical line.

EXAMPLE **10**   Exploring slopes of horizontal and vertical lines   Figure 21 shows a horizontal line, $y = 4$, and a vertical line, $x = -6$. Use the indicated coordinates to calculate the slope of each line.

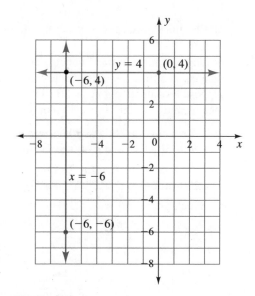

Figure 21

Solution   The slope of the horizontal line, $y = 4$, through $(-6, 4)$ and $(0, 4)$ is

$$\frac{y_2 - y_1}{x_2 - x_1} = \frac{4 - 4}{0 - (-6)} = \frac{0}{6} = 0$$

The change in y is zero, so the slope of $y = 4$ is zero.

The slope of the vertical line, $x = -6$, through $(-6, -6)$ and $(-6, 4)$ is

$$\frac{y_2 - y_1}{x_2 - x_1} = \frac{4 - (-6)}{-6 - (-6)} = \frac{4 + 6}{-6 + 6} = \frac{10}{0}$$

Because the fraction has a zero denominator, the slope of the vertical line, $x = -6$, is undefined.   ●

Summary of Slope

- An increasing function has a positive slope.
- A decreasing function has a negative slope.
- A horizontal line has a zero slope.
- A vertical line has an undefined slope.

Slope has many applications. In construction, slope describes the steepness of a roof. In highway design, the slope of a road is measured by *grade*. An 8% grade, or 8/100 slope, is considered quite steep. In geology, the steepness of a layer of rock is called *dip*. Dip is based on the angle down from horizontal. In economics, the steepness of a cost curve is called *marginal cost*. In investments, the change in stock prices from minute to minute is now "watched" by computer programs, and changes of specified size can lead to an automatic buy or sell order.

PARALLEL LINES   We now consider the slopes of parallel lines.

Parallel Lines

If two lines are parallel, their slopes are equal.

If the slopes of two lines are equal, the lines are parallel.

Parallel lines have special meaning in applications.

EXAMPLE   Writing cost functions and finding slopes: car rental costs   The total cost of renting a car is a function of the basic charge, insurance, and number of miles driven. Suppose a compact car costs $0.10 per mile plus $30 in basic charges and insurance. A mid-size car costs $50 in basic charges and insurance plus $0.10 per mile.

a. Write a cost function for the compact car rental, and use it to find the cost for a 200-mile trip and a 400-mile trip.

b. Write a cost function for the mid-size car rental, and use it to find the cost for a 200-mile trip and a 400-mile trip.

c. Which graph in Figure 22 represents costs for the compact car?

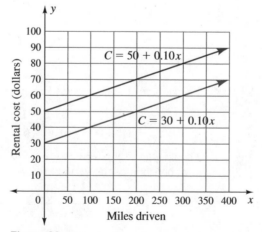

Figure 22

d. What is the slope of each line? What is the meaning of the slope?

e. Why are the lines parallel?

Solution
a. The compact car costs \$30 + \$0.10 per mile for x miles, or $30 + 0.10x$ dollars. The 200-mile trip costs $30 + 0.10(200) = 50$, or \$50. The 400-mile trip costs $30 + 0.10(400) = 70$, or \$70.

b. The mid-size car costs \$50 + \$0.10 per mile for x miles, or $50 + 0.10x$ dollars. The 200-mile trip costs $50 + 0.10(200) = 70$, or \$70. The 400-mile trip costs $50 + 0.10(400) = 90$, or \$90.

c. In Figure 22, the bottom line is for the compact, and the top line is for the mid-size car.

d. The slope of the graph for the compact car is given by

$$\text{Slope} = \frac{\text{change in output}}{\text{change in input}} = \frac{\$70 - \$50}{(400 - 200)\ \text{mi}}$$

$$= \frac{\$20}{200\ \text{mi}} = \frac{\$0.10}{\text{mi}}$$

The slope of the graph for the mid-size car is given by

$$\text{Slope} = \frac{\text{change in output}}{\text{change in input}} = \frac{\$90 - \$70}{(400 - 200)\ \text{mi}}$$

$$= \frac{\$20}{200\ \text{mi}} = \frac{\$0.10}{\text{mi}}$$

The slopes are both \$0.10 per mile—the same for each car.

e. The lines are parallel because their slopes—the cost per mile—are the same.

●

Think about it 4: What is the meaning of the vertical axis intercepts in Example 11?

PERPENDICULAR LINES   Perpendicular lines intersect at a right angle. There are two ways lines can be perpendicular.

First, the lines can be parallel to the horizontal and vertical axes.

Slopes of Perpendicular Lines I

> Two lines are perpendicular if one has a zero slope and the other has an undefined slope.

Second, the lines can be perpendicular if their slopes are *negative reciprocals*.

Slopes of Perpendicular Lines II

> If $\dfrac{a}{b}$ is the slope of one line, $\dfrac{-b}{a}$ is the slope of a perpendicular line.

EXAMPLE **12**   Identifying parallel and perpendicular lines by their slope   Use Figure 23 to find the slopes of these line segments:

a. *AB*   **b.** *BC*   **c.** *CD*   **d.** *AD*

e. *AF*   **f.** *BE*   **g.** *DG*   **h.** *CH*

State which pairs are parallel and which are perpendicular.

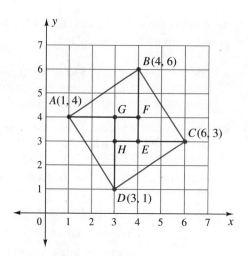

Figure 23

Solution   **a.** The slope of AB is

$$\frac{y_2 - y_1}{x_2 - x_1} = \frac{6 - 4}{4 - 1} = \frac{2}{3}$$

b. The slope of BC is

$$\frac{y_2 - y_1}{x_2 - x_1} = \frac{3 - 6}{6 - 4} = \frac{-3}{2}$$

c. The slope of CD is

$$\frac{y_2 - y_1}{x_2 - x_1} = \frac{3 - 1}{6 - 3} = \frac{2}{3}$$

d. The slope of AD is

$$\frac{y_2 - y_1}{x_2 - x_1} = \frac{1 - 4}{3 - 1} = \frac{-3}{2}$$

e. The slope of AF is

$$\frac{y_2 - y_1}{x_2 - x_1} = \frac{4 - 4}{4 - 1} = \frac{0}{3} = 0$$

f. The slope of BE is

$$\frac{y_2 - y_1}{x_2 - x_1} = \frac{6 - 3}{4 - 4} = \frac{3}{0}, \text{ undefined}$$

g. The slope of DG is

$$\frac{y_2 - y_1}{x_2 - x_1} = \frac{4 - 1}{3 - 3} = \frac{3}{0}, \text{ undefined}$$

h. The slope of CH is

$$\frac{y_2 - y_1}{x_2 - x_1} = \frac{3 - 3}{6 - 3} = \frac{0}{3} = 0$$

There are four sets of parallel line segments: AB and CD, AD and BC, BE and DG, AF and CH. There are eight ways to pair the perpendicular line segments: AB and BC, AD and CD, AB and AD, BC and CD, AF and BE, AF and DG, BE and CH, CH and DG.   ●

Think about it 5: What is the name of the shape *ABCD*? Explain your reasoning.

Before we summarize the slopes of parallel and perpendicular lines, consider Example 13.

EXAMPLE **13**   Investigating negative reciprocals   What is the product of $\frac{2}{3}$ and $-\frac{3}{2}$? What is the product of all pairs of negative reciprocals, $\frac{a}{b}$ and $-\frac{b}{a}$?

Solution

$$\frac{2}{3} \cdot -\frac{3}{2} = -\frac{6}{6} = -1$$

$$\frac{a}{b} \cdot -\frac{b}{a} = -\frac{ab}{ab} = -1$$

The product of negative reciprocals is -1.   ●

The slopes of parallel and perpendicular lines can be summarized as follows, where **nonvertical** means *not vertical*.

Parallel and Perpendicular Lines

> Two lines are parallel if and only if their slopes are equal.
>
> Nonvertical lines are perpendicular if and only if their slopes multiply to -1. The slopes are then negative reciprocals.
>
> A horizontal line and a vertical line are perpendicular.

Note: The logic for "*a* and *b*" and "*a* or *b*" is presented in Section 1.0. Another logical phrase is "*a* if and only if *b*." This statement indicates that if "if *a* then *b*" is true, then so is "if *b* then *a*." The statement "two lines are parallel if and only if their slopes are equal" means two things:

If two lines are parallel, then their slopes are equal.

If the slopes of two lines are equal, then the lines are parallel.

ANSWER BOX

Warm-up 1. 0.75 **2.** 0 **3.** undefined **4.** $-\frac{3}{2}$ **5.** $\frac{2}{3}$ **6.** 0.1 **Think about it 1:** No. If a graph has two intersections with the vertical axis, it has two outputs for one input—that is, two *y*-values for $x = 0$. In that case, it cannot be a function. **Think about it 2:** A vertical line would fail the vertical-line test. For each input *x* there is not exactly one output *y*. **Think about it 3:** We assume that the one price is typical for each month. We assume that the price does not vary during the month. We assume that the prices show the general trend for the year. A different price on other days could result in variations in conclusions. **Think about it 4:** The intercepts are the basic charges (without the per-mile cost) for the rental cars. **Think about it 5:** The corners of *ABCD* are right angles because the sides are perpendicular. If we count on the grid, we find that the lengths of the sides of the four triangles (*ABF*, *BCE*, *CDH*, *DAG*) are identical. Thus, the sides *AB*, *BC*, *CD*, and *AD* are equal, and the figure *ABCD* is a square.

EXERCISES 2.2

In Exercises 1 to 4, name the x- and y-intercept.

1.

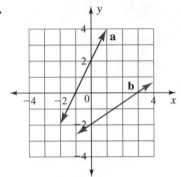

2.

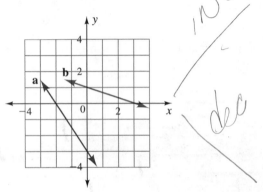

3.

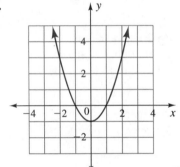

4.

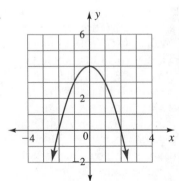

Which of the functions in Exercises 5 to 10 are linear functions?

5. $f(x) = \frac{5}{9}(x - 32)$

6. $f(x) = \frac{9}{5}x + 32$

7. $f(x) = 2\pi x$

8. $f(x) = \pi x^2$

9. $f(x) = x^2 + 2x$

10. $f(x) = 3x - x^2$

Match each description in Exercises 11 to 14 with one of the lines graphed in Exercises 1 and 2.

11. This function decreases less rapidly than the other one.

12. This function increases less rapidly than the other one.

13. This function increases more rapidly than the other one.

14. This function decreases more rapidly than the other one.

Match the descriptions in Exercises 15 and 16 with the graphs in Exercises 3 and 4.

15. This function increases for $x < 0$ and decreases for $x > 0$.

16. This function decreases for $x < 0$ and increases for $x > 0$.

Sketch a graph with lines or line segments to fit each description in Exercises 17 to 22.

17. The graph decreases only.

18. The graph increases only.

19. The graph has zero slope.

20. The graph increases for $x < 0$ and decreases for $x > 0$.

21. The graph decreases for $x < 0$ and has zero slope for $x > 0$.

22. The graph has zero slope for $x < 0$ and increases for $x > 0$.

In Exercises 23 to 34, find the slope of a line that passes through the given points.

23. $(4, 5)$ and $(-2, 7)$

24. $(3, -5)$ and $(6, 2)$

25. $(-3, -5)$ and $(5, -1)$

26. $(-4, -2)$ and $(6, -4)$

27. $(5, 0)$ and $(0, -3)$

28. $(0, 4)$ and $(-5, 0)$

29. $(-2, 0)$ and $(0, -3)$

30. $(4, 0)$ and $(0, 3)$

31. $(4, -2)$ and $(4, 0)$

32. $(-2, 2)$ and $(2, 2)$

33. $(-2, -3)$ and $(4, -3)$

34. $(-2, 5)$ and $(-2, 2)$

Use the graph below for Exercises 35 and 36. The graph shows stock prices for ten consecutive trading days.

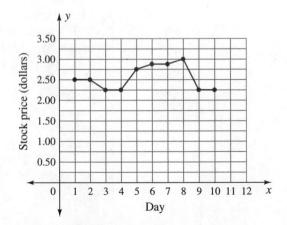

35. a. What is the slope of the line segment between day 3 and day 4?

b. Between day 5 and day 6, is the slope positive or negative?

c. What is the slope of the line between day 9 and day 10?

d. Which segment represents a faster drop in price: day 2 to 3 or day 8 to 9?

e. On how many segments was the stock price function decreasing?

36. a. What is the slope of the line segment between day 2 and day 3?

b. Between day 8 and day 9, is the slope positive or negative?

c. Which segment has a steeper slope: day 4 to 5 or day 7 to 8?

d. On how many segments was the stock price function increasing?

e. On how many segments was the stock price function constant?

In Exercises 37 to 42, decide if the table defines a linear function. If the function is linear,
a. Find the slope and the units on the slope.
b. Name or guess the x-intercept point and its meaning in the problem setting.
c. Name or guess the y-intercept point and its meaning.

37.

Sales, x (dollars)	Sales Tax, y (dollars)
10	0.60
20	1.20
30	1.80
40	2.40

Sales tax table

38.

Tickets, x	Cost, y (dollars)
1	4.00
2	8.00
3	12.00
4	16.00

Economy theater tickets

39.

Trips, x	Value, y (dollars)
0	20.00
1	19.25
2	18.50
3	17.75

Mass transit ticket value

40.

Time, t (seconds)	Height, y (meters)
0	100
1	95.1
2	80.4
3	55.9
4	21.5

Rock falling from 100-meter cliff

41.

Time, t (seconds)	Height, y (feet)
0	6.0
0.5	46.0
1.0	77.9
1.5	101.8
2.0	117.6

Throwing a ball

42.

Cones, x	Value, y (dollars)
0	15.00
1	13.50
2	12.00
3	10.50

Ice cream parlour gift certificate

43. Find the slopes of the line segments connecting the points in the figure on the next page. What can you conclude about the line segments?

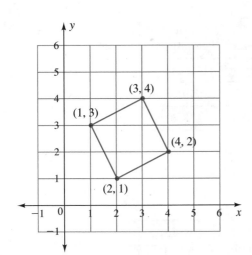

44. Find the slopes of the line segments connecting the points in the figure. What can you conclude about the line segments?

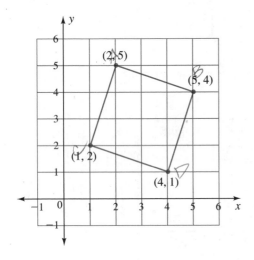

45. Draw the diagonals in the figure in Exercise 43. Use the points below to calculate the slopes for the diagonals. What do you observe?

    a. (2, 1) and (3, 4)   **b.** (1, 3) and (4, 2)

46. Draw the diagonals in the figure in Exercise 44. Use the points below to calculate the slopes for the diagonals. What do you observe?

    a. (2, 5) and (4, 1)   **b.** (1, 2) and (5, 4)

47. Suppose we know the *x*-intercept, *a*, and *y*-intercept, *b*, of a line. Describe how we might use the intercepts *a* and *b* to find the slope of the line. *Hint:* Look for a pattern in Exercises 27 to 30, and find the slope using $(a, 0)$ and $(0, b)$.

48. How might we use the slope to tell if a rectangle is formed by a four-sided figure on a rectangular coordinate graph? Try your description first on $(-2, 0)$, $(0, 5)$, $(2, 0)$, and $(0, -3)$, and then on $(-1, -1)$, $(1, -2)$, $(3, 2)$, and $(1, 3)$.

Projects

49. *Gasoline Prices, 1920 to 1980.* The average gasoline prices in Portland, Oregon, from 1920 to 1980 were as follows: 1920, \$0.27; 1930, \$0.21; 1940, \$0.21; 1950, \$0.28; 1960, \$0.34; 1970, \$0.35; 1980, \$1.19. Prices are per gallon and include local, state, and federal taxes.

    a. Make a graph for the price of gasoline from 1920 to 1980.

    b. Calculate the slope for each segment.

    c. In which decade was the rate of change greatest?

    d. What unit does the slope have? What does it mean?

50. *Gasoline Prices, 1970 to 1980.* The decade from 1970 to 1980 saw significant changes in the price of gasoline: 1970, \$0.35; 1971, \$0.35; 1972, \$0.35; 1973, \$0.38; 1974, \$0.51; 1975, \$0.55; 1976, \$0.57; 1977, \$0.61; 1978, \$0.67; 1979, \$0.86; 1980, \$1.19. (Price includes local, state, and federal taxes.)

    a. Make a graph for the price of gasoline from 1970 to 1980.

    b. When was the price constant?

    c. Calculate the slope of each segment.

    d. In which year was the rate of change greatest?

    e. Research question: What two events caused the price rises during this decade?

51. *Trends in Gasoline Prices, during the Year.* The following are the prices in dollars paid by the author for the first purchase each month of unleaded gasoline:

	Jan.	Feb.	Mar.	April	May	June
1995	1.22	1.18	1.13	1.19	1.24	1.22
1998	1.33	1.18	1.08	1.12	1.19	1.20

	July	Aug.	Sept.	Oct.	Nov.	Dec.
1995	1.25	1.30	1.22	1.28	1.28	1.27
1998	1.15	1.16	1.16	1.16	1.16	1.20

    a. Copy the graph in Figure 18 (page 103), and change it as needed to permit graphing of the 1995 and 1998 data. To graph with a calculator, place the months, in the form of numbers 1 to 12, in a first list and the three years' data in three other lists.

    b. Discuss the similarities and differences among the graphs for 1985, 1995, and 1998. What might explain these patterns?

    c. Show and explain any calculations you make.

52. *Trigonometry and Slope.* In construction, staircase and ramp slopes are often described in terms of degrees. For example, the preferred steepness of a staircase in a home is from 30° to 35° up from horizontal, whereas for an access ramp it is 7°. We use the [TAN] key on a calculator to change an angle in degrees into a slope. The calculator must be set to degree measure.

If your graphing calculator is correctly set, [TAN] 45 = 1. If you get [TAN] 45 = 1.619775 or [TAN] 45 = 0.85408, the calculator is not set to degree measure. Change to degrees under MODE. On a nongraphing calculator, you may need to change to degrees with a [DRG] key and enter the angle before pressing [TAN].

Use a calculator to find the slopes of these settings:

a. A 35° staircase

b. A 30° staircase

c. A 7° ramp

To change slope or percent grade to degrees, use [2nd] [TAN]. Make sure the calculator is set to degree measure. Verify that [2nd] [TAN] 1 = 45. If you get 0.785398 or 50, the calculator is not set to degree measure. On a nongraphing scientific calculator, you may need to enter 1 [INV] [TAN] to obtain 45.

Find the angle measures for these:

d. A very steep road grade: 8% or 0.08

e. The freeway grade descending northbound from Siskiyou Summit on Interstate 5 in southern Oregon: 6%

f. As of 1973, the steepest standard gauge railroad gradient by adhesion, achieved by the Guatemalan State Electric Railway between the Samala River Bridge and Zunil: 1:11

Research question: Find the steepness of Lombard Street in San Francisco, Lookout Mountain in Chattanooga, or a well-known grade in your city or a nearby locality. Write it as both a percent grade and an angle.

MID–CHAPTER ❷ TEST

Exercises I to 8 are to be answered true or false. Explain why each false statement is not true.

1. A decreasing function is higher on the left than on the right.

2. The slope calculation for a vertical line has a zero in the numerator.

3. The closer to horizontal the line, the greater the change in output for a given change in input.

4. A constant slope implies a linear function.

5. No change in output with any change in input gives a zero slope.

6. A graph is a function if it passes the vertical-line test.

7. Perpendicular lines have slopes that multiply to 1.

8. Functions include graphs with equations $x = c$, where c is a constant.

Complete the statements in Exercises 9 to 12.

9. The set of numbers $x \geq 0$ is called _____.

10. The set of inputs to a function is called the _____.

11. A vertical line has a(n) _____ slope.

12. The ordered pair describing the intersection of a graph and the vertical axis is written _____.

13. Evaluate the function $f(x) = 3x - 5$ for the indicated inputs.

a. $f(1)$ **b.** $f(3)$ **c.** $f(-5)$

d. $f(a)$ **e.** $f(a + b)$

14. Evaluate the function $f(x) = x^2 - x$ for the inputs listed in Exercise 13.

For each graph in Exercises 15 to 17, answer the following questions:
a. What is the domain?
b. What is the range?
c. Does the graph describe a function?

15.

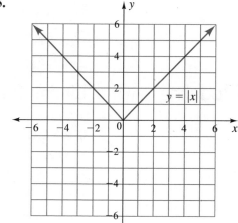

$y = |x|$

16.

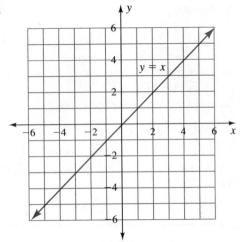

17. circle

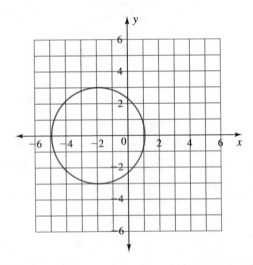

18. Find the slopes of the lines passing through these pairs of points. Which lines are parallel? Which are perpendicular?

a. $(2, -5)$ and $(3, -1)$ **b.** $(-2, 1)$ and $(2, 0)$

c. $(-4, 2)$ and $(-2, 1)$ **d.** $(-3, -3)$ and $(-2, -1)$

e. $(-3, 0)$ and $(-2, 2)$ **f.** $(-4, 3)$ and $(-4, -1)$

19. Following are prices paid by the author from March 1990 through March 1991 for gasoline: March, $1.05; April, $1.04; May, $1.06; June, $1.20; July, $1.13; Aug., $1.11; Sept., $1.26; Oct., $1.33; Nov., $1.36; Dec., $1.38; Jan., $1.35; Feb. $1.18; March, $1.02.

a. Make a graph. Label the axes clearly.

b. Calculate the slope for the steepest positively sloped segment.

c. Calculate the slope for the flattest segment.

d. Calculate the slope for the steepest negatively sloped segment.

e. Between which months was the change greatest?

f. What is the meaning of the slope in terms of the units on the axes?

g. World events bonus: What happened between August and February that explains the changes in price?

20. The basic charge for renting a car for three days is $45.59 per day. The company adds a fuel charge of $0.149 per mile, and then a 7.25% sales tax. On top of this, a $0.97 per day vehicle license fee is added. The total paid is $159.03. How many miles was the car driven? Solve with a guess-and-check table.

21. Calls on a cell phone cost $16.45 for the first 30 minutes and $0.29 for each additional minute. Write equations for the total cost, with conditions on the input.

22. For next-morning delivery over a 1000-mile distance, a shipping company charges $45 for the first pound and $2 per pound thereafter. Draw a graph showing the total cost of shipping 0 to 6 pounds. Show on the graph how to find the weight of a shipment costing $53.

2.3 Modeling with a Linear Function

OBJECTIVES

- Find a linear equation using the slope-intercept equation.
- Find a linear equation using the point-slope equation.
- Find a linear equation using the line of best fit.
- Find a linear equation using calculator regression.

IF WE HAVE REASON to believe that the data in a problem are linearly re-lated, we can *find the equation or function.* This process is called **modeling with a linear function.** To *model* means to give shape or form to information.

This section introduces four ways to obtain the equation describing a linear function. A fifth way, with sequences, is discussed in Section 2.4. We use information about slopes and intercepts or slopes and ordered pairs to find linear equations from word problems or graphs. Finally, we fit a line to data by hand or with a calculator. Each of the ways requires different information and has different applications.

Slope-Intercept Linear Equation

All linear functions may be simplified to a common form, the slope-intercept linear equation.

Slope-Intercept Linear Equation

The **slope-intercept equation** of a straight line is

$$y = mx + b$$

where m is the slope and b is the y-intercept.

We have written many equations in this form already. In Example 1, we return to linear functions introduced in Section 2.2.

EXAMPLE **1**   Finding slope and y-intercept from an equation   Name the slope and y-intercept in each of these equations, and tell the meaning of each in the problem setting.

a. Photocopy card: $y = 10 - 0.10x$, where x = number of copies and y = value in dollars

b. Chocolate bar profit: $y = 0.75x - 15$, where x = number of bars and y = profit in dollars

c. Horizontal line: $y = 4$

d. Car rental cost: $y = 0.10x + 30$, where x = number of miles and y = cost in dollars

Solution   As needed, we rewrite the equation to fit $y = mx + b$.

a. We rewrite $y = 10 - 0.10x$ as $y = -0.10x + 10$. The slope is $-\$0.10$ per photocopy; the y-intercept is $10, the initial cost of the card.

b. The slope is $0.75 profit per bar; the y-intercept is $15, the cost of the display.

c. $y = 4$ is the same as $y = 0x + 4$; $m = 0$, $b = 4$.

d. The slope is $0.10 per mile; the y-intercept is $30, the cost of insurance. ●

The process in Example 2 is the opposite of the process in Example 1. The information is given in numbers only so that you can compare the equations.

EXAMPLE **2**   Finding the equation given the slope and y-intercept   Use the slope and y-intercept to write the linear equation for each of these facts.

a. $m = 5, b = \frac{1}{2}$ $y = 5x + \frac{1}{2}$
b. $m = \frac{1}{2}, b = 5$ $y = \frac{1}{2}x + 5$
c. $m = 0, b = 5$ $y = 0x + 5$
d. $m = 5, b = 0$ $y = 5x$

Solution   We substitute the numbers for m and b into $y = mx + b$.

a. $y = 5x + \frac{1}{2}$
b. $y = \frac{1}{2}x + 5$
c. $y = 0x + 5$, or $y = 5$
d. $y = 5x + 0$, or $y = 5x$ ●

Think about it 1: Solve $y = mx + b$ for b. What does this equation give?

To find the equation of a line using the slope-intercept equation, we use the slope formula and then $b = y - mx$, obtained from $y = mx + b$.

Using the Slope-Intercept Equation

1. Find the slope of the line through (x_1, y_1) and (x_2, y_2):

$$m = \frac{y_2 - y_1}{x_2 - x_1}$$

2. Find the y-intercept of the line:

$$b = y_1 - mx_1$$

3. Place m and b into $y = mx + b$.

EXAMPLE **3**   Finding a linear equation   Find the equation of a line through $(-4, 1)$ and $(0, -5)$.

Solution   We use the steps outlined above.

Step 1: The slope is

$$m = \frac{y_2 - y_1}{x_2 - x_1} = \frac{-5 - 1}{0 - (-4)} = -\frac{6}{4} = -\frac{3}{2}$$

Step 2: The y-intercept is $b = y - mx$. We let $(x, y) = (-4, 1)$ and $m = -\frac{3}{2}$ from part a.

$$b = 1 - \left(-\frac{3}{2}\right)(-4) = 1 - 6 = -5$$

Step 3: We place m and b into the slope-intercept equation.

$$y = mx + b$$
$$y = -\frac{3}{2}x + (-5)$$
$$y = -\frac{3}{2}x - 5$$ ●

Think about it 2: Why was it not necessary to calculate the y-intercept in step 2?

In Examples 4 and 5, we review the role of the slope and y-intercept in the orientation and position of a graph.

EXAMPLE **4**   Comparing graphs: catering a wedding reception   Suppose a caterer has three menus, costing $18 per person, $21 per person, and $25 per person. Write and graph the total cost equations for a $450 wedding cake and each menu choice. What effect does the menu choice have on the graph?

Solution   The cost equations are

$$y = 18x + 450, \qquad y = 21x + 450, \qquad \text{and} \qquad y = 25x + 450$$

All three graphs pass through the same y-intercept, $450. As shown in Figure 24, the higher the cost per person, the steeper the graph.

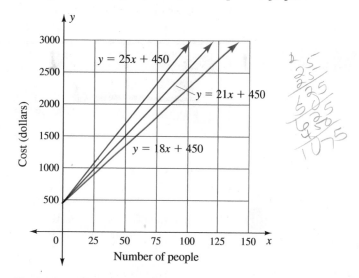

Figure 24

We changed m, the slope, and the orientation, or steepness, of the graph changed. The line became steeper as the price per person increased.   ●

EXAMPLE **5**   Comparing graphs: more catering   Suppose a caterer has three different cakes but only one menu price, $18 per person. The three cakes cost $450, $550, and $800. Write and graph the total cost equations. What effect does the cost of the cake have on the graphs?

Solution   The cost equations are

$$y = 18x + 450, \qquad y = 18x + 550, \qquad \text{and} \qquad y = 18x + 800$$

The graphs are parallel and pass through different y-intercepts. As shown in Figure 25, the higher the cost of the cake, the higher the y-intercept.

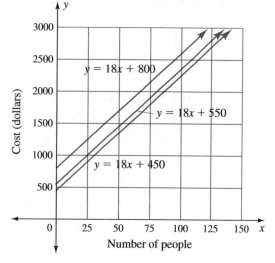

Figure 25

We changed b, the y-intercept, and the position of the graph changed. It moved higher as the price of the cake increased. The graphs are parallel. ●

Point-Slope Equation

Although the slope-intercept equation can be used for any nonvertical linear data, finding the y-intercept is sometimes difficult. In such cases, using the point-slope equation may be easier.

Point-Slope Equation

The **point-slope equation** is $$y - y_1 = m(x - x_1)$$ where m is the slope and (x_1, y_1) is a point on the line.

Example 6 suggests how mathematicians might have discovered the point-slope equation.

EXAMPLE **6**

Using slope to build a linear equation: long-distance telephone costs   It costs $2.16 to make an 11-minute long-distance call and $3.30 to make a 17-minute long-distance call.

a. Explain how to assign the variables x and y to the problem setting. Write the information as two ordered pairs.

b. Draw a graph, and connect the data points with a straight line.

c. Use the two points to find the slope.

d. Find another expression for the slope: Use one point, such as (17, 3.30), and (x, y) in the slope formula. Set this expression equal to the slope in part c. Solve for the $y = mx + b$ form.

e. Does the y-intercept from the graph agree with the one from the equation in part d?

Solution   a. The cost of a call is a function of the length of the call. Because y is a function of x, we let cost in dollars be y and time in minutes be x. The ordered pairs are (11, 2.16) and (17, 3.30).

b. The graph is shown in Figure 26.

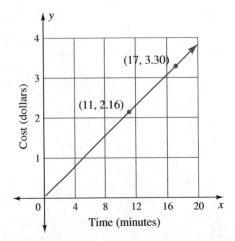

Figure 26

c. We find the slope as follows:

$$m = \frac{y_2 - y_1}{x_2 - x_1} = \frac{3.30 - 2.16}{17 - 11} = \frac{1.14}{6} = \frac{0.19}{1}, \text{ or } \$0.19 \text{ per min}$$

d. Using the point (17, 3.30) and (x, y) in the slope formula, we have

$$m = \frac{y_2 - y_1}{x_2 - x_1} = \frac{y - 3.30}{x - 17}$$

The two slope formulas are equal, so we set them equal and solve for y:

Student Note: The second step shows the point-slope form of the linear equation.

$$\frac{y - 3.30}{x - 17} = \frac{0.19}{1}$$ Multiply both sides by $(x - 17)$.

$$y - 3.30 = 0.19(x - 17)$$ Distribute 0.19 over $(x - 17)$.

$$y - 3.30 = 0.19x - 3.23$$ Add 3.30 to both sides.

$$y = 0.19x - 3.23 + 3.30$$ Simplify.

$$y = 0.19x + 0.07$$

e. The y-intercept is $0.07. The graph agrees, showing a small positive intercept. ●

Think about it 3: Use the slope, $m = 0.19$, and the data point (11, 2.16) to find the y-intercept by substituting into $y = mx + b$. Does your result agree with the y-intercept found in Example 6?

Using the Point-Slope Equation

1. Find the slope:

$$m = \frac{y_2 - y_1}{x_2 - x_1}$$

2. Substitute the slope and one point (x_1, y_1) into $y - y_1 = m(x - x_1)$.

3. Solve for y.

EXAMPLE **7**   Using the point-slope equation   Find the equation of the line passing through (4, 6) and (6, 3).

Solution   **Step 1:** The slope is

$$\frac{y_2 - y_1}{x_2 - x_1} = \frac{3 - 6}{6 - 4} = -\frac{3}{2}$$

Step 2: We substitute the slope, $-\frac{3}{2}$, and (4, 6) into $y - y_1 = m(x - x_1)$:

$$y - 6 = -\frac{3}{2}(x - 4)$$

Step 3: We solve for y:

$$y - 6 = -\frac{3}{2}x + 6$$
$$y = -\frac{3}{2}x + 12$$ ●

The point-slope equation is particularly useful in finding equations of lines parallel or perpendicular to another line.

EXAMPLE **8**   Finding the equation of a line through a given point and perpendicular to another line   Find the equation of a line passing through (3, 1) perpendicular to the line in Example 7, $y = -\frac{3}{2}x + 12$.

Solution   We follow the steps for using the point-slope equation.

Step 1: We find the slope from the equation $y = -\frac{3}{2}x + 12$: $m = -\frac{3}{2}$. The slope of a perpendicular line is the negative reciprocal of $-\frac{3}{2}$ and is $m = \frac{2}{3}$.

Step 2: We substitute the slope, $\frac{2}{3}$, and the point $(3, 1)$ into the point-slope equation.

$$y - y_1 = m(x - x_1)$$
$$y - 1 = \frac{2}{3}(x - 3)$$

Step 3: We solve for y.

$$y - 1 = \frac{2}{3}x - 2$$
$$y = \frac{2}{3}x - 1$$

●

Think about it 4: The data in Examples 7 and 8 are from the material on perpendicular lines in Example 12 of Section 2.2, page 110. Explain why the graph in Figure 27 suggests that the equations are reasonable.

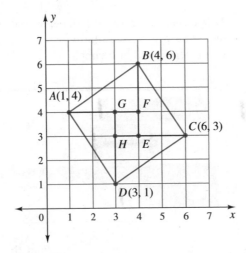

Figure 27

Fitting an Equation to Data

Up to this point, our work with linear equations has involved finding an equation when we know that such an equation exists. In these settings, we need only two points to find the equation. A more common situation, however, is when we observe that data are forming a pattern and must determine whether it is reasonable to assume that the pattern is linear. If the pattern is linear, we then need to "fit" an equation to the data. With multiple points, there may be more than one possible equation.

We look at two methods. The first method is finding a **line of best fit,** where we observe linearity from a graph and *pick two reasonable points to use in finding the equation.* In the second method, **linear regression,** we use the graphing calculator to simultaneously check the data for linearity and find the equation.

LINE OF BEST FIT   The watts and lumens data from packages of light bulbs (see Figure 28 and Table 16) may or may not be related. Watts are a unit of power, and lumens give the rate of light transmission. We investigate a possible relationship in Example 9.

Watts	Lumens
60	855
75	1170
100	1710
250	4500

Figure 28

Table 16

EXAMPLE **9**   Fitting a line with the slope-intercept equation: watts and lumens   Numbers of watts and lumens are given in Table 16.

a. Graph the watts and lumens as input-output pairs. Do they form a pattern?

b. Fit an equation to the data.

c. Try your equation on $x = 25$ watts and $x = 40$ watts.

d. Discuss whether or not the equation is a reasonable description of the data and what you know about light bulbs.

Solution   **a.** The graph in Figure 29 shows the data to be approximately linear. A ruler held on edge along the data passes near each point.

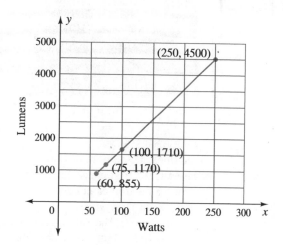

Figure 29

b. Because the data form almost a straight line, we can find the line from any two ordered pairs, using the slope-intercept equation $y = mx + b$. Suppose we use (75, 1170) and (250, 4500):

$$\text{Slope} = \frac{\Delta y}{\Delta x} = \frac{4500 - 1170}{250 - 75} = \frac{3330}{175} \approx \frac{19 \text{ lumens}}{1 \text{ watt}}$$

We substitute the slope, $m = 19$, and one of the ordered pairs used in finding the slope, (75, 1170), into $b = y - mx$:

$$b = 1170 - 19(75)$$
$$b = -255$$

Then we substitute the slope and y-intercept into $y = mx + b$:

$$y = 19x - 255$$

c. At $x = 25$ watts, $y = 19(25) - 255 = 220$ lumens.
At $x = 40$ watts, $y = 19(40) - 255 = 505$ lumens.

d. The y-intercept, -255, is surprising. One might expect the y-intercept to be $(0, 0)$, as zero watts should have a zero rate of light transmission. If we let $y = 0$ and solve for x, we find the x-intercept:

$$0 = 19x - 255 \qquad \text{Add 255 to both sides.}$$
$$255 = 19x \qquad \text{Divide both sides by 19.}$$
$$x \approx 13$$

According to the equation, at the x-intercept, $x \approx 13$, the lumen count is zero, implying that a 13-watt bulb has a zero rate of light transmission. However, tiny holiday lights, at 3 to 5 watts, produce light. We can only conclude that the relationship between watts and lumens is linear for the interval between 60 and 250 watts. ●

Caution: The 60-watt to 250-watt interval for linearity in Example 9 reminds us that, when using an equation obtained from data, we must restrict predictions to inputs near the given data set.

Summary: Finding the Line of Best Fit

1. Graph the data to see whether they have a linear shape.
2. Hold a ruler on edge over the data. Move the ruler so that it passes through the two points and as close as possible to the remaining points.
3. Calculate the equation of the line through the two chosen points.
4. Graph the equation to see that it is reasonably close to the data.

 LINEAR REGRESSION   The graphing calculator statistical functions use a feature called linear regression, or LinReg, to fit linear equations to data. As with the line of best fit, we first check for linearity and then obtain the equation of the line.

The calculator calculates the **coefficient of correlation** to *determine how well the data fit a straight line.* The coefficient of correlation is denoted r. If $r = 1$ or $r = -1$, the data are perfectly linear. If $r = 0$, the data are not linear. The inequality $-1 \leq r \leq 1$ describes the entire range for r. Figure 30 shows the range for the coefficient of correlation. When the coefficient of correlation is close to -1 or $+1$, we may have a good fit to a line. How close r must be to -1 or $+1$ depends on the data. For this course, look at the graph of the data and use your best judgment. In a statistics course, you will learn more about the coefficient of correlation and apply linear regression accordingly.

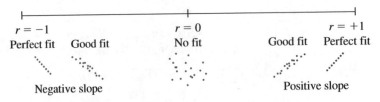

Figure 30   Coefficient of correlation

Figure 30 also shows that a negative coefficient of correlation matches with a negatively sloped line and a positive coefficient of correlation indicates a positively sloped line.

EXAMPLE

Fitting a linear equation to data   Enter the four data points from Table 16 with the STATISTICS option on a graphing calculator. Calculate the equation of the line with linear regression, and record the equation as well as the coefficient of correlation, r.

Solution   Calculator linear regression gives $y = 19x - 251$. The coefficient of correlation is $r \approx 0.999\,718$. If $r = 1$ or $r = -1$, the data fit a line perfectly. These data appear to fit a line nearly perfectly.   ●

Although the calculator equation has a different y-intercept than the equation from the points $(75, 1170)$ and $(250, 4500)$ in Example 9, the difference is not significant, as we can see when the data points and both equations are graphed together, as in Figure 31.

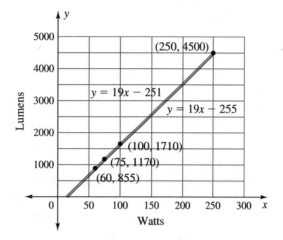

Figure 31

When there are exactly two data points, the calculator linear regression feature gives the same line through the points as the equations $y = mx + b$ and $y - y_1 = m(x - x_1)$. Practice doing linear regression on the graphing calculator with the data from Examples 3, 6, 7, and 9.

APPLICATIONS   Linear functions play such an important role in other subjects that there are many special forms for linear equations. In introductory statistics, the line is written $y = a + bx$. This form is an option in calculator regression. In accounting, a linear equation may be written $y = Vx + F$, where V is variable cost per unit and F is fixed cost. In economics, the linear equation has many different forms, depending on the topic under discussion. Two examples of linear equations from economics are $y = a + bx_t$ and $C = a + bY_d$.

Graphing Calculator Technique:
Linear Regression

Choose the STATISTICS option.

Clear prior lists.

Enter the new data.

Select linear regression, usually LINREG, from the calculation options.

Specify the location of lists of data, usually L_1 and L_2.

To view the data and the line, enter the equation of the line in $\boxed{Y =}$ and plot the data as described in your calculator manual.

Note: On some calculators, the coefficient of correlation, r, is not automatically listed by the regression. Look in your manual for instructions on how to turn on this option.

ANSWER BOX

Warm-up 1. $b = -3$ **2.** $\frac{0.19}{1}$ **3.** $\approx \frac{19}{1}$ **4.** $y = -\frac{3}{2}x + 12$ **Think about it 1:** Subtract mx from each side of $y = mx + b$ to obtain $b = y - mx$. This equation gives the y-intercept of the graph of $y = mx + b$. **Think about it 2:** The ordered pair $(0, -5)$ is a y-intercept point. From that we know that $b = -5$. **Think about it 3:** $y = mx + b$, $2.16 = 0.19(11) + b$, $b = 0.07$, as before. It does not matter which ordered pair we use in finding the equation. **Think about it 4:** The graph of the equation $y = -\frac{3}{2}x + 12$ passes through $B(4, 6)$ and $C(6, 3)$ and would intersect the y-axis near $b = 12$. The graph of the equation $y = \frac{2}{3}x - 1$ passes through points $C(6, 3)$ and $D(3, 1)$ and would intersect the y-axis near $b = -1$.

EXERCISES 2.3

Name the slope and vertical axis or y-intercept for the equations in Exercises 1 to 10.

1. Amount of sales tax: $y = 0.055x$
2. Total cost for a pizza party: $y = 3.50x$
3. Total cost of a party at a wave pool: $y = 3.00x + 10$
4. Total cost of a party at a bowling alley: $y = 5.50x + 15$
5. Circumference is a function of radius: $C = 2\pi r$
6. Circumference is a function of diameter: $C = \pi d$
7. Friction force is a function of normal force N: $F = \mu N$
8. Distance traveled is a function of time: $d = 55t$
9. Consumption (spending) is a function of income Y: $C = a + bY$
10. Total cost is a function of the number of units produced: $y = Vx + F$

Build a linear equation from the information given in Exercises 11 to 20.

11. Slope is 8; y-intercept is -4.
12. Slope is -4; y-intercept is $\frac{1}{2}$.
13. Slope is $\frac{1}{2}$; y-intercept is -8.
14. Slope is 0; y-intercept is -4.
15. Slope is -2; y-intercept is 0.
16. Slope is $\frac{1}{4}$; y-intercept is -2.
17. $(3, 6)$ and $(0, -2)$ are on the line.
18. $(-3, 2)$ and $(0, 4)$ are on the line.
19. $(-2, 4)$ and $(5, -3)$ are on the line.
20. $(4, -3)$ and $(-2, 1)$ are on the line.

21. According to a newspaper medical columnist, aerobic exercise is safe and effective when your pulse rate is between 50 and 70% of your maximum pulse rate. Your maximum pulse rate is found by subtracting your age from 220.*
    a. Which is correct: Pulse rate is a function of age, or age is a function of pulse rate?
    b. In what range is your safe exercise pulse rate?
    c. Write a formula that describes the lower (50%) safe exercise pulse rate.
    d. Write a formula that describes the upper (70%) safe exercise pulse rate.
    e. What range of pulse rates is safe for a 50-year-old?
    f. For what age is the pulse rate between 95 and 133?
22. A budget checking account costs $5.00 per month for the first 15 checks and $0.50 per check for each additional check. Zero is the minimum balance. The total cost per month is a function of the number x of checks written.
    a. What is the cost for 10 checks?
    b. What is the cost for 25 checks?
    c. Write the cost function for 0 to 15 checks.
    d. What is the rate of change, or slope, of the function for 0 to 15 checks?
    e. What is the y-intercept?
    f. What is the meaning, if any, of the y-intercept?
    g. Write the cost function for more than 15 checks.
    h. What is the rate of change, or slope, of the function for more than 15 checks?

*From *To Your Good Health,* Dr. Paul Donohue, © 1992, North American Syndicate.

Exercises 23 and 24 refer to accounting, where total cost, C, is related to variable cost per unit, V, and fixed cost, F, by the equation C = Vx + F.

23. The bank charges 2.5% of the amount of the loan (called *points*) plus $300 in fees. What is the fixed cost? What is the variable cost per dollar borrowed? What is the cost function to borrow x dollars? (This cost does not include the interest paid on the loan!)

24. Suppose blue jeans cost $17.80 per pair plus a $250 order handling fee. What is the fixed cost? What is the variable cost per pair? What is the cost function to buy x pairs of jeans?

In Exercises 25 to 30, find the equations of lines through the given points.

25. (2, 5) and (5, 4)   **26.** (1, 2) and (4, 1)

27. (5, 4) and (4, 1)   **28.** (2, 5) and (1, 2)

29. (4, 1) and (2, 5)   **30.** (4, 1) and (1, 2)

In Exercises 31 and 32, total cost, C, is related to variable cost per unit, V, and fixed cost, F, by the equation C = Vx + F.

31. Maria, owner of the Runner's Outlet, orders 250 pairs of shoes for $11,000. Three months later, under the same purchase plan, she orders 300 pairs of shoes for $13,185. What is the cost function? What is the fixed cost? What is the variable cost per pair?

32. Yuan predicts that manufacturing 10,000 calculators will cost $125,900, while manufacturing 15,000 calculators will cost $177,350. What is the manufacturing cost function? What is the fixed cost? What is the variable cost per unit?

33. Let $(C, F) = (x, y)$. Use the coordinates (100, 212) and (0, 32) to derive the temperature conversion formula for Fahrenheit in terms of Celsius.

34. Let $(F, C) = (x, y)$. Use the coordinates (212, 100) and (32, 0) to derive the temperature conversion formula for Celsius in terms of Fahrenheit.

35. Find the equation of the line between (60, 855) and (250, 4500). Compare these results with those in part b of Example 9.

36. Find the equation of the line between (100, 1710) and (250, 4500). Compare these results with those in part b of Example 9.

In Exercises 37 to 46, find the equation of a line fitting the information given.

37. Parallel to $2x + 3y = 6$ and passing through the origin

38. Parallel to $3x - 2y = 6$ and passing through the origin

39. Perpendicular to $y = -\frac{1}{2}x + 3$ and passing through the origin

40. Perpendicular to $y = \frac{3}{4}x - 2$ and passing through the origin

41. Perpendicular to $y = \frac{5}{8}x - 3$ and passing through (2, 3)

42. Perpendicular to $y = -\frac{1}{4}x + 2$ and passing through (3, 1)

43. Parallel to $4x - 3y = 12$ and passing through (−2, 1)

44. Parallel to $3x + 4y = 12$ and passing through (−1, 3)

45. Perpendicular to $5x - 2y = 8$ and passing through (2, −1)

46. Perpendicular to $2x + 5y = 15$ and passing through (1, 4)

47. For the NCAA basketball tournament, the #1 team is matched with the #16 team in a region, the #2 team is matched with the #15 team, the #3 team with the #14 team, and so forth. Let x be the ranking of the first team in the match-up and y be the ranking of the opponent. Is the match-up a linear function? If so, give its equation.

48. Large eggs weigh a minimum of 50 grams each and cost $0.99 per dozen. Extra-large eggs weigh a minimum of 56 grams each and cost $1.09 per dozen. Jumbo eggs weigh a minimum of 63 grams each and cost $1.19 per dozen. Draw a graph for the cost of the eggs as a function of weight. Fit a linear equation.

49. Hand-packed ice cream costs $2.75 for 12 ounces, $4.75 for 24 ounces, and $7.95 for 50 ounces. Draw a graph for cost as a function of weight. Fit a linear equation.

50. Income eligibility guidelines for free weatherization services by a local utility are based on the number of people in the household and annual income: (1, $16,680), (2, $19,080), (3, $21,420), (4, $23,820), (5, $25,740). Fit a linear equation.

51. Sale prices in 1984 for round diamond pendants set in 14K gold are listed below. Let $x =$ carat value and $y =$ price. Find an equation for the line of best fit by hand or with calculator regression.

Size in Carats	0.05	0.10	0.15	0.20	0.25	0.30	0.50
Price	$39	$79	$99	$159	$229	$299	$495

52. Describe how you can use the slope a/b of a line and an ordered pair (c, d) for a point on the line to find an ordered pair for another point on the line. Your answer can include sentences, a picture, and/or an expression.

Projects

53. *Choosing a Method.* Match one of the four methods of finding a linear equation with each of the reasons in parts a to h.

Method 1: slope-intercept equation, $y = mx + b$

Method 2: point-slope equation, $y - y_1 = m(x - x_1)$

Method 3: line of best fit

Method 4: calculator regression

a. Your instructor gives you permission to use the calculator on any problem.

b. You need to find a parallel or perpendicular line through another point.

c. You need to find a parallel or perpendicular line through the origin.

d. The data set is large and not easy to graph.

e. The data set is easy to graph and points do not all lie in a straight line.

f. Two ordered pairs are provided.

g. The rate of change of the input and the initial, or starting, value of the function are given.

h. One of two ordered pairs is a *y*-intercept.

Write any other reasons you can think of for choosing one method over another.

54. *More Watts and Lumens*

a. Look at other bulbs of the same brand as those in Example 9 to find out if their lumen data are the same. If not, calculate a new linear equation and find $f(40)$ and $f(25)$.

b. Find 40-watt and 25-watt packages of the same brand light bulbs as in part a, and compare the lumens with the calculated lumens in part a.

c. Talk with a physics instructor to obtain other relevant information, such as the expected behavior of the data near zero.

Alternative to a and b: Find another brand light bulb with the same watts, and repeat Example 9 for the other brand. Note whether the lumen data (for light output) are the same for each brand.

55. *Checking Account Costs and Computer Spreadsheets.* Create a spreadsheet to compare personal checking account costs. Design the spreadsheet in four columns. The first column shows the number of checks written per month. The next three columns show the total monthly cost at Bank A (column 2), Bank B (column 3), and Credit Union C (column 4). Make the spreadsheet long enough to find for how many checks each option is the cheapest.

- Bank A charges $7.00 per month. You may write an unlimited number of checks each month. There is no minimum balance required in the account.

- Bank B charges $1.25 per month for a debit card and $0.50 per card use. The debit card is used in place of checks. The amount of the purchase is deducted from your account at the point of purchase when you use the debit card. There is no minimum balance required in the account.

- Credit Union C charges $3.00 per month for a budget account, and you can write 15 checks per month; there is a $0.50 charge for each additional check written. There is no minimum balance.

2.4 Special Functions

OBJECTIVES

- Find a linear equation from a sequence.
- Find a specified term of a sequence based on a linear function.
- Identify and graph constant and identity functions.
- Identify and graph absolute value functions.
- Find the domain and range of special functions.

WARM-UP

What is the next number in each pattern?

1. 1, 3, 5, 7, 9 ...

2. 3, 6, 9, 12, 15

3. 4, 7, 10, 13, ...

4. 1, 3, 6, 10, ...

5. 6, 2, −2, −6, −10

$\mathbf{T}$HIS SECTION INTRODUCES the sequence function, which gives a fifth way to obtain the equation describing a linear function. The section also introduces three special functions: the constant function, identity function, and absolute value function.

Number Patterns

FINDING DIFFERENCES   In Section 1.1, we looked at number patterns and their rules. In Example 1, we explore some ideas about the number patterns in the Warm-up. We will find the *difference*, the answer obtained by subtracting adjacent numbers in the pattern.

EXAMPLE **1**   Exploration: finding differences   For each pattern in the Warm-up, tell if the pattern is increasing or decreasing, and find the differences between the numbers in the patterns. Then describe how the patterns are the same and different.

a. 1, 3, 5, 7, . . .

b. 3, 6, 9, 12, . . .

c. 4, 7, 10, 13, . . .

d. 1, 3, 6, 10, . . .

e. 6, 2, −2, −6, . . .

Solution   **a.** 1, 3, 5, 7, . . .   Number pattern
∨ ∨ ∨
2, 2, 2, . . .   Difference between numbers

b. 3, 6, 9, 12, . . .   Number pattern
∨ ∨ ∨
3, 3, 3, . . .   Difference between numbers

c. 4, 7, 10, 13, . . .   Number pattern
∨ ∨ ∨
3, 3, 3, . . .   Difference between numbers

d. 1, 3, 6, 10, . . .   Number pattern
∨ ∨ ∨
2, 3, 4, . . .   Difference between numbers

e. 6, 2, −2, −6, . . .   Number pattern
∨ ∨ ∨
−4, −4, −4, . . .   Difference between numbers

The first three patterns are increasing. A pattern is said to have a **common difference** if *a constant answer is obtained when adjacent numbers in the pattern are subtracted.* The first pattern has a common difference of 2; the next two have a common difference of 3. The fourth pattern is increasing and has differences that change. The last pattern decreases by 4; it has a constant difference of −4 between each term. In the Warm-up, it is the differences that let us predict the next number in the pattern.   ●

SEQUENCES AND ORDERED PAIRS   The formal name for number patterns is sequences. A **sequence** is *a function with the natural numbers (or positive integers) as its set of inputs (domain).* Sequences are given the special notation a_n, instead of $f(x)$. The subscript n denotes the natural numbers and reminds us that the inputs are natural numbers. The inputs are not shown when the sequence is written. *The numbers in a sequence are* called **terms of a sequence.** Instead of writing $f(1)$, $f(2)$, and $f(3)$, we talk about the terms a_1, a_2, and a_3.

Once we know that the natural numbers are the inputs to the rule for the sequence (function), we can write the sequence as ordered pairs.

Sequences and Ordered Pairs

> To build ordered pairs from a sequence:
>
> 1. Write the position of the number in the sequence (first, second, third, ...) as the first number, n, in the ordered pair (n, a_n).
> 2. Write the number from the sequence as the second number, a_n, in the ordered pair, (n, a_n).

EXAMPLE **2** Changing sequences to ordered pairs Write the first four number patterns in Example 1 as ordered pairs.

Solution **a.** $a_n = \{1, 3, 5, 7, \ldots\}$ is written $(1, 1), (2, 3), (3, 5), (4, 7), \ldots$

b. $a_n = \{3, 6, 9, 12, \ldots\}$ is written $(1, 3), (2, 6), (3, 9), (4, 12), \ldots$

c. $a_n = \{4, 7, 10, 13, \ldots\}$ is written $(1, 4), (2, 7), (3, 10), (4, 13), \ldots$

d. $a_n = \{1, 3, 6, 10, \ldots\}$ is written $(1, 1), (2, 3), (3, 6), (4, 10), \ldots$ ●

GRAPHING SEQUENCES The sequences in Examples 1 and 2 are of two types. Looking at their graphs will reveal more about them.

Graphing a Sequence

> Graph sequences by graphing their ordered pairs.
>
> 1. Place n on the horizontal axis instead of x.
> 2. Place a_n on the vertical axis instead of y or $f(x)$.

EXAMPLE **3** Graphing sequences Graph the first four sequences in Example 1 by graphing their ordered pairs. Comment on the shapes formed by the graphs.

Solution The sequences $a_n = \{1, 3, 5, 7, \ldots\}$ and $a_n = \{3, 6, 9, 12, \ldots\}$ are shown in Figure 31. The sequences $a_n = \{4, 7, 10, 13, \ldots\}$ and $a_n = \{1, 3, 6, 10, \ldots\}$ are shown in Figure 32.

Student Note: Compare the graphs by holding the edge of a ruler along each set of dots.

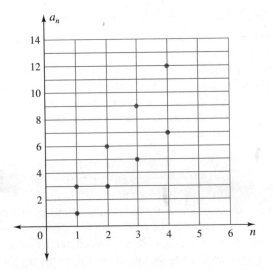

Figure 31

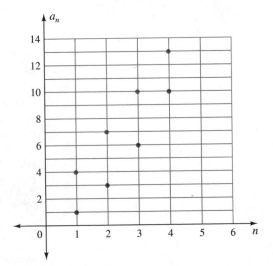

Figure 32

The points for the first three sequences lie in a straight line, which suggest that they are related to linear functions. The points for the fourth sequence do not lie in a straight line.   ●

Think about it 1: Why don't we connect the points in the graph of a sequence?

FINDING RULES FOR SEQUENCES   We can use the common difference to find the rule for a sequence whose points lie on a straight line.

EXAMPLE **4**

Finding the rule   Use the following process to find the rule for $a_n = \{1, 3, 5, 7, \ldots\}$.

a. The first four terms are given. Find the fifth term in the sequence.

b. Always starting with the first term, $a_1 = 1$, how many differences of 2 must we add to get the second term, $a_2 = 3$? The third term, $a_3 = 5$? The fourth term, $a_4 = 7$?

c. Use the number of differences to predict the fifth term, a_5.

d. How is the number of differences related to the term number?

e. What is a_n for this sequence?

Solution   **a.** The fifth term is 9.

b. We add 1 difference to get the second term, a_2:

$$1 + 2 = 3$$

We add 2 differences to get the third term, a_3:

$$1 + 2 + 2 = 5$$

We add three differences to get the fourth term, a_3:

$$1 + 2 + 2 + 2 = 7$$

c. According to the pattern in part b, the fifth term should be 1 plus four 2s:

$$1 + 2 + 2 + 2 + 2 = 9$$

d. The number of differences is 1 less than the term number: $n - 1$.

e. For this sequence, a_n is the first term plus $(n - 1)$ differences of 2:

$$a_n = 1 + (n - 1) \cdot 2$$
$$a_n = 1 + 2n - 2$$
$$a_n = 2n - 1$$

●

> To find a linear equation from a sequence rule, replace n with x and a_n with y.

EXAMPLE **5**

Comparing a rule with a linear equation   What linear equation matches the sequence $a_n = 2n - 1$? Graph the equation with the sequence.

Solution   When we replace n with x and a_n with y, $a_n = 2n - 1$ becomes $y = 2x - 1$. The graph is shown in Figure 33.

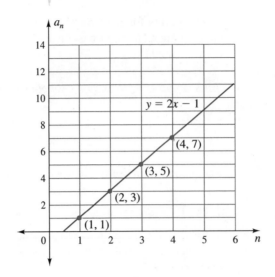

Figure 33

Example 4 suggests the following formula for finding the rule for the nth term of a sequence:

Rule for the nth Term of a Sequence

> When there is a common difference, the nth term of a sequence is given by
>
> $$a_n = a_1 + (n - 1)d$$
>
> where a_1 is the first term, n is the number of terms, and d is the common difference between the terms.

EXAMPLE **6** Finding a rule and its matching linear equation Find a rule for the sequence $a_n = \{6, 2, -2, -6, \ldots\}$. Find the matching linear equation.

Solution From Example 1, we have a common difference of -4. The first term is 6.

$a_n = a_1 + (n - 1)d$ Substitute for d and a_1.
$a_n = 6 + (n - 1)(-4)$ Simplify the right side.
$a_n = 6 - 4n + 4$ Add like terms.
$a_n = 10 - 4n$

The linear equation is $y = 10 - 4x$ or $y = -4x + 10$.

Think about it 2: What is the role of the common difference in the linear equation? What meaning, if any, does 10 have in the sequence?

FINDING A TERM The formula $a_n = a_1 + (n - 1)d$ is useful for both finding a specific term and writing the rule.

EXAMPLE **7** Finding a specific term and finding a rule Predict the rule for the sequence $a_n = \{3, 6, 9, 12, \ldots\}$. Use the formula for a_n to find the 20th term of the sequence and the rule for the sequence.

Solution The numbers 3, 6, 9, and 12 are multiples of 3. We predict that the rule will be $a_n = 3n$. The common difference is 3. The first term is $a_1 = 3$.

The 20th term has $n = 20$:

$$a_n = a_1 + (n - 1)d$$
$$a_{20} = 3 + (20 - 1) \cdot 3$$
$$a_{20} = 3 + 19 \cdot 3 = 60$$

In the rule, we keep n in the formula and replace a_1 and d:

$$a_n = 3 + (n - 1) \cdot 3$$
$$a_n = 3 + 3n - 3 = 3n$$

●

Think about it 3: Show that the rule for 4, 7, 10, 13, ... is $a_n = 3n + 1$.

A sequence with a common difference is a special linear function called an **arithmetic sequence.** (The number pattern in this section that is not an arithmetic sequence, 1, 3, 6, 10, ... , is a quadratic sequence; we will look at quadratic sequences in Section 3.2.)

Graphing Calculator Technique:
Listing an Arithmetic Sequence

Go to TABLE SET-UP. Place the first term, a_1, of the arithmetic sequence under the table minimum, and place the common difference, d, under the table step or table change, Δx. A listing of the table will place the arithmetic sequence in the X column. Practice this procedure with the arithmetic sequences in the Warm-up.

The Constant Function

We now consider a special linear function: the constant function.

EXAMPLE **8**

Exploring the constant function: university tuition   Suppose you attend a university with a fixed tuition of \$8000 per semester. What is the cost for taking 10, 20, and 30 credit hours during the semester? Draw a graph of tuition as a function of number of credit hours taken. What do you observe about the graph?

Solution   The tuition cost remains \$8000, regardless of the number of credit hours taken. Thus, the graph in Figure 34 passes through the points (10, 8000), (20, 8000), and (30, 8000). The graph is a horizontal line.

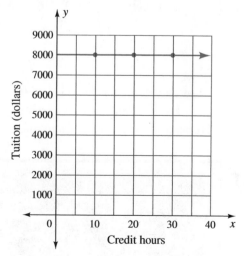

Figure 34

●

   The fact that the university's tuition is the same no matter how many credit hours are taken makes the graph a constant function.

Constant Function

> A **constant function** has a fixed output for all inputs.

In Example 9, we calculate the slope and find the y-intercept, domain, and range for a constant function.

EXAMPLE **9** Finding features of a constant function

a. List the ordered pairs from Table 17, and find the corresponding points in Figure 35.

b. Calculate the slope of this constant function.

c. What is the y-intercept for $f(x) = 3$?

d. What are the domain and range for the function?

Input x	Output $f(x) = 3$
-4	3
-2	3
0	3
5	3

Table 17

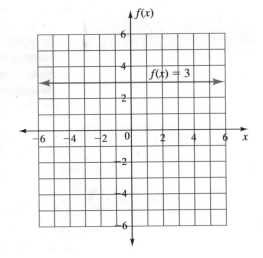

Figure 35 Constant function

Solution a. The listed coordinates are $(-4, 3)$, $(-2, 3)$, $(0, 3)$, and $(5, 3)$. The points all lie on the horizontal line.

b. The slope between $(-2, 3)$ and $(5, 3)$ is given by

$$\frac{y_2 - y_1}{x_2 - x_1} = \frac{3 - 3}{5 - (-2)} = \frac{0}{7} = 0$$

The slope is zero.

c. The y-intercept for the graph of $f(x) = 3$ is $y = 3$.

d. The set of inputs, or domain, for the function is all real numbers, $\mathbb{R}$. The set of outputs, or range, is a constant, 3. ●

Properties of a Constant Function

> Every horizontal line is a constant function.
>
> The slope of the constant function is zero.
>
> The constant function $f(x) = c$ intersects the vertical axis at $(0, c)$.
>
> The domain (unless limited by an application) of the constant function is the set of all real numbers, $\mathbb{R}$.
>
> The range of the constant function $f(x) = c$ is the y-intercept value: the single number c.

Think about it 4: What are the domain and range for university tuition in Example 8?

Identity Function

Have you heard the phrases "what you see is what you get" and "garbage in, garbage out"? In mathematics, the identity function behaves according to the principle "what you put in is what you get out." Each output of the identity function is identical to the corresponding input.

Identity Function

> The identity function is $f(x) = x$. The **identity function** has each output equal to its corresponding input.

The graph of $f(x) = x$ is the line $y = x$; see Table 18 and Figure 36.

Input x	Output $f(x) = x$
-2	-2
-1	-1
0	0
1	1

Table 18

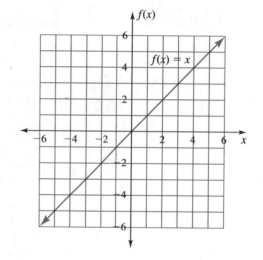

Figure 36 Identity function

EXAMPLE **10**   Identifying constant and identity functions Which situations describe identity functions? Which describe constant functions?

a. Input: Number of local telephone calls made for a $21.95 per month fee
  Output: Total cost of local telephone service

b. Input: Charge balance on a credit card
  Output: The payment due in order to avoid paying interest or finance charges

c. Input: The amount of money put into a "no charge" change machine
  Output: The amount of money received in change from the machine

d. Input: Number of cups of coffee consumed at a restaurant with a $3 "bottomless cup"
  Output: Total cost of the coffee

Solution   **a.** Constant function; the total cost is fixed for any number of local calls.
  b. Identity function; the payment due equals the charge balance.
  c. Identity function; the money put in equals the change received.
  d. Constant function; the $3 cost is fixed for all the coffee you can drink. ●

The graph of the identity function (see Figure 36) extends infinitely to the left and to the right. The set of inputs, or domain, is the set of all real numbers,

ℝ. Because the outputs to the identity function exactly match the inputs, the range is also the set of all real numbers, ℝ. The domain and/or range may be limited for certain applications.

Properties of the Identity Function

> The graph of the identity function is the line $y = x$.
>
> The identity function passes through the origin, (0, 0).
>
> Both the domain and the range (unless limited by an application) of the identity function are the set of all real numbers, ℝ.

Absolute Value Function

To close this section, we examine the absolute value function.

Absolute Value Function

> The **absolute value function** gives the distance a number is from zero on the number line. The function is written $f(x) = |x|$.

EXAMPLE **11** Finding the domain and range   What are the domain and range for the absolute value function, $f(x) = |x|$?

Solution   The absolute value function is shown in Table 19 and Figure 37.

Input, x	Output, $f(x) = |x|$
−5	5
−3	3
−1	1
0	0
1	1
3	3
5	5

Table 19

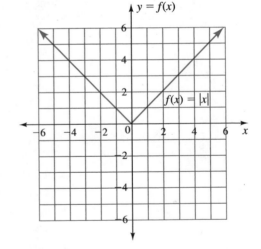

Figure 37

Any real number is a defined input, so the domain is the set of all real numbers, ℝ.

There are no negative outputs, so the range is positive or zero: $f(x) \geq 0$. ●

 Because the absolute value symbol can be used as a grouping symbol, expressions inside the absolute value must be placed in parentheses when a graphing calculator function is used.

EXAMPLE **12** Exploring absolute value on a calculator   Match the absolute value expressions (a, b, c, and d) with the expressions needed on a graphing calculator (1, 2, 3, and 4).

a. $|x + 2|$ **1.** $1 \div (\text{abs}(x) + 2)$

b. $|x| + 2$ **2.** $\text{abs}(x + 2)$

c. $\dfrac{1}{|x + 2|}$ **3.** $\text{abs}(x) + 2$

                  4. $1 \div (\text{abs}(x + 2))$

d. $\dfrac{1}{|x| + 2}$

Solution   See the Answer Box.

Graphing Calculator Technique: Absolute Value	An absolute value key, ABS, is on the keyboard of some calculators; on others, absolute value is listed under CATALOG.

ANSWER BOX

Warm-up: 1. 9 **2.** 15 **3.** 16 **4.** 15 **5.** −10 **Think about it 1:** The inputs to a sequence are natural numbers, not real numbers; it would not make sense for a term number to be a fraction or decimal. **Think about it 2:** The common difference becomes the slope, because the inputs are the natural numbers 1 unit apart. In the linear equation, which intersects the y-axis at the point $(0, 10)$, 10 is the y-intercept. It is 4 units from $a_1 = 6$. If defined, $a_0 = 10$ would be the next number to the left in the sequence.

$$10, 6, 2, -2, -6, \ldots \quad \text{Sequence}$$
$$\vee \;\; \vee \;\; \vee \;\; \vee$$
$$-4, -4, -4, -4, \ldots \quad \text{Differences}$$

Because the inputs to sequences are the natural numbers, the sequence stops at $a_1 = 6$, but the matching linear equation continues to the left. Some people like to find a_0 and then use it as the y-intercept for the linear equation. **Think about it 3:** $a_1 = 4$ and $d = 3$, so $a_n = 4 + (n - 1) \cdot 3 = 4 + 3n - 3 = 3n + 1$. **Think about it 4:** Because tuition is paid only for taking a positive number of credits, the domain must be restricted to positive numbers. Additional information about inputs (whether there is an upper limit on the number of credits, whether the university offers half-credit courses, etc.) may be needed to identify the highest domain number and to decide if fractions or decimals are permitted in the domain. The range is constant: $8000. **Example 12: a.** 2 **b.** 3 **c.** 4 **d.** 1

EXERCISES

In Exercises 1 and 2, find the differences in each sequence. What might be the next term?

1. a. 8, 4, 2, 1, ...

    b. 8, 4, 0, −4, ...

    c. −13, −5, 3, 11, 19, ...

    d. 2, −4, 6, −8, 10, ...

2. a. 6, −1, −8, −15, ...

    b. 4, 8, 16, 32, ...

    c. 4, 8, 12, 16, ...

    d. 7, 2, −3, −8, ...

3. In Exercise 1, model parts b and c with a rule using a_n and n. Match each nth term expression with one of these equations:

$$y = 8x - 21$$
$$y = -4x + 12$$
$$y = 8x + 4$$

4. In Exercise 2, model parts a, c, and d with a rule using a_n and n. Match each nth term expression with one of these equations:

$$y = -7x + 13$$
$$y = -5x + 12$$
$$y = 4x$$

For Exercises 5 to 12:

a. Name the first term and the common difference.

b. Find the tenth term.

c. Find the rule in terms of the nth term, a_n.

d. Rewrite the rule as a linear equation, $y = mx + b$.

5. $-1, 3, 7, 11, 15, \ldots$ 19, 23, 27, 31, 35

6. $4, 9, 14, 19, 24, \ldots$ 29, 34, 39, 44, 49

7. $11, 17, 23, 29, 35, \ldots$

8. $-1, 2, 5, 8, 11, \ldots$

9. $9, 11, 13, 15, 17, \ldots$ 19, 21, 23, 25, 27

10. $2, 0, -2, -4, -6, \ldots$

11. $7, 4, 1, -2, -5, \ldots$ -8, -11, -14, -17, -20

12. $1, -3, -7, -11, -15, \ldots$

13. Describe how to find out if a sequence can be modeled with a straight line.

14. Explain the relation between the common difference in a sequence and the slope of its linear equation.

15. You borrow $400 and agree to pay $100 each month. Suppose $20 of the $100 payment is interest and $80 is for the loan repayment. Write a sequence showing the amount left to repay on the loan after each payment. How long will it take to pay off the loan? What will be the total interest paid? What percent is the total interest paid of the $400 borrowed?

16. *Halley's Comet.* Halley's comet appeared in view of the Earth in the following years:

1456, 1531, 1607, 1682, 1758, 1835, 1910, 1986

Discuss the dates in terms of sequences. Variations are due to the relative position of the Earth in its orbit about the sun and the relative position (and, hence, gravitational attraction) of Jupiter.

a. Predict the next two years in which Halley's comet will be visible.

b. Fit a line to the dates with linear regression. Record the equation.

c. Using 1986 as the table minimum and -75.8 as the change in the table, estimate how close to the years 1000 A.D. and 1 A.D. the comet appeared.

d. Research a connection between Mark Twain (Samuel L. Clemens) and Halley's comet. A collegiate dictionary may be a sufficient source.

In Exercises 17 to 20, graph the functions. State the domain and range of each.

17. $f(x) = 2$

18. $f(x) = -3$

19. $f(x) = -2$

20. $f(x) = \pi$

In Exercises 21 to 28, which situations describe identity functions? Which describe constant functions? Explain your reasoning.

21. Input: Number of visits to a county fair on a $15 season pass.
Output: Total cost for visits to the fair

22. Input: Amount of money loaned without interest
Output: Amount of money paid back, if loan is repaid in full

23. Input: Number of horseback rides per month
Output: Monthly payment of $100 for unlimited rides

24. Input: Number of helpings eaten
Output: Total cost for one person of eating at a restaurant that advertises "All You Can Eat for $10.95"

25. You follow the rule "Eat an apple a day to keep the doctor away."
Input: Days since you started to follow the rule
Output: Number of apples eaten each day

26. You follow the rule "Eat an apple a day to keep the doctor away."
Input: Days since you started to follow the rule
Output: Total number of apples eaten

27. Input: Money put into a postage machine at the post office
Output: Total value of the stamps and change received

28. Input: The key you want to press on the computer keyboard
Output: The character appears that you wanted to enter

Graph the equations in Exercises 29 to 36. Give the domain and range of each.

29. $y = |x + 3|$

30. $y = |x - 3|$

31. $y = |x| + 3$

32. $y = |x| - 3$

33. $y = -|x - 3|$

34. $y = -|x + 3|$

35. $y = -|x| - 3$

36. $y = -|x| + 3$

For Exercises 37 to 39, use the graph in the figure below.

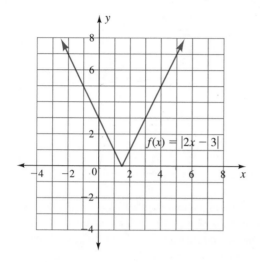

37. Solve these equations from the graph.

  a. $|2x - 3| = 5$ **b.** $|2x - 3| = 3$

  c. $|2x - 3| = 0$ **d.** $|2x - 3| = -2$

38. Solve these equations from the graph.

  a. $|2x - 3| = 6$ **b.** $|2x - 3| = 7$

  c. $|2x - 3| = 8$ **d.** $|2x - 3| = -4$

39. a. What is the slope of the left side of the graph of $f(x) = |2x - 3|$?

  b. What is the slope of the right side?

  c. What is the y-intercept of $f(x) = |2x - 3|$? What input gives the y-intercept?

  d. Write the equation of the piece of the graph containing $(0, 3)$. State the equation as subject to $x \leq 1.5$.

  e. Find $f(1.5)$ with your equation in part d.

  f. Write the equation of the piece of the graph containing $(3, 3)$. State the equation as subject to $x \geq 1.5$.

g. Find $f(1.5)$ with your equation in part f.

For Exercises 40 to 42, use the graph in the figure below. Extend the graph as needed.

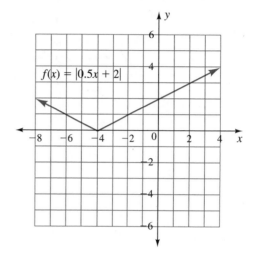

40. Solve these equations from the graph.

  a. $|0.5x + 2| = 2$ **b.** $|0.5x + 2| = 0$

  c. $|0.5x + 2| = -2$ **d.** $|0.5x + 2| = 5$

41. Solve these equations from the graph.

  a. $|0.5x + 2| = 3$ **b.** $|0.5x + 2| = -1$

  c. $|0.5x + 2| = 1$ **d.** $|0.5x + 2| = 4$

42. a. What is the slope of the left side of the graph of $f(x) = |0.5x + 2|$?

  b. What is the slope of the right side?

  c. What is the y-intercept of the function?

  d. Write the equation of the piece of the graph containing $(0, 2)$. State the equation as subject to $x \geq -4$.

  e. Find $f(-4)$ with your equation in part d.

  f. Write the equation of the piece of the graph containing $(-6, 1)$. State the equation as subject to $x \leq -4$.

  g. Find $f(-4)$ with your equation in part f.

43. Solve these absolute value equations and inequalities from the graph shown in the figure.

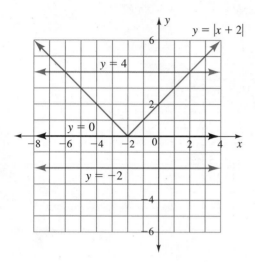

a. $|x + 2| = 0$ **b.** $|x + 2| > 0$

c. $|x + 2| < 4$ **d.** $|x + 2| > -2$

e. $|x + 2| < -2$ **f.** $|x + 2| = 3$

g. $|x + 2| > 3$

44. Solve these absolute value equations and inequalities from the graph shown in the figure.

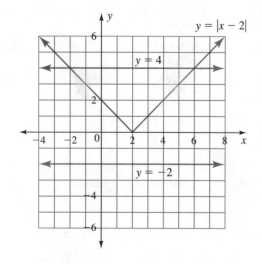

a. $|x - 2| = 0$ **b.** $|x - 2| \geq 0$

c. $|x - 2| > 4$ **d.** $|x - 2| < -2$

e. $|x - 2| > -2$ **f.** $|x - 2| = 3$

g. $|x - 2| < 3$

Projects

45. *Probability Function, I.* The set of outcomes from rolling a die (one of a pair of dice) is $\{1, 2, 3, 4, 5, 6\}$.

a. Toss a die 120 times and record the outcomes in a table similar to the one below, where n = possible outcome from rolling the die, $E(n)$ = expected number of times out of 120 that you think the outcome n will occur, $f(n)$ = actual number of times the outcome occurred, and Percent = $f(n) \div 120$.

n	$E(n)$	$f(n)$	Percent
1			
2			
3			
4			
5			
6			

b. *Probability* is the number of ways something can happen divided by the number of possible outcomes. What is the probability of obtaining any particular number on a single roll of a die?

c. Make a graph showing n (the outcome or face turned up) on the horizontal axis and both Percent and Probability on the vertical axis. Plot ordered pairs $(n, \text{Percent})$ from the table. Plot ordered pairs $(n, \text{Probability})$ from part b.

d. What function has a graph similar to the probability graph? How are the graphs the same? How are they different?

46. *Probability Function, II.* Suppose you have two dice, each with sides numbered from 1 to 6.

a. Complete the first table to show all the sums from adding the numbers facing up on the two dice.

	1	2	3	4	5	6
1	2	3	4	5		
2	3	4	5			
3	4					
4	5					
5						
6						

b. Use the table above to complete the outcomes column of the table below.

Sum, n	Number of Outcomes	Probability Function, $P(n)$
2	1	
3		
4		
5		
6		
7		
8		
9		
10		
11		
12		

c. What is the total of the outcomes column in the second table?

d. The *probability of an outcome n, P(n)*, is the number of ways the outcome n can happen divided by the number of possible outcomes. Unless otherwise directed, we write probability as a fraction. The $P(n)$ column describes the probability function for obtaining the sum n when rolling two dice. Write the probability fraction for the sum of two dice in the $P(n)$ column of the second table. What is the sum of this column?

e. Draw a graph of the probability function for the sum of two dice. Use n as input and $P(n)$ as output.

f. Which mathematical function has a graph similar to that of $P(n)$? How are the graphs the same? How are they different?

47. *Multiplication Table.* Each column and row in the multiplication table may be considered a sequence of numbers.

Describe the common difference in each of these rows:

a. the first row

b. the second row

c. the tenth row

d. the nth row

1	2	3	4	5	6	7	8	9	10	11	12
2	4	6	8	10	12	14	16	18	20	22	24
3	6	9	12	15	18	21	24	27	30	33	36
4	8	12	16	20	24	28	32	36	40	44	48
5	10	15	20	25	30	35	40	45	50	55	60
6	12	18	24	30	36	42	48	54	60	66	72
7	14	21	28	35	42	49	56	63	70	77	84
8	16	24	32	40	48	56	64	72	80	88	96
9	18	27	36	45	54	63	72	81	90	99	108
10	20	30	40	50	60	70	80	90	100	110	120
11	22	33	44	55	66	77	88	99	110	121	132
12	24	36	48	60	72	84	96	108	120	132	144

Multiplication table

What is a_n for these sequences?

e. 2, 4, 6, 8, 10, . . .

f. 3, 6, 9, 12, 15, . . .

g. 4, 8, 12, 16, 20, . . .

Photocopy the multiplication table, and shade the following sequences with different colors or types of shading. Are they arithmetic? Guess a rule in terms of a_n for each sequence. The positions in the table may help, especially after you find the rule for the sequence 1, 4, 9, 16, 25, ...

h. 1, 4, 9, 16, 25, . . .

i. 2, 6 12, 20, 30, . . .

j. 3, 8, 15, 24, . . .

k. The Warm-up pattern 1, 3, 6, 10, . . .
(*Hint:* Each term of the sequence in part k is half the corresponding term of the sequence in part i.)

CHAPTER ❷ SUMMARY

Vocabulary

For definitions and page references, see the Glossary/Index.

absolute value function	coefficient of correlation	condition	domain
arithmetic sequence	common difference	conditional function	dot graph
		constant function	function
		decreasing function	identity function

increasing function

line of best fit

linear equation in one variable

linear equation in two variables

linear function

linear regression

modeling with a linear function

non-negative numbers

nonvertical

point-slope equation

range

relevant domain

relevant range

rule for the nth term of a sequence

sequence

slope

slope-intercept equation

step graph

subscripts

terms of a sequence

vertical-line test

x-intercept

y-intercept

Concepts

See the list of vocabulary for important definitions.

2.0 Writing and Solving Linear Equations

Guess and check may provide a solution to a problem, or it may help you write an equation needed to solve a problem.

Restrictions on inputs or conditions on equations may result in graphs that change direction, dot graphs, or step graphs.

2.1 Functions

We may describe a function with a graph, a written rule, a listing of ordered pairs, or an equation.

When graphing a function, place the inputs on the horizontal axis and the outputs on the vertical axis.

A relationship is identified as a function from its ordered pairs, its table (one input for each output), or its graph (passing the vertical-line test).

Functions are not limited to mathematical relationships.

Functions may be visualized as putting numbers or objects through a function machine or as mapping one set of numbers or objects onto another set.

We evaluate functions by substituting the input for x into the function described by $f(x)$.

We can describe relationships between outputs to functions with the notation such as $f(a)$ and $f(b)$: in $f(x) = x^2, f(-2) = f(2)$.

The domain and range of a mathematical function may be limited in application settings to a relevant domain and relevant range.

2.2 Linear Functions

Graphs cross the axes at intercept points $(a, 0)$ and $(0, b)$. These intercepts, a and b, have special significance in applications. In function notation, a is found by solving $f(x) = 0$ and b is found by $f(0)$.

Vertical lines, with equation $x = c$, are the only linear equations that are not also linear functions, $f(x) = mx + b$.

A function is a linear function if and only if its slope is constant. To find the units on slope, divide the units on the output axis by the units on the input axis.

We describe the change in a function's graph as *increasing* or *decreasing*.

If the slope of a line is positive, the function is an increasing function; if the slope of a line is negative, the function is a decreasing function.

A horizontal line has a zero slope.

A vertical line has an undefined slope.

Parallel lines have the same slope.

Perpendicular lines have slopes that are negative reciprocals—their slopes multiply to -1.

The logical phrase "if and only if" means that both "if a then b" and "if b then a" are true.

2.3 Modeling with a Linear Function

To build an equation with the slope-intercept equation $y = mx + b$, find the slope, m, and the y-intercept, b.

To build an equation with the point-slope equation, substitute one point, (x_1, y_1), and the slope, m, into $y - y_1 = m(x - x_1)$.

To build an equation with a line of best fit, graph the data, position a ruler on the graph so that it passes through two points and reasonably close to the remaining points, and then use the two points and the slope-intercept or point-slope equation to find the equation for the line.

To find an equation with calculator regression, enter the data, set a window appropriate to the data, plot the data to see which regression is most appropriate, and choose the regression from the statistical options. Check by graphing the regression equation with the data.

2.4 Special Functions

Sequences with a constant difference between terms may be described with either an nth term expression or a linear equation.

The terms, a_n, of a sequence of numbers may be paired with the natural numbers to form ordered pairs, (n, a_n).

A constant function has a horizontal graph and an equation $y = c$, where c is a real number.

The graph of the identity function is the line $y = x$.

The absolute value function describes the distance a number is from zero on the number line. Expressions in an absolute value function must be placed in parentheses when they are entered into a calculator.

CHAPTER ② REVIEW EXERCISES

Find words in the Vocabulary list that match each of the descriptions in Exercises 1 to 16. If more than one answer is needed, the number in parentheses tells how many are needed.

1. Used to find out if a graph is a function

2. The numbers in a sequence

3. A graph for which the inputs are only integers

4. A graph for which the inputs are rounded off before the outputs are found.

5. Limits on inputs due to an application setting

6. Limits on outputs due to an application setting

7. Tells if a sequence can be described by a linear equation

8. Used to find a linear equation on a graphing calculator (2)

9. A function where the output exactly matches the input

10. A function with only one input (regardless of input)

11. A function with a positive output for any real-number input

12. Numbers that are zero or positive

13. Numbers that are small and placed below and to the right of letters or variables

14. Number or expression that describes the rate of change of a graph

15. Ways to find a linear equation (5)

16. Functions that are rising, falling, or horizontal when viewed left to right (3)

Use guess and check in a table to solve the problems in Exercises 17 and 18.

17. Body mass index is calculated by multiplying weight in pounds by 704.5 and then dividing by the square of height in inches. A healthy range for body mass index is 19 to 24. To the nearest pound, for what weight will a 5-foot 5-inch person have an index of 19? An index of 24?

18. The figure shows an angle of x degrees in a circle with radius r inches. The length of the arc opposite angle x is S, in inches. To find S, multiply π, r, and x and then divide by 180. In a circle with radius 10 inches, for what angle, x, will S be approximately 15.7 inches? In the same circle, for what x will S exactly equal the radius, 10 inches?

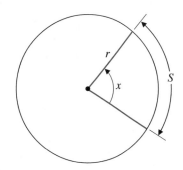

For each situation in Exercises 19 to 24, define variables, write an equation, and say whether the graph will be a dot graph, a step graph, or a straight line. Test your equation with a number to see that it makes sense in the problem setting.

19. The total cost, with a 7% sales tax, of a purchase

20. The total cost of a party at an ice rink that charges $85 for up to ten people and $4.75 for each additional person

21. The total cost of a party at a roller skating ring that charges $40 for up to 10 children and $4.50 for each additional child

22. The total cost of a long-distance telephone call at $0.26 for the first minute or part thereof and $0.19 per minute for each additional minute or part thereof

23. The total cost of a long-distance telephone call at $1.08 for the first 2 minutes or part thereof and $0.63 for each additional minute or part thereof

24. The sales tax on a purchase at an 8% tax rate

In Exercises 25 to 28, answer the following questions:

a. What is the domain?

b. What is the range?

c. Does the graph describe a function?

25. ellipse

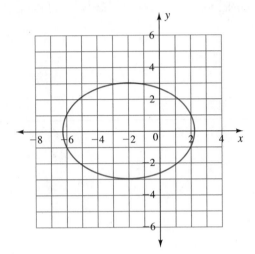

26.

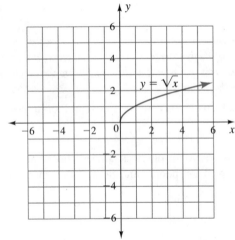

27.

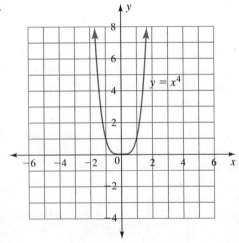

28.

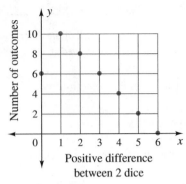

Positive difference between 2 dice

29. Use the graph of $f(x)$ in the figure to evaluate the expressions and answer the questions. The graph does not intersect the x-axis.

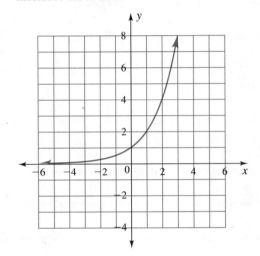

a. $f(0)$ **b.** $f(2)$ **c.** $f(-1)$

d. $f(3)$ **e.** $f(1)$

f. For what inputs is $f(x) \geq 2$?

g. For what inputs is $f(x) < 0$?

h. For what inputs is $f(x) \leq 4$?

i. What is the y-intercept?

j. What is the set of possible inputs (domain) for $f(x)$?

k. What is the set of possible outputs (range) for $f(x)$?

30. Use the graph of $f(x)$ in the figure on page 145 to evaluate the expressions and answer the questions.

a. $f(0)$ **b.** $f(1)$ **c.** $f(-2)$

d. $f(3)$ **e.** $f(-1)$

f. For what inputs is $f(x) \geq 16$?

g. For what inputs is $f(x) < 0$?

h. For what inputs is $f(x) < 4$?

i. What is the y-intercept?

j. What is the set of possible inputs (domain) for $f(x)$?

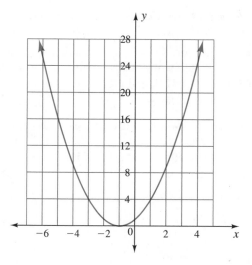

k. What is the set of possible outputs (range) for $f(x)$?

31. In the figure in Exercise 30, what inputs x give these function values?

    a. $f(x) = 0$ **b.** $f(x) = 9$

    c. $f(x) = 16$ **d.** $f(x) = 4$

    e. $f(x) = -2$

32. In the figure in Exercise 29, what inputs x give these function values?

    a. $f(x) = 1$ **b.** $f(x) = 3$

    c. $f(x) = \frac{1}{2}$ **d.** $f(x) = 4$

    e. $f(x) = -2$

33. Use the table to find the indicated values.*

Altitude, x (kilometers)	Summer Temperature, $f(x)$ (°C)
0	15.7
0.5	14.5
1.0	12.0
1.5	10.0
2.0	7.5
3.0	2.4
4.0	−3.0
5.0	−8.9
6.0	−15.1

    a. $f(5 \text{ km})$ **b.** $f(3 \text{ km})$

    c. x, if $f(x) = -3°C$ **d.** x, if $f(x) = 12°C$

    e. $f(0 \text{ km})$

34. Use the table to find the indicated values.*

Altitude, x (kilometers)	Winter Temperature, $f(x)$ (°C)
0	0.7
0.5	0.0
1.0	−1.3
1.5	−3.0
2.0	−4.7
3.0	−9.3
4.0	−15.0
5.0	−21.2
6.0	−28.1

    a. $f(5 \text{ km})$ **b.** $f(3 \text{ km})$

    c. x, if $f(x) = -15°C$ **d.** x, if $f(x) = 0°C$

    e. $f(0 \text{ km})$

In Exercises 35 to 42, evaluate the function $f(x) = 2x^2 - 3x + 1$ for the indicated inputs.

35. $f(1)$ **36.** $f(2)$

37. $f(0.5)$ **38.** $f(1.5)$

39. $f(-2)$ **40.** $f(-0.5)$

41. $f(\square)$ **42.** $f(-a)$

43. Find the equation of a line perpendicular to $y = 3x + 4$ through

    a. the origin   **b.** $(3, 4)$

44. Find the equation of a line parallel to $y = \frac{2}{3}x - 4$ through

    a. the origin   **b.** $(4, 3)$

Make a table in which the columns are headed Ordered Pairs, Slope, Equation, Horizontal or Vertical, x-intercept, y-intercept. Complete the table for each set of ordered pairs in Exercises 45 to 52.

45. $(1, 3)$ and $(2, 2)$

46. $(2, 3)$ and $(2, 2)$

47. $(-3, 3)$ and $(0, 0)$

48. $(-3, 3)$ and $(3, 3)$

49. $(-1, -1)$ and $(1, 3)$

50. $(2, 3)$ and $(-3, 4)$

51. $(-3, 3)$ and $(2, 3)$

52. $(0, 4)$ and $(-1, -1)$

*Data in Exercises 33 and 34 are from Handbook of Chemistry and Physics, 36th ed. (Chemical Rubber Publishing Company, Cleveland, Ohio, © 1954), p. 3093.

53. In Exercises 45, 47, 49, and 51, are any of the lines formed by the ordered pairs parallel? Perpendicular?

54. In Exercises 46, 48, 50, and 52, are any of the lines formed by the ordered pairs parallel? Perpendicular?

55. Is the graph of the function in Exercise 28 linear? Explain your reasoning.

56. Is the graph of the function in Exercise 33 linear? Explain your reasoning.

In Exercises 57 to 60:

a. Write an equation.

b. Indicate the slope and y-intercept, and explain what, if anything, each means.

57. Marcia pays a sales tax of 6.5% on a price x. What is the tax paid?

58. Don's party at a restaurant costs $10.95 per person. What is the total cost?

59. An annual inspection for a Beechcraft Skipper airplane costs $500 plus $45 per hour for any repairs needed. Parts are extra. What is the total cost without parts?

60. David's minimum monthly cellular phone bill is $19 plus $0.20 per minute in send mode. Long-distance charges are extra. What is his monthly bill without long-distance charges?

61. The cost of plastic notebook binding combs is a function of their diameter. For 20 combs of the given diameter, the cost is as follows: $\frac{1}{4}''$, $1.99; $\frac{3}{8}''$, $2.49; $\frac{1}{2}''$, $2.99; and $\frac{5}{8}''$, $4.49. (*Hint:* Round prices to the next highest cent.)

   a. Graph and find the equation of a line of best fit.

   b. Use your equation to predict the cost of 1-inch-diameter combs.

   c. Suggest why the actual cost ($10.99) is different from your prediction.

62. The number of sheets of paper held by a plastic notebook binding comb is a function of its diameter. For a comb of the given diameter, the number of sheets is as follows: $\frac{3}{4}''$, 150 sheets; $1''$, 200 sheets; $1\frac{1}{2}''$, 290 sheets; $2''$, 425 sheets.

   a. Graph and find the equation of a line of best fit.

   b. Use your equation to predict the number of sheets of paper held by a $\frac{1}{4}$-inch-diameter comb.

   c. Suggest why the actual number of sheets (20) is different from your prediction.

In Exercises 63 and 64, use calculator regression to find a linear equation.

63. Time and total score for a computer solitaire game are given by the following data points: (125, 6286), (122, 6391), (118, 6573), and (117, 6618). Let $x =$ time and $y =$ total score.

64. The bonus points awarded for the times in the solitaire game in Exercise 63 are as follows: (125, 5600), (122, 5705), (118, 5915), and (117, 5950). Let $x =$ time to complete a game (in seconds) and $y =$ bonus points.

In Exercises 65 to 68, what is the next number in each sequence?

65. 8, 15, 22, 29, 36, . . .

66. 17, 14, 11, 8, 5, . . .

67. 32, 16, 8, 4, 2, . . .

68. 2, 10, 50, 250, . . .

In Exercises 69 and 70, list the first four terms of the sequence described by each rule.

69. $a_n = 4n - 3$

70. $a_n = 5n + 1$

In Exercises 71 to 73, what is the rule for each sequence?

71. 18, 16, 14, 12, 10, . . .

72. 31, 27, 23, 19, 15, . . .

73. $-5, -2, 1, 4, 7, . . .$

In Exercises 74 to 79, write an equation. Identify each as an increasing, decreasing, or constant function.

74. A ski resort advertises a monthly pass for $350. What is the total cost to Jason if x is the number of ski trips he makes each month?

75. The annual fee for an athletic club is $350. No charge is made for each use of the facility. What is the total annual cost if Chae Hyok uses the club x times?

76. A restaurant advertises "Anything You Want to Eat for $5.95 per Pound." What is the total cost to Kellie if x is the number of pounds she eats?

77. The annual fee for an athletic club is $350. Each use costs an additional $5. What is the total yearly cost Chamique pays if she uses the club x times?

78. Paul pays $350 in advance on his account at the athletic club. Each time he uses the club, $5 is deducted from the account. What value remains in his account after x visits to the club?

79. Kristen buys a $35 mass transportation ticket. Each ride costs $1.50, deducted automatically from the ticket when she leaves the station. What value remains on the ticket after x rides?

80. Graph $y = |x - 1|$. From your graph, solve these equations.

a. $|x - 1| = 4$ **b.** $|x - 1| = 2$

c. $|x - 1| = 0$ **d.** $|x - 1| = -2$

81. Graph the conditional equations

$y = 1 - x$ for $x < 1$

and

$y = x - 1$ for $x \geq 1$

In Exercises 82 to 89, give the domain and range.

82. $y = 365$ **83.** $y = 365x$

84. $y = |x - 1|$ **85.** $y = |x| - 1$

86. $y = x$ **87.** $y = |x + 1|$

88. $y = |x| + 1$ **89.** $y = 1$

CHAPTER ❷ TEST

1. Make a table for the total cost of 1 to 6 transcripts if the cost is \$5 (total) for up to two copies and \$2 for each additional copy. Graph your function. Explain why the points should or should not be connected.

2. Which are functions?

a.

Input: Name	Output: Sport
Ali	Basketball
Ali	Golf
Tarom	Volleyball
Dean	Badminton

b.

Input: Name	Output: Registration Day
Terrie	Sept. 1
Jessica	Sept. 3
Dianne	Sept. 1
Hee-Jin	Sept. 2

c.

Input	Output
2	3
3	3
4	3
5	3

d.

Input	Output
2	4
2	6
3	7
4	3

3. If $f(x) = 3x^2 - 2x - 4$, find

a. $f(-2)$ **b.** $f(0)$

c. $f(2)$

4. a. Find the slope of the line passing through $(-2, 4)$ and $(5, 2)$.

b. Find the equation of the line in part a.

c. Find the slope of a line parallel to the line in part a.

d. Find the equation of a line through $(2, -1)$ and perpendicular to the line in part a.

5. Fill in the missing word or choose the correct word from those given in brackets.

a. The slope of a horizontal line is _____ .

b. A line that falls from left to right has a [negative, zero, positive] slope and is said to be a(an) [increasing, decreasing] function.

c. If the slope of a graph between all pairs of points is constant, the graph is a _____ function.

d. A horizontal linear graph is also called a _____ function.

e. Linear or arithmetic sequences have a _____ difference between terms.

f. The set of inputs to a sequence function is the _____ .

6. Alicia's taxi ride cost \$2.50 plus \$7.00 per mile.

a. Write a linear equation to describe the total cost of an x-mile ride.

b. What is the slope?

7. Heidi's long-distance telephone call cost \$0.13 for 1 minute. A 19-minute call to the same location on another day cost \$2.11.

a. Identify reasonable inputs and outputs, and define the variables.

b. Write the data as two ordered pairs.

c. Fit a linear equation to the data.

8. Sale prices in 1984 for solitaire diamond engagement rings set in 14K gold are listed below. Let x = points and y = price. One carat = 100 points. Fit a linear equation with calculator regression.

Size in Points	10	20	25	33	50	100
Price	$99	$179	$249	$299	$499	$999

9. Give the next number in each sequence. Identify which are linear (arithmetic) sequences.

   a. 10, 18, 26, 34, 42, _____

   b. 4, 10, 18, 28, 40, _____

   c. 3, 9, 27, 81, _____

   d. −16, −9, −2, 5, 12, _____

In Exercises 10 to 15, sketch a graph to fit the equation or description.

10. $y = |x| - 3$

11. $y = x$

12. College tuition is $300 per credit up to 12 credits, with no additional charge for up to a total maximum load of 24 credits.

13. College tuition is $250 per credit. Maximum load is 24 credits.

14. Filling an 8-inch-tall cylindrical bucket with a garden hose, with the depth of the water in the bucket as a function of time. Explain your graph.

15. The total monthly cost for Ki to ride the bus if a monthly bus pass costs $35 and he rides the bus x times.

CUMULATIVE REVIEW OF CHAPTERS 1 AND 2

1.

Input x	Input y	Output xy	Output $x + y$	Output $x - y$
−2	4			
−3	7			
2		−6		
−3		6		
−1			−7	
	−2		−7	
	−2		1	
2			−7	

In Exercises 2 and 3, match each expression with one of the given words.

2. Choose from
   whole numbers, quotient, difference, natural numbers, rational numbers

   a. The answer to a subtraction problem

   b. The set of numbers {1, 2, 3, 4, ...}

   c. The set of numbers that may be written in the form a/b, where a and b are integers and $b \neq 0$

   d. The answer to a division problem

   e. The set of numbers {0, 1, 2, 3, 4, ...}

3. Choose from
   factors, sets, opposites, reciprocals, factoring

   a. Two numbers, n and $-n$, that add to zero

   b. Two numbers or expressions, a and b, that are multiplied to obtain the product ab

   c. Two numbers, n and $1/n$, that multiply to 1

   d. Removing a common factor from two or more terms

   e. Collections of objects or numbers

Simplify the expressions in Exercises 4 to 9.

4. $3[4 - 2(5 - 8) - 6]$

5. $a(b + c) - b(a + c) + c(a - b)$

6. πr^2 for $r = 7.5$ feet

7. $\frac{1}{4}\pi d^2$ for $d = 15$ feet

8. $\frac{ac - bc}{c}$

9. $\frac{16 + 21x}{6}$

Solve the equations in Exercises 10 and 11.

10. $3x + 12 = 5x + 3$ 11. $15 - 4x = 5(6 - x)$

Write and solve an equation in Exercises 12 and 13.

12. Five more than the quotient of a number and 8 is 2.

13. The product of 3 and a number is 15 more than the number.

14. Sketch a graph of $y = 2x - 5$ and $y = 4 - x$. Solve the equations in parts a to d from your graph.

   a. $2x - 5 = 4 - x$ b. $2x - 5 > 0$

   c. $4 - x \leq 0$ d. $2x - 5 < 4 - x$

e. Solve the inequality in part b with algebraic notation.

f. Solve the inequality in part c with algebraic notation.

g. Solve the inequality in part d with algebraic notation.

Complete the sentences in Exercises 15 to 20 to make true statements. Choose from

subtraction to addition of the opposite number, a product to a sum, division to multiplication by the reciprocal, a sum to a product, $x \geq 0, y \geq 0$

15. The multiplication $a(b + c)$ changes _____ .

16. Outputs are zero or positive if we require _____ .

17. Inputs are zero or positive if we require _____ .

18. Factoring $ab + ac$ changes _____ .

19. To subtract real numbers, we may change _____ .

20. To divide real numbers, we may change _____ .

21. *Body Mass Index.* Solve the formulas

$$19 \leq \frac{x(704.5)}{68^2} \quad \text{and} \quad \frac{x(704.5)}{68^2} \leq 24$$

to find the healthy weight range in pounds for a 5-foot 8-inch person.

22. Suppose coffee is $5.98 for each of the first two 1-pound cans and $9.98 for each additional can.

a. Make a table and graph for the total cost of 1 to 5 cans.

b. Discuss whether or not the points should be connected.

c. Does the total cost equation describe a function? If so, what is the domain of the function?

In Exercises 23 to 26, find $f(-1)$, $f(0)$, $f(1)$, and $f(2)$.

23. $f(x) = x + 2$ **24.** $f(x) = 2x$

25. $f(x) = x^2$ **26.** $f(x) = 3 - x$

What are the slope and the equation for each pair of points in Exercises 27 and 28?

27. $(4, -3)$ and $(-6, 1)$ **28.** $(-2, -7)$ and $(2, -1)$

29. What is the equation of a line passing through the origin that is perpendicular to $y = 3x - 5$?

In Exercises 30 to 33, give the next pair of coordinates for each list, and describe the function with a rule in either words or symbols.

30. $(0, 0)$, $(1, 1)$, $(2, 4)$, $(3, 9)$

31. $(0, 0)$, $(1, 1)$, $(2, 2)$, $(3, 3)$

32. $(0, 0)$, $(1, 2)$, $(2, 4)$, $(3, 6)$

33. $(0, 4)$, $(1, 4)$, $(2, 4)$, $(3, 4)$

In Exercises 34 and 35, write an equation.

34. Elise's car rental costs $60 plus $0.07 for each mile driven. What is the total cost of the rental?

35. Marianne's conference exhibit costs $500 plus $65 per worker at the exhibit. What is the total cost of the exhibit?

36. Find an equation to fit these data: $(-4, -20)$, $(5, -15)$, $(14, -10)$, $(23, -5)$, $(41, 5)$.

37. What kind of function is the wind-chill apparent temperature for a 0-mph wind, graphed in the figure?

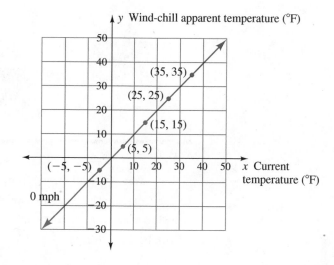

3

Quadratic Functions: Applications and Solutions to Equations

Figure 1

The clavadistas (cliff divers) from La Quebrada at Acapulco, Mexico, dive from a height of 35 meters above the water and need to move horizontally away from the cliff, as shown on the left in Figure 1. The path of their dives, a parabola, may be described with equations in this chapter.

Olympic platform divers attempt to keep their line of motion as vertical as possible from the end of the board to the water's surface, as shown on the right in Figure 1. The path of their dives is vertical, not parabolic. However, the position of the diver with respect to time may be modeled by the parabolas in this chapter.

This chapter introduces quadratic functions; we graph, apply, and solve quadratic equations. Within the material, we review polynomial operations, square roots, and the Pythagorean theorem.

3.0 Basic Operations on Polynomials

OBJECTIVES

- Identify a polynomial and types of polynomials.
- Add and subtract polynomials.
- Multiply monomials, binomials, and trinomials.
- Factor trinomials.
- Find common monomial factors.

WARM-UP

List all the pairs of positive integer factors of these numbers.

Example: $28 = 1 \cdot 28 = 2 \cdot 14 = 4 \cdot 7$

1. 60
2. 12
3. 24
4. 80
5. 100
6. 180

Use the distributive property to multiply these expressions.

7. $x(x^2 + 6x + 9)$
8. $3x(2x + 3)$
9. $x(x^2 - 2xy + y^2)$
10. $3x(4x^2 + 2x - 3)$

IN THIS SECTION, we review polynomials and their addition, subtraction, multiplication, and factorization. The multiplying and factoring are presented in table form. The table method is likely to be new to you.

Polynomials

Lindsey has designed a parabolic arched entry for the front of a museum (Figure 2). She needs to calculate its equation so that she can write specifications for the height at uniform intervals across the entry. To create the equation, she draws the parabola to scale on graph paper (Figure 3). She uses a basic equation form for a parabola, $y = a(x - x_1)(x - x_2)$. The variables x_1 and x_2 are x-intercepts, and a may be found by substituting for x and y an ordered pair for a point on the parabola.

Figure 2

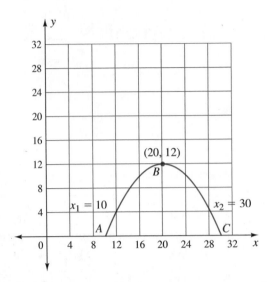

Figure 3

The equation $y = a(x - x_1)(x - x_2)$ is an example of a polynomial equation in factored form. The right side of the equation contains three factors a, $(x - x_1)$, and $(x - x_2)$, hence the name *factored form*. After numbers are placed in the equation for x_1 and x_2, the factors in a polynomial are usually multiplied out, leaving a sum of terms. A polynomial is defined by the nature of these terms.

Definition of Polynomial

> A **polynomial** is an expression written as a sum of terms. The terms in a polynomial are limited to a number or a product of a number and one or more variables having non-negative integers as exponents.

Recall that non-negative means zero or positive. Polynomials include expressions such as $x^2 - y^2$, πr^2, 5, and $x^3 + 8$.

All polynomials have a finite number of terms. Common polynomials are named for the number of terms they contain. *One-, two-, and three-term polynomial expressions* are called **monomials, binomials,** and **trinomials,** respectively.

EXAMPLE **1**   Identifying polynomials   Identify the expression on the right side of each equation as a monomial, a binomial, a trinomial, or not a polynomial.

  a. Bridge-cable equation: $y = \frac{4}{125}x^2$

  b. Square root equation: $y = \sqrt{x}$

  c. Perimeter of a rectangle formula: $P = 2l + 2w$

  d. Vertical motion equation: $h = \frac{1}{2}gt^2 + v_0t + h_0$

  e. Absolute value function: $y = |x|$

  f. $y = x^{-2} + 4x^{-1} + 4$

Solution   **a.** $\frac{4}{125}x^2$ is a monomial; the x is raised to the exponent 2, a positive integer.

  b. $\sqrt{x}$ is not a polynomial; neither zero nor a positive integer exponent will give the square root of x.

  c. $2l + 2w$ is a binomial in two variables; the l and the w are both raised to the exponent 1.

  d. $\frac{1}{2}gt^2 + v_0t + h_0$ is a trinomial in the variable t.

e. The absolute value of x, $|x|$, is not a polynomial; it cannot be created with a zero or positive exponent.

f. $x^{-2} + 4x^{-1} + 4$ has negative exponents and thus is not a polynomial. ●

Addition and Subtraction of Polynomials

We add and subtract polynomials by combining like terms. As mentioned in Chapter 1, *like terms* are expressions with identical variable factors:

$3x^2$, $\frac{1}{2}x^2$, and $0.25x^2$ are like terms.

$4xy$, $-xy$, and $2.5xy$ are like terms.

$-2x^2y$, $-\frac{1}{4}x^2y$, and $6.3x^2y$ are like terms.

EXAMPLE **2**   Adding like terms   Add like terms in these polynomial expressions.

a. $4x + 3y - 6x + 2y$ **b.** $3x^2 - \frac{1}{2}x^2 + 0.25x^2$

c. $4xy - xy + 2.5xy$ **d.** $-2x^2y - \frac{1}{4}x^2y + 6.3x^2y$

e. $3xy + 5yx - yz$

Solution   **a.** $4x + 3y - 6x + 2y = -2x + 5y$

b. $3x^2 - \frac{1}{2}x^2 + 0.25x^2 = 3x^2 - 0.5x^2 + 0.25x^2 = 2.75x^2$

c. Because $-xy$ means $-1xy$, we may write

$$4xy - 1xy + 2.5xy = 5.5xy$$

d. $-2x^2y - \frac{1}{4}x^2y + 6.3x^2y = -2x^2y - 0.25x^2y + 6.3x^2y = 4.05x^2y$

e. There are only two like terms. The terms $3xy$ and $5yx$ contain the same variables; $-yz$ does not. Thus, we write

$$3xy + 5yx - yz = 8xy - yz$$ ●

In Example 3, it may be necessary to apply the distributive property to multiply out expressions before adding like terms.

EXAMPLE **3**   Multiplying and adding like terms   Add or subtract these expressions, as indicated.

a. $(3x - 4y) + (3y - 4x)$

b. $x(x^2 + 6x + 9) + 3(x^2 + 6x + 9)$

c. $3x(2x + 3) - 2(2x + 3)$

d. $x(2x + 1) - 6(2x + 1)$

e. $x(x^2 - 2xy + y^2) - y(x^2 - 2xy + y^2)$

Solution   We multiply with the distributive property and then rearrange the terms to place like terms together.

a. $(3x - 4y) + (3y - 4x) = 3x - 4x - 4y + 3y$
$$= -1x - 1y$$

b. $x(x^2 + 6x + 9) + 3(x^2 + 6x + 9) = x^3 + 6x^2 + 9x + 3x^2 + 18x + 27$
$$= x^3 + 6x^2 + 3x^2 + 9x + 18x + 27$$
$$= x^3 + 9x^2 + 27x + 27$$

c. $3x(2x + 3) - 2(2x + 3) = 6x^2 + 9x - 4x - 6$
$$= 6x^2 + 5x - 6$$

d. $x(2x + 1) - 6(2x + 1) = 2x^2 + x - 12x - 6$
$$= 2x^2 - 11x - 6$$

e. $x(x^2 - 2xy + y^2) - y(x^2 - 2xy + y^2)$
$$= x^3 - 2x^2y + xy^2 - x^2y + 2xy^2 - y^3$$
$$= x^3 - 2x^2y - x^2y + xy^2 + 2xy^2 - y^3$$
$$= x^3 - 3x^2y + 3xy^2 - y^3 \quad \bullet$$

Note that Example 3 uses these exponent facts:

$$x \cdot x = x^2 \quad \text{and} \quad x \cdot x \cdot x = x^2 \cdot x = x^3$$

Also note that the answers are written in *descending order of exponents*. When there are two variables, the terms are arranged in descending order of exponents on the variable occurring first in the alphabet.

Multiplication of Polynomials

The algebraic capabilities of newer calculators allow us to leave much of the tedious work to the calculator, once we understand the structure of multiplying (and factoring). Because of this technological development, the approach to multiplication (and factorization) presented here is based on tables. If you know how to multiply (and factor), be patient while some of your classmates try to understand what it is all about, some for the first time. If you look carefully, you will find some patterns that you have never noticed before and you may improve your understanding of the basic concepts.

The distributive property shows how to multiply a monomial times a binomial or trinomial. To multiply a binomial times another binomial or a trinomial, use the distributive property twice.

Multiplication of Two Binomials

> To multiply the binomials $(ax + b)(cx + d)$, multiply $ax(cx + d)$ and add to this product the product $b(cx + d)$.

The following table depicts the multiplication more visually.

Multiply	cx	$+d$
ax	acx^2	$+adx$
$+b$	$+bcx$	$+bd$

The product is
$$(ax + b)(cx + d)$$
$$= acx^2 + (ad + bc)x + bd$$

A common mental shortcut for multiplying binomials is the FOIL method: take the product of the "first terms," $ax \cdot cx$; add the product of the "outside terms," $ax \cdot d$; add the product of the "inside terms," $b \cdot cx$; and then add the product of the "last terms," $b \cdot d$. The disadvantage of the FOIL method is that it is limited to the product of binomials. The table method may be used for any product of polynomials; see Example 6 for a product of a binomial and a trinomial. Another advantage of the table method is that it organizes the products so that all the terms and patterns are easy to see.

In Figure 3, the x-intercepts of the parabolic arch were $x_1 = 10$ and $x_2 = 30$. We may assume the units are feet. The equation of the parabola becomes

$$y = a(x - x_1)(x - x_2) = a(x - 10)(x - 30)$$

Example 4 shows that the equation of the parabola may also be written $y = a(x^2 - 40x + 300)$.

EXAMPLE 4 Multiplying binomials Multiply $(x - 10)(x - 30)$ with a table.

Multiply	x	-10
x	.	
-30		

Solution The table shows that the distributive property is used twice, once for $x(x - 10)$ and once for $-30(x - 10)$. The entries within the white region are the products.

Multiply	x	-10
x	x^2	$-10x$
-30	$-30x$	$+300$

The product is
$(x - 10)(x - 30)$
$= x^2 + \boxed{(-10x) + (-30x)} + 300$
$= x^2 - 40x + 300$ ●

Scan the next three examples and then refer back to them as you explore the questions in Example 8.

EXAMPLE 5 Multiplying binomials Multiply $(2x + 3)(3x - 2)$ in a table.

Solution We place the binomials on the left and across the top of the table:

Multiply	$3x$	-2
$2x$	$6x^2$	$-4x$
$+3$	$+9x$	-6

The product is
$(2x + 3)(3x - 2)$
$= 6x^2 + \boxed{(-4x) + 9x} - 6$
$= 6x^2 + 5x - 6$ ●

EXAMPLE 6 Multiplying a binomial and a trinomial Multiply $(x + 3)(x^2 + 6x + 9)$ in a table.

Solution Although the positions do not matter, in this text we will place the binomial on the left and the trinomial across the top.

Multiply	x^2	$+6x$	$+9$
x	x^3	$+6x^2$	$+9x$
$+3$	$+3x^2$	$+18x$	$+27$

The product is
$(x + 3)(x^2 + 6x + 9)$
$= x^3 + \boxed{6x^2 + 3x^2}$
$\quad + \boxed{9x + 18x} + 27$
$= x^3 + 9x^2 + 27x + 27$

The middle step in the solution above is not needed, as the table groups like terms. We usually write the answer, $x^3 + 9x^2 + 27x + 27$, directly from the table. ●

EXAMPLE 7 Multiplying two trinomials Multiply $(x^2 - 2x - 1)(x^2 + 2x - 1)$ in a table.

Multiply	x^2	$+2x$	-1
x^2			
$-2x$			
-1			

Solution   The table, which has nine entries, shows a clear organization of like terms, four of which add to zero.

Multiply	x^2	$+2x$	-1
x^2	x^4	$+2x^3$	$-1x^2$
$-2x$	$-2x^3$	$-4x^2$	$+2x$
-1	$-x^2$	$-2x$	$+1$

The product is
$$(x^2 - 2x - 1)(x^2 + 2x - 1)$$
$$= x^4 - 6x^2 + 1$$

●

EXAMPLE **8**   *Exploration*   As mentioned, the reason for using table multiplication is to identify patterns. Stop now and examine the tables in Examples 4 and 5. Consider these questions:

a. Where in the table is the product of the first terms?

b. Where are like terms located?

c. Where is the product of the last terms?

d. What do the products of the terms in the diagonals have in common?

Do the patterns extend to Examples 6 and 7?

Solution   When the terms being multiplied are arranged in descending order of the exponent on x, the following patterns appear:

a. The first entry (upper left corner) in the white part of each table is the product of the first terms.

b. Like terms are on a diagonal.

c. The last entry (lower right corner) in the white part is the product of the last terms.

d. The product of the two terms on one diagonal is equal to the product of the two terms on the other diagonal. In Example 4,

$$300(x^2) = 300x^2 \quad \text{and} \quad (-10x)(-30x) = 300x^2$$

In Example 5,

$$6x^2(-6) = -36x^2 \quad \text{and} \quad (-4x)(+9x) = -36x^2$$

In Examples 6 and 7, like terms in adjacent columns are on a diagonal. In Example 6, the diagonal products for adjacent columns are equal. (Can you find a way to make the products equal in Example 7?)   ●

   The pattern in part d above—equal diagonal products—is a pattern that does not show up in other ways of multiplying polynomials. The equal diagonal products will be central to the table-based factoring process.
   In Example 9, we prove equal diagonal products for binomials.

Definition of Proof

A **proof** is a logical argument that demonstrates the truth of a statement.

In algebra, a proof may be demonstrated by obtaining the result with letters instead of numbers. The letters show that the process works for any real number.

EXAMPLE **9**   *Proving equal diagonal products*   Prove that the diagonals of the table for $(ax + b)(cx + d)$ always multiply to the same product. Suppose a, b, c, and d are any real numbers.

Solution   We multiply $(ax + b)$ and $(cx + d)$ to obtain the table entries, and then we multiply the diagonals to compare the products.

Multiply	cx	$+d$
ax	acx^2	$+adx$
$+b$	$+bcx$	$+bd$

The product of one diagonal is $(acx^2)(bd)$, or $abcdx^2$. The product of the other diagonal is $(bcx)(adx)$, or $abcdx^2$. The diagonal products are the same for any real numbers a, b, c, and d.   ●

Factoring Polynomials with Tables

Factoring is the reverse operation to multiplication of polynomials. The basis for this reversible relationship is the *distributive property*. The distributive property allows us to use multiplication to change the product $a(b + c)$ to the sum $ab + ac$. Through factoring, the sum $ab + ac$ is changed back to the product $a(b + c)$.

Trinomial factoring changes the sum of three terms back to a product $(ax + b)(cx + d)$. If an algebraic calculator is not available, the table method provides a systematic way to factor trinomials.

If you can quickly factor polynomial expressions such as $2x^2 - 11x - 6$, continue to rely on your own method. Learn enough about the table method to try a few exercises and help other students. You will find tables useful in completing the square in Section 3.5.

Example 10 refers to the **greatest common factor**, *the largest number that divides evenly into two other numbers.* The greatest common factor of 16 and 24 is 4. The greatest common factor is abbreviated gcf.

EXAMPLE **10**    Factoring a trinomial   Use the patterns from multiplication to enter information from $2x^2 - 11x - 6$ into a table. Use other patterns to build the white part of the table, and then factor to find the binomials on the left and the top of the table.

Solution   We enter the first and last terms directly into the table.

Factor		
	$2x^2$	
		-6

sum $-11x$

product $-12x^2$

The sum of the diagonal is $-11x$, and the product of the diagonal terms is $-12x^2$. Because the two products of the diagonals are equal, we need to look for terms that multiply to $-12x^2$ and add to $-11x$. Without signs, the possible terms are

$$12x \text{ and } 1x \qquad 6x \text{ and } 2x \qquad 4x \text{ and } 3x$$

Only $-12x$ and $+1x$ add to $-11x$, so we can place these two terms *in any order* in the other diagonal of the table.

Factor		
	$2x^2$	$-12x$
	$+1x$	-6

Next, we complete the top row of factors by removing the greatest common factor (gcf) from the first row of the white part of the table. We then use the top row of factors to find the missing factor in the left column. We use the number in the bottom right corner as a check.

Write the gcf here. $\longrightarrow$

Factor	x	-6
$2x$	$2x^2$	$-12x$
$+1$	$+1x$	-6

The result is

$$2x^2 - 11x - 6 = (2x + 1)(x - 6)$$

EXAMPLE **11**

Factoring a trinomial   Factor $12x^2 - 8x - 15$.

Solution   We first enter $12x^2$ and -15 into the table.

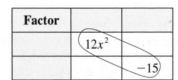

Factor		
	$12x^2$	
		-15

sum $-8x$

product $-180x^2$

The product along the diagonal is $-180x^2$, which must be factored into a pair of terms adding to $-8x$. A list of all factors may not be needed but is included here for completeness:

$1x$ and $180x$	$2x$ and $90x$	$3x$ and $60x$	$4x$ and $45x$
$5x$ and $36x$	$6x$ and $30x$	$10x$ and $18x$	

The factors $-18x$ and $+10x$ give the required sum and are placed in the table.

Factor		
	$12x^2$	$-18x$
	$+10x$	-15

We identify the gcf of the first row of the white part of the table and use it to complete the top row of factors. Next, we find the lower left factor. We then check with the lower right number.

Write the gcf here. $\longrightarrow$

Factor	$2x$	-3
$6x$	$12x^2$	$-18x$
$+5$	$+10x$	-15

The result is $12x^2 - 8x - 15 = (6x + 5)(2x - 3)$.

A graphing calculator may be used to list the factors of the diagonal product. See the Graphing Calculator Technique box.

Graphing Calculator Technique:
Listing Factors

The following process can be used to list the factors of the product ac with the table feature.* It is particularly handy if you have several trinomials, $ax^2 + bx + c$, to factor. Under $\boxed{Y=}$, let

$$Y_1 = A \cdot C \div X$$

$$Y_2 = A \cdot C \div X + X$$

On the HOME, or COMPUTATION, screen,

* Enter the value for a and store it in A.
* Enter the value for c and store it in C.

Going to TABLE SET-UP,

* Enter 1 for table minimum.
* Enter 1 for the change in the table.

Go to TABLE, and scan down the Y_2 column until you find b, the middle term's coefficient. The coefficients for the required factors of ac are in the same row under X and Y_1.

*Thanks to Linda Knauer, Pellissippi State Technical Community College, Knoxville, Tennessee, for this process.

Regardless of the factoring method selected, we must look for a **common monomial factor,** *an expression that divides each term evenly,* before factoring. The monomial $5x$ is the common monomial factor in $20x^2 - 15x$. *If there is a common monomial factor, it should be removed from an expression before any other factoring is done.*

Example 12 illustrates a factorization where the common monomial factor contains the variable x. You may recall a shortcut that could be used to factor the expression in Example 12—more on shortcuts in Section 4.0.

EXAMPLE **12**   *Factoring*   Factor $12x^3 - 75x$.

Solution   First we factor $3x$, the greatest common monomial factor, from each term:

$$12x^3 - 75x = 3x(4x^2 - 25)$$

The second expression, $4x^2 - 25$, may be factored again. We enter $4x^2$ and -25 into the table. This time, the sum of the diagonal terms is zero because the expression $4x^2 - 25$ is missing its middle term. The product of the diagonal terms is $-100x^2$, and its factors, $-10x$ and $10x$, add to zero. We enter them into the table and factor.

Factor	$2x$	$+5$
$2x$	$4x^2$	$+10x$
-5	$-10x$	-25

Thus,

$$12x^3 - 75x = 3x(4x^2 - 25) = 3x(2x - 5)(2x + 5)$$

Think about it: What is the same and what is different about a greatest common factor and a common monomial factor?

The factors in our answers to Examples 10 to 12 are a complete factorization of the original expression. A polynomial expression is *completely factored* when all of the factors are prime. A **prime number** and a **prime factor** are *divisible only by 1 and the number or factor itself.*

Summary: Factoring by Table

To factor $adx^2 + dbx + dc$:

1. First remove common monomial factor d so that $ax^2 + bx + c$ remains:

$$d(ax^2 + bx + c)$$

2. Enter ax^2 and c into one diagonal.

3. Multiply ax^2 and c to obtain the diagonal product.

4. Find factors of the diagonal product that add to bx.

5. Enter the factors into the other diagonal.

6. Factor out the greatest common factor, find the other factors, and write the binomial pair.

7. Multiply factors to check that the white "area" is $ax^2 + bx + c$.

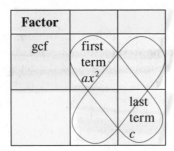

Use the table method in the exercises until you understand the process. Afterward, use guess and check or whatever method seems appropriate.

ANSWER BOX

Warm-up: 1. $1 \cdot 60, 2 \cdot 30, 3 \cdot 20, 4 \cdot 15, 5 \cdot 12, 6 \cdot 10$ **2.** $1 \cdot 12,$ $2 \cdot 6, 3 \cdot 4$ **3.** $1 \cdot 24, 2 \cdot 12, 3 \cdot 8, 4 \cdot 6$ **4.** $1 \cdot 80, 2 \cdot 40, 4 \cdot 20, 5 \cdot 16,$ $8 \cdot 10$ **5.** $1 \cdot 100, 2 \cdot 50, 4 \cdot 25, 5 \cdot 20, 10 \cdot 10$ **6.** $1 \cdot 180, 2 \cdot 90, 3 \cdot 60,$ $4 \cdot 45, 5 \cdot 36, 6 \cdot 30, 9 \cdot 20, 10 \cdot 18, 12 \cdot 15$ **7.** $x^3 + 6x^2 + 9x$ **8.** $6x^2 + 9x$ **9.** $x^3 - 2x^2y + xy^2$ **10.** $12x^3 + 6x^2 - 9x$ **Think about it:** The expressions have almost the same meaning. *Greatest common factor* is more general. A gcf can be numbers or variables, but it can also include square roots or negative powers of variables. For example, $\sqrt{y} + x\sqrt{y}$ has $\sqrt{y}$ as a common factor; $y^{-1} + xy^{-1}$ has y^{-1} as a common factor. Neither $\sqrt{y}$ nor y^{-1} is a monomial. A common monomial factor is a gcf that is also a polynomial expression. The distinction is made in this section to make the directions easier to read.

EXERCISES 3.0

List all the pairs of positive integer factors of the numbers in Exercises 1 and 2.

1. a. 48

  b. 36

  c. 72

2. a. 27

  b. 90

  c. 20

Identify each expression in Exercises 3 to 6 as monomial, binomial, trinomial, or not polynomial.

3. a. $\sqrt{y} + x\sqrt{y}$ **b.** $x^2 + y^2$

4. a. $y^{-1} + xy^{-1}$ **b.** $a\sqrt{3} + b\sqrt{3}$

5. a. $2\pi r$ **b.** $\frac{1}{2}gt^2 + vt + h$

6. a. $x^2y + xy^2 + y^3$ **b.** $|x^2 + y^2|$

Completing the tables in Exercises 7 and 8 will improve your ability to guess and check some sets of factors.

7.

m	n	$m + n$	$m \cdot n$
-3	-5		
		7	12
2	6		
		8	15
		-10	24
		-14	24
-3	-8		
		-4	-12

8.

m	n	$m + n$	$m \cdot n$
1	15		
		10	24
3	8		
		14	24
		2	-24
2	-12		
-3	8		
		-8	-20

In Exercises 9 to 12, add or subtract the polynomials.

9. a. $(x^3 + 2x^2 + 4x) - (2x^2 + 4x + 8)$

  b. $(x^3 - 6x^2 + 9x) - (3x^2 - 18x + 27)$

10. a. $(x^3 - 3x^2 + 9x) + (3x^2 - 9x + 27)$

  b. $(x^3 + 4x^2 + 4x) - (2x^2 + 8x + 8)$

11. a. $(x^3 + 2x^2 + x) + (x^2 + 2x + 1)$

  b. $(a^3 - a^2b + ab^2) + (a^2b - ab^2 + b^3)$

12. a. $(x^3 + x^2 + x) - (x^2 + x + 1)$

  b. $(a^3 + a^2b + ab^2) - (a^2b + ab^2 + b^3)$

Simplify the expressions in Exercises 13 and 14.

13. $x(x^2 - 2xy + y^2) - y(x^2 - 2xy + y^2)$

14. $x(x^2 - xy + y^2) + y(x^2 - xy + y^2)$

15.

Multiply	$2x$	$+3$
$3x$		
-1		

16.

Multiply	$2x$	$+1$
$4x$		
-9		

17.

Multiply	$3x$	$+1$
$3x$		
$+1$		

18.

Multiply	$4x$	$+3$
$4x$		
$+3$		

19.

Multiply	x^2	$-2x$	$+4$
x			
$+2$			

20.

Multiply	x^2	$-6x$	$+9$
x			
-3			

Multiply the expressions in Exercises 21 to 44. Use a table as needed.

21. $(x + 6)(x - 3)$ **22.** $(x - 3)(x + 4)$

23. $(x - 6)(x + 3)$ **24.** $(x - 6)(x + 2)$

25. $(x - 9)(x - 2)$ **26.** $(x + 9)(x + 2)$

27. $(x + 4)(x - 4)$ **28.** $(x - 4)(x - 4)$

29. $(x - 5)(x - 5)$ **30.** $(x + 5)(x - 5)$

31. $(2x - 3)(x + 4)$ **32.** $(2x + 1)(x - 12)$

33. $(2x + 3)(x - 4)$ **34.** $(2x - 3)(x - 4)$

35. $(3x - 1)(3x + 2)$ **36.** $(3x + 4)(2x - 3)$

37. $(x + 2)^2$ **38.** $(x - 1)^2$

39. $(2x - 1)(2x + 1)$ **40.** $(3x + 2)(3x - 2)$

41. $(x - 2)(x^2 + 2x + 4)$

42. $(x + 3)(x^2 - 3x + 9)$

43. $(x^2 - 3x - 2)(x^2 + 3x - 2)$

44. $(x^2 + 4x + 2)(x^2 - 4x + 2)$

Complete the tables in Exercises 45 to 48.

45.

Factor		
	$12x^2$	$+2x$
	$-18x$	-3

46.

Factor		
	$18x^2$	$-15x$
	$+6x$	-5

47.

Factor		
	$3x^2$	$-9x$
	$-4x$	$+12$

48.

Factor		
	$2x^2$	$-6x$
	$+5x$	-15

Factor the expressions in Exercises 49 to 80. Use a table, guess and check, or another method.

49. $x^2 + 7x + 12$ **50.** $x^2 - 9x + 18$

51. $x^2 + x - 12$ **52.** $x^2 - 8x + 12$

53. $x^2 - 7x + 12$ **54.** $x^2 + 4x - 12$

55. $x^2 - 11x - 12$ **56.** $x^2 + 13x + 12$

57. $x^2 + 3x - 28$ **58.** $x^2 - 2x - 15$

59. $6x^2 + 19x + 10$ **60.** $6x^2 + 17x + 10$

61. $6x^2 + 11x - 10$ **62.** $6x^2 + 7x - 10$

63. $6x^2 - 17x + 10$ **64.** $6x^2 - 11x - 10$

65. $15x^2 - x - 6$ **66.** $10x^2 + x - 3$

67. $6x^2 + 13x + 5$ **68.** $10x^2 + 13x - 3$

69. $10x^2 + 61x + 6$ **70.** $10x^2 + 4x - 6$

71. $6x^2 - 32x + 10$ **72.** $15x^2 - 9x - 6$

73. $x^3 + 2x^2 + 4x$ **74.** $2x^2 + 4x + 8$

75. $x^3 - 6x^2 + 9x$ **76.** $3x^2 - 18x + 27$

77. $a^3 - a^2b + ab^2$ **78.** $a^2b - ab^2 + b^3$

79. $a^3 + a^2b + ab^2$ **80.** $a^2b + ab^2 + b^3$

81. Describe how Exercises 17, 18, 37, and 38 are alike. Name other exercises that are like these.

82. Describe how Exercises 27, 39, and 40 are alike. Name other exercises that are like these.

In Exercises 83 to 86, give several numbers that may be used in place of the b in the expression in order to make a trinomial that factors.

83. $x^2 + bx + 12$ **84.** $x^2 + bx - 15$

85. $x^2 + bx - 20$ **86.** $x^2 + bx + 18$

Factor the expressions in Exercises 87 to 94.

87. $3x^2 - 48$ **88.** $15x^2 - 60$

89. $12x^2 - 27$ **90.** $24 - 6x^2$

91. $20x^2 - 45$ **92.** $2x^2 - 18$

93. $28 - 63x^2$ **94.** $18x^2 - 8$

95. Match each definition with the appropriate word. Give an example for each. Choose from the following: square, binomial, monomial, factors, perfect cube, trinomial, exponent.

    a. Two or more numbers or expressions being multiplied

    b. A one-term polynomial

    c. A two-term polynomial

    d. A three-term polynomial

    e. The result of multiplying a number times itself

    f. The result of multiplying three identical factors

Projects

96. *Genetics.* Research the use in genetics of the punnett square and the Hardy-Weinberg principle. Describe both topics in detail, and include in your summary how they relate to polynomial multiplication and the table form of multiplying and factoring.

97. *Powers of Binomials.* Multiply out the powers of binomials in parts a to c.

    a. $(x + 1)^2$ to $(x + 1)^5$

    b. $(x - 1)^2$ to $(x - 1)^5$

    c. $(x + y)^2$ to $(x + y)^5$

    d. Record patterns you find.

    e. Research Pascal's triangle. Relate it to the powers of binomials.

3.1 # The Square Root Function and the Pythagorean Theorem ___

OBJECTIVES

- Explain the difference between square roots and principal square roots.
- Identify a square root function and its domain and range.
- Simplify expressions containing square roots.
- Apply the Pythagorean theorem when appropriate.

WARM-UP

1. List the squares of the numbers from 1 to 16.
2. List the squares of the numbers 20, 25, 30, 40, and 50.
3. List the squares of the numbers $\frac{1}{2}$, $\frac{1}{3}$, and $\frac{1}{4}$.
4. List the squares of the numbers 0.1, 0.2, and 0.3.
5. List the squares of the numbers 1.5, 2.5, 3.5, and 6.5.

IN THIS SECTION, we first do calculator operations and equation solving with square roots. We then practice square root skills with the Pythagorean theorem and related applications.

Square Roots and the Square Root Function

Have you ever looked at the f-stop numbers on a camera lens (Figure 4) and wondered how they were related? The material in this section will help you understand the sequence of f-stops: 1.4, 2, 2.8, 4, 5.6, 8, 11, 16. We begin by examining the meaning of *square root*. We will return to f-stops in Example 6.

Figure 4

SQUARE ROOT   The **square root** of x is *the real number that, multiplied by it-self, produces x.* Each positive number has two real square roots, one positive and one negative. Because $3 \cdot 3 = 9$ and $(-3)(-3) = 9$, 3 and -3 are both square roots of 9. Thus, if $x^2 = 9$, then $x = \pm 3$. The **plus or minus sign,** $\pm$, indicates two numbers, *the positive and the negative of the number that follows the sign.*

In general, when solving equations of the form $x^2 = k$, $k \geq 0$, we have two solutions.

The equation $x^2 = k$, $k \geq 0$, has two solutions:
$$x = +\sqrt{k} \qquad \text{and} \qquad x = -\sqrt{k}$$

EXAMPLE **1**   Finding square roots   For what real numbers are these equations true?

a. $x^2 = 16$ **b.** $x^2 = 0.81$ **c.** $x^2 = 2500$ **d.** $x^2 = 14{,}400$
e. $x^2 = -1$ **f.** $x^2 = -400$

Solution   The numbers that make the equation true are the square roots of the number on the right side of the equation.

a. $x = \pm 4$ **b.** $x = \pm 0.9$ **c.** $x = \pm 50$ **d.** $x = \pm 120$
e. No real number, when squared, gives -1.
f. No real number, when squared, gives -400.   ●

 To solve $x^2 = k$ with a calculator graph, graph $y = x^2$ and $y = k$.

SQUARE ROOT FUNCTION   Did you notice that no square root symbol was used in the last several paragraphs? The square root symbol, $\sqrt{}$, has a special meaning, which we now define.

Definition of Principal Square Root and Square Root Function

The **principal square root** of x is the *positive* real number that, when multiplied by itself, produces x.

The **square root function** is the function that gives the principal square root of a real number. The square root function is written $f(x) = \sqrt{x}$, $x \geq 0$.

EXAMPLE **2**   Finding domain and range   What are the domain and range for $f(x) = \sqrt{x}$? Use a
 graphing calculator or make a table and graph for selected integer inputs between -5 and 5.

Solution   Table 1 and Figure 5 show ordered pairs and a graph for $f(x) = \sqrt{x}$.

The domain for the square root function is $x \geq 0$. The product of two nega-tive numbers is a positive number, and the product of two positive numbers is a positive number. There is no real number that, when multiplied by itself, pro-duces a negative x. Thus, we say that the square root of a negative number is *undefined* in the real numbers. (See Section 4.3 for more on square roots of nega-tive numbers.)

Input x	Output $f(x) = \sqrt{x}$
-5	undefined in $\mathbb{R}$
-3	undefined in $\mathbb{R}$
-1	undefined in $\mathbb{R}$
0	0
1	1
3	1.732
5	2.236

Table 1

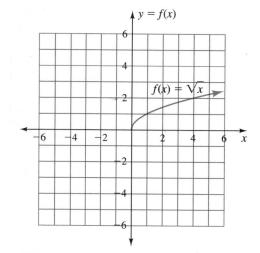

Figure 5

The output column in Table 1 indicates that the square root function has no output for negative inputs. The graph in Figure 5 shows no points to the left of the vertical axis. Both the table and the graph indicate that the function is undefined when $x < 0$.

The range for the square root function is $f(x) \geq 0$. Because $\sqrt{x}$ is the principal or positive square root, the output for $f(x) = \sqrt{x}$ is always non-negative. This is in agreement with both the positive outputs in Table 1 and the first-quadrant graph in Figure 5. ●

Think about it 1: How does the graph in Figure 5 show that $\sqrt{4} = 2$, not ± 2?

SQUARE ROOT SYMBOL   As indicated, the symbol $\sqrt{x}$ refers only to the positive square root and is read "the principal square root of x." Although it is correct to say that 3 and -3 are square roots of 9, it is not correct to write $\sqrt{9} = \pm 3$, because the symbol $\sqrt{}$ stands for the principal, or positive, root.

The *square root symbol* is also known as the **radical sign** (or radical). The *real number under the radical* is called the **radicand.**

$$\underset{\text{radical}}{\nearrow} \sqrt{25} = 5 \underset{\text{radicand}}{\nwarrow} \overset{\text{principal square root of 25}}{\swarrow}$$

In Chapter 1, we defined an irrational number as a number that cannot be written as a rational number a/b, where a and b are integers and $b \neq 0$. Irrational numbers are nonrepeating, nonterminating decimals. Irrational numbers commonly arise when we take square roots.

EXAMPLE **3**

Finding square roots   Find these square roots, using a calculator as needed. Round the decimals to three decimal places. Identify each root as rational or irrational.

a. $\sqrt{2}$ b. $\sqrt{2.25}$ c. $\sqrt{3}$ d. $\sqrt{4}$ e. $\sqrt{4.41}$
f. $\sqrt{5}$ g. $\sqrt{6.25}$ h. $\sqrt{8}$ i. $\sqrt{-9}$ j. $\sqrt{-16}$

Solution   a. $\sqrt{2} \approx 1.414$, irrational   b. $\sqrt{2.25} = 1.5$, rational
c. $\sqrt{3} \approx 1.732$, irrational   d. $\sqrt{4} = 2$, rational
e. $\sqrt{4.41} = 2.1$, rational   f. $\sqrt{5} \approx 2.236$, irrational
g. $\sqrt{6.25} = 2.5$, rational   h. $\sqrt{8} \approx 2.828$, irrational
i. Not defined in the set of real numbers
j. Not defined in the set of real numbers   ●

Think about it 2: Do any of the answers in Example 3 resemble camera f-stops? Look in Example 4 for another clue to the relationships among the numbers.

CALCULATIONS WITH SQUARE ROOTS We now examine the skills you will need to deal with the square roots that arise in working with the Pythagorean theorem and with quadratic equations. We will return to square roots and related properties, operations, and equations in Chapter 6.

When using a calculator to evaluate expressions, follow the order of operations and remember that grouping symbols include fraction bars and radical signs. Wait until the last operation is finished before rounding answers. Unless directed to do otherwise, round decimals to three decimal places.

EXAMPLE **4** Exploring square root expressions Refer to Example 3 to estimate the value of each expression, and then evaluate the expression with a calculator. When you have finished, look for expressions that are equal and try to recall any properties from prior courses that explain why these expressions are equal.

a. $\sqrt{32}$ **b.** $3\sqrt{2}$ **c.** $\sqrt{18}$ **d.** $4\sqrt{2}$

e. $\dfrac{-2 + \sqrt{5}}{2}$ **f.** $\dfrac{3 - \sqrt{2}}{2}$ **g.** $\dfrac{3 + \sqrt{3}}{3}$

Solution **a.** $\sqrt{32}$ is between $\sqrt{25}$ and $\sqrt{36}$, so $\sqrt{32}$ is between 5 and 6.

$$\sqrt{32} \approx 5.657$$

b. $\sqrt{2}$ is between 1 and 2, so $3\sqrt{2}$ is between 3 and 6.

$$3\sqrt{2} \approx 4.243$$

c. $\sqrt{18}$ is between $\sqrt{16}$ and $\sqrt{25}$, so $\sqrt{18}$ is between 4 and 5.

$$\sqrt{18} \approx 4.243$$

d. $\sqrt{2}$ is between 1 and 2, so $4\sqrt{2}$ is between 4 and 8.

$$4\sqrt{2} \approx 5.657$$

The estimations are important because it is easy to incorrectly enter the expressions in parts e, f, and g on a calculator. Use parentheses so that the entire numerator is evaluated before division by the denominator.

e. $\dfrac{-2 + \sqrt{5}}{2}$ is between $\dfrac{-2 + 2}{2} = 0$ and $\dfrac{-2 + 3}{2} = \dfrac{1}{2}$.

$$\dfrac{-2 + \sqrt{5}}{2} = (-2 + \sqrt{5}) \div 2 \approx 0.118$$

f. $\dfrac{3 - \sqrt{2}}{2}$ is between $\dfrac{3 - 1}{2} = 1$ and $\dfrac{3 - 2}{2} = \dfrac{1}{2}$.

$$\dfrac{3 - \sqrt{2}}{2} = (3 - \sqrt{2}) \div 2 = 0.793$$

g. $\dfrac{3 + \sqrt{3}}{3}$ is between $\dfrac{3 + 1}{3} = \dfrac{4}{3}$ and $\dfrac{3 + 2}{3} = \dfrac{5}{3}$.

$$\dfrac{3 + \sqrt{3}}{3} = (3 + \sqrt{3}) \div 3 \approx 1.577$$

We have two pairs of equal answers, implying that $\sqrt{18} = 3\sqrt{2}$ and $\sqrt{32} = 4\sqrt{2}$. The expressions are equal because of the product property of square roots. ●

PRODUCT PROPERTY OF SQUARE ROOTS Both numerical and variable square root expressions may be simplified by using the product property of square roots.

Product Property of Square Roots

$$\sqrt{a \cdot b} = \sqrt{a} \cdot \sqrt{b}$$

where a and b are positive real numbers.

The first two parts of Example 5 show why the expressions in Example 4 are equal.

EXAMPLE **5**   Applying the product property   Use the product property of square roots to simplify these expressions.

a. $\sqrt{32}$ **b.** $\sqrt{18}$ **c.** $\sqrt{48}$ **d.** $\dfrac{3 - \sqrt{18}}{3}$ **e.** $\dfrac{4 + \sqrt{12}}{8}$

Solution   In each expression, we factor the radicand into a perfect square and another number. (If we find the largest square factor possible, the solution is shorter.) Then we take the square root of the square factor.

a. $\sqrt{32} = \sqrt{16 \cdot 2} = \sqrt{16} \cdot \sqrt{2}$
$= 4\sqrt{2}$

b. $\sqrt{18} = \sqrt{9 \cdot 2} = \sqrt{9} \cdot \sqrt{2}$
$= 3\sqrt{2}$

c. $\sqrt{48} = \sqrt{16 \cdot 3} = \sqrt{16} \cdot \sqrt{3}$
$= 4\sqrt{3}$

d. $\dfrac{3 - \sqrt{18}}{3} = \dfrac{3 - \sqrt{9 \cdot 2}}{3} = \dfrac{3 - 3\sqrt{2}}{3} = \dfrac{3(1 - \sqrt{2})}{3}$
$= 1 - \sqrt{2}$

e. $\dfrac{4 + \sqrt{12}}{8} = \dfrac{4 + \sqrt{4 \cdot 3}}{8} = \dfrac{4 + 2\sqrt{3}}{8} = \dfrac{2(2 + \sqrt{3})}{2 \cdot 4}$
$= \dfrac{2 + \sqrt{3}}{4}$ ●

Think about it 3: The 2 and the 4 in the final step in part e of Example 5 do not simplify. Explain why.

Examples 3 and 4 produced several numbers related to f-stop numbers. In Example 6, we summarize the results in a table and determine the rest of the pattern.

EXAMPLE **6**   Finding a pattern: camera f-stops   Complete Table 2, and describe the mathematical pattern behind the f-stops on a camera.

f-stop	Radical form
1.4	$\sqrt{2}$
2	$\sqrt{4} = 2$
2.8	$\sqrt{8} = 2\sqrt{2}$
4	$\sqrt{} = 4$
5.6	$\sqrt{32} = \_\sqrt{}$
8	$\sqrt{} = 8$
11	$\sqrt{} = \_\sqrt{}$
16	$\sqrt{} = 16$

Table 2

Solution The f-stops are decimal forms approximating the square roots of numbers obtained by doubling 2: $\sqrt{2}$, $\sqrt{4}$, $\sqrt{8}$, $\sqrt{16}$, $\sqrt{32}$, $\sqrt{64}$, $\sqrt{128}$, $\sqrt{256}$. When we simplify using the product property of radicals, we obtain an alternation between whole numbers and whole numbers times a square root of 2: $\sqrt{2}$, 2, $2\sqrt{2}$, 4, $4\sqrt{2}$, 8, $8\sqrt{2}$, 16. Researching why these numbers appear as f-stops is Project 90 at the end of this section. ●

The Pythagorean Theorem

Have you ever wondered how a carpenter finds out whether the corners of the forms for a rectangular foundation are right angles? As you will see in Example 10, the carpenter measures the diagonals of the rectangle and looks to see whether they are equal. The Pythagorean theorem gives us a justification for this process.

The Pythagorean theorem relates the length of the *perpendicular sides* (**legs**) of a right triangle to the length of the *longest side* (**hypotenuse**).* The legs and hypotenuse are marked on the right triangle in Figure 6. It is customary to label the sides with lowercase letters and the angles with uppercase letters. Leg a is opposite angle A, leg b is opposite angle B, and the hypotenuse c is opposite right angle C.

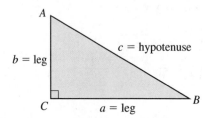

Figure 6

EXAMPLE 7 Exploring the Pythagorean theorem Complete Table 3.

a. In the first three columns of the table, enter the lengths of the sides of the right triangles shown in Figures 7, 8, and 9.

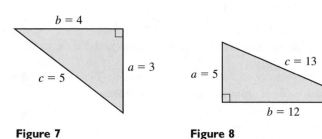

Figure 7   **Figure 8**   **Figure 9**

b. Fill in the remaining four columns.

c. Compare the last two columns.

Triangle	Leg a	Leg b	Hypotenuse c	a^2	b^2	$a^2 + b^2$	c^2
Figure 7							
Figure 8							
Figure 9							

Table 3

*The Greek mathematician and philosopher Pythagoras (ca. 582–500 B.C.) is credited with being the first to record a proof of this relationship for right triangles, but historical documents from China indicate that the relationship was known long before Pythagoras's time.

Solution   Table 4 contains the measures of the sides of three right triangles and the squares of the sides.

Triangle	Leg a	Leg b	Hypotenuse c	a^2	b^2	$a^2 + b^2$	c^2
Figure 7	3	4	5	9	16	25	25
Figure 8	5	12	13	25	144	169	169
Figure 9	6	8	10	36	64	100	100

Table 4

The Pythagorean theorem appears in the last two columns, where

$$a^2 + b^2 = c^2$$ ●

Pythagorean Theorem

> If a triangle is a right triangle, then the sum of the squares of the two shorter sides (legs) is equal to the square of the longest side (hypotenuse).

Note that the theorem refers to *the sum of the squares of the two shorter sides (legs)* and does not specify which is side a and which is side b. Unless the corners of the triangles are labeled with letters, a and b are arbitrarily assigned to the legs—the two perpendicular sides. The letter c is always placed on the hypotenuse—the side opposite the right angle.

EXAMPLE **8**   **Finding missing sides**   Identify the perpendicular sides in the right triangles below. What is the length of the missing side n in each drawing?

a.

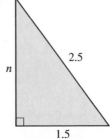

b.

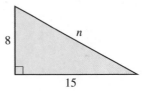

c.

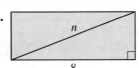

d.

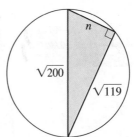

Solution   Because n describes a physical object, $n \geq 0$.

a. The sides labeled n and 1.5 are perpendicular. The hypotenuse is 2.5.

$$a^2 + b^2 = c^2 \qquad \text{Substitute } a = n, b = 1.5, c = 2.5.$$
$$n^2 + 1.5^2 = 2.5^2$$
$$n^2 + 2.25 = 6.25$$
$$n^2 = 4 \qquad \text{Find the square root.}$$
$$n = \pm 2$$

Because $n \geq 0$, the length of the leg is 2 units.

b. The sides labeled 8 and 15 are perpendicular. The hypotenuse is n.

$$a^2 + b^2 = c^2 \qquad \text{Substitute } a = 8, b = 15, c = n.$$
$$8^2 + 15^2 = n^2$$
$$64 + 225 = n^2$$
$$289 = n^2 \qquad \text{Find the square root.}$$
$$\pm 17 = n$$

Because $n \geq 0$, the length of the hypotenuse is 17 units.

c. The sides labeled 3 and 8 are perpendicular. The hypotenuse is n.

$$a^2 + b^2 = c^2 \qquad \text{Substitute } a = 3, b = 8, c = n.$$
$$3^2 + 8^2 = n^2$$
$$9 + 64 = n^2$$
$$73 = n^2 \qquad \text{Find the square root.}$$
$$\pm\sqrt{73} = n$$

Because $n \geq 0$, the length of the hypotenuse is $\sqrt{73}$ units. If the square root is irrational, it is permissible to retain the radical sign or to use a calculator to obtain a decimal approximation. In this case, $n = \sqrt{73} \approx 8.544$, rounded to the nearest thousandth.

d. The sides labeled n and $\sqrt{119}$ are perpendicular, and the hypotenuse is the diameter of the circle.

$$a^2 + b^2 = c^2 \qquad \text{Substitute } a = n, b = \sqrt{119}, c = \sqrt{200}.$$
$$n^2 + \sqrt{119}^2 = \sqrt{200}^2$$
$$n^2 + 119 = 200$$
$$n^2 = 81 \qquad \text{Find the square root.}$$
$$n = \pm 9$$

Because $n \geq 0$, the length of the leg is 9 units.   ●

Changing the position of the *if* and *then* statements in the Pythagorean theorem gives the converse of the Pythagorean theorem. The converse is also true.

Converse of the Pythagorean Theorem	If the sum of the squares of the two shorter sides (legs) is equal to the square of the longest side (hypotenuse), then the triangle is a right triangle.

The converse is used to find out whether a given triangle is a right triangle.

EXAMPLE **9**   Identifying right triangles   Each triangle below is drawn to look like a right triangle. Which *are* right triangles?

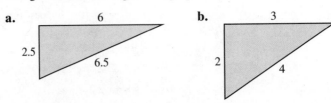

a.   6
   2.5
         6.5

b.   3
   2
         4

c.

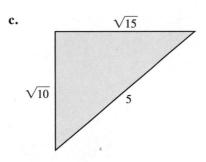

Solution   The $\stackrel{?}{=}$ sign means that we are checking for equality.

a. The shorter sides are 2.5 and 6. The third side is $c = 6.5$.

$$a^2 + b^2 = c^2 \qquad \text{Substitute } a = 2.5, b = 6, c = 6.5.$$
$$2.5^2 + 6^2 \stackrel{?}{=} 6.5^2$$
$$6.25 + 36 \stackrel{?}{=} 42.25$$
$$42.25 = 42.25 \qquad ✔$$

This is a right triangle, with the right angle where the shorter sides meet.

b. The shorter sides are 2 and 3. The third side is $c = 4$.

$$a^2 + b^2 = c^2 \qquad \text{Substitute } a = 2, b = 3, c = 4.$$
$$2^2 + 3^2 \stackrel{?}{=} 4^2$$
$$4 + 9 \stackrel{?}{=} 16$$
$$13 \neq 16$$

This triangle is *not* a right triangle.

c. The shorter sides are $\sqrt{10}$ and $\sqrt{15}$. The third side is $c = 5$.

$$a^2 + b^2 = c^2 \qquad \text{Substitute } a = \sqrt{10}, b = \sqrt{15}, c = 5.$$
$$\sqrt{10}^2 + \sqrt{15}^2 \stackrel{?}{=} 5^2$$
$$10 + 15 \stackrel{?}{=} 25$$
$$25 = 25 \qquad ✔$$

This triangle is a right triangle.   ●

Applications

The Pythagorean theorem has many practical applications. It is a tool for confirming that something, such as a foundation, is built correctly (Example 10). It lets us find distances that cannot be easily measured, such as the length of a support wire to a tower of known height (Example 12). It is vital in proving many concepts in geometry (Example 15). Again, in work with physical objects or geometric shapes, only a positive answer is sensible. Thus, the negative square root will be omitted.

EXAMPLE **10**   Applying the Pythagorean theorem: foundations   Carpenters know that the diagonals of a square or rectangle are equal. They use this fact to check that foundation forms have been built correctly. If the dimensions for a rectangular addition to a house are 16 feet by 20 feet, how long should the diagonal be? (See Figure 10.)

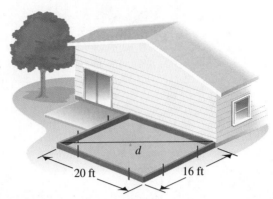

Figure 10

Solution Because the 16 feet and the 20 feet represent the sides of the rectangle and hence are perpendicular, the diagonal will form the hypotenuse c, $c \geq 0$.

$$a^2 + b^2 = c^2 \qquad \text{Substitute } a = 16, b = 20.$$
$$16^2 + 20^2 = c^2$$
$$656 = c^2 \qquad \text{Find the square root.}$$
$$25.6 \approx c$$

The diagonal is approximately 25.6 feet. The carpenter knows that the other diagonal is the hypotenuse of an identical triangle, and so it too must measure 25.6 feet. ●

EXAMPLE **11** Changing feet to inches When the carpenter measures and compares the diagonals in Example 10, what should the 25.6-foot measurement be in feet and inches?

Solution If we multiply 0.6 foot by 12 inches per foot, the feet will be eliminated and the result will be in inches:

$$0.6 \text{ ft} \cdot \frac{12 \text{ in.}}{1 \text{ ft}} = 7.2 \text{ in.}$$

The diagonals should measure 25 feet, 7 inches. We will do similar work with units in Section 5.0. ●

Illustrations for many problems are shown from different views. In Figure 11, the transmission tower is shown first in side view, with three supporting cables. The top view is from the sky, looking down the tower, and shows the positions of the cables attached to the tower. In Example 12, we find the lengths of the cables.

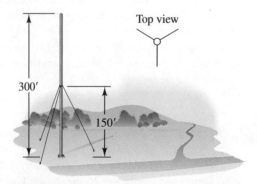

Figure 11

EXAMPLE **12**   Applying the Pythagorean theorem: tower cables   The 300-foot tower in Figure 11 is to be braced by three cables attached halfway up the tower. The cables are to be attached to the ground at a distance from the tower equal to one-fourth the height of the tower. What is the total length of cable needed to brace the tower?

Solution   We assume that the tower is held at a right angle to the ground. The 300-foot tower will be braced at a height of

$$\tfrac{1}{2}(300) = 150 \text{ ft}$$

The cable will be attached to the ground at

$$\tfrac{1}{4}(300) = 75 \text{ ft} \quad \text{from the base of the tower}$$

We label the perpendicular sides of the right triangle 150 ft and 75 ft (see Figure 12). The hypotenuse is labeled c.

$$a^2 + b^2 = c^2 \qquad \text{Substitute } a = 150, b = 75.$$
$$150^2 + 75^2 = c^2$$
$$28{,}125 = c^2 \qquad \text{Find the square root.}$$
$$168 \text{ ft} \approx c$$

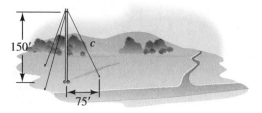

Figure 12

There are three cables, so

$$3(168 \text{ ft}) = 504 \text{ ft}$$

is needed. The person ordering the cable will need to round up to allow some extra for fastening the cable to the ground supports and the tower.   ●

In the next two examples, we apply the Pythagorean theorem to derive information about special triangles. Focus on how the square roots and the Pythagorean theorem are used rather than memorizing the geometric relationships themselves.

ISOSCELES RIGHT TRIANGLES   When a square is cut with a diagonal line (see Figure 13), two identical triangles are formed. These triangles are unique in at least three ways:

1. The angle measures of each triangle are 45°, 45°, and 90°.

2. Each *triangle has two equal sides* and thus is an **isosceles triangle.**

3. Each *triangle has a right angle along with the two equal angles* and thus is an **isosceles right triangle.**

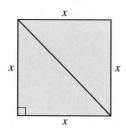

Figure 13

The hypotenuse (here, the diagonal of the square) could be the distance from home plate to second base in baseball or the distance traveled when we cut across an empty block instead of taking the sidewalks around the block.

EXAMPLE **13**   Applying the Pythagorean theorem to geometry   Find the hypotenuse of each isosceles right triangle in Figure 14. Compare the hypotenuse lengths. Assume the units are meters.

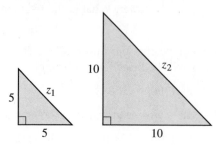

Figure 14

Solution   The equal legs form the base and height of each triangle.
For the smaller triangle,

$$z_1^2 = 5^2 + 5^2$$
$$z_1^2 = 25 + 25$$
$$z_1 = \sqrt{50} \approx 7.071 \text{ m}$$

The hypotenuse of the smaller triangle is approximately 7.071 meters.
For the larger triangle,

$$z_2^2 = 10^2 + 10^2$$
$$z_2^2 = 100 + 100$$
$$z_2 = \sqrt{200} \approx 14.142 \text{ m}$$

The hypotenuse of the larger triangle is approximately 14.142 meters.
The hypotenuse of the larger triangle is exactly twice the hypotenuse of the smaller triangle.   ●

RIGHT TRIANGLES WITHIN EQUILATERAL TRIANGLES   Another important triangle is formed by drawing the height in an equilateral triangle. The **height of a triangle** (also called the **altitude**) is *the perpendicular distance from a vertex (corner) to the opposite side.* The **equilateral triangle** has several special properties:

1. The height of an equilateral triangle is a line of symmetry.
2. The height cuts the triangle into two identical triangles.
3. The two triangles have angle measures 30°, 60°, and 90°, as shown in Figure 15, and are therefore right triangles.

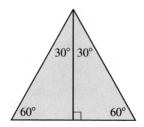

Figure 15

EXAMPLE **14**   Applying the Pythagorean theorem to geometry   If the side of the equilateral triangle in Figure 16 is 2 inches, what is the height of the triangle?

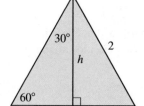

Figure 16

Solution   The height cuts the triangle symmetrically into two identical triangles, so the base of the right triangle is half the side, or 1 inch. From the right triangle, we have

$$a^2 + b^2 = c^2$$ Substitute $a = 1$, $b = h$, $c = 2$.
$$1^2 + h^2 = 2^2$$
$$1 + h^2 = 4$$
$$h^2 = 4 - 1$$
$$h^2 = 3$$ Find the square root.
$$h = \sqrt{3} \approx 1.732 \text{ in.}$$ ●

One of the most powerful uses of algebra is in finding general results or formulas, as required in proving relationships. In Example 15, we use the Pythagorean theorem to solve for the height of any equilateral triangle.

EXAMPLE **15**

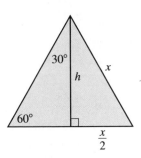

Applying the Pythagorean theorem to geometry Find the height of any equilateral triangle in terms of its side (see Figure 17). Let x be the length of the side of the triangle.

Figure 17

Solution Remember that $h \geq 0$ and $x \geq 0$.

$$a^2 + b^2 = c^2$$ Substitute $a = h$, $b = \dfrac{x}{2}$, $c = x$.

$$h^2 + \left(\frac{x}{2}\right)^2 = x^2$$

$$h^2 + \frac{x^2}{4} = x^2$$ Subtract $\dfrac{x^2}{4}$ from both sides.

$$h^2 = x^2 - \frac{x^2}{4}$$ Find a common denominator.

Student Note: See Section 5.4 for more on this step.

$$h^2 = \frac{4x^2 - x^2}{4}$$ Simplify.

$$h^2 = \frac{3x^2}{4}$$ Find the square root.

$$h = \frac{x\sqrt{3}}{2}, x \geq 0, h \geq 0$$ ●

It is customary to write $\dfrac{\sqrt{3}x}{2}$ rather than $\dfrac{x\sqrt{3}}{2}$. However, carelessly drawn radical signs can cause the x to incorrectly appear under the radical sign. It is recommended that you write variables before the radical unless you are told to do otherwise.

EXERCISES 3.1

In Exercises 1 to 8, for what real numbers are the equations true?

1. $x^2 = 225$ ⒖   **2.** $x^2 = 169$

3. $x^2 = 1.21$ **4.** $x^2 = 0.64$

5. $x^2 = 10000$ **6.** $x^2 = 900$

7. $x^2 = -36$ **8.** $x^2 = -4$

In Exercises 9 to 18, simplify without a calculator.

9. 15^2 **10.** 25^2

11. $\sqrt{49}$ **12.** $\sqrt{144}$

13. 11^2 **14.** 16^2

15. $\sqrt{10^2}$ **16.** $\sqrt{8^2}$

17. $\sqrt{1.96}$ **18.** $\sqrt{0.16}$

In Exercises 19 to 24, use a calculator to simplify. Round to the nearest thousandth.

19. $\sqrt{75}$ **20.** $\sqrt{18}$

21. $\sqrt{40}$ **22.** $\sqrt{20}$

23. $\sqrt{200}$ **24.** $\sqrt{90}$

In Exercises 25 to 30, use a calculator to change the numbers to decimals. Use the product property of square roots to match each expression with an equivalent expression from Exercises 19 to 24.

25. $3\sqrt{2}$ **26.** $2\sqrt{10}$

27. $10\sqrt{2}$ **28.** $5\sqrt{3}$

29. $3\sqrt{10}$ **30.** $2\sqrt{5}$

In Exercises 31 to 36, simplify the radical expressions.

31. $\sqrt{150}$ **32.** $\sqrt{45}$

33. $\sqrt{80}$ **34.** $\sqrt{98}$

35. $\sqrt{32}$ **36.** $\sqrt{300}$

In Exercises 37 to 40, evaluate the expressions using a calculator. Round to the nearest thousandth.

37. $1 + \sqrt{2}$ **38.** $1 - 2\sqrt{2}$

39. $\dfrac{1 - \sqrt{3}}{2}$ **40.** $1 - \sqrt{2}$

In Exercises 41 to 44, evaluate the expressions. Round to the nearest thousandth. Use the product property of square roots to show why the answers are equal to those in Exercises 37 and 40, respectively.

41. $\dfrac{2 + \sqrt{8}}{2}$ **42.** $\dfrac{2 - \sqrt{32}}{2}$

43. $\dfrac{3 - \sqrt{27}}{6}$ **44.** $\dfrac{5 - \sqrt{50}}{5}$

In Exercises 45 to 50, simplify the radical expressions. Evaluate each original expression with a calculator, and compare the result with your answer to verify that you simplified correctly. Round to the nearest thousandth.

45. $\dfrac{4 + \sqrt{8}}{4}$ **46.** $\dfrac{2 - \sqrt{2}}{2}$

47. $\dfrac{2 + \sqrt{2}}{2}$ **48.** $\dfrac{4 - \sqrt{8}}{4}$

49. $\dfrac{4\sqrt{8}}{4}$ **50.** $\dfrac{3\sqrt{6}}{3}$

51. These questions refer to the graphs of $x = y^2$ on page 87 and $y = \sqrt{x}$ on page 165.

   a. How are the graphs alike? How are they different?

   b. List three number facts that show why the graphs have points in common.

   c. What restriction on $y = \sqrt{x}$ indicates why its graph is different from that of $x = y^2$?

   d. Why are there no points on either graph in quadrants 2 and 3?

52. The whole number generated by placing an integer into x^2 is called a perfect square—for example, $2^2 = 2 \cdot 2 = 4$, $4^2 = 4 \cdot 4 = 16$, and $5^2 = 5 \cdot 5 = 25$. The perfect square numbers 4, 16, and 25 end with digits 4, 6, and 5. Name three whole-number perfect squares (if they exist) that satisfy these conditions.

   a. End with the digit 0.   **b.** End with the digit 1.

   c. End with the digit 2.   **d.** End with the digit 3.

   e. What last digits are found on whole-number perfect squares? Summarize your observations.

   f. What last digits are not found on whole-number perfect squares?

In Exercises 53 to 56, solve for the missing side x.

53. a.

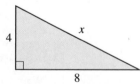

   b.

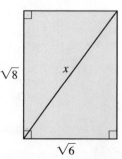

c.

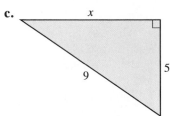

54. a.

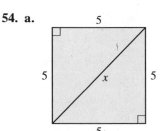

b.

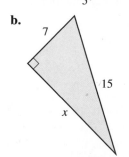

c.

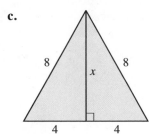

55. a.

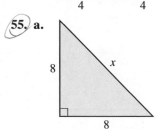

b.

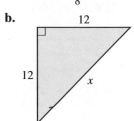

c.

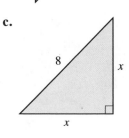

56. a.

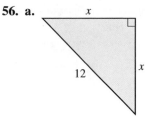

b.

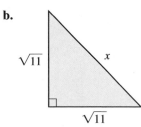

c.

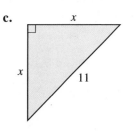

Show which of the sets of numbers in Exercises 57 to 66 could be lengths of sides of right triangles.

57. 12, 16, 20   **58.** 15, 20, 25

59. 4.5, 6, 7.5   **60.** 10.5, 14, 17.5

61. 8, 9, 12   **62.** 11, 12, 16

63. 16, 30, 34   **64.** 11, 60, 61

65. $\sqrt{5}$, $\sqrt{11}$, 4   **66.** $\sqrt{3}$, $\sqrt{6}$, 3

In Exercises 67 to 70, find the diagonal (in feet and inches) of a room with the given dimensions.

67. 12 feet by 12 feet   **68.** 11 feet by 28 feet

69. 13 feet by 19 feet   **70.** 15 feet by 20 feet

71. A support wire for a 50-foot-tall radio antenna is to be fastened halfway up the antenna. The other end of the wire is to be attached to the ground, 16 feet from the base of the tower. Rounding up to the nearest tenth of a foot, how long a wire is needed?

72. A support wire for a 30-foot electric pole is to be fastened two-thirds of the distance up the pole. The other end of the wire is to be attached to the ground, 14 feet from the base of the pole. Rounding up to the nearest tenth of a foot, how long a wire is needed?

A ladder is in a safe position if the ratio of the height it reaches on the wall to the distance of the base from the wall is 4 to 1. To the nearest tenth of a foot, find the length of the ladder in Exercises 73 and 74.

73. The ladder must reach a 16-foot height.

74. The ladder must reach a 20-foot height.

What is the length of the hypotenuse of an isosceles right triangle with sides of the lengths given in Exercises 75 to 80?

75. 20 meters   **76.** 12 meters   **77.** $\sqrt{2}$ feet

78. $\sqrt{8}$ inches   **79.** $\sqrt{18}$ inches   **80.** $\sqrt{32}$ feet

81. Calculate the ratio (by division) of the length of the hypotenuse to the length of the leg for each triangle in Example 13. What do you observe?

82. Find the area of an isosceles right triangle with perpendicular sides of length 5 meters. Find the area of another with perpendicular sides of length 10 meters. Compare the two areas. *Hint:* See Example 13.

What is the height of each of the triangles described in Exercises 83 to 86?

83. Side of the equilateral triangle is 8 inches.

84. Side of the equilateral triangle is 12 inches.

85. Side of the equilateral triangle is 5 inches.

86. Side of the equilateral triangle is 10 inches.

87. Use the results of Example 15 to find the area of an equilateral triangle in terms of its side x by substituting into $A = \frac{1}{2}bh$.

88. Both the Pythagorean theorem and its converse are true, so the two statements can be written as one using the phase *if and only if.* Write such a statement.

Projects

89. *Triangle Investigation.* We have seen that if $a^2 + b^2 = c^2$, the triangle is a right triangle. What can be said about a triangle if $a^2 + b^2 > c^2$? What can be said about a triangle if $a^2 + b^2 < c^2$? Draw several triangles, with sides in centimeters, to illustrate your conclusions.

90. *F-stops.* One set of numbers on a camera lens is the f-stops: 1.4, 2, 2.8, 4, 5.6, 8, 11, 16. Example 6 explained the mathematical sequence generating the numbers. Research the photographic meaning. Why do these numbers appear on the camera?

91. *Dot Paper Areas.* Copy, trace, or draw several grids of dots like that shown below. The 25 dots are 1 unit apart in the horizontal and vertical directions.

      . . . . .
      . . . . .
      . . . . .
      . . . . .
      . . . . .

a. By connecting dots, make squares on the grid. Find 8 squares of different sizes. Your squares must fit inside the 25-dot grids. Label each square with the length of its side (this is where the Pythagorean theorem comes in) and its area.

b. Think of the grid as being the first quadrant in a coordinate graph. Label each side of each square with its slope. Explain how the slopes of the sides of a square are related.

3.2   Quadratic Functions and Solving Quadratic Equations with Tables and Graphs

OBJECTIVES

- Identify a quadratic function by number pattern.
- Identify a quadratic function from the equation form.
- Identify the input variable as well as coefficients *a*, *b*, and *c* in quadratic equations.
- Graph quadratic functions and find the domain and range.
- Find the intercepts, vertex, and axis of symmetry from a graph.
- Solve a quadratic equation with a table or graph.

W A R M - U P

Predict the next number in each set.

1. 1, 4, 9, 16, 25, ...

2. 2, 6, 12, 20, 30, ...

3. 1, 3, 6, 10, 15, ...

Complete the tables in Exercises 4 and 5 by hand. Use a table on a graphing calculator to complete Exercise 6. Round to the nearest tenth.

4.

Input x	Output $f(x) = x^2 + x$
-4	
-3	
-2	
-1	
0	
1	
2	
3	
4	

5.

Input x	Output $f(x) = 4 - x^2$
-3	
-2	
-1	
0	
1	
2	
3	

6.

Input t	Output $f(t) = -16.1t^2 + 6t + 32.8$
0	
0.2	
0.4	
0.6	
0.8	
1.0	
1.2	
1.4	
1.6	
1.8	
2.0	

IN THIS SECTION, we return to the subject of sequences and find patterns for quadratic sequences. The sequences will lead us to defining a quadratic function. In addition, we examine tables and graphs of quadratic functions. The characteristic parabolic shape of the quadratic function, the intercepts, and symmetry permit us to draw the graph with limited information. Tables and graphs allow us to solve quadratic equations using the same skills we use in solving linear and other types of equations.

Patterns and Quadratic Sequences

You should recognize the first sequence in the Warm-up, 1, 4, 9, 16, 25, ..., as the function $f(x) = x^2$, where the set of inputs (domain) is the positive integers. The following Exploration leads to other ways of thinking about the sequence.

For convenience, we call *the first set of differences between consecutive terms* simply **first differences.** Then, *the differences between the differences of consecutive terms* we call the **second differences.**

EXAMPLE **1**   Exploration

a. Find the first and second differences between consecutive terms in the Warm-up sequence 1, 4, 9, 16, 25, ...

b. Comment on any patterns in the differences.

c. Give the next two numbers in the sequence.

d. Can the sequence be described with a linear function?

e. Describe how the number sequence fits the picture in Figure 18.

f. Describe the shapes pictured.

g. Use the descriptions to write the rule for the sequence.

1   4   9   16

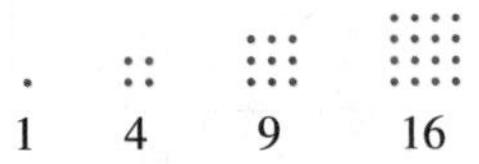

Figure 18

Solution   a. *Sequence:*   1, 4, 9, 16, 25, ...
                                        ∨ ∨ ∨ ∨
            First differences: 3, 5, 7, 9, ...
                                         ∨ ∨ ∨
            Second differences: 2, 2, 2, ...

b. The first differences are odd numbers. The second differences are 2s.

c. The next two first differences are the next two odd numbers: 11 and 13. Thus, the next two numbers in the sequence are

$$25 + 11 = 36 \qquad \text{and} \qquad 36 + 13 = 49$$

d. The first differences are not constant, so the sequence cannot be described by a linear function.

e. The dots in Figure 18 form squares.

f. The sizes of the squares are 1 by 1, 2 by 2, 3 by 3, and 4 by 4.

g. The function is $f(x) = x^2$.   ●

In Example 2, we look at the Warm-up sequence 2, 6, 12, 20, 30, ...

EXAMPLE **2**   Exploration

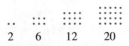

a. Find the first and second differences between consecutive terms in the sequence 2, 6, 12, 20, 30, ...

b. Comment on any patterns in the differences.

c. Give the next two numbers in the sequence.

d. Can the sequence be described with a linear function?

e. Describe how the number sequence fits the picture in Figure 19.

f. Describe the shapes pictured.

2   6   12   20

Figure 19

g. Use the descriptions to write the rule for the sequence.

Solution　**a.** *Sequence*:　　　　　　　2, 6, 12, 20, 30, . . .
　　　　　　　　　　　　　　　　　　∨ ∨ ∨ ∨
　　　　　　First differences:　　　4,　6,　8,　10, . . .
　　　　　　　　　　　　　　　　　　　∨ ∨ ∨
　　　　　　Second differences:　　　　2,　2,　2, . . .

b. The first differences are even numbers. The second differences are 2s.

c. The next two first differences are the next two even numbers: 12 and 14. Thus, the next two numbers in the sequence are

$$30 + 12 = 42 \qquad \text{and} \qquad 42 + 14 = 56$$

d. The first differences are not constant, so the sequence cannot be described by a linear function.

e. The dots in Figure 19 form rectangles.

f. The dimensions of the rectangles are consecutive numbers: 1 by 2, 2 by 3, 3 by 4, and 4 by 5.

g. The function is $f(x) = x(x + 1)$, or $f(x) = x^2 + x$.　　　　　　●

The second differences reveal a numerical distinction between linear and quadratic functions. In Examples 1 and 2, *the sequences have constant second differences* and are called **quadratic sequences.** The name *quadratic* comes from the Latin word *quad*, meaning square, which refers to the exponent in x^2.

Degree of a Polynomial

We define the **degree (*n*) of a polynomial of one variable** to be *the highest power on any one term in the polynomial*. Because they can be written $f(x) = mx^1 + b$ with a highest power of 1, linear functions (see Chapter 2) are also called *polynomial functions of degree 1. Polynomial functions of degree 3*, such as $V = 4\pi r^3 / 3$, are called **cubic functions.**

EXAMPLE ❸　Find the degree of a polynomial　Find the degree of each of these polynomial functions.

a. Volume depends on the side of a cube: $V = s^3$

b. $h(t) = -\frac{1}{2}gt^2 + v_0 t + h_0$

c. $y = x^5 - 5x^4 + 10x^3 - 10x^2 + 5x - 1$

d. Circumference depends on radius: $C = 2\pi r$

Solution　**a.** 3, the highest power on the input variable s

b. 2, the highest power on the input variable t

c. 5, the highest power on the input variable x

d. 1, the highest power on the input variable r　　　　　　●

Quadratic Functions

The rules for the functions in Examples 1 and 2, $f(x) = x^2$ and $f(x) = x^2 + x$, both contain polynomials with the exponent 2 as the highest power. They are *polynomial functions of degree 2*, or **quadratic functions.**

Definition of Quadratic Function

Quadratic functions are polynomial functions of degree 2. They may be written in the form

$$f(x) = ax^2 + bx + c$$

where a, b, and c are real numbers, $a \neq 0$. The input variable is x.

The numbers replacing a, b, and c have considerable impact on the shape and position of the graph of the function. The relationship will be suggested in the next few sections and examined carefully in Section 4.1.

EXAMPLE **4**   Finding the coefficients of terms in quadratic functions   Identify the input variable as well as a, b, and c in these functions.

a. $f(x) = x^2$

b. $f(x) = x(x + 1)$

c. Area of circle: $A = \pi r^2$

d. Approximate length of skid marks (in feet) at s miles per hour on a dry concrete road surface: $L = s^2/24$

e. Height of an object in vertical motion, where g is the acceleration due to gravity, v_0 is the initial velocity, and h_0 is the initial height: $h = -\frac{1}{2}gt^2 + v_0 t + h_0$

f. Another common form for a quadratic equation: $y = 2(x - 1)^2 - 1$

Solution   **a.** In $f(x) = x^2$, the input variable is x. We assume there to be a 1 before the x^2, so $a = 1$. There is nothing to match with the other two terms $bx + c$, so $b = 0$ and $c = 0$.

b. We change $f(x) = x(x + 1)$ to the form $f(x) = x^2 + x$. The input variable is x. Thus, $a = 1$, $b = 1$, and $c = 0$.

c. The area of a circle is a function of radius r, so r is the input variable, $a = \pi$, $b = 0$, and $c = 0$.

d. The length of the skid mark is a function of speed s, so the input variable is s. Thus, $a = \frac{1}{24}$, $b = 0$, $c = 0$.

e. The height is a function of time t. Thus, the input variable is t, $a = -\frac{1}{2}g$, $b = v_0$, and $c = h_0$.

f. We eliminate the parentheses in $y = 2(x - 1)^2 - 1$:

$$y = 2(x^2 - 2x + 1) - 1$$
$$= 2x^2 - 4x + 2 - 1$$
$$= 2x^2 - 4x + 1$$

The input variable is x, $a = 2$, $b = -4$, and $c = 1$.   ●

Graphing Quadratic Functions

Because the graph is a type of *curve* called a **parabola**, it takes more points to make a good graph of a quadratic function than of a linear function. However, identifying particular points on and features of the parabola can reduce the number of points we need for a graph.

INTERCEPTS   Some of the first points we find in graphing a quadratic function are its intercepts with the axes. Recall that the **x-intercept points** are *the points where a graph crosses the x-axis*. Because $y = 0$ everywhere on the x-axis, the

x-intercepts are described as the inputs for which the output is zero, $f(x) = 0$. The **y-intercept point** is *the point where a graph crosses the y-axis*. Because $x = 0$ everywhere on the y-axis, the y-intercept is $f(0)$, the function evaluated at zero.

EXAMPLE **5**   Graphing quadratic functions and finding intercepts

a. Make a table and graph for the quadratic function $f(x) = x^2 + x$. Use integer inputs from -4 to 4.

b. Identify the x- and y-intercept points on the table and graph.

Solution   **a.** The graph in Figure 20 was drawn with the data from Table 5.

Input x	Output $f(x) = x^2 + x$
-4	12
-3	6
-2	2
-1	0
0	0
1	2
2	6
3	12
4	20

Table 5

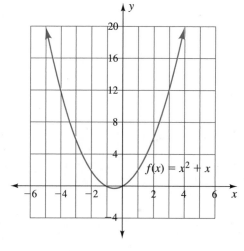

Figure 20

b. The table and the graph show two x-intercept points, $(-1, 0)$ and $(0, 0)$. These ordered pairs satisfy $f(x) = 0$.

$$(-1)^2 + (-1) \stackrel{?}{=} 0 \quad \text{✔}$$

$$(0)^2 + (0) \stackrel{?}{=} 0 \quad \text{✔}$$

The origin, $(0, 0)$, is also the y-intercept point $f(0)$.   ●

AXIS OF SYMMETRY   The parabolic graph of a quadratic function has an **axis, or line, of symmetry**—*a line across which the graph can be folded so that points on one side of the graph match with points on the other side of the graph*. In Example 5, the symmetry is evident in Table 5, where the output numbers appear in pairs as we move up and down on the table from $f(x) = 0$, the lowest output shown. The symmetry is also evident in the graph in Figure 21, in that folding the graph by matching the x-intercepts ($x = -1$ and $x = 0$) would create an axis of symmetry.

EXAMPLE **6**   Using the axis of symmetry to find other ordered pairs   Use the axis of symmetry in Table 5 to predict the output for $x = -5$.

Solution   Symmetry indicates that the 0, 2, 6, 12, 20 pattern for the output numbers in the table will be repeated in both directions. Thus, at $x = -5$, $y = 20$. The ordered pair $(-5, 20)$ lies on the graph.   ●

Think about it 1: Name another function that has a vertical axis of symmetry. Could a function have a horizontal axis of symmetry?

VERTEX   The *highest or lowest point on a parabola* is called the **vertex.** In Figure 21, the vertex is the lowest point, located on the graph between $x = 0$ and $x = 1$. The concept of symmetry allows us to find the coordinates of the vertex for the function $f(x) = x^2 + x$.

EXAMPLE **7**   Finding the vertex

a. Find the equation of the axis of symmetry for $f(x) = x^2 + x$.

b. Find the ordered pair for the vertex.

Solution   **a.** Table 5 in Example 5 shows the x-intercepts $(-1, 0)$ and $(0, 0)$ to have the lowest outputs. The vertex must lie halfway between these two lowest outputs, on the axis of symmetry. The line halfway between $x = -1$ and $x = 0$ is $x = -\frac{1}{2}$, or $x = -0.5$. Thus, the axis of symmetry has the equation $x = -0.5$.

b. To find the vertex, we substitute $x = -0.5$ into the function.

$$f(x) = x^2 + x$$
$$f(-0.5) = (-0.5)^2 + (-0.5)$$
$$= 0.25 - 0.50$$
$$= -0.25$$

The coordinates of the vertex are $(-0.5, -0.25)$, or $\left(-\frac{1}{2}, -\frac{1}{4}\right)$ as fractions.

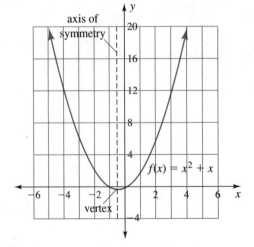

Figure 21

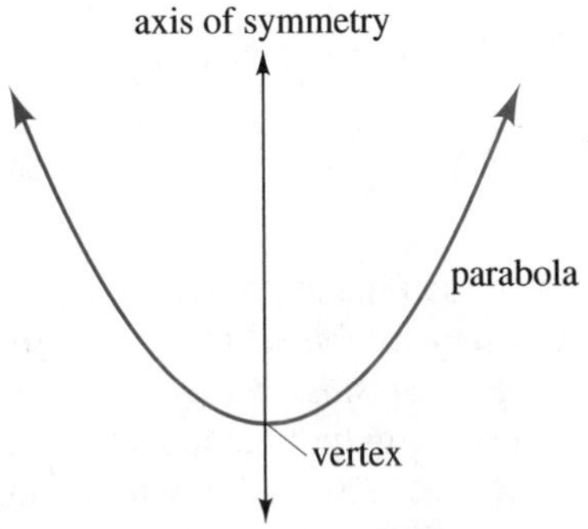

The pairing of the outputs in the table for a quadratic function and the symmetry of the quadratic function graph lead us to conclude that *the axis of symmetry passes through the vertex of a parabola.* The axis of symmetry and vertex are shown in Figure 22.

Figure 22

EXAMPLE **8**   Graphing a quadratic function   Make a table and graph for the quadratic function $f(x) = 4 - x^2$ using the following steps.

a. Start a table with three to four inputs, including $f(0)$. Plot these points and note the y-intercept.

b. Look for patterns that show symmetry, and make other entries in the table without evaluating the function.

c. If possible, identify where $f(x) = 0$ and plot the x-intercept points.

d. Use the x-intercepts or any other two equal outputs to locate the line of symmetry and the vertex.

e. Draw the full parabola.

Solution   **a.** Table 6 shows outputs for inputs 0 to 3. These ordered pairs are plotted in Figure 23.

Input x	Output $f(x) = 4 - x^2$
0	4
1	3
2	0
3	−5

Table 6

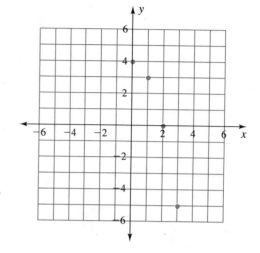

Figure 23

b. The output numbers for $x = 0$ and $x = 1$ are the closest together, so we extend the table to negative x values to look for symmetry.

$$f(-1) = 4 - (-1)^2 = 3$$

We observe that $f(-1) = f(1) = 3$, so we can use the symmetry pattern to extend the table for $f(-2)$ and $f(-3)$. The extended table is shown in Table 7.

c. The x-intercepts, where $f(x) = 0$, are at $x = -2$ and $x = 2$.

Input x	Output $f(x) = 4 - x^2$
−3	−5
−2	0
−1	3
0	4
1	3
2	0
3	−5

Table 7

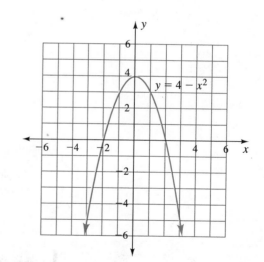

Figure 24

d. Note that only $x = 0$ has 4 as an output. The ordered pair $(0, 4)$ is the vertex. That $x = 0$ is halfway between each pair of inputs with equal outputs confirms that $(0, 4)$ is the vertex. The line of symmetry, $x = 0$, passes through $(0, 4)$, the vertex.

e. The graph of $f(x) = 4 - x^2$ is shown in Figure 24.   ●

Think about it 2: How is the graph of $f(x) = 4 - x^2$ the same as and different from that of $f(x) = x^2 + x$?

The vertex helps identify the range of a quadratic function.

EXAMPLE **9**   Finding domain and range   What are the domain and range for these functions?

a. $f(x) = x^2 + x$ b. $f(x) = 4 - x^2$

Solution   Both functions have any real number as input, so their domains are all real numbers. Both functions have ranges—sets of outputs—that start at their vertex.

a. The graph of $f(x) = x^2 + x$ turns up, with its vertex at the bottom. The vertex for $f(x) = x^2 + x$ is $(-0.5, -0.25)$, so the range is all real numbers larger than -0.25, or $y \geq -0.25$.

b. The graph of $f(x) = 4 - x^2$ turns down, with its vertex at the top. The vertex for $f(x) = 4 - x^2$ is $(0, 4)$, so the range is all real numbers smaller than 4, or $y \leq 4$.   ●

Table 8 lists important points on the graph of a quadratic function.

Name	Definition	Table		Graph
x-intercept points (where $y = 0$ or $f(x) = 0$)	places where graph crosses x-axis	x | y | 0		(_, 0)
y-intercept point (where $x = 0$ or at $f(0)$)	places where graph crosses y-axis	x | y 0 |		(0, _)
vertex	highest or lowest point	The x-coordinate is halfway between two equal outputs.		

Table 8

Solving Quadratic Equations with Tables and Graphs

Tables and graphs of quadratic equations may be used to solve the quadratic equations.

Solving Quadratic Equations $f(x) = n$

> Solve for x by locating the output n in the table for $f(x)$.
> Solve for x by finding the point of intersection of $y = f(x)$ and $y = n$.

EXAMPLE **10**   Solving quadratic equations with a table and graph   Solve these equations with Table 9 and Figure 25.

a. $x^2 + x = 0$ b. $x^2 + x = 6$ c. $x^2 + x = -4$

Input x	Output $f(x) = x^2 + x$
-4	12
-3	6
-2	2
-1	0
0	0
1	2
2	6
3	12
4	20

Table 9

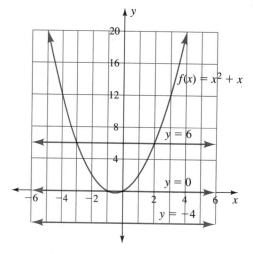

Figure 25

Solution **a.** $x^2 + x = 0$ in two locations in Table 9: at $x = -1$ and $x = 0$. The graph of $y = 0$, the x-axis, intersects the graph of $f(x) = x^2 + x$ twice: at $x = -1$ and $x = 0$.

b. $x^2 + x = 6$ in two locations in Table 9: at $x = -3$ and $x = 2$. The graph of $y = 6$ intersects the graph of $f(x) = x^2 + x$ twice: at $x = -3$ and $x = 2$.

c. The graph of $f(x) = x^2 + x$ does not intersect the line $y = -4$. There are no real-number solutions to the equation $x^2 + x = -4$.   ●

As shown in the chapter opener, many dives require an Olympic diver to jump up as vertically as possible in order to just miss the platform on the way past. The diver's position at any given time (see Figure 26) can be approximated by a quadratic equation called the vertical motion equation.

Figure 26

Vertical Motion Equation

In terms of time t, the height h of an object in vertical motion is

$$h = -\tfrac{1}{2}gt^2 + v_0 t + h_0$$

where g is the acceleration due to gravity (9.81 m/sec² or 32.2 ft/sec²), v_0 is the initial upward velocity (in m/sec or ft/sec), and h_0 is the initial height (in meters or feet).

For a diver, height h is the height above the water and initial velocity v_0 is the rate of upward motion from a jump. Example 11 reminds us that a quadratic equation may have zero, one, or two solutions.

EXAMPLE **11**

Using the vertical motion equation: Olympic diving   A competitor stands on a diving platform 32.8 feet (10 meters) above the water. She jumps from the platform with an initial upward velocity of 6 ft/sec.

a. Write a vertical motion equation describing her height h above the water in terms of time t. Using a calculator table feature, build a table for the equation, with inputs from $t = 0$ to 2 seconds, every two tenths of a second. Graph the equation.

b. Show the equation evaluated at $t = 1.5$ sec.

Use the equation, table, and graph to answer the questions in parts c to g.

c. After how many seconds does she reach the water?

d. After how many seconds is her height above the water 33 feet?

e. After how many seconds is her height above the water 23 feet?

f. After how many seconds is her height above the water 35 feet?

g. What is the maximum height she attains?

Solution **a.** The diver's height at any time is based on $h = -\frac{1}{2}gt^2 + v_0t + h_0$.

$$h = -\frac{1}{2}\left(\frac{32.2 \text{ ft}}{\text{sec}^2}\right)t^2 + \frac{6 \text{ ft}}{\text{sec}}t + 32.8 \text{ ft}$$

$$= -16.1t^2 + 6t + 32.8$$

Because t is in seconds, the seconds in the denominators will be eliminated when the time is substituted into the equation. The final answer will be in feet. Table 10 shows the diver's height at time t. The graph in Figure 27 also shows her height above the water at time t. Remember that the graph does *not* show a picture of the dive because the horizontal axis is time, not distance.

Time (seconds)	Height (feet) $h = -16.1t^2 + 6t + 32.8$
0	32.8
0.2	33.4
0.4	32.6
0.6	30.6
0.8	27.3
1.0	22.7
1.2	16.8
1.4	9.6
1.6	1.2
1.8	−8.6
2.0	−19.6

Table 10

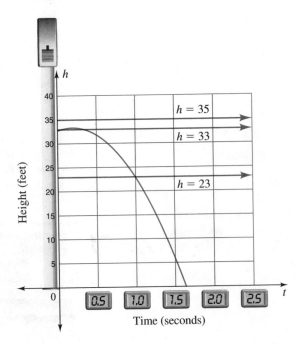

Figure 27

b. The equation for $t = 1.5$ sec is

$$h = -\frac{1}{2}\left(\frac{32.2 \text{ ft}}{\text{sec}^2}\right)(1.5 \text{ sec})^2 + \frac{6 \text{ ft}}{\text{sec}}1.5 \text{ sec} + 32.8 \text{ ft}$$

$$= -\frac{1}{2}(32.2)2.25 \text{ ft} + 6(1.5) \text{ ft} + 32.8 \text{ ft}$$

$$\approx 5.6 \text{ ft}$$

She is about 5 feet above the water after 1.5 seconds.

c. The diver enters the water at $h = 0$; so, using the equation from part a, we have

$$0 = -16.1t^2 + 6t + 32.8$$

A zero output in Table 10 would be located between $t = 1.6$ and $t = 1.8$ sec. The graph crosses the horizontal axis about halfway between 1.5 and 1.75 seconds. A time of 1.65 seconds is a reasonable solution.

d. The equation is $-16.1t^2 + 6t + 32.8 = 33$. There appear to be two times when a height of 33 feet is reached, once as the diver goes up and again as she comes back down. From the table we estimate the two times to be near 0.1 and 0.3 second. In the graph, we seek the inputs for the two points of intersection of the graph and $y = 33$.

e. The equation is $-16.1t^2 + 6t + 32.8 = 23$. Both the table and the graph suggest that 1 second is a reasonable guess for a height of 23 feet.

f. The equation is $-16.1t^2 + 6t + 32.8 = 35$. We see from the graph that the diver never reaches 35 feet. There is no real-number solution to this equation.

g. The maximum height she attains is the vertex. The highest output is about 33.4 feet; however, the vertex is not shown in the table because the output data lack symmetry.   ●

Practice using your graphing calculator intersection feature with the equations in Example 11.

Graphing Calculator Technique: Finding the Intersection of Graphs	Read in your calculator manual about the option for finding the intersection of graphs. You will need to enter the equations into $\boxed{Y=}$ and graph the equations with a window that will show the intersections of the graphs. After selecting the INTERSECTION option, select the graphs. If there is more than one point of intersection, you need to indicate one point at a time. Trace to your chosen point. Repeat for other intersections.

Summary: Solving Quadratic Equations	A quadratic equation $ax^2 + bx + c = 0$ or $ax^2 + bx + c = d$ may have zero, one, or two solutions. • Solve $ax^2 + bx + c = 0$ with a table by finding the input x where $f(x) = y = 0$. • Solve $ax^2 + bx + c = d$ with a table by finding the input x where $f(x) = y = d$. • Solve $ax^2 + bx + c = 0$ with a graph by finding the x-intercepts. • Solve $ax^2 + bx + c = d$ with a graph by finding intersections of $y = d$ and the graph of the corresponding quadratic function.

ANSWER BOX

Warm-up: 1. 36 **2.** 42 **3.** 21 **4.** See Table 5, page 183. **5.** For inputs -4 to 4, the outputs are $-12, -5, 0, 3, 4, 3, 0, -5, -12$. **6.** See Table 10, page 188. **Think about it 1:** There are various possible answers, but $y = |x|$ is one example. A function cannot have a horizontal line of symmetry because the vertical-line test would fail if there were two output values for some inputs. **Think about it 2:** The two graphs are parabolas, each with a line of symmetry. The graph of $f(x) = 4 - x^2$ opens down, whereas the graph of $f(x) = x^2 + x$ opens up.

EXERCISES 3.2

What is the next number in each sequence in Exercises 1 and 2? Identify each as linear, quadratic, or neither.

1. a. 1, 3, 9, 27, 81, 243

   b. 11, 18, 25, 32, 39

   c. 56, 47, 38, 29, 20

   d. -1, 1, 7, 17, 31

2. a. 2, 3, 6, 11, 18

   b. -9, -7, -3, 3, 11

   c. -3, -8, -13, -18, -23

   d. 1, 2, 4, 8, 16

In Exercises 3 and 4, name the degree of the equation.

3. a. $y = 1 - x^3$

   b. $y = x^5 + 1$

   c. $y = x^2 + x + 3$

   d. $y = x - 1$

4. a. $y = 1 - x^2 - x^4$

   b. $y = x + 1$

   c. $y = 5 - x^2$

   d. $y = x^3 + x + 1$

In Exercises 5 and 6, name the input variable and the numbers representing a, b, and c when the equation or formula is written in standard form, $f(x) = ax^2 + bx + c$.

5. a. Surface area of a cylinder of height 3 meters: $f(r) = \pi r^2 + 2\pi r(3 \text{ meters})$

   b. Length of a pendulum: $f(T) = \dfrac{g}{4\pi^2} T^2$

   c. $y = \frac{1}{2}x(x + 1)$

   d. $y = (x - 1)^2$

   e. $y = 2(x + 1)^2 + 2$

6. a. Surface area of a sphere: $f(r) = 4\pi r^2$

   b. $y = \dfrac{(n + 1)(n + 2)}{2}$

   c. $y = x(x + 1)$

   d. $y = (x + 2)^2 + 1$

   e. $y = \frac{1}{2}(x - 1)^2 + 1$

In Exercises 7 and 8, identify the coordinates of the x- and y-intercepts and the vertex from the table.

7.

x	$f(x) = x^2 - 4x + 3$
-2	15
-1	8
0	3
1	0
2	-1
3	0
4	3
5	8

8.

x	$f(x) = x^2 + x - 6$
-3	0
-2	-4
-1	-6
0	-6
1	-4
2	0
3	6
4	14

9. Graph the table in Exercise 7.

10. Graph the table in Exercise 8.

11. Factor $x^2 - 4x + 3$. What do you observe about the relationship between the numbers in the factors and the x-intercepts in Exercise 9?

12. Factor $x^2 + x - 6$. What do you observe about the relationship between the numbers in the factors and the x-intercepts in Exercise 10?

13. From the table in Exercise 7, what is $f(6)$?

14. From the table in Exercise 8, what is $f(-5)$?

In Exercises 15 and 16, solve the equations from the tables in Exercises 7 and 8. Write the answers in set notation.

15. a. $x^2 - 4x + 3 = 8$ **b.** $x^2 - 4x + 3 = 0$

   c. $x^2 - 4x + 3 = 15$ **d.** $x^2 - 4x + 3 = -5$

16. a. $x^2 + x - 6 = 14$ **b.** $x^2 + x - 6 = 0$

   c. $x^2 + x - 6 = -4$ **d.** $x^2 + x - 6 = -10$

17. Refer to the graph of $f(x) = x^2 - 2x - 8$ in the figure.

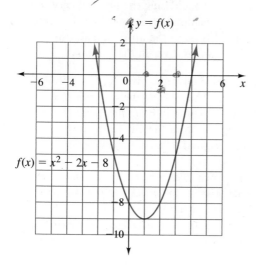

a. Find the x-intercept points.

b. Find the y-intercept point.

c. Find the equation for the axis of symmetry.

d. Find the vertex.

e. Solve $x^2 - 2x - 8 = -9$.

f. Solve $x^2 - 2x - 8 = -5$.

g. Solve $x^2 - 2x - 8 = 0$; compare your answers with those in part a.

h. What is the range for $f(x)$?

i. Factor $x^2 - 2x - 8$. How could you use the factors to get the numbers in part a?

18. Refer to the graph of $f(x) = 2x^2 + x - 1$ in the figure.

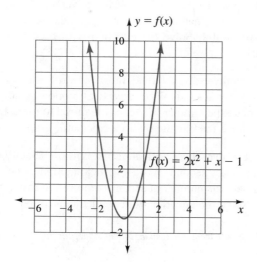

a. Find the x-intercept points.

b. Find the y-intercept point.

c. Find the equation for the axis of symmetry.

d. Find the vertex.

e. Solve $2x^2 + x - 1 = 5$.

f. Solve $2x^2 + x - 1 = 2$.

g. Solve $2x^2 + x - 1 = 0$; compare your answers with those in part a.

h. What is the range for $f(x)$?

i. Factor $2x^2 + x - 1$. How could you use the factors to get the numbers in part a?

19. Make a table and graph $f(x) = -x^2 - 2x + 8$.

a. Find the x-intercept points.

b. Find the y-intercept point.

c. Find the equation for the axis of symmetry.

d. Find the vertex.

e. Solve $-x^2 - 2x + 8 = 0$; compare your answers with those in part a.

f. Solve $-x^2 - 2x + 8 = 8$.

g. Solve $-x^2 - 2x + 8 = 10$.

h. Solve $-x^2 - 2x + 8 = 5$.

i. What is the range for $f(x)$?

j. Factor $-x^2 - 2x + 8$. How could you use the factors to get the numbers in part a?

20. Make a table and graph $f(x) = x^2 + 5x + 6$.

a. Find the x-intercept points.

b. Find the y-intercept point.

c. Find the equation for the axis of symmetry.

d. Find the vertex.

e. Solve $x^2 + 5x + 6 = 6$.

f. Solve $x^2 + 5x + 6 = 2$.

g. Solve $x^2 + 5x + 6 = 0$; compare your answers with those in part a.

h. Solve $x^2 + 5x + 6 = -2$.

i. What is the range for $f(x)$?

j. Factor $x^2 + 5x + 6$. How could you use the factors to get the numbers in part a?

In Exercises 21 and 22, make a table and graph and use the graph to answer the questions.

21. The formula for the surface area of a cylinder is $A = 2\pi r^2 + 2\pi rh$. This cylinder might be a metal storage can (see the figure). Let $h = 6$ in. and $r = 0$ to 6 in.

a. For what radius will the surface area be 200 square inches?

b. For what radius will the surface area be 400 square inches?

c. Why is the radius in part b not double the radius in part a?

22. The formula for the amount of money in an account earning interest compounded yearly for 2 years is $A = P(1 + r)^2$. Let $P = \$1000$ and r be 0 to 20%.

a. For what interest rate, r, will the account contain $1100 at the end of the 2 years?

b. For what interest rate, r, will the account contain $1200 at the end of the 2 years?

c. Find the interest rate needed to have $1400 at the end of 2 years.

Exercises 23 to 26 refer to the vertical motion in Example II.

23. a. What is the meaning of the y-intercept in the graph?

b. Does the part of the parabola to the left of the y-axis have any meaning in the problem setting?

c. Estimate the other time when the curve would be at 23 feet if the parabola were drawn symmetrically. Comment on this result.

d. At what two times is the diver at 32.8 feet?

24. a. Estimate the time taken to travel between the highest point in the dive and the 16-foot height.

b. Estimate the time taken to travel between the 16-foot height and the water surface.

c. The distances in parts a and b are approximately the same. Why is there such a large difference in time?

25. Suppose another Olympic diver has an initial jump velocity $v_0 = 4$ ft/sec. Using the new velocity in the vertical motion equation in Example 11, solve for $h = 0$ to find her total time in the air. Compare this time with that of the original diver in the example.

26. Suppose a third Olympic diver has an initial jump velocity $v_0 = 8$ ft/sec. Using the new velocity in the vertical motion equation in Example 11, solve for $h = 0$ to find her total time in the air. Compare this time with that of the original diver in the example.

The cliff diving from La Quebrada at Acapulco, Mexico, is a popular tourist attraction. The clavadistas (cliff divers) dive from an initial height h_0 of 35 meters above the water (see the figure). In Exercises 27 to 29, assume that $h = 0$ at the water level. In the metric system, the acceleration due to gravity is $g \approx 9.81$ m/sec^2. Round to the nearest tenth of a second.

parabolic path   vertical motion

27. Suppose a diver runs toward the cliff's edge and dives without any upward motion. This means there is no initial upward velocity v_0. Write a vertical motion equation to estimate the number of seconds required to reach the water. Solve the equation with a calculator graph.

28. Suppose a diver dives with an initial upward velocity of 2 meters per second. Write a vertical motion equation to estimate the number of seconds required to reach the water. Solve the equation with a calculator graph.

29. Suppose a diver dives with an initial upward velocity of 1.5 meters per second. Write a vertical motion equation to estimate the number of seconds required to reach the water. Solve the equation with a calculator graph.

Exercises 30 to 40 focus on thinking about quadratic functions and their graphs.

30. If $a = 0$ in $y = ax^2 + bx + c$, what is the resulting equation? Why then does the summary of a quadratic equation state $a \neq 0$?

31. Draw sketches of quadratic functions that explain why quadratic equations can have only 0, 1, or 2 possible solutions.

32. Explain the relationship between the vertex of a parabolic graph and the set of outputs, or range, for the function.

33. The factors of a quadratic expression are $(ax + b)$ and $(cx + d)$. How might the factors be used to find the x-intercepts of $y = (ax + b)(cx + d)$?

34. Describe the relationship between the value of a in $y = ax^2 + bx + c$ and whether the parabola opens up or opens down.

35. If a graph of a quadratic equation has the y-axis as an axis of symmetry and has $x = 4$ as one x-intercept, what will be another x-intercept?

36. If a graph of $f(x)$ has the y-axis as a line of symmetry, what may be said about $f(a)$ and $f(-a)$?

Projects

37. *Positions on a Parabola.* Match each phrase with one of these three positions on a parabola: x-intercept, y-intercept, or vertex. Make a sketch to clarify your answer. Answers may vary depending on your sketch.

a. The highest profit in a parabolic business profit curve

b. The place where a ball on a parabolic path falls to the ground when thrown

c. The start-up costs in a business, where the total cost graph is a parabola

d. The time it takes for a porpoise to re-enter the water after a parabolic leap

e. The down payment in a purchase plan where the total cost graph is a parabola

f. The highest point on a jet of water

g. The number of sales required to change from loss to profit on a parabolic business income curve

h. The maximum height in the parabolic path of a ball

i. The initial height of a golf tee, where the path of the ball is a parabola

j. The lowest point in a parabolic cable on a suspension bridge

k. The initial cost in a parabolic cost curve

l. The point where a diver enters the water, when the curve shows his parabolic path

38. *Estimating Reaction Time.* Turn your hand so that you can catch a ruler between your thumb and fingers. Have someone hold a ruler just above finger level and drop it between your outstretched thumb and fingers (see the figure). Catch the ruler. Use the point on the ruler at which your fingers catch it as an estimate of distance d in inches. Use $d = \frac{1}{2}gt^2$, with $g \approx 32.2$ ft/sec^2, to estimate your reaction time t. (*Hint:* How many inches in a foot?)

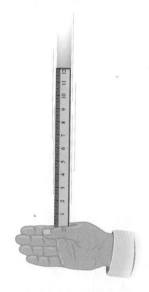

MID–CHAPTER ❸ TEST

1. Simplify the expressions.

a. $x^3 - 6x^2 + 9x - x^2 + 18x - 27$

b. $x^3 + 3x^2 + 9x + (-3x^2 - 9x - 27)$

c. $16a + 4b + c - (9a + 3b + c)$

d. $9a + 3b + c - (4a + 2b + c)$

2. Multiply the expressions.

a. $(x - 5)(x + 5)$ **b.** $(1 - x)(1 - x)$

c. $(1 - x)(x + 3)$ **d.** $(2x + 3)(3 - 2x)$

3. Factor the expressions.

a. $3x^2 + 5x - 12$ **b.** $x^2 - x$

c. $6x^2 + 5x - 4$ **d.** $3x^3 + 6x^2 + 3x$

In Exercises 4 to 9, match the square root or radical expression to one of the following:

$$\sqrt{27}, \quad \sqrt{48}, \quad \sqrt{50}, \quad \sqrt{12}, \quad \sqrt{8}, \quad \sqrt{28}, \quad \sqrt{72}$$

4. $2\sqrt{2}$ **5.** $5\sqrt{2}$ **6.** $4\sqrt{3}$

7. $6\sqrt{2}$ **8.** $3\sqrt{3}$ **9.** $2\sqrt{3}$

10. Solve $x^2 = 16$ and $x = \sqrt{16}$. Explain why the solution sets are not the same.

11. Evaluate with a calculator. Round to the nearest thousandth.

    a. $\dfrac{3 + \sqrt{6}}{3}$ **b.** $\dfrac{\sqrt{30} - 5}{10}$

12. Which of these sets of numbers could be the lengths of the sides of a right triangle?

    a. $\{7.5, 10, 12.5\}$ **b.** $\{8, 15, 17\}$

    c. $\{6, 8, 12\}$

13. Solve for the missing side in each figure.

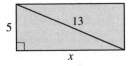

    a.   **b.**

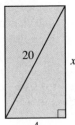

    c.

14. Give the next number in each sequence. Identify each sequence as linear, quadratic, or other.

    a. $-6, -4, 0, 6, 14$

    b. $3, 0, 1, 6, 15$

    c. $8, 11, 14, 17, 20$

    d. $3, 4, 7, 11, 18$

In Exercises 15 and 16, change the equation to standard quadratic form and find the coefficients a, b, and c.

15. $y = \frac{1}{2}n(n + 1) + 1$ **16.** $y = 1000(1 + x)^2$

17. a. Make a table and graph for $A = 6x^2$, the surface area of a cube of side x. Let x be 0 to 10.

    b. If we double the side of the cube, what happens to the surface area?

18. Suppose we stand on the ground, $h_0 = 0$, and throw a ball straight up with an initial velocity of 60 ft/sec.

    a. Write an equation describing the height in terms of time. Use the vertical motion equation, $h = -\frac{1}{2}gt^2 + v_0t + h_0$. Let the acceleration due to gravity be $g = 32$ ft/sec^2.

    b. Make a table and graph for $t = 0, 1, 2, 3, 4$, using a calculator as needed.

    c. What are the x-intercepts? Explain why there are two answers.

    d. Solve the equation for $h = 56$ ft. Use a graph and/or table as needed. Explain your answer(s).

    e. What is the highest point the ball reaches?

3.3 Solving Quadratic Equations with Factors and Square Roots

OBJECTIVES

- Solve a quadratic equation by factoring.
- Find x-intercepts by factoring.
- Apply absolute value in taking square roots.
- Solve linear absolute value equations.
- Solve a quadratic equation by taking the square root.
- Find x-intercepts for $y = ax^2 + c$ with square roots.
- Solve a formula with squares and square roots.

WARM-UP

Multiply these expressions.

1. $(3x + 5)(2x - 3)$

2. $(3x - 2)(4x + 5)$

3. $(2x - 3)(6x + 5)$

4. $(2x + 1)(4x + 9)$

5. $(x + 3)^2$

6. $\frac{100}{12}(0.12x^2 - 2.4x + 9)$

Factor these expressions.

7. $x^2 - x - 6$

8. $x^2 + 6x + 5$

9. $x^2 - 20x + 75$

10. $x^2 - 25y^2$

I N THIS AND THE NEXT SECTION, we continue to look at ways of solving quadratic equations. We use factoring and taking the square root in this section. We formalize the square root method to include absolute value and solving formulas. In the next section, we review squares of binomials and use them to derive the quadratic formula.

A list of solution methods for quadratic equations is given here for your reference.

Summary: Techniques for Solving Quadratic Equations
$$ax^2 + bx + c = 0$$

When $f(x) = ax^2 + bx + c,$

1. Guess numbers for x, and check until the output $f(x)$ is zero.

2. Draw a graph and find x-intercepts, where $f(x) = 0$ (Section 3.2).

3. Make a table and find x, where $f(x) = y = 0$ (Section 3.2).

4. Factor and use the zero product rule (Section 3.3).

5. If $b = 0$, solve for x^2 and take the square root of both sides (Sections 3.1 and 3.3).

6. In $f(x) = d$, if $f(x)$ is a perfect square, take the square root of both sides (Section 3.3).

7. Complete the square (Section 3.4).

8. Use the quadratic formula (Section 3.4).

9. Use the solve feature on a programmable calculator or computer (Section 3.4).

Because students in most advanced courses have calculator or computer technology, the last technique in the list is becoming increasingly important.

Solving by Factoring

In Section 3.2, we solved quadratic equations of the form $ax^2 + bx + c = 0$ with tables and graphs by identifying the x-intercepts. These techniques provided numeric and graphical methods of solving quadratic equations. Factoring provides an algebraic method of solving quadratic equations. The solution-by-factoring method requires an understanding of the zero product rule.

Zero Product Rule

> If the product of two expressions is zero, then either one or the other expression is zero. Symbolically,
>
> $$\text{if } \quad A \cdot B = 0, \quad \text{then either} \quad A = 0 \quad \text{or} \quad B = 0$$

EXAMPLE 1

Exploring the zero product rule Describe the role of zero in these equations and expressions.

a. $A(-5) = 0$ **b.** $5 \cdot B = 0$

c. $(x + 2)(x - 3)$ if $x = 3$ **d.** $(x + 2)(x - 3)$ if $x = -2$

Solution
a. $A = 0$ because the other factor, -5, is not zero.

b. $B = 0$ because the other factor, 5, is not zero.

c. $(3 + 2)(3 - 3) = 5 \cdot 0 = 0$
The product $(x + 2)(x - 3)$ is zero if $x = 3$.

d. $(-2 + 2)(-2 - 3) = 0(-5) = 0$
The product $(x + 2)(x - 3)$ is zero if $x = -2$.

Parts a and b illustrate the fact that, according to the zero product rule, if one factor of $ab = 0$ is not zero, the other factor is zero. Parts c and d illustrate the fact that if one factor is zero, the product is zero. This fact is useful in checking solutions. ●

The zero product rule indicates that if we factor an expression, we may solve for inputs that cause the expression to have a zero output.

EXAMPLE 2

Solving by factoring Solve $x^2 - x - 6 = 0$ by factoring. Compare the solutions with the x-intercepts in Figure 28.

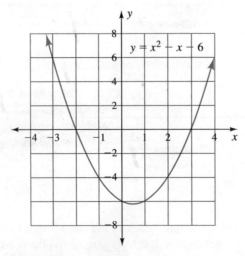

Figure 28

Solution

$$x^2 - x - 6 = 0 \qquad \text{Factor.}$$

$$(x + 2)(x - 3) = 0 \qquad \text{Apply the zero product rule.}$$

$$\text{Either } (x + 2) = 0 \quad \text{or} \quad (x - 3) = 0 \qquad \text{Solve the factor equations.}$$

$$x = -2 \quad \text{or} \qquad x = 3$$

Check: See parts c and d of Example 1.

In Figure 28, the graph of $y = x^2 - x - 6$ crosses the x-axis at $x = -2$ and $x = 3$. The x-intercepts are solutions to the equation when $y = 0$. ●

Writing Two Solutions to an Equation

> In solving by factoring, each factor is the source of a solution. Because x can represent only one input at a time, use the phrasing *either ... or* in the solution.

The *either ... or* phrasing is important. Use it in all solutions by factoring.

EXAMPLE 3 | Solving by factoring   Estimate the solutions to $6x^2 + x - 2 = 0$ from Table 11 and the graph in Figure 29. Solve by factoring.

x	$f(x) = 6x^2 + x - 2$
-1.5	10
-1	3
-0.5	-1
0	-2
0.5	0
1	5
1.5	13

Table 11

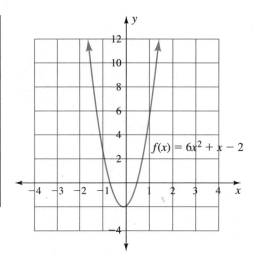

Figure 29

Solution   In the table, $f(x) = 0$ between $x = -1$ and $x = -0.5$ and at $x = 0.5$. The graph agrees with these solutions in that it intersects the x-axis between -1 and -0.5 and again at $x = 0.5$. The solutions to the equation should agree.

$$6x^2 + x - 2 = 0 \qquad \text{Factor.}$$

$$(3x + 2)(2x - 1) = 0 \qquad \text{Apply the zero product rule.}$$

$$\text{Either} \quad 3x + 2 = 0 \quad \text{or} \quad 2x - 1 = 0 \qquad \text{Solve the factor equations.}$$

$$3x = -2 \quad \text{or} \qquad 2x = 1$$

$$x = -\tfrac{2}{3} \quad \text{or} \qquad x = \tfrac{1}{2}$$

The solution $x = -\tfrac{2}{3}$ is between $x = -1$ and $x = -0.5$, as predicted. ●

To use the zero product rule, we must have $f(x) = 0$. Note that $a \cdot b = c$, $c \neq 0$, does *not* imply that $a = c$ or $b = c$. Thus, in the first step of the solution to Example 4, we change the equation to $f(x) = 0$ form.

EXAMPLE

Solving by factoring: museum accessibility A number of parabolic arches surround a museum (see Figure 30). A collector in Greece has offered to loan the museum a large sculpture on the condition that the shipping box be opened inside the controlled environment of the museum. The shipping box is rectangular in cross section and mounted on wheels. According to the design specifications, the equation for the height of the parabolic entry (labeled with letters A, B, and C) is

$$h(x) = -0.12x^2 + 2.4x$$

Both x and $h(x)$ are in feet. The variable x is the distance along the floor from the lower left corner of the entry—point A in both Figure 30 and Figure 31. What width box can pass through the door if the box (including wheels) is 9 feet high?

Figure 30

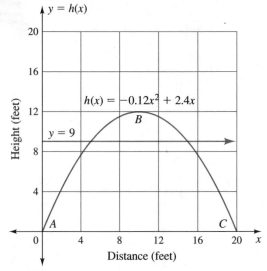

Figure 31

Solution If the height of the box is 9 feet, we need to have $h(x) = 9$. Setting up the equation to factor, we have

$$h(x) = 9$$

$-0.12x^2 + 2.4x = 9$	Multiply both sides by -1 and add 9 on each side.
$0.12x^2 - 2.4x + 9 = 0$	Multiply by 100 to remove decimals.
$12x^2 - 240x + 900 = 0$	Divide by 12, a common factor.
$x^2 - 20x + 75 = 0$	Factor.
$(x - 15)(x - 5) = 0$	Apply the zero product rule.
Either $x - 15 = 0$ or $x - 5 = 0$	Solve the factor equations.
$x = 15$ or $x = 5$	

The coordinates of the arch at the 9-foot height are (5, 9) and (15, 9). There is 10 feet between the coordinates at that height, so a box 9 feet tall and 10 feet wide will fit through the door. ●

Example 5 shows the relationship between the solutions $x = 5$ and $x = 15$ and the x-intercepts in the equations as they are changed in the algebraic solution.

EXAMPLE **5**

Exploring the graphs of equivalent equations The first step in the solution to Example 4 changed $-0.12x^2 + 2.4x = 9$ to $0.12x^2 - 2.4x + 9 = 0$.

Multiplication by 100 and division by 12 transformed the second equation to $x^2 - 20x + 75 = 0$. What is alike and what is different about the graphs $y = -0.12x^2 + 2.4x - 9$ and $y = x^2 - 20x + 75$?

Solution   The graphs are shown in Figure 32. Both are parabolas. The graph of $y = -0.12x^2 + 2.4x - 9$ has its vertex on top, and the parabola opens downward. The graph of $y = x^2 - 20x + 75$ has its vertex on the bottom, and the parabola opens upward. Both have x-intercepts at $x = 5$ and $x = 15$, the solutions found in Example 4.

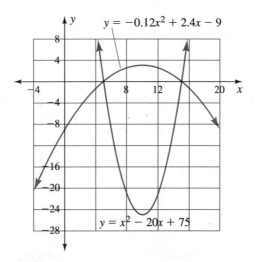

Figure 32   ●

The solutions to an equation in the form $f(x) = 0$ are the x-intercepts of the graph of $f(x)$.

The limitation of factoring as a solution tool is that the quadratic equations in many applications are much too complicated. In Example 4, we transformed the equation into a factorable form. If we had changed the height of the shipping box to 8 or 10 feet, the equation could not have been factored in the same way. Nevertheless, both equations, $-0.12x^2 + 2.4x = 8$ and $-0.12x^2 + 2.4x = 10$, are solvable by calculator table, tracing on a graph, or, as shown in the next section, the quadratic formula.

Absolute Value and Square Roots

In the next example, we review solving equations with tables and graphs while justifying the two solutions to $x^2 = 4$. When we solve $x^2 = 4$, we take the principal square root of both sides and obtain $\sqrt{x^2} = 2$.

EXAMPLE **6**   Exploring the graph of $y = \sqrt{x^2}$ Solve $\sqrt{x^2} = 2$ by table and graph. What other equation could be used to represent the graph of $y = \sqrt{x^2}$?

Solution   Table 12 shows $y = \sqrt{x^2}$ for several inputs. To solve $\sqrt{x^2} = 2$, we look for the value 2 in the second column of Table 12. The value 2 appears twice, so there are two solutions to $\sqrt{x^2} = 2$: $x = -2$ and $x = 2$.

The graphs of $y = 2$ and $y = \sqrt{x^2}$ are shown in Figure 33. The graph of $y = \sqrt{x^2}$ appears to be the same as the graph of the absolute value of x, $y = |x|$. (You may want to check the graph on your own graphing calculator.) The graph shows that $\sqrt{x^2}$ is not x. The graph of $y = x$ is a straight line.

Figure 33 shows the solutions, located at the intersections of the graph of $y = 2$ with the graph of $y = \sqrt{x^2}$. We again conclude that the solutions to $\sqrt{x^2} = 2$ are $x = -2$ and $x = 2$.

x	$y = \sqrt{x^2}$
-3	3
-2	2
-1	1
0	0
1	1
2	2

Table 12

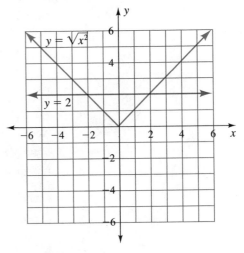

Figure 33

Think about it: Why does a calculator give an error message when evaluating $\sqrt{-2^2}$ in the real numbers? Why is this not a problem in evaluating $f(x) = \sqrt{x^2}$ for $x = -2$?

Because $y = \sqrt{x^2}$ and $y = |x|$ have the same graph, the source of the two solutions in $\sqrt{x^2} = 2$ is the absolute value, not the square root. To understand why the absolute value gives two solutions, we first look at its graph and then use the graph to come up with the formal definition of absolute value. The informal definition of the absolute value of x is the distance x is from zero.

The absolute value graph is made up of two parts. Example 7 explores the equation of each part.

EXAMPLE **7**   Exploring the absolute value graph   Use the absolute value graph in Figure 34 to write the equation and input condition for

a. the left side of the graph   **b.** the right side of the graph

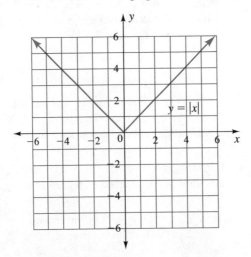

Figure 34

Solution **a.** The left part of the graph of $y = |x|$, in Figure 35(a), has a slope of -1 and a y-intercept at zero. This part is linear for all negative inputs, $x < 0$. The equation is $y = -1x + 0$, or $y = -x$ for $x < 0$.

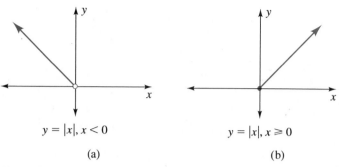

$$y = |x|, x < 0$$
(a)

$$y = |x|, x \geq 0$$
(b)

Figure 35

b. The right part of the graph of $y = |x|$, in Figure 35(b), has a slope of 1 and a y-intercept at zero. This part is linear for all non-negative inputs, $x > 0$. The equation is $y = 1x + 0$, or $y = x$ for $x \geq 0$. ●

The equations in Example 7 lead to the formal definition of absolute value.

Definition of Absolute Value

> The **absolute value** of x is x whenever the input is zero or positive and the opposite of x whenever the input is negative.
> Symbolically, $y = |x|$ is defined as
>
> $$y = \begin{cases} x & \text{if } x \geq 0 \\ -x & \text{if } x < 0 \end{cases}$$
>
> As a function, $f(x) = |x|$ is defined as
>
> $$f(x) = \begin{cases} x & \text{if } x \geq 0 \\ -x & \text{if } x < 0 \end{cases}$$

It is incorrect to think of x as being positive and $-x$ as being negative.

EXAMPLE **8** Finding the opposite of x Evaluate $-x$ for

a. $x = 4$

b. $x = -2$

Solution **a.** For $x = 4$, $-x = -(4) = -4$.

b. For $x = -2$, $-x = -(-2) = 2$.

Note that $-x$ is positive when x is negative. ●

In the definition of absolute value, $|x|$ can be $|w|$, where w is an expression such as $x + 3$. In Example 9, we solve an equation containing the absolute value of an expression.

EXAMPLE **9** Solving absolute value equations   Solve $|x + 3| = 2$. Confirm the results with a graph.

Solution Because the expression inside the absolute value sign may be either positive or negative, we have two outcomes.

If $x + 3$ is positive,

$$x + 3 = 2 \quad \text{and} \quad x = 2 - 3 = -1$$

If $x + 3$ is negative,

$$x + 3 = -2 \quad \text{and} \quad x = -2 - 3 = -5$$

Check: $|-1 + 3| = |2| = 2$ ✔
$|-5 + 3| = |-2| = 2$ ✔

The graphs of $y = |x + 3|$ and $y = 2$ are shown in Figure 36. The line $y = 2$ intersects $y = |x + 3|$ at $x = -1$ and $x = -5$, confirming the results found symbolically.

Student Note: Recall that absolute value is abbreviated ABS or abs on most calculators.

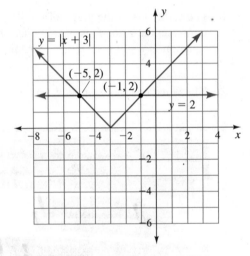

Figure 36

Solving by Taking Square Roots

When $ax^2 + bx + c$ is a perfect square, the quadratic equation $ax^2 + bx + c = d$ can be solved with the technique of *taking the square root of both sides of an equation.*

Consider the equation $x^2 + 6x + 9 = 4$. In solving such an equation, we would normally subtract 4 from both sides to obtain the equivalent equation $x^2 + 6x + 5 = 0$. However, the expression $x^2 + 6x + 9$ is equivalent to $(x + 3)^2$; this type of *two-term expression for the square of a binomial* is called a perfect **binomial square.** A perfect binomial square permits us to apply square roots and absolute value to solve the quadratic equation.

$\sqrt{w^2}$ and $|w|$	In solving equations, $\sqrt{w^2}$ can be replaced by $|w|$.

EXAMPLE **10** Solving by taking the square root of both sides Solve $x^2 + 6x + 9 = 4$.

Solution	$x^2 + 6x + 9 = 4$	Change the left side to the square of $(x + 3)$.
	$(x + 3)^2 = 4$	Take the square root.
	$\sqrt{(x + 3)^2} = \sqrt{4}$	$\sqrt{w^2} = |w|$
	$|x + 3| = 2$	$x + 3$ may equal $+2$ or -2.

Student Note: These steps repeat those in Example 9.

Either $x + 3 = 2$ or $x + 3 = -2$

$x = -1$ or $x = -5$

Check: $(-1)^2 + 6(-1) + 9 \overset{?}{=} 4$ ✔
$(-5)^2 + 6(-5) + 9 \overset{?}{=} 4$ ✔

●

Example 10 illustrates the process we will need in order to derive a general solution to quadratic equations in Section 3.4. As with factoring, if the equation solves quickly with the square root method, use square roots. If not, try another method.

Quotient Property of Square Roots

If you encounter fractions when taking the root, the quotient property of square roots can be useful in simplifying answers.

Quotient Property of Square Roots

For a and b positive and $b \neq 0$,

$$\sqrt{\frac{a}{b}} = \frac{\sqrt{a}}{\sqrt{b}}$$

In Example 11, we solve a quadratic equation $ax^2 + bx + c = 0$, where $b = 0$, by isolating x^2 and taking the square root of both sides.

EXAMPLE **11**   Using the quotient property of square roots   Solve $75x^2 - 3 = 0$.

Solution	$75x^2 - 3 = 0$	Add 3 to both sides.
	$75x^2 = 3$	Divide both sides by 75.
	$x^2 = \frac{3}{75}$	Simplify the fraction to lowest terms.
	$x^2 = \frac{1}{25}$	Take the square root of both sides.
	$\sqrt{x^2} = \frac{\sqrt{1}}{\sqrt{25}}$	Apply the quotient property and $\sqrt{w^2} = |w|$.
	$|x| = \frac{1}{5}$	Inside the absolute value, x is $\frac{1}{5}$ or x is $-\frac{1}{5}$.
	$x = \pm\frac{1}{5}$	

Check: $75\left(\frac{1}{5}\right)^2 - 3 \overset{?}{=} 0$ ✔
$75\left(-\frac{1}{5}\right)^2 - 3 \overset{?}{=} 0$ ✔

●

Solving Formulas Containing Squares and Square Roots

We can apply our work with squares and square roots to formulas. Remember that a solution plan is important in solving formulas.

EXAMPLE **12**   Solving formulas containing squared expressions   Solve the formula for the area of a circle, $A = \pi r^2$, for r. Make an inverse order of operations plan.

Solution   **Plan:** The order of operations on r is to square and then multiply by π. The inverse order of operations is to divide by π and then take the square root.

$$A = \pi r^2 \qquad \text{Divide by } \pi.$$

$$\frac{A}{\pi} = r^2 \qquad \text{Take the square root.}$$

$$\sqrt{\frac{A}{\pi}} = r$$

The absolute value symbol is not needed around r because r represents the radius of a circle and is already positive.

Check: $A \overset{?}{=} \pi\left(\sqrt{\frac{A}{\pi}}\right)^2$ ✔ ●

When we solved $A = \pi r^2$ for r in Example 12, we obtained a radical expression, $r = \sqrt{A/\pi}$. With some restrictions, squares and square roots are inversely related, as are (1) multiplication and division and (2) addition and subtraction.

EXAMPLE **13**   Solving formulas containing square roots: distance to the horizon   On the moon, the approximate distance in miles seen to the horizon from a height h, in feet, is $d = \sqrt{3h/8}$. Solve for h.

Solution   **Plan:** The variable h is multiplied by 3 and divided by 8, and then the square root is applied. The inverse order of operations is to square, multiply by 8, and divide by 3.

$$d = \sqrt{\frac{3h}{8}} \qquad \text{Square both sides.}$$

$$d^2 = \frac{3h}{8} \qquad \text{Multiply by 8.}$$

$$8d^2 = 3h \qquad \text{Divide by 3.}$$

$$\frac{8d^2}{3} = h$$

Check: $d \overset{?}{=} \sqrt{\dfrac{3\left(\dfrac{8d^2}{3}\right)}{8}}$ ✔

Remember that the distance d is positive, so $\sqrt{d^2} = d$ without absolute value. ●

ANSWER BOX

Warm-up: 1. $6x^2 + x - 15$ **2.** $12x^2 + 7x - 10$ **3.** $12x^2 - 8x - 15$
4. $8x^2 + 22x + 9$ **5.** $x^2 + 6x + 9$ **6.** $x^2 - 20x + 75$
7. $(x - 3)(x + 2)$ **8.** $(x + 5)(x + 1)$ **9.** $(x - 5)(x - 15)$
10. $(x - 5y)(x + 5y)$ **Think about it:** According to the order of operations, we first square the 2 in $\sqrt{-2^2}$, then apply the negative or opposite sign to get $\sqrt{-4}$, and then take the square root. The square root of a negative number is not a real number, hence the error message. In evaluating $\sqrt{x^2}$ for $x = -2$, we square the -2 and so the result is positive: $\sqrt{(-2)^2} = \sqrt{4} = 2$.

EXERCISES 3.3

In Exercises 1 to 28, choose your method of solving the equations. However, do at least six of the equations in two different ways: factoring and taking square roots.

1. $x^2 - x - 6 = 0$
2. $x^2 + x - 12 = 0$

3. $2x^2 + x - 1 = 0$
4. $2x^2 + x - 3 = 0$

5. $3x^2 - 48 = 0$
6. $2x^2 - 98 = 0$

7. $x^2 + 4x + 4 = 9$
8. $x^2 - 6x + 9 = 25$

9. $x^2 - 10x + 25 = 36$
10. $x^2 + 8x + 16 = 4$

11. $x^2 + 8x + 16 = 9$
12. $x^2 + 2x + 1 = 16$

13. $x^2 - 4x + 4 = 16$
14. $x^2 - 8x + 16 = 100$

15. $4x^2 - 1 = 0$
16. $9x^2 - 4 = 0$

17. $4x^2 + 4x + 1 = 0$
18. $25x^2 + 10x + 1 = 9$

19. $4x^2 - 16 = 0$
20. $12x^2 - 3 = 0$

21. $x^2 = 225$
22. $x^2 = 169$

23. $x^2 - 121 = 0$
24. $x^2 - 144 = 0$

25. $3x^2 - 48 = 0$
26. $9x^2 - 81 = 0$

27. $5x^2 - 45 = 0$
28. $12x^2 - 48 = 0$

In Exercises 29 to 32, solve. Round to the nearest tenth.

29. $16x^2 - 48 = 0$
30. $8x^2 - 56 = 0$

31. $9x^2 - 45 = 0$
32. $12x^2 - 36 = 0$

33. In solving Example 10, we changed the equation $(x + 3)^2 = 4$ to $|x + 3| = 2$.

  a. Sketch $y = (x + 3)^2 - 4$ and $y = |x + 3| - 2$ for $-8 \le x \le 4$ and $-5 \le y \le 10$.

  b. Compare the x-intercepts.

  c. How are the x-intercepts related to the solutions of the original equation?

34. The equation in Example 11, $75x^2 - 3 = 0$, is equivalent to $x^2 - \frac{3}{75} = 0$.

  a. Sketch $y = 75x^2 - 3$ and $y = x^2 - \frac{3}{75}$ for $-1 \le x \le 1$ and $-1 \le y \le 1$.

  b. Compare the x-intercepts.

  c. How are the intercepts related to the solutions of the original equation?

In Exercises 35 to 40, name the x-intercepts for each equation without graphing the equation.

35. $y = (x - 4)(x + 3)$
36. $y = (x - 2)(x + 5)$

37. $y = (2x - 5)(x + 3)$
38. $y = (3x + 1)(x - 4)$

39. $y = (3x - 4)(2x + 1)$

40. $y = (2x - 3)(4x + 1)$

41. In many applications, the area of a circle is given in terms of diameter instead of radius. This is because the diameter of a cylinder in a combustion engine, the diameter of a log, or the diameter of a highway culvert is more accurately measured than the radius.

  a. The area of a circle in terms of diameter is $A = \pi d^2 / 4$. Solve for diameter in terms of area.

  b. Evaluate $\pi/4$, and round to four decimal places. Observe the location of the four digits on the calculator keyboard. What is it about the digits in $\pi/4$ that makes $A = \pi d^2 / 4$ an easy formula to use?

For Exercises 42 to 44, use the formulas in Exercise 41.

42. A round drainage pipe (culvert) under a highway needs to have a cross-sectional area of 255 square inches. What diameter culvert is needed?

43. The cross-sectional area of a water pipe is 5026 square inches. To the nearest tenth, what is the diameter of the pipe in inches and in feet? (This pipe was estimated to supply 60 gallons of water per person per day to a city of 1 million inhabitants in 1909.)

44. A household water supply pipe has a cross-sectional area of 0.7854 square inch. What is the diameter of the pipe?

45. The surface area of a spherical balloon is $A = 4\pi r^2$. To the nearest tenth, what is the radius of the balloon when the surface area is 100 square inches?

46. The surface area of a cube with side length x is $A = 6x^2$. To the nearest tenth, what is the edge of the cube if the area is 100 square inches?

47. a. For $f(x) = |x - 2|$, find $f(-3)$, $f(-1)$, $f(0)$, $f(1)$, $f(3)$, and $f(5)$. Graph $f(x)$. Enclose the $x - 2$ in parentheses.

  b. Use function notation to say that 0 and 4 give the same output to $f(x)$.

  c. Describe the line of symmetry of $f(x) = |x - 2|$.

48. a. For $f(x) = |x + 2|$, find $f(-3)$, $f(-1)$, $f(0)$, $f(1)$, $f(3)$, and $f(5)$. Graph $f(x)$. Enclose the $x + 2$ in parentheses.

  b. Use function notation to say that -3 and -1 give the same output to $f(x)$.

  c. Describe the line of symmetry of $f(x) = |x + 2|$.

Graph the expressions in Exercises 49 to 52 with a graphing calculator, and record the graphs. Parentheses are needed with both types of expressions.

49. $y = \sqrt{(x - 4)^2}$ and $y = |x - 4|$

50. $y = \sqrt{(x + 1)^2}$ and $y = |x + 1|$

51. $y = \sqrt{(x-1)^2}$ and $y = |x-1|$

52. $y = \sqrt{(x+4)^2}$ and $y = |x+4|$

Solve the equations in Exercises 53 to 60.

53. $|x| = 4$

54. $|x| = 8$

55. $|x+2| = 3$

56. $|x-3| = 4$

57. $|x-5| = 2$

58. $|x+4| = 3$

59. $|x-4| = 2$

60. $|x+1| = 4$

61. Explain the difference between the graphs $f(x) =$ abs $x - 2$ and $f(x) =$ abs $(x - 2)$. Which is the same as $f(x) = |x - 2|$?

62. Explain how to find the x-intercepts from an equation that has been factored, such as $y = (ax + b)(cx + d)$.

Match the expressions in Exercises 63 to 68 with their simplified form. Choose from

$$a\sqrt{b},\ a^2\sqrt{b},\ |a|\sqrt{b},\ \sqrt{ab},\ ab,\ b\sqrt{a},\ |b|\sqrt{a},\ b^2\sqrt{a}$$

63. $\sqrt{b^2a}$, b is positive or zero

64. $\sqrt{a^2b}$, a is any real number

65. $\sqrt{a^2b}$, a is positive or zero

66. $\sqrt{ab^4}$

67. $\sqrt{a^4b}$

68. $\sqrt{ab^2}$, b is any real number

69. The vertical motion formula for the height of an object is $h = -\frac{1}{2}gt^2 + v_0t + h_0$. If an object is dropped, v_0 is zero. Show that the length of time required for the object to hit the ground when dropped from a height h_0 is

$$t = \sqrt{\frac{2h_0}{g}}$$

(*Hint:* When the object hits the ground, the height h will be zero.)

In Exercises 70 to 72, use the vertical motion formula from Exercise 69. The acceleration due to gravity is $g \approx 9.81$ m/sec^2, or $g \approx 32.2$ ft/sec^2. Round to the nearest tenth.

70. What is the time needed for an object dropped from 555 feet to reach the ground? (555 feet is the height of the Washington Monument in our nation's capital.)

71. What is the time needed for an object dropped from 1368 feet to reach the ground? (1368 feet is the height of one tower of the World Trade Center in New York City.)

72. What is the time needed for a metal bolt dropped from the top of the Warsaw (Poland) Radio mast to hit the ground? The mast is 646.4 meters high.

For Exercises 73 and 74, note that in auto accident investigations, the speed r of a car in miles per hour is estimated by measuring the length of the skid marks. Round to the nearest whole number.

73. For a wet concrete road and skid marks of length L, measured in feet, the formula for speed is $r = \sqrt{12L}$.

   a. Solve the formula for L.

   b. If a car is traveling at 55 miles per hour, how long will the skid marks be?

   c. If a car made a 100-foot skid mark, how fast was it traveling?

74. For a dry concrete road and skid marks of length L, measured in feet, the speed formula is $r = \sqrt{24L}$.

   a. Solve the formula for L.

   b. If a car is traveling at 45 miles per hour, how long will the skid marks be?

   c. If a car made a 100-foot skid mark, how fast was it traveling?

Project

75. *Museum Door.* Use a calculator table or graph to answer these questions related to Example 4.

   a. What is the width of the museum arch at 8 feet above the floor?

   b. What is the width of the museum arch at 10 feet above the floor?

   c. Give the cross-sectional area of boxes of height 8, 9, and 10 feet that could pass through the museum arch.

   d. We assume that the wheels supporting the shipping box cause only a minimal reduction in the cross-sectional area of the box. What dimensions (height and width) give the largest area for a box with rectangular cross section that we could fit through the parabolic arch? (To verify your results, guess and check other heights and widths.)

3.4 Solving Quadratic Equations with the Quadratic Formula

OBJECTIVES

- Identify squares of binomials.
- Use completing the square to form the squares of binomials.
- Derive the quadratic formula from $ax^2 + bx + c = 0$.
- Use the quadratic formula to solve equations.
- Use a calculator or solve program to solve quadratic equations.

WARM-UP

Multiply these binomials.

1. $(x - 2)(x - 2)$ **2.** $(n + 3)(n + 3)$

3. $\left(x - \frac{1}{2}\right)^2$ **4.** $(x + b)^2$

5. $(2ax + b)(2ax + b)$

6. $\left(x + \dfrac{b}{2a}\right)^2$

Factor each expression into the square of a binomial.

7. $t^2 - 2t + 1$ **8.** $x^2 - 6x + 9$

9. $a^2 - 5a + 6.25$ **10.** $x^2 + 3x + 2.25$

Fill in the missing parts.

11. $(x + 2)^2 = (\quad)(\quad) = x^2 + 4x + \underline{\quad}$

12. $(x - 5)^2 = (\quad)(\quad) = x^2 - \underline{\quad}x + \underline{\quad}$

13. $(x - \underline{\quad})^2 = x^2 - 8x + 16$

14. $(x + \underline{\quad})^2 = x^2 + 12x + \underline{\quad}$

15. $(x + \underline{\quad})^2 = x^2 + 5x + \underline{\quad}$

HISTORICAL NOTE

Dr. Grace Hopper (1907 to 1992), a U.S. Navy officer and driving force behind the programming language COBOL, is credited with creating the word *debug* when she removed a dead moth that had been causing problems in a computer circuit and recorded in the computer logbook that she had "debugged" the computer.

IN THE FIRST HALF of this section, we examine two concepts leading to the derivation of the quadratic formula: squaring binomials and completing the square. In the second half of this section, the quadratic formula is introduced as the general algebraic method for solving quadratic equations, and we apply it to finding the golden ratio.

One of the reasons quadratic equations have been popular for modeling real-world phenomena is that they may be solved with a formula. Recently, the widespread use of computer software to solve equations has lessened the dependence on such a formula. Nevertheless, both for historical perspective and because the techniques shown here—squaring binomials and completing the square—are useful in other mathematical applications, we will derive and apply the general solution to quadratic equations, the quadratic formula.

Squaring Binomials

When we multiply the binomial $(x + y)$ by itself, the *square of the binomial gives a three-term expression* known as a **perfect square trinomial:** $(x + y)^2 = x^2 + 2xy + y^2$. We may think of the square of a binomial in terms of the multiplication table for $(x + y)(x + y)$, as shown below, or in terms of the area of a square with sides $x + y$, as in Figure 37.

Multiply	x	$+y$
x	x^2	$+xy$
$+y$	$+xy$	$+y^2$

$(x + y)^2 = x^2 + 2xy + y^2$

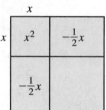

binomial square

Figure 37

The   sum of the diagonal terms, $xy + xy$, is $2xy$, the middle term of $x^2 + 2xy + y^2$.

Completing the Square

In Example 1, we find what must be added to an expression to build a binomial square.

EXAMPLE **1**   Exploration: finding binomial squares   What must be added to the expressions in the squares in order to obtain binomial squares? Draw the resulting binomial square, and describe it in symbols.

a.

	n	
n	n^2	$+3n$
	$+3n$	

binomial square

b.

	x	
x	x^2	$-\frac{1}{2}x$
	$-\frac{1}{2}x$	

binomial square

Solution   We start by finding the factors, or sides of the square.

a. In Figure 38, the factors (sides) are both $n + 3$. These factors produce a 9 in the lower right corner, so a 9 would be added to make a square. The square is

$$n^2 + 6n + 9 = (n + 3)^2$$

b. In Figure 39, the factors are both $x - \frac{1}{2}$. These factors produce a $\frac{1}{4}$ in the lower right corner, so $\frac{1}{4}$ would be added to make a square. The square is

$$x^2 - x + \tfrac{1}{4} = \left(x - \tfrac{1}{2}\right)^2$$

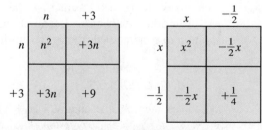

Figure 38   **Figure 39**

Definition of Completing the Square

> **Completing the square** is the process of finding the term to add to $x^2 + nx$ to build a perfect square trinomial.

The diagonal terms (lower left to upper right) in a square are equal and may be found by taking half of the middle term in the original trinomial.

EXAMPLE **2**   Completing the square with a square   Complete the square on these expressions. Draw the resulting binomial square and describe it in symbols.

a. $a^2 - 5a$ **b.** $x^2 + bx$

Solution   **a.** Half of 5 is $\frac{5}{2}$, so $a^2 - 5a = a^2 - \frac{5}{2}a - \frac{5}{2}a$.

To complete the square, we enter $a^2 - \frac{5}{2}a - \frac{5}{2}a$ into Figure 40. In Figure 41, we factor to find the sides of the square.

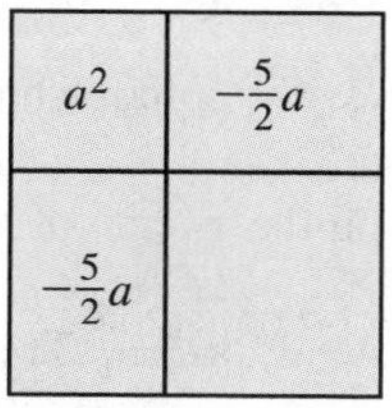

a^2	$-\frac{5}{2}a$
$-\frac{5}{2}a$	

Figure 40

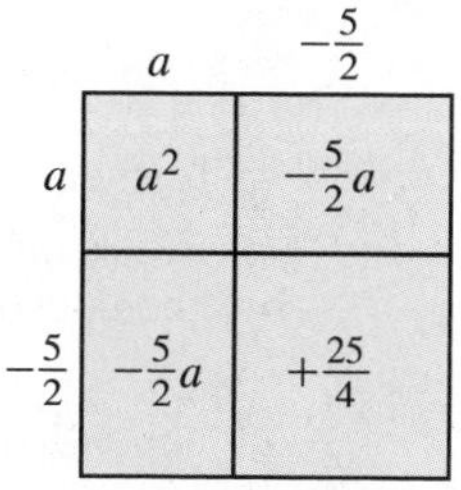

	a	$-\frac{5}{2}$
a	a^2	$-\frac{5}{2}a$
$-\frac{5}{2}$	$-\frac{5}{2}a$	$+\frac{25}{4}$

Figure 41

The square of $-\frac{5}{2}$ is $\frac{25}{4}$, which is added to make a square. The binomial square is

$$a^2 - 5a + \tfrac{25}{4} = \left(a - \tfrac{5}{2}\right)^2$$

b. Half of b is $\dfrac{b}{2}$, so $x^2 + bx = x^2 + \dfrac{bx}{2} + \dfrac{bx}{2}$.

To complete the square, we enter $x^2 + \dfrac{bx}{2} + \dfrac{bx}{2}$ into Figure 42. In Figure 43, we factor to find the sides of the square.

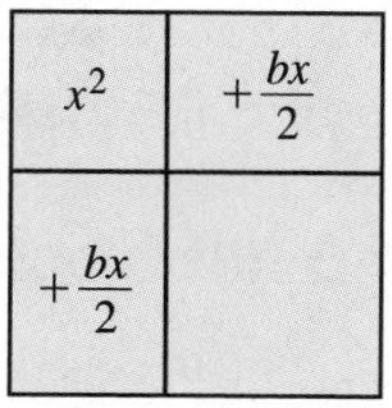

x^2	$+\frac{bx}{2}$
$+\frac{bx}{2}$	

Figure 42

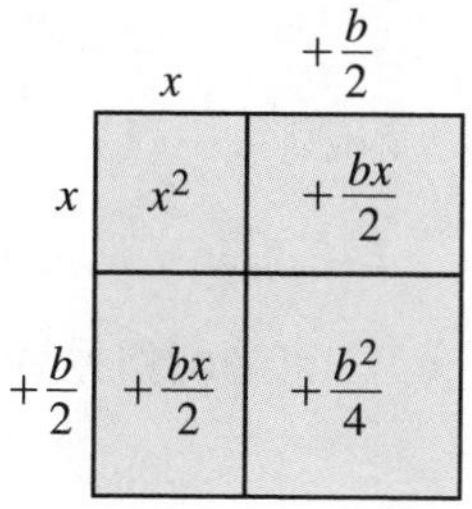

	x	$+\frac{b}{2}$
x	x^2	$+\frac{bx}{2}$
$+\frac{b}{2}$	$+\frac{bx}{2}$	$+\frac{b^2}{4}$

Figure 43

The square of $\dfrac{b}{2}$ is $\dfrac{b^2}{4}$, which is added to make a square. The square is

$$x^2 + bx + \frac{b^2}{4} = \left(x + \frac{b}{2}\right)^2$$

Completing the Square

> To complete the square on an expression $x^2 + nx$, we add the square of half the coefficient on the x term, or $\left(\dfrac{n}{2}\right)^2$.
>
> $$x^2 + nx + \left(\dfrac{n}{2}\right)^2 = \left(x + \dfrac{n}{2}\right)^2$$

EXAMPLE **3**   Completing the square Complete the square of these expressions and then check with a binomial square figure.

  a. $x^2 + 3x$, with decimals

  b. $x^2 + \dfrac{bx}{a}$, with fractions

Solution   **a.** The coefficient on the x term is 3, so to complete the square we add the square of 1.5: $1.5^2 = 2.25$. The binomial square is

$$x^2 + 3x + 2.25 = (x + 1.5)^2$$

To check, we draw the square shown in Figure 44.

b. The coefficient on the x term of $x^2 + \dfrac{bx}{a}$ is $\dfrac{b}{a}$, so to complete the square we

add the square of $\dfrac{b}{2a}$, or $\dfrac{b^2}{4a^2}$. The binomial square is

$$x^2 + \dfrac{bx}{a} + \dfrac{b^2}{4a^2} = \left(x + \dfrac{b}{2a}\right)^2$$

To check, we draw the square shown in Figure 45.

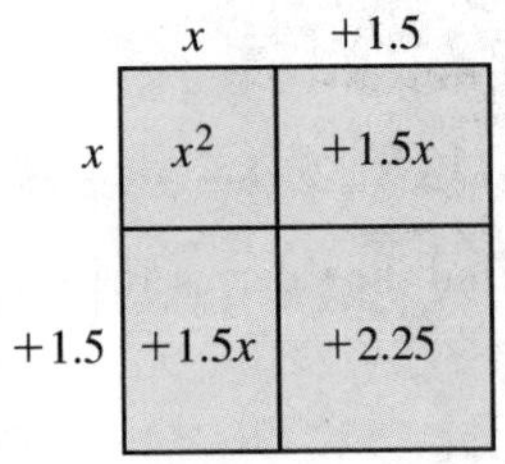

Student Note:
$$\dfrac{1}{2} \cdot \dfrac{b}{a} = \dfrac{b}{2a}$$

Figure 44 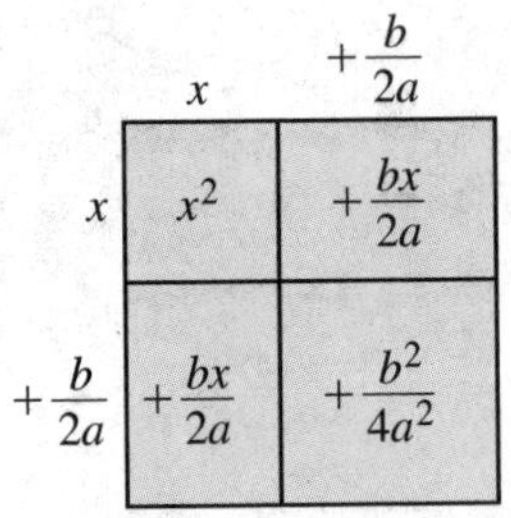   **Figure 45**

Completing the Square in an Equation

To complete the square in an equation, we build a binomial square on one side. We keep the equation balanced by adding the square of half the coefficient of x to both sides.

EXAMPLE **4**   Completing the square in an equation Find what must be added to each side of the quadratic equation $x^2 - 12x = -11$ to create a binomial square on the left. Draw a binomial square figure as needed. Solve the resulting equation.

Solution   We complete the square on $x^2 - 12x$. As shown in Figure 46, $\left(\frac{-12}{2}\right)^2 = (-6)^2 = 36$ is needed to obtain a binomial square.

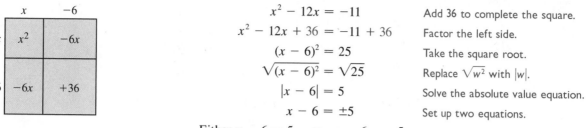

Figure 46

$$x^2 - 12x = -11 \qquad \text{Add 36 to complete the square.}$$
$$x^2 - 12x + 36 = -11 + 36 \qquad \text{Factor the left side.}$$
$$(x - 6)^2 = 25 \qquad \text{Take the square root.}$$
$$\sqrt{(x - 6)^2} = \sqrt{25} \qquad \text{Replace } \sqrt{w^2} \text{ with } |w|.$$
$$|x - 6| = 5 \qquad \text{Solve the absolute value equation.}$$
$$x - 6 = \pm 5 \qquad \text{Set up two equations.}$$

Either $x - 6 = 5$ or $x - 6 = -5$
$x = 11$ or $x = 1$

The purpose of introducing the completing the square process is to derive a general rule, the quadratic formula, not to solve equations by completing the square. The following summary indicates how to complete the square in an equation when $a \neq 1$. This same process will be followed in deriving the quadratic formula in Example 5.

Summary: Completing the Square

For the quadratic equation $ax^2 + bx + c = 0$

1. Subtract c from both sides.

2. Divide all terms by a, the coefficient of x^2.

3. a. Find half the coefficient of the x term to get $\dfrac{b}{2a}$.

   b. Square the result to get $\left(\dfrac{b}{2a}\right)^2$.

   c. Add $\left(\dfrac{b}{2a}\right)^2$ to both sides of the equation.

4. Write the left side as the square of a binomial.

The Quadratic Formula

Major work on the conditions under which an equation can be solved was written out during the night of May 29, 1832, by a 20-year-old Frenchman, Evariste Galois. Galois knew that any early morning duel by pistols would bring his death. Thus, the night before, he prepared his will and wrote extensively about what would be his last mathematical theories. Galois's brilliant work was as significant in solving equations for the next hundred years as the numerical methods employed by calculators and computers are in solving equations today.

The quadratic formula is historically important because it permits solving any quadratic equation. In Example 5, we derive the quadratic formula by completing the square on the quadratic equation $ax^2 + bx + c = 0$.

EXAMPLE **5**   Deriving the quadratic formula   The steps in solving $ax^2 + bx + c = 0, a \neq 0$, for x by completing the square are shown below. Describe what was done at each step.

$$ax^2 + bx + c = 0$$

Step 1 $$ax^2 + bx = -c$$

Step 2 $$x^2 + \frac{b}{a}x = -\frac{c}{a}$$

Step 3 $$x^2 + \frac{b}{a}x + \left(\frac{b}{2a}\right)^2 = \left(\frac{b}{2a}\right)^2 - \frac{c}{a}$$

Step 4 $$\left(x + \frac{b}{2a}\right)^2 = \frac{b^2}{4a^2} - \frac{c}{a}$$

Step 5 $$\left(x + \frac{b}{2a}\right)^2 = \frac{b^2}{4a^2} - \frac{4ac}{4a^2}$$

Step 6 $$\left(x + \frac{b}{2a}\right)^2 = \frac{b^2 - 4ac}{4a^2}$$

Step 7 $$\sqrt{\left(x + \frac{b}{2a}\right)^2} = \frac{\sqrt{b^2 - 4ac}}{2a}$$

Step 8 $$\left|x + \frac{b}{2a}\right| = \frac{\sqrt{b^2 - 4ac}}{2a}$$

Step 9 $$x + \frac{b}{2a} = \pm \frac{\sqrt{b^2 - 4ac}}{2a}$$

Step 10 $$x = -\frac{b}{2a} \pm \frac{\sqrt{b^2 - 4ac}}{2a}$$

Step 11 $$x = \frac{-b \pm \sqrt{b^2 - 4ac}}{2a}$$

Solution   Step 1: Subtract c from both sides.
Step 2: Divide both sides by a, $a \neq 0$.
Step 3: The coefficient of x is b/a. Square half of b/a, and add the result to both sides.
Step 4: Write the left side as the square of a binomial. (See also part b of Example 3.)
Step 5: Set up a common denominator for the right side.
Step 6: Add the fractions on the right side.
Step 7: Take the square root on both sides, applying the quotient rule on the right side.
Step 8: Replace $\sqrt{\left(x + \frac{b}{2a}\right)^2}$ with $\left|x + \frac{b}{2a}\right|$.
Step 9: Solve the absolute value equation.
Step 10: Subtract $b/2a$ from both sides.
Step 11: Add the fractions on the right side.   ●

Because a, b, and c are not specified, the process in Example 5 has solved *all* quadratic equations for x and leads to the quadratic formula.

Quadratic Formula

> If an equation can be written $ax^2 + bx + c = 0$, $a \neq 0$, then its solutions, if they exist, are
>
> $$x = \frac{-b \pm \sqrt{b^2 - 4ac}}{2a}$$

EXAMPLE **6**   Applying the quadratic formula Use the quadratic formula to solve these
   equations.

a. $3x^2 + 7x + 2 = 0$ **b.** $4x^2 = 5x + 1$ **c.** $2x^2 + 4x = -3$

Solution **a.** In $3x^2 + 7x + 2 = 0$, $a = 3$, $b = 7$, and $c = 2$.

$$x = \frac{-b \pm \sqrt{b^2 - 4ac}}{2a} = \frac{-7 \pm \sqrt{7^2 - 4(3)(2)}}{2(3)}$$

$$= \frac{-7 \pm \sqrt{49 - 24}}{6}$$

$$= \frac{-7 \pm \sqrt{25}}{6}$$

The solutions are rational numbers:

$$x = \frac{-7 + 5}{6} = \frac{-2}{6} = \frac{-1}{3} \quad \text{and} \quad x = \frac{-7 - 5}{6} = \frac{-12}{6} = -2$$

b. The equation $4x^2 = 5x + 1$ must be rearranged to equal zero. In $4x^2 - 5x - 1 = 0$, $a = 4$, $b = -5$, and $c = -1$.

$$x = \frac{-b \pm \sqrt{b^2 - 4ac}}{2a} = \frac{-(-5) \pm \sqrt{(-5)^2 - 4(4)(-1)}}{2(4)}$$

$$= \frac{-(-5) \pm \sqrt{25 + 16}}{8}$$

$$= \frac{5 \pm \sqrt{41}}{8}$$

The solutions are irrational numbers:

$$x = \frac{5 + \sqrt{41}}{8} \approx 1.425 \quad \text{and} \quad x = \frac{5 - \sqrt{41}}{8} \approx -0.175$$

c. The equation $2x^2 + 4x = -3$ must be rearranged to equal zero. In $2x^2 + 4x + 3 = 0$, $a = 2$, $b = 4$, and $c = 3$.

$$x = \frac{-b \pm \sqrt{b^2 - 4ac}}{2a} = \frac{-4 \pm \sqrt{4^2 - 4(2)(3)}}{2(2)}$$

$$= \frac{-4 \pm \sqrt{16 - 24}}{4}$$

$$= \frac{-4 \pm \sqrt{-8}}{4}$$

There are no real-number solutions because the square root of a negative number is not a real number. ●

In each part of Example 6, the answer had a different form. Part a had rational solutions. Part b had irrational solutions. Part c had no real-number solution. In Section 4.3, after complex numbers are introduced, we will solve problems like the one in part c. We now compare the graphs related to the equations in Example 6.

EXAMPLE **7** Comparing quadratic graphs and numbers of solutions   Graph these equations, and identify the solutions to $y = 0$ on the graphs.

a. $y = 3x^2 + 7x + 2$ **b.** $y = 4x^2 - 5x - 1$
c. $y = 2x^2 + 4x + 3$

Solution The solutions to $y = 0$ are the *x*-intercepts.

a. The graph in Figure 47 has two *x*-intercepts. This confirms the solutions in Example 6: $3x^2 + 7x + 2 = 0$ at $x = -\frac{1}{3}$ and $x = -2$.

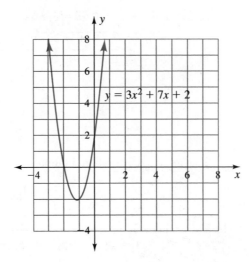

Figure 47

b. The graph in Figure 48 has two x-intercepts. This confirms the solutions in Example 6: $4x^2 - 5x - 1 = 0$ near $x = 1\frac{1}{2}$ and between $x = -\frac{1}{2}$ and $x = 0$.

c. The graph in Figure 49 does not cross the x-axis, so it has no x-intercepts. This confirms that the equation $2x^2 + 4x + 3 = 0$ has no real-number solutions.

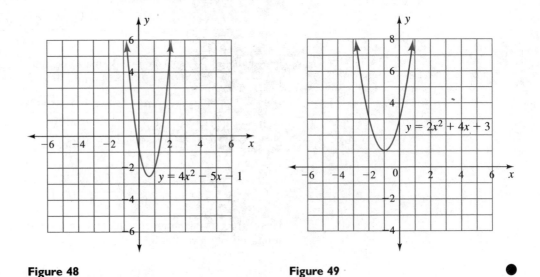

Figure 48   **Figure 49**   ●

The Golden Rectangle

To balance the many applications of algebra in science and engineering, this section closes with an application of quadratic equations to art. Included are two related concepts: ratios and proportions.

The study of beauty in structure has been of interest to artists and architects since before the building of the Parthenon in Greece in the fifth century B.C.

The Greeks believed that *rectangles with certain dimensions,* called **golden rectangles,** were more beautiful than other rectangles. The rectangle drawn around the Parthenon with its upper triangular structure intact, as shown in Figure 50, is a golden rectangle. The dimensions the Greeks had in mind were those that made it possible to remove a square from the rectangle and leave a smaller rectangle with sides in the same ratio as in the original rectangle.

Figure 50

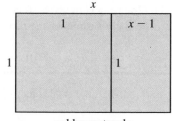

golden rectangle

Figure 51

Recall that a **ratio** is a *comparison of two numbers.* The ratio of two quantities a and b may be written as a fraction $\left(\dfrac{a}{b}\right)$ or in words (a to b).

The ratio of length to width in the rectangle in Figure 51 is x to 1. A square with sides of 1 unit has been marked, leaving a smaller rectangle. For the smaller rectangle, the ratio of length to width is 1 to $x - 1$. In order for the larger rectangle to be a golden rectangle, the two ratios need to be equal, creating a proportion:

$$\frac{x}{1} = \frac{1}{x - 1}$$

Definition of Proportion

Two equal ratios form a **proportion:**

$$\frac{a}{b} = \frac{c}{d}$$

where $b \neq 0$ and $d \neq 0$.

We clear the denominators in a proportion by cross multiplying.

Cross Multiplication Property

The proportion $\dfrac{a}{b} = \dfrac{c}{d}$ implies that $a \cdot d = b \cdot c$.

In Example 8, we solve the proportion for the x that creates a golden rectangle.

EXAMPLE **8**

Applying the quadratic equation   Cross multiply to eliminate the denominators in the proportion $\dfrac{x}{1} = \dfrac{1}{x - 1}$. Solve the resulting quadratic equation, and confirm the solutions with a graphing calculator.

Solution   When we apply cross multiplication to this proportion, we obtain a quadratic equation that does not factor:

$$\frac{x}{1} = \frac{1}{x - 1}$$
$$x(x - 1) = 1 \cdot 1$$
$$x^2 - x = 1$$
$$x^2 - x - 1 = 0$$

In the equation $x^2 - x - 1 = 0$, $a = 1$, $b = -1$, and $c = -1$.

$$x = \frac{-b \pm \sqrt{b^2 - 4ac}}{2a} = \frac{-(-1) \pm \sqrt{(-1)^2 - 4(1)(-1)}}{2(1)} = \frac{1 \pm \sqrt{1 + 4}}{2}$$

The solutions are

$$x = \frac{1 + \sqrt{5}}{2} \approx 1.618 \quad \text{and} \quad x = \frac{1 - \sqrt{5}}{2} \approx -0.618$$

When graphed, $y = x^2 - x - 1$ has x-intercepts at $x \approx 1.618$ and $x \approx -0.618$. Because x represents the length of a rectangle, the negative solution is discarded. ●

Some graphing calculators have the quadratic formula built in under POLYNO-MIAL. If your calculator does not, writing your own program may be an option. As a model, follow the sample program in the Graphing Calculator Technique box. Look at your calculator manual for details on the locations of the programming commands and keys.

Graphing Calculator Technique:
Quadratic Formula

The following sample program gives the vertex and real-number solutions for a quadratic equation.

[PRGM] QUAD
⟨Prompt⟩ A, B, C

$-B/(2A)$ [STO] V

Use the negative key, not subtraction.

⟨Disp⟩ "VERTEX X Y", V, $AV^2 + BV + C$
$B^2 - 4AC$ [STO] D
⟨If⟩ D < 0

⟨Goto⟩ 1
$(-B + \sqrt{(D)})/(2A)$ [STO] R

Use the negative key, not subtraction.

$(-B - \sqrt{(D)})/(2A)$ [STO] S

Use the negative key, then subtraction after B.

⟨Disp⟩ "SOLUTIONS", R, S
⟨Goto⟩ 2
⟨Lbl⟩ 1
⟨Disp⟩ "NO REAL SOLUTIONS"
⟨Lbl⟩ 2
⟨Stop⟩

The program as shown will work on the TI-83 calculator. Some calculators will require a multiplication sign between variables: $A*V^2$, $B*V$, and $A*C$.

ANSWER BOX

Warm-up: 1. $x^2 - 4x + 4$ **2.** $n^2 + 6n + 9$ **3.** $x^2 - x + \frac{1}{4}$
4. $x^2 + 2bx + b^2$ **5.** $4a^2x^2 + 4abx + b^2$ **6.** $x^2 + bx/a + b^2/4a^2$
7. $(t - 1)^2$ **8.** $(x - 3)^2$ **9.** $(a - 2.5)^2$ **10.** $(x + 1.5)^2$
11. $(x + 2)(x + 2) = x^2 + 4x + 4$ **12.** $(x - 5)(x - 5) = x^2 - 10x + 25$ **13.** $(x - 4)^2 = x^2 - 8x + 16$ **14.** $(x + 6)^2 = x^2 + 12x + 36$ **15.** $(x + 2.5)^2 = x^2 + 5x + 6.25$

EXERCISES

Multiply out the expressions in Exercises 1 to 8.

1. $(x + 3)^2$ **2.** $(x + 5)^2$

3. $(x - 6)^2$ **4.** $(x - 7)^2$

5. $\left(x - \frac{1}{2}\right)^2$ **6.** $\left(x + \frac{1}{4}\right)^2$

7. $(2x - 3)^2$ **8.** $(3x + 1)^2$

Write each expression in Exercises 9 to 16 as a binomial square.

9. $x^2 - 4x + 4$ **10.** $x^2 - 10x + 25$

11. $x^2 + 8x + 16$ **12.** $x^2 + 4x + 4$

13. $x^2 + 18x + 81$ **14.** $x^2 + 6x + 9$

15. $x^2 - 16x + 64$ **16.** $x^2 - 2x + 1$

Complete the squares in Exercises 17 to 20. Write the factors and the perfect square trinomial described by each completed square. In cases where there is more than one correct answer, give both.

17.

18.

19.

20.

Complete the square in Exercises 21 to 26.

21. $x^2 + 4x$ **22.** $x^2 - 16x$

23. $x^2 - 18x$ **24.** $x^2 + 22x$

25. $x^2 - 7x$ **26.** $x^2 + 9x$

Complete the square on the equations in Exercises 27 to 30. Identify the square of the binomial that you obtain on the left side.

27. $x^2 + 12x = 13$

28. $x^2 - 8x = -7$

29. $x^2 - 11x = -18$

30. $x^2 + 13x = -12$

Instead of using $ax^2 + bx + c = 0$, some calculator solve programs identify the quadratic coefficients with subscripts, using $a_2x^2 + a_1x + a_0 = 0$. Arrange each equation so that a_2 is positive, and then identify a_2, a_1, and a_0 in the equations in Exercises 31 to 36.

31. $3x - 4x^2 = 5$ **32.** $4x - 5 = 3x^2$

33. $4 = 5 - x^2$ **34.** $4 - 5x = x^2$

35. $5 - 3x^2 = 4$ **36.** $3x^2 = 4 - 5x^2$

In Exercises 37 to 42, change the right side of each formula to $ax^2 + bx + c$ form. Identify the input variable, and then identify a, b, and c.

37. Surface area of a closed base cone with slant side 4: $A = \pi r(4 + r)$

38. Surface area of a cylinder of height 4: $A = 2\pi r^2 + 8\pi r$

39. Rectangular numbers: $r = n(n + 1)$

40. Triangular numbers 4: $s = \frac{1}{2}n(n + 1)$

41. Compound interest for two years: $A = P(1 + r)^2$

42. Purchasing power with $r\%$ inflation: $A = P(1 - r)^2$

In Exercises 43 to 48, evaluate the quadratic formula expressions without a calculator. Identify a, b, and c, and write the original equation.

43. $x = \dfrac{-(-4) - \sqrt{(-4)^2 - 4(1)(-5)}}{2(1)}$

44. $x = \dfrac{-(-4) + \sqrt{(-4)^2 - 4(1)(-12)}}{2(1)}$

45. $x = \dfrac{-5 + \sqrt{5^2 - 4(2)(-12)}}{2(2)}$

46. $x = \dfrac{-5 - \sqrt{5^2 - 4(7)(-2)}}{2(7)}$

47. $x = \dfrac{-1 + \sqrt{1 - 4(3)(-4)}}{2(3)}$

48. $x = \dfrac{-(-5) - \sqrt{(-5)^2 - 4(6)(1)}}{2(6)}$

In Exercises 49 to 58, use factoring or the quadratic formula to solve the equations. If there are real-number solutions, indicate whether they are rational or irrational. Confirm your answers with a graphing calculator. (Hint: To learn the quadratic formula quickly, write it each time you use it.)

49. $x^2 + 6x - 8 = 0$

50. $x^2 + 3x - 12 = 0$

51. $x^2 + 6x + 8 = 0$

52. $x^2 + 11x - 12 = 0$

53. $2x^2 + 3x = -1$

54. $5x^2 + 3x = -3$

55. $2x^2 = -2 - 3x$

56. $2x^2 = 2 - 3x$

57. $5x = 3 - 3x^2$

58. $5x - 3 = -2x^2$

Solve the equations in Exercises 59 to 62 using the quadratic formula. The equations are from the art museum parabolic arch formula. Round to the nearest thousandth.

59. $-0.12x^2 + 2.4x = 8$

60. $-0.12x^2 + 2.4x = 9$

61. $-0.12x^2 + 2.4x = 10$

62. $-0.12x^2 + 2.4x = 11$

63. Returning to the 10-meter platform diver, recall that the height of the diver at time t is $h = -\frac{1}{2}gt^2 + v_0 t + h_0$, where g is the acceleration due to gravity (32.2 ft/sec^2, or 9.81 m/sec^2), v_0 is initial velocity, and h_0 is initial height above the water. The 10-meter platform is approximately 32.8 feet above the water. Set up and solve an equation to find the time taken for the diver to do each of the following. Round to the nearest tenth.

   a. Reach the water with an initial velocity of 6 ft/sec

   b. Reach a 23-foot height with an initial velocity of 6 ft/sec

   c. Start with a 6-ft/sec initial velocity and reach a 33-foot height

   d. Start with a 6-ft/sec initial velocity and reach a 35-foot height

   e. Reach the water with an initial velocity of 4 ft/sec

   f. Reach the water with an initial velocity of 3 m/sec

   g. Start with a 2-m/sec initial velocity and reach a 5-meter height

In Exercises 64 to 67, simplify the expressions where possible. Why can't some of the expressions be simplified?

64. $\sqrt{16x^2}$

65. $\sqrt{16 + x^2}$

66. $\sqrt{4x^2 + 9y^2}$

67. $\sqrt{4x^2 - 4x + 1}$

68. An alternative derivation of the quadratic formula is shown below. Describe what was done at each step and write the last step.

$$ax^2 + bx + c = 0$$
$$4a^2x^2 + 4abx + 4ac = 0$$
$$4a^2x^2 + 4abx = -4ac$$
$$4a^2x^2 + 4abx + b^2 = b^2 - 4ac$$
$$(2ax + b)^2 = b^2 - 4ac$$
$$\sqrt{(2ax + b)^2} = \pm\sqrt{b^2 - 4ac}$$
$$2ax + b = \sqrt{b^2 - 4ac} \text{ or } 2ax + b = -\sqrt{b^2 - 4ac}$$
$$2ax = -b + \sqrt{b^2 - 4ac} \text{ or } 2ax = -b - \sqrt{b^2 - 4ac}$$

Projects

69. Binomial Squares. Use the figure to help prove the Pythagorean theorem by showing that $a^2 + b^2 = c^2$ is true for any a, b, and c that satisfy the conditions of the theorem. The conditions are that a and b are legs of a right triangle and c is the hypotenuse. We are given that $\triangle PQR$ is a right triangle.

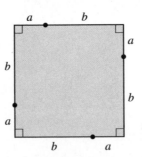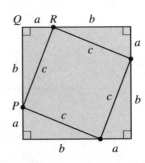

   a. What is the area of the first square, which has $a + b$ on each side?

   b. Write the area of each of the five parts in the second figure, and then add the areas.

   c. Because the two figures are both squares with sides $a + b$, set the expressions for their areas equal and show that $a^2 + b^2 = c^2$.

70. *Pythagorean Theorem and Solving Quadratic Equations.* Prove that the numbers 3, 4, and 5 are the *only consecutive positive integers* that satisfy the Pythagorean theorem.

   a. Let x, $x + 1$, and $x + 2$ be any three consecutive numbers. Write the Pythagorean theorem with this consecutive number notation, and simplify.

   b. Find the solutions to the resulting equation by graphing and factoring.

   c. Explain why the set {3, 4, 5} is the only possible solution.

71. *Generalizing the Quadratic Formula.* Apply the quadratic formula; follow the suggested steps.

   a. In $2\square^2 + 3\square - 5 = 0$, solve for $\square$.

   b. In $2\square^2 + 3\square - 2 = 0$, solve for $\square$.

   c. In $x^4 - 2x^2 + 1 = 0$, substitute A for x^2; solve for A and then for x.

   d. In $x^4 - 1 = 0$, substitute A for x^2; solve for A and then for x.

   e. In $16x^4 - 8x^2 + 1 = 0$, solve for x^2 and then for x.

   f. In $x^4 + 5x^2 - 36 = 0$, solve for x^2 and then for x.

   g. What can you conclude about the quadratic formula and equations in parts d to f?

72. *Golden Ratio.* The *golden ratio* is $(1 + \sqrt{5})/2$ to 1. Any rectangle with this length to width ratio is a golden rectangle. The golden ratio is sometimes given its own Greek letter, ϕ (phi); ϕ is an irrational number like π (pi). The golden ratio is closely related to the Fibonacci sequence: 1, 1, 2, 3, 5, 8, 13, 21, 34, ...

   a. Use differences to find the next ten terms of the Fibonacci sequence.

   b. Describe in words how the sequence is formed.

   c. Divide consecutive terms of the sequence, a_n/a_{n-1}, up to 233/144. Describe what happens as n gets larger.

   d. Extension: Write a spreadsheet to calculate the first 100 terms in the Fibonacci sequence and the ratios of consecutive terms.

73. *Fibonacci and Lucas Sequence Research.* In a report on these sequences, include how the Fibonacci sequence is related to pine cones and sunflowers.

3.5   Solving Quadratic Inequalities with Tables and Graphs ———

OBJECTIVES

- Write inequalities in interval notation.
- Solve inequalities with a table or a graph.
- Describe the solutions to quadratic inequalities relative to x-intercepts.
- Solve application problems involving inequalities.

WARM-UP

Solve $f(x) = 0$. Without graphing $f(x)$, find the vertex from the x-intercepts.

1. $f(x) = x^2 - 3x - 10$
2. $f(x) = 2x^2 - 3x - 5$
3. $f(x) = x^2 - 2$
4. $f(x) = -16.1x^2 + 400x - 1493$

IN THIS SECTION, we write inequalities as intervals. We use tables and graphs to aid us in solving quadratic inequalities.

Interval Notation

We now consider interval notation, which you may find easier to use than inequality notation in writing sets of solutions.

An **interval** is *a set containing all the numbers between its endpoints as well as one endpoint, both endpoints, or neither endpoint.* Interval notation indicates the inclusion or exclusion of endpoints by the use of brackets or parentheses. Brackets, [ ], indicate that the endpoints are included in the set. Parentheses, ( ), are used when the endpoints are excluded from the set. We may mix brackets and parentheses in one interval, as shown in Table 13.

Inequality	Interval	Words	Number Line
$2 \leq x \leq 5$	[2, 5]	Set of numbers between 2 and 5, including 2 and 5	
$2 \leq x < 5$	[2, 5)	Set of numbers between 2 and 5, including 2	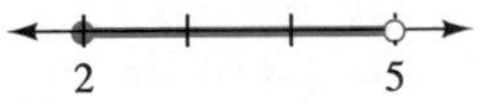
$2 < x \leq 5$	(2, 5]	Set of numbers between 2 and 5, including 5	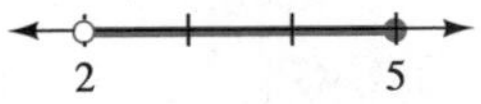
$2 < x < 5$	(2, 5)	Set of numbers between 2 and 5	

Table 13

To write inequalities such as $x > 5$ or $x \leq 2$ with interval notation, we need an **infinity sign, ∞**. **Infinite** means *without bound*. We draw arrows on the ends of a line or axis to indicate that the line goes on without bound. A positive sign before the infinity sign means that a horizontal line extends infinitely to the right; a negative sign indicates that it extends infinitely to the left (see Figure 52).

Figure 52

Because infinity is not a number, it cannot be included in a set of numbers. As shown in Table 14, we always use parentheses next to an infinity sign.

Inequality	Interval	Words	Number Line
$x \geq 5$	[5, $+\infty$)	Set of numbers greater than 5, including 5	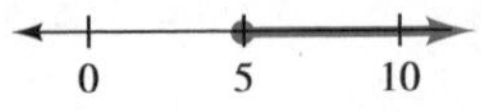
$x < 2$	($-\infty$, 2)	Set of numbers less than 2	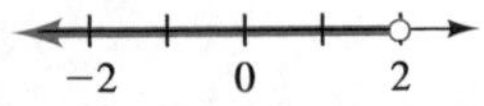
$-\infty < x < +\infty$	($-\infty$, $+\infty$)	Set of all real numbers, $\mathbb{R}$	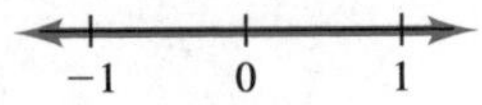

Table 14

EXAMPLE 1 Writing intervals Complete this chart.

	Inequality	Interval	Words	Number Line
a.	$-3 < x \le 4$			
b.			Set of numbers greater than -3, including -3	
c.				
d.	$x < 2$			
e.			Set of numbers less than 4 or greater than 8	

Solution

	Inequality	Interval	Words	Number Line
a.	$-3 < x \le 4$	$(-3, 4]$	Set of numbers greater than -3 and less than 4, including 4	
b.	$x \ge -3$	$[-3, +\infty)$	Set of numbers greater than -3, including -3	
c.	$x \le -2$	$(-\infty, -2]$	Set of numbers less than -2, including -2	
d.	$x < 2$	$(-\infty, 2)$	Set of numbers less than 2	
e.	$x < 4$ or $x > 8$	$(-\infty, 4)$ or $(8, +\infty)$	Set of numbers less than 4 or greater than 8	

Using a Table or Graph to Solve Inequalities

There are many times when we are interested in finding as a solution a set of numbers, instead of a single number. For example, to earn a passing grade we need any score over 70%. For paying our bills, any amount in our checking account over the amount owed (plus the minimum balance) is satisfactory. We describe these sets of numbers with inequalities or intervals. In Example 2, we use inequalities to state the solution to a problem related to emergency rescue flares.

EXAMPLE 2 Solving inequalities with a table and graph: height of an emergency flare An emergency rescue flare is fired straight up through a layer of fog 1500 feet thick. The flare's height in terms of time is found from the vertical motion equation $h = -\frac{1}{2}gt^2 + v_0t + h_0$. Suppose h_0 is the height of the hand holding the flare gun, 7 feet, and the acceleration due to gravity is $g = 32.2$ ft/sec^2. The flare is fired with an initial velocity v_0 of 400 ft/sec.

a. Set up an equation describing the height of the flare.

b. Make a table for the height every 4 seconds until the flare returns to the ground.

c. Graph the equation for the height of the flare and the equation for the fog layer, $y = 1500$ ft.

d. Write an inequality that tells when the height of the flare will be above 1500 feet.

e. From the table and graph estimate the time interval during which the flare will be visible to a plane flying above the fog.

f. Check that a point on the time interval satisfies the inequality in part d.

Solution   **a.** The equation is $h = -16.1t^2 + 400t + 7$.

b. Table 15 shows the height every 4 seconds, to 28 seconds.

c. Figure 53 shows the graph of the equation and $y = 1500$ ft.

Time (seconds)	Height (feet)
0	7
4	1349
8	2177
12	2489
16	2285
20	1567
24	333
28	−1415

Table 15

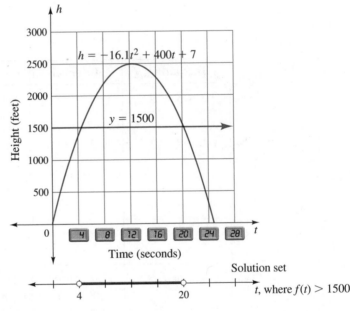

Figure 53

d. The flare will be above 1500 feet for $-16.1t^2 + 400t + 7 > 1500$.

e. The table indicates that the flare will be above 1500 feet between $t \approx 4$ and $t \approx 20$ sec, or $4 < t < 20$. From the graph, we observe the height of the flare and confirm our estimate, the interval (4, 20) seconds. The flare will be visible for about 16 seconds.

f. *Check:* To check our interval, we select an arbitrary time between 4 and 20 and substitute it into our inequality. Suppose $t = 10$.

$$-16.1t^2 + 400t + 7 > 1500$$
$$-16.1(100) + 400(10) + 7 \overset{?}{>} 1500$$
$$-1610 + 4000 + 7 \overset{?}{>} 1500$$
$$2397 > 1500 \quad ✔$$

At $t = 10$, the flare is at 2397 feet and above 1500 feet, so our interval is reasonable.

With a graphing calculator, we can improve on our time estimates from both the table and the graph. We can adjust the table set-up to 1-second intervals between inputs and find a closer estimate. Similarly, we can trace and zoom in on the graph to find a closer estimate of the intersection between the graph of the height of the flare and $y = 1500$. A third possibility is shown in Example 3, where we use an equivalent inequality (one having the same solution set).

EXAMPLE **3** Solving inequalities with the quadratic formula and a graph: more on flare height Solve $-16.1t^2 + 400t + 7 > 1500$ by examining the graph of an equivalent inequality.

a. Subtract 1500 from both sides of the inequality.

b. Graph the resulting inequality.

c. Find the horizontal intercepts of the new inequality.

d. Find the interval for which the new inequality is above zero.

Solution **a.** If we subtract 1500 from both sides of our inequality, we obtain an equivalent inequality with the same solution set as the original.

$$-16.1t^2 + 400t + 7 > 1500$$
$$-16.1t^2 + 400t + 7 - 1500 > 1500 - 1500$$
$$-16.1t^2 + 400t - 1493 > 0$$

By subtracting 1500 from both sides, we have lowered the original height of the flare function by 1500 feet. The time interval when the flare is above the horizontal axis is the time interval when the flare is above the fog.

b. The graph of $h = -16.1t^2 + 400t - 1493$ is shown in Figure 54.

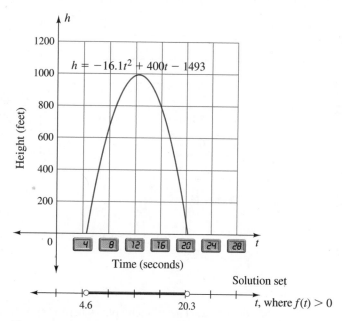

Figure 54

c. We use the quadratic formula to find the intercepts:

$$x = \frac{-400 \pm \sqrt{400^2 - 4(-16.1)(-1493)}}{2(-16.1)}$$

$$= \frac{-400 \pm \sqrt{63850.8}}{-32.2}$$

$$x \approx 4.6 \text{ sec} \quad \text{or} \quad x \approx 20.3 \text{ sec}$$

d. The graph is above the horizontal axis when $4.6 < t < 20.3$. Because we solved an equivalent inequality, we know that the flare is above the clouds for the times $4.6 < t < 20.3$, or for about 15.7 seconds. ●

EXAMPLE    Solving inequalities with tables, graphs and x-intercepts   Build a table and graph for the function $f(x) = x^2 - 3x - 10$, with inputs on the interval $x = -4$ to $x = 7$. Using inequalities and graphs, describe where

a. $f(x) = 0$ **b.** $f(x) > 0$ **c.** $f(x) < 0$

Solution   Table 16 contains data for $f(x) = x^2 - 3x - 10$, and Figure 55 contains the graph.

Input x	Output $f(x) = x^2 - 3x - 10$
−4	18
−3	8
−2	0
−1	−6
0	−10
1	−12
2	−12
3	−10
4	−6
5	0
6	8
7	18

Table 16

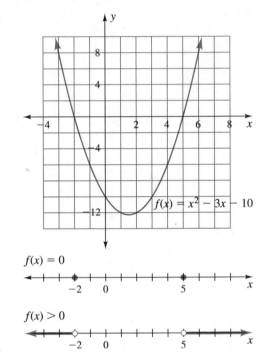

Figure 55

a. The x-intercepts, where $f(x) = 0$, are $x = -2$ and $x = 5$. In Figure 55, the top number line shows the x-intercepts.

b. Table 16 indicates that the output is positive, $f(x) > 0$, for $x < -2$ and for $x > 5$. The graph is above the x-axis when $x < -2$ and $x > 5$. This indicates where $f(x) > 0$ because the x-axis is $y = 0$. Thus,

$$x^2 - 3x - 10 > 0 \qquad \text{for } x < -2 \quad \text{or} \quad x > 5$$

The middle number line in Figure 55 shows the solutions: the intervals $(-\infty, -2)$ and $(5, +\infty)$.

c. For inputs between $x = -2$ and $x = 5$, the table output is negative, $f(x) < 0$. The graph is below the x-axis for x between -2 and 5. This indicates that $f(x) < 0$. Thus,

$$x^2 - 3x - 10 < 0 \qquad \text{for } -2 < x < 5$$

The bottom number line in Figure 55 shows the solutions: the interval $(-2, 5)$. ●

EXAMPLE **5** Solving inequalities with tables, graphs, and x-intercepts Build a table and a graph for $f(x) = 2x^2 - 3x - 5$ for the integers on the interval $[-3, 4]$. Find the intercepts, and determine for what inputs

a. $2x^2 - 3x - 5 < 0$ **b.** $2x^2 - 3x - 5 \leq 0$

Solution Table 17 contains data for $f(x) = 2x^2 - 3x - 5$. The graph is in Figure 56.
     Table 17 shows only one x-intercept, $x = -1$. The graph in Figure 56 shows two intercepts. To solve the inequality $2x^2 - 3x - 5 < 0$, we need both x-intercepts. We could use the quadratic formula, but the factors of $2x^2 - 3x - 5$ are $(x + 1)$ and $(2x - 5)$, so the other x-intercept is where $2x - 5 = 0$, or $x = 2.5$.

Input x	Output $f(x) = 2x^2 - 3x - 5$
-3	22
-2	9
-1	0
0	-5
1	-6
2	-3
3	4
4	15

Table 17

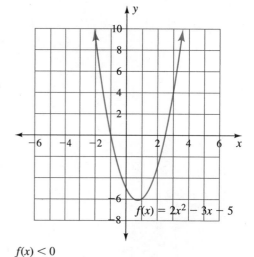

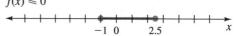

Figure 56

a. The inequality $2x^2 - 3x - 5 < 0$ is true for all numbers between $x = -1$ and $x = 2.5$, or $-1 < x < 2.5$. The solution is shown on the top number line in Figure 56.

b. To solve $2x^2 - 3x - 5 \leq 0$, we include the intercepts on our interval because the intercepts themselves make the inequality true. We replace the $<$ signs with $\leq$ to include the intercepts: $-1 \leq x \leq 2.5$. The solution is shown on the bottom number line in Figure 56. ●

Summary: Solving Inequalities

For inequalities of the form $f(x) > 0, f(x) \geq 0, \ldots,$

1. Find the x-intercepts, where $f(x) = 0$. The x-intercepts are the endpoints or boundaries for the solution set to the inequalities.

   a. For $f(x) \geq 0$ or $f(x) \leq 0$, the endpoints are in the solution set.

   b. For $f(x) > 0$ or $f(x) < 0$, the endpoints are not in the solution set.

2. For $f(x) > 0$ or $f(x) \geq 0$, find where the output is above 0 or the graph is above the x-axis. Describe the solution set as the corresponding inputs.

3. For $f(x) < 0$ or $f(x) \leq 0$, find where the output is below 0 or the graph is below the x-axis. Describe the solution set as the corresponding inputs.

At times, we want to describe regions in the coordinate plane with quadratic inequalities. After graphing the equation corresponding to the inequality, we can use test points to locate regions to be shaded, as we did in Section 1.4.

EXAMPLE 6 Finding regions described by inequalities Match the description with the shaded region in Figure 57, 58, or 59. Use test points as needed.

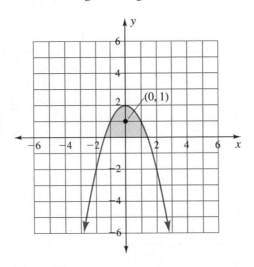

Figure 57

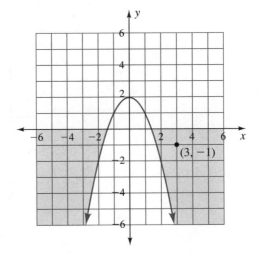

Figure 58

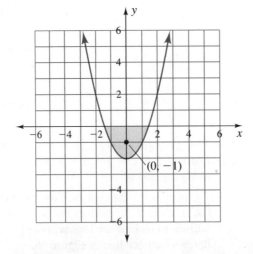

Figure 59

a. $y \leq 0$ and $y \geq x^2 - 2$ b. $y \geq 0$ and $y \leq -x^2 + 2$

c. $y \leq 0$ and $y \geq -x^2 + 2$

Solution   a. The inequality $y \leq 0$ represents the region on or below the x-axis. The inequality $y \geq x^2 - 2$ represents the region above the curve $y = x^2 - 2$. The region that makes both of these inequalities true is shaded in Figure 59. As a check, we see that the point $(0, -1)$ makes both inequalities true.

b. The inequality $y \geq 0$ represents the region on or above the x-axis. The inequality $y \leq -x^2 + 2$ represents the region below the curve $y = -x^2 + 2$. The region that makes both of these inequalities true is shaded in Figure 57. As a check, we see that the point $(0, 1)$ makes both inequalities true.

c. The inequality $y \leq 0$ represents the region on or below the x-axis. The inequality $y \geq -x^2 + 2$ represents the region above the curve $y = -x^2 + 2$. The region that makes both of these inequalities true is shaded in Figure 58. As a check, we see that the point $(3, -1)$ makes both inequalities true.   ●

In the Exercises, we will find areas of regions related to applications.

Applications

COST AND REVENUE CURVES   As indicated earlier, solutions described by inequalities are common. On business graphs that show costs and revenues (sales or income), all inputs giving revenues above costs create profit. Those inputs giving costs greater than revenues create losses.

In Example 7, we look at cost and revenue curves for a seasonal product.

EXAMPLE 7

Solving inequality problems from a graph: cost and revenue curves   In Figure 60, the revenue curve gives sales volume in dollars for the business year starting July 1 and ending the following June 30. The cost curve shows the costs incurred during the year to make those sales. The intercepts for the cost curve are $x = 0$ and $x = 365$.

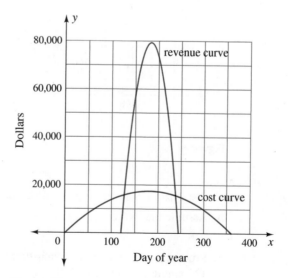

Figure 60

a. Estimate the time period (day ____ to day ____) during which revenue is received.

b. Estimate the time period (day ____ to day ____) during which revenue is greater than cost.

c. Assume each curve is a parabola, and estimate its vertex.

Solution **a.** The revenue curve intersects the *x*-axis from approximately day 120 to day 245. Thus, revenue is received on the interval $120 < x < 245$.

b. The revenue curve is above the cost curve from approximately day 125 to day 240, or on the interval $125 < x < 240$.

c. The parabolic curves appear to be symmetric to the beginning and ending of the business year, so the *x*-coordinate of the vertex for each curve is $\frac{365}{2}$, or about 182. For the revenue curve, the vertex is approximately (182, 80,000). For the cost curve, the vertex is approximately (182, 17,500).   ●

Think about it 1: What season of the year is the revenue curve describing?

SUPPLY AND DEMAND CURVES In agricultural planning, those months for which food demand curves are above food supply are months in which food must be imported. In Example 8, we look at supply (production) and demand (consumption) for an agricultural product.

EXAMPLE **8**   Solving inequality problems from a graph: agricultural supply and demand curves   Figure 61 shows a tropical country's consumption of rice and its twice yearly production of rice. The production (harvest) takes a month to complete.

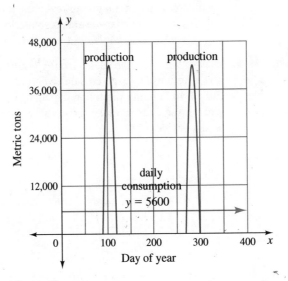

Figure 61

a. For what days in the year is production greater than consumption?

b. Assume the two curves are parabolas. Estimate the vertices.

Solution **a.** Production is greater than consumption from approximately day 90 to day 120 and day 270 to day 300.

b. The first vertex is at approximately (105, 42,000). The second vertex is at approximately (285, 42,000).   ●

Think about it 2: Rice stores easily between harvests. What would it mean to this tropical country if the food represented in Figure 61 could not be stored for future use?

ANSWER BOX

Warm-up: 1. $x = 5$, $x = -2$; $(1.5, -12.25)$ **2.** $x = 2.5$, $x = -1$; $(0.75, -6.125)$ **3.** $x = \pm\sqrt{2}$; $(0, -2)$ **4.** $x \approx 4.6$, $x \approx 20.3$; $\approx(12.4, 991.5)$ **Think about it 1:** If we count from July 1, it appears that the revenue curve describes the winter shopping season, between October and March. **Think about it 2:** If the food could not be stored, it would need to be sold or exported immediately, and the demand (consumption needs) during the rest of the year would have to be met by imports.

EXERCISES 3.5

Complete this chart for Exercises 1 to 12.

	Inequality	Interval	Words	Number Line
1.	$-3 < x < 5$	$(-3, 5)$	Set of numbers greater than -3 and less than 5	
2.			Set of numbers greater than -1 and less than 3, including 3	
3.	$-4 < x \le 2$	$(-4, 2]$	Set of numbers greater than -4 and less than 2, including 2	
4.			Set of numbers greater than -2 and less than 4, including -2 and 4	
5.	$x > 5$	$(5, \infty)$	Set of #'s greater than 5	
6.				
7.	$x < -2$	$(\infty, -2)$	Set of #'s less than -2	
8.	$x \ge -2$		S	
9.	$x \le -3$	$(-\infty, -3]$	Set less than inc. -3	
10.		$(-\infty, 3)$		
11.	$x \ge 4$	$[4, +\infty)$	great inclu 4	
12.		$(-1, +\infty)$		

13. In Example 5, for what inputs x is $2x^2 - 3x - 5 > 0$?

14. For what inputs is $2x^2 - 3x - 5 \ge 0$?

In Exercises 15 to 36, find for what inputs each inequality is true. Use any method (graphs, tables, or factoring) to find the x-intercepts.

15. $x^2 - 5x + 6 > 0$ **16.** $x^2 - 5x + 6 \leq 0$

17. $x^2 - 5x - 14 \leq 0$ **18.** $x^2 - 5x + 4 < 0$

19. $x^2 - 5x - 6 \geq 0$ **20.** $x^2 - 5x - 24 > 0$

21. $x^2 + 6x + 9 > 0$ **22.** $x^2 - 6x + 9 < 0$

23. $4x^2 - 4x + 1 < 0$ **24.** $4x^2 + 4x + 1 \geq 0$

25. $x^2 + 2x - 15 < 0$ **26.** $x^2 - 8x + 15 \geq 0$

27. $x^2 + 10x - 24 \geq 0$ **28.** $x^2 - 10x + 24 < 0$

29. $x^2 - 3x - 18 \leq 0$ **30.** $x^2 + 7x - 18 > 0$

31. $x^2 + 17x - 18 > 0$ **32.** $x^2 + 9x + 18 \leq 0$

33. $x^2 + 8x + 16 \geq 0$ **34.** $x^2 - 10x + 25 < 0$

35. $x^2 - 14x + 49 < 0$ **36.** $x^2 + 12x + 36 > 0$

Estimate answers for Exercises 37 to 42 using the following information: The curved line in the figure shows the path of a kicked ball. The ball is kicked with an initial velocity of 60 ft/sec, at an angle of 45° relative to the ground. The straight line is the path the ball would follow if there were no gravity.

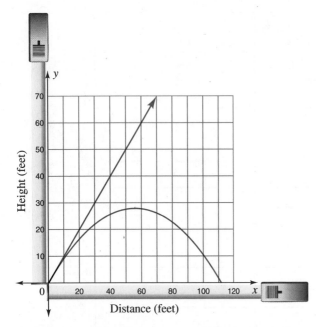
Distance (feet)

37. The ball is to pass over a horizontal bar 10 feet high. For what horizontal distances will the ball go over the bar?

38. The ball may be knocked to the ground by the opposing team if it is below 10 feet. For what horizontal distance is the ball in danger of being knocked down?

39. What is the highest point reached by the ball?

40. Does the graph show the length of time the ball is in the air? Explain your reasoning.

41. If there were no gravity, what would be the height of the ball when it was horizontally 50 feet from the kicker?

42. For what horizontal distances is the ball over 20 feet high?

43. From their truck, an inspection team can inspect suspension bridge cable within 20 feet of the road surface. The parabolic cable has its vertex at the road surface in the center of the bridge. Its equation is $y = \frac{4}{125}x^2$. For what horizontal interval on the 80-foot-long bridge can the inspection be accomplished from the truck?

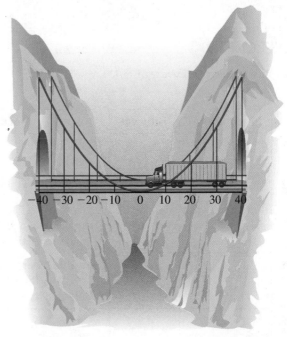

44. The inspection team in Exercise 43 obtains a new truck with a hydraulic basket that can carry a worker high enough to inspect up to 30 feet above the roadway. For what horizontal interval can the inspection now be accomplished?

Refer to Example 7 for Exercises 45 to 49.

45. What type of product might have revenue (sales) and costs as described?

46. Between what calendar dates do the sales occur?

47. What date is the 182nd day from July 1?

48. What date is the 120th day from July 1?

49. The areas under the graphs give the total revenue and total cost for the year. To approximate the area for this portion of a parabola, use the formula $A = \frac{2}{3}bh$, where b is the estimated distance between the x-intercepts and h is the perpendicular distance from the vertex to the x-axis.

   a. Use the area to estimate the total revenue for the year.

b. Use the area to estimate the total costs for the year.

c. Will there be an overall profit (total revenue − total cost > 0) for the year?

Refer to Example 8 for Exercises 50 to 52.

50. The average consumption of rice in this country is 5600 metric tons per day. What is the consumption for the year? What part of the graph represents the yearly rice consumption?

51. The area under the two parabolic graphs gives the total rice production for the year. Find the total production. (*Hint:* Approximate the areas with $A = \frac{2}{3}bh$, where b is the estimated distance between the x-intercepts and h is the perpendicular distance from the vertex to the x-axis.)

52. If annual consumption is larger than annual production, the country must import rice. Estimate the amount of rice imported by comparing the answers to Exercises 50 and 51.

Match each of the descriptions in Exercises 53 to 58 with one of the graphs a–f.

c.

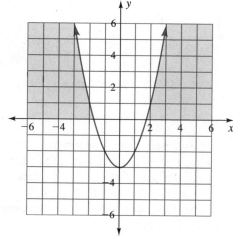

d.

a.

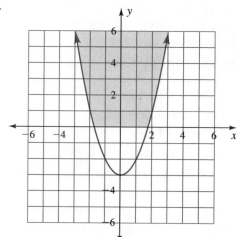

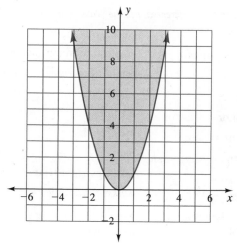

b.

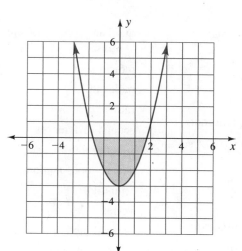

e.

f.

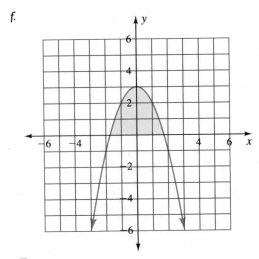

53. $y \le x^2$

54. $y \ge x^2$

55. $y \ge 0$ and $y \le -x^2 + 3$

56. $y \le 0$ and $y \ge x^2 - 3$

57. $y \ge 0$ and $y \ge x^2 - 3$

58. $y \ge 0$ and $y \le x^2 - 3$

59. Why are there no points that make $y < 0$ and $y \ge x^2$ true?

60. Why are there no points that make $y \le -x^2$ and $y > 0$ true?

Projects

61. ***Distinguishing Intervals from Coordinates.*** The expression in parentheses (2, 4) could describe an interval or an ordered pair (coordinate point), because both use the same notation. Identify which meaning is intended in each of these examples. (*Hint:* Sometimes drawing a picture is helpful.)

a. The postage doubled for packages with weight (2, 4).

b. (2, 4) makes the equation $y = 2x$ true.

c. The graphs $y = x + 2$ and $y = x^2$ cross at (2, 4) and (−1, 1).

d. The graph of $y = x^2 - 6x + 8$ is below the x-axis for (2, 4).

e. Make an appointment in the afternoon, (2, 4).

f. The line $y = 4$ passes through (2, 4) and is parallel to the x-axis.

g. The solution to the equation $0 = x - 3$ is in (2, 4).

h. A solution to the equation $y = 3x - 2$ is (2, 4).

62. ***Retail Sales***

a. Sketch a graph showing cost (of purchasing the summer clothing inventory) and income (from sales of the clothes) for a clothing retailer. Let the horizontal axis be the 365 days of the year. Assume the cost curve is parabolic. Make up appropriate dollar amounts, and describe your assumptions about the dates you place along the horizontal axis. Describe sales of the clothes with whatever type of graph you wish, but explain your reasoning.

b. Estimate the total cost of purchasing the inventory by finding the area under the parabola using the formula in Exercise 49.

c. Estimate the total income from sales of the clothes. Describe your technique.

d. Indicate whether the retailer had a profit or a loss.

e. Make up and answer two inequality questions based on your graph.

CHAPTER ❸ SUMMARY

Vocabulary

For definitions and page references, see the Glossary/Index.

absolute value

axis (line) of symmetry

binomial

binomial square

common monomial factor

completing the square

converse of the Pythagorean theorem

cubic function

degree of a polynomial of one variable

equilateral triangle

first differences

golden rectangle

greatest common factor

height of a triangle (altitude)

hypotenuse

infinite

infinity sign

interval

isosceles right triangle

isosceles triangle

leg

monomial

parabola

perfect square trinomial

plus or minus sign

polynomial

prime factor

prime number

principal square root

product property of square roots

proof

proportion

Pythagorean theorem

quadratic formula

quadratic function

quadratic sequence

quotient property of square roots

radical sign

radicand

ratio

second differences

square root

square root function

trinomial

vertex

vertical motion equation

x-intercept points

y-intercept point

zero product rule

Concepts

3.0 Basic Operations on Polynomials:

Name a polynomial according to the number of terms it has.

Arrange the terms in a polynomial in descending order of exponents on the alphabetically first variable.

To factor $adx^2 + dbx + dc$ by table:

1. First remove common monomial factor d so that $ax^2 + bx + c$ remains:

$$d(ax^2 + bx + c)$$

2. Enter ax^2 and c into one diagonal.

3. Multiply ax^2 and c to obtain the diagonal product.

4. Find factors of the diagonal product that add to bx.

5. Enter these factors into the other diagonal.

6. Factor out the greatest common factor, find the other factors, and write the binomial pair.

7. Multiply factors to check that the inner "area" is $ax^2 + bx + c$.

Factor		
gcf	first term ax^2	diagonal sum bx
	last term c	diagonal product acx^2

3.1 Square Root Function and Pythagorean Theorem

$x^2 = n$ has two solutions, $x = \pm\sqrt{n}$.

$x = \sqrt{n}$ has one solution, the principal square root of n.

The square root of a negative number is undefined in the set of real numbers.

3.2 Solving Quadratic Equations with Tables and Graphs

See Table 18 for important points on the graph of a quadratic function:

Quadratic functions are polynomial functions of degree 2. They may be written in the form $y = ax^2 + bx + c$, where a, b, and c are real numbers, $a \neq 0$.

A sequence of numbers is the output of a quadratic function if the second differences are constant. A sequence of numbers is the output to a linear equation if the first differences are constant.

A quadratic equation $ax^2 + bx + c = 0$ or $ax^2 + bx + c = d$ may have no, one, or two real-number solutions.

When solving $ax^2 + bx + c = d$ with a graph, find the intersections of $y = d$ and the graph of the corresponding quadratic function. When solving $ax^2 + bx + c = d$ with a table, find the input x where the output $f(x) = y = d$.

3.3, 3.4 Solving Quadratic Equations with Factors, Square Roots, and the Quadratic Formula

The following are nine techniques for solving quadratic equations $ax^2 + bx + c = 0$ when $f(x) = ax^2 + bx + c$.

1. Guess numbers for x, and check until the output $f(x)$ is zero.

2. Draw a graph and find x-intercepts, where $f(x) = 0$ (Section 3.2).

3. Make a table and find x, where $f(x) = y = 0$ (Section 3.2).

4. Factor and use the zero product rule (if the product of two expressions is zero, then either one or the other expression is zero) (Section 3.3).

5. If $b = 0$, solve for x^2 and take the square root of both sides. The equation $x^2 = k$, $k \geq 0$ has two solutions: $x = +\sqrt{k}$ and $x = -\sqrt{k}$ (Section 3.3).

6. In $f(x) = d$, if $f(x)$ is a perfect square, take the square root of both sides (Section 3.3).

Name	Definition	Table		Graph
x-intercept points (where $y = 0$)	places where graph crosses x-axis	x | y / (blank) | 0		(_, 0)
y-intercept point (where $x = 0$)	places where graph crosses y-axis	x | y / 0 | (blank)		(0, _)
vertex	highest or lowest point	The x-coordinate is halfway between two equal outputs.		

Table 18

7. Complete the square. To complete the square in a quadratic equation $ax^2 + bx + c = 0$, $a \neq 0$, subtract c from both sides, divide all terms by a, and add $(b/2a)^2$ to both sides (Section 3.4).

8. Use the quadratic formula (Section 3.4).

9. Use the solve feature on a programmable calculator or computer (Section 3.4).

3.5 Solving Quadratic Inequalities with Tables and Graphs

Interval notation may look like the coordinates of a point. Read carefully when you see (a, b) to see whether the reference is to a coordinate point (a, b) or an interval (a, b) describing the set $a < x < b$.

See Table 19 for symbols used in inequalities and intervals.

To solve inequalities of the form $f(x) > 0$, $f(x) \geq 0$, $f(x) < 0$, $f(x) \leq 0$:

1. Find the x-intercepts, where $f(x) = 0$. The x-intercepts are the endpoints or boundaries for the solution set to the inequalities.

   a. For $f(x) \geq 0$ or $f(x) \leq 0$, the endpoints are in the solution set.

   b. For $f(x) > 0$ or $f(x) < 0$, the endpoints are not in the solution set.

2. For $f(x) > 0$ or $f(x) \geq 0$, find where the output is above 0 or the graph is above the x-axis. Describe the solution set as the corresponding inputs.

3. For $f(x) < 0$ or $f(x) \leq 0$, find where the output is below 0 or the graph is below the x-axis. Describe the solution set as the corresponding inputs.

Inequality Symbol	Interval Notation	Word Meaning	Line Graph Notation
$<$	$(\,,\,)$	is less than	small circle or $(\,,\,)$
$>$	$(\,,\,)$	is greater than	small circle or $(\,,\,)$
$=$		is equal to	dot
$\leq$	$[\,,\,]$	is less than or equal to	dot or $[\,,\,]$
$\geq$	$[\,,\,]$	is greater than or equal to	dot or $[\,,\,]$
$+\infty$	$,\,+\infty)$	positive infinity	$\rightarrow$
$-\infty$	$(-\infty,\,$	negative infinity	$\leftarrow$

Table 19

CHAPTER ③ REVIEW EXERCISES

1. Identify each of the following expressions as monomial, binomial, trinomial, or not a polynomial.

   a. $x^3 - y^3$ **b.** $\sqrt{x} - \sqrt{y}$

   c. $x^2 + x^{-1}$ **d.** $x\sqrt{3} + \sqrt{3}$

   e. $3x^3 - 3x^2 + 3x$ **f.** $|x + 1|$

2. Simplify.

   a. $(x^2 + 3x) + (x^3 - 4x^2 - 5x)$

   b. $(x^2 + 3x) - (x^3 - 4x^2 - 5x)$

   c. $x^2 - x + 1 - x^2 + x + 1$

   d. $x^2 - x + 1 - (x^2 + x + 1)$

   e. $14 - 6(x + 3)$

   f. $12 - 7(x - 2)$

3. Multiply these expressions.

   a. $(x - 3)(x + 3)$

   b. $(2x - 5)(2x - 5)$

   c. $(x - 1)(x^2 + x + 1)$

   d. $(n + 4)(n + 4)$

   e. $(2x - 3)(3x + 4)$

   f. $(x^2 + x - 1)(x^2 - x - 1)$

4. Multiply these expressions.

   a. $(x - 5)(x - 5)$

   b. $(x + 2)(x^2 - 2x + 4)$

   c. $(n - 2)(n + 2)$

   d. $(3x + 4)(3x + 4)$

e. $(3x - 2)(4x + 3)$

f. $(x^2 - 3x + 3)(x^2 + 3x - 3)$

5. Factor these expressions.

a. $x^2 + 3x - 4$ **b.** $2x^2 - 3x$

c. $2x^2 + x - 3$ **d.** $9x^2 + 12x + 4$

e. $x^2 - x$ **f.** $3x^2 + 6x + 3$

6. Factor these expressions.

a. $x^2 + x$ **b.** $3x^2 - x - 4$

c. $2x^2 - 9x + 10$ **d.** $3x^2 + 10x + 8$

e. $x^2 + x - 6$ **f.** $4x^2 - 16x + 16$

Simplify the radical expressions in Exercises 7 to 10.

7. a. $\sqrt{75}$ **b.** $\sqrt{8}$

c. $\sqrt{32}$

8. a. $\sqrt{50}$ **b.** $\sqrt{18}$

c. $\sqrt{72}$

9. a. $\dfrac{3 - 3\sqrt{6}}{3}$ **b.** $\dfrac{3 + 3\sqrt{6}}{3}$

10. a. $\dfrac{3 + \sqrt{18}}{3}$ **b.** $\dfrac{3 + \sqrt{12}}{2}$

11. Show that each of the following sets of numbers could be the lengths of the sides of a right triangle.

a. $\{7.5, 10, 12.5\}$ **b.** $\{18, 24, 30\}$

c. $\{\sqrt{5}, \sqrt{8}, \sqrt{13}\}$ **d.** $\{4, \sqrt{20}, 6\}$

12. Solve for x.

a.

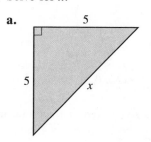

b.

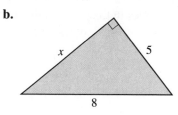

c.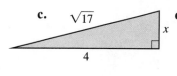

d.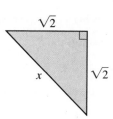

e.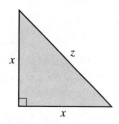

f.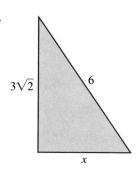

13. Find the length of the hypotenuse z of an isosceles right triangle with legs x (see the figure). Write z in terms of x. Divide the expression for z by x to compare the lengths, hypotenuse to leg.

14. Solve for x.

a. $x^2 = 225$ **b.** $x = \sqrt{225}$

c. $x^2 = 17$ **d.** $x = \sqrt{17}$

e. $|x + 3| = 2$ **f.** $|x - 5| = 3$

15. True or false: $\sqrt{x}$ can be negative. Explain how your answer can be shown by the graph of $y = \sqrt{x}$.

In Exercises 16 to 19, give the next number in each sequence and identify each sequence as linear, quadratic, or neither.

16. $-6, -4, 0, 6, 14$

17. $-3, -2, 3, 12, 25$

18. $-10, -4, 2, 8, 14$

19. $-6, -2, 2, 6, 10$

20. a. Make a table of values for $f(x) = x^2 + x - 12$. Use inputs -2 to $+2$.

b. Identify the x-intercepts.

c. Find the equation for the axis of symmetry.

  d. What is the vertex?

  e. Graph the data.

  f. Solve $x^2 + x - 12 = -6$.

  g. Solve $x^2 + x - 12 = 0$.

  h. Solve $x^2 + x - 12 = -10$.

21. a. Make a table of values for $f(x) = x^2 - x - 6$. Use inputs -2 to $+2$.

  b. Identify the x-intercepts.

  c. Find the equation for the axis of symmetry.

  d. What is the vertex?

  e. Graph the data.

  f. Solve $x^2 - x - 6 = 6$.

  g. Solve $x^2 - x - 6 = 0$.

  h. Solve $x^2 - x - 6 = -7$.

In Exercises 22 and 23, make a table and a graph.

22. Surface area of spheres with radius 0 to 10 inches: $A = 4\pi r^2$

23. Decline in purchasing power due to inflation rate r per year over 2 years: $A = P(1 - r)^2$. Let r be 0 to 8%, with income of $P = \$20,000$.

24. The period of a pendulum is $T = 2\pi\sqrt{L/g}$. Solve the formula for L, the length of the pendulum.

25. An approximation to the perimeter of an ellipse is $C = 2\pi\sqrt{\dfrac{a^2 + b^2}{2}}$. Solve the formula for a.

In Exercises 26 to 33, solve the equations in three ways: by factoring, with the quadratic formula, and by graphing.

26. $x^2 + 3x - 4 = 0$

27. $x^2 - 5x + 6 = 0$

28. $x^2 + 16 = 8x$

29. $x^2 = 6x - 9$

30. $9x^2 + 12x + 4 = 0$

31. $3x^2 + x - 4 = 0$

32. $2x^2 + x = 3$

33. $9x^2 = 4$

34. How many real-number solutions does the equation $5x^2 + 7x = -4$ have? Explain in terms of the graph and again using the quadratic formula.

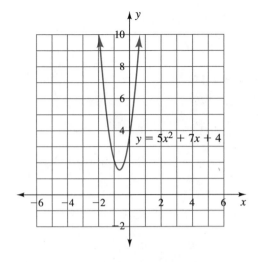

35. Suppose we stand on the ground, $h_0 = 0$, and throw a ball straight up with an initial velocity of 72 ft/sec.

  a. Write an equation describing height in terms of time. Use the vertical motion equation, $h = -\frac{1}{2}gt^2 + v_0t + h_0$. Let the acceleration due to gravity be $g = 32$ ft/sec^2.

  b. What are the x-intercepts? Explain why there are two answers.

  c. Solve the equation for $h = 80$ ft. Use a graph and/or table as needed. Explain your answer(s).

  d. What is the highest point the ball reaches?

Quadratic equations appear in a rather unusual form in equilibrium problems from first-year chemistry. In solving Exercises 36 to 39 for x, remember that A/B = C is equivalent to A = B · C. Because of physical constraints on the problems, the input variable has been restricted. Solve for x. Round to the nearest thousandth.

36. $\dfrac{x^2}{1.000 - x} = 4,\ 0 \le x \le 1$

37. $\dfrac{x^2}{2.000 - x} = 12,\ 0 \le x \le 2$

38. $\dfrac{(3.000 + 2x)^2}{(1.000 - x)(2.000 - x)} = 42,\ 0 \le x \le 1$

39. $\dfrac{(2x)^2}{(2.000 - x)(1.000 - x)} = 24,\ 0 \le x \le 1$

40. Zoning codes in a mid-size city require that a business sign in a residential neighborhood be no greater than 1.5 square feet in area. What are the dimensions for a rectangular sign with sides in the golden ratio (1.618 to 1)? Calculate the length and width of the sign in inches. Round to the nearest tenth.

41. Identify what was done in each step in solving $x^2 - x - 1 = 0$ for x by completing the square. (*Note:*

The coefficient of x^2 is 1, so no division is necessary.)

$$x^2 - x - 1 = 0$$
$$x^2 - x = 1$$
$$x^2 - 1x + \tfrac{1}{4} = 1 + \tfrac{1}{4}$$
$$\left(x - \tfrac{1}{2}\right)^2 = 1 + \tfrac{1}{4}$$
$$\left(x - \tfrac{1}{2}\right)^2 = \tfrac{5}{4}$$
$$\sqrt{\left(x - \tfrac{1}{2}\right)^2} = \sqrt{\tfrac{5}{4}}$$
$$\left|x - \frac{1}{2}\right| = \frac{\sqrt{5}}{2}$$
$$x - \frac{1}{2} = \pm\frac{\sqrt{5}}{2}$$
$$x = \frac{1}{2} \pm \frac{\sqrt{5}}{2}$$

Either $x \approx 1.618$ or $x \approx -0.618$.

In Exercises 42 to 48, find the set of inputs that satisfy each inequality. Use any method to find the x-intercepts. Round to the nearest thousandth.

42. $x^2 + 6x - 8 \le 0$

43. $x^2 + 6x - 7 \ge 0$

44. $1 - 4x^2 \le 0$

45. $2x^2 - 5x - 3 > 0$

46. $3x^2 + 5x - 3 > 19$

47. $-4x^2 + 5x + 2 < -7$

48. From their truck, an inspection team can inspect suspension bridge cable within 20 feet of the road surface. The equation of the parabolic suspension cable is $y = 0.0375(x - 20)^2 + 10$. For what horizontal interval on the 40-foot-long bridge can the inspection be accomplished from the truck?

In Exercises 49 to 51, indicate true or false.

49. The vertex of the graph of a quadratic function is the location of the maximum or minimum y-value.

50. The x-coordinate of the vertex of the graph of a quadratic function is halfway between the x-intercepts of the graph.

51. The vertex of the parabola lies on the axis of symmetry of the graph.

52. Fill in the blank: The solutions to the quadratic equation, $f(x) = 0$, give the _____ on the graph.

Exercises 53 and 54 involve evaluating polynomials. A uniformly loaded cable is parabolic in shape. The length of the cable, L, is approximated by

$$L = a\left|1 + \frac{8}{3}\left(\frac{d}{a}\right)^2 - \frac{32}{5}\left(\frac{d}{a}\right)^4 + \frac{256}{7}\left(\frac{d}{a}\right)^6 - \cdots\right|$$

where d is the sag in feet or meters and a is the span across the parabola in feet or meters. Temperature variations are ignored. Round answers to the nearest ten thousandth.

53. Suppose the span is 500 feet and the sag is 30 feet.

    a. Find the length of the cable when two terms are used inside the absolute value.

    b. Find the length of the cable when three terms are used.

    c. Find the length of the cable when four terms are used.

54. Suppose the span is 200 meters and the sag is 10 meters.

    a. Find the length of the cable when two terms are used inside the absolute value.

    b. Find the length of the cable when three terms are used.

    c. Find the length of the cable when four terms are used.

55. Match each of the descriptions on page 238 with one of Graphs 1 to 4.

Graph 1:

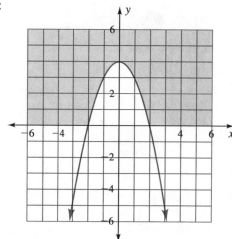

Graph 2:

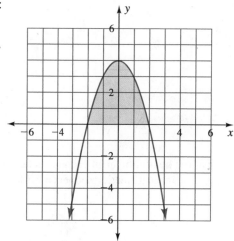

Graph 3:

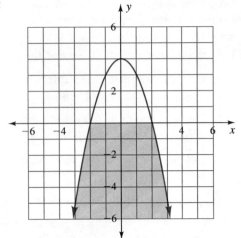

Graph 4:

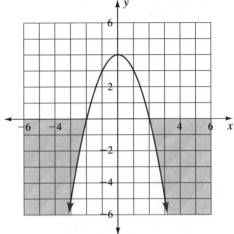

a. $y < 0, y < -x^2 + 4$ **b.** $y > 0, y > -x^2 + 4$

c. $y < 0, y > -x^2 + 4$ **d.** $y > 0, y < -x^2 + 4$

CHAPTER ③ TEST

1. State the quadratic formula for solving $ax^2 + bx + c = 0$.

2. Multiply: $(3x - 4)(4x - 3)$.

3. Multiply: $(x - 3)(x^2 + 3x + 9)$.

4. Factor $2x^2 - 7x + 6$.

5. Solve $0.04x^2 - 169 = 0$ for x.

6. Simplify $\sqrt{98}$.

7. Simplify $\dfrac{4 - 2\sqrt{6}}{2}$.

8. Solve for x:

a.

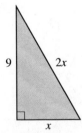

b.

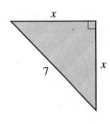

9. Give the next number in each sequence. Identify which are linear (arithmetic) sequences. Identify which are quadratic.

a. $-6, 2, 12, 24, 38$ **b.** $23, 15, 7, -1, -9$

c. $2, 8, 14, 20, 26$ **d.** $25, 17, 10, 4, -1$

10. Solve for x: $|x - 2| = 1$.

11. Below is a table for $y = x^2 + 5x - 6$. Use the table and reasoning to find *all* the solutions to the following equations. Show clearly how you use the table and any numbers you add to it.

x	y
-8	18
-6	0
-4	-10
-2	-12
0	-6
2	8
4	30

a. $x^2 + 5x - 6 = -10$

b. $x^2 + 5x - 6 = 8$

c. $x^2 + 5x - 6 = 0$

d. $x^2 + 5x - 6 = -20$

e. Find the equation for the axis of symmetry.

f. What are the coordinates of the vertex?

12. Solve the following inequalities.

    a. $x^2 + 5x - 6 \geq 0$

    b. $x^2 + 5x - 6 \leq 0$

    c. $x^2 + 5x - 6 > 0$

13. Solve $2x^2 - 7x + 6 = 0$ by factoring.

14. Solve $2x^2 - 7x + 6 = 0$ with the quadratic formula. Show all your steps.

15. a. Make a table and a graph for the annual salary A if r is the percent raise per year for 2 years and P is the starting salary in the formula $A = P(1 + r)^2$. Suppose the starting salary is \$20,000. Let r be 0 to 10%.

    b. For what percent raise will the salary be \$22,898 after 2 years?

c. Extend your graph; for what percent raise will the salary be \$30,000?

16. The velocity (speed) of a falling object, without air resistance, is $v = \sqrt{2gs}$. Solve for s, the distance fallen.

17. A 10-year-old cuts across an empty square city block rather than taking the sidewalk around the block. If she takes 212 steps to cut diagonally across the block, approximately how many steps would she have to take if she stayed on the sidewalk?

18. Compare the solutions to $\sqrt{x^2} = 2$ and $x = \sqrt{4}$. Include a graph with your explanation.

4

Quadratic Functions: Special Topics

Figure 1

The water jets sent up by the fireboat in Figure 1 form several different parabolas. The parabolas may be described by quadratic functions of the form $f(x) = ax^2 + bx + c$. After we review special products of polynomials and their graphs in Section 4.0, we build quadratic equations from data in Section 4.1. In Section 4.2, we investigate the role of a, b, and c in controlling the shape of a parabola such as the water jet. Solving equations with no real-number solution is introduced in Section 4.3. We continue with graphs and translations of the vertex as we examine the vertex form of the quadratic equation in Section 4.4 and its relation to the maximum and minimum values in Section 4.5.

Special Products of Binomials and Higher Order Polynomials and Their Graphs

OBJECTIVES

- Factor perfect square trinomials.
- Factor differences of squares.
- Identify graphs related to squares of binomials and differences of squares.
- Factor sums and differences of cubes.
- Solve higher order polynomial equations with graphs.

WARM-UP

Multiply these expressions

1. $(x - 3)(x - 3)$
2. $(x - 3)(x + 3)$
3. $(x + 3)(x + 3)$
4. $(x - 3)(x^2 + 3x + 9)$
5. $(x - 3)(x^2 - 6x + 9)$
6. $(a + b)(a - b)$
7. $(a - b)(a - b)$
8. $(a + b)(a + b)$
9. $(a + b)(a^2 - ab + b^2)$
10. $(a + b)(a^2 + 2ab + b^2)$

I N THIS SECTION, we review two special products from multiplication of binomials: squares of binomials and differences of squares. We examine the graphs of the expressions, and then we solve higher order polynomial equations with graphs.

Special Products

SQUARES OF BINOMIALS   The products in the Warm-up contain several patterns. One of the patterns is the perfect square trinomial, obtained by squaring a binomial.

As we saw in Section 3.4, the **square of a binomial** (also known as a **binomial square**) is

$$(a + b)(a + b) \quad \text{or} \quad (a + b)^2$$

We may interpret the square of a binomial as the area of a square with sides $a + b$, as shown in Figure 2. The product $(a + b)^2$ simplifies to $a^2 + 2ab + b^2$, a **perfect square trinomial.** The middle term of $a^2 + 2ab + b^2$, $2ab$, is the sum of the diagonal terms in the square, $ab + ab$.

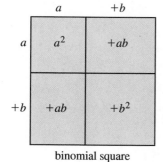

binomial square

Figure 2

EXAMPLE **1**   Identifying squares of binomials   Which of the exercises in the Warm-up contain squares of binomials?

Solution   See the Answer Box.

In Example 2, we solve quadratic equations based on the squares of binomials and compare the graphs to that of $y = x^2$ (Figure 3).

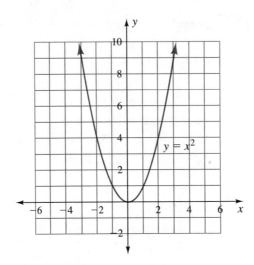

Figure 3

EXAMPLE 2 Exploring solutions and graphs Solve each quadratic equation by factoring and by graphing. Describe the relationship between the solutions and the graph.

a. $x^2 + 6x + 9 = 0$ **b.** $x^2 - 6x + 9 = 0$

Solution **a.**

$x^2 + 6x + 9 = 0$	Factor.
$(x + 3)(x + 3) = (x + 3)^2 = 0$	Apply the zero product rule.
Either $x + 3 = 0$ or $x + 3 = 0$	Solve the factor equations.
$x = -3$ or $x = -3$	

There is one solution, $x = -3$. The graph of $y = x^2 + 6x + 9$ is shown in Figure 4. The solution to $x^2 + 6x + 9 = 0$ is the x-intercept point, $(-3, 0)$. The solution is also the vertex of the graph. The graph of $y = (x + 3)^2$ is shifted 3 units to the left of that of $y = x^2$.

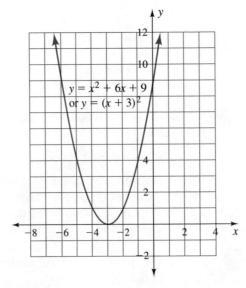

Figure 4

b. $x^2 - 6x + 9 = 0$ Factor.

           $(x - 3)(x - 3) = (x - 3)^2 = 0$ Apply the zero product rule.

Either $x - 3 = 0$ or $x - 3 = 0$ Solve the factor equations.

           $x = 3$ or $x = 3$

There is one solution, $x = 3$. The graph of $y = x^2 - 6x + 9$ is shown in Figure 5. The solution to $x^2 - 6x + 9 = 0$ is the x-intercept point, $(3, 0)$. The solution is also the vertex of the graph. The graph of $y = (x - 3)^2$ is shifted 3 units to the right of that of $y = x^2$.

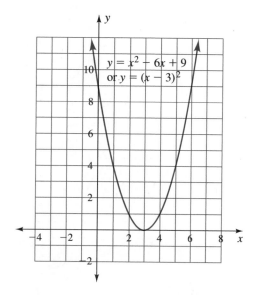

Figure 5   ●

Think about it: How can we tell, from the equations, the location of the y-intercept?

Each of the equations in Example 2 contains a binomial square. Although there is only one number solving each original equation, we say the solution is a **double root**. *When an $(x - r)$ factor appears twice, the solution to $x - r = 0$ is* a **double root.** Each graph is a **horizontal shift,** *a movement left or right,* from the graph of $y = x^2$.

Graphs and Solutions of Equations with Binomial Squares

> The graph of $y = (x - r)^2$ has its vertex on the x-axis at $(r, 0)$.
>
> The graph of $y = (x - r)^2$ is shifted horizontally r units from that of $y = x^2$.
>
> The solution to $(x - r)^2 = 0$ is $x = r$, a double root.

EXAMPLE **3**   Solving equations containing squares of binomials

a. Solve $x^2 - 2x + 1 = 0$ by factoring.

b. Predict the location of the graph of $y = x^2 - 2x + 1$ relative to that of $y = x^2$, its vertex, and the x- and y-intercepts.

Solution   **a.** $x^2 - 2x + 1 = 0$ Factor.

$(x - 1)(x - 1) = (x - 1)^2 = 0$ Apply the zero product rule.

Either $x - 1 = 0$ or $x - 1 = 0$ Solve the factor equations.

$x = 1$ or $x = 1$

The one solution is a double root, $x = 1$, shown as the x-intercept in Figure 6.

b. The graph of $y = (x - 1)^2$ is shifted 1 unit to the right of that of $y = x^2$. The vertex is at $(1, 0)$ and is the only x-intercept point. The y-intercept point is $(0, 1)$.

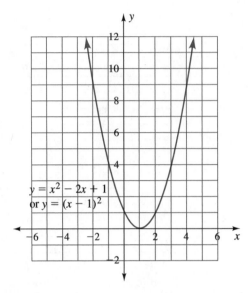

$y = x^2 - 2x + 1$
or $y = (x - 1)^2$

Figure 6   ●

As mentioned, if $y = (x - r)^2$, the vertex of the parabola will be on the x-axis at $x = r$. The graph of $y = (x - r)^2$ will be shifted r units from that of $y = x^2$. Describing the direction of this shift, whether to the left or to the right, is the object of Exercises 35 to 38 at the end of this section.

DIFFERENCES OF SQUARES   The product $(a + b)(a - b) = a^2 - b^2$ is called a **difference of squares.**

EXAMPLE **4**   Identifying differences of squares   Which of the Warm-up exercises contain differences of squares?

Solution   See the Answer Box.   ●

In Example 5, we examine solutions to equations containing differences of squares and their graphs.

EXAMPLE **5**   Exploring solutions and graphs   Solve these quadratic equations by factoring and graphing. Describe the relationship between the solutions to parts a and b and the graphs of $y = x^2 - 9$ and $y = x^2 - 1$, respectively. Describe the positions of the graphs relative to that of $y = x^2$.

a. $x^2 - 9 = 0$ **b.** $x^2 - 1 = 0$

Solution   **a.** $x^2 - 9 = 0$ Factor.

$(x + 3)(x - 3) = 0$ Apply the zero product rule.

Either $x + 3 = 0$ or $x - 3 = 0$ Solve the factor equations.

$x = -3$ or $x = 3$

There are two solutions, $x = -3$ and $x = 3$. The graph of $y = x^2 - 9$ is shown in Figure 7. The solutions are the x-intercept points, $(-3, 0)$ and $(3, 0)$. The vertex of the graph is on the y-axis. The graph is shifted down 9 units from that of $y = x^2$.

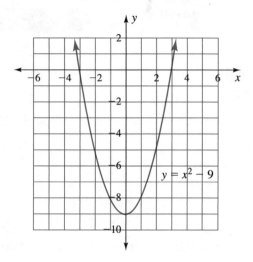

Figure 7

b.
$$x^2 - 1 = 0 \qquad \text{Factor.}$$
$$(x + 1)(x - 1) = 0 \qquad \text{Apply the zero product rule.}$$
Either $\quad x + 1 = 0 \quad$ or $\quad x - 1 = 0 \qquad$ Solve the factor equations.
$$x = -1 \quad \text{or} \qquad x = 1$$

There are two solutions, $x = -1$ and $x = 1$. The graph of $y = x^2 - 1$ is shown in Figure 8. The solutions are the x-intercept points, $(-1, 0)$ and $(1, 0)$. The vertex of the graph is on the y-axis. The graph is shifted down 1 unit from that of $y = x^2$.

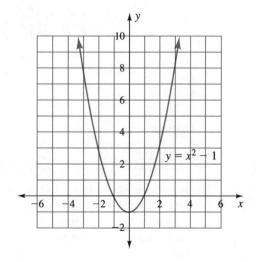

Figure 8

The graphs in Example 5 are **vertical shifts,** *movements up or down* from the graph of $y = x^2$.

Graphs and Solutions of Equations
with Differences of Squares

For c, a positive number,

The graph of $y = x^2 - c^2$ has its vertex on the y-axis at $(0, -c^2)$.

The graph of $y = x^2 - c^2$ is shifted vertically c^2 units from that of $y = x^2$.

The two solutions to $x^2 - c^2 = 0$ are $x = c$ and $x = -c$.

EXAMPLE **6**

Solving equations containing differences of squares

a. Solve $x^2 - 4 = 0$ by factoring.

b. Predict the location of the graph of $y = x^2 - 4$, its vertex, and the x- and y-intercepts.

Solution **a.**

$$x^2 - 4 = 0 \qquad \text{Factor.}$$
$$(x - 2)(x + 2) = 0 \qquad \text{Apply the zero product rule.}$$
$$\text{Either} \quad x - 2 = 0 \quad \text{or} \quad x + 2 = 0 \qquad \text{Solve the factor equations.}$$
$$x = 2 \quad \text{or} \qquad x = -2$$

There are two solutions, $x = 2$ and $x = -2$, shown as the x-intercepts in Figure 9.

b. The graph of $y = x^2 - 4$ is a parabola with its vertex 4 units down the y-axis from the origin. The x-intercept points of $y = x^2 - 4$ are $(-2, 0)$ and $(2, 0)$. The y-intercept point is also the vertex, $(0, -4)$.

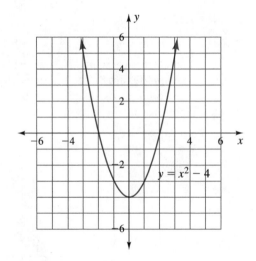

Figure 9 ●

Polynomials of Degree Greater Than 2

The main topic of Chapter 3 and this chapter is quadratic equations and quadratic expressions. However, higher power equations and expressions are included because we need to keep in mind that they all are part of the larger family of polynomial equations. Many of the properties of quadratic equations hold true for all polynomial equations.

CUBIC EXPRESSIONS When *the highest exponent in a polynomial is 3,* we say the polynomial is a **cubic expression.** Familiarity with the cubes of numbers will be helpful in the following discussion, so Example 7 provides review in evaluating a cubic function.

EXAMPLE 7   Graphing a cubic function   Make a table and graph for $f(x) = x^3$, with integer inputs in the interval from -5 to 5.

Solution   The table appears in Table 1 and the graph in Figure 10.

Student Note: Because $x^3 = x \cdot x \cdot x$, $x = 0$ is a triple root for the equation $x^3 = 0$.

Input x	Output $f(x) = x^3$
-5	-125
-4	-64
-3	-27
-2	-8
-1	-1
0	0
1	1
2	8
3	27
4	64
5	125

Table 1

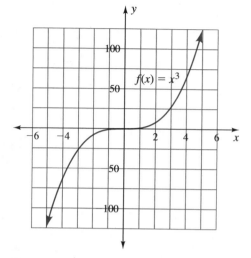

Figure 10

The scale on the y-axis in Figure 10 is large because the cubes of numbers increase rapidly. In the graphs that follow, we stay close to the origin, so the outputs are smaller.

EXAMPLE 8   Graphing cubic polynomials   Multiply these expressions, and graph their products. Compare their graphical positions to that of $y = x^3$.

a. $(x - 2)(x^2 + 2x + 4)$ **b.** $(x + 1)(x^2 - x + 1)$

Solution   **a.**

Multiply	x^2	$+2x$	$+4$
x	x^3	$+2x^2$	$+4x$
-2	$-2x^2$	$-4x$	-8

b.

Multiply	x^2	$-x$	$+1$
x	x^3	$-x^2$	$+x$
$+1$	$+x^2$	$-x$	$+1$

The products are

$$(x - 2)(x^2 + 2x + 4) = x^3 - 8 \quad \text{and} \quad (x + 1)(x^2 - x + 1) = x^3 + 1$$

The graphs of $y = x^3 - 8$ and $y = x^3 + 1$ are shown in Figures 11 and 12, respectively. The graph of $y = x^3 - 8$ is shifted 8 units below that of $y = x^3$. The graph of $y = x^3 + 1$ is shifted 1 unit above that of $y = x^3$.

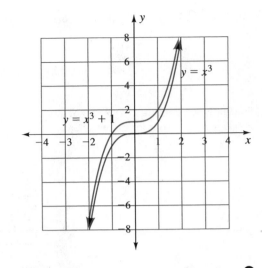

Figure 11

Figure 12

 The graphs of the functions $f(x) = x^3 \pm c$ are vertical shifts of the graph of $f(x) = x^3$. Verify the vertical shifts in Example 8 on a graphing calculator by using the up and down cursor keys to compare the outputs.

SUMS AND DIFFERENCES OF CUBES   The expression $a^3 + b^3$ is called a **sum of cubes,** and the expression $a^3 - b^3$ is called a **difference of cubes.** The factors creating the sums and differences of cubes may be summarized with these products.

Sum and Difference of Cubes

The product $(x + y)(x^2 - xy + y^2)$ gives the sum of cubes, $x^3 + y^3$.

The product $(x - y)(x^2 + xy + y^2)$ gives the difference of cubes, $x^3 - y^3$.

If you have difficulty remembering the second factor for a sum or difference of cubes, you can use a table to find it. Observe that the tables in Example 9, like the products in Example 8, contain two rows and three columns.

EXAMPLE **9**   Factoring cubic expressions   Use a table to find the second factor in each of these expressions.

a. $x^3 - 27 = (x - 3)(\underline{\quad\quad})$
b. $a^3 + b^3 = (a + b)(\underline{\quad\quad})$

Solution   **a.** We enter the factor $x - 3$ on the left side of the table, and we enter x^3 and -27 as the first and last entries inside the table.

Factor			
x	x^3		
-3			-27

Then we find the remaining factors and table entries. We start with the factor above x^3 and then the table entry below x^3. Note that the diagonal terms must add to zero because the expression $x^3 - 27$ contains no middle terms.

Factor	x^2	$+3x$	$+9$
x	x^3	$+3x^2$	$+9x$
-3	$-3x^2$	$-9x$	-27

The factorization is $x^3 - 27 = (x - 3)(x^2 + 3x + 9)$.

b. We enter the factor $a + b$ on the left side of the table, and we enter a^3 and b^3 as the first and last entries inside the table.

Multiply			
a	a^3		
$+b$			$+b^3$

Then we find the remaining factors and table entries. We start with the factor above a^3 and then the table entry below a^3. Note that the diagonal terms must add to zero because the expression $a^3 + b^3$ contains no middle terms.

Multiply	a^2	$-ab$	$+b^2$
a	a^3	$-a^2b$	$+ab^2$
$+b$	$+a^2b$	$-ab^2$	$+b^3$

The factorization is $a^3 + b^3 = (a + b)(a^2 - ab + b^2)$.   ●

GENERAL POLYNOMIAL EQUATIONS   Another name for a sum or difference of cubes is *polynomial of degree 3* or *third-degree polynomial,* because the term with the highest exponent has an exponent of 3. The **degree** of a polynomial is given by *the exponent on the term with the highest exponent.* For a quadratic expression, the degree is 2 because the x^2 is the term with the highest exponent. For the **general polynomial equation** of the form

$$y = a_n x^n + a_{n-1}x^{n-1} + a_{n-2}x^{n-2} + \cdots + a_1 x^1 + a_0$$

the degree is n, $n \geq 0$.

In Example 10, we explore the number of solutions for an equation of degree 3.

EXAMPLE **10**

Finding solutions   Use Figure 13 to solve these third-degree equations. Tell how many real-number solutions there are to each equation, and explain why.

a. $x^3 - 3x^2 = 6$ **b.** $x^3 - 3x^2 = 0$ **c.** $x^3 - 3x^2 = -2$

d. $x^3 - 3x^2 = -4$ **e.** $x^3 - 3x^2 = -6$

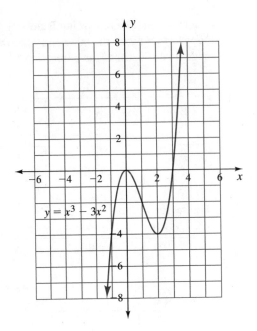

Figure 13

Solution   Figure 14 shows the solutions to the five equations.

   a. There is one solution, as the horizontal line $y = 6$ intersects the graph of $y = x^3 - 3x^2$ once: at $x \approx 3.5$.

   b. There are two solutions, as the x-axis, $y = 0$, intersects the graph twice: at $x = 0$ and $x = 3$.

   c. There are three solutions, as the line $y = -2$ intersects the graph three times: at $x \approx -0.7$, $x = 1$, and $x \approx 2.7$.

   d. There are two solutions, as the line $y = -4$ intersects the graph twice: at $x = -1$ and $x = 2$.

   e. There is one solution, as the line $y = -6$ intersects the graph once: at $x \approx -1.2$.

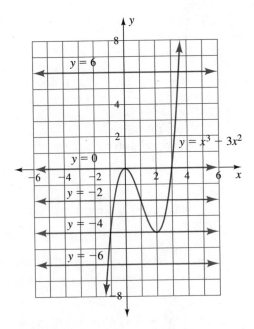

Figure 14

Recall that quadratic equations, which are of degree 2, have 0, 1, or 2 solutions. In Example 10, the cubic equations, or third-degree polynomial equations, had at most 3 real-number solutions, the same number as their degree.

Degree and Number of Solutions

> The degree of a polynomial equation indicates the maximum number of real-number solutions.

ANSWER BOX

Warm-up: **1.** $x^2 - 6x + 9$ **2.** $x^2 - 9$ **3.** $x^2 + 6x + 9$ **4.** $x^3 - 27$
5. $x^3 - 9x^2 + 27x - 27$ **6.** $a^2 - b^2$ **7.** $a^2 - 2ab + b^2$
8. $a^2 + 2ab + b^2$ **9.** $a^3 + b^3$ **10.** $a^3 + 3a^2b + 3ab^2 + b^3$
Example 1: Exercises 1, 3, 7, and 8 **Think about it:** When we substitute $x = 0$ into the equation, the result is $y = c$. The point $(0, c)$ is the y-intercept point. **Example 4:** Exercises 2 and 6

EXERCISES 4.0

Multiply the expressions in Exercises 1 to 10.

1. $(x - 3)(x + 4)$

2. $(2x + 1)(x - 12)$

3. $(2x + 3)(x - 4)$

4. $(x - 3)(x - 3)$

5. $(x + 3)(x + 3)$

6. $(x - 8)(x + 8)$

7. $(x + 6)(x - 6)$

8. $(x - 4)(x - 4)$

9. $(2x - 5)^2$

10. $(x + 5)^2$

11. Which of Exercises 1 to 10 contain binomial squares?

12. Using a graphing calculator, graph the expression in Exercise 10 under $\boxed{Y=}$ (suggested window: $[-10, 10]$ for x, $[-20, 20]$ for y). Compare the position of the graph with that of $y = x^2$. What is the x-intercept point?

13. Describe how you would find the graph of $y = (x - 4)^2$ given the graph of $y = x^2$.

14. Describe how to find the x-intercepts of $y = (2x - 5)^2$ from the equation.

In Exercises 15 to 24, fill in numbers that make the expressions equal to binomial squares.

15. $x^2 + \underline{\hspace{0.5cm}} x + 16 = (x + 4)^2$

16. $x^2 + \underline{\hspace{0.5cm}} x + 9 = (x + 3)^2$

17. $x^2 - \underline{\hspace{0.5cm}} x + 49 = (x - \underline{\hspace{0.5cm}})^2$

18. $x^2 - \underline{\hspace{0.5cm}} x + \underline{\hspace{0.5cm}} = (x - \frac{1}{2})^2$

19. $x^2 - 5x + \underline{\hspace{0.5cm}} = (x - 2.5)^2$

20. $x^2 - 3x + \underline{\hspace{0.5cm}} = (x - 1\frac{1}{2})^2$

21. $x^2 + 7x + \underline{\hspace{0.5cm}} = (x + \underline{\hspace{0.5cm}})^2$

22. $x^2 + 9x + \underline{\hspace{0.5cm}} = (x + \underline{\hspace{0.5cm}})^2$

23. $x^2 - 24x + \underline{\hspace{0.5cm}} = (x - \underline{\hspace{0.5cm}})^2$

24. $x^2 + bx + \underline{\hspace{0.5cm}} = (x + \underline{\hspace{0.5cm}})^2$

Multiply the expressions in Exercises 25 to 27.

25. $(x - 4)(x + 4)$

26. $(x - 5)(x + 5)$

27. $(2x - 3)(2x + 3)$

28. Which of Exercises 1 to 10 have answers that are differences of squares?

29. Using a graphing calculator, graph the expression in Exercise 25 under $\boxed{Y=}$ (suggested window: $[-10, 10]$ for x, $[-25, 25]$ for y). Compare the position of the graph with that of $y = x^2$. Describe the x-intercept point.

30. Describe how you would find the graph of $y = (x - 8)(x + 8)$ given the graph of $y = x^2$.

Factor the differences of squares in Exercises 31 to 34.

31. $x^2 - 144$

32. $25x^2 - 64$

33. $0.25x^2 - 0.01$

34. $0.01x^2 - 0.36$

In Exercises 35 to 38, consider these results: If $y = (x - 3)^2$, the vertex of the parabola will be on the x-axis at $x = 3$. If $y = (x + 3)^2 = [x - (-3)]^2$, the vertex will be on the x-axis at $x = -3$.

35. a. Describe how to find the vertex for $y = (x - r)^2$.

     b. Apply your rule to $y = (x + 1)^2$.

     c. Apply your rule to $y = (x - 4)^2$.

36. a. Describe how to find the vertex in terms of s and t for $y = (sx - t)^2$.

 b. Apply your rule to $y = (2x + 1)^2$.

 c. Apply your rule to $y = (3x - 4)^2$.

37. a. Describe the shift of $y = x^2$ as left or right in terms of the value of r in $y = (x - r)^2$.

 b. Apply your rule to $y = (x + 2)^2$.

 c. Apply your rule to $y = (x - 1)^2$.

38. a. Describe the shift of $y = x^2$ as left or right in terms of the values of s and t in $y = (sx - t)^2$, $s > 0$.

 b. Apply your rule to $y = (2x - 1)^2$.

 c. Apply your rule to $y = (2x + 3)^2$.

Multiply the expressions in Exercises 39 to 44.

39. $(x + 1)(x^2 - x + 1)$

40. $(x - 1)(x^2 - 2x + 1)$

41. $(x - 3)(x^2 - 6x + 9)$

42. $(x + 2)(x^2 + 4x + 4)$

43. $(x + 1)(x^2 - 2x + 1)$

44. $(x - 1)(x^2 + x + 1)$

45. The expressions $a^3 + b^3$ and $a^3 - b^3$ are the sum and difference of cubes, respectively. Which of Exercises 39 to 44 contain a sum or difference of cubes?

Use the tables in Exercises 46 and 47 to answer the questions.

46. $x^3 + 64$ equals $(x + 4)$ multiplied by what trinomial?

Factor			
x	x^3		
$+4$			$+64$

47. $x^3 - 125$ equals $(x - 5)$ multiplied by what trinomial?

Factor			
x	x^3		
-5			-125

Factor the expressions in Exercises 48 and 49.

48. a. $x^3 + 27$ **b.** $x^3 + 8$ **c.** $x^3 - 64$

49. a. $x^3 + 1$ **b.** $x^3 + 125$ **c.** $x^3 - 1000$

In Exercises 50 to 53,
a. Identify the expression in the second set of parentheses as a difference of squares (ds), a sum of squares (ss), a perfect square trinomial (pst), or none of these.
b. Factor the expression in the second set of parentheses.

c. Solve the equation.
d. From your solutions, predict where the graph of the left side crosses the x-axis and how many times.

50. $(x - 3)(x^2 - 6x + 9) = 0$

51. $(x + 1)(x^2 + 2x + 1) = 0$

52. $(x - 2)(x^2 - 4) = 0$

53. $(x + 1)(x^2 - 1) = 0$

54. Compare $y = (x + 2)^3$ with $y = x^3 + 8$ on a graphing calculator. Are they the same? Compare each to $y = x^3$.

55. Compare $y = (x + 1)^3$ with $y = x^3 + 1$ on a graphing calculator. Are they the same? Compare each to $y = x^3$.

56. Match the terms with their definitions, and give an example of each. Choose from the following:
square of a binomial, sum of cubes,
perfect square trinomial, difference of squares,
difference of cubes.

 a. The subtraction of one cubic expression from another

 b. The square of a two-term expression

 c. The subtraction of one squared expression from another

 d. The addition of one cube to another

 e. A square that is also a three-term polynomial

What is the greatest possible number of real-number solutions to an equation $f(x) = c$ for each function in Exercises 57 to 60?

57. $f(x) = x^2 - 4x^5$ **58.** $f(x) = 3x + 2x^2 - x^4$

59. $f(x) = 2x^3 - 2x^2$ **60.** $f(x) = 4x^5 - x^3 + x$

In Exercises 61 to 64, write an output $y = n$ such that $f(x) = n$ has the listed number of real-number solutions, if possible. It may be helpful to draw the horizontal line representing each output $y = n$.
a. no solutions   b. 1 solution   c. 2 solutions
d. 3 solutions   e. 4 solutions

61. $f(x) = x^3 - 4x^2 + 4x - 1$

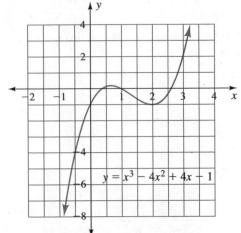
$y = x^3 - 4x^2 + 4x - 1$

62. $f(x) = -x^3 - 2x^2 + x - 1$

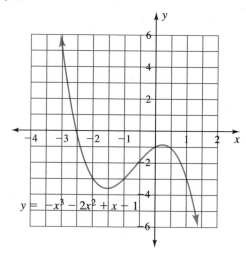

$y = -x^3 - 2x^2 + x - 1$

63. $f(x) = -x^4 - 3x^3 + x - 1$

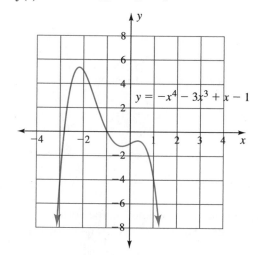

$y = -x^4 - 3x^3 + x - 1$

64. $f(x) = x^4 - 3x^2 + x - 1$

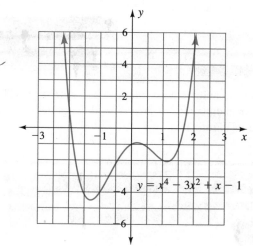

$y = x^4 - 3x^2 + x - 1$

65. Graph and trace to find the solution set to $f(x) = 0$ in Exercise 61. Round to the nearest thousandth.

66. Graph and trace to find the solution set to $f(x) = 0$ in Exercise 62. Round to the nearest thousandth.

67. Explain what feature of the graphs of $y = x^3 + 1$ and $y = x^2 + 1$ shows why $x^3 + 1 = 0$ has a real-number solution while $x^2 + 1 = 0$ does not.

68. Explain what feature of the graphs of $y = x^2 + 4$ and $y = x^2 - 4$ shows why $x^2 - 4 = 0$ has two real-number solutions while $x^2 + 4 = 0$ has none.

69. State and solve four equations from the figure in Exercise 63.

Projects

70. *Cubic Function Differences*

a. Use a graphing calculator table or spreadsheet to evaluate these cubic functions for $x = 1, 2, 3, 4, 5$. List outputs in a sequence. Take the difference between terms, repeat as necessary, and discuss patterns in the differences.

$$y = x^3$$
$$y = 2x^3 + x$$
$$y = 0.5x^3 + x^2$$

b. Evaluate these fourth-degree functions for $x = 1, 2, 3, 4, 5, 6$. List outputs in a sequence. Take the difference between terms, repeat as necessary, and discuss patterns in the differences.

$$y = x^4$$
$$y = 2x^4 + x^2$$
$$y = 0.5x^4 + x^3$$

c. How are the patterns in parts a and b like and unlike linear and quadratic sequences? Predict a relationship between the nth row of differences and a polynomial equation of degree n.

d. Use your observations to find which two of these three sequences are from cubic polynomials.

3, 10, 29, 66, 127

3, 9, 27, 81, 243

−4, 3, 22, 59, 120

71. *Rotating Liquids Research*

a. Fill a jar slightly less than half full of water. Place the jar at the center of an old phonograph turntable, a pottery wheel, or a revolving tray for serving food. What shape does the water form as the base is turned? How does the shape change as the speed of rotation changes?

b. In recent years, the rotating concept was adapted to vats of molten glass, which were cooled as they rotated in order to create large reflecting telescope lenses that required little grinding. The same idea is also behind using rotating vats of mercury to form liquid telescope reflectors. Research and write a report on one of these techniques.

4.1   Modeling Quadratic Functions

OBJECTIVES

- Use x-intercepts and one other point to find a quadratic function.
- Use a table and differences to find a quadratic function.
- Use calculator regression to find a quadratic function.

WARM-UP

What is the rule for each of these quadratic sequences? (*Hint:* See Examples 1 and 2, Section 3.2.)

1. $1, 4, 9, 16, 25, \ldots$

2. $2, 6, 12, 20, 30, \ldots$

3. $1, 3, 6, 10, 15, \ldots$

Simplify these expressions.

4. $0.5x(x + 1)$ **5.** $x(x + 1)$

6. $(x + 1)(x + 2)$ **7.** $2(x - 1)^2 - 1$

Answer the following questions for the quadratic function $f(x) = ax^2 + bx + c$.

8. What is $f(0)$?

9. What is $f(0)$ on the graph of $f(x)$?

10. What is $f(1)$?

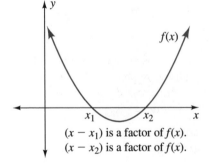

$(x - x_1)$ is a factor of $f(x)$.
$(x - x_2)$ is a factor of $f(x)$.

Figure 15

IN THIS SECTION, we build quadratic equations of the form $y = ax^2 + bx + c$ from intercepts and a point, from sequences, and with calculator regression.

Building an Equation from x-intercepts

To build an equation, we need to find its parameters. A **parameter** is *a letter representing a number that changes the orientation and position of the graph.* The slope m and y-intercept b in the linear equation $y = mx + b$ are parameters. The letters a, b, and c in $y = ax^2 + bx + c$ are all parameters.

Our first method of building an equation is with solutions or x-intercepts. The relationship among the solutions to an equation, its factors, and the graph is given by the factor theorem, illustrated in Figure 15.

Factor Theorem

Suppose that $f(x)$ is a polynomial function and x_1 is a constant. If $x - x_1$ is a factor of the polynomial function, then $x = x_1$ is a solution to the equation $f(x) = 0$. Also, if $x = x_1$ is a solution to the equation $f(x) = 0$, then $x - x_1$ is a factor of $f(x)$.

If we know the x-intercepts (say, x_1 and x_2), we can write the quadratic equation as $y = a(x - x_1)(x - x_2)$.

In Example 1, we use intercepts to derive the equation for the parabolic arches in the museum entryway from Chapter 3.

EXAMPLE **1**

Finding a quadratic equation: museum accessibility, again   The parabolic arches for the art museum, shown again in Figure 16, must be built so that A is $(0, 0)$, B is $(10, 12)$, and C is $(20, 0)$. The dimensions are in feet. Find the quadratic equation satisfying the given coordinates.

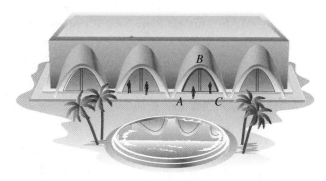

Figure 16

Solution
$$y = a(x - x_1)(x - x_2) \quad\quad x_1 \text{ and } x_2 \text{ are } x\text{-intercepts.}$$
$$y = a(x - 0)(x - 20)$$
$$y = a(x^2 - 20x)$$

We substitute the point $(10, 12)$ for x and y in the equation and simplify:

$$12 = -100a$$

Then we solve for a:

$$a = -\frac{12}{100} = -0.12$$

The equation is $y = -0.12(x^2 - 20x)$, or $y = -0.12x^2 + 2.4x$.   ●

In Example 1, the placement of the origin at point A made it possible for points A and C to be x-intercept points. Well-reasoned placement of axes in a problem situation can simplify both calculating information and finding equations.

Building an Equation from a Sequence

In order to build a quadratic equation from a sequence, we must make some observations about the differences between the terms in the sequence. In Examples 2 and 3, we find first and second differences for three sequences and review the rules for the sequences.

EXAMPLE **2**

Reviewing sequences and their rules   Find the first and second differences for these sequences. What is the rule for each sequence? See Section 3.2 as needed.

a. 1, 4, 9, 16, 25, 36   **b.** 2, 6, 12, 20, 30, 42

Solution **a.** *Sequence:*   1, 4, 9, 16, 25, 36
                        ⋁ ⋁ ⋁ ⋁ ⋁
        First differences: 3, 5, 7, 9, 11
                        ⋁ ⋁ ⋁ ⋁
        Second differences: 2, 2, 2, 2

The rule is x^2.

b. *Sequence:* 2, 6, 12, 20, 30, 42
∨ ∨ ∨ ∨ ∨
First differences: 4, 6, 8, 10, 12
∨ ∨ ∨ ∨
Second differences: 2, 2, 2, 2

The rule is $x(x + 1)$, or $x^2 + x$.

In Example 3, we look at a sequence related to a sequence in Example 2.

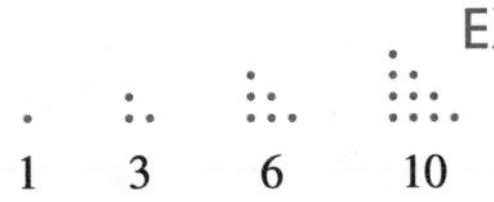

EXAMPLE **3**   Finding an equation from a related function   The first few terms of the sequence 1, 3, 6, 10, 15, 21,...are shown in Figure 17.

a. Find the first and second differences.

b. How is the sequence related to 2, 6, 12, 20, 30, 42,...?

1   3   6   10

Figure 17

Solution   **a.** *Sequence:* 1, 3, 6, 10, 15, 21,...
∨ ∨ ∨ ∨ ∨
First differences: 2, 3, 4, 5, 6,...
∨ ∨ ∨ ∨
Second differences: 1, 1, 1, 1,...

The first differences are consecutive integers. The second differences are 1s. The sequence has a constant second difference, as in earlier examples.

b. Each term of the sequence 1, 3, 6, 10, 15, 21,...is half the corresponding term of the sequence 2, 6, 12, 20, 30, 42,...in Example 2. Likewise, the triangular shapes shown in Figure 17 contain half the dots in the rectangles formed by $x(x + 1)$, as shown in Figure 18.

It is reasonable that the function description is half that of part b in Example 2:

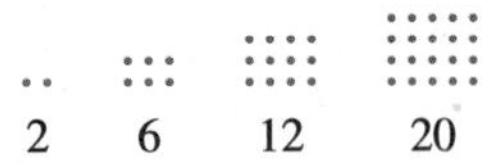

$$f(x) = \tfrac{1}{2}x(x + 1) = \tfrac{1}{2}x^2 + \tfrac{1}{2}x$$

2   6   12   20

Figure 18

FINDING *a*   In Example 4, we find a relationship between the differences and the parameter *a*, the coefficient of x^2.

EXAMPLE **4**   Finding *a*   Use the results from Examples 2 and 3 to find the relationship between the second difference and the sequence or function rule.

Solution   In part a of Example 2, the second difference was 2, and

$$f(x) = 1x^2$$

In part b of Example 2, the second difference was 2, and

$$f(x) = 1x^2 + 1x$$

In Example 3, the second difference was 1, and

$$f(x) = \tfrac{1}{2}x^2 + \tfrac{1}{2}x$$

In each case, the coefficient on the x^2 term, *a*, is half the second difference. The first function has only one term, so the impact of the second difference on the *x* term is not apparent.

From Example 4, we conclude that *the parameter a, the coefficient of x^2, is half the second differences.*

FINDING c When we evaluate $f(x) = ax^2 + bx + c$ for $x = 0$, we obtain

$$f(0) = a(0)^2 + b(0) + c$$
$$f(0) = 0 + 0 + c$$
$$f(0) = c$$

Because the domain for a sequence is the set of natural numbers, $1, 2, 3, \ldots$, the terms of the sequence can be written: $f(1)$ as the first term, $f(2)$ as the second term, $f(3)$ as the third term, and so forth. Although $f(0)$ is not defined for a sequence, the term prior to the first term in a sequence might be considered $f(0)$. Thus, we can find the $f(0)$ term (that is, c) by extending the pattern backwards.

EXAMPLE **5**   Finding c Use differences to find the term prior to the first term in the sequence $-1, 1, 7, 17, 31, 49, \ldots$.

Solution   We write out the sequence, the first differences, and the second differences. After entering another 4 in front of the second differences, we subtract that 4 from 2 to obtain -2 in the first differences. Then we subtract that -2 from -1 to obtain 1 in the sequence.

Sequence: $\quad\quad\quad$ 1, -1, 1, 7, 17, 31, 49, ...
$\quad\quad\quad\quad\quad\quad\quad\quad\quad$ V V V V V V
First differences: $\quad\quad$ -2, 2, 6, 10, 14, 18, ...
$\quad\quad\quad\quad\quad\quad\quad\quad\quad\quad$ V V V V V
Second differences: $\quad$ 4, 4, 4, 4, 4, ...

The term prior to $f(1) = -1$ is $f(0) = 1$. Thus, $c = 1$ for this sequence. ●

Student Note: Project 35 deals with finding the quadratic equation from number patterns in a multiplication table.

The parameter c is the term before the first term, found by working backwards through the second and first differences.

FINDING b We can find parameter b by substituting parameters a and c and any term, such as $f(1)$, from the sequence into $f(x) = ax^2 + bx + c$.

EXAMPLE **6**   Finding b Find b in the rule for the sequence $-1, 1, 7, 17, 31, 49, \ldots$, and then write the rule for the sequence.

Solution   From Example 5, $f(1) = 1$, $c = 1$, and the second difference is 4. The second difference shows that $a = \frac{1}{2}(4) = 2$. We let $x = 1$ in $f(x) = ax^2 + bx + c$:

$$f(1) = a(1)^2 + b(1) + 1 \quad\quad \text{Substitute } f(1) = -1, c = 1, a = 2.$$
$$-1 = 2(1)^2 + b(1) + 1 \quad\quad \text{Simplify.}$$
$$-1 = 2 + b + 1 \quad\quad\quad\quad \text{Solve for } b.$$
$$-4 = b$$

We then substitute a, b, and c into the general quadratic equation:

$$y = 2x^2 - 4x + 1$$

Check: When we enter the equation into $\boxed{\text{Y} =}$ and set up a calculator table with natural numbers as inputs, the table outputs match the sequence. ●

Finding an Equation from a
Sequence: Difference Method

> To find a, b, and c in $f(x) = ax^2 + bx + c$ from a quadratic sequence:
>
> 1. List the sequence, and calculate the first and second differences.
>
> 2. Find a, given that the constant second difference is $2a$.
>
> 3. Work backwards from the second difference row to find c, the $f(0)$ term before the first term of the sequence, $f(1)$.
>
> 4. Let $x = 1$ in $f(x)$. Substitute for the first term, $f(1)$, and coefficients a and c. Solve for b.
>
> 5. Write the equation $y = ax^2 + bx + c$, substituting for a, b, and c.
>
> To check, place the equation in $\boxed{Y=}$ on a graphing calculator, set the natural numbers as inputs on the table feature, and confirm that the sequence is given as table outputs.

EXAMPLE **7**

Finding a quadratic equation from a sequence   Find the rule for the sequence 4, 2, -2, -8, -16, ...

Solution   **Step 1:**   *Sequence:*

$$4,\ 2,\ -2,\ -8,\ -16, \ldots$$
$$\vee \quad \vee \quad \vee \quad \vee$$

First differences: $-2,\ -4,\ -6,\ -8, \ldots$
$$\vee \quad \vee \quad \vee$$

Second differences: $-2,\ -2,\ -2, \ldots$

Step 2:   The constant second difference is -2, so $a = \frac{1}{2}(-2) = -1$.

Step 3:   Working backwards through the differences to find $f(0) = c$, we find that the prior first difference is 0. The $f(0)$ term is then 4, the same as the first term. Thus, $c = 4$.

Step 4:   We know that the first term, $f(1)$, is 4 and that $a = -1$ and $c = 4$. We let $x = 1$ in $f(x)$:

$$f(1) = a(1)^2 + b(1) + c \qquad \text{Let } f(1) = 4, a = -1, c = 4.$$
$$4 = -1(1)^2 + b(1) + 4 \qquad \text{Simplify and solve for } b.$$
$$1 = b$$

Step 5:   The equation is $y = -x^2 + x + 4$.

Check: When we enter the equation into $\boxed{Y=}$ and set up a calculator table with natural numbers as inputs, the table outputs match the sequence.   ●

The difference method has been presented here because it is a numerical approach to finding equations. The problem with using the difference method is that it requires exact data from a sequence generated by a quadratic.

Finding a Quadratic Equation with a Calculator

We now consider a calculator method for finding a quadratic function. The calculator method may be used on approximate or experimental data. We first enter the data into the statistical lists on a graphing calculator. Then we choose the calculator option for **quadratic regression,** or *fitting a quadratic equation.*

Graphing Calculator Technique: Fitting a Quadratic Equation

> Read about regression in your calculator manual.
>
> Enter the data into lists—say L_1 and L_2.
>
> Choose the calculation of quadratic regression.
>
> Run the regression.
>
> Record the values of a, b, and c for $f(x) = ax^2 + bx + c$.

EXAMPLE 8

Fitting an equation with a calculator: falling rock   A small rock is dropped from the top of a 576-foot abandoned mine shaft. (See Figure 19.) Its approximate distance from the bottom each second after release (assuming no air resistance) is given in Table 2. Graph the data and fit an appropriate equation.

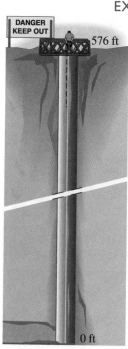

Figure 19

Time after Release (seconds)	Distance from Bottom (feet)
0	576
1	560
2	512
3	432
4	320

Table 2

Solution   The graph in Figure 20 shows the data from Table 2. The graph suggests one side a parabola, so we fit a quadratic equation.

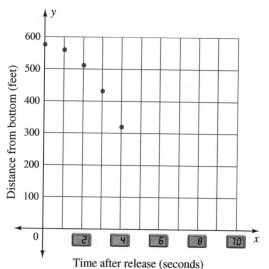

Figure 20

We enter the data into two lists in a graphing calculator. After choosing quadratic regression, we run the regression. The calculator gives $a = -16$, $b = 0$, and $c = 576$. Substituting these values into $f(x) = ax^2 + bx + c$ yields

$$f(x) = -16x^2 + 0x + 576$$

Because our input variable is time, we would usually write the equation as a function of t,

$$f(t) = -16t^2 + 576$$

●

Look at the number of choices for calculating an equation from data on a graphing calculator. Many calculators give ten or more choices of functions. Computer programs may give 50, 200, or more choices of functions. To use technology, you need to have a reasonable idea as to the type of function. Here is a summary of what you have learned thus far, to help you make a selection.

Summary: Modeling Functions

Linear functions (use linear regression):

- The graph of the data lies in a straight line or an approximately straight line.
- The data form a sequence with positive integer inputs, and the first differences are constant.

Quadratic functions (use quadratic regression):

- The graph of the data approximates a parabola, turning up or down.
- The data form a sequence with positive integer inputs, and the second differences are constant.

ANSWER BOX

Warm-up: 1. x^2 **2.** $x(x + 1)$ **3.** $\frac{1}{2}x(x + 1)$ **4.** $0.5x^2 + 0.5x$
5. $x^2 + x$ **6.** $x^2 + 3x + 2$ **7.** $2x^2 - 4x + 1$ **8.** c **9.** y-intercept
10. $a + b + c$

EXERCISES 4.1

In Exercises 1 to 4, find a if the graph of the equation is to pass through the given point.

1. The graph of $y = a(x + 1)(x + 5)$ passes through $(-4, 3)$.

2. The graph of $y = a(x + 1)(x + 5)$ passes through $(-2, 6)$.

3. The graph of $y = a(x + 1)(x + 5)$ passes through $(0, 7.5)$.

4. The graph of $y = a(x + 1)(x + 5)$ passes through $(-2, -3)$.

In Exercises 5 to 16, use the factored equation method to write a quadratic equation passing through the points. Check with a quadratic regression.

5. $(3, 0), (-2, 0), (-3, 12)$

6. $(3, 0), (-2, 0), (0, -6)$

7. $(0, 12), (3, 0), (-2, 0)$

8. $(-2, 0), (1, 18), (3, 0)$

9. $(-3, 0), (5, 0), (3, 6)$

10. $(-2, 0), (-1, 10), (4, 0)$

11. $(-2, 0), (2, -24), (5, 0)$

12. $(-2, 12), (-4, 0), (2, 0)$

13. Parabolic arch: $(10, 0), (30, 0), (20, 12)$

14. Parabolic arch: $(0, 0), (10, 15), (20, 0)$

15. A parabolic suspension bridge cable is to pass through $(-25, 20), (0, 0)$, and $(25, 20)$.

16. A parabolic concrete bridge support is to pass through $(-50, 0), (0, 30)$, and $(50, 0)$. (See the figure, page 261.)

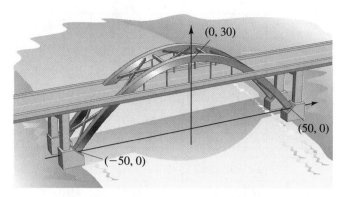

(0, 30)
(50, 0)
(−50, 0)

17. A parabolic entry design has the equation $y = -\frac{3}{40}(x^2 - 50x + 400)$. Return the equation to the form $y = a(x - x_1)(x - x_2)$, and then explain how to find the vertex on the parabola.

18. A parabolic entry design has the equation $y = -0.025x^2 + 2x - 30$. Change the equation to the form $y = a(x - x_1)(x - x_2)$, and then explain how to find the vertex on the parabola.

In Exercises 19 to 32, use first and second differences to find out whether each sequence may be described with a linear function, a quadratic function, or neither. If a linear function may be used, fit an equation, $y = mx + b$. If a quadratic function may be used, fit an equation with the difference method. Check with calculator regression.

19. 7, 16, 27, 40, 55, . . .   **20.** 1, 8, 27, 64, 125, . . .

21. 4, 8, 12, 16, 20, 24, . . .   **22.** 4, 12, 24, 40, 60, . . .

23. 8, 27, 50, 77, 108, . . .   **24.** 12, 23, 34, 45, 56, . . .

25. 5, 12, 21, 32, 45, . . .   **26.** 7, 14, 21, 28, 35, . . .

27. 13, 25, 37, 49, 61, . . .

28. 36, 44, 50, 54, 56, . . .

29. 1, 1, 2, 3, 5, 8, . . .

30. 26, 35, 44, 53, 62, . . .

31. 9, 16, 21, 24, 25, . . .

32. 3, 4, 7, 11, 18, . . .

33. Graph the data from the table below on a calculator. Explain why quadratic regression may be appropriate. Fit an equation. What common formula generates the data?

Input	Output
2	12.57
3	28.27
4	50.27
5	78.54

Projects

34. *Differences.* First consider a quadratic function:

a. Set up a row containing the outputs from the table in the form of a sequence. Below this row, calculate the first and second differences.

x	$f(x) = ax^2 + bx + c$
1	$a + b + c$
2	$4a + 2b + c$
3	$9a + 3b + c$
4	$16a + 4b + c$
5	$25a + 5b + c$

b. Are the second differences constant?

c. Is the lead term of the row of first differences $3a + b$? How might this help you write the equation?

Next consider a linear function:

d. Make a table with inputs 1, 2, 3, 4, and 5 and outputs
$$y = ax + b$$

e. Set up a row containing the outputs in the form of a sequence. Below this row, calculate the first and second differences.

f. Explain how the results may be used to write a linear equation.

Finally, consider a cubic function:

g. Make a table with inputs 1, 2, 3, 4, and 5 and outputs
$$y = ax^3 + bx^2 + cx + d$$

h. Set up a row containing the outputs in the form of a sequence. Below this row, calculate the first, second, and third differences.

i. Explain how the results may be used to write a cubic equation for any cubic sequence.

35. *Multiplication Table.* Make two photocopies of the multiplication table on page 262. On the first photocopy, do parts a to e.

a. Using two different colors or shading patterns, highlight the numbers in the sequences for x^2 and $x(x + 1)$ from Examples 2 and 3.

b. Exercises 19 to 32 contain six quadratic equations. Using a different color or shading pattern for each, highlight the sequences for any four of them.

c. Describe the orientation or location of quadratic sequences in the multiplication table.

d. Choose another quadratic sequence from the multiplication table. Using differences, show that it is indeed quadratic.

e. What is the rule for 1, 4, 9, 16, 25, ...? Use the rule to describe a rule for 2, 5, 10, 17, 26, ...

On the second photocopy, do parts f to i.

f. Locate 5, 10, 15, 20, 25, ... on the multiplication table. What is the rule for this sequence? Use the rule to describe a rule for 6, 11, 16, 21, 26, ...

g. Exercises 19 to 32 contain five linear sequences. Using a different color or shading pattern for each, highlight the sequences for any three of them. (*Hint:* See part f if you can't find a sequence.)

h. What can you say about the rules for rows and columns in the multiplication table?

i. Choose another linear sequence from the multiplication table. Give its equation.

1	2	3	4	5	6	7	8	9	10	11	12
2	4	6	8	10	12	14	16	18	20	22	24
3	6	9	12	15	18	21	24	27	30	33	36
4	8	12	16	20	24	28	32	36	40	44	48
5	10	15	20	25	30	35	40	45	50	55	60
6	12	18	24	30	36	42	48	54	60	66	72
7	14	21	28	35	42	49	56	63	70	77	84
8	16	24	32	40	48	56	64	72	80	88	96
9	18	27	36	45	54	63	72	81	90	99	108
10	20	30	40	50	60	70	80	90	100	110	120
11	22	33	44	55	66	77	88	99	110	121	132
12	24	36	48	60	72	84	96	108	120	132	144

4.2 The Role of a, b, and c in Graphing Quadratic Functions

OBJECTIVES

- Determine the effect of a on the graph of $f(x) = ax^2$.
- Determine the effect of b on the graph of $f(x) = x^2 + bx$.
- Determine the effect of c on the graph of $f(x) = x^2 + c$.

WARM-UP

1. Complete the table.

x	$-x^2$	$-x$	$(-x)^2$
-4			
-2			
1			
3			
5			

2. What other expression would give the same numbers as appear in the last column?

3. Explain why the columns $-x^2$ and $(-x)^2$ are different. If you do not believe they are, graph $y = -x^2$ and $y = (-x)^2$ and compare the graphs.

4. Complete the table.

x	x^2	$2x^2$	$\frac{1}{2}x^2$
-2			
-1			
0			
1			
2			

5. Describe how the 2 makes the $2x^2$ column different from the x^2 column.

6. Describe how the $\frac{1}{2}$ makes the $\frac{1}{2}x^2$ column different from the x^2 column.

$\mathbf{I}$N THIS SECTION, we examine the graphs of quadratic equations in more detail. We consider the role played by the parameters *a*, *b*, and *c* in determining the shape and position of the graph of a quadratic equation. Familiarity with these parameters will be helpful in applications with equations.

Review of Graphing a Quadratic Equation

Have you ever wondered why the game of golf requires so many clubs? The length of a golf club, its weight, and the angle of the striking face are designed to produce a particular path of a golf ball, as shown in Figure 21. Selection of the right club and consistency in its use are prerequisites to playing the game of golf well.

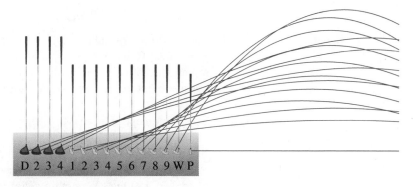

Figure 21   Golf clubs

Example 1 reviews several ideas about a quadratic function and its parabolic graph. The parabolic model approximating the path of a golf ball requires trigonometric coefficients, so we will not show how the function in Example 1 was obtained (as we did for the height-as-a-function-of-time curves for the Olympic divers).

EXAMPLE  Reviewing important features of the graph of a quadratic function: path of a golf ball Suppose we use the quadratic equation $f(x) = -0.0005812x^2 + 0.1763x + 15$ to model the path of a golf ball hit upward at a 10° angle from a tee (starting point) 15 feet above the fairway. The equation includes a 169-ft/sec initial velocity.

a. Make a table and graph for $f(x)$ using 50-foot increments for x.
b. Identify the x-intercepts, $f(x) = 0$, and their meaning, if any.
c. Identify the y-intercept and its meaning as the initial value, $f(0)$.
d. Estimate the vertex from the table and graph. Describe how we could calculate the vertex more exactly.
e. Identify the axis of symmetry in the table and on the graph. Write the equation of the axis of symmetry.

Solution   a. The table is shown in Table 3 and the graph in Figure 22. The path of the ball is really flatter than it appears, because the scales on the x- and y-axes in Figure 22 are not the same.

x	$f(x)$
0	15
50	22.4
100	26.8
150	28.4
200	27.0
250	22.8
300	15.6
350	5.5
400	-7.5

Table 3 $f(x) =$ $-0.0005812x^2 +$ $0.1763x + 15$

$y = f(x)$

$f(x) = -0.0005812x^2 + 0.1763x + 15$

Height (feet)

Distance traveled (feet)

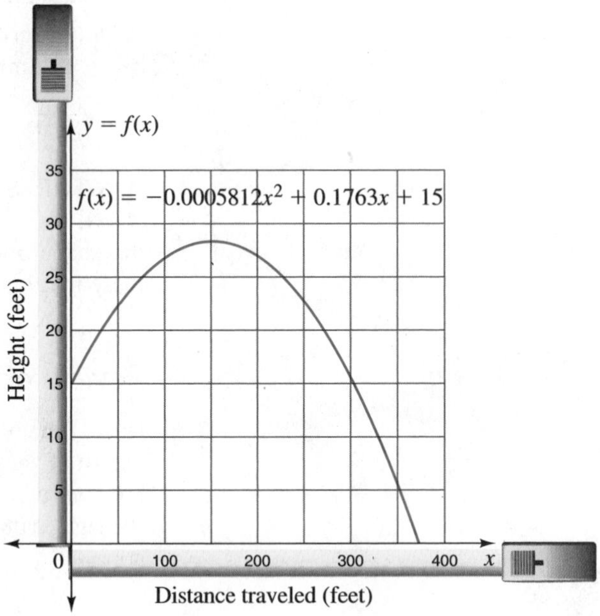

Figure 22

b. The quadratic formula with $a = -0.0005812$, $b = 0.1763$, and $c = 15$ gives $x \approx -69.3$ ft and $x \approx 372.6$ ft as the x-intercepts. Because the function represents the path of a ball, only positive inputs and outputs are reasonable. The positive x-intercept, $x \approx 372.6$ ft, gives the position where the ball strikes the ground.

c. The y-intercept is $(0, 15)$, or $f(0) = 15$ ft. The y-intercept represents the initial height of the ball at the tee.

d. The table outputs 22.4 and 22.8, 26.8 and 27.0 are almost equal and approximate the symmetry expected for a quadratic equation. We estimate the vertex to be near 150 feet. We obtain a similar result from the graph. A better estimate of the x-coordinate of the vertex is halfway between the x-intercepts found in part a, or their average:

$$x = \frac{-69.3 + 372.6}{2} \approx 151.7$$

The output at $x \approx 151.7$ ft is

$$f(x) = -0.0005812x^2 + 0.1763x + 15$$
$$f(151.7) = -0.0005812(151.7)^2 + 0.1763(151.7) + 15$$
$$\approx 28.4 \text{ ft}$$

Thus, the vertex is approximately $(151.7, 28.4)$.

e. The axis of symmetry passes through the vertex. Its equation is $x \approx 151.7$.

The Shape and Position of Quadratic Function Graphs

How high the golf ball reaches and how far it flies before hitting the ground are determined by how hard and at what angle the ball is hit, as well as how high the ball is initially. Establishing the connection of these factors to a, b, and c in the equation $y = ax^2 + bx + c$ is beyond the level of this course, but we can see some of the relationships by exploring some examples.

As mentioned earlier, the letters a, b, and c in $y = ax^2 + bx + c$ are called parameters because they determine the shape, orientation, and position of the parabola. The parameters stay the same within one equation while the variables x and y change.

ROLE OF PARAMETER a Examples 2 and 3 investigate the role of the parameter a on the graph.

EXAMPLE **2**   Exploring the sign on a Use the Warm-up and Example 1 to answer these questions.

a. How does the negative sign in front of the x^2 change the numbers in the table from those generated by x^2?

b. How does the graph of $y = -x^2$ compare to that of $y = x^2$?

c. Test your conclusions with the equation and graph in Example 1.

Solution   **a.** The outputs from the expression $-x^2$ have the opposite sign of the outputs from x^2.

b. The graph of $y = -x^2$ is upside down compared with the graph of $y = x^2$ (see Figure 23). The negative sign changes the orientation of the parabola, from opening up for $y = x^2$ to opening down for $y = -x^2$.

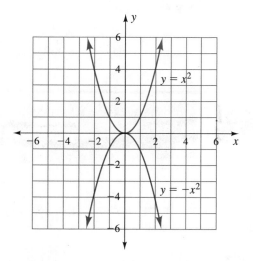

Figure 23

c. The equation in Example 1 has a negative sign on the x^2 term, and the graph opens down.

Example 2 suggests the following conclusion.

The Sign on *a* and the
Direction of the Parabola

> A positive *a* on the x^2 term causes the parabola to open up, ∪. A negative *a* on the x^2 term causes the parabola to open down, ∩.

Example 3 examines the steepness of parabolic graphs.

EXAMPLE **3**   Exploring the value of *a*   Graph $y = 2x^2$, $y = x^2$, and $y = \frac{1}{2}x^2$, and compare the graphs.

Solution   The graph of $y = 2x^2$ in Figure 24 is the steepest. The graph of $y = \frac{1}{2}x^2$ is flatter than the other two. The numbers 2 and $\frac{1}{2}$ appear to change the shape of the parabola.

$$—y = x^2 \quad —y = 2x^2 \quad —y = \frac{1}{2}x^2$$

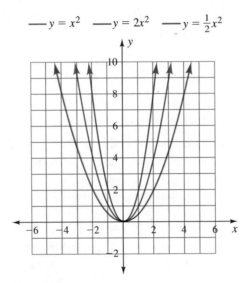

Figure 24

Example 3 suggests the following conclusion.

The Size of *a* and the
Steepness of the Parabola

> The coefficient on the x^2 term, the parameter *a*, controls the shape of the graph. If *a* is larger than 1, the graph is steeper than the graph of $y = x^2$. If *a* is between zero and 1, the graph is flatter than the graph of $y = x^2$.

ROLE OF PARAMETER *b*   Example 4 investigates one effect of the parameter *b* on the graph of $y = x^2 + bx$.

EXAMPLE **4**   Exploring *b*   Graph $y = x^2$, $y = x^2 + 1x$, $y = x^2 + 2x$, and $y = x^2 + 3x$. How does the *b* in $y = x^2 + bx$ change the graph from that of $y = x^2$?

Solution   The graphs are shown in Figure 25. The graphs are the same shape as that of $y = x^2$, but their vertices have moved away from the origin. As *b* increases, the vertices move in a left and downward direction.

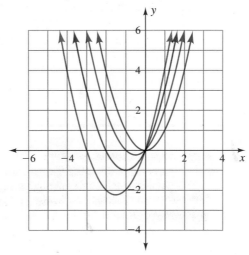

Figure 25

The Role of *b*

> The parameter *b* in $y = x^2 + bx$ contributes to a change in the position of the vertex of the parabola from that of $y = x^2$.

Although our conclusion that *b* changes the position of the vertex is correct, both *a* and *c* also affect the position of the vertex. No simple generalization about *b* is possible.

ROLE OF PARAMETER *c*   Example 5 investigates the impact of the parameter *c* on the graph.

EXAMPLE **5**   Exploring *c*   Use Table 4 and the graph in Figure 26 to answer these questions.

a. How does adding a constant to x^2 change the graph? What does the parameter *c* represent in the graph?

b. Test your conclusion with the function in Example 1.

Input x	Output x^2	Output $x^2 + 2$	Output $x^2 - 1$
-2	4	6	3
-1	1	3	0
0	0	2	-1
1	1	3	0
2	4	6	3

Table 4

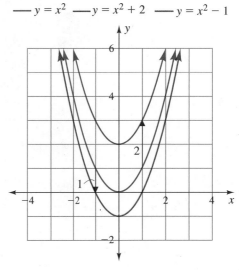

Figure 26

Solution   **a.** The graph of $y = x^2 + 2$ is shifted up 2 units from that of $y = x^2$. The graph of $y = x^2 - 1$ is 1 unit lower than that of $y = x^2$. In $y = x^2 + 2$, the y-intercept is 2. In $y = x^2 - 1$, the y-intercept is -1. Adding a number to x^2 shifts the graph vertically. The number also becomes the y-intercept.

b. In Example 1, the function $f(x) = -0.0005812x^2 + 0.1763x + 15$ has 15 as its y-intercept.   ●

Example 5 suggests the following conclusion.

The Role of c

> The parameter c in $y = ax^2 + bx + c$ is the output when $x = 0$. The co-ordinate $(0, c)$ is the y-intercept point of the graph of a quadratic function.
>
> If the equation is of the form $y = ax^2 + c$, then $|c|$ is the distance (number of units) the parabola $y = ax^2$ is shifted vertically (up or down) and the vertex is $(0, c)$.

Application: Fireboat

Our conclusions can be applied to the path of the water jet coming from the cannons on the fireboats shown in Figure 1 (on page 240) and Figure 27.

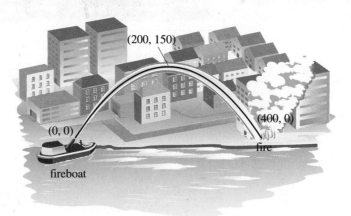

Figure 27

EXAMPLE **6**   Predicting parameters: fireboat water jet

a. What are the x-intercept points, y-intercept point, and vertex of the water spray between the boat and the fire in Figure 27?

b. Assume that the path of the water is parabolic. Predict the sign on a and the value of c in an equation that fits the path. Find the equation of the path of the water spray.

Solution   **a.** If the origin $(0, 0)$ is placed at the water cannon, as shown in Figure 27, the x-intercept points of the water spray graph represent the starting point of the water as well as the position at which the water returns to the same level as the cannon, $(400, 0)$. The y-intercept point is also at the origin. The vertex of the water spray is at its highest point, $(200, 150)$.

b. The graph appears parabolic. The parabola turns down, so we expect a negative coefficient for x^2. The curve passes through the origin, so $c = 0$. We let (x, y) be $(200, 150)$ in $y = a(x - 0)(x - 400)$ and solve for a. We substitute a back into the equation and obtain $y = -0.00375x^2 + 1.5x + 0$.   ●

Student Note: Check your results with quadratic regression.

The placement of the origin in any given problem is arbitrary. Figure 27 shows the origin at the cannon because the angle of the cannon controls the trajectory of the water and, consequently, the height of the water spray.

Think about it: What parameters (a, b, and/or c) change if we place the origin directly below the cannon, at water level?

ANSWER BOX

Warm-up:

1.

x	$-x^2$	$-x$	$(-x)^2$
-4	-16	4	16
-2	-4	2	4
1	-1	-1	1
3	-9	-3	9
5	-25	-5	25

2. x^2 **3.** $-x^2$ is the opposite of x^2, whereas $(-x)^2$ is the square of the opposite of x and equals x^2.

4.

x	x^2	$2x^2$	$\frac{1}{2}x^2$
-2	4	8	2
-1	1	2	$\frac{1}{2}$
0	0	0	0
1	1	2	$\frac{1}{2}$
2	4	8	2

5. The 2 doubles the x^2. **6.** The $\frac{1}{2}$ takes half of x^2. **Think about it:** When we move the origin down, the parameter c changes from 0 to the distance between the cannon and the water.

EXERCISES 4.2

1. In Example 1, what part of the equation tells us that the path of the golf ball is a parabola that turns down?

2. In Example 1, how does the graph indicate that the coefficient on the x^2 term is negative?

In Exercises 3 and 4, match the situation described with one of the following equations.

$$y = -0.16x^2 + 0.09x, \qquad y = \frac{4}{125}x^2$$

Explain how you found the correct equation.

3. The equation of a parabolic cable supporting a bridge

4. The equation of the path of a cliff diver

5. Graph $y = \frac{1}{2}x^2$ and $y = \frac{1}{2}x^2 + 3$. Draw arrows to show the shift in the graph of $y = \frac{1}{2}x^2$ to the position of the second graph.

6. Graph $y = \frac{1}{2}x^2$ and $y = \frac{1}{2}x^2 - 2$. Draw arrows to show the shift in the graph of $y = \frac{1}{2}x^2$ to the position of the second graph.

7. Graph $y = 2x^2$ and $y = 2x^2 - 3$. Draw arrows to show the shift in the graph of $y = 2x^2$ to the position of the second graph.

8. Graph $y = 2x^2$ and $y = 2x^2 + 1$. Draw arrows to show the shift in the graph of $y = 2x^2$ to the position of the second graph.

Eight quadratic functions, with different function names, are listed below. Find which functions satisfy the conditions in Exercises 9 to 14. It is not necessary to graph the functions.

$$f(x) = 2x^2 + 2x - 3 \qquad g(x) = 0.5x^2 + 3x - 2$$

$$h(x) = \tfrac{1}{2}x^2 + 2x \qquad j(x) = 2(x - 1)^2$$

$$k(x) = -2x^2 - 1 \qquad m(x) = \frac{x^2 + 2x}{2}$$

$$p(x) = -\tfrac{1}{2}(x + 1)^2 \qquad q(x) = \frac{(x + 1)(x + 2)}{2}$$

9. Which of the functions have the same coefficient on x^2 as $g(x)$ does?

10. Which of the functions have the same coefficient on x^2 as $f(x)$ does?

11. Which of the functions have a parabolic graph that turns up?

12. Which of the functions have a parabolic graph that passes through the origin?

13. Which of the functions have a parabolic graph that has $(0, 2)$ as its y-intercept?

14. Which of the functions have a parabolic graph that turns down?

In Exercises 15 to 22, identify the function whose graph will make a steeper parabola.

15. $f(x) = 2x^2$ or $g(x) = \pi x^2$

16. $g(x) = \pi x^2$ or $h(x) = 3x^2$

17. $j(x) = 4x^2$ or $g(x) = \pi x^2$

18. $k(x) = 3.5x^2$ or $g(x) = \pi x^2$

19. $f(x) = \tfrac{2}{3}x^2$ or $h(x) = \tfrac{3}{4}x^2$

20. $g(x) = 2.299x^2$ or $h(x) = 2.32x^2$

21. $p(x) = 2.41x^2$ or $q(x) = 2.288x^2$

22. $r(x) = \tfrac{3}{8}x^2$ or $s(x) = \tfrac{1}{3}x^2$

23. Choose the correct word, *steeper* or *flatter*, to complete this sentence: A coefficient on x^2 larger than 1 gives a graph ____ than the graph of $y = x^2$.

24. Choose the correct word, *larger* or *smaller*, to complete this sentence: A positive coefficient ____ than 1 gives a graph flatter than the graph of $y = x^2$.

25. In $y = mx + b$, which parameter, m or b, controls the steepness of a linear graph?

26. In $y = mx + b$, which parameter, m or b, controls the y-intercept of a linear graph?

27. Graph $y = x^2$, $y = x^2 - 1x$, and $y = x^2 - 2x$. What is the vertex of each graph? How do the graphs change from that of $y = x^2$?

28. Graph $y = -x^2$, $y = -x^2 + 1x$, and $y = -x^2 + 2x$. What is the vertex of each graph? How do the graphs change from that of $y = -x^2$?

29. Sketch graphs of $y = x^2$, $y = x^2 + 2$, and $y = x^2 - 1$ on one set of axes.

30. Sketch graphs of $y = x^2$, $y = x^2 - 2$, and $y = x^2 + 1$ on one set of axes.

31. A graph has the same shape as $y = x^2$. Its vertex is $(0, -3)$. What is its equation? (There are two possibilities.)

32. A graph has the same shape as $y = x^2$. Its vertex is $(0, 4)$. What is its equation? (There are two possibilities.)

33. Suppose the origin in the fireboat problem (Example 6) is placed 20 feet directly below the cannon at the water line.

    a. What are the coordinates for the cannon, the highest point on the spray of water, and the fire?

    b. Predict the new equation of the water spray. Confirm your solution by fitting a quadratic equation through the three points.

34. Suppose the origin in the fireboat problem (Example 6) is placed at the fire and the remaining coordinates are adjusted to be in the second quadrant or on the negative x-axis.

    a. What are the coordinates for the cannon, the highest point on the spray of water, and the fire?

    b. Predict the new equation of the water spray. Confirm your solution by fitting a quadratic equation through the three points.

Project

35. *Vertex Patterns.* Graph all of the following quadratic equations on one pair of axes.

$$y = x^2 + 1x$$
$$y = x^2 + 3x$$
$$y = x^2 - 4x$$
$$y = x^2 - 2x$$

Now graph all of these quadratic equations on another pair of axes:

$$y = 2x^2 + 1x$$
$$y = 2x^2 + 3x$$
$$y = 2x^2 - 4x$$
$$y = 2x^2 - 2x$$

With a colored pencil, mark the vertex of each parabola. List the vertices. Connect the vertices of the parabolas with a smooth curve. Fit an appropriate curve to the set of vertices with your calculator.

Finally, generalize your conclusions by completing these statements:

a. For curves of the form $y = ax^2 + bx$ with the coefficient a held constant, changing b will change the location of the ___ of each parabola.

b. All equations intersect at ___ .

c. The vertices of the parabolas $y = ax^2 + bx$, if a is constant, lie on a smooth curve described by the equation ___ .

4.3 Complex Numbers and Graphs of Functions

OBJECTIVES

- Use the discriminant, $b^2 - 4ac$, to identify the number and type of roots of quadratic equations.
- Change square roots of negative numbers into complex number notation.
- Identify complex conjugates.
- Add, subtract, and multiply complex numbers.
- Multiply and simplify binomials containing irrational and complex numbers.
- Write complex solutions to quadratic and other equations.

WARM-UP

1. Simplify these expressions.

   a. $(-1)^0$ **b.** $(-1)^1$ **c.** $(-1)^3$

   d. $(-1)^2$ **e.** $(-1)^8$ **f.** $(-1)^4$

2. Multiply and simplify these expressions.

   a. $\sqrt{2} \cdot \sqrt{32}$ **b.** $\sqrt{8} \cdot \sqrt{2}$ **c.** $\sqrt{3} \cdot \sqrt{27}$ **d.** $\sqrt{20} \cdot \sqrt{5}$

3. Look for perfect square factors and simplify, using $\sqrt{a \cdot b} = \sqrt{a} \cdot \sqrt{b}$.

   Example: $\sqrt{52} = \sqrt{4 \cdot 13} = \sqrt{4} \cdot \sqrt{13} = 2\sqrt{13}$

   a. $\sqrt{12}$ **b.** $\sqrt{24}$

   c. $\sqrt{45}$ **d.** $\sqrt{8}$

THROUGHOUT OUR WORK with quadratic equations, the intersection of a parabolic graph with the x-axis gave the solutions to $ax^2 + bx + c = 0$. This section answers the question *What solutions are obtained when the graph does not touch the x-axis?* The imaginary unit and the complex number system will be introduced, as well as operations with complex numbers.

The Discriminant

In Example 1, we examine three quadratic equations, one of which has no real-number solutions. The example shows how we can predict the number of real-number solutions a quadratic equation may have.

EXAMPLE   *Finding numbers of solutions*   Graph each equation. Comment on the position of the graph relative to the x-axis. Use the quadratic formula to find the solution to each equation when $f(x) = 0$. Comment on the number and types of solutions.

  a. $f(x) = x^2 + 2x + 1$ **b.** $f(x) = x^2 + 2$ **c.** $f(x) = x^2 - 2x$

Solution   The equations are graphed in Figure 28.

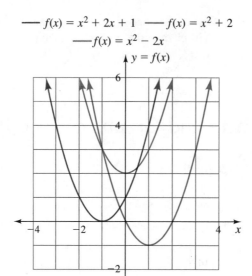

Figure 28

a. The graph of $f(x) = x^2 + 2x + 1 = (x + 1)^2$ has one x-intercept: $x = -1$, a double root. The solution to $x^2 + 2x + 1 = 0$ with the quadratic formula is

$$x = \frac{-b \pm \sqrt{b^2 - 4ac}}{2a} = \frac{-2 \pm \sqrt{4 - 4 \cdot (1) \cdot (1)}}{2(1)} = \frac{-2 \pm 0}{2} = -1$$

In this problem, $\sqrt{b^2 - 4ac} = 0$, so the $\pm$ has no impact on the answer. Thus, there is only one real-number solution: $x = -1$. The solution is a double root.

b. The graph of $f(x) = x^2 + 2$ has no x-intercepts. The solution to $x^2 + 2 = 0$ with the quadratic formula is

$$x = \frac{-b \pm \sqrt{b^2 - 4ac}}{2a} = \frac{-0 \pm \sqrt{0 - 4 \cdot (1) \cdot (2)}}{2(1)} = \frac{-0 \pm \sqrt{-8}}{2}$$

In this problem, $\sqrt{b^2 - 4ac} = \sqrt{-8}$ is undefined in the real numbers, so there are no real-number solutions.

c. The graph of $f(x) = x^2 - 2x$ has two x-intercepts: $x = 0$ and $x = 2$. The solution to $x^2 - 2x = 0$ with the quadratic formula is

$$x = \frac{-b \pm \sqrt{b^2 - 4ac}}{2a} = \frac{-(-2) \pm \sqrt{4 - 4 \cdot (1) \cdot (0)}}{2(1)} = \frac{2 \pm 2}{2}$$

In this problem, $\sqrt{b^2 - 4ac} = 2$, so the $\pm$ gives two real-number solutions: $x = 0$ and $x = 2$.   ●

The square root portion of the quadratic formula, $\sqrt{b^2 - 4ac}$, controls the nature of the solutions. The expression $b^2 - 4ac$ is called the **discriminant.** The following statements summarize the role of $b^2 - 4ac$ in the number of real-number solutions to $ax^2 + bx + c = 0$ and the graph of $f(x) = ax^2 + bx + c$.

Finding the Number of Solutions

For quadratic equations in the form $ax^2 + bx + c = 0,$

- If $b^2 - 4ac$ is positive, there are two real-number solutions. The graph of $f(x)$ passes through the x-axis twice.
- If $b^2 - 4ac$ is zero, there is one real-number solution, a double root. The graph of $f(x)$ touches the x-axis once.
- If $b^2 - 4ac$ is negative, there are no real-number solutions. The graph of $f(x)$ does not touch the x-axis.

EXAMPLE **2**   Predicting the number of solutions   Use the discriminant to predict the number of real-number solutions to $f(x) = 0$ for these equations.

a. $f(x) = \frac{4}{125}x^2$ **b.** $f(x) = -0.00375x^2 + 1.5x$

c. $f(x) = (x - 2)^2$ **d.** $f(x) = x^2 + 1$

Solution   **a.** In $f(x) = \frac{4}{125}x^2$, $a = \frac{4}{125}$, $b = 0$, and $c = 0$.

$$b^2 - 4ac = 0^2 - 4\left(\frac{4}{125}\right)(0) = 0$$

Because the discriminant is zero, there is one real-number solution, a double root.

b. In $f(x) = -0.00375x^2 + 1.5x$, $a = -0.00375$, $b = 1.5$, and $c = 0$.

$$b^2 - 4ac = (1.5)^2 - 4(-0.00375)(0) = 2.25$$

The discriminant is positive, so there are two real-number solutions to $f(x) = 0$.

c. In $f(x) = (x - 2)^2$, we must first multiply out $(x - 2)^2$. We obtain

$$f(x) = x^2 - 4x + 4$$

so $a = 1$, $b = -4$, and $c = 4$.

$$b^2 - 4ac = (-4)^2 - 4(1)(4) = 0$$

The discriminant is zero, so there is one real-number solution to $f(x) = 0$.

d. In $f(x) = x^2 + 1$, $a = 1$, $b = 0$, and $c = 1$.

$$b^2 - 4ac = (0)^2 - 4(1)(1) = -4$$

The discriminant is -4, causing the quadratic formula to contain $\sqrt{-4}$. Hence, there are no real-number solutions to $f(x) = 0$. ●

The Imaginary Unit

Although the square roots of negative numbers cannot be graphed on the rectangular coordinate axes, mathematicians have nevertheless investigated the properties of these numbers. In order to do so, they defined a new symbol, the imaginary unit i.

Definition of Imaginary Unit

$$i = \sqrt{-1} \quad \text{or} \quad i^2 = -1$$

where i is the **imaginary unit**.

The imaginary unit is important in the study of alternating current in electricity. Because the letter i is already used for something else in electronics, the imaginary unit is j, or the j-operator, as in $z = R + j\omega L$ or $z = R - j/\omega C$. In mathematics, however, we traditionally use i for the imaginary unit.

Our properties of radicals apply only to the real numbers. To work with square roots containing negative numbers, we first change them to products of a real number and the imaginary unit.

$$\sqrt{-a} = i\sqrt{a}$$

if a is real and positive.

EXAMPLE **3**   Changing square root notation to imaginary units   Change these numbers into the product of a real number and the imaginary unit.

a. $\sqrt{-4}$ **b.** $\sqrt{-8}$ **c.** $\sqrt{-32}$

Solution   **a.** $\sqrt{-4} = i\sqrt{4} = i \cdot 2 = 2i$
b. $\sqrt{-8} = i\sqrt{8} = i\sqrt{4 \cdot 2} = i\sqrt{4}\sqrt{2} = 2i\sqrt{2}$
c. $\sqrt{-32} = i\sqrt{32} = i\sqrt{16}\sqrt{2} = 4i\sqrt{2}$ ●

Although placing the i after the radical sign is acceptable, it may lead to errors if the i appears to be under the radical. When a number or letter is multiplying a radical, place it in front of the radical.

Where convenient, the square roots of negatives may be written with decimals:

$$\sqrt{-8} \approx 2.828i \quad \text{and} \quad \sqrt{-32} \approx 5.657i$$

The Complex Number System

When we combine *the real numbers with the imaginary unit,* we create the **complex number system.** Complex numbers may be written $a + bi$, where a and b are any real numbers, including zero. In a complex number $a + bi$, a is the real part and bi is the imaginary part. All real numbers a are complex numbers with a zero imaginary part: $a + 0i$.

EXAMPLE **4**   Writing numbers in complex number notation   Write the following as complex numbers.

a. $5i$ **b.** $\sqrt{-27}$ **c.** 15

d. $\dfrac{6 + 8i}{2}$ **e.** $\dfrac{\pm\sqrt{-8}}{2}$ **f.** $\dfrac{-4 \pm \sqrt{-8}}{4}$

Solution   **a.** $5i = 0 + 5i$
b. $\sqrt{-27} = i\sqrt{27} = i\sqrt{9}\sqrt{3} = 3i\sqrt{3}$, or $0 + 3i\sqrt{3}$
c. $15 = 15 + 0i$

d. $\dfrac{6 + 8i}{2} = \dfrac{6}{2} + \dfrac{8i}{2} = 3 + 4i$

e. $\dfrac{\pm\sqrt{-8}}{2} = \dfrac{\pm i\sqrt{8}}{2} = \dfrac{\pm i\sqrt{4}\sqrt{2}}{2}$

$= \dfrac{\pm i \cdot 2 \cdot \sqrt{2}}{2} = \pm i\sqrt{2}$, or $0 \pm i\sqrt{2}$

f. $\dfrac{-4 \pm \sqrt{-8}}{4} = -\dfrac{4}{4} \pm \dfrac{i\sqrt{8}}{4} = -1 \pm \dfrac{i\sqrt{4}\,\sqrt{2}}{4}$

$$= -1 \pm \dfrac{2i\sqrt{2}}{4} = -1 \pm \dfrac{i\sqrt{2}}{2}$$

The two expressions may be written separately: $-1 + \dfrac{i\sqrt{2}}{2}$ and $-1 - \dfrac{i\sqrt{2}}{2}$.

The real part of each answer is -1. The decimal form may be simpler: $-1 \pm 0.707i$. ●

Complex Conjugates

Complex-number solutions always occur in pairs, such as $x = i\sqrt{2}$ and $x = -i\sqrt{2}$ or $x = -1 + \dfrac{i\sqrt{2}}{2}$ and $x = -1 - \dfrac{i\sqrt{2}}{2}$. These pairs are given a special name—complex conjugates.

Definition of Complex Conjugates

> If a and b are real numbers, expressions of the form $a + bi$ and $a - bi$ are **complex conjugates.**

Complex conjugates are complex numbers with the same real-number part and opposite imaginary parts.

EXAMPLE **5**   Writing complex conjugates   Name the complex conjugate for each expression.

a. i **b.** $4 + i$ **c.** $5 - 2i$

Solution   *Hint:* To obtain a complex conjugate, replace i with $-i$ or $-i$ with i. See the Answer Box. ●

Think about it: What part of the quadratic formula shows that the complex solutions to a quadratic equation always appear as complex conjugates?

We now turn to operations with complex numbers. Look for special results when we add, subtract, or multiply complex conjugates.

Operations with Complex Numbers

We define addition and subtraction of complex numbers as follows.

Addition of Complex Numbers

> $(a + bi) + (c + di) = (a + c) + (b + d)i$

Subtraction of Complex Numbers

> $(a + bi) - (c + di) = (a - c) + (b - d)i$

Addition and subtraction with complex numbers are similar to the operations with polynomials. Both operations may be thought of as adding like terms. We add and subtract like numbers, real to real and imaginary to imaginary.

EXAMPLE **6** Adding and subtracting complex numbers   Add or subtract these complex numbers. If any are complex conjugates, note anything special about the results.

a. $(2 + 3i) + (2 - 3i)$ **b.** $(3 - 5i) - (3 + 5i)$

c. $(4i + 6) - (7i + 2) - (3i - 5)$

Solution   **a.** $(2 + 3i) + (2 - 3i) = (2 + 2) + (3 - 3)i = 4 + 0i = 4$
These numbers are complex conjugates. Their sum is a real number.

b. $(3 - 5i) - (3 + 5i) = (3 - 3) + (-5 - 5)i = 0 - 10i = -10i$
The difference of these complex conjugates has only an imaginary part.

c. $(4i + 6) - (7i + 2) - (3i - 5) = 6 - 2 - (-5) + (4 - 7 - 3)i$
$$= 9 - 6i$$

●

B efore we define multiplication with complex numbers, we return to our definition of the imaginary unit and note the statement $i^2 = -1$.

Definition of Imaginary Unit

$$i = \sqrt{-1} \quad \text{or} \quad i^2 = -1$$

where i is the **imaginary unit.**

The $i^2 = -1$ is used to simplify after multiplying complex numbers.

Multiplication of Complex Numbers

$$(a + bi)(c + di) = ac + adi + bci + bdi^2$$
$$= ac - bd + (ad + bc)i$$

The multiplication of complex numbers is similar to the multiplication of binomials $(a + b)(c + d)$. Try a favorite method, or use a table as shown in Example 7.

EXAMPLE **7** Multiplying complex numbers   Multiply these complex numbers. If any are complex conjugates, note anything special about the products.

a. $(3 + 4i)(3 + 4i)$ **b.** $(3 + 4i)(3 - 4i)$

c. $(3 - 4i)(3 - 4i)$ **d.** $(\sqrt{3} + i)(\sqrt{3} - i)$

Solution   We multiply each in a table.

a.

Multiply	3	$+4i$
3	9	$+12i$
$+4i$	$+12i$	$+16i^2$

$$(3 + 4i)(3 + 4i) = 9 + 24i + 16i^2 = 9 + 24i - 16$$

because $i^2 = -1$. Thus, $(3 + 4i)(3 + 4i) = -7 + 24i$.

Student Note: In parts b and d, the diagonal terms add to zero.

b.

Multiply	3	$+4i$
3	9	$+12i$
$-4i$	$-12i$	$-16i^2$

$$(3 + 4i)(3 - 4i) = 9 - 16i^2 = 9 - 16(-1) = 9 + 16 = 25$$

These numbers are complex conjugates. Their product is a real number.

c.

Multiply	3	$-4i$
3	9	$-12i$
$-4i$	$-12i$	$+16i^2$

$$(3 - 4i)(3 - 4i) = 9 - 24i + 16i^2 = 9 - 24i - 16 = -7 - 24i$$

d.

Multiply	$\sqrt{3}$	$+i$
$\sqrt{3}$	3	$+i\sqrt{3}$
$-i$	$-i\sqrt{3}$	$-i^2$

$$(\sqrt{3} + i)(\sqrt{3} - i) = 3 - i^2 = 3 - (-1) = 4$$ ●

Solving Quadratic and Higher Order Equations

We now return to solving equations when the solution set may contain complex numbers.

QUADRATIC EQUATIONS   The quadratic equations in Example 8 have complex conjugates as solutions.

EXAMPLE 8   Solving equations with complex solutions   Solve these quadratic equations.

a. $x^2 + 1 = 0$

b. $x^2 + 2x + 5 = 0$

Solution   **a.** $x^2 + 1 = 0$

$$x^2 = -1$$ Take the square root of both sides.

$$\sqrt{x^2} = \sqrt{-1}$$ Replace $\sqrt{-1}$ with i and $\sqrt{x^2}$ with the absolute value of x.

$$|x| = i$$

$$x = \pm i$$

b. In $x^2 + 2x + 5 = 0$, $a = 1$, $b = 2$, and $c = 5$.

$$x = \frac{-b \pm \sqrt{b^2 - 4ac}}{2a} = \frac{-2 \pm \sqrt{2^2 - 4(1)(5)}}{2(1)}$$

$$= \frac{-2 \pm \sqrt{-16}}{2}$$

$$= \frac{-2}{2} \pm \frac{i\sqrt{16}}{2}$$

$$= -1 \pm \frac{4i}{2} = -1 \pm 2i$$ ●

HIGHER ORDER EQUATIONS   In Section 4.0, we saw that the degree of a polynomial equation indicates the maximum number of real solutions. If we include complex solutions and allow for the repetitions from double roots, then all quadratic equations have two solutions, all cubic equations have three solutions, all fourth-degree equations have four solutions, and so forth. It is important to note that *only the real-number solutions appear on a rectangular coordinate graph.*

In the next two examples, we use factoring to obtain both real and complex solutions to equations with degree higher than 2.

EXAMPLE **9**   Solving equations with complex solutions

a. Use a graph to discuss the number of real solutions to $x^4 - 1 = 0$.

b. Solve $x^4 - 1 = 0$ algebraically, and indicate the number of real- and complex-number solutions.

Solution   **a.** Two real solutions to $x^4 - 1 = 0$, $x = 1$ and $x = -1$, appear as x-intercepts in the graph of $y = x^4 - 1$ (see Figure 29).

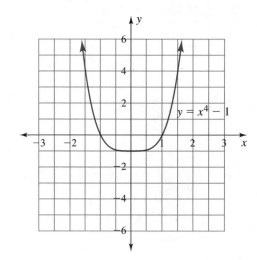

Figure 29

b.
$$x^4 - 1 = 0 \qquad \text{Factor.}$$
$$(x^2 - 1)(x^2 + 1) = 0 \qquad \text{Factor } (x^2 - 1).$$
$$(x - 1)(x + 1)(x^2 + 1) = 0 \qquad \text{Apply the zero product rule.}$$
Either $\quad x - 1 = 0, \quad x + 1 = 0, \quad$ or $\quad x^2 + 1 = 0$
$$x = 1, \qquad x = -1, \qquad x = +i, \qquad x = -i$$

The latter two solutions are from part a of Example 8. There are four solutions altogether: two real and two complex.   ●

It is useful to combine use of the quadratic formula and factoring in solving higher order polynomial equations. The first expression in Example 10 has been partially factored. Use the factorization and the quadratic formula to show that the degree of the equation indicates the number of solutions.

EXAMPLE **10**

Solving equations with complex solutions

a. Use a graph to discuss the number of real solutions and the number of complex solutions to $x^3 - 1 = 0$.

b. Solve $x^3 - 1 = 0$ algebraically, where $x^3 - 1 = (x - 1)(x^2 + x + 1)$.

Solution   **a.** Only one x-intercept appears in the graph of $y = x^3 - 1$ (see Figure 30). We anticipate finding only one real-number solution among the three possible solutions.

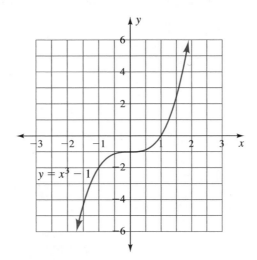

Figure 30

b.
$$x^3 - 1 = 0$$
$$(x - 1)(x^2 + x + 1) = 0 \qquad \text{Apply the zero product rule.}$$
Either $x - 1 = 0$ or $x^2 + x + 1 = 0$

The solution to the first equation is $x = 1$. We need the quadratic formula to solve the second factor equation.

$$x = \frac{-b \pm \sqrt{b^2 - 4ac}}{2a} = \frac{-1 \pm \sqrt{1^2 - 4(1)(1)}}{2(1)} = \frac{-1 \pm \sqrt{-3}}{2} = \frac{-1}{2} \pm \frac{i\sqrt{3}}{2}$$

There are three solutions: one real number, $x = 1$, and two complex numbers,

$$x = -\frac{1}{2} + \frac{i\sqrt{3}}{2} \qquad \text{and} \qquad x = -\frac{1}{2} - \frac{i\sqrt{3}}{2}$$

We can write the answers in decimal form:

$$x = 1, \qquad x \approx -0.5 + 0.866i, \qquad \text{and} \qquad x \approx -0.5 - 0.866i \qquad \bullet$$

ANSWER BOX

Warm-up: 1. a. 1 **b.** -1 **c.** -1 **d.** 1 **e.** 1 **f.** 1 **2. a.** 8 **b.** 4 **c.** 9 **d.** 10 **3. a.** $2\sqrt{3}$ **b.** $2\sqrt{6}$ **c.** $3\sqrt{5}$ **d.** $2\sqrt{2}$ **Example 5: a.** $-i$ **b.** $4 - i$ **c.** $5 + 2i$ **Think about it:** The square root is preceded by $\pm$, so each time the square root is nonzero, there will be two solutions. When complex, the solutions will be of the form $a \pm bi$, complex conjugates.

EXERCISES ④③

Use the discriminant to predict the number and type of solutions to the equations in Exercises 1 to 10.

1. $4x^2 + 7x - 2 = 0$ **2.** $2x^2 + 3x - 5 = 0$

3. $5x^2 + 4x + 1 = 0$ **4.** $x^2 + 2x + 5 = 0$

5. $4x^2 + 4x + 1 = 0$ **6.** $9x^2 - 6x + 1 = 0$

7. $2x^2 + 6x + 5 = 0$ **8.** $2x^2 + 2x + 5 = 0$

9. $\frac{1}{2}x^2 - 3 = 0$ **10.** $\frac{1}{2}x^2 - 5 = 0$

In Exercises 11 to 14, change each to an expression containing a product of a real number and the imaginary unit. Leave answers in either decimal or simplified radical form.

11. a. $\sqrt{-20}$ **b.** $\sqrt{-40}$ **c.** $\sqrt{-72}$

12. a. $\sqrt{-63}$ **b.** $\sqrt{-75}$ **c.** $\sqrt{-80}$

13. a. $\sqrt{-3}$

    b. $\sqrt{-54}$

    c. $\sqrt{-27}$

14. a. $\sqrt{-12}$

    b. $\sqrt{-48}$

    c. $\sqrt{-24}$

Change each expression in Exercises 15 to 18 to a complex number.

15. a. 16

    b. $\sqrt{-16}$

    c. $\sqrt{36} + \sqrt{-6}$

16. a. $\sqrt{25}$

    b. $\sqrt{-11}$

    c. $8 - \sqrt{-49}$

17. a. $\dfrac{6 + \sqrt{-12}}{2}$

    b. $\dfrac{2 - \sqrt{-24}}{4}$

18. a. $\dfrac{5 - \sqrt{-50}}{5}$

    b. $\dfrac{3 + \sqrt{-18}}{6}$

In Exercises 19 to 26, give the conjugate of each complex number.

19. $2 + 3i$ **20.** $4 - 2i$

21. $i + 1$ **22.** $2i + 2$

23. $2i - 3$ **24.** $3 - i$

25. $4 - 2i$ **26.** $3 + 2i$

Simplify the expressions in Exercises 27 to 36.

27. $2i + 3i - 4i - 5i + 6i + 7i - 8i - 9i$

28. $3i - (-5i)$

29. $6 + 4i - 7i + 5 - 12i$

30. $a + bi - a + bi$

31. $-a + bi - (a + bi)$

32. $a - bi + (bi - a)$

33. $3 - 4(5 - i)$

34. $3i - 4(5 - i)$

35. $4i - 3(5 - i)$ **36.** $4 - 3(5 - i)$

Multiply the conjugate expressions in Exercises 37 to 40. Compare and explain the results.

37. a. $(4 - 3i)(4 + 3i)$

    b. $(3 + 4i)(3 - 4i)$

38. a. $(5 + 12i)(5 - 12i)$

    b. $(12 - 5i)(12 + 5i)$

39. a. $(\sqrt{2} + i)(\sqrt{2} - i)$

    b. $(1 - i\sqrt{2})(1 + i\sqrt{2})$

40. a. $(3 + i\sqrt{2})(3 - i\sqrt{2})$

    b. $(\sqrt{2} + 3i)(\sqrt{2} - 3i)$

Multiply the expressions in Exercises 41 to 44.

41. $(x + i)(x - i)$

42. $(x + 2i)(x - 2i)$

43. $(x - 3i)(x + 3i)$

44. $(x - 4i)(x + 4i)$

In Exercises 45 to 58, solve for all solutions, real or complex.

45. $x^2 - 8 = 0$ **46.** $x^2 - 27 = 0$

47. $x^2 + 2x + 4 = 0$

48. $x^2 + x + 1 = 0$

49. $x^2 - 4x + 8 = 0$

50. $x^2 - 3x + 6 = 0$

51. $x^2 + 3x = 4$ **52.** $x^2 - 3x = 10$

53. $x^2 - 4x + 4 = 0$ **54.** $x^2 + 6x + 9 = 0$

55. $(x + 1)^2 + 1 = 0$ **56.** $(x - 2)^2 + 1 = 0$

57. $2x^2 + 10x + 13 = 0$ **58.** $13x^2 + 2x + 2 = 0$

Solve for all solutions, real or complex, in Exercises 59 to 64. Some factoring hints are given.

59. $x^3 + 8 = 0$, where $x^3 + 8 = (x + 2)(x^2 - 2x + 4)$

60. $x^4 - 16 = 0$

61. $x^4 - 81 = 0$

62. $x^3 + 1 = 0$, where $x^3 + 1 = (x + 1)(x^2 - x + 1)$

63. $x^3 + 3x^2 + 2x = 0$

64. $x^3 + 3x^2 + x = 0$

In Exercises 65 to 68, tell whether the statement is true or false. Explain your response.

65. It is possible to predict the type of solution (real or complex) from the quadratic equation.

66. It is possible to add two complex numbers and obtain a real number.

67. It is possible to multiply two complex numbers and obtain a real number.

68. If the graph of a quadratic equation $y = f(x)$ has no x-intercepts, then solutions to $f(x) = 0$ are complex.

For Exercises 69 to 71, choose the phrase below that best completes each sentence.

a. then the quadratic equation has one real-number solution, a double root.

b. then the quadratic equation has two real-number solutions.

c. then the quadratic equation has two complex-number solutions.

69. If $b^2 - 4ac$ is negative, _____ .

70. If $b^2 - 4ac$ is zero, _____ .

71. If $b^2 - 4ac$ is positive, _____ .

Which of graphs a–f match each condition in Exercises 72 to 74?

a. 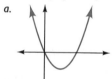 b.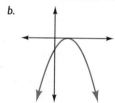

c. 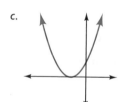 d.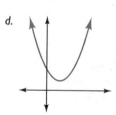

e. 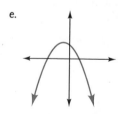 f.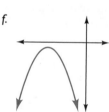

72. $b^2 - 4ac < 0$ **73.** $b^2 - 4ac > 0$

74. $b^2 - 4ac = 0$

Projects

75. Complex Cubes. Multiply the expressions. What do you observe about the results?

a. $(1 + i\sqrt{3})^3$

b. $(-1 - i\sqrt{3})^3$

c. $\left(-\dfrac{3}{2} + \dfrac{3i\sqrt{3}}{2}\right)^3$

d. $\left(-\dfrac{1}{2} + \dfrac{i\sqrt{3}}{2}\right)^3$

76. Complex Conjugates. What conclusions can be drawn about these operations with complex conjugates?

a. $(a + bi) + (a - bi)$

b. $(a + bi) - (a - bi)$

c. $(a + bi)(a - bi)$

The complex conjugate is used to eliminate a complex number from a denominator. Multiply the top and bottom of each of the following fractions by the complex conjugate of the denominator.

d. $\dfrac{1}{3i}$ **e.** $\dfrac{2}{i}$ **f.** $\dfrac{3}{-2i}$

g. $\dfrac{1}{2 + 3i}$ **h.** $\dfrac{3}{2 + i}$ **i.** $\dfrac{4}{3 - 2i}$

77. Rationalizing Denominators. Over the years, some unusual "division" methods have been developed to simplify operations. Long division by irrational numbers, such as $\sqrt{3} \approx 1.73205$, was considered too cumbersome, so mathematicians developed a technique to make the division easier. Suppose we wish to divide 4 by $\sqrt{3}$. If we write this division as a fraction, we have $4/\sqrt{3}$. To simplify the division, we multiply by $\sqrt{3}/\sqrt{3} = 1$. This multiplication gives

$$\frac{4}{\sqrt{3}} \cdot \frac{\sqrt{3}}{\sqrt{3}} = \frac{4\sqrt{3}}{\sqrt{9}} = \frac{4\sqrt{3}}{3}$$

(Although modern calculators make division by irrational numbers of no consequence, by hand the division by 3 is much easier.)

This technique is called *rationalizing the denominator*. Rationalize these fractions by multiplying top and bottom by the denominator.

a. $\dfrac{3}{\sqrt{5}}$ **b.** $\dfrac{2\sqrt{2}}{\sqrt{3}}$ **c.** $\dfrac{2\sqrt{2}}{3\sqrt{8}}$

d. $\dfrac{3\sqrt{5}}{\sqrt{2}}$ **e.** $\dfrac{6}{\sqrt{3}}$ **f.** $\dfrac{5\sqrt{3}}{\sqrt{15}}$

g. Exercise 76 deals with division by complex numbers. How is rationalizing the denominator similar to division by a complex number? How is rationalizing the denominator different from division by a complex number?

78. Powers of i. Use $i^1 = i$ and $i^2 = -i$ to investigate the powers of i. Look for a pattern. Use your pattern to predict i^{16}, i^{24}, i^{35}, i^{46}, and i^{97}. (*Hint:* $i^{11} = -i$ and $i^{32} = i$.)

MID–CHAPTER ④ TEST

1. Multiply, and give the name of any special expressions formed.

   a. $(2x + 3)(2x - 3)$

   b. $(2x - 3)(2x - 3)$

   c. $(2x + 3)(2x + 3)$

   d. $(3x - 2)(2x + 3)$

2. **a.** Multiply $(x - 2)(x^2 + 2x + 4)$.

   b. Factor $x^3 + 27$.

3. Explain how the graph of $f(x) = (x - 2)(x^2 + 2x + 4)$ is like and different from the graph of $f(x) = x^3$.

4. Compare the graph of $y = (x + 2)^2$ with that of $y = x^2$. Discuss vertices and intercepts.

5. The graph of what quadratic equation has -3 and 4 as x-intercepts and passes through $(-2, 3)$?

6. Identify each sequence as linear, or quadratic, or neither, and then find the function for each.

   a. $15, 8, 3, 0, -1, \ldots$

   b. $4, 9, 14, 19, 24, \ldots$

   c. $-3, 4, 11, 18, 25, \ldots$

   d. $1, -1, 1, -1, 1, \ldots$

7. Use calculator regression to fit a function to this table. Explain your choice of functions.

x	$f(x)$
2	19.620
3	44.145
4	78.480
5	122.625

8. Describe how the -2 in $y = -2x^2$ changes the graph of $y = x^2$.

9. Explain how the -4 in $y = x^2 - 4$ changes the position, vertex, and intercepts from those of $y = x^2$.

10. Write as complex numbers:

    a. $\sqrt{-16}$ **b.** 4 **c.** $3 + \sqrt{-4}$ **d.** $\sqrt{-56}$

11. Multiply, and simplify using properties of complex numbers.

    a. $(3 - 2i)(3 + 2i)$ **b.** $(2 - i)(2 + i)$

    c. $(x - i)(x + i)$ **d.** $(2 - 3i)(2 + 3i)$

12. **a.** Compare your answers to parts a and d in Problem 11. What is unusual about the answers?

    b. Find the products

    $$(1 + 2i)(1 - 2i) \quad \text{and} \quad (2 + i)(2 - i)$$

13. Evaluate the discriminant to find how many real-number solutions exist to $f(x) = 0$, where $f(x) = -x^2 + 2x + 3$. Find the solutions.

14. Find all real- or complex-number solutions to $f(x) = 0$, where $f(x) = (x - 2)(x^2 + 2x + 4)$.

⬤
4.4

Shifts, Vertex Form, and Applications of Quadratic Functions

OBJECTIVES

- Identify horizontal or vertical shifts of $y = x^2$.

- Use the vertex form to predict the graph of a quadratic equation.

- Find the vertex from a quadratic equation in vertex form.

- Find the equation of a parabola from its vertex and one other point.

- Use completing the square to change a quadratic equation to vertex form.

WARM-UP

1. Complete the table.

x	x^2	$(x-2)^2$	$(x+1)^2$
-2			
-1			
0			
1			
2			

2. Calculate these binomial squares.
   a. $(x-2)^2$ b. $(x+1)^2$
   c. $(x-1.5)^2$ d. $(2x+1)^2$
3. What must be added to make each of these expressions the square of a binomial?
   a. x^2+6x b. x^2-5x

HAVE YOU EVER tried to use a drinking fountain when the water flow failed to reach an appropriate height? Suppose, for sanitary and comfortable drinking, the vertex of the parabola formed by the water flowing from a drinking fountain needs to be horizontally 5 inches from the nozzle and vertically 6 inches above the nozzle (see Figure 31). In this section, we explore how knowing the vertex helps describe an equation for the water flow.

We begin by shifting the graphs of quadratic equations horizontally and vertically. We then find the vertex of a parabola from the vertex form of a quadratic equation and complete the square on a quadratic equation to change the equation into vertex form.

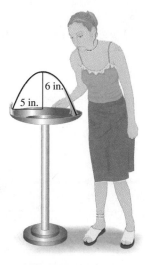

Figure 31

Horizontal Shifts

In Section 4.0, we graphed equations containing the squares of binomials, such as $y=(x+3)^2$ and $y=(x-3)^2$, and observed the following.

Horizontal Shift

> If the quadratic equation can be written as $y=(x-h)^2$, the vertex is at $(h,0)$. The graph of $y=(x-h)^2$ is *shifted horizontally h units* from the graph of $y=x^2$.

The formal name given to the horizontal shift of $y=x^2$ to $y=(x-h)^2$ is **translation,** *shifting a graph parallel to one of the axes.*

EXAMPLE

Predicting horizontal shifts   Predict the graphs for the following equations and then check by graphing.

   a. $y=(x+1)^2$ b. $y=(x-2)^2$

Solution   **a.** Because $a = 1$, the graph of $y = (x + 1)^2 = x^2 + 2x + 1$ is the same shape as that of $y = x^2$. When we let $x + 1 = 0$, we obtain $x = -1$, the x-coordinate of the vertex. The vertex $(-1, 0)$ is 1 unit to the left of the origin. Thus, the graph of $y = (x + 1)^2$ is shifted 1 unit to the left of that of $y = x^2$, as shown in Figure 32.

b. Because $a = 1$, the graph of $y = (x - 2)^2 = x^2 - 4x + 4$ is the same shape as that of $y = x^2$. When we let $x - 2 = 0$, we obtain $x = 2$, the x-coordinate of the vertex. The vertex $(2, 0)$ is 2 units to the right of the origin. Thus, the graph of $y = (x - 2)^2$ is shifted 2 units to the right of that of $y = x^2$, as shown in Figure 33.

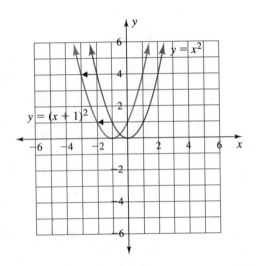

Figure 32

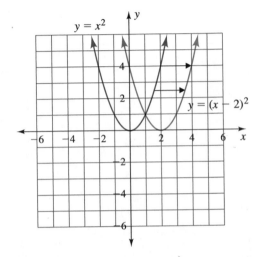

Figure 33

Think about it 1: For $(x - h)^2$ or $(x + h)^2$, tell how you might know whether the graph shifts to the left or to the right.

Vertical Shifts

In Section 4.0, we graphed equations containing differences of squares, such as $y = x^2 - 1$ and $y = x^2 - 4$. In Section 4.2, we extended the graphing to equations containing either the addition or the subtraction of a constant, $y = x^2 \pm c$, and observed the following.

Vertical Shift

> If the quadratic equation can be written as $y = x^2 + c$, the vertex is at $(0, c)$. The parabola for $y = x^2 + c$ is *shifted vertically $|c|$ units* from the graph of $y = x^2$.

EXAMPLE **2**   Predicting vertical shifts   Predict the graphs for the following equations, and then check by graphing.

  a. $y = x^2 + 1$ **b.** $y = x^2 - 2$

Solution   **a.** Because $a = 1$, the graph of $y = x^2 + 1$ is the same shape as that of $y = x^2$. The equation shows 1 added to x^2. The vertex $(0, 1)$ is 1 unit above the origin. Thus, the graph of $y = x^2 + 1$ is shifted 1 unit above that of $y = x^2$, as shown in Figure 34.

b. Because $a = 1$, the graph of $y = x^2 - 2$ is the same shape as that of $y = x^2$. The equation shows 2 subtracted from x^2. The vertex $(0, -2)$ is 2 units below the origin. Thus, the graph of $y = x^2 - 2$ is shifted 2 units below that of $y = x^2$, as shown in Figure 35.

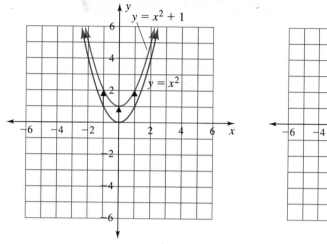

Figure 34   **Figure 35**

Think about it 2: For $x^2 + c$ or $x^2 - c$, tell how you might know whether the graph shifts up or down.

Quadratic Equations in Vertex Form

In Example 1, we obtained a horizontal shift from equations containing a binomial square. In Example 2, we obtained a vertical shift from equations having a constant c added or subtracted. In Example 3, we explore the graph that results from adding a constant to a quadratic equation containing a binomial square.

EXAMPLE **3**   Combining horizontal and vertical shifts   Graph $y = (x - 2)^2 + 3$, and describe how the graph might be obtained from $y = x^2$.

Solution   The graph of $y = (x - 2)^2 + 3$ is the same shape as that of $y = x^2$ but with the vertex at $x = 2$ and $y = 3$ (see Figure 36). The vertex shifted to the right because of the $(x - 2)^2$, but then the vertex shifted up 3 units because 3 is added to $(x - 2)^2$.

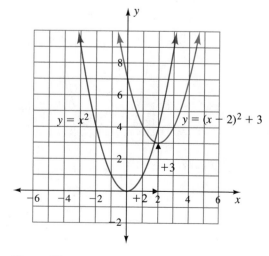

Figure 36

Quadratic equations written as in Example 3 are said to be in vertex form.

Definition of Vertex Form

The **vertex form of a quadratic equation** is $y = a(x - h)^2 + k$, where the vertex coordinates are (h, k).

The letter a is the same as in $y = ax^2 + bx + c$. The formula shows a horizontal and vertical translation, or shift, of the vertex point $(0, 0)$ on $y = x^2$ to a new vertex (h, k), as shown in Figure 37.

Student Note: The signs on h and k are not known, so Figure 37 represents any shift from (0, 0) to (h, k).

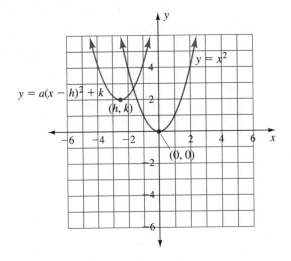

Figure 37

EXAMPLE **4**

Graphing and finding the vertex from vertex form

a. Use shifts to graph $y = (x + 1)^2 - 3$.

b. Find the vertex from the vertex form of this quadratic equation.

Solution

a. From our work with shifts, we set $x + 1 = 0$ and solve for x. We obtain $x = -1$, so the curve first shifts to a vertex at $(-1, 0)$. The -3 added to $(x + 1)^2$ shifts the curve down 3 units. Thus, the vertex is at $(-1, -3)$. The graph of $y = (x + 1)^2 - 3$ is the same shape as that of $y = x^2$, but with the vertex at $x = -1$ and $y = -3$ (see Figure 38).

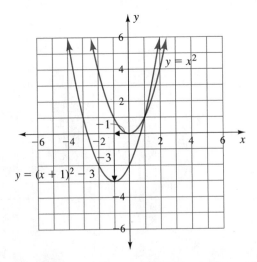

Figure 38

b. We change the equation to $y = (x - h)^2 + k$ form:

$$y = (x + 1)^2 - 3$$
$$y = (x - (-1))^2 + (-3)$$

Since $h = -1$ and $k = -3$, the vertex is at $(-1, -3)$.   ●

BUILDING QUADRATIC EQUATIONS IN APPLICATIONS   In Examples 5 and 6, we use the vertex form to find equations for application settings. In these applications, we are given the vertex and one other point and need to find the equation for the parabola.

Building a Quadratic Equation from Its Vertex and One Other Point	1. Start with the vertex formula, $y = a(x - h)^2 + k$. 2. Using the vertex (h, k), substitute for h and k in $y = a(x - h)^2 + k$. 3. Using the ordered pair for another point on the graph, substitute for x and y. 4. Solve for a. 5. Substitute a and the vertex into the vertex formula.

EXAMPLE **5**   Building an equation from a vertex and one other point: a water fountain's stream
Suppose a parabolic stream of water from a drinking fountain has a vertex at $(5, 6)$ in inches and origin, $(0, 0)$, at the nozzle where the water emerges. Fit an equation to the stream of water.

Solution

$y = a(x - h)^2 + k$ Substitute the vertex, $(h, k) = (5, 6)$, into the vertex formula.

$y = a(x - 5)^2 + 6$ Because the origin, $(0, 0)$, lies on the curve and makes the equation true, substitute $x = 0$, $y = 0$ into the equation.

$0 = a(0 - 5)^2 + 6$ Solve for a.

$-6 = 25a$

$-\frac{6}{25} = a$

$a = -0.24$

Now we substitute $a = -0.24$ and the vertex into the vertex formula.

$$y = -0.24(x - 5)^2 + 6$$

The coefficient a is negative, in agreement with the path of the water.   ●

EXAMPLE **6**

Building an equation from a vertex and one other point: suspension bridge cable
Suppose the lowest point on a parabolic cable for a suspension bridge is to be 10 feet above the roadbed. One support for the cable is 20 feet away (horizontally) and 25 feet above the roadbed.

a. Sketch a coordinate picture of the situation, using any convenient origin.

b. Fit an equation to the cable, using the coordinates in your picture.

Solution   **a.** A sketch of the cable is shown in Figure 39. Other locations for the origin are possible.

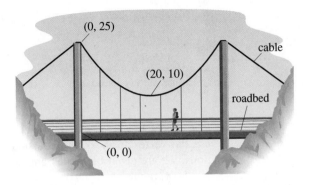

Figure 39

b. The vertex of the cable is at $(20, 10)$. One other point on the cable is at the support $(0, 25)$.

$y = a(x - h)^2 + k$ Substitute the vertex, $(h, k) = (20, 10)$, into the vertex formula.

$y = a(x - 20)^2 + 10$ Substitute the other point, $(0, 25)$.

$25 = a(0 - 20)^2 + 10$ Solve for a.

$15 = a(400)$

$\frac{15}{400} = a$

$a = 0.0375$

Now we substitute a and the vertex into the vertex formula.

$y = 0.0375(x - 20)^2 + 10$

The coefficient a is positive, in agreement with the upward direction of the parabolic cable.   ●

Other methods for finding the equations in Examples 5 and 6 will be considered in the exercises.

Think about it 3: How will the equation in Example 6 change if the origin is placed at the top of the left end of the parabola?

COMPLETING THE SQUARE TO CHANGE TO VERTEX FORM   We can change a quadratic equation into vertex form by completing the square on the equation in standard form, $y = ax^2 + bx + c$.

If $a \neq 1$, we divide both sides by a. Solving for y is then somewhat complicated, so we are not going to do it at this time.

If $a = 1$, first we apply the process used in Section 3.4 to complete the square on a quadratic equation. Then we change to the less general vertex form, $y = (x - h)^2 + k$.

Changing a Quadratic Equation to
Vertex Form, $a = 1$

1. Write the equation as $y = x^2 + bx + c$.
2. Subtract the constant term, c, from both sides.
3. Complete the square on the right, adding the required number to both sides.
4. Change to squared form on the right.
5. Change to vertex form, $y = (x - h)^2 + k$.

The vertex (h, k) can now be read from the equation.

EXAMPLE **7**

Changing a quadratic equation to vertex form   Complete the square to find the vertex form of the equation, and identify the vertex. Check by graphing the original equation.

a. $y = x^2 + 6x + 5$

b. $y = x^2 - 5x - 1$

Solution **a.**

$$y = x^2 + 6x + 5$$ Subtract the constant term, 5, from each side.

$$y - 5 = x^2 + 6x$$ Complete the square on the right.

$$y - 5 + 9 = x^2 + 6x + 9$$ Change to squared form on the right.

$$y + 4 = (x + 3)^2$$ Change to vertex form.

$$y = (x + 3)^2 - 4$$

The vertex is at $(-3, -4)$. The graph is shown in Figure 40.

b.

$$y = x^2 - 5x - 1$$ Add 1 to each side.

$$y + 1 = x^2 - 5x$$ Complete the square on the right.

$$y + 1 + (2.5)^2 = x^2 - 5x + (2.5)^2$$ Simplify.

$$y + 7.25 = x^2 - 5x + 6.25$$ Change to squared form on the right.

$$y + 7.25 = (x - 2.5)^2$$ Change to vertex form.

$$y = (x - 2.5)^2 - 7.25$$

The vertex is at $(2.5, -7.25)$. The graph is shown in Figure 40.

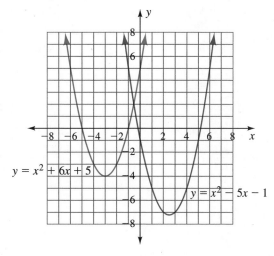

Figure 40

ANSWER BOX

Warm-up: 1.

x	x^2	$(x - 2)^2$	$(x + 1)^2$
-2	4	16	1
-1	1	9	0
0	0	4	1
1	1	1	4
2	4	0	9

2. a. $x^2 - 4x + 4$ **b.** $x^2 + 2x + 1$ **c.** $x^2 - 3x + 2.25$
d. $4x^2 + 4x + 1$ **3. a.** 9 **b.** 6.25 **Think about it 1:** For $(x - h)^2$, the
vertex is at $x - h = 0$, or $x = h$. The graph shifts to the vertex
position. Thus, for $(x - h)^2$, the shift is to the right; for $(x + h)^2$, the
shift is to the left. **Think about it 2:** For $x^2 + c$, the vertex is at
$(0, c)$, so the shift is up. For $x^2 - c$, which is equal to $x^2 + (-c)$, the
vertex is at $(0, -c)$, so the shift is down. **Think about it 3:** Moving
the origin up 25 units will place the part of the parabola shown below
the x-axis. The vertex will be 25 units lower than before. The equation
is $y = 0.0375(x - 20)^2 + 10 - 25$, or $y = 0.0375(x - 20)^2 - 15$.

EXERCISES 4.4

*Each of the graphs in Exercises 1 to 8 has the same steepness as the
graphs of $y = x^2$ and $y = -x^2$. Identify the equation of each graph.
Check with a graphing calculator.*

1.

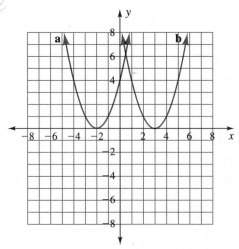

2.

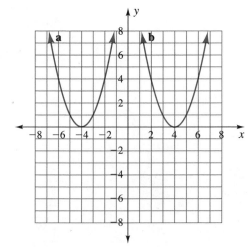

(3.)

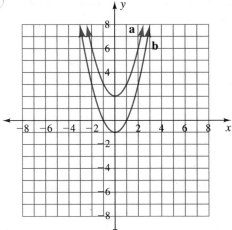

4.

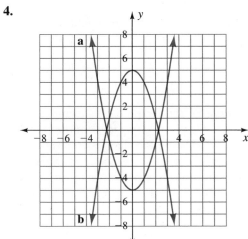

(5.)

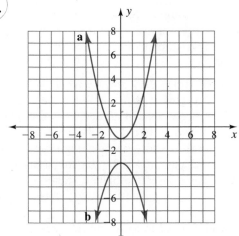

6.

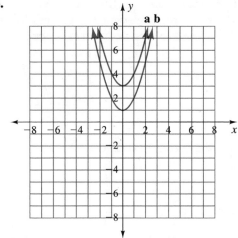

7.

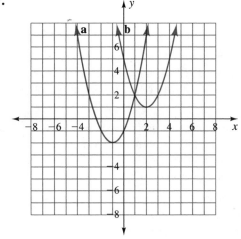

8.

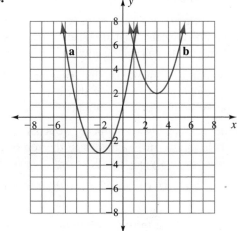

In Exercises 9 to 16, describe how the graph of each equation might be obtained from $y = x^2$. Use complete sentences, and describe all shifts or translations. Sketch your graph, and check with a graphing calculator.

9. $y = (x + 2)^2$

10. $y = (x - 3)^2$

11. $y = (x - 4)^2$

12. $y = (x + 4)^2$

13. $y = (x - 3)^2 + 4$

14. $y + 3 = (x + 2)^2$

15. $y - 3 = (x + 4)^2$

16. $y = (x - 4)^2 - 3$

Identify the vertex of each equation in Exercises 17 to 22.

17. $y = (x - 1)^2$

18. $y = (x + 2)^2$

19. $y = (x + 3)^2 + 4$

20. $y = (x - 4)^2 + 3$

21. $y = (x - 2)^2 - 3$

22. $y = (x + 1)^2 - 2$

In Exercises 23 to 28, find the equation of the parabola, given the vertex V and another point P.

23. $V(3, 1)$, $P(-1, 3)$

24. $V(2, -1)$, $P(-2, 3)$

25. $V(-1, -1)$, $P(1, 7)$

26. $V(-2, 1)$, $P(-3, -1)$

27. $V(3, -2)$, $P(5, -10)$

28. $V(-3, 2)$, $P(-2, 4)$

29. *A Water Fountain's Stream.* In Example 5, we substituted one point, $(0, 0)$, and the vertex, $(5, 6)$, into the vertex formula to find the equation of the stream of water.

    a. Use the symmetry property of parabolas to find a third point on the stream of water, and use quadratic regression on the three points to find the equation of the stream of water.

    b. Multiply out the square in the equation from Example 5, and simplify to show that it is the same equation as the regression equation.

30. *Suspension Bridge Cable.* In Example 6, we substituted one point and the vertex into the vertex formula to find the equation of the cable.

    a. Use the symmetry property of parabolas to find a third point on the cable, and use quadratic regression on the three points to find the equation of the cable.

    b. Multiply out the square in the equation in Example 6, and simplify to show that it is the same equation as the regression equation.

31. *Suspension Bridge Cable.* Suppose we change the location of the origin in Example 6 to the top of the support tower, currently labeled $(0, 25)$.

    a. Write the resulting coordinates for the other two labeled points.

    b. Use the vertex form to find the resulting equation.

    c. Compare your equation with that found in Example 6, and explain the difference in the equations.

32. *Suspension Bridge Cable, Again.* Suppose we change the location of the origin in Example 6 to the vertex of the cable, currently labeled $(20, 10)$.

    a. Write the resulting coordinates for the other two labeled points.

    b. Use the vertex form to find the resulting equation.

    c. Compare your equation with that found in Example 6, and explain the difference in the equations.

33. *Water Jet Equation.* A jet of water is to reach a height of 100 feet when it is a horizontal distance of 30 feet from its origin. Find the equation of the jet of water.

34. *Another Water Jet Equation.* A jet of water is to reach a height of 200 feet when it is a horizontal distance of 40 feet from its origin. Find the equation of the jet of water.

In Exercises 35 to 42, change the equations to vertex form by completing the square. Check by graphing the original and vertex forms with a graphing calculator.

35. $y = x^2 + 10x + 25$

36. $y = x^2 + 4x + 4$

37. $y = x^2 - 6x + 9$

38. $y = x^2 - 10x + 25$

39. $y = x^2 + 10x + 30$

40. $y = x^2 + 4x - 5$

41. $y = x^2 - 6x + 8$

42. $y = x^2 - 10x + 20$

Projects

43. *Absolute Value Shifts.* Graph the following equations on a graphing calculator. Compare the positions of the pairs of graphs. How do the changes in position compare with the movements of parabolas in this section? Write a quadratic equation that shows the same change in position with $y = x^2$.

    a. $y = |x|$ and $y = |x + 2|$

    b. $y = |x|$ and $y = |x| + 2$

    c. $y = |x|$ and $y = |x| - 2$

    d. $y = |x|$ and $y = |x - 2|$

44. *Measuring a Fountain's Stream.* Using the source of water as the origin, measure the stream of water coming from a drinking fountain turned on full force. Find its equation.

45. *A Summer or Warm Day Investigation.* In writing, describe an experiment that will solve the following problem. List the equipment needed. Check your plan with your instructor before doing the project.

    Find the equations for water coming from a garden hose held at various angles. If you can measure the height, what angle gives the highest stream of water? What angle gives the greatest distance reached by the stream of water?

    Include a record of 15 to 20 angles, the heights, and the distance reached by the water at each angle. Plot the angles on the x-axis and the distance on the y-axis.

46. *Temperature and Water Volume.* In the table, the input is the temperature of water, and the output is the volume filled by the water at that temperature. Increased pressure keeps the water liquid.

Input: Temperature (°C)	Output: Volume (cc)
−5	1.00070
−4	1.00055
−3	1.00042
−2	1.00031
−1	1.00021
0	1.00013
1	1.00007
2	1.00003
3	1.00001
4	1.00000
5	1.00001
6	1.00003
7	1.00007
8	1.00012
9	1.00019
10	1.00027

Volume of water near freezing

Note: The mass of 1 cubic centimeter of water at 4 degrees Celsius (°C) is 1 gram. Each of the volumes listed in the table represents 1 gram of water heated or cooled to the given temperature.

Source: Data reprinted with permission of CRC Press, Boca Raton, Florida, from the *Handbook of Chemistry and Physics,* 36th Edition, 1954–55, Chemical Rubber Publishing Company, 2310 Superior Ave. N.E., Cleveland, Ohio, page 1963.

Answer the following questions to help you understand the relationship between temperature and water volume.

a. Where is the smallest volume in the table?

b. Does the volume get larger or smaller as the temperature moves away from 4°C?

c. Does the volume of water change the same amount for each degree change?

d. For the data given, where does the greatest change take place?

e. Graph the 16 data points. Discuss how you set the scale on the axes.

f. Set up an equation in $y = a(x - h)^2 + k$ form. Guess and check the coefficient a that makes the parabola fit the data. Describe your process. (*Hint:* Where is the vertex in the table?)

g. Apply calculator regression to all 16 data points. Compare the calculator results with your equation in part f.

4.5 Solving Minimum and Maximum Problems

OBJECTIVES

- Find the vertex from the graph, intercepts, and quadratic formula.
- Solve minimum and maximum application problems.

WARM-UP

Use the quadratic formula to find the x-intercepts of each function. Simplify and reduce any fractions, but leave your answers in the form

$$x = -\frac{b}{2a} \pm \frac{\sqrt{b^2 - 4ac}}{2a}$$

1. $f(x) = x^2 + 6x - 16$ **2.** $f(x) = x^2 - 2x$

3. $f(x) = x^2 - 2x - 8$ **4.** $f(x) = x^2 + 4x - 21$

5. $f(x) = 2x^2 + 3x - 4$

THIS SECTION FOCUSES on finding a formula for the vertex, calculating the vertex, and solving minimum and maximum problems.

Finding the Vertex from the Graph and Intercepts

Have you ever looked up facts in the *Guinness Book of Records*? This book is filled with facts about greatest and smallest values. In mathematics applications, the vertex is the position on parabolic graphs where the **maximum** (*greatest*) and **minimum** (*smallest*) values are obtained.

Suppose we wanted to find the maximum height reached by the Olympic competitor in Example 11 of Section 3.2. The equation for her height at time t is $h = -16.1t^2 + 6t + 32.8$. This equation does not lend itself to easily finding the vertex from intercepts (Section 3.2) or completing the square (Section 4.4).

Although Table 5 shows the diver's height at time t, it lacks the symmetry of outputs needed to determine a vertex. We estimate the vertex to be near 33.4 feet. The graph in Figure 41 also shows her height above the water at t seconds. Duplicating the graph on a graphing calculator would permit us to zoom and trace to find the vertex.

Input	Output
0	32.8
0.2	33.4
0.4	32.6
0.6	30.6
0.8	27.3
1.0	22.7
1.2	16.8
1.4	9.6
1.6	1.2
1.8	−8.6

Table 5

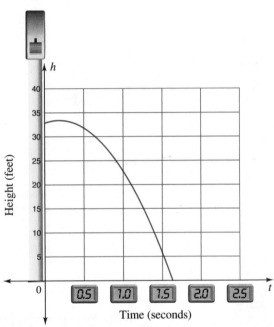

Figure 41

When the equation is known, we can apply the quadratic formula, find the x-intercepts, and, because of the symmetry of the graph, locate the x-coordinate of the vertex halfway between the x-intercepts.

EXAMPLE **1**   Finding the vertex: Olympic diver

a. Find the horizontal axis intercepts for the quadratic function $h(t) = -16.1t^2 + 6t + 32.8$.

b. Use the horizontal axis intercepts to find the vertex.

c. Compare the vertex with the table estimate of the maximum height.

Solution   a. We want to find t when $h = 0$. The coefficients are $a = -16.1$, $b = 6$, and $c = 32.8$.

$$t = \frac{-b \pm \sqrt{b^2 - 4ac}}{2a} = \frac{-6 \pm \sqrt{36 - 4(-16.1)(32.8)}}{2(-16.1)}$$

$$= \frac{-6 \pm \sqrt{2148.32}}{-32.2}$$

$t \approx -1.3$ or $t \approx 1.6$, rounded to the nearest tenth

b. The sum of the intercepts divided by 2 gives the time coordinate of the vertex:

$$t = \frac{-1.3 + 1.6}{2} \approx 0.2$$

The height coordinate of the vertex is

$$h = f(0.2) = -16.1(0.2)^2 + 6(0.2) + 32.8 \approx 33.4$$

Rounding to the nearest tenth, we have for the vertex (0.2, 33.4).

c. The table estimate of the maximum height, $h = 33.4$ ft, was correct to the nearest tenth.   ●

Finding the Vertex from the Quadratic Formula

The quadratic formula in the form

$$x = -\frac{b}{2a} \pm \frac{\sqrt{b^2 - 4ac}}{2a}$$

not only shows why the vertex is symmetrically placed between the x-intercepts but also gives a formula for the x-coordinate of the vertex. In Figure 42, the distance between the axis of symmetry and each x-intercept is $\sqrt{b^2 - 4ac}/2a$. The axis of symmetry, which passes through the vertex V, has equation $x = -b/2a$. This same expression is also the x-coordinate of the vertex. Because the x-coordinate is $x = -b/2a$, the y-coordinate of the vertex may be described with function notation $y = f(-b/2a)$.

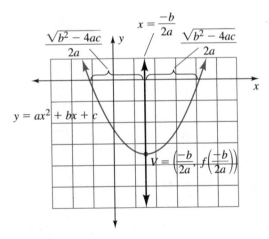

Figure 42

Finding the Vertex	For the function $f(x) = ax^2 + bx + c$,

• The **x-coordinate of the vertex** is a portion of the quadratic formula, $x = -b/2a$.
• The **y-coordinate of the vertex** is the output, $f(-b/2a)$.

EXAMPLE 2   Finding the vertex   Find the vertex for the diver's equation, using $x = -b/2a$ to find the horizontal coordinate (time t) of the vertex.

Solution   For $t = -b/2a$, $a = -16.1$ and $b = 6$. Then $t = -6/(2(-16.1)) \approx 0.2$. The height, the y-coordinate of the vertex, is $f(0.2) \approx 33.4$ ft. We have for the vertex (0.2, 33.4), as before.   ●

EXAMPLE **3**   Finding the vertex

a. Find the x-intercepts of $f(x) = x^2 + 6x - 16$.   **b.** Find the vertex.

Solution   **a.** We obtain the x-intercepts by evaluating the quadratic formula:

$$x = -\frac{b}{2a} \pm \frac{\sqrt{b^2 - 4ac}}{2a} = -\frac{6}{2(1)} \pm \frac{\sqrt{36 - 4(1)(-16)}}{2(1)}$$

$$= -3 \pm 5$$

$$x = -8 \quad \text{or} \quad x = 2$$

b. The x-intercepts are each 5 units from the axis of symmetry, $x = -3$, as shown in Figure 43. The x-coordinate of the vertex is $x = -3$. [The same result is obtained using $x = -b/2a = -6/2(1) = -3$.] We substitute $x = -3$ into $f(x)$ to find the y-coordinate of the vertex:

$$f(x) = x^2 + 6x - 16$$

$$f(-3) = (-3)^2 + 6(-3) - 16$$

$$= -25$$

The vertex is $(-3, -25)$ (see Figure 43).

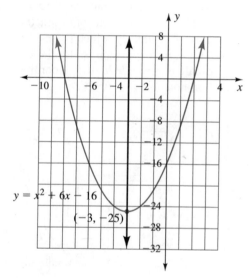

Figure 43

The advantage of the formula method of finding the vertex is that it does not require graphing or finding the x-intercepts. As a result, it may be used with quadratic equations whose graphs have no x-intercepts, as in Example 4.

EXAMPLE **4**   Finding the vertex

a. Find the x-intercepts of $f(x) = 2x^2 + 3x + 4$.

b. Find the vertex of $f(x)$.

Solution   **a.** The results of the quadratic formula are as follows:

$$x = -\frac{b}{2a} \pm \frac{\sqrt{b^2 - 4ac}}{2a} = -\frac{3}{2(2)} \pm \frac{\sqrt{9 - 4(2)(4)}}{2(2)}$$

$$= -\frac{3}{4} \pm \frac{\sqrt{-23}}{4}$$

$$= -\frac{3}{4} \pm i\frac{\sqrt{23}}{4}$$

The solutions are complex numbers, so there are no x-intercepts. See Figure 44.

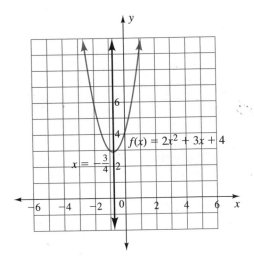

Figure 44

b. From the vertex formula, the x-coordinate of the vertex is

$$x = -\frac{b}{2a} = -\frac{3}{4}$$

We substitute $x = -\frac{3}{4}$ into $f(x)$ to find the y-coordinate of the vertex:

$$f(x) = 2x^2 + 3x + 4$$
$$f\left(-\tfrac{3}{4}\right) = 2\left(-\tfrac{3}{4}\right)^2 + 3\left(-\tfrac{3}{4}\right) + 4$$
$$= 2\tfrac{7}{8}$$

The vertex is $\left(-\tfrac{3}{4}, 2\tfrac{7}{8}\right)$, or $(-0.75, 2.875)$.

Thus far, we have the following results about the vertex of a parabola.

Summary: Vertex of the Parabolic Graph of $y = ax^2 + bx + c$

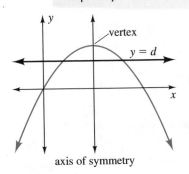

Figure 45

- Once symmetry is observed in an input-output table, the vertex can be located at the highest or lowest output.
- The vertex is the highest or lowest point on a parabolic graph (Figure 45).
- The vertex lies on the axis of symmetry of a parabolic graph (Figure 45).
- The x-coordinate of the vertex is the midpoint between the x-intercepts [solutions to $f(x) = 0$].
- The x-coordinate of the vertex is the midpoint of any horizontal segment, $y = d$, that intersects the parabola in two points (Figure 45).
- The vertex is (h, k) when we have completed the square on a quadratic equation and changed it into vertex form,

$$y = a(x - h)^2 + k$$

- The vertex has the ordered pair (x, y), where $x = -\dfrac{b}{2a}$ and $y = f\left(\dfrac{-b}{2a}\right)$.

Applications

PATH OF A BALL   In Example 5, we work with a parabola that represents the path of a ball. In this example, the horizontal and vertical axes are both labeled in feet.

EXAMPLE **5**

Finding the vertex: path of a soccer ball   Orlando kicks a soccer ball at a 45° angle relative to the ground. If we ignore wind and air resistance, the path the ball follows is given by the quadratic equation

$$y = -\frac{g}{v^2}x^2 + x$$

The input x is the horizontal distance traveled by the ball, and the output y is the vertical position. The constant g is the average acceleration due to gravity, 32.2 ft/sec², and v is the initial velocity from the kick. Suppose the ball is kicked with a 80-ft/sec velocity.

a. What is the maximum height reached by the soccer ball?

b. What horizontal distance will it travel before returning to the ground?

Solution   **a.** We substitute $v = 80$ ft/sec and $g = 32.2$ ft/sec² into the given quadratic equation and simplify to identify a, b, and c.

$$y = -\frac{g}{v^2}x^2 + x = -\frac{(32.2)}{80^2}x^2 + x \approx -0.005x^2 + x$$

The coefficient of x^2, $a = -0.005$, is rounded to the nearest thousandth. The equation is graphed in Figure 46.

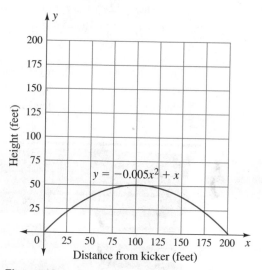

Figure 46

Because $a \approx -0.005$, $b = 1$, and $c = 0$, the x-coordinate of the vertex, in feet, is

$$x = -\frac{b}{2a} = -\frac{1}{2(-0.005)} = 100 \text{ ft}$$

The y-coordinate of the vertex is the maximum height reached by the ball. This height is $f(-b/2a)$, or $f(100 \text{ ft})$:

$$f(100) = -0.005(100)^2 + 1(100) = 50 \text{ ft}$$

These results look much more accurate than they are. Remember that the coefficient a was rounded.

b. Because the vertex, (100, 50), lies on the axis of symmetry and one x-intercept is 0, the second x-intercept is twice the x-coordinate of the vertex, or 200 feet. Of course, the quadratic formula would give the x-intercepts, but the equation factors easily.

$$0 = -0.005x^2 + 1x \qquad \text{Factor.}$$
$$0 = x(-0.005x + 1) \qquad \text{Apply the zero product rule.}$$

Either $\quad x = 0 \quad$ or $\quad (-0.005x + 1) = 0 \qquad$ Solve the resulting equations.

$$x = 0 \quad \text{or} \qquad x = \frac{1}{0.005} = 200 \text{ ft}$$

Figure 46 also shows that the ball will travel 200 feet. ●

Think about it 1: What is the meaning of the solution $x = 0$ in Example 5?

MAXIMUM AREA   Example 6 reminds us that we may have applications in which solving two equations in a system of equations leads to a quadratic equation.

EXAMPLE **6**

Finding the vertex   Mei has 80 feet of wire bird-netting to use around her vegetable garden to keep out slugs and cats. The garden is to be rectangular (Figure 47). What length and width give the maximum area for the available netting?

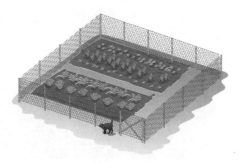

Figure 47

Solution   The perimeter of the garden is given by $P = 2l + 2w$. The perimeter is 80 feet. Substituting $P = 80$ ft and solving for l yields

$$80 = 2l + 2w$$
$$l = 40 - w$$

The area of the garden is given by $A = lw$. Substituting $l = 40 - w$ for l in the area formula yields

$$A = lw$$
$$A = (40 - w)w$$
$$A = 40w - w^2$$

This is a quadratic equation, so its vertex will give the maximum area as a function of width. The vertex is located where

$$w = -\frac{b}{2a} = -\frac{40}{2(-1)} = 20 \text{ ft}$$

Thus,

$$A = 40w - w^2 = 40(20) - 20^2 = 400 \text{ ft}^2$$

The width is 20 feet, and from $l = 40 - w$ we find that the length is also 20 feet. Thus, the maximum area rectangular garden with a fixed perimeter of 80 feet is a square. ●

In Example 6, you may be uncomfortable with the idea that the answer is a square when the problem asked for a rectangle. Keep in mind that a square is a rectangle with four equal sides. *Rectangle* is the more generic name, as you may explore in Project 31.

TRANSITION CURVES   Have you ever driven over a hill so fast that the motion seemed to lift you out of your car seat? Civil engineers try to avoid such a possibility when they calculate the *transition curve,* the roadbed design over a hill or between two hills. Transition curves based on quadratic equations permit comfortable and safe travel. The design of the transition curve takes into consideration the slopes of the hills and the elevation of the highway from some reference point. The vertex of a transition curve between two hills is the lowest point, and it indicates the site for the storm water drain.

Although we do not derive the equation for a transition curve until Chapter 7, we apply it in Example 7 to find the drain position—the minimum point.

EXAMPLE **7**   Finding the vertex: roadbed transition curve   A roadbed is to pass through point *G* on one hill and point *H* on the next hill. A side view of the required roadbed is shown in Figure 48. There is to be a 3% downgrade slope at *G* and a 2% upgrade slope at *H*. Suppose point *G* has coordinates (0, 1000), with units in feet. Point *H* has an *x*-coordinate of 1500 feet because it is 1500 feet away from point *G*.

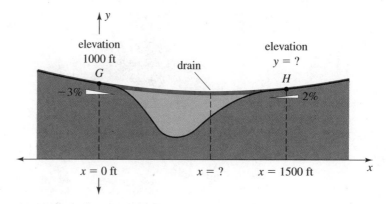

Figure 48

The roadbed transition curve to meet these requirements is given by

$$y = \frac{1}{60000}x^2 - \frac{3}{100}x + 1000$$

a. Find the *y*-coordinate for point *H* in Figure 48.

b. Find the vertex (storm drain location) of the parabolic roadbed.

c. Use a calculator to complete Table 6 and confirm that your answers to parts a and b are correct.

x (feet)	y (feet)
0	
300	
600	
900	
1200	
1500	

Table 6   Roadbed coordinates

Solution **a.** To find the elevation at point H, we place $x = 1500$ in the transition curve equation.

$$y = \tfrac{1}{60000}x^2 - \tfrac{3}{100}x + 1000 = \tfrac{1}{60000}(1500)^2 - \tfrac{3}{100}(1500) + 1000$$
$$= 992.5 \text{ ft}$$

(It may be necessary to place parentheses around the fractions in a calculator.)

b. The x-coordinate of the vertex is $x = -b/2a$.

$$x = -\frac{b}{2a} = -\frac{-\tfrac{3}{100}}{2\left(\tfrac{1}{60000}\right)} = \tfrac{3}{100} \cdot \tfrac{60000}{2} = 900$$

The y-coordinate of the vertex is $f(900)$.

$$y = \tfrac{1}{60000}x^2 - \tfrac{3}{100}x + 1000 = \tfrac{1}{60000}(900)^2 - \tfrac{3}{100}(900) + 1000$$
$$= 986.5 \text{ ft}$$

c. We place the transition curve in the calculator as Y_1 and evaluate for inputs in the table. The symmetry in outputs in Table 7 is at the input 900. Thus, the vertex of the quadratic equation, the location of the storm water drain, is at (900, 986.5). The output at $x = 1500$ agrees with our results in part a.

x (feet)	y (feet)
0	1000
300	992.5
600	988
900	986.5
1200	988
1500	992.5

Table 7 Roadbed coordinates

●

Think about it 2: How long a "water puddle" would form if the storm drain were placed 1.5 feet in elevation above the vertex? Assume the water has nowhere else to run.

It may be helpful to think of the key features of a quadratic function in terms of inputs x and outputs y, as shown in Table 8. The coordinates of the vertex are $(-b/2a, f(-b/2a))$. The x-intercepts are where $y = f(x) = 0$ and may be found in a variety of ways, usually with the quadratic formula. The y-intercept is where $x = 0$; in function notation, the y-intercept is $f(0)$.

	x	y
Vertex	$-b/2a$	$f(-b/2a)$
x-intercept	Find with quadratic formula.	0
y-intercept	0	$f(0)$

Table 8 Key features of graphs of quadratic functions, $f(x) = ax^2 + bx + c$

ANSWER BOX

Warm-up: **1.** -3 ± 5 **2.** 1 ± 1 **3.** 1 ± 3 **4.** -2 ± 5 **5.** $-\dfrac{3}{4} \pm \dfrac{\sqrt{41}}{4}$

Think about it 1: The solution $x = 0$ when height $y = 0$ means that the ball started on the ground before it was kicked.

Think about it 2: If the water had nowhere else to run, it would back up to the 600- and 1200-foot marks relative to the x-axis (see Table 7). The puddle would be 600 feet long.

EXERCISES

In Exercises 1 to 4, use factoring to find the x-intercepts, and then use the intercepts to find the vertex for each equation.

1. $y = x^2 - 2x$

2. $y = x^2 - 2x - 8$

3. $y = -x^2 - 4x + 21$

4. $y = x^2 + 4x - 21$

Use

$$x = -\frac{b}{2a} \quad and \quad y = f\left(-\frac{b}{2a}\right)$$

to find the vertex of each quadratic equation in Exercises 5 to 10.

5. $f(x) = x^2 + 8x + 15$

6. $f(x) = x^2 - 6x + 8$

7. $f(x) = x^2 - 4x + 5$

8. $f(x) = x^2 + 4x + 7$

9. $f(x) = 2x^2 - x - 3$

10. $f(x) = 2x^2 + 5x - 3$

Exercises 11 to 16 refer to Example 5. Round answers to the nearest tenth.

11. Suppose Rosa kicks a soccer ball at a 45° angle with an initial velocity of 90 ft/sec. Predict the maximum height of the ball and the horizontal distance traveled by the ball before it returns to the ground.

12. Suppose Arne kicks a soccer ball at a 45° angle with an initial velocity of 60 ft/sec. Predict the maximum height of the ball and the horizontal distance traveled by the ball before it returns to the ground.

13. A signal flare with a parabolic flight path needs to be seen over a 500-foot-high obstacle. Assume the parabolic path is at a 45° angle. Will it reach sufficient height if it is fired with an initial velocity of 200 ft/sec? What is the highest point it reaches?

14. Use guess and check, a table, or a graph to determine the initial velocity required to reach 500 feet in Exercise 13.

15. Sketch a set of three graphs that show the path of a ball (or other object) hit from (0, 0) at a 45° angle with initial velocities of 50, 100, and 200 ft/sec. Use the graphs to finish these statements:

   a. When the initial velocity is doubled, the height of the vertex above the ground is _____ .

   b. When the initial velocity is doubled, the distance the ball travels is _____ .

16. Set up three tables showing the coordinates of the vertex and x-intercepts for the path of a ball (or other object) hit from (0, 0) at a 45° angle with initial velocities of 40, 80, and 160 ft/sec. Use the tables to finish the statements in parts a and b in Exercise 15.

17. The answer in Example 6 was a square when the perimeter was 80 feet. What nonrectangular shape would make a larger area than the square? What is its area?

18. Suppose Jiri builds a pen shaped as an equilateral triangle. If the perimeter of the pen is 80 feet, what is the area? The formula for the area of an equilateral triangle is $A = x^2\sqrt{3}/4$, where x is the length of one side. Compare the area with the results in Example 6 and Exercise 17.

19. Miguel has 60 feet of deer fencing for a garden next to his house. The garden is to be rectangular, with the house forming one side as shown in the figure. What is the largest possible rectangular area? What length and width give the largest possible area?

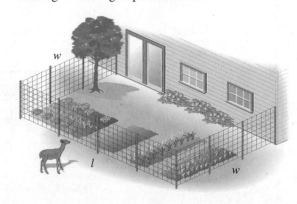

20. Rita has 90 feet of fencing for a corral next to her barn. The corral is to be rectangular, with the barn forming one side, similar to the garden in Exercise 19. What is the largest possible rectangular area? What length and width give the largest possible area?

In Exercises 21 and 22, a ball is thrown straight up. Its height above ground level relative to the time in the air behaves according to the formula $h = -0.5gt^2 + v_0t + h_0$, where g is acceleration due to gravity (32.2 ft/sec^2), v_0 is the initial throwing velocity, and h_0 is the initial height from which the ball is thrown.

21. What is the maximum height for a ball thrown with an initial velocity of 40 ft/sec? Assume the ball leaves the hand at 5.5 feet.

22. What is the maximum height for a ball thrown with an initial velocity of 30 ft/sec? Assume the ball leaves the hand at 4.5 feet.

In Exercises 23 to 26, the road grades connecting two hills are designed with a parabolic curve. Engineers need to find the minimum point for a storm drain.

23. In the figure, point K has a 4% downgrade slope and point L has a 3% upgrade slope. The horizontal distance between points K and L is 2000 feet. The elevation of point K is 700 feet. The equation for the road grade is

$$y = \frac{7}{400,000}x^2 - \frac{1}{25}x + 700$$

Find the elevation of point L and the location for the storm drain.

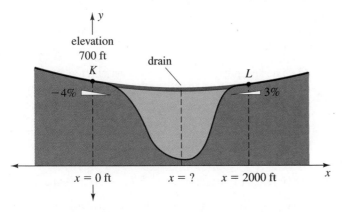

24. In the figure in Exercise 23, suppose point K has a 3% downgrade slope and point L has a 5% upgrade slope. The horizontal distance between points K and L is 2000 feet. The elevation of point K is 700 feet. The new equation for the road grade is

$$y = \frac{1}{50,000}x^2 - \frac{3}{100}x + 700$$

Find the elevation of point L and the location for the storm drain. How did the change in slopes change the equation and the location of the drain from those of Exercise 23?

25. Point M in the figure has a 5% downgrade slope, and point N has a 6% upgrade slope. The horizontal distance between point M and point N is 2400 feet. The elevation of point N is 1200 feet. The equation for the road grade is

$$y = \frac{11}{480,000}x^2 - \frac{1}{20}x + 1188$$

Find the elevation of point M and the location for the storm drain.

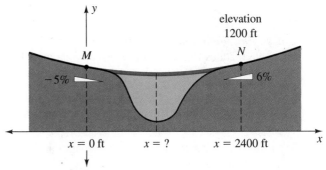

26. In the figure in Exercise 25, suppose point M has a 4% downgrade slope and point N has a 2% upgrade slope. The horizontal distance between point M and point N is 2400 feet. The elevation of point N is 1200 feet. The new equation for the road grade is

$$y = \frac{1}{80,000}x^2 - \frac{1}{25}x + 1224$$

Find the elevation of point M and the location for the storm drain. How did the change in slopes change the equation and the location of the drain from those of Exercise 25?

27. For the highway roadbed in Example 7, the transition curve between the hills was given by

$$y = \frac{1}{60000}x^2 - \frac{3}{100}x + 1000$$

with a storm drain at the vertex, at (900, 986.5) feet. Suppose the storm drain plugs and water backs up onto the roadway to a depth of $\frac{1}{10}$ foot, or an elevation of 986.6 feet. Solve the inequality

$$\frac{1}{60000}x^2 - \frac{3}{100}x + 1000 < 986.6$$

to see how long a "puddle" is formed by the water.

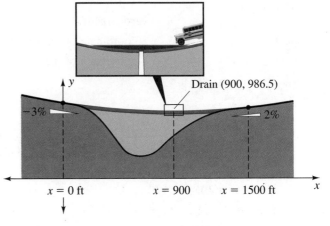

28. Repeat Exercise 27 but with water accumulating to a depth of 3 inches, or 0.25 foot, on the roadway.

29. Describe how to determine the second x-intercept if we know one x-intercept $(x, 0)$ and the vertex (h, k).

30. Substitute $x = -b/2a$ into $y = ax^2 + bx + c$ and find a formula for the y-coordinate of the vertex. Simplify your formula with a common denominator. Where have you seen parts of this formula before?

Project

31. *Classifying Four-Sided Figures.* The text following Example 6 states that a square is a rectangle with four equal sides. Make a chart that shows the relationships among the four-sided figures defined below.

   rectangle: a four-sided figure with opposite sides equal and one right angle

   square: a four-sided figure with one right angle and all sides equal

   trapezoid: a four-sided figure with exactly one pair of parallel sides

   rhombus or diamond: a four-sided figure with all sides equal and no right angle

   quadrilateral: any four-sided figure

   parallelogram: a four-sided figure with opposite sides parallel

   kite: a four-sided figure with two pairs of adjacent sides equal and none parallel

A sketch of each figure may help. Observe that the definitions give the minimum requirements for the figure, not all its properties. Your chart should look somewhat like the bird chart below, with the most generic shape on top and the most specialized shapes toward the bottom.

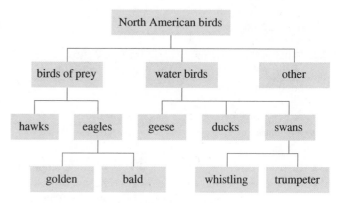

CHAPTER 4 SUMMARY

Vocabulary

For definitions and page references, see the Glossary/Index.

binomial square	maximum
complex conjugate	minimum
complex number system	parameter
cubic expression	perfect square trinomial
degree	quadratic regression
difference of cubes	square of a binomial
difference of squares	sum of cubes
discriminant	translation
double root	vertex form of a quadratic equation
factor theorem	
general polynomial equation	vertical shift
horizontal shift	x-coordinate of the vertex
imaginary unit	y-coordinate of the vertex

Concepts

4.0 Special Products

For equations containing binomial squares,

1. The graph of $y = (x - r)^2$ has its vertex on the x-axis at $(r, 0)$.

2. The graph of $y = (x - r)^2$ is shifted horizontally r units from that of $y = x^2$.

3. The solution to $(x - r)^2 = 0$ is $x = r$, a double root.

For equations containing differences of squares, where c is a positive number,

1. The graph of $y = x^2 - c^2$ has its vertex on the y-axis at $(0, -c^2)$.

2. The graph of $y = x^2 - c^2$ is shifted vertically c^2 units from that of $y = x^2$.

3. The two solutions to $x^2 - c^2 = 0$ are $x = c$ and $x = -c$.

Sum of cubes: $x^3 + y^3 = (x + y)(x^2 - xy + y^2)$

Difference of cubes: $x^3 - y^3 = (x - y)(x^2 + xy + y^2)$

The graph of $y = x^3 \pm c$ is shifted vertically from that of $y = x^3$.

The degree of a polynomial equation $f(x) = 0$ indicates the maximum number of real-number solutions it can have.

4.1 Modeling Quadratic Functions

Fit a linear function to data if the graph of the data is a straight line or an approximately straight line or if the data form a sequence with constant common (or first) differences. Use linear regression to find linear functions.

To find a quadratic equation with x-intercepts and one other point, use $y = a(x - x_1)(x - x_2)$.

Use the difference method to find a, b, and c in $f(x) = ax^2 + bx + c$ from a quadratic sequence:

1. List the sequence, and calculate the first and second differences.

2. Find a, given that the constant second difference is $2a$.

3. Work backwards from the second difference row to find c, the $f(0)$ term before the first term, $f(1)$.

4. Use $f(1)$, $x = 1$, a, and c to find b in $f(x) = ax^2 + bx + c$.

Fit a quadratic function to data if the data form a sequence with constant second differences or if the graph of the data has a parabolic or approximately parabolic shape. Use quadratic regression to find quadratic functions.

4.2 The Role of a, b, and c

The parameters a, b, and c in $f(x) = ax^2 + bx + c$ interact to control the shape, position, and orientation of the graph of $f(x)$.

A positive a on the x^2 term causes the parabola to open up, $\cup$. A negative a on the x^2 term causes the parabola to open down, $\cap$.

The coefficient on the x^2 term, the parameter a, controls the steepness of the graph. If a is larger than 1, the graph is steeper than the graph of $y = x^2$. If a is between zero and 1, the graph is flatter than the graph of $y = x^2$.

The parameter b in $y = x^2 + bx$ contributes to a change in the position of the vertex of the parabola from that of $y = x^2$.

The parameter c in $y = ax^2 + bx + c$ is the output when $x = 0$. The coordinate $(0, c)$ is the y-intercept point of the graph of a quadratic function. If the equation is of the form $y = ax^2 + c$, then c is the distance (number of units) the parabola $y = ax^2$ is shifted vertically (up or down).

4.3 Complex Numbers

The degree of a polynomial equation indicates the maximum number of real-number solutions or the total number of real-number and complex-number solutions.

Only the real numbers appear on a rectangular coordinate graph.

If the discriminant $b^2 - 4ac$ is positive, there are two real-number solutions to the quadratic equation $ax^2 + bx + c = 0$. The graph of $f(x) = ax^2 + bx + c$ passes through the x-axis twice.

If $b^2 - 4ac$ is zero, there is one real-number solution, a double root. The graph of $f(x)$ touches the x-axis once.

If $b^2 - 4ac$ is negative, there are two complex-number solutions. The graph of $f(x)$ does not touch the x-axis. If n is a positive number, then $\sqrt{-n} = i\sqrt{n}$, where i is the imaginary unit.

Addition of complex numbers:

$$(a + bi) + (c + di) = (a + c) + (b + d)i$$

Subtraction of complex numbers:

$$(a + bi) - (c + di) = (a - c) + (b - d)i$$

Multiplication of complex numbers:

$$(a + bi)(c + di) = ac + adi + bci + bdi^2$$
$$= ac - bd + (ad + bc)i$$

4.4 Vertex of a Quadratic Function

Once symmetry is observed in an input-output table, the vertex can be located at the highest or lowest output.

The vertex is the highest or lowest point on a quadratic graph (see Figure 49).

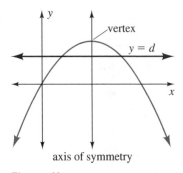

Figure 49

The vertex lies on the axis of symmetry of a quadratic graph (see Figure 49).

The x-coordinate of the vertex is the midpoint between the x-intercepts [solutions to $f(x) = 0$].

The x-coordinate of the vertex is the midpoint of any horizontal segment that intersects the parabola in two points.

The vertex is (h, k) when we have completed the square on a quadratic equation and changed it into vertex form,

$$y = a(x - h)^2 + k$$

4.5 Minimum and Maximum Problems

For the function $f(x) = ax^2 + bx + c$, the x-coordinate of the vertex is a portion of the quadratic formula, $x = -b/2a$. The y-coordinate of the vertex is the output, $f(-b/2a)$.

Table 9 gives key features of graphs of quadratic functions, $f(x) = ax^2 + bx + c$.

	x	y
Vertex	$-b/2a$	$f(-b/2a)$
x-intercept	Find with quadratic formula.*	0
y-intercept	0	$f(0)$

Table 9

*For other ways to find x-intercepts, see the Chapter 3 Summary on pages 233 and 234.

CHAPTER 4 REVIEW EXERCISES

Multiply the expressions in Exercises 1 to 6. Identify the product as a perfect square trinomial, difference of squares, or other.

1. $(x - 6)(x - 2)$

2. $(x - 12)(x + 1)$

3. $(x + 12)(x + 1)$

4. $(2x - 3)(x + 4)$

5. $(x - 2)^2$

6. $(2x + 1)(2x - 1)$

In Exercises 7 to 10, fill in numbers that make these expressions binomial squares.

7. $x^2 - \underline{\hspace{0.5cm}}x + 225 = (x - \underline{\hspace{0.5cm}})^2$

8. $x^2 + 10x + \underline{\hspace{0.5cm}} = (x + \underline{\hspace{0.5cm}})^2$

9. $x^2 + 20x + \underline{\hspace{0.5cm}} = (x + \underline{\hspace{0.5cm}})^2$

10. $x^2 + \underline{\hspace{0.5cm}}x + 169 = (x + \underline{\hspace{0.5cm}})^2$

Factor the expressions in Exercises 11 to 18.

11. $x^2 - 49$

12. $4x^2 - 9$

13. $4x^2 - 4x + 1$

14. $x^2 + 24x + 144$

15. $x^2 + 16$

16. $x^2 - 225$

17. $4x^2 + 12x + 9$

18. $x^2 + 36$

Multiply the expressions in Exercises 19 to 22.

19. $(a - b)(a^2 - 2ab + b^2)$

20. $(a - b)(a^2 + ab + b^2)$

21. $(a + b)(a^2 - ab + b^2)$

22. $(a + b)(a^2 + 2ab + b^2)$

23. Which of Exercises 19 to 22 contains an expression equal to $(a - b)^3$?

24. Which of Exercises 19 to 22 contains an expression equal to $(a + b)^3$?

Factor the expressions in Exercises 25 to 30.

25. $a^3 - 8$

26. $x^3 - 125$

27. $x^3 + 27$

28. $a^3 + 1$

29. $x^3 - 1000$

30. $a^3 + 64$

In Exercises 31 to 34, write an output $y = n$ such that $f(x) = n$ has the number of real-number solutions indicated, if possible. Use the graph and draw the horizontal line representing each output.

a. no solutions   b. 1 solution

c. 2 solutions   d. 3 solutions

e. 4 solutions

31. $f(x) = 2x^2 - 8x + 4$

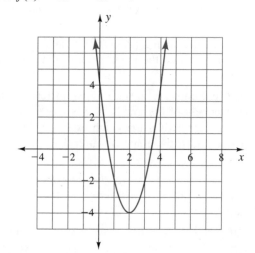

32. $f(x) = -2x^2 - 4x - 1$

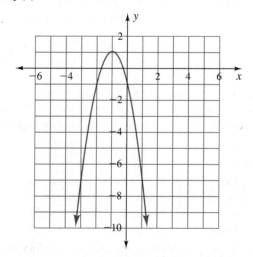

33. $f(x) = x^3 + 2x^2 - 5x - 6$

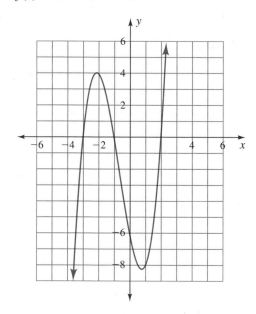

34. $f(x) = 2x^4 + x^3 - 9x^2 - 4x + 4$

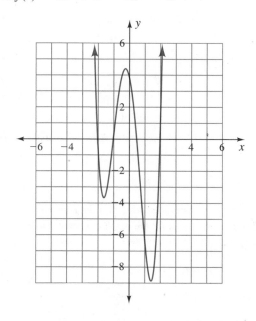

Use the graphs in Exercises 31 to 34 to solve the equations in Exercises 35 to 38. Estimate x, as needed, to the nearest 0.5.

35. a. $2x^2 - 8x + 4 = 0$

  b. $2x^2 - 8x + 4 = 4$

36. a. $-2x^2 - 4x - 1 = 0$

  b. $-2x^2 - 4x - 1 = -7$

37. a. $x^3 + 2x^2 - 5x - 6 = 0$

  b. $x^3 + 2x^2 - 5x - 6 = -6$

38. a. $2x^4 + x^3 - 9x^2 - 4x + 4 = 0$

  b. $2x^4 + x^3 - 9x^2 - 4x + 4 = -10$

In Exercises 39 and 40, find at least one quadratic equation whose graph passes through the x-axis at the given points.

39. $x = 4$ and $x = -2$

40. $x = -3$ and $x = 1$

In Exercises 41 and 42, find a quadratic equation whose graph passes through the given points.

41. (1, 3) and the intercepts in Exercise 39

42. (−1, −4) and the intercepts in Exercise 40

Find out whether the sequences in Exercises 43 to 50 are linear, quadratic, or neither. Fit an equation to the linear or quadratic sequences.

43. 6, 15, 28, 45, 66

44. −1, 3, 13, 29, 51

45. −1, 2, 5, 8, 11

46. 8, 2, −4, −10, −16

47. 15, 12, 7, 0, −9

48. 16, 15, 8, −11, −48

49. 7, 9, 5, −11, −45

50. 24, 21, 16, 9, 0

51. The table below gives the boiling point of water at various elevations above sea level. The differences in boiling points are due to atmospheric pressure.

Elevation above Sea Level (feet), x	Temperature (°C), y
0	100
5,280 (Denver)	95
14,500 (Mt. Whitney)	85
29,500 (Mt. Everest)	71

Boiling point of water

Data from G. Tyler Miller and David G. Lygre, *Chemistry*, Wadsworth Publishing Co., 1991, p. 128.

a. Fit a linear function to the data.

b. Fit a quadratic function to the data.

c. Graph the functions with the data, and discuss which you think is a better description.

d. Use both functions to predict the boiling point of water near the Dead Sea, 1312 feet below sea level.

52. The table below gives the surface area of a sphere for various radii.

Radius	Surface Area of Sphere
1	12.57
2	50.27
3	113.10
4	201.06

a. Fit a linear function to the data.

b. Fit a quadratic function to the data.

c. Discuss which you think is a better description. (*Hint:* Use differences.)

d. Use both functions to predict the surface area of a sphere for a radius of 10. Which is correct?

53. How does the graph of $y = 3x^2$ differ from that of $y = x^2$?

54. How does the graph of $y = -x^2$ differ from that of $y = x^2$?

55. How does the graph of $y = x^2 - 9$ differ from that of $y = x^2$?

56. How does the graph of $y = x^2 + 4$ differ from that of $y = x^2$?

57. How does the graph of $y = x^2 - 9$ show that $x^2 - 9 = 0$ has two real-number solutions while that of $y = x^2 + 9$ shows that $x^2 + 9 = 0$ has no real-number solutions?

58. How does the graph of $y = (x + 4)^2$ indicate that $(x + 4)^2 = 0$ has one real-number solution?

59. Explain why the left side of $(x + 4)^2 = 0$ has two factors and yet the equation has only one real-number solution. What do we call this solution?

Evaluate $b^2 - 4ac$ for the quadratic equations in Exercises 60 to 64. Indicate the number and type of solutions to $f(x) = 0$ and the number of x-intercepts in the graph of f(x).

60. $f(x) = -2x^2 + x + 3$

61. $f(x) = -1x^2 + 2x - 1$

62. $f(x) = 2x^2 + 3x + 4$

63. $f(x) = 3x^2 + 2x + 1$

64. $f(x) = x^2 + 6x + 9$

For Exercises 65 to 67, choose the phrase that best completes each sentence.

a. then the quadratic equation has one real root.

b. then the quadratic equation has two real roots.

c. then the quadratic equation has no real roots.

65. If $b^2 - 4ac$ is negative, _____.

66. If $b^2 - 4ac$ is zero, _____.

67. If $b^2 - 4ac$ is positive, _____.

Which description matches each statement in Exercises 68 to 70?

a. The graph of the quadratic equation just touches the x-axis.

b. The graph of the quadratic equation intersects the x-axis twice.

c. The graph of the quadratic equation does not intersect the x-axis.

68. $b^2 - 4ac < 0$ 69. $b^2 - 4ac > 0$

70. $b^2 - 4ac = 0$

Change each radical expression in Exercises 71 and 72 to a product of a real number and the imaginary unit.

71. a. $\sqrt{-16}$ b. $\sqrt{-50}$

72. a. $\sqrt{-64}$ b. $\sqrt{-18}$

In Exercises 73 and 74, simplify the complex numbers.

73. $\dfrac{8 + 6i}{2}$ 74. $\dfrac{6 + 3i}{3}$

In Exercises 75 and 76, multiply and simplify the expressions.

75. a. $(4 + 3i)(4 + 3i)$ b. $(4 - 3i)(4 + 3i)$

c. $(3 - 4i)(3 + 4i)$

76. a. $(1 - 3i)(1 + 3i)$ b. $(3 - i)(3 + i)$

c. $(1 + 3i)(1 + 3i)$

In Exercises 77 and 78, find all solutions to f(x) = 0, real or complex.

77. $f(x) = (x - 3)(x^2 + 3x + 9)$

78. $f(x) = (x + 2)(x^2 - 2x + 4)$

In Exercises 79 to 82, use any method to find the x-intercepts, if they exist, and then find the vertex for the graph of each equation.

79. $f(x) = x^2 - x - 2$

80. $f(x) = x^2 - x - 6$

81. $y = 2x^2 - 4x + 5$

82. $y = 3x^2 - 6x + 5$

Use the graph of $y = x^2$ and shifts to graph the equations in Exercises 83 to 86.

83. $y = (x - 3)^2 + 2$

84. $y = (x + 3)^2 + 1$

85. $y = (x + 2)^2 - 1$

86. $y = (x - 4)^2 - 2$

In Exercises 87 and 88, complete the square to find the vertex form of the equation.

87. $y = x^2 + 4x + 9$

88. $y = x^2 - 6x + 6$

89. Use the graph and equation in the figure to answer these questions.

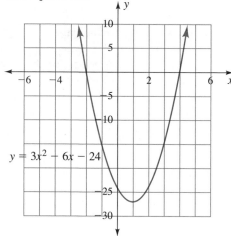

$y = 3x^2 - 6x - 24$

a. What are the solutions to $y = 0$?

b. What is the equation of the axis of symmetry of the graph?

c. What is the vertex of the graph?

90. Use the graph and equation in the figure to answer these questions.

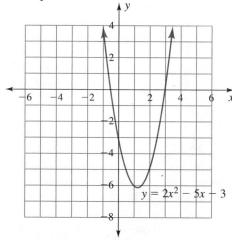

$y = 2x^2 - 5x - 3$

a. What are the solutions to $y = 0$?

b. What is the equation of the axis of symmetry of the graph?

c. What is the vertex of the graph?

91. A jet of water is to reach a maximum height of 100 feet when it is a horizontal distance of 30 feet from its origin. Find the equation of the jet of water.

92. A jet of water is to reach a maximum height of 75 feet when it is a horizontal distance of 25 feet from its origin. Find the equation of the jet of water.

93. Suppose an emergency rescue flare is fired into the air at a 45° angle. The path of the flare is

$$y = -\frac{g}{v^2}x^2 + x$$

The input x is the horizontal distance traveled by the flare, and the output y is the vertical position. The constant g is the average acceleration due to gravity, 32.2 ft/sec^2, and v is the initial velocity of the flare in feet per second. Find the vertex of the path of the flare for each of these initial velocities of the flare.

a. 132 ft/sec (the speed of a very fast baseball pitch)

b. 220 ft/sec (the speed of a golf club in the hands of a professional golfer)

c. 400 ft/sec

d. 500 ft/sec

94. A wrestling mat has a perimeter of 157 feet and has the maximum possible area for a rectangular shape. Write the area as a function of only the width, and show that the vertex of the graph of the area function gives the maximum area.

95. A boxing ring has a perimeter of 80 feet and has the maximum possible area for a rectangular shape. Write the area as a function of only the width, and show that the vertex of the graph of the area function gives the maximum area.

96. In Exercise 92, over what horizontal distance will the jet of water be over 60 feet?

97. In Exercise 91, over what horizontal distance will the jet of water be below 50 feet?

CHAPTER ④ TEST

1. Multiply $(x - 3)(x^2 - 6x + 9)$.

2. Factor these:

a. $x^2 - 49$

b. $x^3 - 27$

3. Compare the graph of $y = x^2 + 2$ with the graph of $y = x^2$. Discuss intercepts and vertices.

4. Explain how the negative sign in $y = -x^2$ changes its graph from that of $y = x^2$.

5. Which of the two equations

$$y = x - \frac{32.2}{2500}x^2 \quad \text{or} \quad y = \frac{4}{125}x^2$$

could describe the path of a ball through the air? Explain.

6. Give the next number in each sequence. Identify which are linear (arithmetic) sequences; give the rule. Identify which are quadratic; fit a rule.

   a. $-6, 2, 12, 24, 38$

   b. $23, 15, 7, -1, -9$

   c. $2, 8, 14, 20, 26$

   d. $25, 17, 10, 4, -1$

7. Evaluate the discriminant for $f(x) = -1x^2 + 2x + 3$. Indicate the number and type of solutions to $f(x) = 0$ and the number of x-intercepts in the graph of $f(x)$.

8. Change the radical expression to a product of a real number and the imaginary unit.

   a. $\sqrt{-36}$ **b.** $\sqrt{-75}$

9. Simplify these complex numbers.

   a. $\dfrac{5 + 10i}{5}$ **b.** $\dfrac{8 + 4i}{2}$

10. Multiply and simplify the expressions.

    a. $(1 + 5i)(1 - 5i)$ **b.** $(5 - i)(5 + i)$

    c. $(1 - 5i)(1 - 5i)$

11. Find all solutions to $f(x) = 0$, real or complex, for $f(x) = (x - 1)(x^2 + x + 1)$.

12. Use the graph and equation in the figure to answer these questions.

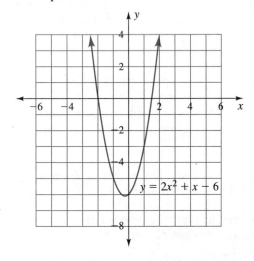

    a. What are the solutions to $y = 0$?

    b. What is the equation of the axis of symmetry of the graph?

    c. What is the vertex of the graph?

    d. For what inputs is $2x^2 + x - 6 > 0$?

    e. Give a value of y such that $y = 2x^2 + x - 6$ has no real-number solutions.

    f. Between what two values of y will $y = 2x^2 + x - 6$ have two negative real-number solutions?

13. Use any method to find the x-intercepts, if they exist, and then find the vertex for the graph of $f(x) = x^2 - x - 2$.

14. Quadratic equations can model the path of water in a drinking fountain, the path of a ball thrown across a field, a cable in a suspension bridge, or the roadbed of a highway between two hills. Describe how the x-intercepts, the y-intercept, and the vertex of a graph relate to one of these applications.

15. Complete the square to change $y = x^2 + 4x - 1$ into vertex form.

16. Describe the graph of $y = (x - 3)^2 - 1$ in terms of shifts of $y = x^2$.

17. A parabola has its vertex at $(-1, -8)$ and contains the point $(2, 19)$. Find the equation.

18. Let the input be the length of the side of the square, and let the output be the number of toothpicks needed to build the square design. The squares for inputs 1, 2, and 3 are shown in the figure. Make an input-output table for inputs 1 to 5. What kind of function describes the outputs? Find an equation to describe the outputs.

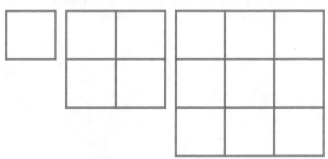

19. You have 15 feet of fencing and wish to make a rectangular kennel for your dog. You build the kennel next to your garage so that you need to enclose only three sides, as shown in the figure.

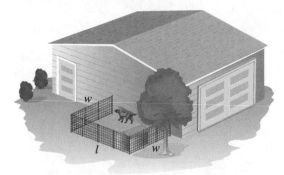

    a. Write an equation that describes the area y in terms of the width w and the length l.

    b. Eliminate l from your area equation with a substitution.

    c. Find the vertex of the graph of the area function, and explain how it shows the maximum area inside the kennel.

CUMULATIVE REVIEW OF CHAPTERS 1 TO 4 _____

1. Simplify $8 - 3(x - 2) - (x - 3)$.

2. Solve $3x - 3(2x - 1) = 27$.

3. Find a linear function through the origin that is parallel to $2y = 3x + 4$.

4. Suppose the amount earned is a function of hours worked and wage per hour. Write an equation and solve it for each situation.

   a. How many hours will Lynn have to work at $5.50 per hour to match what Tamika earns working 40 hours at $8.00 per hour?

   b. How many hours will Shea have to work at $6.00 per hour to pay Kyra for a 3-hour automobile repair at $35 per hour?

5. Solve for n:

   a. $c = nt^2$ **b.** $s = 2a + d(n - 1)$

6. Solve for x: $\dfrac{x + 7}{2} > x + 6$.

7. Show whether these lengths of sides form a right triangle: 4.5, 20, 20.5.

8. Evaluate $f(x) = 3x^2 - 5x + 2$ for each of the following.

   a. $f(-1)$ **b.** $f(0)$

   c. $f(2)$ **d.** $f(x + 1)$

9. Give the next number in each pattern. Identify whether the sequence is linear or quadratic and fit an equation to each.

   a. 5, 12, 21, 32, 45, ...   **b.** 8, 15, 22, 29, 36, ...

   c. 6, 12, 18, 24, 30, ...   **d.** 4, 15, 30, 49, 72, ...

10. Find the equation of a parabola passing through the set of points.

   a. $(3, 0)$, $(-4, 0)$, $(2, -4)$

   b. Vertex $(2, 4)$ and point $(-3, 2)$

11. Use the graph below of $f(x) = x^2 - 2x - 3$ to solve the following.

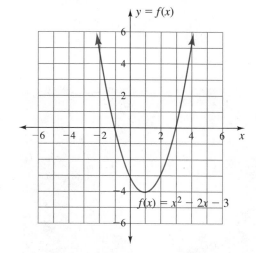

   a. $x^2 - 2x - 3 = 0$ **b.** $x^2 - 2x - 3 \le 0$

   c. $x^2 - 2x - 3 \ge -3$ **d.** $x^2 - 2x - 3 < 5$

12. What term must be added to $x^2 + 3x$ to make a perfect square trinomial?

13. Solve these equations:

   a. $x^2 + 6x + 10 = 0$

   b. $5x^2 + x - 4 = 0$

   c. $7x^2 + 4x = 3$

   d. $6x^2 + 3 = 11x$

14. Multiply.

   a. $(5x - 3)(3x - 5)$

   b. $(3 - i)(3 + i)$

15. Factor.

   a. $16 - x^2$

   b. $x^3 - 64$

5

Rational Functions and Variation

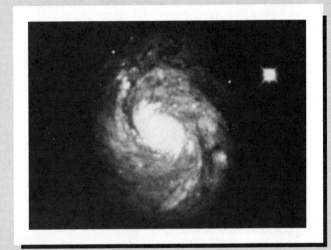

Figure 1   Galaxy NGC 1068 (M77)

Figure 2   Hurricane Felix, August 14, 1995

Both the galaxy in Figure 1 and the hurricane in Figure 2 are in rotational motion, and both have a rate of turning that is dependent on how compact their structure is relative to the center of rotation. For the galaxy, the more densely packed the stars are toward the center of the galaxy, the faster the galaxy rotates. For the hurricane, the smaller the "eye" is, the closer the cloud mass and moisture are to the center of rotation and the faster the accompanying winds. The relationship between position relative to the center of rotation and rate of turning is described as an inverse variation. Inverse variation is discussed in Section 5.2.

The chapter opens with a review of rational numbers and direct variation in Sections 5.0 and 5.1. Sections 5.3, 5.4, and 5.5 examine algebraic skills with rational expressions. The chapter closes with a discussion of solving rational equations in Section 5.6.

5.0 Review of Rational Numbers and Unit Analysis _____

OBJECTIVES

- Identify rational numbers.
- Find equivalent rational numbers.
- Find the least common denominator.
- Perform operations with rational numbers.
- Simplify expressions containing units, and set up a unit analysis.

WARM-UP

Add, subtract, multiply, and divide $\frac{5}{6}$ and $\frac{2}{3}$. Assume that the operations are to be done on the numbers in the order in which they appear.

IN THIS SECTION, we review operations with rational numbers in fraction notation, and we return to the unit analysis method of working with units of measurement.

Rational Numbers

Definition of Rational Numbers

Rational numbers are the set of numbers that may be written as the quotient of two integers,

$$\frac{a}{b}, \quad b \neq 0$$

A rational number may be written in many different forms. For example, the number *one-half* may be written in fraction notation as $\frac{1}{2}$, in decimal notation as 0.5, or in scientific notation as 5×10^{-1}. We will return to scientific notation in Section 6.0.

EQUIVALENT FRACTIONS   When we add or subtract fractions, we change to fractions with common denominators, using the concept of an equivalent fraction. The fractions b/c and ab/ac are said to be *equivalent* because they simplify to the same fraction, b/c. To find equivalent fractions, we multiply the fraction by a/a, or 1.

Equivalent Fraction Property

For all real numbers, $a \neq 0$, $c \neq 0$,

$$\frac{b}{c} = \frac{b}{c} \cdot \frac{a}{a} = \frac{ab}{ac}$$

EXAMPLE **1**   Finding equivalent fractions   Change each fraction to an equivalent fraction with the indicated denominator.

a. $\dfrac{4}{5} = \dfrac{}{15}$ **b.** $\dfrac{5}{3} = \dfrac{}{21}$

Solution   **a.** We multiply $\frac{4}{5}$ by $\frac{3}{3}$ to obtain 15 in the denominator:

$$\frac{4}{5} = \frac{4 \cdot 3}{5 \cdot 3} = \frac{12}{15}$$

b. We multiply $\frac{5}{3}$ by $\frac{7}{7}$ to obtain 21 in the denominator:

$$\frac{5}{3} = \frac{5 \cdot 7}{3 \cdot 7} = \frac{35}{21}$$

●

LEAST COMMON DENOMINATORS   The **least common denominator** is *the smallest number divisible by all the denominators under consideration.* Finding the least common denominator is important in operations with fractions because it keeps the expressions as simple as possible.

One way of finding the least common denominator is to list the multiples of each denominator until a common multiple is found.

EXAMPLE **2**   Finding a least common denominator with multiples   Find the least common denominator for $\frac{1}{300}$ and $\frac{1}{500}$.

Solution   We list the multiples of each denominator:

Student Note: These denominators will appear in Section 5.4.

> *Multiples of* 300:   300, 600, 900, 1200, **1500,** 1800
> *Multiples of* 500:   500, 1000, **1500,** 2000

The first multiple that both lists have in common is 1500. Thus, the least common denominator is 1500.   ●

As we will see in Section 5.4, listing multiples for the denominators of algebraic fractions is not practical. A more useful method of finding the least common denominator in algebra is to use prime factors.

Least Common Denominator (LCD)

To find the least common denominator:

1. List the prime factors of each denominator.

2. Compare the lists of prime factors.

3. a. If the lists have no common factors, the LCD is the product of the denominators.

b. If the lists have common factors, write each factor the highest number of times it appears in any one denominator. The LCD is the product of these factors.

EXAMPLE **3**   Finding the least common denominator with factors   Find the least common denominator for $\frac{1}{500}$ and $\frac{1}{300}$.

Solution   To find the least common denominator, we start with a list of prime factors of each denominator:

$$500 = 2 \cdot 2 \cdot 5 \cdot 5 \cdot 5$$
$$300 = 2 \cdot 2 \cdot 3 \cdot 5 \cdot 5$$

The least common denominator needs to be a number divisible by both denominators. Hence, the denominator needs to be divisible by two factors of 2, one factor of 3, and three factors of 5. The product of these factors, $2 \cdot 2 \cdot 3 \cdot 5 \cdot 5 \cdot 5$, is the least common denominator: 1500.   ●

Operations with Rational Numbers

In Example 4, we review operations with fractions. Look carefully at the steps, because these same steps are employed when we work with fractions containing algebraic expressions.

EXAMPLE **4**   Adding and subtracting fractions   Add and subtract $\frac{1}{6}$ and $\frac{1}{10}$.

Solution   In addition and subtraction, we find a common denominator and change the fractions to equivalent fractions with that denominator. We factor the denominators: $6 = 2 \cdot 3$ and $10 = 2 \cdot 5$. The factors needed in the least common denominator are 2, 3, and 5, so the required denominator is $2 \cdot 3 \cdot 5 = 30$.

Student Note: Adding across is common in this chapter.

$$\frac{1}{6} + \frac{1}{10} = \frac{1 \cdot 5}{2 \cdot 3 \cdot 5} + \frac{1 \cdot 3}{2 \cdot 5 \cdot 3} = \frac{5}{30} + \frac{3}{30} = \frac{8}{30} = \frac{4}{15}$$

$$\frac{1}{6} - \frac{1}{10} = \frac{1 \cdot 5}{2 \cdot 3 \cdot 5} - \frac{1 \cdot 3}{2 \cdot 5 \cdot 3} = \frac{5}{30} - \frac{3}{30} = \frac{2}{30} = \frac{1}{15}$$

Addition and Subtraction of Fractions

> 1. Find a common denominator, if needed.
> 2. Change each fraction to an equivalent fraction with the common denominator.
> 3. Add or subtract the numerators.

EXAMPLE **5**   Multiplying and dividing fractions   Multiply and divide $\frac{1}{6}$ and $\frac{1}{10}$.

Solution   No common denominator is needed for multiplication:

$$\frac{1}{6} \cdot \frac{1}{10} = \frac{1}{60}$$

To divide, we change the division to multiplication by the reciprocal of the second fraction:

$$\frac{1}{6} \div \frac{1}{10} = \frac{1}{6} \cdot \frac{10}{1} = \frac{10}{6} = \frac{2 \cdot 5}{2 \cdot 3} = \frac{5}{3}$$

Multiplication and division of fractions follow these rules.

Multiplication of Fractions

$$\frac{a}{b} \text{ times } \frac{c}{d} = \frac{ac}{bd}, \quad b \neq 0, d \neq 0$$

Division of Fractions

$$\frac{a}{b} \text{ divided by } \frac{c}{d} = \frac{a}{b} \text{ times } \frac{d}{c} = \frac{ad}{bc}, \quad b \neq 0, c \neq 0, d \neq 0$$

Operations with Units

In working with units, it is helpful to remember the following concepts.

Simplifying Expressions with Units

- Multiplication by 1 does not change number values:

  $$a \cdot 1 = a$$

- The number 1 appears in many forms—numbers, variables, and units:

  $$\frac{25}{25}, \quad \frac{a}{a}, \quad \frac{m}{m}, \quad \frac{\text{feet}}{\text{foot}}, \quad \frac{\text{pounds}}{\text{pound}}$$

- We can simplify within a product of fractions as well as within a single fraction:

  $$\frac{\cancel{3}}{4} \cdot \frac{5}{\cancel{3}} = \frac{5}{4}, \quad \frac{\$7.50}{1 \text{ hour}} \cdot \frac{8 \text{ hours}}{1 \text{ day}} = \frac{\$60}{1 \text{ day}}$$

EXAMPLE 6 Simplifying expressions containing units Simplify.

a. $\dfrac{50 \text{ m}^3}{25 \text{ m}}$

b. $\dfrac{300 \text{ foot·pounds}}{10 \text{ pounds}}$

c. $\dfrac{30 \text{ degree·days}}{6 \text{ days}}$

d. $\dfrac{9.81 \text{ m}}{\text{sec}^2} \cdot 5 \text{ sec}$

Solution a. $\dfrac{50 \text{ m}^3}{25 \text{ m}} = \dfrac{25 \cdot 2 \text{ m·m·m}}{25 \text{ m}} = \dfrac{2 \text{ m}^2}{1}$

Student Note: Engineers and mechanics use the foot-pound.

b. $\dfrac{300 \text{ foot·pounds}}{10 \text{ pounds}} = \dfrac{30 \cdot 10 \text{ foot·pounds}}{10 \text{ pounds}} = 30 \text{ feet}$

The foot·pound is a unit of measure resulting from the product of feet and pounds. It is preferable to use a dot to show the multiplication rather than the customary dash (foot-pound).

c. $\dfrac{30 \text{ degree·days}}{6 \text{ days}} = \dfrac{5 \cdot 6 \text{ degree·days}}{6 \text{ days}} = 5 \text{ degrees}$

The degree·day is used by public utilities to describe changes from day to day in power consumption.

d. $\dfrac{9.81 \text{ m}}{\text{sec}^2} \cdot 5 \text{ sec} = \dfrac{49.05 \text{ m}}{\text{sec}}$

Here we have the speed that results from a 5-second acceleration due to gravity. Units involving the division of distance by time are usually changed to a *rate,* such as meters per second. ●

In Example 7, units of measure are used in solving problems and evaluating formulas. In this example, **evaluate** means *to substitute numbers and units in place of the variables in expressions or formulas.* The word *per* means division, so a unit following *per* belongs in the denominator.

EXAMPLE 7 Working with units Answer the questions or evaluate the formulas with the given information. Show units of measure.

a. The distance traveled in a car is given by $D = rt$. Suppose r is 55 miles per hour and t is 3 hours. What is the distance traveled?

b. Total pay is given by wage times hours worked. Suppose the wage is $5.10 per hour and the number of hours worked is 40 hours per week. Find total pay.

c. The distance an object falls is given by $h = \frac{1}{2}gt^2$. Suppose g is 32.2 feet per second squared and t is 2 seconds. What is the distance fallen?

d. Total calories from protein is the number of calories per gram times the number of grams consumed. There are 4 calories per gram of protein. Suppose a 2.1-ounce Baby Ruth® bar contains 4 grams of protein. How many calories are in the protein?

Solution   **a.** $D = rt = \dfrac{55 \text{ miles}}{1 \text{ hour}} \cdot 3 \text{ hours} = 165 \text{ miles}$

b. Total pay $= \dfrac{\$5.10}{1 \text{ hour}} \cdot \dfrac{40 \text{ hours}}{1 \text{ week}} = \$204$ per week

c. $D = \dfrac{1}{2}gt^2 = \dfrac{1}{2}\left(\dfrac{32.2 \text{ ft}}{\text{sec}^2}\right)(2 \text{ sec})^2 = 64.4 \text{ ft}$

d. $\dfrac{4 \text{ calories}}{1 \text{ gram of protein}} \cdot 4 \text{ grams of protein} = 16 \text{ calories}$ ●

Unit Analysis

Unit analysis is *a method for changing from one unit of measure to another or for changing from one rate to another.* A key idea in unit analysis is that facts such as 12 inches = 1 foot are formed into a fraction worth 1. The numerator and denominator represent different ways of expressing the same measure. That is,

$$\frac{12 \text{ inches}}{1 \text{ foot}} = 1, \qquad \frac{5280 \text{ feet}}{1 \text{ mile}} = 1, \qquad \text{and} \qquad \frac{60 \text{ seconds}}{1 \text{ minute}} = 1$$

UNIT TO UNIT   To set up a unit analysis, we arrange the facts into fractions so that each unit of measure appears once in a numerator and once in a denominator. Where possible, we use the four problem-solving steps: understand, plan, do, and check.

EXAMPLE **8**   Changing from one unit to another   Convert 10 miles into inches.

Solution   **Understand**: The first step is to read the problem and identify the question. Here we start with miles and want to end with inches. We list the units-of-measure facts that relate the starting and ending units:

    12 inches = 1 foot
    5280 feet = 1 mile

Plan: We will start with 10 miles (marked in yellow) and arrange the facts as fractions so that other units are eliminated and only inches remain.
Do:

$$\frac{10 \text{ mi}}{1} \cdot \frac{5280 \text{ ft}}{1 \text{ mi}} \cdot \frac{12 \text{ in.}}{1 \text{ ft}} = 633,600 \text{ in.}$$

Check: We must make sure that all facts are correctly written, the unwanted units are eliminated, and the answer is reasonable. As we might expect, we obtain a large number of inches in a mile.   ●

Unit Analysis

---

1. Identify the units of measure to be changed, and identify the units needed in the answer.
2. List facts that contain the starting units of measure and the ending units. List facts needed to relate the starting and ending units.
3. Write the starting units. Using your list of facts, set up a product of fractions so that each unit of measure appears once in the numerator and once in the denominator.
4. Use $a/a = 1$ to eliminate the unwanted units of measure, and then calculate with the numbers.

EXAMPLE **9**

Solution

Changing from one unit to another   A rectangular floor is 13 feet by 18 feet. Find its area in square yards.

Understand: The area is the product of width and length. Our unit fact is

$$3 \text{ feet} = 1 \text{ yard}$$

Plan: We multiply to find the area in square feet and use the fact that 3 feet = 1 yard twice to obtain square yards.

Do:

$$\text{Area} = \frac{13 \text{ ft}}{1} \cdot \frac{18 \text{ ft}}{1} \cdot \frac{1 \text{ yd}}{3 \text{ ft}} \cdot \frac{1 \text{ yd}}{3 \text{ ft}} = \frac{13 \cdot \overset{2}{18} \text{ yd}^2}{\underset{1}{9}} = 26 \text{ yd}^2$$

Check: We can change the room size to yards first and then find the area.

$$13 \text{ ft} \cdot \frac{1 \text{ yd}}{3 \text{ ft}} = \frac{13}{3} \text{ yd}$$

$$18 \text{ ft} \cdot \frac{1 \text{ yd}}{3 \text{ ft}} = 6 \text{ yd}$$

$$\frac{13}{\underset{1}{3}} \text{ yd} \cdot \frac{\overset{2}{6}}{1} \text{ yd} = 26 \text{ yd}^2$$

If you change units in your head, the method suggested here can help you put your mental process into written form.

RATE TO RATE   A **rate** is *a comparison of a quantity of one unit to a quantity of another unit.* In Example 10, we change from one rate to another. Changing rates usually means changing two units at a time.

EXAMPLE **10**

Solution

Changing rates: snail's pace   Convert a snail's pace, 6 inches per minute, to miles per hour.

Understand: Here we start with inches per minute and want to end with miles per hour. We need to change inches to miles and minutes to hours. The facts are

$$12 \text{ inches} = 1 \text{ foot}$$
$$5280 \text{ feet} = 1 \text{ mile}$$
$$60 \text{ minutes} = 1 \text{ hour}$$

Plan: We start with a fraction containing inches over minutes and arrange the facts as fractions so that other units are eliminated and only miles and hours remain.

Do:

$$\frac{6 \text{ in.}}{1 \text{ min}} \cdot \frac{1 \text{ ft}}{12 \text{ in.}} \cdot \frac{1 \text{ mi}}{5280 \text{ ft}} \cdot \frac{60 \text{ min}}{1 \text{ hr}} = \frac{6 \cdot 1 \cdot 1 \cdot 60}{1 \cdot 12 \cdot 5280 \cdot 1} \frac{\text{mi}}{\text{hr}}$$

$$\approx 0.00568 \text{ mile per hour}$$

Check: We must make sure that all facts are correctly written, the unwanted units are eliminated, and the answer is reasonable. As we might expect, the snail's pace is slower than 1 mile per hour, so the answer is reasonable.   ●

When changing both units in a rate, it is convenient to put all fractions that change the denominator units to the left of the given rate and all fractions that change the numerator units to the right. This placement of the fractions puts like units close together. In Example 10, we might have written

$$\frac{60 \text{ min}}{1 \text{ hr}} \cdot \frac{6 \text{ in.}}{1 \text{ min}} \cdot \frac{1 \text{ ft}}{12 \text{ in.}} \cdot \frac{1 \text{ mi}}{5280 \text{ ft}}$$

Think about it: Why can we arrange the unit fractions in any order?

PROBLEM SOLVING   We apply the unit analysis approach to problems with units by setting up rates or facts as fractions, as shown in Example 11.

EXAMPLE **11**

Solving problems with units: volume of shower water   An energy-efficient shower-head permits a flow of 2 gallons per minute. How many gallons of water are used in a 30-day month by a person taking a 5-minute shower every other day?

Solution

Understand: We start by listing key phrases describing the facts:

2 gallons per minute

30 days per month

5 minutes per shower

1 shower every 2 days

Plan: We are looking for gallons per month, so we start with a fraction containing gallons on the top and look for facts that eliminate all other units except months.

Do:

$$\frac{2 \text{ gallons}}{1 \text{ minute}} \cdot \frac{5 \text{ minutes}}{\text{shower}} \cdot \frac{1 \text{ shower}}{2 \text{ days}} \cdot \frac{30 \text{ days}}{1 \text{ month}} = \frac{2(5)(30)}{2} \frac{\text{gallons}}{\text{month}}$$

$$= 150 \text{ gallons per month}$$

Check: We must confirm that all facts are correctly written and that unwanted units are eliminated.   ●

ANSWER BOX

Warm up: $\frac{3}{2}, \frac{1}{6}, \frac{5}{9}, \frac{5}{4}$. **Think about it:** The commutative property of multiplication allows us to multiply in any order.

EXERCISES 5.0

In Exercises 1 to 4, change the fractions to equivalent fractions with the indicated denominator.

1. a. $\dfrac{3}{5} = \dfrac{}{35}$ **b.** $\dfrac{5}{8} = \dfrac{}{48}$ **c.** $\dfrac{8}{3} = \dfrac{}{48}$

2. a. $\dfrac{5}{9} = \dfrac{}{54}$ **b.** $\dfrac{7}{4} = \dfrac{}{36}$ **c.** $\dfrac{5}{9} = \dfrac{}{36}$

3. a. $\dfrac{3}{8} = \dfrac{}{40}$ **b.** $\dfrac{5}{6} = \dfrac{}{54}$ **c.** $\dfrac{5}{4} = \dfrac{}{32}$

4. a. $\dfrac{6}{5} = \dfrac{}{45}$ **b.** $\dfrac{3}{4} = \dfrac{}{28}$ **c.** $\dfrac{5}{8} = \dfrac{}{56}$

Find the least common denominator for each pair of fractions in Exercises 5 and 6.

5. a. $\dfrac{5}{6}, \dfrac{3}{10}$ **b.** $\dfrac{4}{15}, \dfrac{7}{10}$ **c.** $\dfrac{1}{48}, \dfrac{1}{32}$

6. a. $\dfrac{2}{15}, \dfrac{6}{25}$ **b.** $\dfrac{3}{14}, \dfrac{4}{21}$ **c.** $\dfrac{1}{27}, \dfrac{1}{36}$

In Exercises 7 to 14, add, subtract, multiply, and divide each pair of fractions in the order shown.

7. $\frac{1}{4}$ and $\frac{1}{10}$ **8.** $\frac{1}{8}$ and $\frac{1}{10}$

9. $\frac{3}{4}$ and $\frac{5}{6}$ **10.** $\frac{2}{3}$ and $\frac{5}{6}$

11. $1\frac{1}{3}$ and $2\frac{1}{2}$ **12.** $2\frac{1}{4}$ and $1\frac{2}{3}$

13. $2\frac{1}{5}$ and $1\frac{1}{4}$ **14.** $1\frac{3}{4}$ and $2\frac{1}{7}$

Simplify the expressions in Exercises 15 to 22.

15. $\dfrac{186 \text{ cm}^2}{6 \text{ cm}}$ **16.** $\dfrac{125 \text{ ft}^3}{25 \text{ ft}}$

17. $\dfrac{36 \text{ in.}}{1728 \text{ in}^3}$ **18.** $\dfrac{54 \text{ yd}^3}{9 \text{ yd}}$

19. $\dfrac{1500 \text{ foot·pounds}}{25 \text{ feet}}$ **20.** $\dfrac{1024 \text{ degree·gallons}}{128 \text{ degrees}}$

21. $\dfrac{1200 \text{ kilowatt hours}}{24 \text{ hours}}$ **22.** $\dfrac{4800 \text{ kilowatt hours}}{60 \text{ kilowatts}}$

23. The distance an object falls when dropped is given by $d = \frac{1}{2}gt^2$, where $g \approx 32.2$ ft/sec^2. Find the distance fallen for the times given below. Show the units clearly in your work.

   a. $t = 1$ sec   **b.** $t = 3$ sec

   c. $t = 9$ sec

24. The distance an object falls when dropped is given by $d = \frac{1}{2}gt^2$, where $g \approx 9.81$ m/sec^2. Find the distance fallen for the times given below. Show the units clearly in your work.

   a. $t = 1$ sec   **b.** $t = 2$ sec

   c. $t = 3$ sec

25. The speed of a falling object is given by $s = gt$, where t is the time during which the object falls. Suppose $g = 9.81$ m/sec^2. Find the speed for each of these times.

   a. $t = 1$ sec   **b.** $t = 2$ sec

   c. $t = 4$ sec

26. The speed of a falling object is given by $s = gt$, where t is the time during which the object falls. Suppose $g = 32.2$ ft/sec^2. Find the speed for each of these times.

   a. $t = 1$ sec   **b.** $t = 2$ sec

   c. $t = 4$ sec

In Exercises 27 to 40, arrange the listed facts into a unit analysis and solve the problem.

27. 300 milliliters is how many liters?
1 liter = 1000 milliliters

28. 160 pounds is how many kilograms?
2.2 pounds is 1 kilogram.

29. 25 feet is how many meters?
1 foot = 12 inches
1 meter is 39.37 inches.

30. 55 gallons is how many pints?
1 gallon = 4 quarts
2 pints = 1 quart

31. 150 cubic feet is how many cubic yards?
1 yard = 3 feet

32. 48 square inches is how many square feet?
1 foot = 12 inches

33. 100 cubic inches is how many cubic feet?
12 inches = 1 foot

34. 1600 square centimeters is how many square meters?
1 meter = 100 centimeters

35. 200 milliliters of water is how many grams?
1 kilogram is 1000 milliliters of water.
1000 grams = 1 kilogram

36. 55 feet per second is how many miles per hour?
1 mile = 5280 feet
1 minute = 60 seconds
1 hour = 60 minutes

37. 40 miles per hour is how many feet per second?
1 mile = 5280 feet
1 minute = 60 seconds
1 hour = 60 minutes

38. One gallon for five miles is how many dollars per day of driving?
1 hour to travel 55 miles
1 gallon is $1.35.
1 driving day is 10 hours.

39. 240 milliliters in 12 hours is how many microdrops per minute?
60 microdrops = 1 milliliter
1 hour = 60 minutes

40. Prescription dosage for young children may be based on their age relative to 150 months, as 150 months is considered "adult" for many prescriptions. If the adult dosage is 500 milligrams, how many milligrams should a 1-year-old infant receive?
1 year = 12 months

In Exercises 41 to 44, use these food values to find the calories from each source and the total calories per serving: 9 calories per gram of fat, 4 calories per gram of carbohydrate, and 4 calories per gram of protein. Show the units clearly in your work.

41. A $\frac{3}{4}$-cup serving of Wheat Chex®

   a. 1 gram of fat

   b. 41 grams of carbohydrate

   c. 5 grams of protein

42. A $\frac{1}{2}$ cup serving of Campbell's® cream of chicken soup

   a. 8 grams of fat

   b. 11 grams of carbohydrate

   c. 3 grams of protein

43. A 1.5-ounce serving of Nissin Top Ramen®

   a. 8 grams of fat

   b. 27 grams of carbohydrate

   c. 4 grams of protein

44. An 11.5-ounce can of V8 100% Vegetable Juice®

   a. 0 grams of fat

   b. 15 grams of carbohydrate

   c. 2 grams of protein

Identify the missing operation sign (add, subtract, multiply, or divide) in Exercises 45 to 48.

45. a. $\frac{3}{5} \boxed{} \frac{2}{7} = \frac{6}{35}$ **b.** $\frac{3}{5} \boxed{} \frac{2}{7} = \frac{21}{10}$

46. a. $\frac{2}{7} \boxed{} \frac{3}{5} = \frac{10}{21}$ **b.** $\frac{3}{5} \boxed{} \frac{2}{7} = \frac{11}{35}$

47. a. $\frac{3}{5} \boxed{} \frac{2}{7} = \frac{31}{35}$ **b.** $\frac{5}{7} \boxed{} \frac{2}{3} = \frac{1}{21}$

48. a. $\frac{7}{5} \boxed{} \frac{2}{3} = \frac{11}{15}$ **b.** $\frac{7}{3} \boxed{} \frac{2}{5} = \frac{14}{15}$

Indicate whether the solutions to Exercises 49 to 54 are right or wrong. If wrong, explain what was done wrong. If right, state whether the method would always work and what its disadvantage is.

49. $\frac{3}{4} + \frac{2}{3} = \frac{3 + 2}{4 \cdot 3} = \frac{5}{12}$

50. $\frac{3}{5} - \frac{2}{4} = \frac{3 - 2}{5 - 4} = \frac{1}{1}$

51. $\frac{5}{6} \div \frac{1}{2} = \frac{5 \div 1}{6 \div 2} = \frac{5}{3}$

52. $\frac{3}{8} + \frac{1}{4} = \frac{3(4) + 8(1)}{8 \cdot 4} = \frac{12 + 8}{32} = \frac{20}{32} = \frac{5}{8}$

53. $\frac{5}{6} + \frac{1}{3} = \frac{5 + 1}{6 + 3} = \frac{6}{9} = \frac{2}{3}$

54. $\frac{4}{5} \div \frac{2}{3} = \frac{4 \div 2}{5 \div 3} = \frac{2}{\frac{5}{3}} = \frac{6}{5}$

Projects

55. *Oil Spill.* The 1989 Exxon Valdez oil spill in Alaska released 11 million gallons of oil into Prince William Sound. Over how many square miles would this much oil spread if it were uniformly the thickness of a sheet of 20-pound photocopy paper? (Useful facts: 1 gallon is 231 cubic inches, 500 sheets of 20-pound photocopy paper is 2 inches thick, 1 mile is 5280 feet.)

56. *Driveway Remodeling.* A homeowner wants to fill an inclined driveway and reverse the incline to drain toward the street instead of toward the basement of the house. The driveway is to be 12.5 feet wide. The other dimensions are shown in the figure. How many cubic yards of concrete must be ordered for this remodeling project?

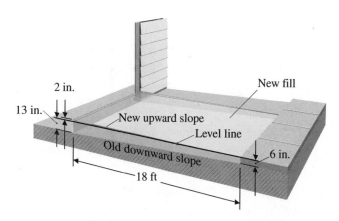

57. *Entry Remodeling.* A homeowner wants to build a short sidewalk, staircase, and landing. The dimensions are shown in the figure. How many cubic yards of concrete must be ordered for this remodeling project?

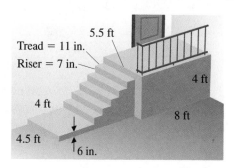

58. *Patterns in Fractions.* To change a rational number from a fraction to a decimal, we divide the numerator by the denominator.

a. Find the decimal for each of these fractions. Identify which are terminating decimals and which are repeating decimals.

$$\frac{1}{7}$$
$$\frac{1}{9}$$
$$\frac{1}{5}$$
$$\frac{1}{8}$$
$$\frac{4}{25}$$
$$\frac{1}{11}$$
$$\frac{2}{3}$$
$$\frac{3}{20}$$

$$\frac{2}{7}$$
$$\frac{4}{9}$$
$$\frac{1}{20}$$
$$\frac{3}{16}$$
$$\frac{7}{50}$$
$$\frac{2}{11}$$
$$\frac{5}{6}$$
$$\frac{4}{5}$$

b. List the denominators for the fractions that form terminating decimals. What do the denominators have in common?

c. What repeating decimal pattern is formed by the ninths $\left(\frac{1}{9}, \frac{2}{9}, \frac{3}{9}, \frac{4}{9}, \text{etc.}\right)$?

d. Add the repeating decimals for $\frac{8}{9}$ and $\frac{1}{9}$. Add the fractions $\frac{8}{9}$ and $\frac{1}{9}$. What do you observe?

e. What repeating decimal pattern is formed by the elevenths $\left(\frac{1}{11}, \frac{2}{11}, \frac{3}{11}, \frac{4}{11}, \text{etc.}\right)$?

5.1   Ratios, Proportions, and Direct Variation

OBJECTIVES

- Find when ratios are equal.
- Solve word problems and similar triangle problems with proportions.
- Find when linear data represent a direct variation.
- Translate word expressions into linear, quadratic, or joint variation.

WARM-UP

Perform the following operations, leaving the answers in simplest form.

1. $\dfrac{-2}{3} \cdot \dfrac{3}{4}$ **2.** $\dfrac{14}{15} \cdot \dfrac{6}{7}$

3. $\dfrac{15 \text{ cm}}{2 \text{ hr}} \cdot \dfrac{24 \text{ hr}}{1 \text{ day}}$ **4.** $225 \text{ mi} \div \dfrac{45 \text{ mi}}{1 \text{ hr}}$

5. $\dfrac{5}{-8} \div \dfrac{1}{2}$ **6.** $\dfrac{-10}{15} \div \dfrac{1}{3}$

I N THIS SECTION, we use ratios, proportions, and linear equations to examine the concept of direct variation. We discuss why proportions cannot be applied to all linear situations, and we consider quadratic and joint variation. The emphasis is on vocabulary and new ways of describing familiar relationships. Understanding variation is important for reading and writing about relationships.

Ratios

A **ratio** is *the quotient of two quantities.* Ratios compare like or unlike quantities. Recall that the slope of a linear equation is the ratio of rise to run, or change in y to change in x.

***Think about it 1*:** The fraction $\frac{5}{0}$ is undefined, as is the slope of a vertical line. Give examples of situations where a ratio of 5 to 0 is meaningful.

If two ratios divide to the same decimal or simplify to the same fraction, they are equivalent.

Definition of Equivalent Ratios	Two ratios are **equivalent ratios** if they simplify to the same number.

EXAMPLE

Exploring pi   The number pi is the ratio of the circumference to the diameter of a circle. Many arithmetic textbooks use the fraction $\frac{22}{7}$ as an approximation to pi. Find out whether $\frac{22}{7}$ and $\frac{2218}{706}$ are equivalent ratios. Compare them to the value of pi programmed into your calculator.

Solution   See the Answer Box.

Simplifying ratios that contain only numbers or units is exactly like simplifying fractions: We look for common factors in the numerator and denominator and eliminate those factors based on the fact that $a/a = 1$.

If a ratio contains units of measure of the same type (length, mass, capacity), *the units should be made the same before the ratio is simplified or is compared with another ratio.* The facts in Table 1 may be useful in simplifying ratios of units. More facts are listed on the inside cover.

1000 milliliters = 1 liter	16 ounces = 1 pound
1000 grams = 1 kilogram	1 yard = 36 inches
100 centimeters = 1 meter	1 yard = 3 feet
1000 meters = 1 kilometer	1 mile = 5280 feet
	16 tablespoons = 1 cup
	4 cups = 1 quart
	4 quarts = 1 gallon

Table I   Measurement facts

EXAMPLE 2

Simplifying ratios   Use unit analysis to eliminate units, and simplify.

a. 150 centimeters to 2 meters

b. 2500 grams to 1 kilogram

c. 2 tablespoons to 1 gallon

d. n inches to m feet

e. x cups to y quarts

Solution   **a.** $\dfrac{150 \text{ cm}}{2 \text{ m}} \cdot \dfrac{1 \text{ m}}{100 \text{ cm}} = \dfrac{150}{200} = \dfrac{3}{4}$

b. $\dfrac{2500 \text{ g}}{1 \text{ kg}} \cdot \dfrac{1 \text{ kg}}{1000 \text{ g}} = \dfrac{2500}{1000} = \dfrac{5}{2}$

c. $\dfrac{2 \text{ tbsp}}{1 \text{ gal}} \cdot \dfrac{1 \text{ cup}}{16 \text{ tbsp}} \cdot \dfrac{1 \text{ qt}}{4 \text{ cups}} \cdot \dfrac{1 \text{ gal}}{4 \text{ qt}} = \dfrac{2}{256} = \dfrac{1}{128}$

d. $\dfrac{n \text{ in.}}{m \text{ ft}} \cdot \dfrac{1 \text{ ft}}{12 \text{ in.}} = \dfrac{n}{12m}$

e. $\dfrac{x \text{ cups}}{y \text{ qt}} \cdot \dfrac{1 \text{ qt}}{4 \text{ cups}} = \dfrac{x}{4y}$

Proportions

The main floor of the house in Figure 3 is 70 inches above sidewalk level. A slope ratio of 1 foot to 8 feet makes a steep but adequate wheelchair ramp (see Figure 4). We need proportions to find the number of horizontal feet needed for the ramp.

Figure 3

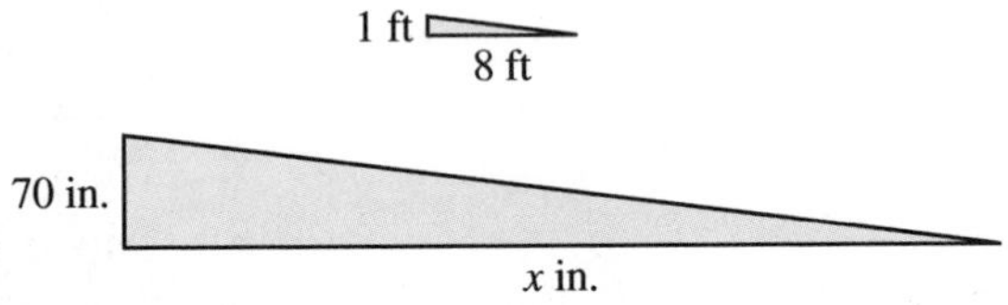

1 ft
8 ft

70 in.

x in.

Figure 4

Two equal ratios form a **proportion**

$$\frac{a}{b} = \frac{c}{d}, \quad b \neq 0, d \neq 0$$

To find whether two ratios are equal or to solve for a variable in a proportion, we use the cross multiplication property.

Cross Multiplication Property

The proportion $\dfrac{a}{b} \times \dfrac{c}{d}$ implies that $a \cdot d = b \cdot c.$

Changing the proportion to the equation $a \cdot d = b \cdot c$ is called cross multiplication because lines drawn between the parts being multiplied form an X-shaped cross.

In Examples 3 to 5, we set up proportions in problem settings. We look for phrases that give the relevant facts and then rewrite the phrases into a proportion.

EXAMPLE **3** Using a proportion: wheelchair ramp How many horizontal feet of ramp are needed to rise 70 inches above sidewalk level? Use a slope ratio of 1 foot to 8 feet.

Solution   **Phrases**: The slope ratio of rise to run is 1 to 8.

The ramp ratio is 70 inches to x inches.

Proportion: We set up a proportion based on the slope, rise over run:

$$\frac{1}{8} = \frac{70 \text{ in.}}{x}$$

$$1x = 560 \text{ in.}$$

We next change from inches to feet:

$$x = 560 \text{ in.} \cdot \frac{1 \text{ ft}}{12 \text{ in.}} = 46\tfrac{2}{3} \text{ ft}$$

The ramp must run nearly 47 feet around the house in order to rise from the sidewalk to the main floor.   ●

PROPORTIONS AND UNITS   Proportions have many gardening applications.

EXAMPLE **4**   Applying proportions to plant food preparation   To prepare Speedy Grow® plant food, we add 2 tablespoons of Speedy Grow to a gallon of water. How many tablespoons are needed for a 60-gallon nursery tank?

Solution   **Phrases**: 2 tablespoons for one gallon

x tablespoons for 60 gallons

Proportion:

$$\frac{2 \text{ tbsp}}{x} = \frac{1 \text{ gal}}{60 \text{ gal}}$$

The facts appear in the proportion in the same order as in the phrases.

$$(1 \text{ gal})(x) = (2 \text{ tbsp})(60 \text{ gal})$$

$$x = \frac{2 \cdot 60 \text{ tbsp} \cdot \text{gal}}{1 \text{ gal}}$$

$$x = 120 \text{ tbsp}$$

120 tablespoons of Speedy Grow are needed for the 60-gallon tank.   ●

Think about it 2: List three other proportions with solutions $x = 120$ tbsp that can be made from the data in Example 4.

We often combine unit analysis work with proportional settings.

EXAMPLE **5**   Using unit analysis and proportions: more plant food preparation   Speedy Grow is sold in a 1-quart container. The quart container will prepare how many gallons of mixture? How many quarts will be needed to prepare a mixture for the 60-gallon tank in Example 4?

Solution   **Phrases**: 2 tablespoons for 1 gallon

1 quart for x gallons

We change quarts into tablespoons with unit analysis:

$$\frac{1 \text{ qt}}{1 \text{ container}} \cdot \frac{4 \text{ cups}}{1 \text{ qt}} \cdot \frac{16 \text{ tbsp}}{1 \text{ cup}} = \frac{64 \text{ tbsp}}{1 \text{ container}}$$

Proportion:

$$\frac{2 \text{ tbsp}}{64 \text{ tbsp}} = \frac{1 \text{ gallon}}{x}$$

$$(2 \text{ tbsp})x = (1 \text{ gal})(64 \text{ tbsp})$$

$$x = 32 \text{ gal}$$

The quart container will mix with 32 gallons of water. Because the tank in Example 4 holds 60 gallons of water, 2 quarts of Speedy Grow will be sufficient to prepare a mixture for the tank. ●

SIMILAR TRIANGLES   Proportions are commonly found in geometric applications, particularly those involving similar triangles. **Similar triangles** *have the same shape but different sizes*. Figure 5 shows two similar triangles.

The similar triangles in Figure 5 are arranged to show how the sides of the smaller triangle and the sides of the larger triangle compare. When triangles are arranged in this way, we say that the corresponding sides are lined up. **Corresponding sides** are *the sides that are in the same position compared to the other parts of the triangles*.

Corresponding angles are *the angles that are in the same position compared to the other parts of the triangles*. As we change the size of similar triangles, the angles stay the same size. Thus, similar triangles have corresponding angles that are equal.

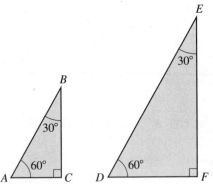

Figure 5

EXAMPLE **6**   Finding corresponding angles and sides   Identify the corresponding angles and sides in Figure 5.

Solution   The angles in the two triangles are the same, so the corresponding angles are *A* and *D*, at 60°; *B* and *E*, at 30°; and *C* and *F*, at 90°. Side *AB* corresponds to side *DE*, side *AC* to *DF*, and side *BC* to *EF*. ●

A s we enlarge a triangle by, say, doubling one side, the other sides grow proportionately and also double.

Corresponding sides of similar triangles are proportional.

EXAMPLE **7**   Writing proportions for sides of similar triangles   Write three different proportions showing that the ratios of the sides of the similar triangles in Figure 6 are the same.

Solution

$$\frac{4}{8} = \frac{10}{20}$$ Cross multiplication gives 80 each way.

$$\frac{4}{8} = \frac{12}{24}$$ Cross multiplication gives 96 each way.

$$\frac{20}{10} = \frac{24}{12}$$ Cross multiplication gives 240 each way.   ●

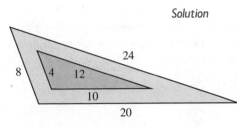

Figure 6

Think about it 3: What other proportions can be written from the triangles in Figure 6?

EXAMPLE **8**   Applying proportions to similar triangles   Use proportions to find the lengths of the indicated sides in Figure 7.

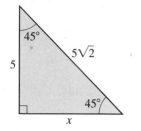

Figure 7

Solution To find size z in Figure 7, we use the proportion

$$\frac{5\sqrt{2}}{5} = \frac{z}{3}$$

We solve the proportion by cross multiplying:

$$\frac{5\sqrt{2}}{5} = \frac{z}{3} \qquad \text{Cross multiply.}$$

$$3 \cdot 5\sqrt{2} = 5 \cdot z \qquad \text{Divide by 5.}$$

$$3\sqrt{2} = z$$

Both side x and side y are missing. We may use the Pythagorean theorem to find x and then a proportion to find y. To find x, we use

$$a^2 + b^2 = c^2 \qquad \text{Substitute the lengths of the sides.}$$

$$x^2 + 5^2 = (5\sqrt{2})^2 \qquad \text{Simplify square root expressions.}$$

$$x^2 + 25 = 50 \qquad \text{Subtract 25 from both sides.}$$

$$x^2 = 25 \qquad \text{Take the square root.}$$

$$x = 5$$

To find y, we use the proportion

$$\frac{5}{5} = \frac{3}{y}$$

Thus, $y = 3$. ●

Think about it 4: From Example 8, what geometry fact can we guess about the lengths of sides opposite equal angles in a triangle?

OVERLAPPING SIMILAR TRIANGLES In many situations, similar triangles overlap. In Figure 8, the light from a streetlamp casts a shadow, b_2, beyond a person of height h_2. The triangle with the shadow and person as sides is similar to the triangle with the ground, b_1, and the streetlamp, h_1, as sides.

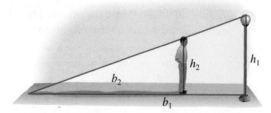

Figure 8

EXAMPLE **9** Identifying proportions in overlapping triangles: shadows Describe the relationship among sides $b_1, b_2, h_1,$ and h_2 in Figure 8. Write a sentence and a proportion.

Solution The bases b_1 and b_2 and heights h_1 and h_2 are corresponding sides of similar triangles and are proportional. Thus,

$$\frac{h_1}{b_1} = \frac{h_2}{b_2}$$ ●

EXAMPLE **10** Identifying proportions in overlapping triangles: more shadows Use a proportion to find the height h of the streetlamp shown in Figure 9.

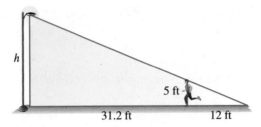

Figure 9

Solution   In this figure, the similar triangles overlap. We build a proportion with the ratios of height to base:

$$\frac{h}{(31.2 + 12)\ \text{ft}} = \frac{5\ \text{ft}}{12\ \text{ft}} \qquad \text{Cross multiply.}$$

$$12h\ \text{ft} = 5(43.2)\ \text{ft}^2 \qquad \text{Divide by 12 ft.}$$

$$h = 18\ \text{ft}$$

The streetlamp is 18 feet in height.   ●

Direct Variation

In Example 11, we return to the data from Examples 4 and 5.

EXAMPLE **11**   Exploring the plant food preparation data   Make a graph of the data in Table 2, derived from the proportions in Examples 4 and 5. Describe the type of function, intercepts, slope, and equation. What is the ratio of output to input?

Gallons of Water	Tablespoons of Speedy Grow
1	2
60	120
32	64

Table 2   Speedy Grow mixtures

Solution   Figure 10 shows the data points, placed in a graph and connected from left to right. The data lie on a straight line.

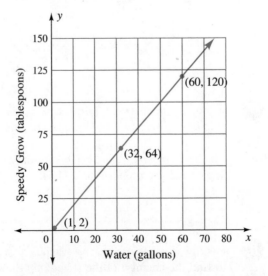

Figure 10

The graph passes through the origin, (0, 0), because 0 gallons of water will require 0 tablespoons of Speedy Grow. Thus, both the x- and the y-intercept are at the origin.

$$\text{Slope} = \frac{\Delta y}{\Delta x} = \frac{120 - 64 \text{ tbsp}}{60 - 32 \text{ gal}} = \frac{56 \text{ tbsp}}{28 \text{ gal}}$$

$$= 2 \text{ tbsp per gal}$$

The ratio of outputs to inputs, y/x, is

$$\frac{y}{x} = \frac{120}{60} = \frac{64}{32} = \frac{2 \text{ tbsp}}{1 \text{ gal}}$$

Using $y = mx + b$, we have the equation $y = 2x$. ●

In Example 11, the ratio of outputs to inputs is the same as the slope. This equality creates a special situation called **direct variation.** When *the ratio of outputs to inputs is constant,* we say that the data *vary directly.* In Example 11, we say that the number of tablespoons of Speedy Grow *varies directly* with the number of gallons of water. The *constant ratio, y/x,* is known as the **constant of variation** (or **constant of proportionality**).

There are two common ways of writing direct variation. The *ratio form* for direct variation is $y/x = k$. The *function form* for direct variation is $f(x) = kx$ or $y = kx$. The following box summarizes.

Direct Variation
(Direct Proportion)

> If there is a constant k such that $y = kx$ for all x, we say y *varies directly as x.* Also, if there is a constant k such that $y/x = k$, we say that y *is directly proportional to x.* The constant k is the constant of variation or constant of proportionality.

EXAMPLE **12**   Describing variation   Describe the situation as a direct variation, and identify the constant of proportionality in these linear settings.

a. The circumference of a circle is pi times the diameter.

b. A can of corn costs $0.69. The output is the total cost of x cans.

Solution   **a.** $C = \pi d$; C varies directly with d. The constant of proportionality is π.

b. Cost varies directly with the number of cans purchased. The constant of proportionality is $0.69 per can. ●

The constant of variation in $y = kx$ is identified by the letter k, not m, for two reasons:

1. The constant of variation also appears in direct variation for nonlinear functions. Nonlinear functions exhibiting direct variation include $f(x) = kx^2$ and $f(x) = k\sqrt{x}$.

2. Not all linear functions vary directly.

Proportions and Direct Variation

You should not assume that all linear functions represent direct variation. *Linear data must be proportional in order to vary directly.*

> For data to be proportional, the ratio of the output to input, y/x, for each data pair must be the same.

EXAMPLE Identifying proportional data   Which of these linear data sets represent proportional relationships?

    a. Two cans of refried beans cost 98 cents. Fifteen cans of refried beans cost $7.35.

    b. One Friday, a long-distance call costs $2.10 for 5 minutes. Another Friday at the same time, the call costs $4.44 for 14 minutes.

    c. Repairing a 12-foot length of sidewalk takes 12 hours. The same contractor takes 17 hours to repair a 32-foot length of sidewalk.

    d. One 15-minute phone call costs $12.45. Another call to the same place at the same time of day costs $6.64 for 8 minutes.

Solution **a.** $\dfrac{\$0.98}{2 \text{ cans}} = \$0.49$ per can; $\dfrac{\$7.35}{15 \text{ cans}} = \$0.49$ per can

       The data form equal ratios. A proportion is appropriate.

    b. $\dfrac{\$2.10}{5 \text{ min}} = \$0.42$ per minute; $\dfrac{\$4.44}{14 \text{ min}} \approx \$0.32$ per minute

       The data do not form equal ratios.

    c. $\dfrac{12 \text{ hr}}{12 \text{ ft}} = 1$ hour per foot; $\dfrac{17 \text{ hr}}{32 \text{ ft}} \approx 0.53$ hour per foot

       The data do not form equal ratios.

    d. $\dfrac{\$12.45}{15 \text{ min}} = \$0.83$ per minute; $\dfrac{\$6.64}{8 \text{ min}} = \$0.83$ per minute

       The data form equal ratios. A proportion is appropriate. ●

Think about it 5: How is the ratio of outputs to inputs, y/x, different from the slope ratio?

    We may identify direct variation or proportionality by graphing linear data.

EXAMPLE Identifying direct variation with a graph   The data from Example 13 are graphed in the figures below. Use the answers from Example 13 to tell which are graphs showing direct variation and which are graphs of nonproportional linear data, and then identify the differences between the two.

a.

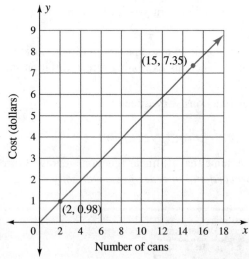

b.

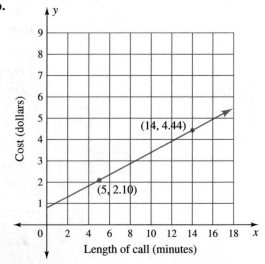

c.

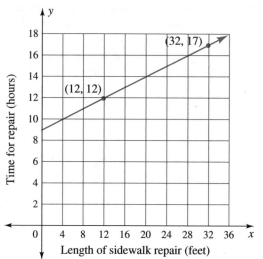

Length of sidewalk repair (feet)

d.

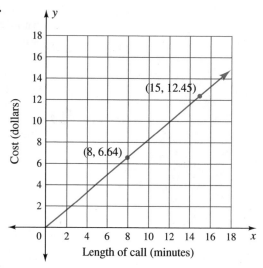

Length of call (minutes)

Solution Parts a and d of Example 13 contain sets of proportional data. Their graphs (a and d above) pass through the origin. The rules for these graphs can be written as direct variations.

Parts b and c of Example 13 do not deal with proportional data. Their graphs (b and c above) do not pass through the origin. The rules for these graphs cannot be written as direct variations. ●

Think about it 6: When a linear graph passes through the origin, why is the ratio of outputs to inputs, y/x, the same as the slope?

We have two ways to find whether linear data vary directly:

1. The ratio of outputs to inputs is the same for all data pairs.
2. The line connecting data points passes through the origin.

When we want to focus on the type of function, we describe *direct variation* as **linear variation.**

Linear Variation	If there is a constant k such that $y = kx$, we say *y varies linearly as x*. The ratio of outputs to inputs is the same for all ordered pairs, and the line connecting the data points passes through the origin.

Quadratic Variation

Given the relationship between direct variation and linear equations, $y = mx$, you should not be surprised that there is a similar relationship between variation and quadratic equations.

Quadratic Variation	If there is a constant k such that $y = kx^2$, we say *y varies with the square of x*. The constant k is the constant of variation.

The function $f(x) = kx^2$ describes quadratic variation. In **quadratic variation,** the output varies with the square of the input, so *the ratio of the output and the square of the input is constant, $y/x^2 = k$.*

The area of a circle provides one example of quadratic variation. The area varies with the square of the radius, $A = \pi r^2$, or with the square of the diameter, $A = (\pi/4)d^2$.

EXAMPLE 15   Applying quadratic variation: area of a circle   Find the area of the circle associated with each of these diameters.

   a. The diameter of one automobile engine cylinder bore is 83 millimeters.

   b. The diameter of the head of a pin is $\frac{1}{8}$ inch.

Solution   **a.** $A = (\pi/4)d^2 \approx 0.7854(83 \text{ mm})^2 \approx 5411 \text{ mm}^2$

       b. $A = (\pi/4)d^2 \approx 0.7854(0.125 \text{ in.})^2 \approx 0.0123 \text{ in}^2$ ●

The motion of a falling object also behaves with quadratic variation.

EXAMPLE 16   Applying quadratic variation: distance an object falls   The distance an object falls in t seconds is $d = \frac{1}{2}gt^2$. The constant g is the acceleration due to gravity, $g \approx 32.2 \text{ ft/sec}^2$.

   a. Write a sentence describing the variation.

   b. Make a table for d, the distance fallen, with t as input. Let t be on the interval 0 to 8 seconds.

   c. Graph the data from the table.

Solution   **a.** The distance fallen varies with the square of the time in motion.

b.

Time, t (seconds)	Distance, d (feet)
0	0
1	16.1
2	64.4
3	144.9
4	257.6
5	402.5
6	579.6
7	788.9
8	1030.4

Distance fallen

c.

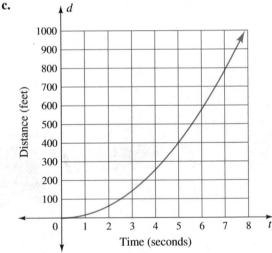

●

Joint Variation

Joint variation describes the product of two or more input variables.

Joint Variation   If there is a constant k such that $y = kwx$, we say y *varies jointly with w and x.* The constant k is the constant of variation.

In joint variation, there may be multiple inputs contributing to an output.

EXAMPLE **17**   *Applying joint variation*   Translate each of these variations into an equation. Identify the constant of variation in each setting.

a. The volume of a cylinder varies jointly with pi times the square of the radius of the base and the height.

b. The sales tax on a purchase of apples varies jointly with the price c per pound and number w of pounds purchased. The sales tax rate is 7%.

Solution   **a.** $V = \pi r^2 h$; the constant of variation is π.

Student Note: The variables in part a are in the customary order.

b. $T = 0.07cw$; the constant of variation is the tax rate, 0.07.   ●

In Example 18, we find the constant of variation for the volume of a pyramid. The **volume of a pyramid** *varies jointly with the area of its base and its height,*

$$V = kBh$$

EXAMPLE **18**

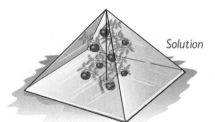

Applying joint variation: volume of a gardening pyramid   A clear plastic pyramid is used to cover tomatoes in a garden and speed their growth (Figure 11). The plastic cover has a square base of 2.5 feet on each side and a height of 3 feet. The volume of the pyramid is 6.25 cubic feet. Find the constant of variation for the volume of a pyramid.

Solution   The volume is $V = kBh$, where B is the area of the base and h is the height. For the plastic cover,

$$6.25 = k(2.5^2)3$$
$$6.25 = k(6.25)3$$
$$k = \tfrac{1}{3}$$

Figure II

The volume formula for a pyramid is $V = \tfrac{1}{3}Bh$.   ●

ANSWER BOX

Warm-up: 1. $-\tfrac{1}{2}$ **2.** $\tfrac{4}{5}$ **3.** 180 cm/day **4.** 5 hr **5.** $-\tfrac{5}{4}$ **6.** -2
Think about it 1: A sports team may have a 5 to 0 ratio of wins to losses. A score in soccer might be 5 to 0. **Example 1:** $\tfrac{22}{7} \approx 3.142\ 857$ and $\tfrac{2218}{706} \approx 3.141\ 643$. The ratios do not have the same decimal value, so they are not equal. Both ratios may be used to approximate the constant pi (π), but neither is exactly pi. Rounded to fourteen decimal places, pi is 3.141 592 653 589 79. **Think about it 2:** Three other possible proportions are $\dfrac{2\ \text{tbsp}}{1\ \text{gal}} = \dfrac{x}{60\ \text{gal}}$, $\dfrac{1\ \text{gal}}{60\ \text{gal}} = \dfrac{2\ \text{tbsp}}{x}$, and $\dfrac{1\ \text{gal}}{2\ \text{tbsp}} = \dfrac{60\ \text{gal}}{x}$.
Each proportion cross multiplies to 2 tbsp · 60 gal = 1 gal · x tbsp, as required. The solution to each is $x = 120$ tbsp.
Think about it 3: Some other proportions are $\tfrac{8}{4} = \tfrac{20}{10}$, $\tfrac{24}{12} = \tfrac{8}{4}$, and $\tfrac{10}{20} = \tfrac{12}{24}$. **Think about it 4:** The lengths of sides that are opposite equal angles in a triangle are equal. **Think about it 5:** The slope is the change in y over the change in x, so two ordered pairs are required to calculate it. The constant of variation is the ratio of the x- and y-coordinates of a single point. **Think about it 6:** In proportional data, the ratio y/x is the same as the slope because the two ordered pairs on the line are (0, 0), the origin, and (x, y):

$$\text{Slope} = m = \frac{y_2 - y_1}{x_2 - x_1} = \frac{y - 0}{x - 0} = \frac{y}{x}$$

EXERCISES

Simplify the expressions in Exercises I to 4.

1. a. $\dfrac{\frac{1}{2}\text{ foot}}{2\text{ inches}}$ **b.** $\dfrac{3000\text{ grams}}{6\text{ kilograms}}$ **c.** $\dfrac{32\text{ ounces}}{6\text{ pounds}}$

2. a. $\dfrac{1\text{ foot}}{4\text{ inches}}$ **b.** $\dfrac{2\text{ meters}}{150\text{ centimeters}}$ **c.** $\dfrac{1500\text{ meters}}{1\text{ kilometer}}$

3. a. $\dfrac{300\text{ milliliters}}{30\text{ liters}}$ **b.** $\dfrac{2\text{ years}}{180\text{ months}}$ **c.** $\dfrac{40\text{ minutes}}{\frac{1}{4}\text{ hour}}$

4. a. $\dfrac{2\text{ liters}}{300\text{ milliliters}}$ **b.** $\dfrac{4\text{ years}}{150\text{ months}}$ **c.** $\dfrac{12\text{ minutes}}{\frac{1}{2}\text{ hour}}$

In Exercises 5 to 8, round to the nearest thousandth.

5. The diameter of the golf hole shown below is $4\frac{1}{2}$ inches. The diameter of a golf ball is 1.68 inches.

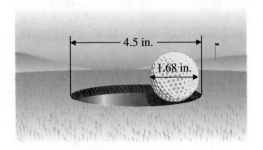

a. Find the ratio of the diameter of the hole to the diameter of the golf ball.

b. Find the ratio of the cross-sectional area of the hole to the cross-sectional area of the ball.

6. The diameter of a basketball hoop is 45 centimeters. The diameter of a basketball is 24.8 centimeters.

a. Find the ratio of the diameter of the hoop to the diameter of the basketball.

b. Find the ratio of the cross-sectional area of the hoop to the cross-sectional area of the ball.

7. The diameter of the head of a pin is 1.6 millimeters.

a. Find the area of the head of the pin.

b. What is the ratio of the area of one square centimeter to the area of the head of a pin?

8. The diameter of one machine bolt is $\frac{1}{2}$ inch and of a second bolt is $\frac{1}{4}$ inch.

a. Find the ratio of the area of the circular cross section of the larger bolt to that of the smaller bolt.

b. Explain why the $\frac{1}{2}$-inch bolt is considered more than twice as strong as the $\frac{1}{4}$-inch bolt.

In Exercises 9 and 10, suppose that when a bale of peat moss is opened and loosened, it expands to three times its original volume. The bale originally contains 4 cubic feet of peat moss.

9. How many bales of peat moss will need to be opened and loosened to fill a truck tank 8 feet by 12 feet by 4 feet?

10. How many bales of peat moss will be needed to cover to a depth of 2 inches a garden plot that is 15 feet by 40 feet?

Solve the proportions in Exercises II and I2. Round to the nearest tenth.

11. a. $\dfrac{3}{x}=\dfrac{5}{14}$ **b.** $\dfrac{4}{25}=\dfrac{x}{15}$ **c.** $\dfrac{7}{24}=\dfrac{16}{x}$

12. a. $\dfrac{4}{13}=\dfrac{x}{7}$ **b.** $\dfrac{x}{15}=\dfrac{8}{35}$ **c.** $\dfrac{24}{x}=\dfrac{32}{9}$

13. An access ramp is to have a rise-to-run ratio of 1 to 12. How long a horizontal distance (in feet) is needed for the ramp to rise 15 inches?

14. An access ramp covers a horizontal distance of 60 feet. Its slope is 1 to 12. What is its vertical rise?

In Exercises I5 and I6, find the missing sides for the similar triangles shown.

15.

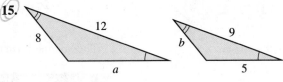

16.

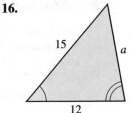

In Exercises 17 to 22, set up proportions to find the lengths of the sides marked with a letter in the figure.

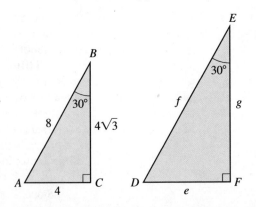

17. Let $e = 6$.

18. Let $f = 10$.

19. Let $g = 10$.

20. Let $g = 8$.

21. Let $e = \sqrt{5}$.

22. Let $f = 2\sqrt{3}$.

In Exercises 23 and 24, set up proportions and/or apply the Pythagorean theorem to find the lengths of the sides marked with a letter.

23.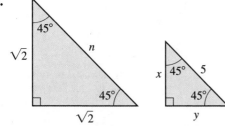

24.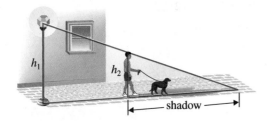

The ratios of base to height for similar triangles are equal. With the help of the figure below, set up proportions to describe the situations in Exercises 25 to 28 and answer the questions. Round your answers to the nearest tenth.

25. A 25-foot streetlight casts light past a 5-foot-tall person, causing a shadow. The person is standing 18 feet from the base of the streetlight. How long is the shadow?

26. A small tree has been planted 10 feet from a streetlight. The streetlight is 20 feet tall. The light creates a tree shadow 16 feet long. What is the height of the tree?

27. A gorilla is walking away from a streetlight. At the instant the gorilla is 12 feet from the base of the streetlight, the gorilla's shadow is 5 feet long. The streetlight is 20 feet tall. What is the height of the gorilla?

28. A 5-foot-tall parking meter is located 16 feet from the base of a 20-foot streetlight. How long is the parking meter's shadow?

In Exercises 29 to 34,

a. Identify variables for input and output. For each situation, write a linear equation, $y = mx$ or $y = mx + b$.

b. State whether the relationship is a direct variation ($y = kx$).

c. Explain the meaning of the nonzero y-intercept for those situations in which the data are not proportional.

29. A 10-pound weight stretches a spring 4 inches. A 15-pound weight stretches the same spring 6 inches.

30. A first-class letter costs $0.55 for 2 ounces and $1.01 for 4 ounces.

31. Tuition and fees are $185 for 2 credit hours and $345 for 4 credit hours.

32. A bicyclist travels 15 miles in 1 hour and 45 miles in 3 hours.

33. Tax on a $16 purchase is $1.20. Tax on a $30 purchase is $2.25.

34. Making 3 dozen cookies takes 2 hours. Making 6 dozen cookies takes 3 hours.

In Exercises 35 to 40, find the constant of variation k, and indicate the units of measure associated with k. Then answer the question.

35. The distance a car travels varies directly with the time traveled. Data to use:

(time, distance) = (0, 0), (1, 55), (2, 110), (3, 165)

How far will the car travel in 8 hours? Assume distance is in miles and time is in hours.

36. The amount of money earned varies directly with the time worked. Data to use:

(time, earnings) = (8, 45.20), (10, 56.50), (12, 67.80)

How much will be earned in 40 hours? Assume time is in hours and earnings is in dollars.

37. The total cost of compact discs varies directly with the number purchased. Data to use:

(number, total cost)
= (4, 57.96), (6, 86.94), (8, 115.92)

What is the cost of 10 discs?

38. The cost of attending a concert varies directly with the number of tickets purchased. Data to use:

(number, total cost) = (5, 195), (7, 273), (9, 351)

What is the cost of 10 tickets?

39. The weight of a pancake is proportional to the square of its radius. A 7-inch-diameter pancake weighs 4.5 ounces. What will a 3-inch-diameter pancake of the same thickness weigh?

40. The surface area of a sphere is proportional to the square of the radius. Data to use:

$$(\text{radius, area}) = (3, 113.1), (4, 201.06), (5, 314.16)$$

What is the surface area of a sphere with an 8-inch radius?

41. Use the data provided in parts a and b of Example 13 to write linear equations. Give an interpretation for any nonzero y-intercepts.

42. Use the data provided in parts c and d of Example 13 to write linear equations. Give an interpretation for any nonzero y-intercepts.

Complete each sentence in Exercises 43 to 46 by choosing the correct phrase or equation from within the parentheses.

43. Linear equations that can be written as $y = mx$ (are/are not) linear variations.

44. Linear equations that can be written as $y = mx + b$, $b \neq 0$ (are/are not) linear variations.

45. If a set of data is proportional, then its equation is of the form ($y = mx$/$y = mx + b$, $b \neq 0$).

46. If a linear set of data is not proportional, then its equation is of the form ($y = mx$/$y = mx + b$, $b \neq 0$).

In Exercises 47 and 48, identify the constant of variation.

47. **a.** $A = \frac{1}{2}bh$

   b. $A = lw$

   c. $V = \frac{1}{3}\pi r^2 h$

48. **a.** $I = \$1000rt$

   b. $D = rt$

   c. $V = \pi r^2 h$

49. Use the results from Example 18 to estimate the original volume of the Great Pyramid of Cheops (see the figure). The square base of the pyramid had an area of approximately 53,095 square meters, and the original height was approximately 146.6 meters.

In Exercises 50 to 57, translate the sentence into a variation equation with constant of variation k. Research, as needed, to find the exact constant of variation.

50. The circumference of a circle varies directly with the diameter.

51. The circumference of a circle varies directly with the radius.

52. The area of a circle varies directly with the square of the radius.

53. The area of a square varies with the square of its side.

54. The area of a rectangle varies jointly with length and width.

55. Distance traveled varies jointly with rate and time.

56. The volume of a cylinder varies jointly with the height and the square of the radius.

57. The surface area of a cube varies with the square of the side of the cube.

58. Multiply both sides of the proportion $\dfrac{a}{b} = \dfrac{c}{d}$ by the common denominator bd, and simplify to show that the cross multiplication property is true.

59. The area of a circle in terms of radius is $A = \pi r^2$. Substitute $d/2$ for r, and show that $A = \pi d^2/4$. Show also that $A \approx 0.7854d^2$.

In Exercises 60 to 63, explain what has been done wrong in each solution.

60. Simplify the ratio 6 ft/2 yd:

$$\frac{6 \text{ ft}}{2 \text{ yd}} = \frac{3}{1}$$

61. Simplify the ratio 3 qt/2 gal:

$$\frac{3 \text{ qt}}{2 \text{ gal}} = \frac{3 \text{ qt}}{2 \text{ gal}} \cdot \frac{4 \text{ qt}}{1 \text{ gal}} = \frac{12}{2} = \frac{6}{1}$$

62. Solve the proportion $\dfrac{4}{15} = \dfrac{6}{x}$:

$$\frac{4}{15} = \frac{6}{x} = \frac{24}{15x}$$

63. Solve the proportion $\dfrac{4}{15} = \dfrac{6}{x}$:

$$\frac{4}{15} = \frac{6}{x} = \frac{90}{4x}$$

Projects

64. ***Astronomy and the Small Angle Equation.*** The small angle equation is

$$\frac{\alpha}{206,265 \text{ sec}} = \frac{d}{D}$$

where α is an arc angle measured in seconds (see the figure). One degree of arc angle is 60 minutes of arc, or 3600 seconds of arc. The distance from the viewer to the object is D. The diameter of the object is d.

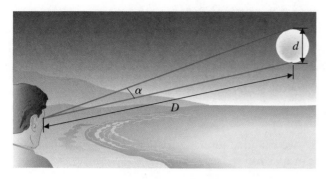

a. The moon has a diameter of approximately 3480 kilometers and is at a distance of about 384,000 kilometers from Earth. Find α, the viewing angle at which we see the moon.

b. Design a tool for sighting objects and estimating the arc angle.

c. Use your tool to estimate the arc angle for five distant objects of known size (such as a parked compact car 16 feet long). Use the formula to predict how far away the car is.

d. The small angle equation comes from a proportion based on the ratio between the 360 degrees in a circle and the circumference of a circle, $2\pi r$. In this situation, the radius of a circle is D, so the circumference is $2\pi D$ (see the figure below).

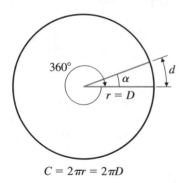

$$C = 2\pi r = 2\pi D$$

The small angle equation may be found by starting with

$$\frac{360°}{\text{circumference}} = \frac{\text{angle in seconds}}{\text{length of distant object}}$$

$$\frac{360°}{2\pi D} = \frac{\alpha \text{ seconds}}{d}$$

Rearrange the above equation into the same form as the small angle equation. It is necessary to apply unit analysis to change degrees into seconds and to obtain the constant 206,265.

65. ***Making Mountains out of Mole Hills.*** Assume that mole hills and mountains are both shaped as cones, as shown in the figure. The formula for the volume of a cone is $V = \frac{1}{3}\pi r^2 h$. Suppose that Fuji-San (a mountain in Japan) and a mole hill are of similar proportions. Find how many mole hills it would take to make the mountain.

5.2 Inverse Variation and Related Graphs

OBJECTIVES

• Identify variation as direct or inverse.

• Find the constant of variation.

• Use inverse variation formulas in applications.

• Graph inverse variation equations.

• Identify the behavior of a graph near an input that gives an undefined output.

• Find the axis of symmetry in inverse variation graphs.

I N THIS SECTION, we examine inverse variation, inverse proportions, and the behavior of the graphs of inverse equations.

Inverse Variation

The photographs on the chapter opening page show a hurricane and a galaxy. The pictures were chosen because of the similarity in form of these two different objects. As vastly different as a hurricane and a galaxy of stars are, both have a rate of turning that is dependent on how compact their structure is relative to the center of rotation. The smaller the hurricane's "eye," the closer the cloud mass and moisture are to the center of rotation and the faster the accompanying winds. In the case of a galaxy, the more densely packed the stars are toward the center of the galaxy, the faster the galaxy rotates. We describe the relationship between the position of hurricane clouds and galaxy stars relative to their centers and their rate of turning as an inverse variation.

Think about it 1: How is an ice skater's spin similar to the motion of a hurricane or galaxy?

In Example 1, we use numbers to investigate an inverse variation setting.

EXAMPLE   Exploring driving speed   Suppose Indy car driver Lyn St. James has 120 miles to drive—how long will it take her? Record your data in a table. Graph your table data with rate on the horizontal axis and time on the vertical axis.

Solution   The time it takes for St. James to drive 120 miles depends on how fast she drives. In an Indy car street race, the 120 miles might take only an hour (see Table 3). At congested freeway speeds, 2 hours might be reasonable. At slower city traffic speeds, the trip may take 4 hours.

Rate (miles per hour)	Time (hours)	Distance (miles)
120	1	120
60	2	120
40	3	120
30	4	120
20	6	120
15	8	120
10	12	120

Table 3

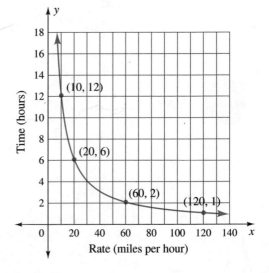

Figure 12

Because the distance is constant, both the rate and the time change. As the rate in miles per hour increases, the time in hours to travel 120 miles decreases. This relationship is shown most clearly by the graph in Figure 12.

In more formal language, we say that the time needed to drive 120 miles varies inversely with the rate. We write this relationship as

$$r \cdot t = 120 \qquad \text{or} \qquad t = \frac{120}{r}$$ ●

Inverse Variation

> If there is a constant k such that $xy = k$, we say y *varies inversely with x.* The constant k is the constant of variation.

There are at least three ways to write inverse variation. We may write inverse variation as a function: $f(x) = k/x$. We may write $y = k/x$, with the constant of variation in a ratio. The third variation equation, $xy = k$, places the emphasis on the constant product of x and y. In Example 1, the distance, 120 miles, is the constant of variation. Look for the constant of variation in Example 2.

EXAMPLE **2**   Identifying inverse variation: lottery jackpot   Suppose the current lottery jackpot is $24 million. Susumu has a winning ticket but hears that there may be up to 24 winners.

a. Make a table and a graph of the number of winning tickets and the prize received for each winning ticket.

b. Write an equation describing the fact that the amount won varies inversely with the number of winning tickets.

c. Identify the constant of variation.

Solution   **a.** The number of winning tickets, x, and the prize received, y, are given in Table 4 and are graphed in Figure 13.

Number of Tickets	Prize (dollars)	Lottery Jackpot (dollars)
1	24 million	24 million
2	12 million	24 million
4	6 million	24 million
6	4 million	24 million
12	2 million	24 million
24	1 million	24 million

Table 4   Lottery jackpot with different numbers of winning tickets

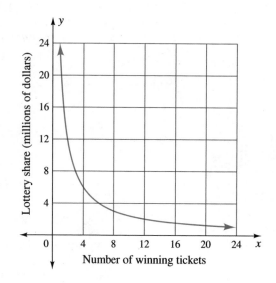

Figure 13

A negative number of winners is meaningless, so we graph only first-quadrant numbers.

b. If x is the number of winning tickets and y is the price received, then the equation is $xy = 24$ million, or $y = 24$ million$/x$.

c. The constant of variation is the $24 million jackpot. The number remains constant regardless of the number of winners. ●

Think about it 2: How many winners would reduce the lottery prize to $1? Where would such a point be located on a graph such as Figure 13, where 2 inches represents 24 winners?

Inverse Variation and Proportions

In Examples 1 and 2, the constants of variation, 120 miles and $24 million, were given. The meaning of the constants was quite clear. In the settings below, the value and meaning of the constant of variation may not be as obvious.

EXAMPLE **3**   Exploring inverse variation: balance   The equipment needed for this exploration is a 12-inch ruler, three coins of the same value, and a pencil.

Place the pencil under the 6-inch position on the 12-inch ruler (see Figure 14). The ruler should balance, with neither end touching the table top.

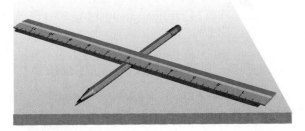

Figure 14

a. Center a stack of two coins at the 4-inch position. Where will the third coin need to be placed to bring the ruler into balance?

b. Repeat part a, placing the two coins at the $3\frac{1}{2}$-inch position.

c. Explain your results.

Solution   **a.** When the third coin is placed 4 inches from the center (at the 10-inch mark), it will balance two coins placed 2 inches from the center (at the 4-inch mark). See Figure 15.

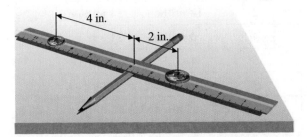

Figure 15

b. When the third coin is placed 5 inches from the center (at the 11-inch mark), it will balance two coins placed $2\frac{1}{2}$ inches from the center.

c. The single coin is placed twice as far from the center of the ruler as the two coins because it weighs half as much. *The distance the coins are placed from the center varies inversely with the number of coins.* ●

Example 4 shows how we can use the data from Example 3 to explain our exploration results and to find a constant of variation.

EXAMPLE **4**   Finding the constant of variation Write an equation relating the numbers and positions of the coins in Example 3. Identify the constant of variation.

Solution   **a.** 2 coins(2 inches from center) = 1 coin(x inches from center)

$$4 \text{ coin·inches} = x \text{ coin·inches}$$
$$x = 4$$

Check: 2 coins(2 in.) $\overset{?}{=}$ 1 coin(4 in.) ✔

The constant of variation k is 4 coin·inches.

b. 2 coins(2.5 inches from center) = 1 coin(x inches from center)

$$5 \text{ coin·inches} = x \text{ coin·inches}$$
$$x = 5$$

Check: 2 coins(2.5 in.) $\overset{?}{=}$ 1 coin(5 in.) ✔

The constant of variation k is 5 coin·inches.

This example illustrates two different inverse variations. ●

When the focus of the problem is on the inverse relationship in the data and not on the constant of variation, we describe the relationship with the phrase *inversely proportional*. The inverse proportion for (x_1, y_1) and (x_2, y_2) is $\dfrac{x_1}{x_2} = \dfrac{y_1}{y_2}$, but it is usually written $x_1 \cdot y_1 = x_2 \cdot y_2$ to emphasize the constant product.

Inverse Proportions

> Quantities are *inversely proportional* if their product, k, is constant. For paired data, such as (x_1, y_1) and (x_2, y_2), being inversely proportional means that $x_1 \cdot y_1 = x_2 \cdot y_2$. The constant product k is called the *constant of proportionality*.

The derivation of $x_1 \cdot y_1 = x_2 \cdot y_2$ appears in Example 7 on page 343. Because either phrasing—*varies inversely* or *inversely proportional*—may be used in application settings, you should be familiar with both sets of vocabulary.

In Example 5, the setting is children on a seesaw rather than coins on a ruler. *The product of one child's weight and distance from the center is equal to the product of the other child's weight and distance from the center.*

EXAMPLE **5**   Finding the constant of proportionality: balancing a seesaw Use an inverse proportion to describe the relationship between the weight and the position of the children in balance on the seesaw in Figure 16. State the constant of proportionality.

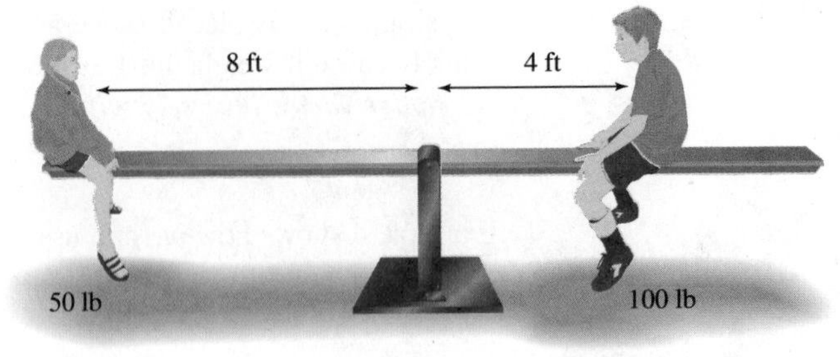

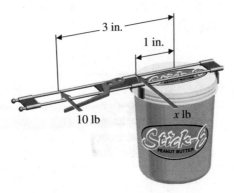

Figure 16

Solution

$$100 \text{ pounds} \cdot 4 \text{ feet} = 50 \text{ pounds} \cdot 8 \text{ feet}$$

$$400 \text{ pound·feet} = 400 \text{ pound·feet}$$

The position on the seesaw is inversely proportional to the weight of the child. The constant of proportionality for these positions is $k = 400$ pound·feet. ●

In Example 6, we use an inverse proportion in another setting. We start by identifying the phrases that relate the data in the problem.

EXAMPLE **6**   Finding the constant of proportionality: loosening a jar lid   The force you need to loosen a jar lid is inversely proportional to the distance your grip is from the center of the lid. It takes 10 pounds of force to open a lid with a gripper handle that places the force 3 inches from the center (see Figure 17). How many pounds of force are needed to open the lid without the gripper if the lid is held 1 inch from the center?

a. Identify the phrases that describe (x_1, y_1) and (x_2, y_2).

b. Construct and solve an inverse proportion.

c. Identify the constant of proportionality.

Figure 17

Solution   **a.** 10 pounds applied at 3 inches is (x_1, y_1).
x pounds applied at 1 inch is (x_2, y_2).

b. $x_1 \cdot y_1 = x_2 \cdot y_2$
10 pounds · 3 inches $= x$ pounds · 1 inch
30 pound·inches $= x$ pound·inches
$x = 30$

It takes 30 pounds of force to open the lid directly.

c. The constant of proportionality is $k = 30$ pound·inches.   ●

The constant product of quantities that vary inversely permits us to derive the inverse proportion formula.

EXAMPLE **7**   Deriving the inverse proportion formula   Suppose the quantities x and y in (x_1, y_1) and (x_2, y_2) vary inversely. Derive the inverse proportion formula, $\dfrac{x_1}{x_2} = \dfrac{y_2}{y_1}$.

Solution   We set up each constant product, $x \cdot y = k$:

$$x_1 \cdot y_1 = k \quad \text{and} \quad x_2 \cdot y_2 = k \qquad \text{Set the left sides of the equations equal.}$$

$$x_1 \cdot y_1 = x_2 \cdot y_2 \qquad \text{Divide both sides by } x_2 y_1.$$

$$\frac{x_1 y_1}{x_2 y_1} = \frac{x_2 y_2}{x_2 y_1} \qquad \text{Simplify.}$$

$$\frac{x_1}{x_2} = \frac{y_2}{y_1}$$

●

Think about it 3: We are accustomed to seeing a direct proportion in the form

$$\frac{x_1}{x_2} = \frac{y_1}{y_2}$$

Is that proportion the same as the proportion that gives the equation $x_1 \cdot y_1 = x_2 \cdot y_2$? Cross multiply to check your answer.

Direct versus Inverse Variation

To distinguish direct variation from inverse variation or a proportion, look for constant ratio or constant products.

Distinguishing Direct and
Inverse Variation

> For data (x_1, y_1) and (x_2, y_2):
>
> Direct variation has a constant ratio, $y/x = k$, the variation constant.
>
> Inverse variation has a constant product, $x_1 y_1 = x_2 y_2 = k$, the variation constant.

EXAMPLE **8**   Identifying direct and inverse variation   Look for a constant ratio or constant product in each table, and identify the table as showing direct or inverse variation or neither. What is the meaning, if any, of the constant of variation?

a.

Altitude (feet)	Cooking Time (minutes)
1000	33
3000	39
5000	45

b.

Size of Glass (ounces)	Number of Glasses
6	$21\frac{1}{3}$
8	16
12	$10\frac{2}{3}$

c.

Vacation Days	Budget ($ per day)
5	200
10	100
15	$66\frac{2}{3}$

d.

Time (hours)	Distance (miles)
2	110
3	165
4	220

Solution **a.**

$$\frac{y}{x} = \frac{33}{1000} \neq \frac{39}{3000} \neq \frac{45}{5000}$$

$$x \cdot y = 1000(33) \neq 3000(39) \neq 5000(45)$$

Neither the ratios nor the products are constant. There is neither direct nor inverse variation here.

b. The product is

$$x \cdot y = 6 \cdot 21\frac{1}{3} = 8 \cdot 16 = 12 \cdot 10\frac{2}{3} = 128$$

$$k = 128$$

There is a constant product. A total of 128 ounces of milk is available (in a gallon of milk, for example). The table shows inverse variation.

c. The product is

$$x \cdot y = 5 \cdot 200 = 10 \cdot 100 = 15 \cdot 66\frac{2}{3} = 1000$$

$$k = 1000$$

There is a constant product. A total of $1000 is available for the vacation. The table shows inverse variation.

d. The ratio is

$$\frac{y}{x} = \frac{110}{2} = \frac{165}{3} = \frac{220}{4} = \frac{55}{1}$$

$$k = \frac{55}{1}$$

The constant ratio in the table is $\frac{55}{1}$. For $k = \frac{55}{1}$, the distance increases by 55 miles with each hour traveled. The table shows direct variation. ●

Inverse Square Variation

EXAMPLE **9**

Exploring light intensity   Shine a flashlight toward a wall from a distance of 1 foot. Next, shine it from a distance of 2 feet. Finally, shine it from a distance of 3 feet. Compare the brightness of the light on the wall in each of the three positions.

Solution   The light becomes significantly dimmer with each foot we move away from the wall. ●

The amount of light energy reaching a surface per unit time is called the *intensity* of the light. Figure 18 shows light striking a surface at three different distances from the source.

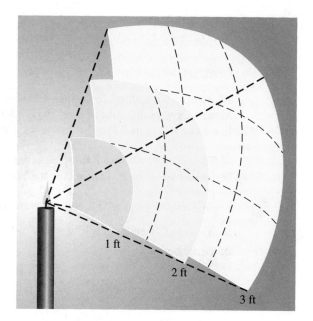

Figure 18

The first square in Figure 18 shows the light striking an area of 1 square foot at a distance of 1 foot from a candle. This same amount of light must spread over 4 square feet at a distance of 2 feet from the candle. Thus, at 2 feet, the intensity of light is $\frac{1}{4}$ of that reaching the original square at 1 foot. The light must spread over 9 square feet at a distance of 3 feet from the candle. Thus, at 3 feet, the intensity of light is $\frac{1}{9}$ of that reaching the original square at 1 foot.

We say that the intensity of light is inversely proportional to the square of the distance from the light source. Thus, the amount of light, or intensity, is described by $y = k/d^2$, an inverse variation in which the input variable is distance squared.

EXAMPLE   Applying inverse square variation: reading light intensity   Nellie places her chair 10 feet from a light source. With a ratio, compare the intensity of light on her reading material to the intensity of light available 1 foot from the light source.

Solution   The light intensity is

$$y = \frac{k}{d^2}$$

At 10 feet, the intensity is

$$y_1 = \frac{k}{d_1{}^2} = \frac{k}{10^2} = \frac{k}{100} = 0.01k$$

At 1 foot, the intensity is

$$y_2 = \frac{k}{d_2{}^2} = \frac{k}{1^2} = \frac{k}{1} = k$$

The ratio of y_1 to y_2 is

$$\frac{y_1}{y_2} = \frac{0.01k}{k} = 0.01 = \frac{1}{100}$$

Thus, the light is only $\frac{1}{100}$ as strong at 10 feet from the light source.   ●

Example 10 shows that the intensity of light diminishes considerably at a short distance from the light source.

Think about it 4: What is the behavior of a graph of light intensity if the distance from the source is on the horizontal axis and light intensity is on the vertical axis?

Inverse Variation Graphs

We return to graphing to further examine inverse variation. The graphs for St. James's driving rate over 120 miles (Figure 12) and for the prize from a $24 million lottery (Figure 13) are portions of a graph called a *hyperbola*.

UNDEFINED VALUES   In Example 11, we examine the behavior of a hyperbola on both sides of an input that makes the equation undefined.

EXAMPLE **11**   Investigating outputs near a zero in the denominator

a. Make tables and a graph of $y = 2/x$ for both negative and positive inputs.

b. Describe the behavior of $y = 2/x$ near $x = 0$ and why it occurs.

Solution **a.** Tables 5 and 6 show outputs for $y = 2/x$ as x gets close to zero. When graphed, the data in each table form one side of the hyperbola (Figure 19).

Input, x	Output, $y = 2/x$
−5	$2/−5 = −0.4$
−4	$2/−4 = −0.5$
−2	$2/−2 = −1.0$
−1	$2/−1 = −2.0$
−0.5	$2/−0.5 = −4.0$
−0.25	$2/−0.25 = −8.0$
−0.1	$2/−0.1 = −20.0$
−0.001	$2/−0.001 = −2000$

Table 5 $y = 2/x$ as x approaches 0 from the left

Input, x	Output, $y = 2/x$
5	$2/5 = 0.4$
4	$2/4 = 0.5$
2	$2/2 = 1.0$
1	$2/1 = 2.0$
0.5	$2/0.5 = 4.0$
0.25	$2/0.25 = 8.0$
0.1	$2/0.1 = 20.0$
0.001	$2/0.001 = 2000$

Table 6 $y = 2/x$ as x approaches 0 from the right

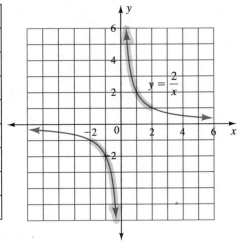

Figure 19

b. The hyperbola is quite steep between $x = −1$ and $x = 0$ and between $x = 0$ and $x = 1$. If we trace the third-quadrant curve from left to right, the inputs x approach zero and the curve turns downward. The curve never touches the y-axis.

If we trace the first-quadrant graph from right to left, the inputs x approach zero and the curve rises. The curve gets closer to but never touches the y-axis. At $x = 0$, the equation becomes $y = \frac{2}{0}$ and is undefined. There is no point on the graph for $x = 0$. ●

The hyperbolic graph of an inverse variation equation, $y = k/x$, is characterized by a nearly vertical graph whenever the denominator approaches zero. This behavior is found in graphs of rational expressions.

Graphical Behavior Near a Zero Denominator	The graph of a rational expression that has been simplified to lowest terms becomes nearly vertical whenever the denominator approaches zero.

Graphing Calculator Note: Calculators are designed to reject an attempt to divide by zero. As you trace along the graph of a curve such as $y = 2/x$, the output will be blank when the input creates a zero denominator. It can take considerable adjustment of the viewing window to locate such an input with the cursor.

VERY LARGE AND VERY SMALL INPUTS   As St. James increases the speed of her car, the time it takes to drive 120 miles drops toward zero. In the lottery example, as the number of winning tickets increases, the prize approaches zero. In Example 12, we look at similar behavior in the graph of $y = 2/x$. We look at the outputs first as the input x approaches negative infinity and then as the input x approaches positive infinity.

EXAMPLE **12**   Investigating outputs for inputs approaching infinity   Describe the behavior of $y = 2/x$ for x smaller than -4 and for x larger than 4.

Solution

Input, x	Output, $y = 2/x$
-4	$2/-4 = -0.5$
-5	$2/-5 = -0.4$
-10	$2/-10 = -0.2$
-100	$2/-100 = -0.02$
-1000	$2/-1000 = -0.002$
$-10,000$	$2/-10,000 = -0.0002$

Table 7 $y = 2/x$ as x approaches $-\infty$

Input, x	Output, $y = 2/x$
4	$2/4 = 0.5$
5	$2/5 = 0.4$
10	$2/10 = 0.2$
100	$2/100 = 0.02$
1000	$2/1000 = 0.002$
10,000	$2/10,000 = 0.0002$

Table 8 $y = 2/x$ as x approaches ∞

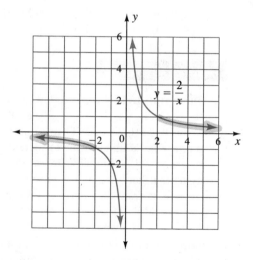

$$y = \frac{2}{x}$$

Figure 20

As Tables 7 and 8 indicate, the expression $2/x$ becomes close to zero as we input numbers larger than 4 and smaller than -4. For these same inputs, the graph in Figure 20 approaches the x-axis.   ●

Figure 20 shows the graph $y = 2/x$ approaching the x-axis. In general, rational expressions approach a horizontal line to the left and right.

Graphical Behavior as x
Goes to $+\infty$ or $-\infty$

> For large x (toward $+\infty$) or small x (toward $-\infty$), the graph of a rational expression approaches a horizontal line.

SYMMETRY   The hyperbola in Figure 21 has two axes of symmetry, shown as dashed lines. One axis of symmetry, $y = x$, passes through the first and third quadrants. The second, $y = -x$, passes through the second and fourth quadrants.

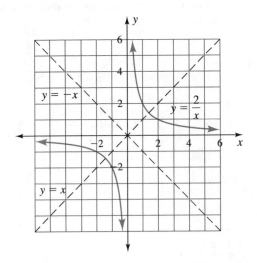

Figure 21

 Graphing Calculator Note: The symmetry in the hyperbola may be difficult to observe because of distortion in the viewing window. You can adjust the display with the square window option.

ANSWER BOX

Warm-up: 1. $1 \cdot 120$, $2 \cdot 60$, $3 \cdot 40$, $4 \cdot 30$, $5 \cdot 24$, $6 \cdot 20$, $8 \cdot 15$, $10 \cdot 12$ **2.** $1 \cdot 64$, $2 \cdot 32$, $4 \cdot 16$, $8 \cdot 8$ **3.** $1 \cdot 400$, $2 \cdot 200$, $4 \cdot 100$, $5 \cdot 80$, $8 \cdot 50$, $10 \cdot 40$, $20 \cdot 20$ **Think about it 1:** When a skater goes into a spin, she pulls in her arms to increase her rate of turning. She pushes her arms back out to slow down. Physicists describe this effect by saying that her angular velocity is inversely proportional to the distance her mass is from the center of rotation. **Think about it 2:**   The lottery prize would be $1 if there were 24 million winners. The point (24 million, $1) would be located 2,000,000 inches to the right, just above the horizontal axis. Use unit analysis to change 2,000,000 inches into an appropriate unit of measure. For a further challenge, calculate an estimate of the distance the graph would be above the x-axis. **Think about it 3:** The proportion $\dfrac{x_1}{x_2} = \dfrac{y_1}{y_2}$ is for direct variation and cross multiplies to $x_1 y_2 = x_2 y_1$. The proportion $\dfrac{x_1}{x_2} = \dfrac{y_2}{y_1}$ is for inverse analysis and gives the cross product $x_1 \cdot y_1 = x_2 \cdot y_2$. **Think about it 4:** Because light intensity decreases as distance increases, we would expect the graph to fall rapidly as we move from an input of 1 foot to an input of 10 feet.

EXERCISES 5.2

Describe each of the situations in Exercises 1 to 6 with an inverse variation equation. Note the units on your variables, and identify the constant of variation.

1. The number of years a landfill can be used varies inversely with the number of cubic yards of trash deposited each year. The landfill has 1 million cubic yards of space.

2. The number of years the world resources of silver will last varies inversely with the number of metric tons used per year. Estimated world supply in 1990 was 420,000 metric tons.

3. The number of years the world resources of copper will last varies inversely with the number of metric tons used per year. Estimated world supply in 1990 was 574 million metric tons.

4. Assuming no interest or other funds are added, the number of years a $100,000 savings account will last after retirement varies inversely with the number of dollars spent per year.

5. The number of points given to each problem on a 100-point test varies inversely with the number of problems on the test.

6. The time required to drill a 1200-foot well varies inversely with the speed of the drill in feet per minute.

In Exercises 7 to 16, identify the data as reflecting direct variation or inverse variation or neither, and identify the constant of variation.

7.

Production	Life Span
160 tons/day	35 years
200 tons/day	28 years

8.

Times to Wear Blouse	Cost per Wearing
2	$26.00
10	$ 5.20

9.

Number of Items	Total Cost
160	$35.00
200	$43.75

10.

Number of Jeans	Total Cost
2	$ 56
10	$280

11.

Rate	Time
15 mi/hr	6 hr
25 mi/hr	3.6 hr

12.

Rate	Distance
15 mi/hr	75 mi
25 mi/hr	125 mi

13.

Number of Workers	Time to Re-roof House
2	6 days
6	2 days

14.

Time of Travel	Distance Traveled
3 hours	90 miles
6 hours	150 miles

15.

Number of Semesters	Loan Amount
2	$2200
3	$3300

16.

Length of Vacation	Total Cost
3 days	$ 540
8 days	$1440

In Exercises 17 to 24, set up and solve an inverse proportion. Identify the constant of variation for each setting.

17. Jehan weighs 35 pounds and sits 6 feet from the balance point of a seesaw. She just balances Dalia, sitting 4 feet from the balance point. How heavy is Dalia?

18. Shu-Ju bicycles 15 miles per hour for 3 hours. Su Lin bicycles the same route in 3 hours and 15 minutes. What is Su Lin's rate?

19. A batch of stew serves each of 80 people a $\frac{3}{4}$-cup serving. How many people may be served 1-cup servings?

20. A landfill will be in service 10 years if each of the 50,000 households throws away a full 90-gallon container per week. If the same households recycle and reduce the throw-away portion to 30 gallons per week, how long will the landfill be in service?

21. The world reserves of lead will last 133 years if 900,000 metric tons are used each year. How many years will the reserves last if 1,300,000 metric tons are used each year?

22. If the entire world production of rice in 1990 had been evenly divided among the world's population, each person would have received about 0.26 kilogram (half a pound) of rice per day. The world population in 1990 was 5.333 billion. How much rice will there be per person per day in 2025 with the same annual production if the population increases to 8.2 billion?

23. If the yellow stripe on a road is 4 inches wide, a gallon of paint will make a stripe 900 feet long. If the stripe is reduced to 3 inches wide, how long a stripe will the gallon paint?

24. At $40 per ticket, an athletic office can sell out its 75,000-seat stadium. In order to maintain the same total revenue, how many tickets would have to be sold at $64 per ticket?

25. If the intensity of light relative to the distance from its source is described by $y = k/d^2$, how many times greater is the intensity of light reaching 3 feet from a light source than that of light reaching 6 feet from the same source?

26. If the intensity of light relative to the distance from its source is described by $y = k/d^2$, how many times greater is the intensity of light reaching 2 feet from a light source than that of light reaching 6 feet from the same source?

27. Divide both sides of $x_1 y_1 = x_2 y_2$ by $x_2 y_1$. Describe why the phrase *inverse proportion* might be appropriate to the result. (*Hint:* Compare the resulting proportion with $x_1/x_2 = y_1/y_2$.)

28. Cross multiply $x_1/x_2 = y_1/y_2$ and compare the result with $x_1 y_1 = x_2 y_2$. How are they the same or different?

29. Graph the inverse square function, $f(x) = 1/x^2$.

   a. What is $f(0)$?

   b. What is the equation of the axis of symmetry?

   c. Describe the behavior of the graph for $x < -3$ and $x > 3$.

30. Graph the reciprocal function, $f(x) = 1/x$.

   a. What is $f(0)$?

   b. What are the equations of the axes of symmetry?

   c. Describe the behavior of the graph for $x < -4$.

   d. Describe the behavior of the graph for $x > 4$.

31. a. Graph $y = 1/x^2$ and $y = 1/x$ on the same axes for x between -5 and 5.

   b. For what inputs will the graph of $y = 1/x$ be above that of $y = 1/x^2$?

   c. For what inputs will the graphs be equal?

   d. Why is $y = 1/x^2$ always above $y = 1/x$ for $x < 0$?

32. Graph $y = 1/x$ and $y = -1/x$ on the same axes. Compare the positions of the graphs.

33. a. Graph $y = 2/x$, $y = 8/x$, and $y = 16/x$ on the same axes.

   b. What is the same about the graphs?

   c. What is different?

   d. What is the constant of variation for each equation?

   e. How does increasing the constant of variation change the graph?

   f. Predict the position of $y = 1/x$ relative to the other graphs.

   g. Predict the position of $y = 32/x$ relative to the other graphs.

   h. Describe the behavior of the graphs as x gets larger than 50.

34. a. Set up coordinate axes to the same scale on both axes. Graph $y = 1/(x + 2)$ and $y = x + 2$.

   b. For what input will $y = 1/(x + 2)$ be undefined?

   c. What happens to the graph of $y = 1/(x + 2)$ near the undefined point?

   d. What are the coordinates of intersection of the two graphs?

   e. What are the axes of symmetry for $y = \dfrac{1}{x + 2}$?

Extended Problems for Extra Practice

35. Dave lives 6 miles from work. He normally drives, but today he is considering taking the time to roller-blade. With the time needed to roller-blade 6 miles as output, make a table and a graph for inputs of 1 mile per hour to 10 miles per hour. What is a reasonable speed for getting to work in a timely fashion?

36. How many U.S. coins of one kind are needed to make $2.00? Make a table and graph, where the input is the value of the coin in cents and the output is the number of that type of coin needed. What is the equation of the graph containing all of the points?

37. a. The legal freeway speed is 65 mph, and you have 20 miles left to travel. How long in minutes will it take you to drive 20 miles at the posted speed?

    b. Make a table for the number of minutes it will take you to drive the 20 miles at 60, 65, 70, 75, and 80 mph.

    c. How fast would you need to drive to cut 5 minutes off the driving time at 65 mph?

38. a. The resistance of a wire of fixed length varies inversely with the square of the diameter of the wire. Use the data in the table to determine the constant of variation between diameter and resistance.

American Wire Gauge Size	Diameter (inches)	Resistance at 20°C per 1000 Feet (ohms)
24	0.0201	25.67
22	0.0254	16.14
20	0.0320	10.15
18	0.0403	6.39
16	0.0508	4.02

    b. Use power regression to fit a curve $y = ax^b$ to the diameter and resistance data.

    c. As the American wire gauge size decreases, the diameter _____ .

39. a. The base unit of a portable telephone is plugged into an electrical outlet with a small box. The box is a transformer that changes 120-volt electricity (V_1) into 12-volt electricity (V_2). The output current (I_2) from the transformer is 100 mA. What is the input current (I_1) when the voltage and current are related by the formula $V_1/V_2 = I_2/I_1$?

    b. Are voltage (V) and current (I) directly or inversely proportional in this problem setting? Show why. (*Hint:* Cross multiply the proportion in part a.)

    c. What is the constant of proportionality?

40. a. The formula $F = GM_1M_2/d^2$ describes the gravitational force F between two masses M_1 and M_2. The masses are a distance d apart. Describe the gravitational force as a joint and an inverse variation.

    b. A 60-kilogram mass on Earth's surface registers a force of 588.6 N when acted on by Earth's gravity. The radius of Earth is 6.378×10^6 meters. The gravitational constant G is 6.672×10^{-11} N·m²/kg². Use the gravitational force equation to find the mass of Earth. (N is a newton, the unit of force in the metric system.)

    c. Suppose the 60-kilogram mass is moved to a distance of 3.92×10^8 meters from Earth's center. This is approximately the distance to the center of the moon. What force, in newtons, acts on the mass at this position? (You can divide newtons by 4.4 to approximate pounds of force.)

Project

41. *Water Reservoirs.* In February 1996, articles appeared in several newspapers and magazines (such as *Science News*) about research by geophysicist Benjamin Fong Chao, who had found that water stored in reservoirs in the Northern Hemisphere had increased the rate of Earth's rotation. Research and write a one-page report on his finding and how it relates to other relationships mentioned in this section.

5.3   Simplification, Multiplication, and Division of Rational Expressions

OBJECTIVES

- Find when a rational expression is not defined.
- Simplify rational expressions.
- Multiply and divide rational expressions.
- Simplify complex fractions.
- Multiply and divide expressions containing units of measure.

W A R M - U P

Factor.

1. $x^2 - 16$	2. $x^2 - 4x - 5$
3. $2x^2 + 9x + 10$	4. $x^2 - 3x$
5. $x^2 + x - 6$	6. $3x^2 + 6x$
7. $x^2 + 4x + 3$	8. $x^2 + 10x + 25$
9. $6x^2 + 4x$	10. $4x^2 - 25$
11. $5 + 4x - x^2$	12. $4 + 3x - x^2$

THIS AND THE NEXT TWO SECTIONS focus on operations with rational expressions. Work with units of measure is included.

EXAMPLE ❶   Working with rational expressions: comparative ages   Table 9 lists comparative ages for a mother and child if the mother is 20 years old when her child is born.

Child's Age, x	Mother's Age, y_1	Difference in Ages, y_2	Ratio of Ages (mother to child), y_3
0	20	20	
1	21	20	
2	22		
3			
4			

Table 9

a. Find the missing values in Table 9.

b. What kind of function is the difference in ages?

c. What result do we get as the first entry in the column for the ratio of the mother's age to the child's age? Why?

d. As the child's age increases, what happens to the ratio of ages? What is the ratio of ages when the child is 50?

e. If the child's age is x, write equations to describe the other columns, y_1, y_2, and y_3.

Solution   a. The mother's age increases by one each year: 20, 21, 22, 23, 24. The difference in ages is always 20. The ratios of ages are $\frac{20}{0}$ (which is undefined), $\frac{21}{1}$, $\frac{22}{2} = 11$, $\frac{23}{3} \approx 7\frac{2}{3}$, and $\frac{24}{4} = 6$.

b. Since the child's age and the mother's age are always 20 years apart, the difference in ages is a constant function, $y = 20$.

c. The first entry in the column for the ratio of ages is undefined because we are dividing by a child's age of zero.

d. The ratio of the ages gets smaller but stays above 1. When the child's age is 50, the ratio is

$$\frac{50 + 20}{50} = \frac{70}{50} = 1.4$$

e. If the child's age is x, $y_1 = 20 + x$, $y_2 = 20$, and $y_3 = (x + 20)/x$.   ●

Think about it 1: When is the ratio of ages 2 to 1? Will the ratio of ages ever reach 1 to 1?

Rational Expressions

The expression on the right in the equation $y_3 = (x + 20)/x$ is a rational expression. A **rational expression** is a *fraction formed by the quotient of polynomials.* A rational expression is undefined if the denominator is zero.

EXAMPLE **2** Finding when a rational expression is undefined For what values of the variables is each expression undefined?

a. $\dfrac{a + 3}{3 - a}$ **b.** $\dfrac{a + b}{a}$

c. $\dfrac{a}{a - b}$ **d.** $\dfrac{x}{x^2 - 3x - 4}$

Solution   **a.** $\dfrac{a + 3}{3 - a}$ has a zero denominator if $a = 3$.

b. $\dfrac{a + b}{a}$ has a zero denominator if $a = 0$.

c. $\dfrac{a}{a - b}$ has a zero denominator if $a = b$.

d. $\dfrac{x}{x^2 - 3x - 4} = \dfrac{x}{(x + 1)(x - 4)}$ has a zero denominator if $x = -1$ or $x = 4$.   ●

Simplification of Rational Expressions

When we simplify a rational expression, we factor the numerator and denominator and remove the common factors, such as a/a. If there are no common factors, the expression cannot be simplified.

EXAMPLE **3** Simplifying rational expressions Simplify the following expressions. Factor as needed. Any variable making a zero denominator must be excluded from possible inputs.

a. $\dfrac{b^2}{2b}, b \neq 0$ **b.** $\dfrac{6xy}{3x^2}, x \neq 0$

c. $\dfrac{x + 2}{(x + 2)(x - 3)}, x \neq -2, x \neq 3$ **d.** $\dfrac{a + a^2}{1 + a}, a \neq -1$

e. $\dfrac{x^2 - 4x - 5}{x^2 + 2x + 1}, x \neq -1$

Solution   **a.** $\dfrac{b^2}{2b} = \dfrac{b \cdot \boxed{b}}{2 \cdot \boxed{b}} = \dfrac{b}{2}$

b. $\dfrac{6xy}{3x^2} = \dfrac{2 \cdot 3 \cdot x \cdot y}{3 \cdot x \cdot x} = \dfrac{2y}{x}$

c. $\dfrac{x + 2}{(x + 2)(x - 3)} = \dfrac{(x + 2)}{(x + 2)(x - 3)} = \dfrac{1}{x - 3}$

d. $\dfrac{a + a^2}{1 + a} = \dfrac{a(1 + a)}{(1 + a)} = a$

e. $\dfrac{x^2 - 4x - 5}{x^2 + 2x + 1} = \dfrac{(x - 5)(x + 1)}{(x + 1)(x + 1)} = \dfrac{x - 5}{x + 1}$ ●

A COMMON ERROR   Example 4 illustrates a common student error: eliminating terms.

EXAMPLE **4**   Finding errors in simplifying   What is wrong with these simplifications? Explain the correct process.

a. $\dfrac{3 + 4}{4} = \dfrac{3 + \cancel{4}}{\cancel{4}}$

b. $\dfrac{x^2 + x - 2}{x + 2} = \dfrac{x^2 + \cancel{x} - \cancel{2}}{\cancel{x} + \cancel{2}}$

Solution   The expressions are not factored, and several *terms* have been eliminated. The correct process is to factor and, if possible, eliminate common *factors* from the numerator and denominator.

a. $\dfrac{3 + 4}{4} = \dfrac{7}{4}$

b. $\dfrac{x^2 + x - 2}{x + 2} = \dfrac{(x - 1)(x + 2)}{(x + 2)} = x - 1$

Placing parentheses around binomials, as in the denominator of part b, may help prevent the error of eliminating terms.   ●

SIMPLIFYING TO −1   Example 5 illustrates various algebraic ways of dividing opposite numbers in problems such as

$$\frac{5 - 8}{8 - 5} = \frac{-3}{3} = -1$$

EXAMPLE **5**   Simplifying rational expressions   Simplify $\dfrac{x - y}{y - x}$.

Solution   There are at least three ways to simplify the expression.

Method 1: We multiply the numerator and denominator of the fraction by −1:

$$\frac{x - y}{y - x} = \frac{(-1)(x - y)}{(-1)(y - x)} = \frac{(-1)(x - y)}{(x - y)} = -1$$

We obtain common factors by distributing −1 in only the denominator.

Method 2: We multiply either the numerator or the denominator by $(-1)(-1)$. Because $(-1)(-1) = 1$, this multiplication will not change the fraction.

$$\frac{x - y}{y - x} = \frac{(-1)(-1)(x - y)}{(y - x)} = \frac{(-1)(y - x)}{(y - x)} = -1$$

We obtain common factors by distributing one of the −1 factors in the numerator.

Method 3: We factor −1 from either the numerator or the denominator:

$$\frac{x - y}{y - x} = \frac{(x - y)}{(-1)(-y + x)} = \frac{(x - y)}{(-1)(x - y)} = \frac{1}{-1} = -1$$ ●

Think about it 2: Simplify the expression in Example 5 two more ways. First, multiply the denominator by $(-1)(-1)$ and simplify as in method 2. Second, factor −1 from the numerator and simplify as in method 3.

It does not matter which method you use to simplify expressions in Example 5. Choose a method that makes sense and consistently gives you the correct result. Following are three important observations about Example 5:

The factors $(x - y)$ and $(y - x)$ are additive inverses or opposites. **Additive inverses** *add to zero.*

A rational expression $\dfrac{x - y}{y - x}$, containing additive inverses in the numerator and denominator, simplifies to -1.

A negative sign on a fraction may be placed in any of three positions—in the numerator, in the denominator, or before the fraction. Thus, these fractions all have the same value:

$$\frac{-a}{b} = \frac{a}{-b} = -\frac{a}{b}, \quad b \neq 0$$

Caution: Expressions such as $x - y$ and $x + y$ may appear to be additive inverses. However, the sum of $x - y$ and $x + y$ is $2x$, not zero. The expressions $x - y$ and $x + y$ are not additive inverses.

EXAMPLE **6**

Simplifying with additive inverses   Which expressions simplify to -1?

a. $\dfrac{5 - x}{x - 5}$ **b.** $\dfrac{2 - x}{x + 2}$

c. $\dfrac{x + 3}{x - 3}$

Solution   We add to show whether the expressions contain additive inverses (opposites).

a. $5 - x + x - 5 = 0$
The numerator and denominator are additive inverses. The expression simplifies to -1.

b. $2 - x + x + 2 = 4$
The numerator and denominator are not additive inverses. No other simplification is possible.

c. $x + 3 + x - 3 = 2x$
The numerator and denominator are not additive inverses. No other simplification is possible.   ●

We can summarize our simplification as follows:

When the numerator and denominator of a rational expression are the same, $\dfrac{a}{a}$, the expression simplifies to 1.

When the numerator and denominator of a rational expression are additive inverses, $\dfrac{a - b}{b - a}$, the expression simplifies to -1.

Multiplication of Rational Expressions

Recall that to multiply fractions we multiply the numerators and the denominators.

If $b \neq 0$ and $d \neq 0$, then

$$\frac{a}{b} \cdot \frac{c}{d} = \frac{a \cdot c}{b \cdot d}$$

As in multiplying fractions, we factor rational expressions and simplify before multiplying. In some cases, no multiplication will need to be done.

Multiplying Rational Expressions

1. Write the product with a single numerator and denominator:

$$\frac{a}{b} \cdot \frac{c}{d} = \frac{a \cdot c}{b \cdot d}$$

2. Factor as needed.
3. Eliminate common factors from the numerator and denominator. *Do not eliminate common terms.*
4. Multiply remaining factors in the numerator and in the denominator.

Your instructor may direct you to leave your answers in factored form to show that no further common factors need to be eliminated.

In Example 7, the expressions contain single terms in the numerator and denominator.

EXAMPLE **7**   Multiplying rational expressions   **Multiply** $\dfrac{5x}{y} \cdot \dfrac{y^2}{10x^2}$, $x \neq 0$, $y \neq 0$.

Solution

$$\frac{5x}{y} \cdot \frac{y^2}{10x^2} = \frac{5xy^2}{10x^2 y} = \frac{5 \cdot x \cdot y \cdot y}{2 \cdot 5 \cdot x \cdot x \cdot y} = \frac{y}{2x}$$

In this solution, we write the numerators and denominators as products, factor them completely, and eliminate common factors.   ●

In Example 8, the expressions contain addition or subtraction operations. To eliminate common factors, we must factor the expressions into the product of a monomial and a binomial, such as $a(b + c)$, or the product of two binomials, $(a + b)(c + d)$.

EXAMPLE **8**   Multiplying rational expressions   **Multiply** $\dfrac{x + 2}{x + 6} \cdot \dfrac{4x^2 - 25}{2x^2 + 9x + 10}$, $x \neq -6$, $-\frac{5}{2}$, -2.

Solution

$$\frac{x + 2}{x + 6} \cdot \frac{4x^2 - 25}{2x^2 + 9x + 10} = \frac{(x + 2)(2x - 5)(2x + 5)}{(x + 6)(x + 2)(2x + 5)} = \frac{2x - 5}{x + 6}$$

We write the two fractions as one fraction in factored form and then eliminate common factors.   ●

Caution: If we multiply the numerators (or denominators) in Example 8, the resulting expression is not easily simplified:

$$\frac{x + 2}{x + 6} \cdot \frac{4x^2 - 25}{2x^2 + 9x + 10} = \frac{4x^3 + 8x^2 - 25x - 50}{2x^3 + 21x^2 + 64x + 60}$$

If your homework solutions contain expressions like the one on the right above, go back to the original exercise statement, factor it, and eliminate common factors from the numerator and denominator before multiplying.

Division of Rational Expressions

The fact that dividing a dollar into 4 equal parts is the same as finding $\frac{1}{4}$ of a dollar reminds us that division by a number a is the same as multiplication by the reciprocal $1/a$:

$$\$1.00 \div 4 = \$1.00 \cdot \tfrac{1}{4} = \$0.25$$

Division of Fractions

To divide, multiply the first fraction by the reciprocal of the second:

$$\frac{a}{b} \div \frac{c}{d} = \frac{a}{b} \cdot \frac{d}{c} = \frac{ad}{bc}, \quad b, c, d \neq 0$$

Rational expressions have the same relationship between multiplication and division as fractions do. Examples 9 and 10 illustrate division of rational expressions and remind us to simplify the expressions by factoring.

EXAMPLE 9   Dividing rational expressions   Divide $\dfrac{a^2 + ab}{a^2 - b^2} \div \dfrac{a}{b}$. Assume no zero denominators.

Solution

$$\frac{a^2 + ab}{a^2 - b^2} \div \frac{a}{b}$$ Change division to multiplication.

$$= \frac{a^2 + ab}{a^2 - b^2} \cdot \frac{b}{a}$$ Factor.

$$= \frac{a(a + b) \cdot b}{(a - b)(a + b) \cdot a}$$ Eliminate common factors.

$$= \frac{b}{a - b}$$ ●

EXAMPLE 10   Dividing rational expressions   Divide $\dfrac{x^2 - 4x - 5}{x - 3} \div \dfrac{5 + 4x - x^2}{x^2 - 9}$. Assume no zero denominators.

Solution

$$\frac{x^2 - 4x - 5}{x - 3} \div \frac{5 + 4x - x^2}{x^2 - 9}$$ Change to multiplication.

$$= \frac{x^2 - 4x - 5}{x - 3} \cdot \frac{x^2 - 9}{5 + 4x - x^2}$$ Factor.

$$= \frac{(x - 5)(x + 1)(x - 3)(x + 3)}{(x - 3)(5 - x)(1 + x)}$$ Eliminate common factors.

$$= \frac{(x - 5)(x + 3)}{(5 - x)}$$ Look for additive inverses.

$$= (-1)(x + 3)$$ ●

Complex Fractions

The technique of changing division to multiplication by a reciprocal may be applied to simplifying complex fractions and expressions containing units of measure.

A fraction that contains a fraction in either the numerator or the denominator is called a **complex fraction.** When the numerator, the denominator, or both are simple fractions, we may replace the fraction bar with a division sign.

EXAMPLE **11**   Simplifying complex fractions   Simplify the complex fraction $\dfrac{\dfrac{a}{b}}{\dfrac{c}{d}}$. Assume no zero denominators.

Solution

$$\frac{\dfrac{a}{b}}{\dfrac{c}{d}} = \frac{a}{b} \div \frac{c}{d} = \frac{a}{b} \cdot \frac{d}{c} = \frac{ad}{bc}$$

We simplify the complex fraction by writing it as a division problem with the longer fraction bar replaced by a division sign. The division is then changed to multiplication by the reciprocal. ●

EXAMPLE **12**   Simplifying complex fractions   Simplify the complex fraction $\dfrac{\dfrac{x+2}{x^2}}{\dfrac{x^2-4}{x}}$.

Solution

$$\frac{\dfrac{x+2}{x^2}}{\dfrac{x^2-4}{x}}$$

Replace the middle fraction bar with division.

$$= \frac{x+2}{x^2} \div \frac{x^2-4}{x}$$

Change to multiplication.

$$= \frac{x+2}{x^2} \cdot \frac{x}{x^2-4}$$

Factor.

$$= \frac{(x+2) \cdot x}{x \cdot x \cdot (x+2)(x-2)}$$

Eliminate common factors.

$$= \frac{1}{x(x-2)}$$ ●

Simplifying units of measure is one of the most useful applications of simplifying complex fractions.

EXAMPLE **13**   Simplifying with units   Predict the units obtained from each expression. Simplify by first writing each as a complex fraction.

a. $\dfrac{\$2.70 \text{ per dozen}}{12 \text{ cookies per dozen}}$

b. $\dfrac{24 \text{ capsules per box}}{4 \text{ capsules per day}}$

Solution **a.** We anticipate that the dozens will be eliminated, leaving the units as dollars per cookie:

$$\frac{\$2.70 \text{ per dozen}}{12 \text{ cookies per dozen}} = \frac{\dfrac{\$2.70}{1 \text{ dozen}}}{\dfrac{12 \text{ cookies}}{1 \text{ dozen}}} = \frac{\$2.70}{1 \text{ dozen}} \cdot \frac{1 \text{ dozen}}{12 \text{ cookies}}$$

$$= \frac{\$2.70}{12 \text{ cookies}} = \$0.225 \text{ per cookie}$$

Our prediction is correct.

b. Following the pattern in part a, we might predict that the capsules will be eliminated, leaving the units as boxes per day:

$$\frac{24 \text{ capsules per box}}{4 \text{ capsules per day}} = \frac{\dfrac{24 \text{ capsules}}{1 \text{ box}}}{\dfrac{4 \text{ capsules}}{1 \text{ day}}} = \frac{24 \text{ capsules}}{1 \text{ box}} \cdot \frac{1 \text{ day}}{4 \text{ capsules}}$$

$$= \frac{24 \text{ days}}{4 \text{ boxes}} = 6 \text{ days per box}$$

Our prediction is not correct. The units are the reciprocal of our prediction. ●

ANSWER BOX

Warm-up: **1.** $(x - 4)(x + 4)$ **2.** $(x - 5)(x + 1)$ **3.** $(2x + 5)(x + 2)$
4. $x(x - 3)$ **5.** $(x + 3)(x - 2)$ **6.** $3x(x + 2)$ **7.** $(x + 1)(x + 3)$
8. $(x + 5)^2$ **9.** $2x(3x + 2)$ **10.** $(2x - 5)(2x + 5)$
11. $(5 - x)(1 + x)$ **12.** $(4 - x)(1 + x)$

Think about it 1: The ratio of ages is 2 to 1 when the mother reaches twice what her age was when the child was born. The ratio will never reach 1 to 1 because the mother will always be 20 years older.

Think about it 2:

$$\frac{x - y}{y - x} = \frac{(x - y)}{(-1)(-1)(y - x)} = \frac{(x - y)}{(-1)(x - y)} = \frac{1}{-1} = -1$$

$$\frac{x - y}{y - x} = \frac{(-1)(-x + y)}{y - x} = \frac{(-1)(y - x)}{(y - x)} = -1$$

EXERCISES

For what numbers will the expressions in Exercises 1 to 6 be undefined?   *Give the additive inverse of each expression in Exercises 7 and 8.*

1. $\dfrac{4 - x}{x + 3}$

2. $\dfrac{x + 3}{2 - x}$

3. $\dfrac{a + 3}{4 - a}$

4. $\dfrac{3 - b}{b^2 + 3b + 2}$

5. $\dfrac{x(x - 3)}{x^2 - 3x + 28}$

6. $\dfrac{x(x + 1)}{x^2 - 2x - 8}$

7. a. $x + y$

   b. $-x + y$

   c. $y - x$

8. a. $-x + y$

   b. $-x - y$

   c. $x - y$

What expression should be placed in the denominator or numerator in Exercises 9 and 10 to make a true statement? State any restrictions on the variables, as needed, to prevent undefined expressions.

9. a. $\dfrac{x-3}{} = 1$

   b. $\dfrac{x+2}{} = -1$

   c. $\dfrac{b-a}{} = 1$

   d. $\dfrac{3-x}{} = -1$

10. a. $\dfrac{x-3}{} = -1$

    b. $\dfrac{x+2}{} = 1$

    c. $\dfrac{}{a-b} = -1$

    d. $\dfrac{3-x}{} = 1$

In Exercises 11 to 20, identify the common factors and simplify. State any restrictions on the variables.

11. a. $\dfrac{16a^2}{10a}$

    b. $\dfrac{2x^2 y}{10xy^2}$

12. a. $\dfrac{12k^2}{4k^3}$

    b. $\dfrac{8xy^2}{2xy}$

13. a. $\dfrac{xy}{2x+y}$

    b. $\dfrac{xy}{2x+xy}$

14. a. $\dfrac{y^2}{6x^2+y}$

    b. $\dfrac{xy}{6x^2+2xy}$

15. a. $\dfrac{3-x}{(x+3)(x-3)}$

    b. $\dfrac{x^2-9}{x^2+5x+6}$

16. a. $\dfrac{x-1}{(1-x)(x+5)}$

    b. $\dfrac{x^2-16}{x^2+3x-4}$

17. a. $\dfrac{2ac+4bc}{4ad+8bd}$

    b. $\dfrac{6x^2+3x}{12x^2-6x}$

18. a. $\dfrac{3ab+3ac}{5b^2+5bc}$

    b. $\dfrac{5x^2-10x}{2x^2-4x}$

19. a. $\dfrac{x^2+x-6}{2-x}$

    b. $\dfrac{x-3}{6-2x}$

20. a. $\dfrac{x^2+x-2}{2x+4}$

    b. $\dfrac{x-4}{12-3x}$

In Exercises 21 and 22, multiply or divide as indicated. Assume no zero denominators.

21. a. $\dfrac{1}{x} \cdot \dfrac{x^2}{1}$

    b. $\dfrac{1}{a} \div \dfrac{a^2 b^2}{1}$

    c. $\dfrac{b}{a} \div \dfrac{a^2}{b^2}$

    d. $\dfrac{x}{y} \div \dfrac{x^3}{y^2}$

22. a. $\dfrac{1}{x} \cdot \dfrac{x^3}{1}$

    b. $\dfrac{1}{b} \div \dfrac{a^2 b^2}{1}$

    c. $\dfrac{a}{b} \cdot \dfrac{b^2}{a^2}$

    d. $\dfrac{x^2}{y^3} \div \dfrac{x}{y}$

In Exercises 23 to 32, multiply or divide as indicated, and simplify. Assume no zero denominators.

23. a. $\dfrac{a^2+7a+12}{a^2-4} \div \dfrac{a^2+4a}{a-2}$

    b. $\dfrac{x^2-2x}{x} \cdot \dfrac{x^2-1}{x^2-3x+2}$

24. a. $\dfrac{b-3}{b^2-4b+3} \div \dfrac{b^2-b}{b-1}$

    b. $\dfrac{x^2-6x+9}{x^2+3x} \div \dfrac{x-3}{x+3}$

25. a. $\dfrac{x^2+3x}{x} \cdot \dfrac{x^2-x-6}{x^2-9}$

    b. $\dfrac{4a^2+4a+1}{4-9a^2} \div \dfrac{4a^2+2a}{3a-2}$

26. a. $\dfrac{b^2+2b+1}{b+1} \cdot \dfrac{b^2}{b^2+b}$

    b. $\dfrac{x^2-4}{x-2} \div \dfrac{1}{x^2-x}$

27. a. $\dfrac{x+2}{x^2-4x+4} \cdot \dfrac{x^2-2x}{x+2}$

b. $\dfrac{a^2 - 5a}{a^2 + 5a} \div \dfrac{a^2 - 10a + 25}{a}$

28. a. $\dfrac{x^2 - 6x + 9}{x^2 + 3x} \div \dfrac{x^2 - 9}{x}$

b. $\dfrac{x^2 - 2x}{3x^2 - 5x - 2} \cdot \dfrac{9x^2 - 4}{9x^2 - 12x + 4}$

29. a. $\dfrac{a^2 - b^2}{a^2 + 2ab + b^2} \cdot \dfrac{a + b}{a - b}$

b. $\dfrac{x^2 - 2xy + y^2}{x^2 - y^2} \cdot \dfrac{x - y}{y - x}$

30. a. $\dfrac{a^2 - 2ab + b^2}{a^2 - b^2} \cdot \dfrac{a + b}{a - b}$

b. $\dfrac{y^2 - x^2}{x^2 + 2xy + y^2} \cdot \dfrac{x + y}{x - y}$

31. a. $\dfrac{a^3 - b^3}{a + b} \cdot \dfrac{a^2 + 2ab + b^2}{a^2 - b^2}$

b. $\dfrac{x^2 + 2xy + y^2}{x^3 + y^3} \cdot \dfrac{x^2 - y^2}{x + y}$

32. a. $\dfrac{a^3 + b^3}{a^2 - b^2} \cdot \dfrac{b - a}{a + b}$

b. $\dfrac{y - x}{x^2 + 2xy + y^2} \cdot \dfrac{x + y}{x^3 - y^3}$

In Exercises 33 and 34, write each phrase as a complex fraction and then divide. Assume that neither a nor b is zero.

33. a. The quotient of $\dfrac{1}{a}$ and $\dfrac{1}{b}$

b. The quotient of $\dfrac{1}{b}$ and a

c. The quotient of $\dfrac{1}{b}$ and b

d. The quotient of $\dfrac{a}{b}$ and $\dfrac{1}{b}$

34. a. The quotient of $\dfrac{1}{b}$ and $\dfrac{1}{a}$

b. The quotient of $\dfrac{a}{b}$ and $\dfrac{1}{a}$

c. The quotient of a and $\dfrac{1}{a}$

d. The quotient of $\dfrac{1}{b}$ and $\dfrac{a}{b}$

Simplify the complex fractions in Exercises 35 to 38.

35. $\dfrac{\dfrac{x^2 - 3x + 2}{x}}{\dfrac{x^2 - 1}{x^2}}$

36. $\dfrac{\dfrac{x^2 - 4}{x^2}}{\dfrac{x^2 + 5x + 6}{x}}$

37. $\dfrac{\dfrac{x - 2}{x^2 + 3x - 4}}{\dfrac{2 - x}{x - 1}}$

38. $\dfrac{\dfrac{x - 1}{x^2 - 9}}{\dfrac{x^2 + 2x - 3}{3 - x}}$

Simplify the expressions in Exercises 39 to 48. The word "per" means division and may be replaced by a fraction bar.

39. $\dfrac{5280 \text{ feet}}{88 \text{ feet per second}}$

40. $\dfrac{93{,}000{,}000 \text{ miles}}{186{,}000 \text{ miles per second}}$

41. $\dfrac{65 \text{ miles per hour}}{15 \text{ miles per gallon}}$

42. $\dfrac{500 \text{ miles per hour}}{200 \text{ gallons per hour}}$

43. $\dfrac{24 \text{ cans per case}}{\$3.98 \text{ per case}}$

44. $\dfrac{250 \text{ vitamins per bottle}}{7 \text{ vitamins per week}}$

45. $\dfrac{8 \text{ stitches per inch}}{\dfrac{1 \text{ foot}}{12 \text{ inches}}}$

46. $\dfrac{140 \text{ heartbeats per minute}}{0.25 \text{ mile per minute}}$

47. $\dfrac{95 \text{ words per minute}}{300 \text{ words per page}}$

48. $\dfrac{440 \text{ cycles per second}}{344 \text{ meters per second}}$

Describe errors in the simplification of the first expression in Exercises 49 to 52.

49. $\dfrac{x^2 + x + 2}{x} = x^2 + 2$

50. $\dfrac{(x - 1)(x - 2)}{2 - x} = x - 1$

51. $\dfrac{x^2 - 2x + 1}{x - 2} = \dfrac{x(x - 2) + 1}{(x - 2)}$

52. $\dfrac{x \cdot x + 2 \cdot 2}{x - 2} = x - 2$

53. When we multiply $a \cdot b \cdot c$, why do we not multiply $a \cdot b$ and $a \cdot c$?

Projects

54. *Comparative Ages.* Return to the equations in part e of Example 1, page 352. Enter the equations into a graphing calculator, with $\boxed{Y=}$.

   a. Look at the columns for Y_1, Y_2, and Y_3 in the graphing calculator table. Does the table agree with that in the example?

   b. What causes the error in the entry for $x = 0$ under Y_3?

   c. Find the child's age when the ratio of ages is 2 to 1.

   d. Move the cursor down the column for Y_3. What number does this column seem to be approaching?

   e. Look at the expression that describes the ratio of ages, Y_3. Change it into two fractions, and simplify if possible. How does the result explain your answer to part d?

   f. Set an appropriate viewing window, and graph the three functions. Use a window setting that gives integers for x as you trace. List the dimensions of your window.

   g. Trace the graphs, and confirm that the values agree with those in Table 9. Name the function formed by each graph.

   h. Trace Y_3 for values of x between 0 and 1. Reset the window as needed. What happens to Y_3 as x gets close to zero?

   i. Trace Y_3 to find the coordinates for $x = 0$. What does the calculator show for coordinates at $x = 0$? Why does this happen?

   j. Trace Y_3 into negative values of x. Do these have any meaning in the problem situation?

55. *Graphs of Rational Functions.* Graph the expressions in Exercises 15a, 15b, and 19a both before and after simplifying. Use parentheses to separate the numerator from the denominator. Then answer the following questions for each expression.

   a. Is there any major difference between the graphs before and after simplifying?

   b. Does the expression give a straight line? If so, what is the slope of the line?

   c. Are there any nearly vertical parts? Where? Why?

   d. Are there any holes in the graphs? To make the holes show up, set a viewing window that gives x in tenths. Where are the holes? Why?

56. *Reciprocal Functions*

   a. List these six functions across the top of a sheet of paper.

$$f(x) = x$$
$$f(x) = x + 3$$
$$f(x) = x - 2$$
$$f(x) = x^2$$
$$f(x) = x^2 - 2$$
$$f(x) = (x - 3)^2$$

   b. Below each $f(x)$, list the reciprocal function $1/f(x)$.

   c. Give the domain of each function.

   d. Give the domain of each reciprocal function.

   e. On graph paper, sketch a graph of each function $f(x)$ and its reciprocal $1/f(x)$. Use two colors, one for $f(x)$ and one for the reciprocal. Use a graphing calculator to assist you. Label key points (intercepts, vertices, axis of symmetry).

   f. Describe what happens to $1/f(x)$ when the outputs for $f(x)$ are near zero.

   g. Describe what happens to $1/f(x)$ when the outputs for $f(x)$ are large.

   h. Write observations about symmetry.

   i. Relative to the original function, describe where the reciprocal function is positive.

   j. Relative to the original function, describe where the reciprocal function is increasing.

MID–CHAPTER ⑤ TEST

Simplify the expressions in Exercises 1 to 5.

1. $\dfrac{32}{24}$

2. $\dfrac{3}{10} + \dfrac{5}{18}$

3. $\dfrac{abc}{aces}$

4. $\dfrac{3}{10} \div \dfrac{5}{18}$

5. $\dfrac{12cd^2}{8c^2d}$

6. $\dfrac{(x-2)(x+3)}{(x+3)}$

7. $\dfrac{a-6}{a^2 - 3a - 18}$

8. $\dfrac{b + bc}{b}$

9. $\dfrac{a^2 b}{c} \cdot \dfrac{c^2}{ab^2}$

10. $\dfrac{ab^2}{c^2} \div \dfrac{ac}{b}$

11. $\dfrac{x^2 - 16}{x^2 + 6x + 8} \cdot \dfrac{x + 2}{x - 2}$

12. $\dfrac{\dfrac{x^2}{x+1}}{\dfrac{x}{x^2-1}}$

13. $\dfrac{x^2 - 16}{x^2 + 6x + 8} \div \dfrac{4 - x}{x + 2}$

14. $\dfrac{2\frac{2}{3} \text{ yards}}{5 \text{ feet}}$

15. $\dfrac{150 \text{ months}}{15 \text{ years}}$

16. For what values of x is the expression shown in Exercise 11 undefined?

17. Change 1,000,000 tablespoons into gallons, using the following facts.

$$1 \text{ quart} = \tfrac{1}{4} \text{ gallon}$$
$$2 \text{ cups} = 1 \text{ pint}$$
$$16 \text{ tablespoons} = 1 \text{ cup}$$
$$1 \text{ quart} = 2 \text{ pints}$$

18. A baby grows from 15 inches at birth to 5 feet 2 inches at age 14 years. What is the rate of growth in miles per hour? (*Hint:* 1 mile = 5280 feet; assume 1 year = 365 days.)

In Exercises 19 and 20, solve the proportions. Round to the nearest hundredth.

19. $\dfrac{a}{15} = \dfrac{12}{35}$

20. $\dfrac{x + 1}{8} = \dfrac{2x - 1}{14}$

21. Solve for the missing sides in these similar triangles.

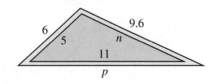

In Exercises 22 and 23, find the equation of the line through the ordered pairs. State whether the inputs and outputs are proportional.

22. $(-4, -6)$, $(10, 15)$

23. (1 minute, $0.60), (3 minutes, $1.50)

24. **a.** What is the constant of proportionality in the figure below?

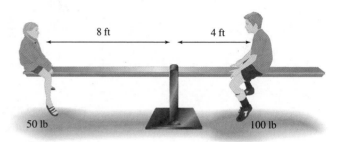

   b. If the 100-pound child is replaced by an 80-pound child, where should the 80-pound child sit?

25. Suppose that the number of cookies per child varies inversely with the number of children present. There are 100 cookies altogether. Describe the situation with an inverse variation equation. Identify the constant of variation.

26. The price of a round pizza varies with the square of its diameter, $p = kd^2$. If a 13-inch pizza costs $10, what will a 16-inch pizza cost?

27. Finishing nails are used for cabinet work and interior wood trim. Suppose that the number of finishing nails per pound varies inversely with the square of their length, $n = k/l^2$. If a pound of 1-inch nails contains 1350 nails, how many 4-inch nails will be in a pound?

5.4 Addition and Subtraction of Rational Expressions

OBJECTIVES

- Find the common denominator for two or more rational expressions.
- Add and subtract rational expressions with like denominators.
- Add and subtract rational expressions with unlike denominators.
- Simplify complex fractions.

WARM-UP

Add or subtract, as indicated.

1. $\frac{1}{5} + \frac{1}{3}$ 2. $\frac{1}{30} + \frac{1}{40}$ 3. $\frac{1}{4} + \frac{1}{4} + \frac{1}{8}$

4. $\frac{7}{12} - \frac{5}{18}$ 5. $\frac{5}{6} - \frac{11}{16}$ 6. $\frac{7}{10} - \frac{5}{18}$

Multiply and simplify these expressions.

7. $\dfrac{6\left(\dfrac{1}{3} - x\right)}{6\left(x + \dfrac{1}{2}\right)}$ 8. $\dfrac{4\left(\dfrac{3}{4} + x\right)}{4\left(x - \dfrac{1}{2}\right)}$ 9. $\dfrac{2x\left(\dfrac{1}{x} - 2\right)}{2x\left(\dfrac{x}{2} + 1\right)}$

IT IS COMMONLY ASSUMED that if a plane travels 100 miles at 500 miles per hour and another 100 miles at 300 miles per hour, the average speed for the whole trip is 400 miles per hour. This is inaccurate reasoning. This section will illustrate the correct answer and show how competitors in a race take advantage of this error in thinking.

Average Speed

EXAMPLE **1** Exploring average trip speed An airplane travels 100 miles at 500 miles per hour and a second 100 miles at 300 miles per hour.

a. What is the total distance traveled?

b. What is the total time traveled?

c. What is the average speed for the whole trip?

Solution a. The total distance is 200 miles.

b. The total time is the time for the first 100 miles plus the time for the second 100 miles.

For the first 100 miles,

$$t = \frac{D}{r} = \frac{100 \text{ mi}}{500 \text{ mph}} = \frac{100 \text{ mi}}{\dfrac{500 \text{ mi}}{1 \text{ hr}}} = 100 \text{ mi} \cdot \frac{1 \text{ hr}}{500 \text{ mi}} = \frac{1}{5} \text{ hr}$$

For the second 100 miles,

$$t = \frac{D}{r} = \frac{100 \text{ mi}}{300 \text{ mph}} = \frac{100 \text{ mi}}{\dfrac{300 \text{ mi}}{1 \text{ hr}}} = 100 \text{ mi} \cdot \frac{1 \text{ hr}}{300 \text{ mi}} = \frac{1}{3} \text{ hr}$$

The total time is

$$\left(\frac{1}{5} + \frac{1}{3}\right) \text{ hr} = \left(\frac{3}{15} + \frac{5}{15}\right) \text{ hr} = \frac{8}{15} \text{ hr}$$

c. The average speed is the total distance divided by the total time. The average speed is

$$\text{Distance} \div \text{time} = 200 \text{ mi} \div \frac{8}{15} \text{ hr}$$

$$= \frac{200}{1} \text{ mi} \cdot \frac{15}{8} \frac{1}{\text{hr}}$$

$$= 375 \text{ miles per hour}$$

The average speed for the 200 miles is 375 miles per hour, slightly slower than the anticipated average of 400 miles per hour. One might think that the average was somehow due to the specific distances of 100 miles. In Example 2, we eliminate that possibility by letting each 100-mile distance be replaced by the variable x.

EXAMPLE

Exploring average trip speed   Find the average speed for a trip where x miles are flown at 500 miles per hour and another x miles are flown at 300 miles per hour.

Solution   The total distance is $x + x = 2x$. The total time, in hours, is $\dfrac{x}{500} + \dfrac{x}{300}$. The factors used to find the least common denominator (LCD) need not be primes if common factors are readily apparent.

$$500 = 5 \cdot 100$$
$$300 = 3 \cdot 100$$
$$\text{LCD} = 3 \cdot 5 \cdot 100 = 1500$$

Adding the total time expression gives

$$\frac{x}{500} + \frac{x}{300} = \frac{x \cdot 3}{5 \cdot 100 \cdot 3} + \frac{x \cdot 5}{3 \cdot 100 \cdot 5}$$
$$= \frac{3x}{1500} + \frac{5x}{1500} = \frac{3x + 5x}{1500} = \frac{8x}{1500} \text{ hr}$$

The average speed is total distance divided by total time:

$$2x \text{ mi} \div \frac{8x}{1500} \text{ hr} = \frac{2x \text{ mi}}{1} \cdot \frac{1500}{8x} \frac{1}{\text{hr}}$$
$$= \frac{3000x}{8x} \frac{\text{mi}}{\text{hr}} = 375 \text{ miles per hour} \qquad \bullet$$

The average speed, 375 miles per hour, does not contain the variable x. Thus, this result is true for any trip where the first half is traveled at 500 miles per hour and the second half at 300 miles per hour. A round trip taken over the same route would be a typical example of such a situation. Projects 56 to 58 explore how hard it is to make up for a slow first half of a race with a faster second half.

Adding and Subtracting Rational Expressions

In order to add the fractions in Examples 1 and 2, we changed to the least common denominator. Recall that the *least common denominator* (LCD) is the smallest number into which both denominators divide evenly.

EQUIVALENT FRACTIONS   To change to the common denominator in an algebraic setting, we factor the denominators, find the LCD, and change to equivalent fractions. When we want to find an equivalent fraction, we multiply the numerator and denominator of the fraction by the same factor.

EXAMPLE

Finding equivalent rational expressions   Change to an equivalent expression with the indicated denominator.

a. $\dfrac{5}{6} = \dfrac{}{54}$

b. $\dfrac{3}{x} = \dfrac{}{2x^2}$

c. $\dfrac{a}{a + 3} = \dfrac{}{2a(a + 3)}$

d. $\dfrac{x}{x + 1} = \dfrac{}{x^2 + 3x + 2}$

Solution   **a.** $\dfrac{5}{6} = \dfrac{5 \cdot 9}{6 \cdot 9} = \dfrac{45}{54}$

b. $\dfrac{3}{x} = \dfrac{3 \cdot 2x}{x \cdot 2x} = \dfrac{6x}{2x^2}$

c. $\dfrac{a}{a + 3} = \dfrac{a \cdot 2a}{2a(a + 3)} = \dfrac{2a^2}{2a(a + 3)}$

d. $\dfrac{x}{x + 1} = \dfrac{x(x + 2)}{(x + 1)(x + 2)} = \dfrac{x(x + 2)}{x^2 + 3x + 2}$

●

LEAST COMMON DENOMINATORS   Although any common denominator may be used to add or subtract fractions, the advantage of using the least common denominator is that the answer is less likely to need simplifying. In the next several examples, we find the LCD and add or subtract rational expressions with variables in the denominator.

EXAMPLE **4**   Adding rational expressions using a least common denominator   Find the LCD, and add:

$$\frac{2}{a^2b} + \frac{5}{abc}, \quad a \neq 0, b \neq 0, c \neq 0$$

Solution   To find the LCD, we factor each denominator:

$$a^2b = a \cdot a \cdot b$$
$$abc = a \cdot b \cdot c$$

The LCD needs to be divisible by both denominators and needs two factors of a as well as one each of b and c. The product of these factors, $a \cdot a \cdot b \cdot c$, gives the LCD, a^2bc.

$$\frac{2}{a^2b} + \frac{5}{abc}$$ Change to equivalent fractions with an LCD.

$$= \frac{2 \cdot c}{a^2b \cdot c} + \frac{5 \cdot a}{abc \cdot a}$$ Multiply.

$$= \frac{2c}{a^2bc} + \frac{5a}{a^2bc}$$ Add numerators.

$$= \frac{2c + 5a}{a^2bc}$$

●

We also use factoring to find the LCD for rational expressions containing binomials or trinomials. Again, an LCD produces answers that are less likely to need simplification.

EXAMPLE **5**   Subtracting rational expressions using an LCD   Find the LCD, and subtract:

$$\frac{x}{x^2 - 2x + 1} - \frac{2}{x^2 - 1}, \quad x \neq 1, -1$$

Solution   To find the LCD, we factor the denominators:

$$x^2 - 2x + 1 = (x - 1)^2 = (x - 1)(x - 1)$$
$$x^2 - 1 = (x - 1)(x + 1)$$

The LCD will be the product of the three factors: $(x - 1)$, $(x - 1)$, and $(x + 1)$.

$$\frac{x}{x^2 - 2x + 1} - \frac{2}{x^2 - 1}$$
Factor the denominators.

$$= \frac{x}{(x - 1)(x - 1)} - \frac{2}{(x - 1)(x + 1)}$$
Set up the LCD and equivalent fractions.

$$= \frac{x(x + 1)}{(x - 1)(x - 1)(x + 1)} - \frac{2(x - 1)}{(x - 1)(x + 1)(x - 1)}$$

Combine the numerators.

$$= \frac{x(x + 1) - 2(x - 1)}{(x - 1)(x + 1)(x - 1)}$$
Apply the distributive property.

$$= \frac{x^2 + x - 2x + 2}{(x - 1)(x + 1)(x - 1)}$$
Add like terms.

$$= \frac{x^2 - x + 2}{(x - 1)(x + 1)(x - 1)}$$

Because $x^2 - x + 2$ does not factor, we know the expression cannot be simplified further. ●

Example 5 illustrates the fact that subtraction problems must be worked carefully because there may be a sign change when numerators are subtracted. The next example also deals with finding the common denominator and changing signs with subtraction.

EXAMPLE **6**   Subtracting rational expressions using an LCD   Find the LCD, and subtract:

$$\frac{3x}{x^2 + 5x + 6} - \frac{3}{2x + 6}, \quad x \neq -3, x \neq -2$$

Solution   To find the LCD, we factor the denominators:

$$x^2 + 5x + 6 = (x + 2)(x + 3)$$
$$2x + 6 = 2(x + 3)$$

The LCD will be the product of the three factors: 2, $(x + 2)$, and $(x + 3)$.

$$\frac{3x}{x^2 + 5x + 6} - \frac{3}{2x + 6}$$
Factor the denominators.

$$= \frac{3x}{(x + 2)(x + 3)} - \frac{3}{2(x + 3)}$$
Set up the LCD with equivalent fractions.

$$= \frac{2 \cdot 3x}{2 \cdot (x + 2)(x + 3)} - \frac{3 \cdot (x + 2)}{2(x + 3) \cdot (x + 2)}$$

Combine the numerators.

$$= \frac{6x - 3(x + 2)}{2(x + 2)(x + 3)}$$
Apply the distributive property.

$$= \frac{6x - 3x - 6}{2(x + 2)(x + 3)}$$
Add like terms.

$$= \frac{3x - 6}{2(x + 2)(x + 3)}$$
Factor the numerator.

$$= \frac{3(x - 2)}{2(x + 2)(x + 3)}$$

There are no common factors in the numerator and denominator, so the expression cannot be simplified. ●

DENOMINATORS CONTAINING ADDITIVE INVERSES   In Section 5.3, we simplified rational expressions containing additive inverses to -1. Here, we add (or subtract) expressions containing additive inverses as denominators.

Additive inverses appear in two different ways in Example 7: first as denominators and then in a rational expression.

EXAMPLE   Adding expressions containing additive inverses   Find the common denominator, and add:

$$\frac{1}{a-1} + \frac{a}{1-a}$$

Solution   The denominators this time are additive inverses, or opposites. We obtain a common denominator, $a - 1$, if we multiply the numerator and denominator of the second expression by -1:

$$\frac{1}{a-1} + \frac{a}{1-a} = \frac{1}{a-1} + \frac{a(-1)}{(1-a)(-1)} \qquad \text{Set up the LCD.}$$

$$= \frac{1}{a-1} + \frac{-a}{a-1} \qquad \text{Combine the numerators.}$$

$$= \frac{1-a}{a-1} \qquad \text{Simplify.}$$

$$= -1 \qquad\qquad\qquad ●$$

If denominators of rational expressions are additive inverses, multiply the numerator and denominator of one expression by -1 to change to a common denominator.

Summary: Adding or Subtracting Rational Numbers

> 1. To find the least common denominator (LCD), factor each denominator.
>   a. If the denominators have no common factors, the LCD is the product of the denominators.
>   b. If the denominators have common factors, the LCD is the product formed by including each factor the highest number of times it appears in any one denominator.
> 2. Convert to a common denominator.
> 3. Add or subtract the numerators, and place the result over the common denominator.
> 4. Factor, and simplify the answer.

Applications

In the next two examples, we explore the rate at which a theater empties when various doors are used. This is an important consideration in theater design and in building codes.

EXAMPLE   Exploring theater exit rates   Suppose a theater has two doors, one of which permits emptying the theater in 4 minutes and the other in 8 minutes. The rate of exit is $\frac{1}{4}$ of the theater capacity per minute for the first door and $\frac{1}{8}$ of the theater capacity per minute for the second door. How much of the theater has been emptied after 1 minute? 2 minutes? 3 minutes?

Solution   Figure 22 shows the portion of the theater that has exited at the end of each minute.

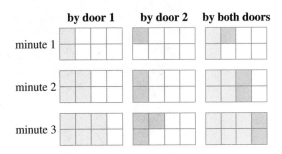

Figure 22

In the first minute, each door permits its portion of the theater to exit:

$$1 \text{ minute}\left(\frac{1}{4} + \frac{1}{8}\right)\left(\frac{\text{capacity}}{\text{minute}}\right) = 1\left(\frac{3}{8}\right) \text{ capacity} = \frac{3}{8} \text{ capacity}$$

At the end of the first minute, $\frac{3}{8}$ of the theater has exited.

In two minutes, each door permits twice the one-minute portion to exit:

$$2 \text{ minutes}\left(\frac{1}{4} + \frac{1}{8}\right)\left(\frac{\text{capacity}}{\text{minute}}\right) = 2\left(\frac{3}{8}\right) \text{ capacity} = \frac{3}{4} \text{ capacity}$$

At the end of the second minute, $\frac{3}{4}$ of the theater has exited.

In three minutes, each door permits three times its portion to exit:

$$3 \text{ minutes}\left(\frac{1}{4} + \frac{1}{8}\right)\left(\frac{\text{capacity}}{\text{minute}}\right) = 3\left(\frac{3}{8}\right) = \frac{9}{8} = 1\frac{1}{8} \text{ capacity}$$

At the end of the third minute, $1\frac{1}{8}$ of the theater has exited. Because full capacity is $\frac{8}{8} = 1$, the theater is empty when $\frac{8}{8}$ of the people have passed through the doors. The theater will become empty during the third minute.   ●

In each step in Example 8, we multiply the time in minutes by the rate of exit per minute to obtain the fraction of the theater capacity that has exited. When the product reaches 1, the theater is empty. The process in Example 8 extends to three or more doors, as we will see in Example 9.

EXAMPLE 9

Exploring theater exit rates   Three doors in a theater empty the theater in 4 minutes, 4 minutes, and 8 minutes, respectively. If all doors are functioning properly, in how many minutes can the theater be emptied? (See Figure 23.)

Solution

$$t \text{ minutes}\left(\frac{1}{4} + \frac{1}{4} + \frac{1}{8}\right)\left(\frac{\text{capacity}}{\text{minute}}\right) = 1 \qquad \text{Set up the LCD.}$$

$$\left(\frac{1 \cdot 2}{4 \cdot 2} + \frac{1 \cdot 2}{4 \cdot 2} + \frac{1}{8}\right)t = 1 \qquad \text{Simplify.}$$

$$\frac{2 + 2 + 1}{8}t = 1 \qquad \text{Simplify.}$$

$$\frac{5t}{8} = 1 \qquad \text{Multiply by 8.}$$

$$5t = 8 \qquad \text{Divide by 5.}$$

$$t = 1.6 \text{ minutes} \qquad ●$$

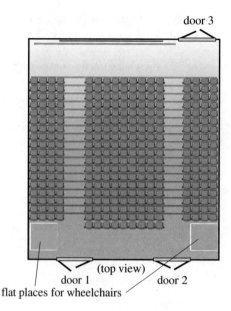

door 3

(top view)

door 1   door 2

flat places for wheelchairs

Figure 23

We can summarize the theater exit process with a formula. The formula uses subscripts to distinguish the amounts of time the different doors require to empty the theater.

If the first door can empty the theater in t_1 minutes, the second in t_2 minutes, and the third in t_3 minutes, then a formula to find the number of minutes t required to empty the theater if all three doors are functioning is

$$t\left(\frac{1}{t_1} + \frac{1}{t_2} + \frac{1}{t_3}\right) = 1$$

This formula is usually divided on both sides by t:

$$\left(\frac{1}{t_1} + \frac{1}{t_2} + \frac{1}{t_3}\right) = \frac{1}{t}$$

Examples 8 and 9 and the formula above illustrate the method for adding rates, one of the most common applications of rational expressions. In the exercises at the end of this section, you will get practice in adding rational expressions as you apply several related formulas.

Common Denominators and Complex Fractions

We close this section by simplifying **complex rational expressions**—*expressions that contain fractions in either the numerator or the denominator or both.* In Example 10, we return to the setting of Examples 1 and 2 and set up a general formula for the average speed.

EXAMPLE **10**   Finding a formula for average trip speed   Find a formula for determining the average speed v for a trip of two equal distances, each of length x, given any two speeds v_1 and v_2.

Solution   The total distance for the trip is $x + x = 2x$. The total time for the trip is the sum of the times for the two parts:

$$t = t_1 + t_2 = \frac{x}{v_1} + \frac{x}{v_2}$$

The average speed is total distance divided by total time, or

$$v = \frac{d}{t} = \frac{2x}{t_1 + t_2} = \frac{2x}{\dfrac{x}{v_1} + \dfrac{x}{v_2}}$$

The average speed is a complex fraction. The fraction bar represents division, so to simplify the fraction we add the denominator fractions together and change to multiplication.

$$v = \frac{2x}{\dfrac{x \cdot v_2}{v_1 \cdot v_2} + \dfrac{x \cdot v_1}{v_2 \cdot v_1}}$$ 　　Set up the LCD.

$$= \frac{2x}{\dfrac{xv_2 + xv_1}{v_1 v_2}}$$ 　　Change division to multiplication.

$$= \frac{2x}{1} \cdot \frac{v_1 v_2}{xv_2 + xv_1}$$ 　　Multiply and factor x from the denominator.

$$= \frac{2xv_1 v_2}{x(v_2 + v_1)}$$ 　　Simplify.

$$= \frac{2v_1 v_2}{v_2 + v_1}$$ 　　Note that there is no x here.

The distance traveled, x, is not part of the average speed formula for a round trip or for two segments of equal length. ●

I n Example 11, the least common denominator of the terms in a rational expression is used to simplify the expression.

EXAMPLE

Simplifying rational expressions by multiplying by the LCD　**Simplify the complex fraction** $\dfrac{\frac{1}{3} - x}{x + \frac{1}{2}}$.

Solution　The common denominator for terms in both the numerator and the denominator expressions is 6. Thus, we multiply both the numerator and the denominator by 6 to clear out the fractions within them:

$$\frac{\frac{1}{3} - x}{x + \frac{1}{2}} = \frac{6\left(\frac{1}{3} - x\right)}{6\left(x + \frac{1}{2}\right)} = \frac{\frac{6}{3} - 6x}{6x + \frac{6}{2}} = \frac{2 - 6x}{6x + 3}$$

After removing denominators from a complex fraction, we factor to determine whether simplification is possible:

$$\frac{2 - 6x}{6x + 3} = \frac{2(1 - 3x)}{3(2x + 1)}$$

There are no common factors, so the expression is simplified. ●

ANSWER BOX

Warm-up: **1.** $\frac{8}{15}$ **2.** $\frac{7}{120}$ **3.** $\frac{5}{8}$ **4.** $\frac{11}{36}$ **5.** $\frac{7}{48}$ **6.** $\frac{19}{45}$
7. $\dfrac{2 - 6x}{6x + 3}$ **8.** $\dfrac{3 + 4x}{4x - 2}$ **9.** $\dfrac{2 - 4x}{x^2 + 2x}$

EXERCISES 5.4

Add or subtract the expressions in Exercises 1 and 2. State any restrictions on the variables.

1. a. $\dfrac{3}{4} - \dfrac{x}{4}$

b. $\dfrac{2}{5x} - \dfrac{7}{5x}$

c. $\dfrac{4}{x+3} - \dfrac{x^2}{x+3}$

d. $\dfrac{2}{x^2+1} - \dfrac{x-1}{x^2+1}$

2. a. $\dfrac{2}{5} + \dfrac{x}{5}$

b. $\dfrac{2}{3x} - \dfrac{5}{3x}$

c. $\dfrac{4}{x+1} + \dfrac{x^2}{x+1}$

d. $\dfrac{2}{x^2-1} - \dfrac{x+1}{x^2-1}$

What is the common denominator in each of the fractions or rational expressions in Exercises 3 and 4?

3. a. $\dfrac{1}{4} + \dfrac{7}{12}$

b. $\dfrac{2}{a} - \dfrac{5}{2a}$

c. $\dfrac{b}{a^2} - \dfrac{c}{ab^2}$

d. $\dfrac{5}{x^2-2x} + \dfrac{3}{x^2-4}$

e. $\dfrac{5x}{x^2-6x+9} - \dfrac{3x}{x^2-9}$

4. a. $\dfrac{5}{18} + \dfrac{7}{24}$

b. $\dfrac{7}{y^2} - \dfrac{2}{y}$

c. $\dfrac{3}{a} - \dfrac{2}{b} + \dfrac{7}{a} - \dfrac{3}{b}$

d. $\dfrac{4}{x-5} - \dfrac{3x}{x^2-25}$

e. $\dfrac{x}{x^2+5x+6} - \dfrac{2x}{x^2+2x}$

5. Add or subtract, as indicated, the expressions in Exercise 3. State any restrictions on the variables.

6. Add or subtract, as indicated, the expressions in Exercise 4. State any restrictions on the variables.

In Exercises 7 to 20, add or subtract the rational expressions, as indicated.

7. $\dfrac{5}{4a} + \dfrac{3}{6b}$

8. $\dfrac{5}{2b} + \dfrac{7}{6a}$

9. $\dfrac{x}{x-3} + \dfrac{1}{x}$

10. $\dfrac{1}{x+2} + \dfrac{x+1}{x-2}$

11. $\dfrac{5x}{x-1} - \dfrac{8+x}{x}$

12. $\dfrac{x}{x+1} - \dfrac{3}{x-1}$

13. $\dfrac{x}{x-3} + \dfrac{3}{x^2-6x+9}$

14. $\dfrac{5x}{x+2} + \dfrac{2}{x^2+4x+4}$

15. $\dfrac{2b}{b^2-1} - \dfrac{3}{1-b}$

16. $\dfrac{1}{1-x} + \dfrac{x^2}{x-1}$

17. $\dfrac{2}{2a+ab} - \dfrac{3}{2b+b^2}$

18. $\dfrac{a}{ac-c^2} - \dfrac{c}{a^2-ac}$

19. $\dfrac{x}{x^2-6x+9} + \dfrac{3}{x^2-3x}$

20. $\dfrac{5}{x^2+x} + \dfrac{x}{x^2-2x-3}$

Exercises 21 to 24 give common formulas that, like the formulas in the theater examples, involve the sum of rates or other rational expressions. Add the fractions on the right side of each formula.

21. Ventilation fans in an attic:

$$\dfrac{1}{t} = \dfrac{1}{t_1} + \dfrac{1}{t_2}$$

22. Days to complete a wheat harvest with two machines:

$$\frac{1}{D} = \frac{1}{D_1} + \frac{1}{D_2}$$

23. Traffic flow through parallel doors:

$$\frac{1}{m} = \frac{1}{m_1} + \frac{1}{m_2}$$

24. Focal distance for a lens in optics:

$$\frac{1}{F} = \frac{1}{f_1} + \frac{1}{f_2}$$

Solve the application problems in Exercises 25 to 32. In Exercises 31 and 32, use the formulas from the reading or from Exercises 21 to 24. Round to the nearest tenth.

25. Two ventilation fans are operating in an attic. One can change the air in 5 hours and the other in 6 hours. Together, will they be able to change the air in 3 hours, as required by code?

26. Norman's corn harvester cuts his crop in 10 days. Karen's can cut the same crop in 8 days. If both harvesters are used together, how many days will be needed to cut Norman's crop?

27. Because of bad weather, Shane averages 45 miles per hour driving from Memphis to Louisville, a distance of 367 miles. On the return trip, he averages 60 miles per hour. What is his total time for the round trip? What is his average rate for the round trip?

28. Heavy traffic slows Luis to an average of 55 miles per hour while driving from Amarillo to Albuquerque, a distance of 284 miles. On the return trip, he averages 65 miles per hour. What was his total time for the trip? What is his average rate for the round trip?

29. What is the average speed for the round trip if Farah drives 55 miles per hour in one direction and 65 miles per hour on the return trip?

30. What is the average speed if Azra runs 18 kilometers per hour for the first half of a race and 22 kilometers per hour for the second half of the race?

31. The Artist-in-Residence theater has one double door and one single door. The larger door permits the audience to leave in 5 minutes. The smaller door permits the audience to leave in 9 minutes. If both doors are used at once, in how many minutes will the theater be empty?

32. The Hole-in-the-Wall theater has three exits. The largest door permits the audience to leave in 6 minutes. The emergency exit permits the audience to leave in 8 minutes. A third exit through the backstage permits the audience to leave in 15 minutes. If all three doors are used at once, in how many minutes will the theater be empty?

In Exercises 33 to 36, add the right side of the formula.

33. Approximating an exponential function:

$$e^x \approx 1 + x + \frac{x^2}{2} + \frac{x^3}{6} + \frac{x^4}{24}$$

34. Approximating a trigonometric function:

$$\cos(x) \approx 1 - \frac{x^2}{2} + \frac{x^4}{24} - \frac{x^6}{720}$$

35. The van der Waal equation for gases in physics:

$$P = \frac{RT}{v - b} - \frac{a}{v^2}$$

36. The x- and y-intercept form of a linear equation:

$$1 = \frac{x}{a} + \frac{y}{b}$$

37. Solve the theater-emptying formula for t. First, add the three fractions on the left side of $\frac{1}{t_1} + \frac{1}{t_2} + \frac{1}{t_3} = \frac{1}{t}$. Next, cross multiply and solve for t. Is the time required to empty the theater the sum of the individual door times?

38. Simplify the average speed formula from Example 10,

$$v = \frac{2x}{\dfrac{x}{v_1} + \dfrac{x}{v_2}}$$

by first multiplying the numerator and denominator by $v_1 v_2$.

In Exercises 37 to 46, simplify the complex fractions to eliminate the fractions from the numerator and the denominator.

39. $\dfrac{7 + 1}{\dfrac{1}{7} + 1}$

40. $\dfrac{4 + 1}{1 + \dfrac{1}{4}}$

41. $a = \dfrac{V}{\dfrac{4}{3}\pi b^2}$

42. $h = \dfrac{V}{\dfrac{\pi}{8}d^2}$

43. $\dfrac{x - \dfrac{x}{2}}{2 + \dfrac{x}{3}}$

44. $\dfrac{\dfrac{x}{3} - x}{3 + \dfrac{x}{2}}$

45. Parallel electric cells: $I = \dfrac{E}{R + \dfrac{r}{2}}$

46. Materials science: $\dfrac{\dfrac{p_2 - p_1}{v_1 - v_2}}{v_1}$

In Exercises 47 to 50, identify the missing operation sign (add, subtract, multiply, or divide).

47. **a.** $\dfrac{a}{b} \;\square\; \dfrac{1}{a} = \dfrac{1}{b}$ **b.** $\dfrac{a}{b} \;\square\; \dfrac{1}{a} = \dfrac{a^2}{b}$

48. **a.** $\dfrac{b}{a} \;\square\; \dfrac{1}{a} = \dfrac{b+1}{a}$ **b.** $\dfrac{b}{a} \;\square\; \dfrac{1}{a} = \dfrac{b}{a^2}$

49. **a.** $\dfrac{1}{a} \;\square\; \dfrac{1}{b} = \dfrac{a+b}{ab}$ **b.** $\dfrac{1}{a} \;\square\; \dfrac{1}{b} = \dfrac{1}{ab}$

50. **a.** $\dfrac{1}{a} \;\square\; \dfrac{1}{b} = \dfrac{b-a}{ab}$ **b.** $\dfrac{1}{b} \;\square\; \dfrac{1}{a} = \dfrac{a}{b}$

51. Calculate the reciprocal of $\frac{2}{3} + \frac{4}{7}$, and show that it is not $\frac{3}{2} + \frac{7}{4}$.

52. Calculate the reciprocal of $\frac{2}{3} + \frac{4}{5}$, and show that it is not $\frac{3}{2} + \frac{5}{4}$.

53. What is the reciprocal of $\dfrac{a}{b} + \dfrac{c}{d}$?

54. What is the reciprocal of $\dfrac{x}{y} - \dfrac{w}{z}$?

55. Describe in your own words how to find the least common denominator for two rational expressions.

Projects

56. *Average Trip Speed.* If you average x miles per hour during the first half of a trip, calculate the speed y in miles per hour needed during the second half of the trip to reach 400 miles per hour as an overall average. (*Hint:* Start with a table.)

57. *Leading a Race.* Arturo and Brahim are the leaders in a 10-kilometer race. Arturo is ahead by a distance x at the halfway point in the race. From prior race results, Arturo knows that Brahim's average speed for the race is likely to be 21 kilometers per hour.

  a. If the time required for the two runners to finish the race from their respective positions is equal, what will the result of the race be?

  b. If Arturo finishes in less time, what will the result of the race be?

  c. Write an expression to describe the time it will take Arturo to finish the race at 20 km/hr.

  d. Write an expression containing x to describe the time it will take Brahim to finish the race.

  e. Subtract the expression in part d from that in part c. Graph the result, with length of the lead on the horizontal axis and time on the vertical axis.

  f. Find the point on the graph that shows the length of the lead so that the race ends in a tie.

  g. Which points show lead lengths that give Arturo a win?

  h. Which points show the lead lengths that give Brahim a win?

58. *Indy Car Challenge.* When the leader reaches the halfway mark of a 500-mile Indy car street race, Michael is x miles behind. A recent pit stop adjustment has improved the performance of Michael's car to allow an average of 165 miles per hour for the rest of the race. The leader is currently averaging 158 miles per hour.

  a. What will determine whether it is possible for Michael to catch the leader?

  b. State the conditions under which Michael wins, ties, and loses.

59. *Lost Keys.* Marsha travels to work by bicycle at 15 miles per hour. Half an hour after she leaves one morning, her husband Jon finds her office keys by the door. He hopes to catch her before she reaches the bike path (at which point there is no way for him to find her). The bike path entrance is 12 miles away.

  a. Suppose Jon travels an average of 35 miles per hour in the car. Show whether he catches her in time.

  b. Write an equation that shows Jon's travel time.

  c. At what average speed must he travel to catch her?

5.5   Division of Polynomials and Related Graphs _____

OBJECTIVES

- Graph selected rational expressions.
- Divide polynomial expressions using long division.
- Find whether a divisor is a factor of the dividend.
- Use long division to factor cubic expressions.

WARM-UP

Divide these numbers with long division.

1. $11\overline{)1331}$ **2.** $35\overline{)7213}$

Subtract these expressions.

3. $2x^2 + 3x$ **4.** $x^3 + 3x^2$
      $-(2x^2 + 2x)$ $-(x^3 + x^2)$

5. $x^3 + 3x^2$ **6.** $x^2 - 3x$
      $-(x^3 - x^2)$ $-(x^2 - 4x)$

I N THIS SECTION, we examine the division of polynomials. Look for general ideas in this section: how division of polynomials with algebra is similar to division with numbers, how division relates to finding factors, and how division verifies factoring formulas. Also look for some unusual features in the graphs.

EXAMPLE Exploring the graph of a rational expression

a. Make a table and a graph for $y = (x^3 + 3x^2 + 3x + 1) \div (x + 1)$.

b. Predict the equation of the graph, and check with calculator regression. Why does the graph have a gap?

Solution   **a.** Table 10 contains values for $(x^3 + 3x^2 + 3x + 1) \div (x + 1)$. The table is symmetric about $x = -1$. Figure 24 contains the graph of the quotient from $y = (x^3 + 3x^2 + 3x + 1) \div (x + 1)$.

b. The graph appears to be a parabola with the same shape as $y = x^2$. The vertex at $(-1, 0)$ means a horizontal shift to the left by 1 unit and suggests the equation $y = (x + 1)^2$. Applying quadratic regression on a calculator, we obtain $y = x^2 + 2x + 1$. When we trace to exactly $x = -1$ on a graphing calculator, there is a blank for y and a *hole in the graph*. This is because the original expression is undefined at $x = -1$.

x	$\dfrac{x^3 + 3x^2 + 3x + 1}{x + 1}$
-4	9
-3	4
-2	1
-1	undefined
0	1
1	4
2	9

Table 10

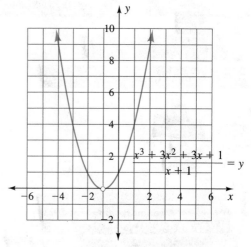

Figure 24

Think about it 1: What is the degree of the numerator? the denominator? the equation of the graph? How do the degrees relate to one another?

Division Vocabulary

In our work in this section, we will use the following words to describe the parts of a division problem:

$$\begin{array}{r} \text{quotient } + \text{ remainder} \\ \hline \text{divisor)}\overline{\text{dividend}} \end{array}$$

When a division problem is written as a fraction, the dividend becomes the numerator and the divisor becomes the denominator:

$$\frac{\text{dividend}}{\text{divisor}}$$

Division with a Zero Remainder

In Example 2, we apply long division to a numerical problem and to the expression in Example 1 to emphasize the similarity between the two division processes.

EXAMPLE ❷

Finding a quotient with long division Divide 1331 by 11, and divide $x^3 + 3x^2 + 3x + 1$ by $x + 1$.

Solution   At each step shown below, compare the numerical long division (in the left column) with polynomial long division (in the right column).

Numerical long division: Divide 1331 by 11.

1. We estimate how many times 11 divides into 13.

2. We enter the estimate, 1, in the quotient.

3. We multiply 11 by the estimate, and place the product below 13.

4. We subtract: $13 - 11 = 2$.

5. We bring down the next number, 3.

$$\begin{array}{r} 1 \\ 11)\overline{1331} \\ \underline{11} \\ 023 \end{array}$$

Now we repeat these same five steps.

1. We estimate how many times 11 divides into 23.

2. We enter the estimate, 2, in the quotient.

Polynomial long division: Divide $x^3 + 3x^2 + 3x + 1$ by $x + 1$.

1. We estimate how many times x divides into x^3.

2. We enter the estimate, x^2, in the quotient.

3. We multiply $x + 1$ by the estimate and place the product below $x^3 + 3x^2$.

4. We subtract: $x^3 + 3x^2 - (x^3 + x^2) = 2x^2$.

5. We bring down the next number, $3x$.

$$\begin{array}{r} x^2 \\ x + 1)\overline{x^3 + 3x^2 + 3x + 1} \\ \underline{-(x^3 + x^2)} \\ 2x^2 + 3x \end{array}$$

Now we repeat these same five steps.

1. We estimate how many times x divides into $2x^2$.

2. We enter the estimate, $2x$, in the quotient.

3. We multiply 11 by the estimate and place the product below 23.

4. We subtract: $23 - 22 = 1$.

5. We bring down the next number, 1.

$$
\begin{array}{r}
12 \\
11\overline{)1331} \\
\underline{11} \\
023 \\
\underline{22} \\
011
\end{array}
$$

1. We estimate how many times 11 divides into 11.

2. We enter the estimate, 1, in the quotient.

3. We multiply 11 by the estimate and place the product below 11.

4. We subtract: $11 - 11 = 0$.

5. The difference is zero and no numbers remain in the dividend, so we stop.

$$
\begin{array}{r}
121 \\
11\overline{)1331} \\
\underline{11} \\
023 \\
\underline{22} \\
011 \\
\underline{11} \\
0
\end{array}
$$

The division indicates that the original expression and the quotient are equivalent:

$$\frac{1331}{11} = 121$$

We check by multiplying the divisor by the quotient:

$$11 \cdot 121 = 1331$$

3. We multiply $x + 1$ by the estimate and place the product below $2x^2 + 3x$.

4. We subtract: $2x^2 + 3x - (2x^2 + 2x) = 1x$.

5. We bring down the next number, 1.

$$
\begin{array}{r}
x^2 + 2x \\
x + 1\overline{)\ x^3 + 3x^2 + 3x + 1} \\
\underline{-(x^3 + x^2)} \\
2x^2 + 3x \\
\underline{-(2x^2 + 2x)} \\
1x + 1
\end{array}
$$

1. We estimate how many times x divides into $1x$.

2. We enter the estimate, 1, in the quotient.

3. We multiply $x + 1$ by the estimate and place the product below $1x + 1$.

4. We subtract: $1x + 1 - (1x + 1) = 0$.

5. The difference is zero and no terms remain in the dividend, so we stop.

$$
\begin{array}{r}
x^2 + 2x + 1 \\
x + 1\overline{)\ x^3 + 3x^2 + 3x + 1} \\
\underline{-(x^3 + x^2)} \\
2x^2 + 3x \\
\underline{-(2x^2 + 2x)} \\
1x + 1 \\
\underline{-(1x + 1)} \\
0
\end{array}
$$

The division indicates that the original expression and the quotient (without the restrictions on x) are equivalent:

$$\frac{x^3 + 3x^2 + 3x + 1}{x + 1} = x^2 + 2x + 1, \quad x \neq -1$$

We check by multiplying the divisor by the quotient:

$$(x + 1)(x^2 + 2x + 1) = x^3 + 3x^2 + 3x + 1 \ \bullet$$

In Example 2, the division came out even—with a zero remainder—leading us to conclude that $x^3 + 3x^2 + 3x + 1$ factors into $(x + 1)(x^2 + 2x + 1)$.

Think about it 2: Write $x^3 + 3x^2 + 3x + 1$ as the product of three factors.

Summary: Dividing Polynomial Expressions with Long Division

- Arrange the divisor and the dividend in descending order of exponents.
- Replace missing terms in the dividend with terms having a zero numerical coefficient.
- Divide, using the five steps:
  1. Estimate how many times the first term of the divisor divides into the first term of the dividend.
  2. Enter the estimate in the quotient.
  3. Multiply the divisor by the estimate, and place the product below the like terms in the dividend.
  4. Subtract.
  5. Bring down the next term.
- Repeat steps 1 to 5, as needed.

In Examples 3 and 4, we repeat the table, graph, and polynomial division process for a rational expression with x^2 and x terms missing in the numerator.

EXAMPLE **3**

Exploring the graph of a rational expression

a. Make a table and a graph for $y = (x^3 - 1) \div (x - 1)$. What happens at $x = 1$?

b. Predict the degree of the resulting equation from the graph, and fit an equation with calculator regression.

Solution

a. Table 11 shows a symmetry of outputs for pairs of inputs. The graph in Figure 25 appears to be a parabola. We can adjust the calculator window for integer inputs to find the hole that corresponds to the undefined value at $x = 1$.

x	$\dfrac{x^3 - 1}{x - 1}$
-3	7
-2	3
-1	1
0	1
1	undefined
2	7
3	13

Table II

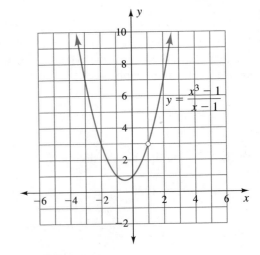

Figure 25

b. The parabolic shape of the graph in Figure 25 suggests a quotient of degree 2, or a quadratic expression. Applying quadratic regression to the data in the table, we obtain a quotient equation $y = x^2 + x + 1$.

Think about it 3: What is the value of $\dfrac{x^3 - 1}{x - 1}$ as x takes on decimal values close to 1?

In Example 4, we do the division on the expression graphed in Example 3. *Note how we adjust the dividend expression in order to carry out the division steps.*

EXAMPLE 4

Solution

Finding a quotient with long division   Divide $x^3 - 1$ by $x - 1$.

1. We estimate how many times x divides into x^3.

2. We enter the estimate, x^2, in the quotient.

3. We multiply $x - 1$ by the estimate and place the product below $x^3 - 1$.

4. We try to subtract $x^3 - x^2$ from $x^3 - 1$.

$$
\begin{array}{r}
x^2 \\
x - 1 \overline{)\ x^3 - 1 } \\
-(x^3 - x^2)
\end{array}
$$

Unfortunately, 1 and x^2 are not like terms and cannot be subtracted.

To provide like terms for the subtraction, we must add terms with a coefficient of zero, $0x^2$ and $0x$, between the x^3 and the 1 in the dividend. We try the first four steps again, with the adjusted expression.

1. We estimate how many times x divides into x^3.

2. We enter the estimate, x^2, in the quotient.

3. We multiply $x - 1$ by the estimate and place the product below $x^3 + 0x^2 + 0x - 1$.

4. We subtract: $x^3 + 0x^2 - (x^3 - x^2) = x^2$.

5. We bring down the next term, $0x$.

$$
\begin{array}{r}
x^2 \\
x - 1 \overline{)\ x^3 + 0x^2 + 0x - 1} \\
-(x^3 - x^2) \\
\hline
+x^2 + 0x
\end{array}
$$

1. We estimate how many times x divides into x^2.

2. We enter the estimate, x, in the quotient.

3. We multiply $x - 1$ by the estimate and place the product below $x^2 + 0x$.

4. We subtract: $x^2 + 0x - (x^2 - x) = x$.

5. We bring down the next term, -1.

$$
\begin{array}{r}
x^2 + x \\
x - 1 \overline{)\ x^3 + 0x^2 + 0x - 1} \\
-(x^3 - x^2) \\
\hline
x^2 + 0x \\
-(x^2 - x) \\
\hline
+x - 1
\end{array}
$$

1. We estimate how many times x divides into x.

2. We enter the estimate, 1, in the quotient.

3. We multiply $x - 1$ by the estimate and place the product below $x - 1$.

4. We subtract: $x - 1 - (x - 1) = 0$.

5. There are no more expressions in the dividend.

$$
\begin{array}{r}
x^2 + x + 1 \\
x - 1 \overline{\smash{)}\ x^3 + 0x^2 + 0x - 1} \\
\underline{-(x^3 - x^2)} \\
x^2 + 0x \\
\underline{-(x^2 - x)} \\
+x - 1 \\
\underline{-(x - 1)} \\
0
\end{array}
$$

The division indicates that the original expression and the quotient with the restriction on x are equivalent:

$$
\frac{x^3 - 1}{x - 1} = x^2 + x + 1, \quad x \neq 1
$$

Check: We check by multiplying the divisor by the quotient:

$$
(x - 1)(x^2 + x + 1) = x^3 - 1 \quad \checkmark
$$

⬤

The division in Example 4 shows that $x^3 - 1$ factors to $(x - 1)(x^2 + x + 1)$. In the examples thus far, we have obtained a zero remainder for each division. *When there is no remainder, we say the divisor is a factor of the dividend.* This idea can be represented visually as follows:

$$
\begin{array}{r}
\text{quotient} + \text{zero} \\
\text{factor} \overline{\smash{)}\ \text{dividend}}
\end{array}
$$

Zero Remainders and Factors

> A zero remainder from division indicates that the divisor is a factor of the dividend.
>
> When the division is written as a fraction, a zero remainder indicates that the numerator and the denominator have a common factor.
>
> A zero remainder indicates that the graph will contain a hole with an x-coordinate that makes a zero denominator in the original fraction.

Division with a Nonzero Remainder

In Examples 5 and 6, we repeat the table, graph, and polynomial division process for an expression that does not come out even after division; it has a nonzero remainder. We start with the division.

EXAMPLE **5** Finding a quotient and remainder with long division   Predict the degree of the quotient, and then divide $x^3 - 2x^2 - 5x + 6$ by $x + 1$.

Solution   The quotient will be of degree 2, a quadratic expression.

$$
\begin{array}{r}
x^2 - 3x - 2 \\
x + 1 \overline{\smash{)}\ x^3 - 2x^2 - 5x + 6} \\
\underline{-(x^3 + x^2)} \\
-3x^2 - 5x \\
\underline{-(-3x^2 - 3x)} \\
-2x + 6 \\
\underline{-(-2x - 2)} \\
8
\end{array}
$$

This time, the remainder is not zero. We have

$$
\frac{x^3 - 2x^2 - 5x + 6}{x + 1} = x^2 - 3x - 2 + \frac{8}{x + 1}, \quad x \neq -1
$$

●

EXAMPLE **6**   Graphing a quotient and remainder Make a table and a graph for $y = (x^3 - 2x^2 - 5x + 6) \div (x + 1)$.

Solution   Table 12 shows $y = (x^3 - 2x^2 - 5x + 6) \div (x + 1)$. The table lacks the symmetry found in the previous two division problems. This may surprise you, because the division gave a quadratic expression. There is an undefined output at $x = -1$ because of a zero denominator, yet instead of being a parabolic graph with a hole, the graph in Figure 26 is nearly vertical near $x = -1$.

x	$\dfrac{x^3 - 2x^2 - 5x + 6}{x + 1}$
-4	23.3
-3	12
-2	0
-1	undefined
0	6
1	0
2	-1.3

Table 12

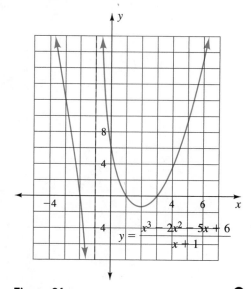

$$y = \frac{x^3 - 2x^2 - 5x + 6}{x + 1}$$

Figure 26

●

Nothing in Table 12 or Figure 26 explains the difference between the shape of the graphs in Examples 3 and 6. The division, on the other hand, does explain why the graph in Figure 26 is nearly vertical at $x = -1$.

In Example 5, we wrote the remainder, 8, over the divisor, $x + 1$. This is a standard form for nonzero remainders:

$$
\text{quotient} + \frac{\text{remainder}}{\text{divisor}}
$$
$$
\text{divisor} \overline{\smash{)}\text{dividend}}
$$

The remainder, 8, over the divisor, $x + 1$, forms a rational expression that is undefined at $x = -1$. The graph of $y = 8/(x + 1)$ has a nearly vertical part as it approaches -1 from either the left or the right (see Figure 27). This nearly vertical behavior causes a change in the parabolic shape that we might expect from the quotient, $x^2 - 3x - 2$, found in Example 5.

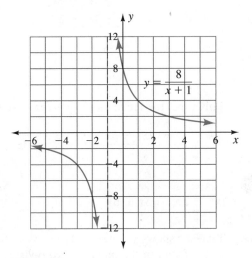

Figure 27

In Example 7, we compare the graph of the quotient $x^2 - 3x - 2$ with that of the quotient and remainder to show the effect of the nonzero remainder.

EXAMPLE **7**   Graphing the quotient together with the quotient and remainder   Graph $y = (x^3 - 2x^2 - 5x + 6) \div (x + 1)$ with $y = x^2 - 3x - 2$.

Solution   In Figure 28, the equations are graphed to the same scale as in Figure 26. In Figure 29, they are graphed with a zoom out. The parabolic shape of the graph seems more clear in Figure 29. The original curve follows the parabola closely except near $x = -1$. The nearly vertical behavior near $x = -1$ is due to the remainder, $8/(x + 1)$.

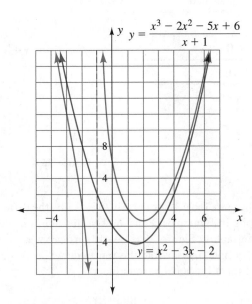

Figure 28

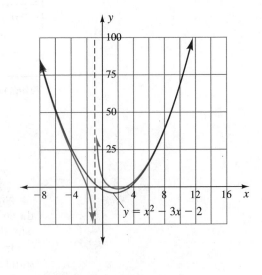

Figure 29

There is a line associated with an undefined point for rational expressions. That line is not important now, but it should be mentioned for completeness. When the number n makes the denominator zero in a rational expression that is in lowest terms (contains no common factors), we say that the line with equation $x = n$ is a *vertical asymptote*. An **asymptote** is *the name given to the line that a graph approaches.* Remember: The vertical asymptote is *not* part of the set of points forming the graph. (See the Graphing Calculator Technique box.) The vertical asymptote, $x = -1$, is a boundary. The asymptote is the dashed line drawn in Figures 27, 28, and 29.

Graphing Calculator Technique:
Lines at Undefined Points

With certain viewing window settings, many graphing calculators draw an almost vertical line on the graph at an undefined point. *This line is an error made by the calculator. The calculator evaluates inputs to the left and right of the undefined point and connects the outputs.*

The line will disappear if the calculator is set on dot rather than connected mode or if the viewing window is adjusted to force the calculator to evaluate the input making the expression undefined.

ANSWER BOX

Warm-up: 1. 121 **2.** $206\frac{3}{35}$ **3.** x **4.** $2x^2$ **5.** $4x^2$ **6.** x
Think about it 1: degree 3; degree 1; degree 2; $3 - 1 = 2$
Think about it 2: $x^3 + 3x^2 + 3x + 1 = (x + 1)(x^2 + 2x + 1)$
$= (x + 1)^3$
Think about it 3: As x gets close to 1, the output gets close to 3.

EXERCISES

In Exercises 1 to 12, the division problems are stated in fraction form, with the denominator containing the divisor. Indicate for what values the expressions are defined, and complete the divisions.

1. $\dfrac{x^2 - 3x - 4}{x - 4}$

2. $\dfrac{x^2 + 3x + 2}{x - 1}$

3. $\dfrac{x^2 - 3x - 4}{x - 3}$

4. $\dfrac{x^2 + 3x + 2}{x}$

5. $\dfrac{x^2 - 3x - 4}{x - 2}$

6. $\dfrac{x^2 + 3x + 2}{x + 1}$

7. $\dfrac{x^2 - 3x - 4}{x - 1}$

8. $\dfrac{x^2 + 3x + 2}{x + 2}$

9. $\dfrac{x^2 - 3x - 4}{x}$

10. $\dfrac{x^2 + 3x + 2}{x + 3}$ **11.** $\dfrac{x^2 - 3x - 4}{x + 1}$

12. $\dfrac{x^2 + 3x + 2}{x + 4}$

13. Factor $x^2 - 3x - 4$. What was special about the quotient when the denominator was a factor of the numerator in Exercises 1 to 11 odd?

14. Factor $x^2 + 3x + 2$. What was special about the quotient when the denominator was not a factor in Exercises 2 to 12 even?

15. Make a table for $f(x) = x^2 - 3x - 4$. Use integer inputs on the interval $[-1, 4]$. How do the function values compare with the remainders in the answers to Exercises 1 to 11 odd?

16. Make a table for $f(x) = x^2 + 3x + 2$. Use integer inputs on the interval $[-4, 1]$. How do the function values compare with the remainders in the answers to Exercises 2 to 12 odd?

17. Use a graphing calculator to graph the expression in Exercise 3. Make a careful sketch of the graph.

   a. Why does the graph have a nearly vertical portion? Where is it located?

   b. Describe the behavior of the graph for $x < -8$ and $x > 8$ compared with that of $y = x$.

   c. Graph the expression in Exercise 1. Why is this graph different from that of the expression in Exercise 3?

   d. Graph the expression in Exercise 5. How is this graph the same as that of the expression in Exercise 3?

18. Use a graphing calculator to graph the expression in Exercise 12. Make a careful sketch of the graph.

   a. Why does the graph have a nearly vertical portion? Where is it located?

   b. Describe the behavior of the graph for $x < -8$ and $x > 8$ compared with that of $y = x - 1$.

   c. Graph the expression in Exercise 4. How is this graph the same as that of the expression in Exercise 12?

   d. Graph the expression in Exercise 8. Why is this graph different from that of the expression in Exercise 12?

Divide the expressions in Exercises 19 to 26. In which exercises is the denominator a factor of the numerator?

19. $\dfrac{x^3 - 1}{x - 2}$ **20.** $\dfrac{x^3 + 3x^2 + 3x + 1}{x + 2}$

21. $\dfrac{x^3 - 1}{x - 1}$ **22.** $\dfrac{x^3 + 3x^2 + 3x + 1}{x + 1}$

23. $\dfrac{x^3 - 1}{x}$ **24.** $\dfrac{x^3 + 3x^2 + 3x + 1}{x}$

25. $\dfrac{x^3 - 1}{x + 1}$ **26.** $\dfrac{x^3 + 3x^2 + 3x + 1}{x - 1}$

27. Make a table for $f(x) = x^3 - 1$ for integer inputs from $x = 2$ to $x = -1$. Compare the outputs with the remainders in Exercises 19 to 25 odd.

28. Make a table for $f(x) = x^3 + 3x^2 + 3x + 1$ for integer inputs on the interval $[-2, 1]$. Compare the outputs with the remainders in Exercises 20 to 26 even.

In Exercises 29 to 36, divide. In which exercises is the denominator a factor of the numerator?

29. $\dfrac{x^3 + 1}{x + 1}$ **30.** $\dfrac{x^4 - 1}{x - 1}$

31. $\dfrac{x^4 + 1}{x + 1}$ **32.** $\dfrac{x^3 + 8}{x + 2}$

33. $\dfrac{x^4 - 4x^3 + 6x^2 - 4x + 1}{x - 1}$

34. $\dfrac{x^4 + 4x^3 + 6x^2 + 4x + 1}{x + 1}$

35. $\dfrac{x^2 + 3x^3 - 1}{x + 1}$ **36.** $\dfrac{x + x^3 - 2x^2}{x + 1}$

37. Describe how it is possible to predict the highest exponent on the quotient from the original rational expression.

38. Give an example of a rational expression with a quotient of degree 4.

In Exercises 39 to 42, use the calculator table feature to locate the undefined position (hole) in the graph. Zoom in with the table, and estimate the ordered pair describing the hole.

39. Exercise 29 (*Hint*: $f(-1.01) \approx 3.03$.)

40. Exercise 30 (*Hint*: $f(0.99) \approx 3.94$.)

41. Exercise 33

42. Exercise 32

43. Use long division, $(x^3 + y^3) \div (x + y)$, to prove the factoring formula for $x^3 + y^3$.

44. Use long division, $(x^3 - y^3) \div (x - y)$, to prove the factoring formula for $x^3 - y^3$.

45. True or false: If the divisor $(x - a)$ is a factor of the dividend, there is a hole in the graph at $x = a$.

46. True or false: If the divisor is a factor of the dividend, the remainder is zero.

47. Find $f(-2)$ and $f(5)$ for $f(x) = x^2 - 3x - 4$. Use long division to find the remainders for division of $x^2 - 3x - 4$ by $(x + 2)$ and $(x - 5)$. What do you observe?

48. Find $f(-5)$ and $f(2)$ for $f(x) = x^2 + 3x + 2$. Use long division to find the remainders for division of $x^2 + 3x + 2$ by $(x + 5)$ and $(x - 2)$. What do you observe?

Fill in the blanks in Exercises 49 and 50.

49. The output at $x = a$, $f(a)$, is the _____ upon division of $f(x)$ by $(x - a)$.

50. When we divide a polynomial by one of its factors, the remainder is _____.

51. Do a quadratic regression on the last three ordered pairs from the table in Example 1. Compare the resulting quadratic equation with the quotient to the division.

52. What is the vertex of the parabola in Example 3? Use $y = a(x - h)^2 + k$ to find the equation of the parabola. Compare your equation with that obtained in the example.

Projects

53. *Factoring Cubes.* List the values of these expressions.

a. 4^3 **b.** 5^3 **c.** 6^3

d. $(0.1)^3$ **e.** $(0.4)^3$ **f.** $(0.5)^3$

g. $(0.01)^3$ **h.** $\left(\frac{1}{4}\right)^3$ **i.** $\left(\frac{1}{5}\right)^3$

j. $\left(\frac{1}{6}\right)^3$

Multiply to find these products. Use tables as needed.

k. $(x - 1)(x^2 + x + 1)$

l. $(x + 2)(x^2 - 2x + 4)$

m. $(3x - 1)(9x^2 + 3x + 1)$

n. $(x - y)(x^2 + xy + y^2)$

o. $(x - y)(x - y)(x - y)$

The sum and the difference of two cubes may be factored using these rules:

$$x^3 + y^3 = (x + y)(x^2 - xy + y^2)$$
$$x^3 - y^3 = (x - y)(x^2 + xy + y^2)$$

Factor these expressions.

p. $x^3 - 8$ **q.** $x^3 + 27$

r. $8x^3 - 64$ **s.** $27a^3 + 1$

t. $y^3 + 0.001$

u. $\frac{1}{64}x^3 + \frac{1}{27}$

v. $8b^3 + 125$

w. $27x^3 - 64$

x. $0.008a^3 + 1$

y. $y^3 + 0.027$

54. *Cubes.* Puzzle 1: Given that $1729 = 12^3 + 1^3$, find two other whole numbers such that $a^3 + b^3 = 1729$.
Puzzle 2: Use the pattern below to do the following.

```
 1
 3   5
 7   9  11
13  15  17  19
21  23  25  27  29
31  33  35  37  39  41
```

a. Find the sum of each row.

b. Build three more rows into the pattern.

c. Find the sum of the new rows.

d. Find the sum of the nth row.

e. Determine how many numbers will appear in the nth row.

f. Why will a quadratic expression generate the last number in each row for $n = 1, 2, 3, 4, \ldots$? Find the expression.

5.6 Solving Rational Equations

OBJECTIVES

- Find the least common denominator of rational expressions in an equation.
- Solve equations containing rational expressions.
- Solve equations graphically.
- Solve application problems related to $\frac{1}{a} + \frac{1}{b} = \frac{1}{c}$.
- Solve application problems containing rational expressions.

WARM-UP

Factor.

1. $x^2 + x - 2$ **2.** $2x^2 + 7x + 3$ **3.** $x^2 - 4x - 77$

Complete the multiplication tables. Summarize by writing the resulting product in simplified form.

4.

Multiply	$\dfrac{1}{3}$	$+\dfrac{1}{4x}$
$12x$		

5.

Multiply	$\dfrac{1}{x-3}$	$+\dfrac{5}{x-2}$
$(x-2)(x-3)$		

Multiply and simplify.

6. $\dfrac{18}{1}\left(\dfrac{x}{2} + \dfrac{x}{9}\right)$

7. $5\left(2w + \dfrac{16w}{5}\right)$

8. $x(x+1)\left(\dfrac{2}{x+1} + \dfrac{3}{x}\right)$

9. $x^2\left(\dfrac{3}{x^2} + \dfrac{2}{x} - 1\right)$

10. $20\left(\dfrac{3}{4} + \dfrac{1}{5}\right)$

11. What happened to the fractions in the expressions in Warm-ups 6 to 10? Why did this happen?

THE PURPOSE OF THIS SECTION is to extend methods of solving equations to **rational equations**—*equations that contain rational expressions.* We solve rational equations by graphing and by one of two algebraic methods: cross multiplication of a proportion or multiplication on both sides of the equation by the least common denominator.

Review of Equation and Formula Solving

We have already solved equations containing fractions, using both symbolic and graphical techniques. Recall that in earlier equation and formula solving, we wrote a plan in which we reversed the order of operations as applied to x. By doing the opposite operations, we were able to solve for x or other variables.

EXAMPLE **1**   Solving equations   Solve $\frac{2}{3}x + 4 = 22$ for x.

Symbolic Solution　**Plan**: Because x is multiplied by $\frac{2}{3}$ and then added to 4, we subtract 4 and divide by $\frac{2}{3}$. Recall that division by $\frac{2}{3}$ is the same as multiplication by $\frac{3}{2}$.

$$\frac{2}{3}x + 4 = 22 \qquad \text{Subtract 4 from each side.}$$
$$\frac{2}{3}x = 18 \qquad \text{Multiply by the reciprocal, } \frac{3}{2}.$$
$$\frac{3}{2} \cdot \frac{2}{3}x = \frac{3}{2}(18) \qquad \text{Simplify.}$$
$$x = 27$$

Check: $\frac{2}{3}(27) + 4 \overset{?}{=} 22$　✔

Graphical Solution　We graph the left side and the right side of $\frac{2}{3}x + 4 = 22$ separately. The graph of $y_1 = \frac{2}{3}x + 4$ is a straight line with a slope of $\frac{2}{3}$ and a y-intercept at 4. The line $y_2 = 22$ is horizontal. The graphical solution to $\frac{2}{3}x + 4 = 22$ is $x = 27$, the x-coordinate of the point of intersection of the graphs (see Figure 30).

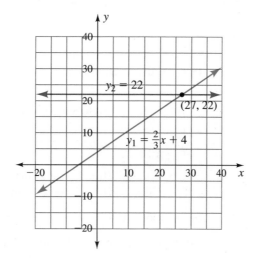

Figure 30　●

We also write a plan in formula problems. Recall that multiplication by $\frac{1}{3}$ is the same as division by 3.

EXAMPLE **2**　*Solving formulas*　Solve the volume-of-a-cone formula for h, the height:

$$V = \frac{1}{3}\pi r^2 h$$

Solution　**Plan**: h is multiplied by πr^2 and divided by 3, so we first multiply by 3 and then divide by πr^2. It is a good idea to identify the h with an arrow or other mark.

$$V = \frac{1}{3}\pi r^2 \overset{\downarrow}{h} \qquad \text{Multiply by 3.}$$
$$3V = \pi r^2 h \qquad \text{Divide by } \pi r^2.$$
$$\frac{3V}{\pi r^2} = h$$

●

In Example 2, we could have divided both sides by $\frac{1}{3}\pi r^2$, but that would have produced a complex fraction. Multiplying by 3 eliminated the fraction and resulted in a simpler expression.

Solving Proportions with Cross Multiplication and Graphs

Many rational equations are proportions or may be written as proportions. In Example 3, we obtain a quadratic equation from the cross multiplication.

EXAMPLE 3   Solving proportions   Solve $\dfrac{7}{x+3} = \dfrac{x-7}{8}$ for x. Assume $x \neq -3$.

Symbolic Solution   This equation is a proportion, and we cross multiply.

$$\frac{7}{x+3} = \frac{x-7}{8} \qquad \text{Cross multiply.}$$

$$(x+3)(x-7) = 7(8) \qquad \text{Multiply the sides.}$$

$$x^2 - 4x - 21 = 56 \qquad \text{Change to } ax^2 + bx + c = 0.$$

$$x^2 - 4x - 77 = 0 \qquad \text{Factor (or use the quadratic formula).}$$

$$(x-11)(x+7) = 0 \qquad \text{Apply the zero product rule.}$$

Either $x = 11$ or $x = -7$

Checking the solution is left as an exercise.

Graphical Solution   We graph the left and right sides of the equation separately,

$$y_1 = \frac{7}{x+3} \qquad \text{and} \qquad y_2 = \frac{x-7}{8}$$

At $x = -3$, y_1 is undefined. The graph turns nearly vertical near $x = -3$. The graph of y_2 is linear with a slope of $\frac{1}{8}$ and a y-intercept at $-\frac{7}{8}$.

The solutions to $\dfrac{7}{x+3} = \dfrac{x-7}{8}$ are $x = -7$ and $x = 11$, the x-coordinates of the intersections of the graphs in Figure 31.

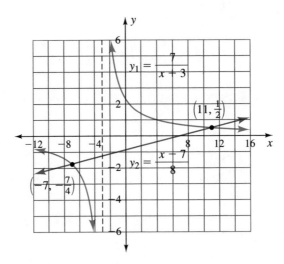

Figure 31

Example 4 shows the importance of checking solutions.

EXAMPLE 4   Solving proportions   Solve $\dfrac{x+1}{x-1} = \dfrac{1}{1-x}$, $x \neq 1$ for x.

Symbolic Solution   When we cross multiply, we find a quadratic equation with two solutions.

$$\frac{x+1}{x-1} = \frac{1}{1-x}$$

> Cross multiply.

$$(1-x) \cdot (x+1) = 1 \cdot (x-1)$$

> Simplify.

$$1 - x^2 = x - 1$$

> Change to $ax^2 + bx + c = 0$.

$$x^2 + x - 2 = 0$$ Factor (or use the quadratic formula).

$$(x-1)(x+2) = 0$$ Apply the zero product rule.

Either $x - 1 = 0$ or $x + 2 = 0$ Solve the factor equations.

$$x = 1 \quad \text{or} \quad x = -2$$

Check: In checking our solutions, we find that $x = -2$ satisfies the equation:

$$\frac{-2+1}{-2-1} \stackrel{?}{=} \frac{1}{1-(-2)} \quad \checkmark$$

However, $x = 1$ gives a zero denominator. As noted in the original problem, $x = 1$ has been excluded from the set of possible inputs.

Graphical Solution   We enter the left and right sides of $\dfrac{x+1}{x-1} = \dfrac{1}{1-x}$ as separate equations:

$$y_1 = \frac{x+1}{x-1} \quad \text{and} \quad y_2 = \frac{1}{1-x}$$

The graphs, shown in Figure 32, intersect only at $x = -2$. At $x = 1$, both sides of the equation have a zero denominator and both curves become nearly vertical. There is no point on either graph corresponding to $x = 1$.

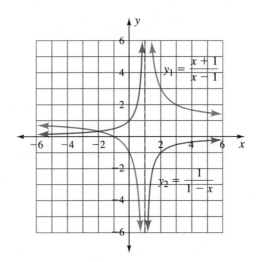

Figure 32   ●

The solution $x = 1$ in Example 4 is an **extraneous root**—*a solution found algebraically that does not satisfy the original equation.* Extraneous roots can also occur when we square both sides of an equation, as we will in Section 6.4. In Exercise 62, we will solve the equation in a different way and avoid the extraneous root.

M any fractional equations, such as those in Section 5.4, may be easily changed to proportions.

EXAMPLE **5**   Solving with a proportion   Solve $\dfrac{3}{4} + \dfrac{1}{5} = \dfrac{1}{x}$ for x. Assume $x \neq 0$.

Symbolic Solution   ***Plan***: By first adding the fractions on the left side, we change the equation to a proportion, and then we solve by cross multiplication.

$$\frac{3}{4} + \frac{1}{5} = \frac{1}{x} \qquad \text{Add the left side.}$$

$$\frac{3 \cdot 5}{4 \cdot 5} + \frac{1 \cdot 4}{5 \cdot 4} = \frac{1}{x} \qquad \text{Simplify.}$$

$$\frac{19}{20} = \frac{1}{x} \qquad \text{Cross multiply.}$$

$$19x = 20(1) \qquad \text{Divide by 19.}$$

$$x = \frac{20}{19}$$

Checking the solution is left as an exercise.

Graphical Solution   We graph the left and right sides separately:

$$y_1 = \frac{3}{4} + \frac{1}{5} = \frac{19}{20} \qquad \text{and} \qquad y_2 = \frac{1}{x}$$

The graph of y_1 is a horizontal line. The graph of y_2 is undefined at $x = 0$ and nearly vertical on either side of the y-axis. The solution to $\dfrac{3}{4} + \dfrac{1}{5} = \dfrac{1}{x}$ is $x = \frac{20}{19}$, the x-coordinate of the point of intersection of the graphs in Figure 33.

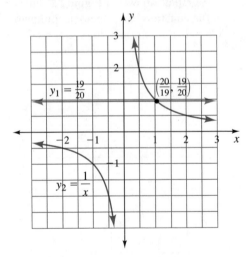

Figure 33

Solving Equations with the Least Common Denominator and Graphs

At least two of the steps in the symbolic solution to Example 5 involve multiplication by numbers related to the denominators. The denominators contain 4, 5, and x. The product of these is $20x$, the least common denominator for all the fractions in the equation. When we multiply both sides of an equation containing denominators by the least common denominator, we eliminate the denominators from each side.

Solving Equations
Containing Denominators

We may eliminate denominators if we multiply both sides of an equation by the least common denominator.

In Example 6, multiplying both sides of the equation by the least common denominator permits us to obtain a simpler equation.

EXAMPLE **6**   Solving with an LCD   Solve $\dfrac{3}{4} + \dfrac{1}{5} = \dfrac{1}{x}$ for x by multiplying both sides of the equation by $20x$, the least common denominator. Assume $x \neq 0$.

Solution

$$\frac{3}{4} + \frac{1}{5} = \frac{1}{x}$$ Multiply by the LCD.

$$20x\left(\frac{3}{4} + \frac{1}{5}\right) = 20x\left(\frac{1}{x}\right)$$ Distribute on the left side.

$$20x\left(\frac{3}{4}\right) + 20x\left(\frac{1}{5}\right) = \frac{20x}{x}$$ Factor.

$$4 \cdot 5x\left(\frac{3}{4}\right) + 4 \cdot 5x\left(\frac{1}{5}\right) = \frac{4 \cdot 5x}{x}$$ Simplify.

$$15x + 4x = 20$$ Add like terms.

$$19x = 20$$ Divide by 19.

$$x = \frac{20}{19}$$

In Example 7, the denominators contain expressions. To eliminate the denominators, we multiply by the least common denominator.

EXAMPLE **7**   Solving with an LCD   Solve $\dfrac{2}{x + 1} + \dfrac{3}{x} = -2, x \neq -1, 0$ for x.

Symbolic Solution

$$\frac{2}{x + 1} + \frac{3}{x} = -2$$

Multiply by the LCD.

$$x(x + 1)\left(\frac{2}{x + 1} + \frac{3}{x}\right) = -2 \cdot x(x + 1)$$

Distribute the LCD.

$$\frac{x(x + 1) \cdot 2}{x + 1} + \frac{x(x + 1) \cdot 3}{x} = -2x(x + 1)$$

Simplify fractions.

$$2x + 3(x + 1) = -2x^2 - 2x$$

Change to $ax^2 + bx + c = 0$.

$$2x + 3x + 3 + 2x^2 + 2x = 0$$ Combine like terms.

$$2x^2 + 7x + 3 = 0$$ Factor (or use the quadratic formula).

$$(2x + 1)(x + 3) = 0$$ Apply the zero product rule.

Either $2x + 1 = 0$ or $x + 3 = 0$ Solve the factor equations.

$$x = -\tfrac{1}{2} \quad \text{or} \quad x = -3$$

Checking the solution is left as an exercise.

Graphical Solution   We graph the left and right sides separately:

$$y_1 = -2 \quad \text{and} \quad y_2 = \frac{2}{x+1} + \frac{3}{x}$$

The graph of y_1 is a horizontal line. The value of y_2 is undefined at both $x = -1$ and $x = 0$, and the graph becomes nearly vertical near those values. As a result, y_2 has three pieces, with breaks at $x = -1$ and $x = 0$. The solutions are $x = -\frac{1}{2}$ and $x = -3$, the x-coordinates of the intersections of the graphs in Figure 34.

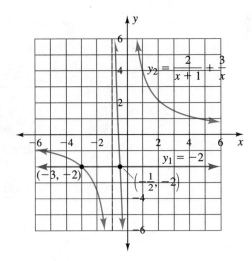

Figure 34   ●

Applications

Example 8 illustrates how a rational equation might be obtained from a problem situation.

EXAMPLE   Setting up a rational equation: movie theater exits   Suppose a movie theater has two exits. One is a single-door emergency exit that, by itself, will permit the theater to empty in 10 minutes. The other is a double-wide door. If building codes require that it be possible to empty the theater in 3 minutes, write an equation for finding the exit time through the double-wide door when it and the emergency exit are both in use.

Solution   Changing exit time to a rate for emptying the theater, we say that during each minute the emergency exit and the double-wide doors permit 1/10 and 1/x of the theater to empty, respectively. The following expressions show the portion of the theater emptied after each minute:

First minute: $1\left(\dfrac{1}{10} + \dfrac{1}{x}\right)$

Second minute: $2\left(\dfrac{1}{10} + \dfrac{1}{x}\right)$

Third minute: $3\left(\dfrac{1}{10} + \dfrac{1}{x}\right)$

To satisfy building codes, the theater must be empty when the expression for the third minute equals 1:

$$3\left(\frac{1}{10} + \frac{1}{x}\right) = 1$$

We divide both sides by 3 to obtain the traditional form of this equation:

$$\frac{1}{10} + \frac{1}{x} = \frac{1}{3}, \quad x \neq 0$$

●

EXAMPLE **9**   Solving with an LCD   Solve $\frac{1}{10} + \frac{1}{x} = \frac{1}{3}, x \neq 0$ for x.

Solution   The least common denominator for the equation is $30x$.

$$\frac{1}{10} + \frac{1}{x} = \frac{1}{3}$$ Multiply by the LCD.

$$30x\left(\frac{1}{10} + \frac{1}{x}\right) = 30x\left(\frac{1}{3}\right)$$ Simplify.

$$3x + 30 = 10x$$ Subtract 3x from both sides.

$$30 = 7x$$ Divide by 7.

$$x = \frac{30}{7} \approx 4.3 \text{ min}$$

Check: Checking with a calculator, we have

$$\frac{1}{10} + \frac{1}{4.3} \approx 0.1 + 0.23 = 0.33 \approx \frac{1}{3} \quad \checkmark$$

If the double-wide door by itself can empty the theater in approximately 4.3 minutes, it can do so together with the emergency exit in 3 minutes.   ●

Think about it: Besides the fact that the equation would be undefined, what would $x = 0$ minutes mean in the problem setting for Example 9?

The equation in Example 9 is a common formula,

$$\frac{1}{a} + \frac{1}{b} = \frac{1}{c}$$

where a = time for one activity, b = time for second activity, c = time for both activities done together. In Example 9, $a = 10$, $c = 3$, and b is unknown. The formula may describe times for filling containers. If a and b are the times needed to fill a child's wading pool with two separate hoses, then c is the time needed to fill it with the two hoses together.

The same formula occurs in applications that seem to have nothing to do with emptying or filling. The formula appears in optics and at least twice in basic electronics. The formula to determine the focal length in lenses is

$$\frac{1}{p} + \frac{1}{q} = \frac{1}{f}$$

The formula for calculating resistance for parallel resistors is

$$\frac{1}{R_1} + \frac{1}{R_2} = \frac{1}{R}$$

and capacitance of condensers in series is given by

$$\frac{1}{C_1} + \frac{1}{C_2} = \frac{1}{C}$$

A similar formula may apply to three doors, pieces of harvest equipment, hoses, lenses, resistors, or condensers. Each additional door, harvester, hose, lens face, resistor, or condenser adds another fraction to the left side.

To close this section, we solve a pair, or system, of equations in which one equation contains a ratio and the solution involves a rational equation. A **system of equations** is *a set of two or more equations to be solved for the values of the variables that satisfy all of the equations.*

EXAMPLE **10**   Solving a system of equations: framing materials   The ratio of width to length of a rectangular picture frame is 5 to 8 (see Figure 35). The amount of framing material—or the perimeter of the frame—is 130 centimeters. Write equations that permit solving for the length and width. Solve the equations.

Figure 35

Solution   The relationships are described with two variables, w and l. The ratio information allows us to build a proportion, $w/l = 5/8$. The perimeter of the rectangle is $2w + 2l = 130$. We solve the perimeter formula for l in terms of w:

$$2w + 2l = 130 \qquad \text{Subtract } 2w.$$

$$2l = 130 - 2w \qquad \text{Divide by 2.}$$

$$l = 65 - w$$

We substitute for l in the proportion and solve for w:

$$\frac{w}{l} = \frac{5}{8} \qquad \text{Substitute for } l.$$

$$\frac{w}{65 - w} = \frac{5}{8} \qquad \text{Cross multiply.}$$

$$8w = 5(65 - w) \qquad \text{Simplify.}$$

$$8w = 325 - 5w \qquad \text{Add } 5w \text{ to each side.}$$

$$13w = 325 \qquad \text{Divide by 13 on each side.}$$

$$w = 25$$

Then we substitute for w:

$$l = 65 - w = 65 - 25 = 40$$

Check both conditions: The ratio 25:40 simplifies to 5:8; the perimeter is $2(25) + 2(40) = 130$ centimeters. ✔

Graphical Solution   The graphs $l = 8w/5$ and $l = (130 - 2w)/2$ in Figure 36 intersect at (25, 40). Thus, for $w = 25$ and $l = 40$, both equations are satisfied.

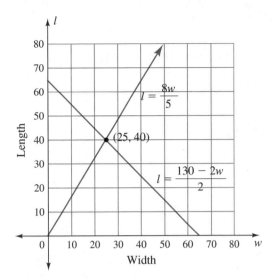

Figure 36

ANSWER BOX

Warm-up: 1. $(x + 2)(x - 1)$ **2.** $(2x + 1)(x + 3)$
3. $(x + 7)(x - 11)$ **4.** $4x + 3$ **5.** $6x - 17$ **6.** $11x$ **7.** $26w$
8. $5x + 3$ **9.** $3 + 2x - x^2$ **10.** 19 **11.** No fractions remain; the factor in front of the parentheses is the least common denominator of the expression and eliminates both denominators. **Think about it:** It would mean that people left the theater instantly, in $t = 0$ time. This is impossible without transporter technology from science fiction.

EXERCISES 5.6

In Exercises 1 to 8, solve for the indicated variable.

1. $\frac{3}{4}x + 5 = 23$ for x

2. $\frac{3}{5}x - 8 = 13$ for x

3. $L = \frac{\pi r \theta}{180}$ for r

4. $A = \frac{(a + b)}{2} \cdot h$ for b

5. $x\left(\frac{1}{10} + \frac{1}{12}\right) = 1$ for x

6. $x\left(\frac{1}{3} + \frac{1}{8}\right) = 1$ for x

7. $\frac{2n + 4}{7} = \frac{3n - 7}{4}$ for n

8. $\frac{4 - 3k}{2} = \frac{3 - 5k}{3}$ for k

9. Why is it not necessary to state any restrictions on the variables in Exercises 5 to 8?

10. In the answer to Exercise 3, we state $\theta \neq 0$. Why do we not also say $\pi \neq 0$?

In Exercises 11 to 15, check the examples by substituting the given number into the equation for the indicated example.

11. $x = 11$ for Example 3

12. $x = -7$ for Example 3

13. $x = \frac{20}{19}$ for Example 5

14. $x = -\frac{1}{2}$ for Example 7

15. $x = -3$ for Example 7

In Exercises 16 to 23, solve with a proportion. Note any restrictions on the variables.

16. $\frac{x + 1}{6} = \frac{x}{9}$

17. $\frac{x + 1}{2} = \frac{x - 3}{1}$

18. $\frac{x - 7}{10} = \frac{-3}{x + 6}$

19. $\frac{x - 8}{5} = \frac{-6}{x + 5}$

20. $\dfrac{x+1}{5} = \dfrac{4}{2x-1}$

21. $\dfrac{x-3}{6} = \dfrac{3}{2x-1}$

22. $\dfrac{x}{4} + \dfrac{x}{6} = 15$

23. $\dfrac{x}{32} + \dfrac{x}{8} = 10$

Multiply the expressions in Exercises 24 to 31, and simplify the results.

24. $12x^2\left(\dfrac{1}{3x^2} + \dfrac{1}{4x}\right)$

25. $6x^2\left(\dfrac{1}{2x^2} - \dfrac{1}{3x}\right)$

26. $9x^2\left(\dfrac{2}{3x^2} + \dfrac{1}{9}\right)$

27. $2x^2\left(\dfrac{1}{2x^2} + \dfrac{3}{x}\right)$

28. $x(x-2)\left(\dfrac{1}{x-2} - \dfrac{3}{x}\right)$

29. $x(x+1)\left(\dfrac{1}{x+1} - \dfrac{2}{x}\right)$

30. $(x-1)(x+2)\left(\dfrac{1}{x-1} + \dfrac{1}{x+2}\right)$

31. $(x+3)(x-1)\left(\dfrac{1}{x+3} + \dfrac{2}{x-1}\right)$

32. Why are there no denominators in the answers in Exercises 24 to 31?

In Exercises 33 to 60, identify any restrictions on the variables. Solve for x using whichever method (proportion, calculator, multiplication by LCD) seems appropriate.

33. $\dfrac{3}{5} + \dfrac{2}{3} = \dfrac{1}{x}$

34. $\dfrac{5}{6} + \dfrac{2}{5} = \dfrac{1}{x}$

35. $\dfrac{1}{3} + \dfrac{1}{x} = \dfrac{1}{2}$

36. $\dfrac{1}{8} + \dfrac{1}{x} = \dfrac{1}{6}$

37. $\dfrac{5}{x} + \dfrac{2}{x} = \dfrac{4}{x}$

38. $\dfrac{1}{x} = \dfrac{1}{2x} + \dfrac{1}{2}$

39. $\dfrac{1}{2} - \dfrac{1}{x} = \dfrac{1}{2x}$

40. $\dfrac{5}{x} - \dfrac{2}{x} = \dfrac{4}{x}$

41. $\dfrac{4}{2x-1} = \dfrac{1}{x-1}$

42. $\dfrac{10}{x+1} = \dfrac{2}{x-1}$

43. $\dfrac{x+1}{2} = \dfrac{15}{x+2}$

44. $\dfrac{1}{x+5} = \dfrac{x+5}{9}$

45. $\dfrac{4}{x-1} = \dfrac{x+2}{10}$

46. $\dfrac{x-2}{3} = \dfrac{6}{x+5}$

47. $\dfrac{18}{x^2} + \dfrac{9}{x} - 2 = 0$

48. $2 + \dfrac{5}{x} - \dfrac{3}{x^2} = 0$

49. $\dfrac{1}{x-2} - 4 = \dfrac{3-x}{x-2}$

50. $\dfrac{4}{x-1} = \dfrac{5-x}{x-1} + 2$

51. $\dfrac{3}{x-1} + \dfrac{5}{2x+2} = \dfrac{3}{2}$

52. $\dfrac{1}{x+1} + \dfrac{1}{x-1} = \dfrac{5}{12}$

53. $\dfrac{x}{6-x} = \dfrac{2}{x-3}$

54. $\dfrac{x}{4} = \dfrac{-1}{x+3}$

55. $\dfrac{1}{x-2} + \dfrac{x-1}{x} = \dfrac{2x+1}{2x}$

56. $\dfrac{1}{x} + \dfrac{x+1}{x+2} = \dfrac{6x-1}{5x}$

57. $\dfrac{x-3}{x} + \dfrac{x-2}{x+3} = \dfrac{6x+1}{8x}$

58. $\dfrac{x-2}{x-1} + \dfrac{x}{x-2} = \dfrac{5x+4}{3(x-1)}$

59. $\dfrac{2}{x-1} + \dfrac{x-2}{x+1} = \dfrac{4x}{3(x+1)}$

60. $\dfrac{1}{x} + \dfrac{x+1}{x+2} = \dfrac{3x+1}{3x}$

61. Solve the equation $\dfrac{7}{x+3} = \dfrac{x-7}{8}$ from Example 3 by multiplying both sides by the least common denominator.

62. Solve the equation $\dfrac{x+1}{x-1} = \dfrac{1}{1-x}$ from Example 4 by multiplying both sides by the least common denominator.

63. Solve $\dfrac{2}{x} + \dfrac{1}{x-1} = 2$ from the graph in the figure.

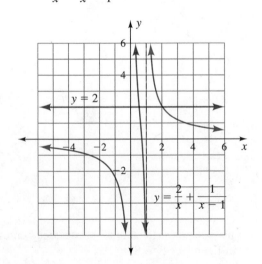

64. Solve $\dfrac{4}{x} + \dfrac{1}{x+3} = -1$ from the graph in the figure.

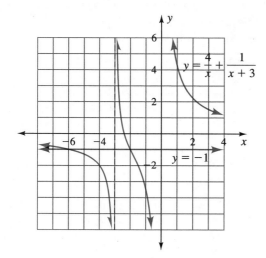

In Exercises 65 to 72, set up equations and solve. Use the formulas from the reading material.

65. One door in a large lecture hall permits a class to exit in 6 minutes. What exit time is needed for a second door in order to drop the exit time to 2 minutes when both doors are available?

66. If $R_1 = 6000$ ohms and $R_2 = 2000$ ohms for parallel resistors in a circuit, find the total resistance R.

67. One hose fills a wading pool in 20 minutes. When both the original and a second hose are used, the filling time drops to 12 minutes. How long would it take the second hose to fill the pool by itself?

68. One ventilation fan changes the air in a house in 7.5 hours. How fast could a second fan vent the house by itself if, working together, the fans meet building code (a complete change of air every 3 hours)?

69. Two resistors are in parallel with $R_1 = 10,000$ ohms and $R_2 = 4000$ ohms. Find the total resistance R.

70. Two capacitors are in series with $C_1 = 0.01$ μF and $C_2 = 0.07$ μF. Find the total capacitance C.

71. Busloads of passengers arrive at a subway station at regular intervals. Ernest runs a toll gate where he can serve a busload in 5 minutes by himself. An exact-change automated gate could serve a busload in 3 minutes. If Ernest and two automated gates were all operating, how many minutes would be needed for each busload of passengers?

72. If a third automated gate were opened in the setting in Exercise 71, how many minutes would be needed for each busload of passengers?

The answers in Exercises 73 to 76 should be rounded to the nearest tenth of a centimeter. The numbers seem weird because they represent relatively accurate dimensions.

73. The ratio of length to width of a 43-gram Hershey® candy bar is 5 to 2. The perimeter is 39.9 centimeters. How long are the sides?

74. The ratio of width to length of a 113-gram Hershey candy bar is 4.6 to 10. The perimeter is 50.6 centimeters. How long are the sides?

75. The ratio of length to width of a 142-gram box of Milk Duds® is 7 to 3. The perimeter is 44.4 centimeters. What are the length and width?

76. The ratio of length to width of a 45.6-gram box of Milk Duds is 8 to 3. The perimeter is 33 centimeters. What are the length and width?

In Exercises 77 to 82, solve for the indicated letter. Assume none of the variables are zero.

77. $\dfrac{1}{a} + \dfrac{1}{b} = \dfrac{1}{c}$ for a **78.** $\dfrac{1}{a} + \dfrac{1}{b} = \dfrac{1}{c}$ for b

79. $\dfrac{1}{a} + \dfrac{1}{b} = \dfrac{1}{c}$ for c **80.** $\dfrac{1}{C} = \dfrac{1}{C_1} + \dfrac{1}{C_2}$ for C_2

81. An electronic formula: $I = \dfrac{E}{R+r}$ for R

82. A refrigeration formula: $\dfrac{W}{Q} = \dfrac{T_2}{T_1} - 1$ for T_1

Exercises 83 to 86 each contain a different incorrect solution to the following problem:

$$\frac{1}{x+2} + \frac{2}{x} = \frac{11}{15}$$

Explain what was done wrong, and correct the work.

83. $15x(x+2)\left(\dfrac{1}{x+2} + \dfrac{2}{x}\right) = \left(\dfrac{11}{15}\right)15x(x+2)$

$$15x + \frac{2}{x} = 11x(x+2)$$

84. $15x(x+2)\left(\dfrac{1}{x+2} + \dfrac{2}{x}\right) = \left(\dfrac{11}{15}\right)15x(x+2)$

$$15x + 2\cdot 15 = 11x(x+2)$$

85. $15x(x+2)\left(\dfrac{1}{x+2} + \dfrac{2}{x}\right) = \left(\dfrac{11}{15}\right)15x(x+2)$

$$\frac{15x(x+2)}{x+2} + \frac{15x(x+2)}{x} = 11x(x+2)$$

86. $15x(x+2)\left(\dfrac{1}{x+2}+\dfrac{2}{x}\right)=\left(\dfrac{11}{15}\right)15x(x+2)$

$\qquad 15x+2\cdot15x+2=11x(x+2)$

Projects

87. ***Scale-Drawing Solutions.*** A scale-drawing solution to problems of the form $\dfrac{1}{x}+\dfrac{1}{y}=\dfrac{1}{z}$ (found in this section and in Section 3.4) has been described by J. E. Thompson in *Arithmetic for the Practical Man* (D. Van Nostrand Company, 1946, p. 237).

As an illustration, we will use this technique to solve the following swimming pool problem: Hazel's hose fills a wading pool in 12 minutes, and Earl's smaller hose fills the wading pool in 18 minutes. If they use both hoses, how many minutes will it take to fill the pool?

On an x-axis, we draw two perpendicular lines to scale, say $AC=12$ centimeters and $BD=18$ centimeters. As shown in the figure (reproduced at a reduced scale), these lines are parallel to the y-axis; they may be any distance AB apart. Each centimeter in the vertical direction represents 1 minute. Next, we draw diagonals AD and BC. We label the point of intersection of AD and BC as E. The number of minutes required for the two hoses to fill the pool together is the vertical distance from the line AB to the point E. Measuring carefully, we determine the length to be 7.2 centimeters; thus, the solution is 7.2 minutes.

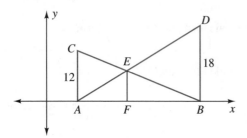

a. Solve Exercise 25 in Section 5.4 using this technique.

b. Solve Exercise 31 in Section 5.4 using this technique.

c. Solve Exercise 65 in this section using a modification of the graphing technique. Remember that Exercise 65 gives one of the times and the total time.

d. Solve Exercise 67 from this section using a modification of the graphing technique.

88. ***Scale-Drawing Proof.*** Prove the scale-drawing solution method in Exercise 87. The proof depends only on similar triangles and substitution.

a. Begin by sketching the figure in Exercise 87 and letting the length of each segment be described by one letter: segment $BD=x$, segment $AC=y$, segment $EF=z$, segment $AB=w$, and segment $BF=v$.

b. Points EFB form a triangle that is similar to triangle CAB. Write a height-to-base proportion for the two triangles. Cross multiply the proportion.

c. Points EFA form a triangle that is similar to triangle DBA. Write a height-to-base proportion for the two triangles. Cross multiply the proportion.

d. Solve the equation in part c for v, and substitute it into the expression in part b.

e. Use a few algebra steps (including common monomial factoring and division by wxz) to obtain $\dfrac{1}{x}+\dfrac{1}{y}=\dfrac{1}{z}$.

89. ***Proportion Proofs.*** If four nonzero numbers a, b, c, and d are in proportion, then several other proportions can be derived from the basic proportion. Prove any four of the five proportion statements below, using accepted equation-solving steps.

a. If $\dfrac{a}{b}=\dfrac{c}{d}$, then $\dfrac{a}{c}=\dfrac{b}{d}$.

b. If $\dfrac{a}{b}=\dfrac{c}{d}$, then $\dfrac{b}{a}=\dfrac{d}{c}$.

c. If $\dfrac{a}{b}=\dfrac{c}{d}$, then $\dfrac{a+b}{b}=\dfrac{c+d}{d}$. (*Hint:* Add 1 to each side.)

d. If $\dfrac{a}{b}=\dfrac{c}{d}$, then $\dfrac{a-b}{b}=\dfrac{c-d}{d}$.

e. If $\dfrac{a}{b}=\dfrac{c}{d}$, then $\dfrac{a+b}{a-b}=\dfrac{c+d}{c-d}$, where $a\neq b$, $c\neq d$. (*Hint:* Use the results in parts c and d.)

CHAPTER **5** SUMMARY

Vocabulary

For definitions and page references, see the Glossary/Index.

additive inverse

asymptote

complex fractions

complex rational
   expressions

constant of proportionality

constant of variation

corresponding angles

corresponding sides

cross multiplication
   property

direct variation

equivalent fraction
   property

equivalent ratios

evaluate

extraneous root

inverse proportion

inverse variation

joint variation

least common denominator

linear variation

proportion

quadrant variation

rate

ratio

rational equations

rational expressions

rational numbers

reciprocal

similar triangles

system of equations

unit analysis

volume of a pyramid

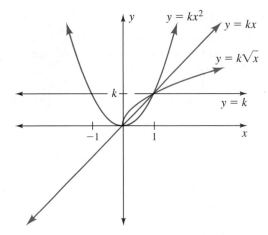

Concepts

5.0 Rational Numbers and Unit Analysis

Division in algebra is usually written as a fraction or rational expression.

Division by zero creates an undefined expression.

To simplify a rational expression, factor the numerator and denominator and eliminate common factors of the form a/a.

Determine whether two ratios are equal by simplifying them to the same ratio, by dividing them to the same decimal, or by cross multiplying to the same number.

To change units with a unit analysis:

1. Identify the units of measure to be changed, and identify the units needed in the answer.

2. List facts that contain the starting units and the ending units. List facts needed to relate the starting and ending units.

3. Write the starting units. Using your list of facts, set up a product of fractions so that each unit of measure appears once in the numerator and once in the denominator.

4. Use $a/a = 1$ to eliminate the unwanted units, and calculate numbers.

To change rates with a unit analysis, follow the steps above. The facts making up the product of fractions may be placed to the left or to the right of the starting rate.

5.1 Ratios, Proportions, Direct Variation

If a ratio contains units of measure of the same type, the units should be made the same before the ratio is simplified or is compared with another ratio.

For data to be proportional, the ratio of the output to the input, y/x, for each data pair must be the same. The graph of a line through proportional data passes through the origin.

Corresponding angles of similar triangles are equal. Corresponding sides of similar triangles are proportional.

5.2 Inverse Variation

The graph of a rational expression that has been simplified to lowest terms becomes nearly vertical whenever the denominator approaches zero.

For data (x_1, y_1) and (x_2, y_2), direct variation has a constant ratio, $x_1/y_1 = x_2/y_2$, and inverse variation has a constant product, $x_1 y_1 = x_2 y_2 = k$.

In Figure 37, the upper graphs illustrate direct variation, $k = \dfrac{y}{x^2} = \dfrac{y}{x} = \dfrac{y}{\sqrt{x}}$. The lower graphs illustrate inverse variation, $k = yx^2 = yx = y\sqrt{x}$.

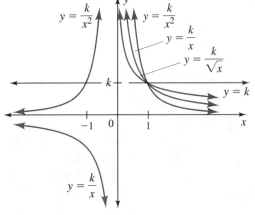

Figure 37

5.3 Simplification, Multiplication, Division

When the numerator and denominator of a fraction are the same, the fraction equals 1.

When the numerator and denominator of a fraction are additive inverses, the fraction equals -1.

To multiply rational expressions, write the product with a single numerator and denominator. Factor as needed. Eliminate common factors from the numerator and denominator. Multiply remaining factors in the numerator and in the denominator.

To divide rational expressions, multiply the first expression by the reciprocal of the second expression.

Simplify complex fractions by writing the fraction as a division.

5.4 Addition and Subtraction

To find the least common denominator, form a product by including each factor the highest number of times it appears in any denominator.

To add or subtract rational numbers, convert to a common denominator, if needed; add or subtract the numerators and place the result over the common denominator; factor and simplify the answer.

Simplify complex fractions by multiplying the numerator and denominator by the least common denominator of all terms.

5.5 Division of Polynomials

To divide algebraic expressions with long division, arrange the dividend into descending order of exponents and replace missing terms with terms having a zero numerical coefficient. Then proceed as follows, repeating the five steps until the degree of the dividend is less than the degree of the divisor:

1. Estimate how many times the divisor divides into the dividend.

2. Enter the estimate in the quotient.

3. Multiply the divisor by the estimate, and place the product below the dividend.

4. Subtract.

5. Bring down the next term in the dividend.

When there is no remainder in a polynomial division, we say the divisor is a factor of the dividend.

A zero remainder indicates that the graph of the expression will contain a hole with an x-coordinate that makes a zero denominator in the original rational expression.

5.6 Solving Rational Equations

When a rational equation is in the form of a proportion,

$\dfrac{a}{b} = \dfrac{c}{d}$, the equation may be cross multiplied to $a \cdot d = b \cdot c$ and then solved.

To eliminate denominators in an equation, multiply each side by the least common denominator.

Table 13 shows the relationships among equations, sequences, and variations.

Linear Equations $y = mx + b$ and $x = c$ Linear Functions $f(x) = mx + b$ Domain: Real numbers	Arithmetic Sequences $a_n = a_1 + (n - 1)d$ Domain: Positive integers Constant first differences	Direct Variation Constant ratio for x and y $\dfrac{y}{x} = k$ or $y = kx$ Linear Variation $y = mx$ Domain: Real numbers
Quadratic Equations $y = ax^2 + bx + c$ Quadratic Functions $f(x) = ax^2 + bx + c$ Domain: Real numbers	Quadratic Sequences Domain: Positive integers Constant second differences	Quadratic Variation $y = kx^2$ Domain: Real numbers
Rational Equations Example: $y = \dfrac{1}{x}$ Rational Functions Example: $f(x) = \dfrac{1}{x}$ Domain: Real numbers, $x \neq 0$ Range: Real numbers, $y \neq 0$		Inverse Variation Constant product for x and y $xy = k$ or $y = \dfrac{k}{x}$ Inverse Square Variation $y = \dfrac{k}{x^2}$

Table 13

CHAPTER ⑤ REVIEW EXERCISES

In Exercises 1 to 4, add, subtract, multiply, and divide the fractions or rational expressions in the order shown.

1. $\frac{2}{3}$ and $\frac{4}{3}$

2. $\frac{3}{4}$ and $\frac{5}{8}$

3. $\dfrac{a}{b}$ and $\dfrac{a}{c}$

4. $\dfrac{a}{c}$ and $\dfrac{c}{b}$

Simplify the expressions in Exercises 5 to 10.

5. a. $\dfrac{27xy^2}{15x^2y}$ b. $\dfrac{2 - x}{x - 2}$ c. $\dfrac{a - ac}{a}$

6. a. $\dfrac{12x^3yz}{39xy^2z^2}$ b. $\dfrac{x - 5}{5 - x}$ c. $\dfrac{ab + b^2}{ab}$

7. a. $\dfrac{3x^2 - 12}{x + 2}$

   b. $\dfrac{(a - b)}{(a + b)(a - b)}$

   c. $8x\left(\dfrac{1}{4x} + \dfrac{1}{2}\right)$

8. a. $\dfrac{21ab + 7b^2}{6a + 2b}$

   b. $\dfrac{(x - 2)(x + 3)}{(x + 3)(x + 2)}$

   c. $x(2 - x)\left(\dfrac{1}{x - 2} + \dfrac{3}{-x}\right)$

9. a. $\dfrac{x^2 - 5x - 6}{x^2 - 4x - 5}$

   b. $\dfrac{4x^2 - 1}{2x^2 + 5x + 2}$

   c. $\dfrac{x^2 + 3x - 4}{x^2 - 16}$

10. a. $\dfrac{x^2 - 7x + 12}{x^2 - 16}$

   b. $\dfrac{2x^2 - 7x + 3}{2x^2 + 7x - 4}$

   c. $\dfrac{3x + 6}{x^2 + 4x + 4}$

In Exercises 11 and 12, what expression is needed in the numerator or denominator to make a true statement?

11. $\dfrac{6 - x}{\rule{2cm}{0.4pt}} = -1$ **12.** $\dfrac{\rule{1.5cm}{0.4pt}}{a - c} = 1$

Simplify the expressions in Exercises 13 and 14.

13. 3 quarts to 4 gallons

14. 5 feet to 5 yards

In Exercises 15 and 16, for what inputs x will the expressions be undefined?

15. $\dfrac{2}{(x - 1)(x + 3)}$ **16.** $\dfrac{-1}{(x + 1)(x + 1)}$

Perform the indicated operations in Exercises 17 to 28. Leave the answers in simplified form.

17. $\dfrac{4y^2}{9x^2} \cdot \dfrac{3x}{8y}$

18. $\dfrac{x + 3}{x^2 - 6x + 9} \div \dfrac{1}{x^2 - 9}$

19. $15x\left(\dfrac{1}{3x} + \dfrac{2}{5x}\right)$

20. $2x(x + 5)\left(\dfrac{3}{2x} - \dfrac{1}{x + 5}\right)$

21. $\dfrac{x^2 + 5x + 4}{x^2 - 16} \cdot \dfrac{2x - 8}{1 - x^2}$

22. $\dfrac{2x^2 + x - 3}{x - 3} \div \dfrac{x^2 - 2x + 1}{3 - x}$

23. $\dfrac{1}{x - 3} + \dfrac{x}{x - 3}$

24. $\dfrac{x}{x - 3} - \dfrac{2}{x + 2}$

25. $\dfrac{x}{x - 1} - \dfrac{1}{1 - x}$

26. $\dfrac{x + 2}{2 - x} - \dfrac{x + 2}{x - 2}$

27. $\dfrac{x - 2}{x^2 - 1} - \dfrac{2}{x - 1}$

28. $\dfrac{1 - x}{x^2 - 4} + \dfrac{2}{x - 2}$

Describe the output behavior of the graph in the figure under the conditions described in Exercises 29 to 32.

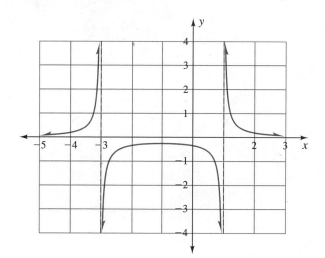

29. As x approaches 1 from the left

30. As x approaches 1 from the right

31. As x approaches -3 from the right

32. As x approaches -3 from the left

33. Change 1,000,000 hours to years. Assume 1 year = 365 days.

34. In 1984, the record for the fastest men's tennis service was 138 miles per hour, a record held by Steve Denton, USA. How long does it take such a service to travel the 78-foot length of the court?

35. a. Suppose a baby grows from 0 to 21 inches in the 9 months before birth. How many miles per hour is this growth rate? Assume 30 days per month.

b. Suppose the growth rate for the 18 years after birth were the same as the growth rate before birth. How tall would the young person be, in inches? in feet?

36. a. A young person grows from 21 inches at birth to 69 inches in 18 years. How many miles per hour is this growth rate? Assume 365 days per year.

b. If the growth rate before birth were the same as that from birth to age 18, how long would it take for the unborn child to reach 21 inches?

In Exercises 37 to 40, solve for the missing sides.

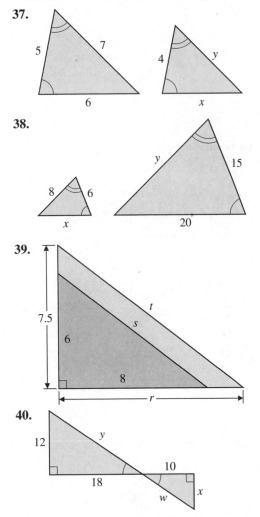

37.

38.

39.

40.

For Exercises 41 to 44, write a linear equation. State whether the relationship is a direct variation (y = kx). Explain the meaning of the y-intercept for those situations where the relationship is not proportional (y = mx + b).

41. A 12-print roll of film costs $9 to develop and print. A 24-print roll of film costs $12. What equation gives the cost in terms of the number of prints?

42. An order of 6 donuts costs $2.99, and another of 12 donuts costs $5.76. What equation gives the cost in terms of the number of donuts?

43. If 40 gallons of aviation fuel costs $74.80 and 65 gallons costs $121.55, what equation describes the cost in terms of the number of gallons?

44. A subscription to *Science News* costs $44.50 for 51 issues or $78 for 102 issues. What equation describes the cost in terms of the number of issues?

Write a formula for each of the sentences in Exercises 45 to 48. Identify your variables, and use k as the constant of proportionality, if needed.

45. The distance you can see from a height varies directly as the square root of the height.

46. The earnings for a week vary jointly as the hours worked and the wage per hour.

47. The volume of a box varies jointly as its length, width, and height.

48. The cost of sending a package varies directly with its weight.

In Exercises 49 to 52, write a formula for each of the sentences, which are based on automobile traffic accident investigation manuals.

49. The skid distance d required to stop a vehicle varies directly as the square of the speed s.

50. In some auto accidents, a car hits a curb, flips into the air, and vaults some distance d through the air to a point of impact with the ground. The speed s of the car at the curb varies directly with the square root of the distance d.

51. The speed s of a car varies directly as the square root of the product of the skid distance d and coefficient of friction factor f.

52. The time required to skid to a stop varies directly as the square root of the length l of the skid mark and inversely as the square root of the coefficient of friction f. Friction is determined by the road surface and weather conditions.

Describe each situation in Exercises 53 to 56 with an inverse variation equation. Identify the constant of variation.

53. An associate's degree from one community college requires a total of 90 credits. If a student takes x credits each quarter, how many quarters does the student need to complete the degree?

54. The output is the number of copies per press required to print 100,000 copies of a book, and the input is the number of printing presses used.

55. The output is the dollars per hour to earn $10,000, and the input is the number of hours worked.

56. A school district allocates $100,000 for textbooks. If the average book price is the input, what equation describes the possible number of books purchased?

57. Mridula starts on a 300-mile trip. Almost immediately, her car is struck from behind, and it takes three hours to get the accident reported and to rent a replacement car.

   a. Explain why her speed, or rate, is

$$r = \frac{300}{t - 3} \text{ hr}$$

   b. In order to investigate the speed she now needs to drive if she is to complete the whole trip in a total of t hours, complete the table for the rate equation.

Total Trip Time (hours)	Rate (miles per hour)
4	
5	
6	
7	
8	60

Rate required to finish trip

   c. The graph in the figure is a portion of a rectangular hyperbola. Explain why the graph is shifted to the right 3 units from the graph for $r = 300/t$.

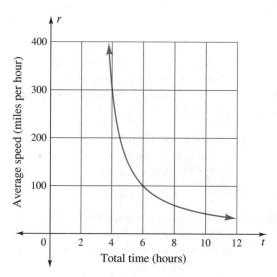

Total time (hours)

58. Referring to part a of Exercise 57, calculate the rate Mridula would need to travel in order to complete her trip in a total of

   a. 10 hours   **b.** 18 hours

   c. 24 hours

59. Referring to part a of Exercise 57, calculate the rate Mridula would need to travel in order to complete her trip in a total of

   a. 3.5 hours   **b.** 3.1 hours

   c. 3.01 hours   **d.** 3 hours

Simplify the complex fractions in the equations or expressions in Exercises 60 to 68.

60. $b = \dfrac{A}{\dfrac{1}{2}h}$ **61.** $h = \dfrac{V}{\dfrac{1}{3}\pi r^2}$

62. $\dfrac{2500 \text{ miles}}{12.5 \text{ miles per gallon}}$

63. $40 \text{ month} \cdot \dfrac{\dfrac{1}{150} \text{ grain}}{150 \text{ month}}$

64. $\dfrac{x + \dfrac{x}{3}}{3 + \dfrac{2}{x}}$ **65.** $\dfrac{x - \dfrac{4}{x}}{1 - \dfrac{2}{x}}$

66. $\dfrac{\dfrac{x^2 - x}{2}}{\dfrac{x^2 - 1}{x}}$

67. Two ventilation fans are operating in an attic. Working alone, one can change the air in 5.5 hours and the other in 7 hours. Together, will they be able to change the air in 3 hours, as required by code?

68. If both the front and the back door are used, a theater can be emptied in 3 minutes. The theater can be emptied using the back door, by itself, in 7 minutes. How rapidly can the theater be emptied using only the front door?

In Exercises 69 to 76, the division problems are stated in fraction form with the denominator containing the divisor. Indicate restrictions on the inputs, and complete the divisions.

69. $\dfrac{x^3 - 3x^2 + 3x - 1}{x}$ **70.** $\dfrac{x^3 + 1}{x}$

71. $\dfrac{x^3 - 3x^2 + 3x - 1}{x - 1}$ **72.** $\dfrac{x^3 + 1}{x + 1}$

73. $\dfrac{x^3 - 3x^2 + 3x - 1}{x - 2}$ **74.** $\dfrac{x^3 + 1}{x + 2}$

75. $\dfrac{x^3 - 3x^2 + 3x - 1}{x - 3}$ **76.** $\dfrac{x^3 + 1}{x + 3}$

77. Graph the expression in Exercise 71. What creates a hole in the graph?

78. Graph the expression in Exercise 73. What creates a nearly vertical line in the graph?

In Exercises 79 to 88, solve for x using whichever method (proportion, calculator, multiplication by LCD) seems appropriate. Exclude inputs that make the expressions undefined.

79. $\dfrac{6}{x} = \dfrac{15}{32}$

80. $\dfrac{x+1}{6} = \dfrac{x-2}{5}$

81. $\dfrac{x}{2} = \dfrac{7}{x+5}$

82. $\dfrac{x-6}{8} = \dfrac{5}{x}$

83. $\dfrac{x-2}{4} = \dfrac{x+1}{3} - 3$

84. $\dfrac{x+2}{4} = \dfrac{x-1}{5}$

85. $\dfrac{1}{x} + \dfrac{5}{x+2} = \dfrac{13}{3x}$

86. $\dfrac{2}{x-4} - \dfrac{1}{x} = \dfrac{11}{3x}$

87. $\dfrac{1}{x-1} + \dfrac{2}{x} = \dfrac{7}{6}$

88. $\dfrac{4}{x} + \dfrac{1}{x-2} = 3$

89. **a.** Fit a linear equation to the coordinate points $(5, 8)$ and $(10, 4)$.

   b. Fit an inverse variation equation to the same points.

   c. Graph the two equations.

   d. How are the graphs the same? How are they different?

CHAPTER ⑤ TEST

Simplify the expressions or perform the indicated operations in Exercises 1 to 18. Exclude any inputs that make the expression undefined.

1. $\dfrac{ab^2c}{a^2bc^2}$

2. $\dfrac{b+3}{b-3}$

3. $\dfrac{3ac}{15ac^2}$

4. $\dfrac{15b^2c^3}{10b^3c}$

5. $\dfrac{x^2-25}{20+x-x^2}$

6. $\dfrac{12-2a}{18+3a-a^2}$

7. $\dfrac{3x^2-7x+2}{2x^2-5x+2}$

8. $\dfrac{2x^2+7x+6}{x^2+4x+4}$

9. $\dfrac{12xy^2}{7y-y^2} \div \dfrac{6x^2}{7-y}$

10. $\dfrac{x^2-9}{x^2+4x+3} \cdot \dfrac{x-3}{x+1}$

11. $\dfrac{x^2+8x+16}{x-2} \cdot \dfrac{x^2-4}{x+4}$

12. $x(x+1)\left(\dfrac{3}{x} + \dfrac{2}{x+1}\right)$

13. $\dfrac{2-x}{x+2} + \dfrac{x+2}{x-2}$

14. $\dfrac{x-2}{x+2} - \dfrac{x}{x^2+2x}$

15. $\dfrac{10000 \text{ cm}^3}{10 \text{ cm}}$

16. $\dfrac{30 \text{ miles per gallon}}{60 \text{ miles per hour}}$

17. $\dfrac{\dfrac{3}{x} - x}{x + \dfrac{2}{x}}$

18. $\dfrac{\dfrac{x^2-3x}{3}}{\dfrac{9-x^2}{x}}$

19. Suppose a baby grows from 0 to 7 pounds in the 9 months before birth. How many ounces per hour is this growth rate? There are 16 ounces in a pound. Assume 30 days per month.

20. A young person grows from 7 pounds to 140 pounds in 18 years. How many ounces per hour is this growth rate? There are 16 ounces in a pound. Assume 365 days per year.

In Exercises 21 and 22, the skid distance in feet for a vehicle varies directly as the square of the speed and inversely as the coefficient of friction f (road surface and weather conditions). Suppose $d = s^2/30f$, where speed is given in miles per hour.

21. What is the skid distance at 45 miles per hour on an asphalt surface with a 0.6 coefficient of friction?

22. A gravel road with a coefficient of friction of 0.4 shows a skidmark of 208 feet. How fast was the car traveling in miles per hour?

23. The volume of a sphere is $V = \frac{4}{3}\pi r^3$. Predict the type of variation in the volume of a sphere. What is the constant of proportionality in the volume of a sphere formula?

24. Triangles *ABC* and *DEF* below are similar. Find the lengths of segments *EF*, *AB*, and *DE*.

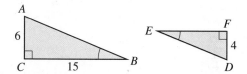

25. Shuhei drives 60 miles per hour for 6 hours. For how many hours must Shinji drive 55 miles per hour to cover the same distance as Shuhei?

    a. What phrase identifies this problem as a direct or an inverse proportion?

    b. Identify the constant of variation.

    c. Write and solve an equation that can be used to find Shinji's travel time.

In Exercises 26 to 30, solve for x using whichever method (proportion, calculator, multiplication by LCD) seems appropriate. Exclude inputs that make the expressions in the equations undefined.

26. $\dfrac{x-2}{10} = \dfrac{2}{x-1}$ **27.** $\dfrac{1}{3} - \dfrac{1}{x} = \dfrac{2x}{3}$

28. $\dfrac{1}{x-4} + \dfrac{1}{x} = \dfrac{10}{3x}$ **29.** $\dfrac{x-1}{9} = \dfrac{x+3}{10}$

30. $\dfrac{3}{x} + \dfrac{2}{x+1} = 4$

In Exercises 31 and 32, the division problems are stated in fraction form with the denominator containing the divisor. Indicate for what values the expressions are defined, and complete the divisions. Graph both expressions; show any holes and nearly vertical portions.

31. $\dfrac{x^3 + 6x^2 + 12x + 8}{x+2}$

32. $\dfrac{x^3 + 6x^2 + 12x + 8}{x+4}$

6

Exponents and Radicals

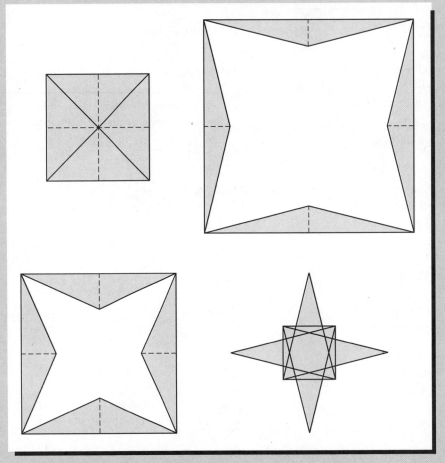

Figure 1

The shapes in Figure 1 are all related by an area formula investigated in a project in this chapter. They represent some surprising results obtained when we apply algebra to a problem setting and disregard the geometric consequences. In this chapter, we examine a variety of exponential and radical expressions.

6.0 Review of Exponents and Scientific Notation

OBJECTIVES

- Simplify expressions containing zero and negative numbers as exponents.
- Find products, quotients, and powers of exponential expressions.
- Change numbers between decimal notation and scientific notation.
- Use a calculator to do operations with scientific notation, and write answers with appropriate significant digits.

WARM-UP

Find the values of the expressions.

1. $2^6, 2^5, 2^4, 2^3, 2^2, 2^1$
2. $3^4, 3^3, 3^2, 3^1$
3. $4^4, 4^3, 4^2, 4^1$
4. $5^4, 5^3, 5^2, 5^1$
5. $10^4, 10^3, 10^2, 10^1$

THIS SECTION REVIEWS exponents and scientific notation.

Zero and Negative Numbers as Exponents

When we apply an exponent to a base, the result is called the power. The number 8 is the power in $2^3 = 8$. We would say that *8 is the 3rd power of 2.*

$$\text{base}^{\text{exponent}} = \text{power}$$

The traditional definition of exponents applies to positive integers used as exponents:

Definition of Positive Integer Exponent

In b^n, the positive integer exponent n indicates the number of times the base b is used as a factor. Thus,

$$b^n = b \cdot b \cdot b \cdot b \cdot \ \cdots \ \cdot b$$

has n factors of b.

This definition is limited to positive integer exponents; it makes no sense if the exponent is zero, negative, or a fraction. We cannot write zero factors of the base, a negative number of factors, or even a fraction of one factor.

Use the following calculator exploration to investigate the possible meanings of $0, -1, -2$, and $1/2$ as exponents.

EXAMPLE **1** Exploring exponents   Use a calculator to investigate the possible meanings of the following expressions. Write answers in both decimal and fraction notation.

$$4^0, (-3)^0, (2.5)^2, 0^0$$
$$2^{-1}, 3^{-1}, 4^{-1}$$
$$2^{-2}, 4^{-2}$$

The exponent key may be $\boxed{\wedge}$, $\boxed{y^x}$, or $\boxed{x^y}$, depending on your calculator.

Solution   The calculator values are listed in Table 1.

Expression	Calculator Value
4^0	1
$(-3)^0$	1
$(2.5)^0$	1
0^0	undefined
2^{-1}	$0.5 = \frac{1}{2}$
3^{-1}	$0.3\overline{3} = \frac{1}{3}$
4^{-1}	$0.25 = \frac{1}{4}$
2^{-2}	$0.25 = \frac{1}{4}$
4^{-2}	$0.0625 = \frac{1}{16}$

Table 1

In spite of the senselessness of these expressions in terms of the positive integer definition of exponents, the calculator readily gives an answer for all table entries except 0^0.   ●

Example 1 suggests some meanings for exponents that are not positive integers. Keep in mind that mathematicians wanted these more unusual exponents to have meanings that would keep the properties for operations consistent with those for positive integer exponents.

Definitions

Zero exponents:

$$b^0 = 1, \quad b \neq 0$$

The expression 0^0 is not defined.

−1 exponents:

$$b^{-1} = \frac{1}{b}, \quad b \neq 0$$

the reciprocal of b. The expression $\frac{1}{0}$ is not defined.

−2 exponents:

$$b^{-2} = \frac{1}{b^2}, \quad b \neq 0$$

the reciprocal of b^2. The expression $\frac{1}{0^2}$ is not defined.

EXAMPLE Simplifying zero exponents   Simplify these exponential expressions.

a. $(a + b)^0$ **b.** $(5x^2y + 2x^{10})^0$

Solution   Recall that 0^0 is undefined, so we must avoid that expression.

a. If a and b do not add to zero, then

$$(a + b)^0 = 1$$

b. If the expression inside the parentheses does not equal zero, then

$$(5x^2y + 2x^{10})^0 = 1$$ ●

EXAMPLE ③ Simplifying with negative exponents   Simplify by using the definitions to rewrite these expressions without negative exponents.

a. 16^{-1} **b.** 9^{-2} **c.** $\left(\dfrac{3}{4}\right)^{-2}$ **d.** $(0.25)^{-1}$

e. 0.5^{-2} **f.** $2a^{-2}, a \neq 0$ **g.** $\dfrac{1}{x^{-2}}, x \neq 0$

Solution   **a.** $16^{-1} = \dfrac{1}{16}$ **b.** $9^{-2} = \dfrac{1}{9^2} = \dfrac{1}{81}$

c. $\left(\dfrac{3}{4}\right)^{-2} = \left(\dfrac{4}{3}\right)^2 = \dfrac{16}{9}$ **d.** $(0.25)^{-1} = \left(\dfrac{1}{4}\right)^{-1} = \dfrac{4}{1}$

e. $0.5^{-2} = \left(\dfrac{1}{2}\right)^{-2} = \left(\dfrac{2}{1}\right)^2 = \dfrac{4}{1}$ **f.** $2a^{-2} = 2 \cdot \dfrac{1}{a^2} = \dfrac{2}{a^2}$

g. $\dfrac{1}{x^{-2}} = \dfrac{1}{\dfrac{1}{x^2}} = 1 \div \dfrac{1}{x^2} = 1 \cdot \dfrac{x^2}{1} = x^2$ ●

The expression x^{-1}, or $1/x$, may be thought of as the reciprocal function of x: $f(x) = 1/x$. Scientific calculators have a function key for reciprocals, labeled $\boxed{x^{-1}}$ or $\boxed{1/x}$. Practice using the reciprocal key by doing parts a to e of Example 3 on a calculator.

We obtain a more general definition of negative integer exponents by combining the reciprocal with the definition of positive integer exponents.

The exponential expression b^{-n} is equivalent to the nth power of the reciprocal of b, $b \neq 0$:

$$b^{-n} = (b^{-1})^n = \left(\dfrac{1}{b}\right)^n$$

For negative exponents on fractions,

$$\left(\dfrac{a}{b}\right)^{-n} = \left(\dfrac{b}{a}\right)^n, \quad a \neq 0, \; b \neq 0$$

In part g of Example 3, the negative exponent was in the denominator. In this case, we have

$$\dfrac{1}{b^{-n}} = \dfrac{1}{\dfrac{1}{b^n}} = 1 \div \dfrac{1}{b^n} = 1 \cdot \dfrac{b^n}{1} = b^n \quad \text{for } b \neq 0$$

B ecause of the nature of the definitions of zero and negative exponents, the properties of powers are still valid.

Product Property of Like Bases	To multiply numbers with like bases, add the exponents: $$b^m \cdot b^n = b^{m+n}$$

Quotient Property of Like Bases	To divide numbers with like bases, subtract the exponents: $$\frac{b^m}{b^n} = b^{m-n}, \ b \neq 0$$

EXAMPLE **4** Simplifying ratios containing exponents The radius of a basketball is about 3.5 times the radius of a baseball. Comparing their volumes with a ratio, we have

$$\frac{\text{Volume of basketball}}{\text{Volume of baseball}} = \frac{\frac{4}{3}\pi(3.5r)^3}{\frac{4}{3}\pi r^3}$$

The volume of the basketball is how many times that of the baseball?

Solution

$$\frac{\text{Volume of basketball}}{\text{Volume of baseball}} = \frac{\frac{4}{3}\pi(3.5r)^3}{\frac{4}{3}\pi r^3} \qquad \text{Line up factors that are alike.}$$

$$= \frac{\frac{4}{3}}{\frac{4}{3}} \cdot \frac{\pi}{\pi}(3.5)^3 \cdot r^{3-3} \qquad \text{Use the quotient property on the variables.}$$

$$= 1 \cdot 1 \cdot (3.5)^3 \cdot r^0 \qquad \text{Use } r^0 = 1.$$

$$\approx 43$$

The volume of the basketball is approximately 43 times that of the baseball. ●

Power Properties of Exponents	To apply an exponent to a number already in exponential form, multiply the exponents: $$(b^m)^n = b^{m \cdot n} \qquad \text{Power property 1}$$ An exponent outside the parentheses applies to all parts of a product or quotient inside the parentheses: $$(a \cdot b)^n = a^n \cdot b^n \qquad \text{Power property 2}$$ $$\left(\frac{a}{b}\right)^n = \frac{a^n}{b^n}, \ b \neq 0 \qquad \text{Power property 3}$$

We use the term *simplify* to indicate that the definitions and properties of exponential expressions are to be used to remove zero or negative exponents, as well as to remove parentheses, to perform operations, and, where possible, to change exponential expressions into numbers or expressions without exponents.

EXAMPLE **5**   Simplifying with properties   Simplify these expressions, replacing expressions with negative exponents with equivalent expressions having positive exponents.

a. $\left(\dfrac{3x}{2}\right)^{-3}$ b. $\left(\dfrac{4x^3}{y^2}\right)^{-2}, y \neq 0$ c. $\dfrac{1}{a^{-3}}, a \neq 0$

d. $\dfrac{b}{a^{-4}}, a \neq 0$ e. $\dfrac{3x^{-2}y}{6x^2y^{-3}}, x \neq 0, y \neq 0$ f. $x^1 \cdot x^n$

g. $2^1 \cdot 2^n$

Solution   a. $\left(\dfrac{3x}{2}\right)^{-3} = \left(\dfrac{2}{3x}\right)^3 = \left(\dfrac{2}{3x}\right)\left(\dfrac{2}{3x}\right)\left(\dfrac{2}{3x}\right) = \dfrac{8}{27x^3}, x \neq 0$

b. $\left(\dfrac{4x^3}{y^2}\right)^{-2} = \left(\dfrac{y^2}{4x^3}\right)^2 = \dfrac{y^{2 \cdot 2}}{4^2x^{3 \cdot 2}} = \dfrac{y^4}{16x^6}, x \neq 0$

c. $\dfrac{1}{a^{-3}} = a^3$

d. $\dfrac{b}{a^{-4}} = b \cdot a^4 = a^4b$

e. $\dfrac{3x^{-2}y}{6x^2y^{-3}} = \dfrac{3 \cdot y \cdot y^3}{2 \cdot 3 \cdot x^2 \cdot x^2} = \dfrac{y^4}{2x^4}, x \neq 0$

f. $x^{1+n} = x^{n+1}$

g. 2^{n+1}

Example 6 reviews powers of 10.

EXAMPLE **6**   Finding powers of 10   Find these powers of 10 without a calculator. Write the expressions in parts d to f as both fractions and decimals.

a. 10^0 b. 10^1 c. 10^2 d. 10^{-1} e. 10^{-2} f. 10^{-3}

Solution   a. $10^0 = 1$ b. $10^1 = 10$

c. $10^2 = 100$ d. $10^{-1} = \frac{1}{10} = 0.1$

e. $10^{-2} = \frac{1}{100} = 0.01$ f. $10^{-3} = \frac{1}{1000} = 0.001$

Observe that the negative powers of 10 give numbers between 0 and 1.   ●

Because powers of 10 are easy to calculate mentally, they became part of scientific notation.

Scientific Notation

Scientific notation is a *short way of writing large and small numbers in which each number is written as a product of a decimal number between 1 and 10 and a power of 10.* Numbers written in scientific notation have two parts, as shown below. The first part is a decimal between 1 and 10; the second part is a power of 10.

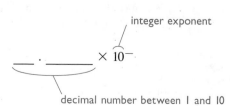

integer exponent

decimal number between 1 and 10

The decimal number is multiplied by the power of 10. The multiplication sign may be written as either a dot or a $\times$.

Because multiplying by 10 only changes the decimal position, writing numbers in scientific notation merely requires moving the decimal point and keeping track of how many places it is moved.

EXAMPLE 7

Changing decimal numbers to scientific notation   Change these numbers into scientific notation. *If a number does not contain a decimal point, you can assume the decimal point is located at the right end of the number.*

a. 2000   **b.** 25,000   **c.** 0.025   **d.** 0.000 025

Solution   **a.** $2000 = 2 \times 10^3$
The decimal point is moved 3 places to create a number between 1 and 10. Because 2000 is larger than 10, the exponent on 10 is positive 3.

b. $25,000 = 2.5 \times 10^4$
The decimal point is moved 4 places and $25,000 > 10$, so the exponent on 10 is positive 4.

c. $0.025 = 2.5 \times 10^{-2}$
The decimal point is moved 2 places. Because 0.025 is between 0 and 1, the exponent on 10 is negative 2.

d. $0.000\,025 = 2.5 \times 10^{-5}$
The decimal point is moved 5 places. Because 0.000 025 is between 0 and 1, the exponent on 10 is negative 5.   ●

There may be some question as to how many zeros to leave on the decimal number. Unless told otherwise, drop all zeros after the last nonzero digit to the right of the decimal point. See also Example 12 on page 414.

EXAMPLE 8

Changing to scientific notation   Change each number to scientific notation.

a. The maximum distance from the sun to Earth is 94,600,000 miles.

b. The minimum distance from the sun to Pluto is 2,756,400,000 miles.

c. The mass of an electron is
0.000 000 000 000 000 000 000 000 000 000 910 9 kilogram.

d. The mass of a bacterium is 0.000 000 000 000 1 kilogram.

Solution   In parts a and b, the distances are numbers larger than 1, so the scientific notation will have a positive exponent on 10.

a. 9.46×10^7 miles   **b.** 2.7564×10^9 miles

In parts c and d, the masses are numbers between 0 and 1, so the scientific notation will have a negative exponent on 10.

c. 9.109×10^{-31} kilogram   **d.** 1×10^{-13} kilogram   ●

> To change scientific notation to regular decimal notation and the reverse, remember that numbers larger than 1 have a positive exponent on the 10 and small numbers between 0 and 1 have a negative exponent on the 10.

EXAMPLE 9

Changing scientific notation to decimal notation   Change these numbers into decimal notation.

a. 3.6×10^2 **b.** 3.6×10^{-3} **c.** 3.6×10^{-1}

Solution   **a.** The exponent on 10 is positive 2. The decimal point is moved 2 places, making a number greater than 3.6:

$$3.6 \times 10^2 = 360$$

b. The exponent on 10 is negative 3. The decimal point is moved 3 places, making a number smaller than 3.6:

$$3.6 \times 10^{-3} = 0.0036$$

c. The exponent on 10 is negative 1. The decimal point is moved 1 place, making a number smaller than 3.6:

$$3.6 \times 10^{-1} = 0.36$$

EXAMPLE

 Changing to decimal notation   Change each number from scientific notation to decimal notation.

a. In 1918, after World War I, the national debt of the United States was 2.6×10^{10} dollars.

b. In 1975, two years after the Vietnam War, the national debt was 5.33×10^{11} dollars.

c. Avogadro's number, 6.022×10^{23} molecules per mole, measures the number of molecules in chemistry.

d. The mass of a neutron is 1.675×10^{-27} kilogram.

Solution   **a.** 26,000,000,000 dollars   **b.** 533,000,000,000 dollars

c. 602,200,000,000,000,000,000,000 molecules per mole

d. 0.000 000 000 000 000 000 000 000 001 675 kilogram

Graphing Calculator Technique:
Scientific and Decimal Notation

Changing to scientific notation: Look for options under [MODE] or [SCI] to change into scientific notation. In this mode, any number entered into the calculator will be changed automatically into scientific notation. The calculator displays for 25,000 and 0.000 025 are in Figure 2(a). The calculator shows E to represent "×10 [∧]."

Changing to decimal notation: When in Normal mode, a graphing calculator will change any number in scientific notation into decimal notation if the decimal number will fit on the display. Otherwise, the number will remain in scientific notation. Use the [EE] key—not [e^x], [10^x], or [∧]—to enter a number in scientific notation. The EE signifies "enter exponent." Graphing calculators permit entering the negative sign before the exponent. A calculator display in Normal mode for 9.46×10^7 and 9.109×10^{-31} is shown in Figure 2(b).

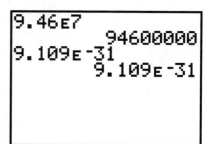

(a)   (b)

Figure 2

Calculators are designed to do operations with scientific notation. It is important to use the $\boxed{\text{EE}}$ key, because using other options may result in errors when division problems are worked.

EXAMPLE

Estimating with scientific notation   Do these problems mentally, and then check the answers with scientific notation on a calculator.

a. $(1.2 \times 10^5)(4.0 \times 10^{-3})$ **b.** $(3.0 \times 10^2)(1.5 \times 10^{-3})$

c. $\dfrac{5.6 \times 10^{-2}}{1.4 \times 10^4}$ **d.** $\dfrac{3.6 \times 10^4}{6.0 \times 10^{-2}}$

Solution
a. $(1.2 \times 10^5)(4.0 \times 10^{-3}) = (1.2)(4) \cdot 10^{5+(-3)} = 4.8 \times 10^2$
We multiply the decimal parts and then add the exponents on the 10.

b. $(3.0 \times 10^2)(1.5 \times 10^{-3}) = (3)(1.5) \cdot 10^{2+-3} = 4.5 \times 10^{-1}$

c. $\dfrac{5.6 \times 10^{-2}}{1.4 \times 10^4} = \dfrac{5.6}{1.4} \cdot 10^{(-2-4)} = 4.0 \times 10^{-6}$

d. $\dfrac{3.6 \times 10^4}{6.0 \times 10^{-2}} = \dfrac{3.6}{6.0} \cdot 10^{4-(-2)} = 0.6 \times 10^6 = 6.0 \times 10^{-1} \times 10^6$
$= 6.0 \times 10^5$ ●

In part d of Example 11, the second step of the solution contains 0.6, a number smaller than 1. Because scientific notation requires a decimal number between 1 and 10, it is necessary to change the decimal point and adjust the exponent on the 10. In this case, the 0.6 is written in scientific notation, and then the resulting exponents on the 10s are combined.

Think about it: What is the calculator answer if you enter the expression $\dfrac{5.6 \times 10^{-2}}{1.4 \times 10^4}$ with these keystrokes: $5.6 \times 10 \boxed{\wedge} \boxed{+/-} 2 \div 1.4 \times 10 \boxed{\wedge} 4$? Why is it incorrect?

As you might expect, scientific notation is common in science. *In many chemistry and physics books, scientific notation is called exponential notation.* In mathematics, however, we use the word *exponential* more generally to describe expressions such as ab^x and the term *scientific notation* to refer specifically to those exponential expressions where a is a number between 1 and 10 and b is 10.

Significant Digits

When writing numbers in scientific notation, we look for significant digits. **Digits** are *the numbers 0 to 9.* **Significant digits** are *all nonzero digits and certain zeros:*

Zeros between nonzero digits (as in 707)

Zeros following a nonzero digit after a decimal point (as in 0.020)

Zeros that are placeholders and marked with an overbar (as in $50\overline{0}0$)

Placeholders are *zeros that are used to properly position the decimal point,* as in 5000 or 0.0025. Placeholders are not significant, except as specified above.

EXAMPLE

Finding significant digits   Identify the significant digits in these numbers, and write the numbers in scientific notation.

a. 2500   **b.** 0.0025   **c.** 2090   **d.** 0.01090   **e.** $25\overline{0}0$

Solution **a.** The zeros after the 5 are placeholders and are omitted; there are two significant digits in 2500.

$$2500 = 2.5 \times 10^3$$

b. The three zeros on the left are placeholders and are omitted; there are two significant digits in 0.0025.

$$0.0025 = 2.5 \times 10^{-3}$$

c. The zero on the right is a placeholder and is omitted; there are three significant digits in 2090.

$$2090 = 2.09 \times 10^3$$

d. The two zeros on the left are placeholders; there are four significant digits in 0.01090. The zero on the right is written only if it is significant, so no overbar is needed.

$$0.01090 = 1.090 \times 10^{-2}$$

e. The overbar above the right zero indicates that it is significant. There are four significant digits.

$$250\overline{0} = 2.500 \times 10^3$$ ●

All significant digits should be shown when numbers are written in scientific notation.

ANSWER BOX

Warm-up: 1. 64, 32, 16, 8, 4, 2 **2.** 81, 27, 9, 3 **3.** 256, 64, 16, 4
4. 625, 125, 25, 5 **5.** 10,000; 1000; 100; 10 **Think about it:** The calculator interprets $1.4 \times 10 \boxed{\wedge} 4$ as the product of two numbers, whereas it interprets $1.4 \boxed{\text{EE}} 4$ as one number. Because the order of operations on a calculator is left to right for multiplication and division, only the 1.4 will remain in the denominator; the 10^4 will be interpreted as being in the numerator.

EXERCISES 6.0

In Exercises 1 to 8, simplify the expressions. Leave no negative or zero exponents.

1. a. $2x^{-1}$ **b.** $\left(\dfrac{y}{x}\right)^{-1}$ **c.** $\left(\dfrac{s}{t}\right)^0$

  d. 0.25^{-1} **e.** $\left(\dfrac{1}{2}\right)^0$

2. a. $3y^{-1}$ **b.** $\left(\dfrac{a}{b}\right)^{-1}$ **c.** $\left(\dfrac{r}{s}\right)^0$

  d. $\left(\dfrac{1}{3}\right)^{-1}$ **e.** 0.5^{-1}

3. a. $\left(\dfrac{m}{n}\right)^{-1}$ **b.** $\left(\dfrac{b}{c}\right)^0$ **c.** $\left(\dfrac{c}{ab}\right)^{-1}$

  d. $\left(\dfrac{3}{4}\right)^0$ **e.** 2.5^{-1}

4. a. $\left(\dfrac{b}{d}\right)^0$ **b.** $\left(\dfrac{y}{x}\right)^{-2}$ **c.** $\left(\dfrac{d}{ef}\right)^{-1}$

  d. 0.05^0 **e.** 0.05^{-1}

5. a. $(2y)^{-2}$ **b.** $2x^{-4}$ **c.** $\left(\dfrac{2x}{y}\right)^{-2}$

  d. $\left(\dfrac{a}{2c}\right)^{-3}$ **e.** $\dfrac{2x^3y^{-2}}{6x^{-1}y^2}$ **f.** $b^1 \cdot b^n$

6. a. $2a^{-2}$ **b.** $(2b)^{-3}$ **c.** $\left(\dfrac{2c}{b}\right)^{-2}$

  d. $\left(\dfrac{c}{2a}\right)^{-3}$ **e.** $\dfrac{3xy^{-3}}{12x^{-2}y^{-1}}$ **f.** $a^1 \cdot a^x$

7. a. $\left(\dfrac{3a^2}{c}\right)^{-3}$ **b.** $\left(\dfrac{a}{c^2}\right)^{-2}$ **c.** $\dfrac{1}{c^{-3}}$

d. $\dfrac{2}{a^{-2}}$ **e.** $\dfrac{12a^3 b^{-2}}{4a^{-1}b^{-5}}$ **f.** $3 \cdot 3^x$

8. a. $\left(\dfrac{2x^2}{y}\right)^{-3}$ **b.** $\left(\dfrac{2b}{c^2 d}\right)^{-2}$ **c.** $\dfrac{1}{a^{-2}}$

d. $\dfrac{3}{x^{-4}}$ **e.** $\dfrac{4a^{-2}b^{-3}}{20ab^{-2}}$ **f.** $5 \cdot 5^x$

9. Explain why $3x^{-2}$ is not $\dfrac{1}{(3x)^2}$.

10. Explain why $\dfrac{2}{x^{-1}}$ is not $\dfrac{x}{2}$.

11. Explain why $5\left(\frac{5}{6}\right)^3$ does or does not equal $6\left(\frac{5}{6}\right)^4$. Show your steps carefully.

12. Sort the following expressions into two sets A and B, where the expressions in set A have variables as bases and the expressions in set B have variables as exponents:

$x^2,\ 2^x,\ 3x^2,\ \frac{1}{2}x^3,\ x^4,\ x^{1/3},\ \left(\frac{1}{2}\right)^x,\ \left(\frac{1}{3}\right)^x,\ \frac{1}{4}x^{1/2},\ x^0,\ x^{-2},$ $(-3)^x,\ x^\pi,\ \pi^x$

13. Compare the statements $x^2 = y^2$ and $2^x = 2^y$. Can we conclude that $x = y$ for one or both or neither? Give an example of numbers that make the original equality true but make the conclusion $x = y$ false.

14. Explain in your own words why $(-3)^4 \neq -3^4$.

15. The equations $2^3 + 3^3 = 35$ and $1^3 + 4^3 = 65$ illustrate sums of cubes. The number 1729 is the smallest number expressible as the sum of two cubes in two different ways. What are the pairs of cubes adding to 1729?

16. Graph $y = x^0$. Describe the graph. Trace the graph. What can you conclude? Is there any x-coordinate for which the function is not defined?

17. Rewrite these formulas, equations, or units of measurement so that they contain no negative signs in the exponents.

   a. Theater emptying: $t^{-1} = t_1^{-1} + t_2^{-1}$

   b. Gravitational constant: $g \approx 9.81$ m·sec^{-2}

   c. Compound interest, continuous compounding: $P = Ae^{-rt}$

   d. Density: $d = 1$ g·cm^{-3}

   e. Tire pressure: $p = 31$ lb·in^{-2}

18. Rewrite these formulas, equations, or units of measurement so that they contain negative exponents instead of division.

   a. Jet aircraft take-off sound at 60 meters:
     $I = 1$ W/m^2

   b. Pressure: $p = 20$ kg/cm^2

   c. Resistance in a parallel circuit: $\dfrac{1}{R} = \dfrac{1}{R_1} + \dfrac{1}{R_2}$

   d. Radioactive decay: $A = \dfrac{A_0}{e^{kt}}$

   e. Speed of spine-tailed swift: $s = 106$ mph

The expressions in Exercises 19 to 22 are simplifications found in a calculus course. Rewrite the expressions by eliminating negative exponents, adding or subtracting fractions, and simplifying the expressions. Remember that $x^{-n} = 1/x^n$.

19. $\dfrac{\dfrac{1}{x+h} - \dfrac{1}{x}}{h}$ **20.** $\dfrac{\dfrac{1}{(x+h)^2} - \dfrac{1}{x^2}}{h}$

21. $x^2(-1)(x+1)^{-2} + \dfrac{2x}{x+1}$

22. $\left(\dfrac{1}{x-1}\right)(2x+2) + (x^2+2x)(-1)(x-1)^{-2}$

Change the numbers in Exercises 23 and 24 to powers of 10.

23. a. 0.001   **b.** $\dfrac{1}{100{,}000}$

   c. 10,000   **d.** 1 hundredth

   e. 1 million   **f.** 1 millionth

24. a. 100,000   **b.** $\dfrac{1}{10000}$

   c. 0.00001   **d.** ten thousand

   e. 1 thousandth   **f.** 1 hundred thousandth

In Exercises 25 and 26, write the numbers in scientific notation.

25. a. 3000   **b.** 350

   c. 350,$\overline{0}$00   **d.** 0.003 50

26. a. 30$\overline{0}$ **b.** 3.50

   c. 0.000 035   **d.** 0.000 000 3

27. How many significant digits are there in each part of Exercise 25?

28. How many significant digits are there in each part of Exercise 26?

Change the numbers in Exercises 29 and 30 to decimal notation.

29. a. The speed of light is 2.9979×10^{10} cm/sec.

   b. The mass of a proton is 1.6726×10^{-24} g.

   c. The charge on an electron is -1.6022×10^{-19} C.

   d. The energy of an electron volt is 1.6022×10^{-19} J.

30. The hydrogen ion (H^+) concentration is used to find pH in chemistry.

   a. The H^+ ion concentration of milk is 2.511×10^{-7} M.

b. The H^+ ion concentration of HCl (hydrochloric acid) is 2.0×10^{-1} M.

c. The H^+ ion concentration of sodium bicarbonate (baking soda) is 3.981×10^{-9} M.

d. The H^+ ion concentration of ammonia is 2.512×10^{-12} M.

Estimate the scientific notation expressions in Exercises 31 to 34, and then check with a calculator.

31. $(4.0 \times 10^{-2})(1.5 \times 10^6)$

32. $(5.0 \times 10^2)(2.5 \times 10^{-4})$

33. $\dfrac{2.4 \times 10^6}{0.3 \times 10^{-2}}$

34. $\dfrac{5.0 \times 10^{-2}}{2.5 \times 10^3}$

35. Many scientific calculators show the decimal part of scientific notation, a space, and the exponent (in large or small print). What number is described by each of these non-graphing calculator displays?

a. 1.25 −3

b. 2.06 2

c. 2.13 10

d. 4.23 −15

36. Refer to the national debt information in Example 10.

a. By how many dollars did the national debt increase between 1918 and 1975?

b. How many times, n, did the national debt double between 1918 and 1975? (*Hint:* $2^n(2.6 \times 10^{10}) = 5.3 \times 10^{11}$.)

c. Divide n into the number of years between 1918 and 1975 to find years required for the debt to double. List the 7 doubling years after 1918.

d. Predict the debt in each year in part c. Compare the growth with the facts. In 1980, the debt was $907.7 billion; in 1985, $1823.1 billion; and in 1990, $3233.3 billion.

37. In science fiction space travel, warp speed is related to the speed of light. If c is the speed of light, 186,000 miles per second, warp speed is the cube of the warp number times the speed of light. Write your answers to the following in scientific notation.

a. Use unit analysis to find the speed of light in miles per hour.

b. Calculate warp 2 in miles per hour.

c. Calculate warp 3 in miles per hour.

d. Calculate warp 14.1 in miles per hour. (This was the highest speed ever reached by the starship Enterprise in the original Star Trek television series.)

38. The Hoover Dam reservoir has a capacity of 2.8253×10^7 acre·feet of water. An acre·foot has an area of one acre (43,560 square feet) and a depth of 1 foot.

a. What is the capacity of the reservoir in cubic feet?

b. To what depth, in feet, would this volume of water cover the state of Rhode Island? Rhode Island is 1045 square miles in land area.

39. What is the largest whole number that your calculator will display in decimal notation without converting to scientific notation?

40. What is the smallest number that your calculator will display in decimal notation without converting to scientific notation?

Projects

41. *Counting Principles.* In the field of probability, there are several methods of counting to determine the likelihood of an outcome. One of these methods is the principle of multiplication. If there are a_1 ways to make the first choice, a_2 ways to make the second choice, and so on, with a_n ways to make the last choice, then the total number of different ways is $a_1 \cdot a_2 \cdot \cdots \cdot a_n$.

Several states have license plates that are three numbers followed by three letters. If there are 10 possible digits for each number and 26 possible digits for each letter, we can use the principle of multiplication to find how many different license plates are possible. The number of possibilities for three numbers followed by three letters is

$$10 \cdot 10 \cdot 10 \cdot 26 \cdot 26 \cdot 26 = 10^3 \cdot 26^3$$
$$= 17,576,000$$

(In reality, several combinations of numbers and letters are not actually used.)

Use the principle of multiplication to answer these questions. Where appropriate, summarize your answer with exponents before multiplying it out.

a. An expensive padlock permits setting a 4-digit number as the combination. If the digits are 0 to 9, how many combinations are possible?

b. A voice-mail code is a 6-digit number, but the first digit cannot be 0. How many different codes are possible?

c. An ice cream parlor offers 3 types of cones, 31 flavors of ice cream, and 5 toppings. How many different cones are possible with one scoop of ice cream and one topping?

d. How many 5-letter words are theoretically possible in English?

e. A balanced coin is tossed 10 times. How many different sequences of heads and tails may result?

f. By 1995, U.S. telephone companies had run out of 3-digit area codes with 1 or 0 as the middle digit. The change to a different middle number created problems for many business telephone networks and caused considerable expense. How many 3-digit area codes are possible if the first digit is nonzero and the second digit is a 1 or a 0? How many 3-digit area codes are possible if the first digit is nonzero and no restrictions are made on the other two digits?

g. Suppose that computer file names are limited to 8 spaces. The first space must be a letter, but all remaining spaces may be a number or letter. How many different file names are possible? (The calculator answer may be in scientific notation.)

h. A lottery has 40 numbered balls. Five balls are selected randomly, one at a time, without replacement of the balls. How many different outcomes are possible?

42. *Combination Locks.* In *Surely You're Joking, Mr. Feynman!* (W. W. Norton and Company, New York, 1985), physicist Richard Feynman described his fascination with file cabinet combination locks while he was working at Los Alamos during World War II. Read the section entitled "Safecracker Meets Safecracker" and summarize his strategy.

43. *Ramanujan and Hardy.* Research the work of Srinivasa Ramanujan, and, if possible, describe the interaction between this mathematician from India and his colleague, Godfrey Hardy (1877–1947), related to the problem in Exercise 15.

6.1   Rational Exponents

OBJECTIVES

- Use a calculator to explore powers with integer bases.

- Evaluate exponential expressions with and without a calculator.

- Apply rational exponents in settings involving compound interest and body fat composition.

WARM-UP

1. List the powers of 2 from 2^{-1} to 2^{10}.

2. List the powers of 3 from 3^{-1} to 3^5.

3. List the powers of 4 from 4^{-1} to 4^5.

4. List the powers of 5 from 5^{-1} to 5^4.

5. List the powers of 6 from 6^{-1} to 6^3.

IN THIS SECTION, we expand our concept of exponents to include rational-number exponents and explore their meaning in terms of factors of a number. With a calculator, we apply rational exponents in the contexts of compound interest and percent body fat.

Exponents That Are Not Integers

In Example 1, we explore solving exponential equations $y = b^x$ with a calculator. Reviewing the powers found in the Warm-up may be helpful in making guesses.

EXAMPLE Exploring powers with a calculator   Solve the exponential equations by guess and check. Use a calculator, as needed. If the exponents are not integers, write them in both fraction and decimal notation.

a. $2^n = 256$ b. $4^n = 256$ c. $16^n = 256$ d. $8^n = 256$

e. $32^n = 256$ f. $4^n = 8$ g. $8^n = 4$ h. $8^n = 16$

i. $4^n = 128$ j. $64^n = 8$ k. $8^n = 128$ l. $8^n = 2$

Solution   See the Answer Box.

Exponential Equation

An exponential equation $y = b^x$ has the variable in the exponent and a constant in the base.

Example 1 suggests that exponents in exponential equations can be fractions and decimals as well as integers. We look at this idea again as we graph the exponential equation $y = 4^x$ in Example 2.

EXAMPLE 2   Solving equations from an exponential graph   Use the numbers from Warm-up Exercise 3 to make a table and a graph of $y = 4^x$. Use the graph to find the solutions to these equations.

a. $4^x = 256$ b. $4^x = 8$ c. $4^x = 128$

Solution   The table for $y = 4^x$ appears in Table 2; the graph is in Figure 3. The horizontal axis is labeled *Exponents* to remind us that the inputs are exponents on a constant base, $b = 4$.

x	4^x
-1	$\frac{1}{4}$
0	1
1	4
2	16
3	64
4	256
5	1024

Table 2 $y = 4^x$

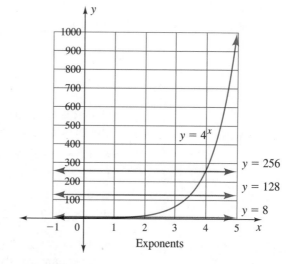

Figure 3

a. The solution to $4^x = 256$ is $x = 4$, the x-coordinate of the intersection of $y = 4^x$ and $y = 256$.

b. The solution to $4^x = 8$ is $x = 1.5$, the x-coordinate of the intersection of $y = 4^x$ and $y = 8$.

c. The solution to $4^x = 128$ is $x = 3.5$, the x-coordinate of the intersection of $y = 4^x$ and $y = 128$.

In the graph of $y = 4^x$ in Figure 3, the input x can take on any value along the x-axis, so the input to an exponential function is the set of all real numbers. This suggests that there are many possible exponents—not only fractions and decimals but also irrational real numbers such as π.

Think about it 1: The number pi, π, is on the interval between what two integers? Use that interval to estimate a low and high value for 4^π. Evaluate 4^π with a calculator.

Rational Exponents

For now, we will focus on exponents that can be written as fractions or repeating decimals—that is, exponents that are rational numbers.

Rational Exponent

> The expression b^x is said to have a rational exponent if the exponent can be written as the quotient of two integers $\dfrac{m}{n}$, where $n \neq 0$.

Our approach to rational exponents will be numerical, involving factors. The next section will introduce a connection between rational exponents and radical functions.

Example 3 shows how factors of numbers can be grouped to suggest rational exponents.

EXAMPLE **3**   Grouping factors   We can arrange the factors of 256 in several ways to obtain different exponential expressions.

$$256 = 2 \cdot 2 \cdot 2 \cdot 2 \cdot 2 \cdot 2 \cdot 2 \cdot 2$$
$$= 2^8$$

The factor 2 appears 8 times, so $2^8 = 256$.

$$256 = (2 \cdot 2) \cdot (2 \cdot 2) \cdot (2 \cdot 2) \cdot (2 \cdot 2)$$
$$= 4^4$$

The 4 sets of the base 4 make 4^4.

$$256 = (2 \cdot 2 \cdot 2) \cdot (2 \cdot 2 \cdot 2) \cdot 2 \cdot 2$$
$$= 8^{2\frac{2}{3}} = 8^{8/3}$$

The $2\frac{2}{3}$ sets of the base 8 make $8^{2\frac{2}{3}}$.

Show the meaning of the exponents in these problems by placing parentheses around sets of 2s equal to the base.

a. $16^2 = 2 \cdot 2 \cdot 2 \cdot 2 \cdot 2 \cdot 2 \cdot 2 \cdot 2 = 256$

b. $32^{1\frac{3}{5}} = 2 \cdot 2 \cdot 2 \cdot 2 \cdot 2 \cdot 2 \cdot 2 \cdot 2 = 256$

c. $64^{1\frac{1}{3}} = 2 \cdot 2 \cdot 2 \cdot 2 \cdot 2 \cdot 2 \cdot 2 \cdot 2 = 256$

d. $128^{1\frac{1}{7}} = 2 \cdot 2 \cdot 2 \cdot 2 \cdot 2 \cdot 2 \cdot 2 \cdot 2 = 256$

Solution   **a.** $16^2 = (2 \cdot 2 \cdot 2 \cdot 2) \cdot (2 \cdot 2 \cdot 2 \cdot 2) = 256$
There are exactly 2 sets of the base 16 in 256.

b. $32^{1\frac{3}{5}} = (2 \cdot 2 \cdot 2 \cdot 2 \cdot 2) \cdot 2 \cdot 2 \cdot 2 = 256$
There are $1\frac{3}{5}$ sets of the base 32 in 256. We may also write $32^{8/5} = 256$.

c. $64^{1\frac{1}{3}} = (2 \cdot 2 \cdot 2 \cdot 2 \cdot 2 \cdot 2) \cdot 2 \cdot 2 = 256$
There are $1\frac{2}{6}$, or $1\frac{1}{3}$, sets of the base 64 in 256. We may also write $64^{4/3} = 256$.

d. $128^{1\frac{1}{7}} = (2 \cdot 2 \cdot 2 \cdot 2 \cdot 2 \cdot 2 \cdot 2) \cdot 2 = 256$
There are $1\frac{1}{7}$ sets of the base 128 in 256. We may also write $128^{8/7} = 256$.

Student Note: After this example, we will tend to use improper fractions, rather than mixed numbers, as exponents.

The Warm-up provided practice in finding powers of common numbers. Knowing the powers of common numbers can help you simplify expressions containing rational exponents. Remember:

$$8^{1/3} = 2 \quad \text{because} \quad 2 \cdot 2 \cdot 2 = 8$$
$$81^{1/4} = 3 \quad \text{because} \quad 3 \cdot 3 \cdot 3 \cdot 3 = 81$$

Keep factors in mind as you think about Examples 4 and 5.

EXAMPLE **4**   Simplifying with rational exponents $1/n$ Simplify these expressions mentally. List the factors of the base, as needed.

a. $27^{1/3}$ **b.** $25^{1/2}$ **c.** $64^{1/3}$ **d.** $64^{1/6}$

      3   5   4   2

Solution   See the Answer Box.   ●

When the exponent contains a number other than 1 in the numerator, we simplify mentally by applying the numerator after the denominator:

$$8^{2/3} = (8^{1/3})^2 = 2^2 = 4$$
$$81^{3/4} = (81^{1/4})^3 = 3^3 = 27$$
$$4^{3/2} = (4^{1/2})^3 = 2^3 = 8$$

EXAMPLE **5**   Simplifying with rational exponents m/n Simplify these expressions.

a. $16^{3/4}$ **b.** $27^{2/3}$ **c.** $81^{5/4}$ **d.** $8^{4/3}$

    8

Solution   See the Answer Box.   ●

Compound Interest

COMPOUND INTEREST, COMPOUNDED ANNUALLY   Compound interest is a common application of exponents on expressions. In the next two examples, we look at interest in wage settings. The examples are included to show the derivation of the formulas for interest compounded annually and compounded twice a year. Then we will work with compound interest using fractional exponents.

EXAMPLE **6**   Finding annual wages   Myrna starts a job at $5.50 per hour. Suppose her hourly wage increases 6% at the end of each year she works.

a. Calculate her wage at the end of each of her first two years.

b. Write a formula for the amount she will earn per hour at the end of n years.

c. Evaluate the formula for $n = 10$.

Solution   The increase is 6% of the wage for each year. Writing 6% as a decimal, we have 0.06.

a. At the end of year 1, Myrna earns

    $5.50 + 0.06(5.50) = \$5.83$ per hr

At the end of year 2, Myrna earns

    $5.83 + 0.06(5.83) \approx \$6.18$ per hr

Student Note: In Examples 6 and 7, we assume that the employer is generous and rounds up to the nearest cent.

b. We identify a pattern in the annual wages by factoring:

    $5.50 + 0.06(5.50) = 5.83$

    $5.50(1 + 0.06) = 5.83$ Year 1

If we substitute the factored expression 5.50(1 + 0.06) into year 2 for 5.83, the expression now contains only the interest rate and the original wage:

$$5.83 + 0.06(5.83) \approx 6.18$$

$$\underline{5.50(1 + 0.06)} + \underline{0.06[5.50(1 + 0.06)]} \approx 6.18$$

The two terms on the left side have been underlined.
We factor [5.50(1 + 0.06)] from each term:

$$[5.50(1 + 0.06)](1 + 0.06) \approx 6.18$$

$$5.50(1 + 0.06)^2 \approx 6.18 \qquad \text{Year 2}$$

The same pattern continues, yielding as the expression for n years

$$5.50(1 + 0.06)^n$$

c. At the end of 10 years, Myrna's wage would be

$$5.50(1 + 0.06)^{10} \approx \$9.85 \text{ per hr} \qquad \bullet$$

The steps in Example 6 show how to calculate annual compound interest.

Annual Compound Interest
(Interest Compounded Once a Year)

The compound interest formula for annual interest is

$$S = P(1 + r)^t$$

where S is the value in the future, P is the value at present, r is the annual rate of interest, and t is the time in years.

Compound Interest, Compounded More Than Once a Year

EXAMPLE **7** Finding annual wages with two raises per year   Suppose Myrna's boss decides to calculate her wage increase so that she receives half the 6% increase every six months.

a. What would her hourly wage be at the end of one year?

b. Write a factored expression in terms of the original wage and percent.

Solution a. At the end of six months, Myrna's wage increases 3%:

$$\$5.50 + 0.03(5.50) \approx 5.67 \text{ per hr}$$

At the end of the second six months, her wage increases another 3%:

$$\$5.67 + 0.03(5.67) \approx 5.84 \text{ per hr}$$

She gains one cent per hour over what she would have gotten if her raise had been an annual 6% raise.

b. The annual rate is 6%, or 0.06. At the end of six months, the expression is

$$5.50 + \frac{0.06}{2}(5.50)$$

At the end of the next six months, it is

$$5.67 + \frac{0.06}{2}(5.67)$$

so we have

$$5.50\left(1 + \frac{0.06}{2}\right) + \frac{0.06}{2}\left[5.50\left(1 + \frac{0.06}{2}\right)\right]$$

As before, we may remove the common factor, $5.50(1 + 0.06/2)$, and obtain a factored expression:

$$\left[5.50\left(1 + \frac{0.06}{2}\right)\right]\left(1 + \frac{0.06}{2}\right) = 5.50\left(1 + \frac{0.06}{2}\right)^2 \qquad \bullet$$

The expression in Example 7 leads us to the more general formula for compound interest:

Compound Interest

If n is the number of times a year interest is calculated (compounded),

$$S = P\left(1 + \frac{r}{n}\right)^{nt}$$

where S is the future value, P is the present value (also known as the principal, or starting amount), r is the annual rate of interest (expressed as a decimal), and t is the number of years.

EXAMPLE 8   Finding annual wages   Find the ending wage earned after 5 years if an employee receives a total 7% annual raise, but the raise is given in two installments over the course of the year. The starting wage is $6.50 per hour.

Solution

$$S = P\left(1 + \frac{r}{n}\right)^{nt}$$

$$= 6.50\left(1 + \frac{0.07}{2}\right)^{(2 \cdot 5)}$$

$$\approx 9.17 \text{ per hr} \qquad \bullet$$

Think about it 2: Why might $9.17 per hour not be the final hourly wage earned?

Compound interest is most often applied to savings accounts, but as Examples 6 to 8 show, it works reasonably well for calculating hourly wages that are increased regularly by a given rate.

EXAMPLE 9   Calculating forgotten savings   Find the amount of money in a forgotten savings account if $1.00 is invested at 7% interest for 200 years. Assume the interest is calculated monthly, $n = 12$.

Solution

$$S = P\left(1 + \frac{r}{n}\right)^{nt}$$

$$= 1.00\left(1 + \frac{0.07}{12}\right)^{(12 \cdot 200)}$$

$$\approx \$1{,}154{,}669.56$$

The answer is rounded down. The dollar has increased to over $1 million. $\bullet$

Think about it 3: When a compound interest expression is entered into a calculator, the exponential factors must be placed in parentheses. Why would the answer to Example 9 change if the calculations were carried out without the parentheses on the factored exponent?

The following calculator box describes a **recursive formula**—*a formula that uses a prior number to calculate the next number.* The answer key, $\boxed{\text{ANS}}$, is used to obtain the answer for use within a calculation.

Graphing Calculator Technique:
Finding Wages with
a Recursive Formula

Use the $\boxed{\text{ANS}}$ key to build a recursive formula for calculating the wages in Example 6, where the starting wage is $5.50 and the interest rate is 6%.

5.50 $\boxed{\text{ENTER}}$

$\boxed{\text{ANS}}$ + 0.06 × $\boxed{\text{ANS}}$ $\boxed{\text{ENTER}}$

Repeat $\boxed{\text{ENTER}}$ to obtain successive results.

COMPOUND INTEREST WITH RATIONAL EXPONENTS   In Examples 6 and 7, we worked with money compounded annually and twice a year. We now combine rational exponents with compound interest by working problems with fractional parts of years, such as $\frac{3}{4}$ year, $1\frac{1}{2}$ years, and $2\frac{2}{3}$ years, and other compounding periods. The compound interest formula is repeated below, for reference:

$$S = P\left(1 + \frac{r}{n}\right)^{nt}$$

EXAMPLE **10**   Finding an account balance with compound interest   Suppose we deposit $1000 at 7% interest.

a. Find the amount of money in the account in $2\frac{1}{2}$ years with weekly compounding.

b. Find the amount of money in the account in $1\frac{2}{3}$ years with weekly compounding.

Solution   Weekly compounding means $n = 52$.

a.
$$S = P\left(1 + \frac{r}{n}\right)^{nt}$$
$$= 1000\left(1 + \frac{0.07}{52}\right)^{52\left(2\frac{1}{2}\right)}$$
$$= 1191.10$$

To evaluate the formula with a calculator, we enter the entire expression into the calculator:

$$1000 \times (1 + 0.07/52)\boxed{\wedge}(52 \times 2.5)$$

This prevents rounding errors and permits us to substitute different numbers into the same expression for subsequent calculations. (Note the use of parentheses around the entire exponents.)

b.
$$S = P\left(1 + \frac{r}{n}\right)^{nt}$$
$$= 1000\left(1 + \frac{0.07}{52}\right)^{52\left(1\frac{2}{3}\right)}$$
$$= 1123.65$$

On a graphing calculator, we use the replay option to retrieve the expression for part a. We move the cursor to the exponent and change the exponent to $(52 \times (1 + 2/3))$. *It is essential that parentheses be placed around the exponential expression in the calculator.*   ●

Caution on Rounding: If you are receiving the interest, then you should drop digits after the hundredth place, because a bank or credit union will never round up and give you extra money. If you are paying the interest, then you should round all numbers up.

Think about it 4: Which way were the answers in Example 10 rounded? Why?

E xample 11 illustrates the use of a calculator to solve a compound interest equation, both by guess and check and with a graph.

EXAMPLE **11** Solving for the exponent on compound interest   To the nearest hundredth of a year, how many years will it take for the $1000 in Example 10 to double to $2000 at 7% interest compounded weekly?

a. Use guess and check with replay.

b. Solve with a graph.

Solution
$$S = P\left(1 + \frac{r}{n}\right)^{nt}$$

$$2000 = 1000\left(1 + \frac{0.07}{52}\right)^{52t}$$

a. We guess and check on a calculator by entering the right side of the equation as an expression and trying different inputs for time. We use x for t. Again the replay feature of the calculator permits us to enter the next guess without retyping the entire expression. At $x = 9.91$ years, the output S will be $2000.17.

b. In solving for t with a graph, we replace t with x and let

$$Y_1 = 2000 \quad \text{and} \quad Y_2 = 1000\left(1 + \frac{0.07}{52}\right)^{52x}$$

Setting a viewing window with x = time in years between 0 and 20 years and y between 0 and 5000, we graph and trace to the point of intersection (Figure 4). We estimate x to be 10 years and zoom in for a better approximation, as needed. The graph confirms the result from guess and check.

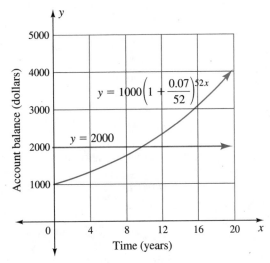

Figure 4

In Example 11, we solved for the exponent . Solving exponential equations for the exponent is a major topic in Chapter 7. For now, we will either guess and check (with or without a calculator) or graph the left and right sides of the equation and solve by tracing and zooming in on the point of intersection.

Body Composition

The closing application introduces a formula with rational exponents in decimal notation.

In his article "Younger than Yesterday" in the May 1990 issue of *Bicycling Magazine,* Steve Johnson describes formulas for estimating the percentage of a person's body composition in fat. The formulas contain expressions with decimal exponents, and they require weight in kilograms (calculated by dividing the weight in pounds by 2.205 pounds per kilogram) and height in centimeters (calculated by multiplying the height in inches by 2.54 centimeters per inch). For men, the percent body fat is

$$\frac{(\text{weight in kg})^{1.2}}{(\text{height in cm})^{3.3}} \cdot 3{,}000{,}000$$

For women, the percent body fat is

$$\frac{(\text{weight in kg})^{1.2}}{(\text{height in cm})^{3.3}} \cdot 4{,}000{,}000$$

The source of the formulas and the units on the number factors are not explained in the article.

EXAMPLE **12**   Evaluating with rational exponents: body fat percentages   Find the percent body fat for these two people:

a. a 5-foot, 8-inch woman weighing 138 pounds

b. a 6-foot, 1-inch man weighing 165 pounds

Solution   **a.** First, we convert to metric units:

$$5 \text{ ft } 8 \text{ in.} = (68 \text{ in.})(2.54 \text{ cm per in.}) = 172.72 \text{ cm}$$

$$138 \text{ lb} = \frac{138 \text{ lb}}{2.205 \text{ lb per kg}} \approx 62.585 \text{ kg}$$

The percent body fat is

$$\frac{(62.585)^{1.2}}{(172.72)^{3.3}} \cdot 4{,}000{,}000 \approx 23.7$$

b. We convert to metric units:

$$6 \text{ ft } 1 \text{ in.} = (73 \text{ in.})(2.54 \text{ cm per in.}) = 185.42 \text{ cm}$$

$$165 \text{ lb} = \frac{165 \text{ lb}}{2.205 \text{ lb per kg}} \approx 74.83 \text{ kg}$$

The percent body fat is

$$\frac{(74.83)^{1.2}}{(185.42)^{3.3}} \cdot 3{,}000{,}000 \approx 17.4$$

●

ANSWER BOX

Warm-up: 1. $\frac{1}{2}$, 1, 2, 4, 8, 16, 32, 64, 128, 256, 512, 1024 **2.** $\frac{1}{3}$, 1, 3, 9, 27, 81, 243 **3.** $\frac{1}{4}$, 1, 4, 16, 64, 256, 1024 **4.** $\frac{1}{5}$, 1, 5, 25, 125, 625 **5.** $\frac{1}{6}$, 1, 6, 36, 216 **Example 1: a.** $n = 8$ **b.** $n = 4$ **c.** $n = 2$ **d.** $n = \frac{8}{3} \approx 2.67$ **e.** $n = \frac{8}{5} = 1.6$ **f.** $n = \frac{3}{2} = 1.5$ **g.** $n = \frac{2}{3} \approx 0.67$ **h.** $n = \frac{4}{3} \approx 1.33$ **i.** $n = \frac{7}{2} = 3.5$ **j.** $n = \frac{1}{2} = 0.5$ **k.** $n = \frac{7}{3} \approx 2.33$ **l.** $n = \frac{1}{3} \approx 0.33$ **Think about it 1:** π is on the interval between 3 and 4, so 4^π is between 4^3 and 4^4, or 64 and 256. $4^\pi \approx 77.880\,233\,65$. **Example 4: a.** 3 **b.** 5 **c.** 4 **d.** 2 **Example 5: a.** 8 **b.** 9 **c.** 243 **d.** 16 **Think about it 2:** If the employer rounded every new wage down to the nearest cent or dollar, the final wage would be lower. Similarly, rounding up might raise the final amount to over $9.17. **Think about it 3:** The calculator would calculate the exponent with only the first number and then multiply the result by the final factor (the time in years). The resulting incorrect formula is $S = P(1 + r/n)^n \cdot t$. **Think about it 4:** Both numbers in Example 10 were rounded down, because they represented an account balance, not a payment.

$27^{\frac{1}{3}}$

EXERCISES 6.1

Use a calculator to guess and check for solutions to the exponential equations in Exercises 1 and 2.

1. a. $27^x = 9$ **b.** $9^x = 27$ **c.** $81^x = 27$

2. a. $25^x = 125$ **b.** $625^x = 125$ **c.** $125^x = 25$

In Exercises 3 and 4, consider this example. When we examine a list of factors, we find different facts, depending on how we choose to group the factors:

$$2^6 = 2 \cdot 2 \cdot 2 \cdot 2 \cdot 2 \cdot 2 = 64$$
$$4^3 = (2 \cdot 2) \cdot (2 \cdot 2) \cdot (2 \cdot 2) = 64$$
$$8^2 = (2 \cdot 2 \cdot 2) \cdot (2 \cdot 2 \cdot 2) = 64$$

Write these expressions with groups of factors to show why they are equal.

3. $81 = 3^4 = 9^2$

4. $625 = 25^2 = 5^4$

5. What exponents make each of the statements below true? Group the 2s in parentheses to show the relationship between the exponent and the base.

a. $4^x = 2 \cdot 2 \cdot 2 \cdot 2 \cdot 2 \cdot 2 \cdot 2 \cdot 2 \cdot 2 \cdot 2 = 1024$

b. $8^x = 2 \cdot 2 \cdot 2 \cdot 2 \cdot 2 \cdot 2 \cdot 2 \cdot 2 \cdot 2 \cdot 2 = 1024$

c. $16^x = 2 \cdot 2 \cdot 2 \cdot 2 \cdot 2 \cdot 2 \cdot 2 \cdot 2 \cdot 2 \cdot 2 = 1024$

d. $32^x = 2 \cdot 2 \cdot 2 \cdot 2 \cdot 2 \cdot 2 \cdot 2 \cdot 2 \cdot 2 \cdot 2 = 1024$

e. $64^x = 2 \cdot 2 \cdot 2 \cdot 2 \cdot 2 \cdot 2 \cdot 2 \cdot 2 \cdot 2 \cdot 2 = 1024$

f. $128^x = 2 \cdot 2 \cdot 2 \cdot 2 \cdot 2 \cdot 2 \cdot 2 \cdot 2 \cdot 2 \cdot 2 = 1024$

g. $256^x = 2 \cdot 2 \cdot 2 \cdot 2 \cdot 2 \cdot 2 \cdot 2 \cdot 2 \cdot 2 \cdot 2 = 1024$

6. What exponent makes each of the statements below true? Group the 3s in parentheses to show the relationship between the base and the exponent.

a. $9^x = 3 \cdot 3 \cdot 3 \cdot 3 \cdot 3 \cdot 3 = 729$

b. $27^x = 3 \cdot 3 \cdot 3 \cdot 3 \cdot 3 \cdot 3 = 729$

c. $81^x = 3 \cdot 3 \cdot 3 \cdot 3 \cdot 3 \cdot 3 = 729$

d. $243^x = 3 \cdot 3 \cdot 3 \cdot 3 \cdot 3 \cdot 3 = 729$

In Exercises 7 to 12, simplify the expressions without a calculator.

7. a. $64^{1/2}$ **b.** $16^{1/4}$ **c.** $32^{1/5}$

8. a. $125^{1/3}$ **b.** $36^{1/2}$ **c.** $64^{1/6}$

9. a. $32^{2/5}$ **b.** $16^{3/2}$ **c.** $36^{3/2}$ **d.** $32^{6/5}$

10. a. $64^{5/6}$ **b.** $64^{2/3}$ **c.** $125^{2/3}$ **d.** $64^{4/3}$

11. a. $16^{5/4}$ **b.** $64^{3/2}$ **c.** $25^{3/2}$ **d.** $81^{3/4}$

12. a. $27^{4/3}$ **b.** $8^{5/3}$ **c.** $81^{3/2}$ **d.** $27^{2/3}$

In Exercises 13 to 16, find numbers that make each statement true.

13. $a^b = b^a$

14. $a^b = a^c$ (b and c are not equal)

15. Suppose $b \neq 0$ and $b \neq 1$. If $b^x = b^y$, then $x = y$.

16. $x^a = y^a$, $x \ne y$.

17. Use Figure 3 (page 419) to describe how to solve these equations from a graph. Estimate the solutions from the graph.

   a. $4^x = 500$

   b. $4^x = 700$

   c. $4^x = 1000$

Simplify the expressions in Exercises 18 and 19.

18. a. $16^{1.5}$ **b.** $36^{1.5}$ **c.** $9^{2.5}$ **d.** $0.01^{0.5}$

19. a. $25^{1.5}$ **b.** $4^{2.5}$ **c.** $16^{2.5}$ **d.** $0.09^{0.5}$

20. Explain how a rational exponent makes it possible to simplify an expression such as $16^{1.5}$ mentally.

21. Common compounding time periods n are given special names. Match each number of compoundings with its appropriate name. Choose from
annual, daily, monthly, quarterly,
semiannual, weekly

   a. once a year

   b. twice a year

   c. 4 times a year

   d. 12 times a year

   e. 52 times a year

   f. 365 times a year

22. The number of compounding periods n per year indicates how often interest is calculated. (Assume 365 days per year.)

   a. If interest were calculated each hour, what n would be used?

   b. If interest were calculated each second, what n would be used?

23. What is the base in the compound interest formula for calculating interest n times a year?

24. What is the base in the compound interest formula for calculating interest once a year?

In Exercises 25 to 28, enter the expressions into a calculator so that the calculator will correctly evaluate them.

25. a. $10,000\left(1 + \dfrac{0.04}{12}\right)^{12 \cdot 3}$

   b. $10,000\left(1 + \dfrac{0.04}{4}\right)^{4 \cdot 3}$

26. a. $10,000\left(1 + \dfrac{0.04}{52}\right)^{52 \cdot 3}$

   b. $10,000\left(1 + \dfrac{0.04}{2}\right)^{2 \cdot 3}$

27. a. $10,000\left(1 + \dfrac{0.08}{12}\right)^{12 \cdot 3}$

   b. $10,000\left(1 + \dfrac{0.08}{4}\right)^{4 \cdot 3}$

28. a. $10,000\left(1 + \dfrac{0.08}{52}\right)^{52 \cdot 3}$

   b. $10,000\left(1 + \dfrac{0.08}{2}\right)^{2 \cdot 3}$

29. Describe a problem setting for each expression given in Exercise 25.

30. Describe a problem setting for each expression given in Exercise 26.

31. Find the final hourly wage if a \$6.00 starting wage is increased 5% each year for 10 years.

32. Find the final hourly wage if a \$6.00 starting wage is increased 2% each year for 10 years.

33. Find the amount of money in an account if \$1000 is deposited at 7% interest compounded quarterly, and the money is left for 5 years.

34. Find the amount of money in an account if \$1000 is deposited at 7% interest compounded monthly, and the money is left for 5 years.

35. Use calculator recursion to extend the calculation started in part a of Example 6 to the third, fourth, and fifth years. Compare your result with that from the formula. Explain any differences in results.

36. Use calculator recursion to extend the calculation started in part a of Example 7 to the second and third years. Compare your result with that from the formula. Explain any differences in results.

In Exercises 37 and 38, suppose that you are working with the compound interest formula and you change from compounding quarterly to compounding monthly.

37. How does the base change?

38. How does the exponent change?

In Exercises 39 to 46, find the amount of money in a savings account for the given time period, rate of interest, and type of compounding. Start with \$1000 in each account.

39. $1\frac{3}{4}$ years at 8%, with monthly compounding

40. $10\frac{1}{2}$ years at 6%, with weekly compounding

41. $2\frac{1}{4}$ years at 8%, with monthly compounding

42. $8\frac{1}{3}$ years at 6%, with weekly compounding

43. $5\frac{1}{3}$ years at 7%, with quarterly compounding

44. 1.5 years at 5%, with monthly compounding

45. 4.75 years at 7%, with quarterly compounding

46. $6\frac{1}{4}$ years at 5%, with monthly compounding

47. Graph the left-hand side of the equation $8000 = 1000(1 + 0.07/52)^{52t}$ as Y_1, and graph the right-hand side as Y_2. Think carefully about an appropriate viewing window. Trace and zoom in as needed to find (to the nearest hundredth) the time t required to raise $1000 to $8000. Give the dimensions of the viewing window (Xmin, Xmax, Ymin, Ymax) where you find your answer.

48. With the formula $1000(1 + 0.07)^t$, use guess and check to find how long it will take for $1000 to exceed $8000. Take your answer to the nearest hundredth of a year.

Exercises 49 to 54 give a range of normal weights for healthy adults. (Data are from the American Medical Association's Encyclopedia of Medicine, *Random House, New York, 1989.) Find the percent body fat for both numbers in the range, using the formulas preceding Example 12. Use the replay feature on your calculator to save time.*

49. Woman: 5 feet 1 inch, 100 to 130 pounds

50. Woman: 5 feet 10 inches, 132 to 165 pounds

51. Man: 6 feet 2 inches, 153 to 193 pounds

52. Woman: 5 feet 5 inches, 112 to 143 pounds

53. Man: 5 feet 4 inches, 118 to 149 pounds

54. Man: 5 feet 7 inches, 128 to 162 pounds

Projects

55. *Practice with Exponential Expressions.* These exponential expressions give practice with a variety of bases and exponents. They are intended to be simplified mentally.

a. $\left(\frac{4}{25}\right)^{1/2}$ **b.** $\left(\frac{1}{8}\right)^{1/3}$ **c.** $\left(\frac{64}{27}\right)^{2/3}$

d. $((64)^{1/2})^{1/3}$ **e.** $\left(\frac{1}{4}\right)^{1/2}$ **f.** $\left(\frac{8}{27}\right)^{1/3}$

g. $\left(\frac{9}{25}\right)^{3/2}$ **h.** $0.09^{3/2}$ **i.** $\left(\frac{1}{32}\right)^{2/5}$

j. $\left(\frac{1}{27}\right)^{2/3}$ **k.** $\left(\frac{1}{64}\right)^{-1/6}$ **l.** $0.04^{3/2}$

m. $\left(\frac{1}{81}\right)^{3/4}$ **n.** $\left(\frac{1}{8}\right)^{5/3}$ **o.** $4^{-1/2}$

p. $32^{-3/5}$ **q.** $8^{-1/3}$ **r.** $\left(\frac{3}{4}\right)^{-2}$

s. $\left(\frac{1}{3}\right)^{-2}$ **t.** $0.008^{2/3}$ **u.** $\left(\frac{9}{25}\right)^{-3/2}$

v. $\left(\frac{27}{8}\right)^{-1/3}$ **w.** $9^{-3/2}$ **x.** $0.027^{2/3}$

56. *Lottery Investment Spreadsheet.* You just won a million-dollar cash lottery. Suppose you deposit the money at 10% simple interest and begin spending at the end of the first year. How many years will the money last if you spend each of the following amounts annually?

a. $500,000

b. $250,000

c. $200,000

d. $150,000

e. $100,000

6.2   Roots and Rational Exponents

OBJECTIVES

- Simplify *n*th root expressions.
- Change expressions from rational exponent to radical notation and vice versa.
- Use properties of powers and radicals to simplify expressions.

WARM-UP

Use the product and quotient properties of square roots to evaluate these square roots. Assume $x \geq 0$, $y \geq 0$.

1. a. $\sqrt{25}$ **b.** $\sqrt{2500}$ **c.** $\sqrt{0.25}$ **d.** $\sqrt{0.0025}$

2. a. $\sqrt{144}$ **b.** $\sqrt{1,440,000}$ **c.** $\sqrt{0.0144}$ **d.** $\sqrt{14,400}$

3. a. $\sqrt{x^2y}$ **b.** $\sqrt{xy^2}$ **c.** $\sqrt{(xy)^2}$ **d.** $\sqrt{x/y^2}$

Use a calculator to evaluate the following, and give your conclusions about 1/2 as an exponent.

4. a. $36^{1/2}$ **b.** $9^{1/2}$ **c.** $256^{1/2}$ **d.** $(-9)^{1/2}$

THE EXPONENT DEFINITIONS in Sections 6.0 and 6.1 are true for different types of functions: power and root functions and exponential functions. Power and root functions have the variable in the base, $y = x^n$. Exponential functions have the variable in the exponent, $y = b^x$. We will study powers and roots for the rest of Chapter 6 and will return to exponential functions in Chapter 7.

In this section, we connect the notation for rational exponents with that for expressions containing roots (radicals). The two different notations are useful in a variety of settings.

Principal Square Roots and nth Roots

Recall the definition of principal square root from Chapter 3:

Principal Square Root

> The **principal square root** of x is the positive real number that, when multiplied by itself, produces x:
>
> $$(\sqrt{x})^2 = x$$
>
> Thus, the principal square root of 25 is 5, not 5 and -5.

RESTRICTIONS   In Section 6.0, we stated that $b^0 = 1$ for $b \neq 0$ because 0^0 is undefined and that $b^{-1} = \dfrac{1}{b}$ for $b \neq 0$ because $\dfrac{1}{0}$ is undefined. Because zero may create undefined (or even false) statements, a first restriction on the material in this section is that we exclude zero as a base; in many general rules, we state that *x is a nonzero real number.*

Chapter 4 introduced the imaginary unit i, the square root of -1. We could include imaginary units here, but to continue to use them with higher roots, we would need trigonometry. Thus, as a second restriction, we will keep the requirement that inputs be positive (and nonzero), for $\sqrt{x}$ *(and all other even roots),* $x > 0$.

n NOTATION AND DEFINITIONS   Another name for the *square root sign,* $\sqrt{}$, is **radical sign.** The small n in the display below represents the index of the radical. The **index of a radical** is *the integer indicating which root is being taken.* For the square root, the index 2 is usually omitted. For other radicals, such as the cube root with index 3, the number n is always shown.

index ⟶ ↘   ↙ ⟶ radical sign

$$n\text{th root} = \sqrt[n]{\text{radicand}}$$

Cube roots, fourth roots, and higher roots have definitions similar to that of square roots. The following table summarizes those definitions.

nth Roots

> If x is a real number and n is a positive integer, then
>
> the **principal square root** of x is the number $\sqrt{x}$ such that
>
> $$(\sqrt{x})^2 = x, \ x \geq 0$$
>
> the **cube root** of x is the number $\sqrt[3]{x}$ such that
>
> $$(\sqrt[3]{x})^3 = x$$

the principal fourth root of x is the number $\sqrt[4]{x}$ such that

$$(\sqrt[4]{x})^4 = x, \ x \geq 0$$

the fifth root of x is the number $\sqrt[5]{x}$ such that

$$(\sqrt[5]{x})^5 = x$$

$\vdots$

the **nth root** of x is the number $\sqrt[n]{x}$ such that

$$(\sqrt[n]{x})^n = x, \ x \geq 0 \text{ if } n \text{ is even}$$

Roots and Negative Numbers

The condition $x \geq 0$ was placed only on the definitions for the even nth roots. Odd roots of negative numbers are defined.

EXAMPLE **1**   Simplifying roots of negative numbers   Simplify, if possible.

a. $\sqrt[3]{-8}$ **b.** $\sqrt[5]{-243}$ **c.** $-\sqrt{16}$ **d.** $\sqrt[4]{-16}$

Solution   **a.** $\sqrt[3]{-8} = -2, (-2)^3 = 8$
b. $\sqrt[5]{-243} = -3, (-3)^5 = -243$
c. $-\sqrt{16} = -4$; this is the opposite of the square root of 16.
d. $\sqrt[4]{-16}$; this is not a real number. There is no real number that can be raised to the fourth power to obtain -16.   ●

Negative Numbers and Roots

> No real number can be raised to any even power to obtain a negative number.
>
> A negative real number raised to any odd power yields a negative number.

Connecting Rational Exponents and Roots

1/2 AS AN EXPONENT   The product property of like bases states that when the bases are the same, we can add the exponents:

$$x^m \cdot x^n = x^{m+n}$$

Example 2 uses the product property to suggest the meaning of 1/2 as an exponent.

EXAMPLE **2**   Exploring the exponent 1/2   Simplify these exponential and radical expressions.

a. $16^{1/2} \cdot 16^{1/2}$ (add exponents)   **b.** $\sqrt{16} \cdot \sqrt{16}$
c. $9^{1/2} \cdot 9^{1/2}$ (add exponents)   **d.** $\sqrt{9} \cdot \sqrt{9}$

Solution   **a.** $16^{1/2} \cdot 16^{1/2} = 16^{(1/2+1/2)} = 16^1 = 16$ **b.** $\sqrt{16} \cdot \sqrt{16} = 4 \cdot 4 = 16$
c. $9^{1/2} \cdot 9^{1/2} = 9^{(1/2+1/2)} = 9^1 = 9$ **d.** $\sqrt{9} \cdot \sqrt{9} = 3 \cdot 3 = 9$ ●

As Example 2 shows, $x^{1/2}$ and $\sqrt{x}$ satisfy the same conditions, so it is natural to define the 1/2 power as meaning the square root.

The Exponent 1/2 and the
Principal Square Root

> A number x raised to the 1/2 power is said to be the principal square root of x.
>
> $$x^{1/2} = x^{0.5} = \sqrt{x}, \; x \geq 0$$

1/n AS AN EXPONENT   It is reasonable to expect the nth root and the $1/n$ exponent to be related in the same way as the square root and the 1/2 exponent.

Rational Exponents 1/n and Radicals

> Let x be a nonzero real number and let n be a positive number. If $\sqrt[n]{x}$ is defined, the rational exponent $1/n$ means
>
> $$x^{1/n} = \sqrt[n]{x}$$
>
> If $1/n$ is a positive rational number, $0^{1/n} = 0$.

EXAMPLE 3   Changing notations   Write each rational exponent expression as a root. Write each root as a rational exponent expression.

a. $x^{1/3}$ b. $\sqrt[6]{x}$ c. $x^{0.25}$ d. $x^{0.2}$

Solution   a. $\sqrt[3]{x}$ b. $x^{1/6}$ c. $x^{0.25} = x^{1/4} = \sqrt[4]{x}$ d. $x^{0.2} = x^{1/5} = \sqrt[5]{x}$ ●

EXAMPLE 4   Changing notation   Change the radicand into exponential notation, and simplify.

a. $\sqrt[3]{27}$ b. $\sqrt[4]{256}$ c. $\sqrt[5]{-243}$ d. $\sqrt[3]{-8}$

Solution   a. $\sqrt[3]{27} = 27^{1/3} = (3^3)^{1/3} = 3^1 = 3$ b. $\sqrt[4]{256} = 256^{1/4} = 4$

c. $\sqrt[5]{-243} = (-243)^{1/5} = -3$ d. $\sqrt[3]{-8} = (-8)^{1/3} = -2$ ●

RATIONAL EXPONENTS m/n   When the exponent contains both a numerator and a denominator, $x^{m/n}$, both a power and a root are implied. The following definition shows the connection.

Rational Exponents m/n and Radicals

> Let x be a nonzero real number. Suppose that m and n are integers with n positive and m/n in lowest terms. Then, if $\sqrt[n]{x}$ is defined in the real numbers, the rational exponent m/n means
>
> $$x^{m/n} = (x^{1/n})^m = (\sqrt[n]{x})^m = \sqrt[n]{x^m}$$
>
> If m/n is a positive rational number, $0^{m/n} = 0$.

EXAMPLE 5   Changing notation   Write each rational exponent expression as a root. Write each root as a rational exponent expression.

a. $x^{2/3}$ b. $\sqrt[5]{x^3}$ c. $\sqrt[3]{x^5}$ d. $x^{0.75}$ e. $x^{0.4}$

Solution   a. $\sqrt[3]{x^2}$ b. $x^{3/5}$ c. $x^{5/3}$

d. $x^{0.75} = x^{3/4} = \sqrt[4]{x^3}$ e. $x^{0.4} = x^{2/5} = \sqrt[5]{x^2}$ ●

In order to use $(\sqrt[n]{b})^m = \sqrt[n]{(b^m)} = b^{m/n}$, the fraction m/n must be in lowest terms and $\sqrt[n]{b}$ must be defined in the real numbers.

EXAMPLE **6**   Changing notation   If possible, write each expression as a base with a rational exponent. Simplify the expression, if possible.

a. $(\sqrt[3]{-5})^3$ b. $(\sqrt[4]{-9})^2$ c. $\sqrt[3]{8^2}$

d. $\sqrt[2]{4^3}$ e. $\sqrt[6]{(-8)^2}$ f. $(\sqrt[6]{-8})^2$

Solution   a. $(\sqrt[3]{-5})^3 = (-5)^{3/3} = (-5)^1 = -5$

b. $(\sqrt[4]{-9})^2$ is not defined in the real numbers.

c. $\sqrt[3]{8^2} = 8^{2/3} = (8^{1/3})^2 = 2^2 = 4$

d. $\sqrt[2]{4^3} = 4^{3/2} = (4^{1/2})^3 = 2^3 = 8$
It is probably easier to observe that $4^3 = 64$ and $\sqrt{64} = 8$.

e. $\sqrt[6]{(-8)^2} = \sqrt[6]{64} = 2$

f. $(\sqrt[6]{-8})^2$ is not defined in the real numbers.   ●

Absolute Value

Recall that placing the absolute value symbol around a variable guarantees a positive output. We extend the concept of the principal square root to square roots of variables with exponents, as follows:

$$\sqrt[n]{x^n} = \begin{cases} x & \text{if } n \text{ is odd} \\ |x| & \text{if } n \text{ is even} \end{cases}$$

EXAMPLE **7**   Simplifying radicals   Simplify, where x and y are any real numbers.

a. $\sqrt{x^4}$ b. $\sqrt{y^2}$ c. $\sqrt{x^8}$ d. $\sqrt[4]{y^4}$

Solution   a. $\sqrt{x^4} = x^2$ b. $\sqrt{y^2} = |y|$ c. $\sqrt{x^8} = x^4$ d. $\sqrt[4]{y^4} = |y|$ ●

In Example 8, we apply the absolute value principle to rational exponents.

EXAMPLE **8**   Simplifying rational exponent expressions   Simplify, where a, b, x, and y are any real numbers.

a. $(a^2)^{\frac{1}{2}}$ b. $(b^5)^{\frac{3}{5}}$ c. $(x^4 \cdot x^2)^{\frac{2}{3}}$ d. $(y^6)^{\frac{3}{2}}$

Solution   We look at the denominator in the exponent to see whether absolute value is needed.

a. $(a^2)^{\frac{1}{2}} = a^{\frac{2}{1} \cdot \frac{1}{2}} = a^{\frac{2}{2}} = |a|$
The denominator is 2. The exponent $\frac{2}{2}$ contains an even root, so the absolute value symbol is needed.

b. $(b^5)^{\frac{3}{5}} = b^{\frac{5}{1} \cdot \frac{3}{5}} = b^3$
The denominator is 5. No absolute value symbol is needed on b, because the root $\left(\text{or } \frac{1}{5} \text{ exponent}\right)$ is odd.

c. $(x^4 \cdot x^2)^{\frac{2}{3}} = (x^6)^{\frac{2}{3}} = x^{\frac{6}{1} \cdot \frac{2}{3}} = x^4$
The denominator is 3. No absolute value symbol is needed on a, because the root $\left(\text{or } \frac{1}{3} \text{ exponent}\right)$ is odd.

d. $(y^6)^{\frac{3}{2}} = y^{\frac{6}{1} \cdot \frac{3}{2}} = |y^9|$
The denominator is 2. The exponent $\frac{3}{2}$ contains an even root, so the absolute value symbol is needed.   ●

Absolute value symbols are not needed if the directions indicate that the variables represent positive numbers or if the root is odd.

Properties of Powers and Radicals

The properties of expressions with rational exponents are given in the following box, with selected equivalent radical notation.

Properties of Powers and Radicals

If m and n are rational numbers, p and q are integers, and each power is defined in the real numbers, the following properties hold.

Product property:

$$b^m \cdot b^n = b^{m+n}$$

Quotient property:

$$\frac{b^m}{b^n} = b^{m-n}, \ b \neq 0$$

Power properties:

$$(b^m)^n = b^{m \cdot n} \qquad \sqrt[p]{\sqrt[q]{b}} = \sqrt[pq]{b}, \ b > 0$$

$$(a \cdot b)^m = a^m \cdot b^m \qquad \sqrt[p]{a \cdot b} = \sqrt[p]{a} \cdot \sqrt[p]{b}, \ a > 0, \ b > 0$$

$$\left(\frac{a}{b}\right)^m = \frac{a^m}{b^m}, \ b \neq 0 \qquad \sqrt[p]{\frac{a}{b}} = \frac{\sqrt[p]{a}}{\sqrt[p]{b}}, \ a > 0, \ b > 0$$

(The properties may be written with x and y instead of a and b.)

EXAMPLE **9**   Applying properties   Simplify, using the properties in the box.

a. $a\sqrt{b} \cdot a\sqrt{b}, \ b > 0$

b. $\sqrt{x}(-\sqrt{x}), \ x > 0$

c. $\sqrt[4]{x} \cdot \sqrt[4]{x^3}, \ x > 0$

d. $\sqrt[3]{b}(\sqrt[3]{b})^2$

Solution   In parts a to c, the variables b and x are positive, so $\sqrt[n]{b}$ and $\sqrt[n]{x}$ are defined in real numbers.

a. $a\sqrt{b} \cdot a\sqrt{b} = a \cdot a \cdot \sqrt{b^2} = a^2 b$

b. $\sqrt{x}(-\sqrt{x}) = -\sqrt{x^2} = -x$

c. Using rational exponents, we have

$$\sqrt[4]{x} \cdot \sqrt[4]{x^3} = x^{1/4} x^{3/4} = x^{1/4+3/4} = x$$

Using radicals yields

$$\sqrt[4]{x} \cdot \sqrt[4]{(x^3)} = \sqrt[4]{(x \cdot x^3)} = \sqrt[4]{x^4} = x$$

No absolute value symbol is needed, because x is positive.

d. Using rational exponents, we have

$$\sqrt[3]{b}(\sqrt[3]{b})^2 = b^{1/3} \cdot b^{2/3} = b^{1/3+2/3} = b$$

Using radicals yields

$$\sqrt[3]{b}(\sqrt[3]{b})^2 = \sqrt[3]{b} \cdot \sqrt[3]{b^2} = \sqrt[3]{(b \cdot b^2)} = \sqrt[3]{b^3} = b$$

No absolute value symbol is needed because of the odd root.   ●

In Example 10, operations are carried out on fractions.

EXAMPLE

Applying properties   Simplify. Assume the variables represent positive numbers.

a. $\sqrt[3]{x^2} \cdot \sqrt[2]{x}$ **b.** $a^{1/2} \cdot a^{1/3}$ **c.** $\dfrac{x^{2/3}}{x^{1/2}}$

d. $\dfrac{\sqrt[5]{x^4}}{\sqrt[3]{x^2}}$ **e.** $\sqrt[2]{\sqrt[3]{b}}$ **f.** $(8 \cdot y)^{2/3}$

Solution   In parts a through d, common denominators are needed in order to add or subtract the exponents.

a. $\sqrt[3]{x^2} \cdot \sqrt[2]{x} = x^{2/3} \cdot x^{1/2} = x^{2/3+1/2} = x^{4/6+3/6} = x^{7/6}$

b. $a^{1/2} \cdot a^{1/3} = a^{1/2+1/3} = a^{3/6+2/6} = a^{5/6}$

c. $\dfrac{x^{2/3}}{x^{1/2}} = x^{2/3-1/2} = x^{4/6-3/6} = x^{1/6}$

d. $\dfrac{\sqrt[5]{x^4}}{\sqrt[3]{x^2}} = \dfrac{x^{4/5}}{x^{2/3}} = x^{4/5-2/3} = x^{12/15-10/15} = x^{2/15}$

e. $\sqrt[2]{\sqrt[3]{b}} = (b^{1/3})^{1/2} = b^{1/6}$

The fractions in the exponents are multiplied, so no common denominator is needed.

f. $(8 \cdot y)^{2/3} = 8^{2/3} \cdot y^{2/3} = (\sqrt[3]{8})^2 \cdot y^{2/3} = 2^2 \cdot y^{2/3} = 4y^{2/3}$ ●

It is generally easier to change radical notation to exponential notation before carrying out the operations. If a root can be easily recognized, as in part f of Example 10, then changing back to a root is useful.

As suggested in the solution to Example 10, there are times when it is easier to work with radicals and not change to rational exponents. Here is a checklist for applying operations and simplifying radical expressions.

Simplified Form for Radical Expressions

> A radical expression is simplified if
>
> **1.** All possible roots have been taken.
> **2.** The exponents in the radicand are smaller than the index.
> **3.** There are no fractions under the radical sign.
> **4.** There are no radicals in the denominator (more on this in Section 6.3).

In Example 11, we simplify radical expressions without changing to rational exponents.

EXAMPLE **11**

Simplifying radicals   Simplify these expressions. Assume all variables represent positive numbers.

a. $\sqrt[4]{2} \cdot \sqrt[4]{8}$ **b.** $\dfrac{\sqrt[4]{32}}{\sqrt[4]{2}}$ **c.** $\sqrt[4]{x} \cdot \sqrt[4]{x^3}$ **d.** $\sqrt[4]{\dfrac{x^6}{x^2}}$

Solution   **a.** $\sqrt[4]{2} \cdot \sqrt[4]{8} = \sqrt[4]{2 \cdot 8} = \sqrt[4]{16} = 2$

b. $\dfrac{\sqrt[4]{32}}{\sqrt[4]{2}} = \sqrt[4]{\dfrac{32}{2}} = \sqrt[4]{16} = 2$

c. $\sqrt[4]{x} \cdot \sqrt[4]{x^3} = \sqrt[4]{x \cdot x^3} = \sqrt[4]{x^4} = x$

d. $\sqrt[4]{\dfrac{x^6}{x^2}} = \sqrt[4]{x^4} = x$ ●

When an exponent inside a radical is larger than the index, factor the expression so as to remove multiples of the factor. For example, if $x \geq 0$,

$$\sqrt{x^3} = \sqrt{x^2 \cdot x} = \sqrt{x^2} \cdot \sqrt{x} = x\sqrt{x}$$

In Example 12, we simplify without changing to rational exponents.

EXAMPLE **12**

Simplifying radicals   Simplify these expressions. Assume all variables represent positive numbers.

a. $\sqrt{50x^7}$ **b.** $\sqrt[4]{a^7}$ **c.** $\sqrt[3]{-27x^4y^5z^6}$ **d.** $\sqrt[4]{a^4b^8c^{10}}$

Solution   **a.** $\sqrt{50x^7} = \sqrt{2 \cdot 25 \cdot x^6 \cdot x} = 5x^3\sqrt{2x}$

b. $\sqrt[4]{a^7} = \sqrt[4]{a^4 \cdot a^3} = a\sqrt[4]{a^3}$

c. $\sqrt[3]{-27x^4y^5z^6} = \sqrt[3]{(-3)^3 x^3 \cdot x \cdot y^3 \cdot y^2 \cdot z^6} = -3xyz^2\sqrt[3]{xy^2}$

The factors may be spread out within one radical sign, as in parts a, b, and c, or they may be placed in separate radical signs, as in part d.

d. $\sqrt[4]{a^4b^8c^{10}} = \sqrt[4]{a^4}\sqrt[4]{b^8}\sqrt[4]{c^8}\sqrt[4]{c^2} = ab^2c^2\sqrt[4]{c}$

The fourth root of c^2 simplifies to the square root of c.   ●

Think about it: Although some older graphing calculators do not have an nth root key, the equivalence between $b^{1/n}$ and $\sqrt[n]{b}$, when $\sqrt[n]{b}$ is defined, makes it possible to evaluate these roots on such a calculator. Simplify each expression twice with a calculator. The first time, use the nth root key $\boxed{\sqrt[n]{x}}$; the second time, use the exponent key $\boxed{\wedge}$ or $\boxed{y^x}$ with a fraction exponent. Describe the result. Indicate any expressions that are not real; explain why.

a. $\sqrt[2]{9}$ **b.** $\sqrt[4]{16}$ **c.** $(\sqrt[3]{-5})^3$ **d.** $(\sqrt[4]{9})^4$ **e.** $(\sqrt{-4})^2$

ANSWER BOX

Warm-up: 1. a. 5 **b.** 50 **c.** 0.5 **d.** 0.05 **2. a.** 12 **b.** 1200 **c.** 0.12 **d.** 120 **3. a.** $x\sqrt{y}$ **b.** $y\sqrt{x}$ **c.** xy **d.** $1/y \cdot \sqrt{x}$ **4. a.** 6 **b.** 3 **c.** 16 **d.** not a real number ($3i$ in the complex number system) **Think about it: a.** 2 $\boxed{\sqrt[n]{}}$ 9; 9 $\boxed{\wedge}$ (1/2); 3; the principal square root **b.** 4 $\boxed{\sqrt[n]{}}$ 16; 16 $\boxed{\wedge}$ (1/4); 2; the principal fourth root **c.** (3 $\boxed{\sqrt[n]{}}$ (−5)) $\boxed{\wedge}$ 3); (−5) $\boxed{\wedge}$ (3/3); −5; odd roots of negative numbers are real numbers. **d.** (4 $\boxed{\sqrt[n]{}}$ 9) $\boxed{\wedge}$ 4; 9 $\boxed{\wedge}$ (4/4); 9; the fourth power of the principal fourth root **e.** not a real number; $\sqrt{-4}$ is undefined in the real numbers.

EXERCISES **6.2**

In Exercises 1 and 2, evaluate the square roots and explain patterns within the answers. Use a calculator only as needed.

1. a. $\sqrt{0.4}$ **b.** $\sqrt{4}$

   c. $\sqrt{40}$ **d.** $\sqrt{400}$

   e. $\sqrt{4000}$

2. a. $\sqrt{0.09}$ **b.** $\sqrt{0.9}$

   c. $\sqrt{9}$ **d.** $\sqrt{90}$

   e. $\sqrt{900}$

Use the patterns from Exercises 1 and 2 to evaluate the square roots in Exercises 3 and 4.

3. a. $\sqrt{400,000}$ **b.** $\sqrt{4,000,000}$

   c. $\sqrt{0.04}$ **d.** $\sqrt{0.000004}$

4. a. $\sqrt{9,000,000}$

   b. $\sqrt{90,000}$

   c. $\sqrt{0.00009}$

   d. $\sqrt{0.0009}$

Simplify the expressions in Exercises 5 to 10 using the product and quotient properties for square roots.

5. $\sqrt{100n}$ **6.** $\sqrt{1000n}$

7. $\sqrt{0.01n}$ **8.** $\sqrt{0.0001n}$

9. $\sqrt{\dfrac{n}{100}}$ **10.** $\sqrt{\dfrac{n}{10000}}$

Evaluate, if possible, the expressions in Exercises 11 to 18.

11. a. $\sqrt[3]{64}$ **b.** $\sqrt[5]{32}$

12. a. $\sqrt[3]{-125}$ **b.** $\sqrt[7]{-128}$

13. a. $\sqrt[3]{-27}$ **b.** $\sqrt[4]{625}$

14. a. $\sqrt[5]{-243}$ **b.** $\sqrt[4]{81}$

15. a. $\sqrt{-4}$ **b.** $\sqrt[4]{-16}$

16. a. $\sqrt[3]{-8}$ **b.** $-\sqrt[3]{1000}$

17. a. $-\sqrt[4]{10000}$ **b.** $\sqrt[3]{-1000}$

18. a. $\sqrt[6]{-64}$ **b.** $\sqrt{-25}$

In Exercises 19 to 22, write each rational expression as a root. Write each root as an expression with a rational exponent.

19. a. $x^{1/6}$ **b.** $\sqrt[3]{x}$ **c.** $x^{0.5}$

20. a. $x^{0.125}$ **b.** $x^{1/7}$ **c.** $\sqrt[5]{x}$

21. a. $x^{3/2}$ **b.** $\sqrt[4]{x^3}$ **c.** $x^{0.8}$

22. a. $\sqrt[4]{x^5}$ **b.** $x^{1.5}$ **c.** $x^{5/6}$

In Exercises 23 to 26, change the rational exponents and simplify.

23. a. $(\sqrt[3]{-8})^5$ **b.** $\sqrt[3]{27^2}$

24. a. $\sqrt[3]{125^2}$ **b.** $(\sqrt[3]{64})^2$

25. a. $\sqrt[4]{16^2}$ **b.** $(\sqrt[4]{-64})^2$

26. a. $(\sqrt[3]{-27})^2$ **b.** $(\sqrt[4]{-16})^2$

Simplify in Exercises 27 and 28. Assume the variables represent any real number.

27. a. $\sqrt{y^2}$ **b.** $\sqrt{z^8}$ **c.** $\sqrt{x^6}$

28. a. $\sqrt{y^4}$ **b.** $\sqrt{z^6}$ **c.** $\sqrt{x^2}$

Simplify in Exercises 29 and 30. Assume the variables represent any real number. State any restrictions.

29. a. $(x^4)^{1/2}$ **b.** $(x^{2/3})^3$ **c.** $(y^6)^{1/6}$

30. a. $(y^6)^{1/3}$ **b.** $(y^{3/4})^4$ **c.** $(x^4)^{1/4}$

Simplify in Exercises 31 to 34. State any restrictions.

31. a. $\sqrt{2}\sqrt{18}$ **b.** $-\sqrt{3}\sqrt{27}$

c. $\sqrt{2}\sqrt{8}$ **d.** $\sqrt[2]{\sqrt[3]{64}}$

32. a. $-\sqrt{12}\sqrt{3}$ **b.** $\sqrt{32}\sqrt{2}$

c. $\sqrt{24}\sqrt{6}$ **d.** $\sqrt[3]{\sqrt[2]{64}}$

33. a. $\dfrac{\sqrt[4]{243}}{\sqrt[4]{3}}$ **b.** $\sqrt[2]{\dfrac{27}{3}}$

c. $\sqrt[3]{16}\cdot\sqrt[3]{4}$ **d.** $\dfrac{\sqrt[4]{64}}{\sqrt[4]{4}}$

34. a. $\sqrt[3]{9}\cdot\sqrt[3]{3}$ **b.** $\sqrt[3]{\dfrac{81}{3}}$

c. $\dfrac{\sqrt[4]{1024}}{\sqrt[4]{4}}$ **d.** $\dfrac{\sqrt[5]{64}}{\sqrt[5]{2}}$

Simplify the expressions in Exercises 35 and 36, and indicate any restrictions on the inputs a and b if the outputs are to be real numbers.

35. a. $\sqrt{a}\sqrt{a^3}$

b. $\sqrt[3]{b}\sqrt[3]{b^2}$

36. a. $\sqrt[3]{b^4}\sqrt[3]{b^2}$

b. $\sqrt[2]{a^3}\sqrt[2]{a^3}$

In Exercises 37 to 42, change to exponential notation where appropriate. Simplify the expressions. Assume x > 0, y > 0.

37. a. $x^{3/4}x^{1/3}$ **b.** $x^{3/4}x^{2/3}$

c. $x^{-1/3}$ **d.** $\sqrt[3]{x^2}\sqrt[4]{x}$

38. a. $(x^{\frac{3}{4}})^{\frac{1}{3}}$ **b.** $x^{2/3}x^{1/2}$

c. $x^{-3/4}$ **d.** $\sqrt[4]{x^3}\sqrt[2]{x^3}$

39. a. $\sqrt[3]{x}\sqrt[2]{x^3}$ **b.** $\dfrac{x^{3/4}}{x^{1/3}}$

c. $(x^{\frac{3}{4}})^{\frac{2}{3}}$

40. a. $\sqrt[4]{x}\sqrt[2]{x^5}$ **b.** $\dfrac{x^{2/3}}{x^{3/5}}$

c. $(x^{\frac{2}{5}})^{\frac{3}{2}}$

41. a. $\sqrt[2]{\sqrt[3]{x^2}}$ **b.** $\left(\dfrac{16x^4}{y^8}\right)^{\frac{3}{4}}$

c. $\dfrac{x^{-1/2}}{x^{3/2}}$

42. a. $\sqrt[3]{\sqrt[2]{x}}$ **b.** $\left(\dfrac{25x^5}{xy^4}\right)^{\frac{1}{2}}$

c. $\dfrac{x^{2/3}}{x^{-1/3}}$

Simplify Exercises 43 and 44 with properties of radicals. Assume all variables represent positive numbers.

43. a. $\sqrt[3]{16}$ **b.** $\sqrt{75a^5}$ **c.** $\sqrt[4]{32x^9}$

d. $\sqrt[5]{-32a^4b^5c^6}$ **e.** $\sqrt[3]{x^4y^8z^9}$

44. a. $\sqrt[3]{81}$ **b.** $\sqrt{18x^3}$ **c.** $\sqrt[5]{64b^{12}}$

d. $\sqrt[4]{16x^3y^4z^5}$ **e.** $\sqrt[3]{x^4y^6z^{10}}$

45. *Circular velocity,* V_{circ}, is the velocity a space vehicle needs in order to keep circling around a planet and not be pulled back to the planet by gravity. *Escape velocity,* V_{esc}, is the velocity a space vehicle needs in order to leave circular orbit around a planet and "escape" the gravitational attraction of that planet. Circular velocity is given by

$$V_{circ} = \sqrt{\frac{GM}{R}} \text{ for } G = 6.67 \times 10^{-11} \text{ N·m}^2\text{·kg}^{-2}$$

where G is Newton's gravitational constant, M is the mass of the planet, and R is the distance of the orbiter from the center of the planet. Escape velocity is given by

$$V_{esc} = \sqrt{2} \cdot V_{circ}$$

Make sure the units are correct in your work. N represents the newton, a unit of force: $1 \text{ N} = 1 \text{ kg·m/sec}^2$. Be aware of significant digits in your final answer.

Find the circular velocity of a space vehicle orbiting the planet at the given distance.

a. Mercury: $R = 3.24 \times 10^6$ m, $M = 3.30 \times 10^{23}$ kg

b. Jupiter: $R = 8.14 \times 10^7$ m, $M = 1.90 \times 10^{27}$ kg

Find the escape velocity of a space vehicle orbiting the planet at the given distance.

c. Earth: $R = 7.18 \times 10^6$ m, $M = 5.98 \times 10^{24}$ kg

d. Mars: $R = 4.19 \times 10^6$ m, $M = 6.44 \times 10^{23}$ kg

46. *Terminal velocity* is the maximum falling speed attained with air resistance. The terminal velocity V of a sky diver depends on the position taken during the dive. Calculate V for each of the following three dive positions.* Use

$$V = \sqrt{\frac{32.2 \frac{\text{ft}}{\text{sec}^2}}{A}}$$

a. $A = 5.18 \times 10^{-4}$ ft^{-1} in fetal position.

b. $A = 7.75 \times 10^{-4}$ ft^{-1} in nose-dive position (arms back along body)

c. $A = 10.4 \times 10^{-4}$ ft^{-1} in horizontal position (arms and legs spread outward from body)

47. Does the expression $\sqrt{(-x)}$, $x < 0$ give a real number? Why or why not?

48. A student tries to use the product property $\sqrt{a \cdot b} = \sqrt{a} \cdot \sqrt{b}$ to take the square root of b^2 in $\sqrt{b^2 - 4ac}$. Explain why the student is wrong.

*The data are for 135-pound female sky diver Jennifer Phillips of Orange, Massachusetts. From William Ralph Bennett, Jr., *Scientific and Engineering Problem-Solving with the Computer*, Prentice-Hall, Englewood Cliffs, NJ, 1976.

49. Explain why $5 \cdot 5^n \neq 25^n$.

50. Under what conditions is no absolute value symbol needed on x in $\sqrt{x^2} = x$?

51. Why is no absolute value symbol needed on x^2 in $\sqrt{x^4} = x^2$?

Projects

52. *Product of Radicals Investigation.* The table below gives several expressions to help you investigate the product of radicals with the same radicand.

a. Calculate each expression, and record the values in the table. Indicate in the center column whether the expressions in the row are equal.

Expression 1	Equal (=) or Not Equal ($\neq$)	Expression 2
$\sqrt[5]{5} \cdot \sqrt[2]{5} =$		$\sqrt[5]{5} =$
$\sqrt[3]{5} \cdot \sqrt[2]{5} =$		$\sqrt[6]{5} =$
$\sqrt[3]{2} \cdot \sqrt[3]{2} =$		$\sqrt[6]{2} =$
$\sqrt[3]{2} \cdot \sqrt[3]{2} =$		$\sqrt[9]{2} =$

The examples in the table show that neither of the following equations is true:

$$\sqrt[p]{x} \cdot \sqrt[q]{x} = \sqrt[p+q]{x}$$
$$\sqrt[p]{x} \cdot \sqrt[q]{x} = \sqrt[p \cdot q]{x}$$

b. Use the equivalence between $\sqrt[n]{x}$ and $x^{1/n}$ (if $\sqrt[n]{x}$ is defined) to find a correct formula for $\sqrt[p]{x} \cdot \sqrt[q]{x}$. You can probably see from your results why this formula is not generally stated in a textbook.

c. Evaluate the expressions in the first column using your formula.

d. Verify that your formula also works for $\sqrt{2} \cdot \sqrt{2} = 2$.

53. *The Generalized Distributive Property.* The power properties—such as $(a \cdot b)^n = a^n \cdot b^n$ and its corresponding radical property $\sqrt{ab} = \sqrt{a}\sqrt{b}$—look somewhat like the distributive property of multiplication over addition, $n(a + b) = na + nb$. The power properties do indeed illustrate a generalized distributive property.

Match the descriptions of the distributive property in a to h with one of the following expressions:

$$(b - c) \div a \qquad a(b + c)$$
$$a(b - c)$$
$$(bc)^a \qquad \sqrt[a]{\frac{b}{c}}$$
$$(b + c) \div a$$
$$\sqrt[a]{(bc)} \qquad \left(\frac{b}{c}\right)^a$$

Show the distribution from each expression. Each statement applies only if the operations are defined, so assume all the usual conditions on the variables.

a. Multiplication distributes over addition.

b. Multiplication distributes over subtraction.

c. Division on the right distributes over addition.

d. Division on the right distributes over subtraction.

e. Exponents distribute over multiplication.

f. Exponents distribute over division.

g. Roots distribute over multiplication.

h. Roots distribute over division.

Simplify each expression, if possible. Which illustrate the generalized distributive property?

i. $2(x - 2)$ j. $2(3x)$

k. $(x + y)^2$

l. $\sqrt{\dfrac{x}{y}}, x \geq 0, y > 0$

m. $a \div (b + c)$

n. $a(b + c)$

o. $\frac{1}{2}(3 - x)$

p. $(x \cdot y)^2$

q. $\sqrt{x - y}, x - y \geq 0$

MID–CHAPTER **6** TEST

Write the expressions in Exercises 1 and 2 another way.

1. $x^{1/2}, x \geq 0$

2. $\dfrac{1}{x}, x \neq 0$

Simplify the expressions in Exercises 3 to 5.

3. $\dfrac{a^{-1}b^2}{c^0}$

4. $\dfrac{a^2b^{-1}}{c^{-2}}$

5. $a \cdot a^x$

6. Simplify.

a. $\left(\dfrac{2x}{3}\right)^{-2}$

b. $\dfrac{x}{y^{-2}}$

7. a. Change 4.3×10^{-3} to decimal notation.

b. Change 0.000123 to scientific notation.

c. Simplify $\dfrac{3.6 \times 10^{-2}}{1.2 \times 10^3}$.

d. How many significant digits are in 0.060?

e. Describe two examples of zero as a placeholder.

8. Find the amount of money in a savings account if $1000 is deposited at 7% interest compounded weekly (52 times a year) for

a. 5 years

b. $2\frac{3}{4}$ years

9. Use guess and check to solve each exponential equation for x.

a. $4^x = 64$

b. $3^x = 27$

c. $\left(\frac{1}{2}\right)^x = \frac{1}{8}$

d. $\pi^x = 1$

10. Why is a 2% wage increase twice a year better than a 4% wage increase once a year?

11. Simplify, and indicate any restrictions on the inputs n needed in order to have real-number answers.

a. $\sqrt{4n}$

b. $\sqrt{40n^2}$

c. $\sqrt{400n}$

d. $\sqrt{4000n}$

In Exercises 12 and 13, change each expression to radical notation. Simplify.

12. a. $8^{2/3}$

b. $125^{1/3}$

13. a. $16^{3/4}$

b. $(a^3)^{2/3}$

14. Simplify.

a. $\sqrt[3]{-64}$

b. $\sqrt[4]{16y^4}$

15. Simplify. Assume variables represent only positive numbers.

a. $\sqrt{9x^5y^2}$

b. $\sqrt[3]{24x^5y^6}$

6.3 More Operations with Radicals

OBJECTIVES

- Add and subtract radicals.
- Multiply two- and three-term radical expressions.
- Identify the conjugate of a radical expression.
- Rationalize the numerator or denominator of a radical expression.

WARM-UP

Solve for x using the quadratic formula. Leave the answer in radical notation.

1. $x^2 + 4x + 2 = 0$ **2.** $x^2 - 2x - 2 = 0$

THIS SECTION DEVELOPS skills in working with radical expressions. Although most sections in this text note applications outside mathematics, sometimes the most appropriate applications of a mathematics skill are within mathematics itself. Skills in this section are used to check solutions to equations, justify patterns, and prove geometric results.

The Quadratic Formula Revisited

We start our consideration of skills with radical expressions by returning to a main source of radical expressions: solving quadratic equations, $ax^2 + bx + c = 0$, with the quadratic formula,

$$x = \frac{-b \pm \sqrt{b^2 - 4ac}}{2a}$$

EXAMPLE **1**   Reviewing quadratic equation solutions   Solve these quadratic equations. Use the quadratic formula, and leave the answers in radical notation.

a. $x^2 + 4x + 2 = 0$ **b.** $x^2 - 2x - 2 = 0$

Solution   **a.** There are two real-number solutions to $x^2 + 4x + 2 = 0$:

$$x = \frac{-b \pm \sqrt{b^2 - 4ac}}{2a} = \frac{-4 \pm \sqrt{16 - 4(1)(2)}}{2(1)}$$

$$= \frac{-4 \pm \sqrt{8}}{2} = \frac{-4 \pm 2\sqrt{2}}{2} = -2 \pm \sqrt{2}$$

b. There are two real-number solutions to $x^2 - 2x - 2 = 0$:

$$x = \frac{-b \pm \sqrt{b^2 - 4ac}}{2a} = \frac{-(-2) \pm \sqrt{4 - 4(1)(-2)}}{2(1)}$$

$$= \frac{2 \pm \sqrt{12}}{2} = \frac{2 \pm 2\sqrt{3}}{2} = 1 \pm \sqrt{3} \qquad \bullet$$

To check our solutions to part a of Example 1, we would substitute $x = -2 + \sqrt{2}$ into $x^2 + 4x + 2$:

$$(-2 + \sqrt{2})^2 + 4(-2 + \sqrt{2}) + 2 \overset{?}{=} 0$$

We would then square $(-2 + \sqrt{2})$, multiply $4(-2 + \sqrt{2})$, add like terms, and verify that the sum is zero. A similar approach is needed to check part b of Example 1. The next few examples summarize the operations with radical expressions that permit us to carry out these operations. You will be asked to check the solutions to Example 1 in the exercise set.

Addition and Subtraction with Radical Expressions

Just as we can add and subtract like terms in polynomials, we can add and subtract similar radicals. **Similar radicals** have *identical indices and radicands.*

EXAMPLE **2**   Adding and subtracting radical expressions   Add or subtract, if possible.

a. $2\sqrt{3} + 3\sqrt{3}$ b. $\sqrt{3} + \sqrt{2}$ c. $2\sqrt{3} - 2\sqrt{3}$

d. $4\sqrt[3]{5} + 6\sqrt[3]{5}$ e. $ab\sqrt{c} + 2ab\sqrt{c}$ f. $\sqrt[3]{x^2 y} + \sqrt[3]{xy}$

Solution   **a.** The terms both contain $\sqrt{3}$ and may be added:

$$2\sqrt{3} + 3\sqrt{3} = (2 + 3)\sqrt{3} = 5\sqrt{3}$$

Using the distributive property, we factor $\sqrt{3}$ from both terms. The radical may be placed after the parentheses to prevent confusion as to whether the expression in the parentheses is under the radical.

b. $\sqrt{3} + \sqrt{2}$ cannot be added because the radicands are different.

c. The radicals contain identical radicands, so

$$2\sqrt{3} - 2\sqrt{3} = 0$$

d. $4\sqrt[3]{5} + 6\sqrt[3]{5} = (4 + 6)\sqrt[3]{5} = 10\sqrt[3]{5}$

e. The expressions have identical variable factors:

$$ab\sqrt{c} + 2ab\sqrt{c} = (1 + 2)ab\sqrt{c} = 3ab\sqrt{c}$$

f. In $\sqrt[3]{x^2 y} + \sqrt[3]{xy}$, the x-variables have different exponents; the radicals cannot be added.   ●

As noted in Section 6.2, *a square root is simplified if it contains no perfect square factor. Other radicals with index n are simplified if they contain no perfect nth power factor.*

In Example 3, we factor and simplify the radical expressions. In doing so, we obtain similar radicals.

EXAMPLE **3**

Adding and subtracting radical expressions   Add or subtract by changing to similar radicals.

a. $\sqrt{27} + \sqrt{3}$ b. $\sqrt{4x} + \sqrt{x}, x \ge 0$

c. $a\sqrt{b} + a\sqrt{9b}, b \ge 0$ d. $x\sqrt[3]{2} - x\sqrt[3]{16}$

Solution   **a.** $\sqrt{27} = \sqrt{9 \cdot 3} = 3\sqrt{3}$, so

$$\sqrt{27} + \sqrt{3} = 3\sqrt{3} + 1\sqrt{3} = 4\sqrt{3}$$

b. $\sqrt{4x} = \sqrt{4 \cdot x} = 2\sqrt{x}$, so

$$\sqrt{4x} + \sqrt{x} = 2\sqrt{x} + 1\sqrt{x} = 3\sqrt{x}, x \ge 0$$

c. $a\sqrt{9b} = a\sqrt{9}\sqrt{b} = 3a\sqrt{b}$, so

$$a\sqrt{b} + a\sqrt{9b} = a\sqrt{b} + 3a\sqrt{b} = 4a\sqrt{b}, b \ge 0$$

d. $x\sqrt[3]{16} = x\sqrt[3]{8 \cdot 2} = x\sqrt[3]{8} \cdot \sqrt[3]{2} = 2x\sqrt[3]{2}$, so

$$x\sqrt[3]{2} - x\sqrt[3]{16} = x\sqrt[3]{2} - 2x\sqrt[3]{2} = -x\sqrt[3]{2}$$ ●

Multiplication of Expressions with Two or More Terms

Multiplying radical expressions is similar to multiplying polynomials. We use the distributive property and addition of like terms to find the product of two or more terms.

EXAMPLE **4**

Multiplying radical expressions   Multiply these expressions.

a. $4(-2 + \sqrt{2})$ b. $(2 - \sqrt{3})(2 - \sqrt{3})$ c. $(-2 + \sqrt{2})^2$

Solution **a.** We apply the distributive property:

$$4(-2 + \sqrt{2}) = -8 + 4\sqrt{2}$$

b. Using a table, we have

Multiply	2	$-\sqrt{3}$
2	4	$-2\sqrt{3}$
$-\sqrt{3}$	$-2\sqrt{3}$	$+\sqrt{9}$

The two terms $-2\sqrt{3}$ and $-2\sqrt{3}$ add to $-4\sqrt{3}$. Thus,

$$(2 - \sqrt{3})(2 - \sqrt{3}) = 4 - 4\sqrt{3} + 3 = 7 - 4\sqrt{3}$$

c. $(-2 + \sqrt{2})^2 = (-2 + \sqrt{2})(-2 + \sqrt{2}) = 4 - 2\sqrt{2} - 2\sqrt{2} + \sqrt{2}^2$
$$= 4 - 4\sqrt{2} + 2 = 6 - 4\sqrt{2}$$ ●

The answers in parts b and c of Example 4 both contain two terms. Products of radical expressions often simplify to fewer terms than do products of polynomials. Such simplification also may occur when variables appear in the expression, as shown in part b of Example 5.

EXAMPLE **5** Multiplying radical expressions Multiply and simplify these expressions.

  a. $(a + \sqrt{b})(a + \sqrt{b}), b \geq 0$ **b.** $(x - \sqrt[3]{2})(x^2 + x\sqrt[3]{2} + (\sqrt[3]{2})^2)$

Solution **a.** $(a + \sqrt{b})(a + \sqrt{b}) = a^2 + a\sqrt{b} + a\sqrt{b} + (\sqrt{b})^2$
$$= a^2 + 2a\sqrt{b} + b, b \geq 0$$

b. $(x - \sqrt[3]{2})(x^2 + x\sqrt[3]{2} + (\sqrt[3]{2})^2)$ requires a large table:

Multiply	x^2	$+x\sqrt[3]{2}$	$(\sqrt[3]{2})^2$
x	x^3	$+x^2\sqrt[3]{2}$	$x(\sqrt[3]{2})^2$
$-\sqrt[3]{2}$	$-x^2\sqrt[3]{2}$	$-x(\sqrt[3]{2})^2$	$-(\sqrt[3]{2})^3$

Four terms in the table add to zero. Thus,

$$(x - \sqrt[3]{2})(x^2 + x\sqrt[3]{2} + (\sqrt[3]{2})^2) = x^3 - (\sqrt[3]{2})^3 = x^3 - 2$$ ●

CONJUGATES Section 4.3 introduced the term *complex conjugates* to describe $a + bi$ and $a - bi$. The term **real-number conjugates** describes *expressions written $a + b$ and $a - b$*. Real-number conjugates include numbers such as $a + \sqrt{b}$ and $a - \sqrt{b}$ or $\sqrt{a} + \sqrt{b}$ and $\sqrt{a} - \sqrt{b}$. When the number contains one radical sign, the change in sign is on the term containing the radical sign.

EXAMPLE **6** Finding conjugates Write the conjugate for each of these real numbers.

  a. $2 - \sqrt{5}$ **b.** $-2 + \sqrt{3}$ **c.** $0 - \sqrt{3}$ **d.** $\sqrt{2} + \sqrt{7}$

Solution **a.** $2 + \sqrt{5}$ **b.** $-2 - \sqrt{3}$ **c.** $0 + \sqrt{3}$ **d.** $\sqrt{2} - \sqrt{7}$ ●

Think about it 1: What is another possible conjugate for part d of Example 6? Why?

In Example 7, we observe an important result from multiplying conjugates.

EXAMPLE **7**   Multiplying conjugates Simplify these expressions, and look for patterns in the answers.

> **a.** $(3 + \sqrt{5})(3 - \sqrt{5})$
>
> **b.** $(\sqrt{5} + \sqrt{2})(\sqrt{5} - \sqrt{2})$
>
> **c.** $(\sqrt{x} - \sqrt{2})(\sqrt{x} + \sqrt{2}), x \geq 0$
>
> **d.** $(\sqrt{x} + \sqrt{y})(\sqrt{x} - \sqrt{y}), x \geq 0, y \geq 0$

Solution   **a.** $(3 + \sqrt{5})(3 - \sqrt{5})$ may be multiplied mentally or with a table.

Multiply	3	$+\sqrt{5}$
3	9	$+3\sqrt{5}$
$-\sqrt{5}$	$-3\sqrt{5}$	$-\sqrt{25}$

The terms $-3\sqrt{5}$ and $+3\sqrt{5}$ add to zero, so

$$(3 + \sqrt{5})(3 - \sqrt{5}) = 9 - \sqrt{25} = 9 - 5 = 4$$

b. $(\sqrt{5} + \sqrt{2})(\sqrt{5} - \sqrt{2}) = 5 - \sqrt{5}\sqrt{2} + \sqrt{5}\sqrt{2} - \sqrt{4} = 5 - 2 = 3$

c. $(\sqrt{x} - \sqrt{2})(\sqrt{x} + \sqrt{2}) = \sqrt{x^2} - \sqrt{2x} + \sqrt{2x} - \sqrt{4} = x - 2$

d. $(\sqrt{x} + \sqrt{y})(\sqrt{x} - \sqrt{y}) = \sqrt{x^2} - \sqrt{xy} + \sqrt{xy} - \sqrt{y^2} = x - y$

None of the answers contain a radical sign. The number expressions simplify to a single rational number.   ●

Think about it 2: Why are no absolute values needed in part d of Example 7?

The products in Example 7 suggest a property of conjugates.

Conjugate Products

> The product of real-number conjugates is a rational number.

Rationalizing Numerators and Denominators

The fact that conjugates multiply to a rational number (hence eliminating the square roots or radical signs) means that conjugates can be used to eliminate a radical from the numerator or denominator of a fraction. **Rationalizing** the numerator or denominator is the name given to *the process of multiplying both the numerator and the denominator by a number that eliminates radicals from one of these positions in the fraction.*

EXAMPLE **8**   Exploring rationalization

> **a.** Multiply the numerator and denominator of $\dfrac{1}{2 + \sqrt{5}}$ by $2 - \sqrt{5}$.
>
> **b.** Multiply the numerator and denominator of $\dfrac{3 - \sqrt{7}}{2}$ by $3 + \sqrt{7}$.

Solution

a. $\dfrac{1}{(2 + \sqrt{5})} \cdot \dfrac{(2 - \sqrt{5})}{(2 - \sqrt{5})} = \dfrac{2 - \sqrt{5}}{4 - 2\sqrt{5} + 2\sqrt{5} - \sqrt{25}}$

$= \dfrac{2 - \sqrt{5}}{4 - 5} = \dfrac{2 - \sqrt{5}}{-1} = -2 + \sqrt{5}$

b. $\dfrac{(3 - \sqrt{7})}{2} \cdot \dfrac{(3 + \sqrt{7})}{(3 + \sqrt{7})} = \dfrac{9 + 3\sqrt{7} - 3\sqrt{7} - \sqrt{49}}{2(3 + \sqrt{7})}$

$= \dfrac{9 - 7}{2(3 + \sqrt{7})} = \dfrac{2}{2(3 + \sqrt{7})} = \dfrac{1}{3 + \sqrt{7}}$

In part a of Example 8, we eliminated the radical from the denominator. In part b of Example 8, we eliminated the radical from the numerator.

Rationalizing

> To rationalize a one-term expression, multiply by a number whose product with the radical makes an exact root.
>
> To rationalize $\sqrt{x}$, multiply by $\sqrt{x}$.
> To rationalize $\sqrt[3]{x}$, multiply by $\sqrt[3]{x^2}$.
> To rationalize $\sqrt[3]{x^2}$, multiply by $\sqrt[3]{x}$.
>
> To rationalize a two-term expression, multiply by the conjugate.
>
> To rationalize $a + \sqrt{b}$, multiply by $a - \sqrt{b}$.
>
> To rationalize $\sqrt{a} + \sqrt{b}$, multiply by either $\sqrt{a} - \sqrt{b}$
> or $-\sqrt{a} + \sqrt{b}$.

EXAMPLE **9**   Rationalizing denominators   Rationalize the denominators in these fractions.

a. $\dfrac{2}{\sqrt{3}}$ **b.** $\dfrac{1}{\sqrt[3]{a}}$

c. $\dfrac{3}{4 - \sqrt{2}}$ **d.** $\dfrac{1}{a - \sqrt{b}}$, $a \neq \sqrt{b}$, a and b not both zero

Solution

a. $\dfrac{2}{\sqrt{3}} \cdot \dfrac{\sqrt{3}}{\sqrt{3}} = \dfrac{2\sqrt{3}}{\sqrt{9}} = \dfrac{2\sqrt{3}}{3}$

b. $\dfrac{1}{\sqrt[3]{a}} \cdot \dfrac{\sqrt[3]{a^2}}{\sqrt[3]{a^2}} = \dfrac{\sqrt[3]{a^2}}{\sqrt[3]{a^3}} = \dfrac{\sqrt[3]{a^2}}{a}$

c. $\dfrac{3}{4 - \sqrt{2}} \cdot \dfrac{4 + \sqrt{2}}{4 + \sqrt{2}} = \dfrac{3(4 + \sqrt{2})}{16 + 4\sqrt{2} - 4\sqrt{2} - \sqrt{4}} = \dfrac{3(4 + \sqrt{2})}{14}$

Because the numerator and denominator contain no common factors, the fraction cannot be simplified.

d. $\dfrac{1}{(a - \sqrt{b})} \cdot \dfrac{(a + \sqrt{b})}{(a + \sqrt{b})} = \dfrac{a + \sqrt{b}}{a^2 + a\sqrt{b} - a\sqrt{b} - \sqrt{b^2}} = \dfrac{a + \sqrt{b}}{a^2 - b}$,

$a^2 \neq b$, $b > 0$, a and b not both zero

Rationalizing of the denominator is needed in order to perform a division with radical expressions. The division is considered complete when the denominator no longer contains a radical.

The following checklist may be helpful in determining whether a radical expression has been simplified.

Simplified Radical Expressions

1. For a radical with index 2, all perfect square factors have been removed.

2. For a radical with index 3, all perfect cube factors have been removed.

3. The index and exponent of factors in the radicand have no common factors.

4. Exponents in the radicand are smaller than the index.

5. Radicals have been eliminated from the denominator.

6. All possible operations have been performed.

We close this section with a geometry proof that uses rationalization.

EXAMPLE **10**   *Applying rationalization in a proof*   Prove that a right angle is formed by two line segments connecting a point (x, y) on a circle to the endpoints of the diameter of the circle.

a. Find the slopes of the two line segments.
b. Find y in terms of x and r, the radius of the circle.
c. Substitute for y, and show that the slope of one segment is the negative reciprocal of the slope of the other segment.

Solution   The line segments will form a right angle if they are perpendicular. We prove perpendicularity by showing that the slopes of the segments are negative reciprocals.

a. We place a circle of radius r with its center at the origin (Figure 5). The endpoints of the diameter lie at $A(r, 0)$ and $B(-r, 0)$. We place point P on the circle at (x, y).

$$\text{Slope } \overline{AP} = \frac{y_2 - y_1}{x_2 - x_1} = \frac{y - 0}{x - r} = \frac{y}{x - r}$$

$$\text{Slope } \overline{BP} = \frac{y_2 - y_1}{x_2 - x_1} = \frac{y - 0}{x - (-r)} = \frac{y}{x + r}$$

b. To write y in terms of x and r, we apply the Pythagorean theorem to the triangle formed by the radius to point P (Figure 6). The right triangle shown in the figure has base x, height y, and hypotenuse r. Thus,

$$x^2 + y^2 = r^2$$
$$y^2 = r^2 - x^2$$
$$y = \sqrt{r^2 - x^2}$$

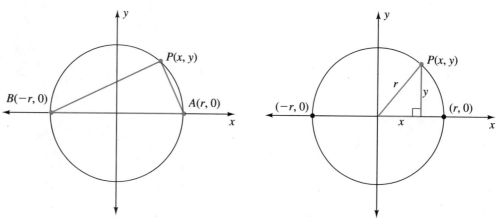

Figure 5   **Figure 6**

c. We start with the slope of AP and relate it to the slope of BP:

$$\text{Slope } \overline{AP} = \frac{y}{x - r}$$ Substitute $\sqrt{r^2 - x^2}$ for y.

$$= \frac{\sqrt{r^2 - x^2}}{x - r}$$ Because $r^2 - x^2$ factors, rationalize the numerator.

$$= \frac{\sqrt{r^2 - x^2}}{x - r} \cdot \frac{\sqrt{r^2 - x^2}}{\sqrt{r^2 - x^2}}$$ Simplify.

$$= \frac{r^2 - x^2}{(x - r)\sqrt{r^2 - x^2}}$$ Factor the numerator.

$$= \frac{(r - x)(r + x)}{(x - r)\sqrt{r^2 - x^2}}$$ Simplify.

$$= \frac{(-1)(r + x)}{\sqrt{r^2 - x^2}}$$ Replace $\sqrt{r^2 - x^2}$ by y.

$$= \frac{(-1)(x + r)}{y}$$ This is the negative reciprocal of the slope of BP.

Because the two segments are perpendicular, a right angle is formed by the segments connecting a point on the circle to the endpoints of the diameter. ●

ANSWER BOX

Warm-up: 1. $x = -2 \pm \sqrt{2}$ **2.** $x = 1 \pm \sqrt{3}$ **Think about it 1:** If we had written $\sqrt{7} + \sqrt{2}$, then $\sqrt{7} - \sqrt{2}$ could be the conjugate. Either $\sqrt{7} - \sqrt{2}$ or $\sqrt{2} - \sqrt{7}$ will work to rationalize $\sqrt{2} + \sqrt{7}$.
Think about it 2: $x \geq 0$ and $y \geq 0$ are stated. $\sqrt{x^2} = |x|$ when x is any real number. For $x \geq 0$, $\sqrt{x^2} = x$. The same applies to $\sqrt{y^2}$.

EXERCISES 6.3

In Exercises 1 to 24, add or subtract like terms. Assume all variables represent positive numbers.

1. $3\sqrt{2} - 4\sqrt{2} + 7\sqrt{2}$

2. $4\sqrt{5} + 5\sqrt{5} - 2\sqrt{5}$

3. $\sqrt{20} + \sqrt{45}$

4. $\sqrt{50} + \sqrt{32}$

5. $\sqrt{75} + \sqrt{27}$

6. $\sqrt{98} + \sqrt{8}$

7. $\sqrt{9x} + \sqrt{16x} - \sqrt{x}$

8. $\sqrt{25x} + \sqrt{4x} + \sqrt{2.25x}$

9. $\sqrt{0.01x} + \sqrt{49x} - \sqrt{16x}$

10. $\sqrt{121x} - \sqrt{81x} + \sqrt{9x}$

11. $a\sqrt{b} + a\sqrt{c}$

12. $a\sqrt{b} + c\sqrt{b}$

13. $\sqrt{ab} + 2\sqrt{ab}$

14. $\sqrt{ab} + \sqrt{ac}$

15. $3\sqrt[3]{x} - \sqrt[3]{x}$

16. $5\sqrt[4]{x} - \sqrt[5]{x}$

17. $3\sqrt[4]{x} - 3\sqrt[2]{x}$

18. $5\sqrt[4]{x} - \sqrt[4]{x}$

19. $\sqrt[3]{8x} + \sqrt[3]{27x}$

20. $\sqrt[3]{64x} - \sqrt[3]{27x}$

21. $a\sqrt[4]{81ab} - \sqrt[4]{a^5b}$

22. $b\sqrt[3]{125ab} + \sqrt[3]{ab^4}$

23. $\sqrt[4]{16x^4y} + x\sqrt[4]{y}$

24. $x\sqrt[5]{32xy} + \sqrt[5]{x^6y}$

In Exercises 25 to 28, multiply and simplify.

25. a. $(2 - \sqrt{3})(2 + \sqrt{3})$

  b. $(5 - \sqrt{2})(5 - \sqrt{2})$

26. a. $(3 - \sqrt{2})(3 - \sqrt{3})$

  b. $(5 + \sqrt{2})(5 - \sqrt{2})$

27. a. $(x - \sqrt{5})(x + \sqrt{5})$

  b. $(x - \sqrt{7})(x - \sqrt{7})$

28. a. $(x - \sqrt{3})(x + \sqrt{3})$

  b. $(x - \sqrt{3})(x - \sqrt{3})$

Multiply the expressions in Exercises 29 to 32. Assume $a \geq 0$, $b \geq 0$, and $x \geq 0$.

29. a. $(\sqrt{x} + 3)^2$

   b. $(\sqrt{a} - 3)^2$

30. a. $(2 - \sqrt{a})^2$

   b. $(\sqrt{a} - \sqrt{b})(\sqrt{a} + \sqrt{b})$

31. a. $(1 - \sqrt{a})^2$

   b. $(\sqrt{a} - \sqrt{b})(\sqrt{a} - \sqrt{b})$

32. a. $(\sqrt{x} - 2)^2$

   b. $(\sqrt{2a} - 1)^2$

33. Show that $x = -2 - \sqrt{2}$ satisfies $x^2 + 4x + 2 = 0$.

34. Show that $x = -2 + \sqrt{2}$ satisfies $x^2 + 4x + 2 = 0$.

35. Show that $x = 1 + \sqrt{3}$ satisfies $x^2 - 2x - 2 = 0$.

36. Show that $x = 1 - \sqrt{3}$ satisfies $x^2 - 2x - 2 = 0$.

37. The golden ratio is the first of the following two expressions:

$$\frac{1 + \sqrt{5}}{2} \quad \text{and} \quad \frac{1 - \sqrt{5}}{2}$$

   a. Multiply the two expressions. What do you observe?

   b. Evaluate each expression with a calculator.

   c. What is the reciprocal of each expression? What do you observe?

38. Substitute the expressions $x = \dfrac{1 + \sqrt{5}}{2}$ and $x = \dfrac{1 - \sqrt{5}}{2}$ into the equation $x^2 - x - 1 = 0$ to check that they are solutions.

In Exercises 39 and 40, multiply the expressions.

39. $(x + \sqrt[3]{2})(x^2 - x\sqrt[3]{2} + \sqrt[3]{4})$

40. $(x - \sqrt[3]{3})(x^2 + x\sqrt[3]{3} + \sqrt[3]{9})$

In Exercises 41 and 42, give the conjugate.

41. a. $3 + \sqrt{2}$ **b.** $3 - \sqrt{a}$

   c. $a - \sqrt{b}$ **d.** $\sqrt{2} - \sqrt{3}$

42. a. $2 - \sqrt{11}$ **b.** $1 + \sqrt{b}$

   c. $c - \sqrt{a}$ **d.** $5 + \sqrt{3}$

Rationalize the denominators in Exercises 43 to 50.

43. a. $\dfrac{4}{\sqrt{5}}$ **b.** $\dfrac{8}{\sqrt{6}}$

44. a. $\dfrac{3}{\sqrt{6}}$ **b.** $\dfrac{5}{\sqrt{15}}$

45. a. $\dfrac{a}{\sqrt{c}}, c > 0$ **b.** $\dfrac{a}{\sqrt{a}}, a > 0$

46. a. $\dfrac{c}{\sqrt{c}}, c > 0$ **b.** $\dfrac{x}{\sqrt{y}}, y > 0$

47. $\dfrac{4}{7 - \sqrt{5}}$

48. $\dfrac{6}{5 - \sqrt{2}}$

49. $\dfrac{x}{x - \sqrt{y}}, x \neq \sqrt{y}$, x and y not both zero

50. $\dfrac{y}{y - \sqrt{x}}, y \neq \sqrt{x}$, x and y not both zero

In Exercises 51 and 52, multiply the numerator and denominator by a number or variable that makes a perfect cube under the radical in the denominator. Further simplify, as needed.

51. a. $\dfrac{1}{\sqrt[3]{2}}$ **b.** $\dfrac{2}{\sqrt[3]{4}}$ **c.** $\dfrac{3}{\sqrt[3]{3}}$

52. a. $\dfrac{2}{\sqrt[3]{2}}$ **b.** $\dfrac{1}{\sqrt[3]{3}}$ **c.** $\dfrac{3}{\sqrt[3]{9}}$

53. Return to the solution of Example 10. Use steps similar to those in part c to show that the slope of $\overline{BP}$ is the negative reciprocal of the slope of $\overline{AP}$:

$$\text{Slope } \overline{AP} = \frac{y}{x - r} \quad \text{and} \quad \text{Slope } \overline{BP} = \frac{y}{x + r}$$

54. A circle with diameter d and center at $(d/2, 0)$ is shown in the figure.

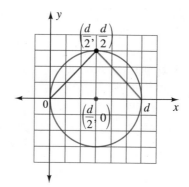

   a. Find the slope of the line segment connecting $(d/2, d/2)$ with the origin $(0, 0)$.

   b. Find the slope of the line segment connecting $(d/2, d/2)$ with the end of the diameter $(d, 0)$.

   c. Comment on what you observe about the slopes and the angle between them.

55. Apply the Pythagorean theorem to the lengths of the sides of the triangle shown on page 448, to show that the *distance formula* between points A and B is given by

$$d = \sqrt{(x_2 - x_1)^2 + (y_2 - y_1)^2}$$

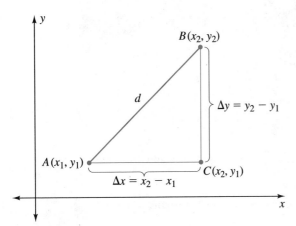

In Exercises 56 to 6l, find the slope and the distance between the given points. Use the distance formula from Exercise 55.

56. $(2, 5)$ and $(6, -2)$

57. $(4, -3)$ and $(-6, 3)$

58. $(-1, 5)$ and $(4, -3)$

59. $(6, 7)$ and $(-3, -4)$

60. (a, b) and (c, d)

61. $\left(\dfrac{a}{2}, \dfrac{b}{2}\right)$ and (a, b)

62. Hero's (or Heron's) formula permits us to find the area of a triangle when we know only the lengths of its sides. The formula is named for a Greek mathematician of the 1st century A.D. or earlier. To find the area, we calculate

$$A = \sqrt{s(s - a)(s - b)(s - c)}$$

where $s = \dfrac{a + b + c}{2}$. The s is called the *semiperimeter* because it is half the perimeter of the triangle. Calculate the area of these triangles.

a. $a = 4, b = 6, c = 3$

b. $a = 5, b = 6, c = 7$

c. $a = 3, b = 4, c = 5$

d. $a = 5, b = 12, c = 13$

e. $a = 8, b = 15, c = 17$

f. Which three triangles in parts a to e are right triangles? How do you know? Calculate their area in another way. Hero's formula is particularly useful in land measure, because it is easy to measure the three sides of a triangular plot of ground.

Projects

63. *Nested Squares.* The design in the figure is a set of squares, nested one inside another. Each inner square has its corners at the midpoints of the sides of the outer square.

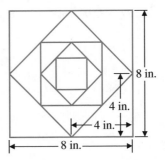

a. Build a table for the squares shown in the figure, with length of side as input and area as output.

b. How many squares would you need to draw for the smallest to have an area of 1 square inch?

c. How many squares would you have altogether if the smallest had an area of $\frac{1}{64}$ square inch?

d. If each of the squares in part c were made from separate paper, what would be the total area covered by all the squares?

64. *Drawing on Graph Paper.* Suppose you have a sheet of graph paper where each square is one unit on a side. Explain how, using only the graph paper, you could draw a line $\sqrt{5}$ units long. Repeat for lines of length $\sqrt{10}$, $\sqrt{13}$, $\sqrt{40}$, and $\sqrt{45}$ units.

65. *Minimum Cost.* This problem may be solved by table, guess and check, graph, or spreadsheet.

A telephone company is replacing copper cable with fiber-optic cable between city A and city B (see the figure). The land above line BC is swamp, and the land along the line between B and C is farm land. It costs \$1000 per foot to put the cable through farm land and \$1500 per foot to put it through swamp.

The distance between B and C is 12 miles. The distance between A and C is 5 miles. There are 5280 feet in a mile. The line AC is perpendicular to BC. Round answers to the nearest thousand dollars.

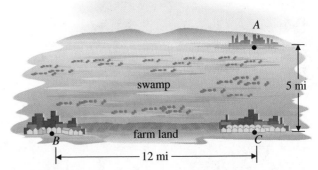

a. What is the cost of a direct route between A and B through the swamp?

b. What is the cost of installing the cable from A to C to B?

c. What is the cost of installation from A to C_1 to B, where C_1 is 1 mile from C along the line between B and C?

d. What is the cost of installation from A to C_2 to B, where C_2 is 2 miles from C along the line between B and C?

e. What is the cost of installation from A to C_n to B, where C_n is x miles from C along the line between B and C?

f. Using a graph or table, find the minimum possible cost of installation. With what point along the line BC will the minimum cost be associated?

66. *Placing a Water Pump.* Toni, a Peace Corps volunteer, has one water pump at P and two villages at A and B that need water supplied (see the figure). One village is 10 kilometers from the river, and the other village is 20 kilometers from the river. The horizontal distance along the river bank is 20 kilometers. (A spreadsheet or graphing calculator may be useful for this project.)

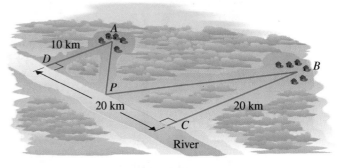

a. Where on the river bank should she place the pump to minimize the amount of pipe needed to deliver water to the two villages?

b. If the villages were equal distances from the river, where should the pump be placed?

6.4 Inverse Functions: Solving Power and Root Equations

OBJECTIVES

- Find the inverse of a function from a table, from a graph, and from an equation.
- Identify power and root functions.
- Solve a power equation by taking the nth root of both sides.
- Solve a radical equation by taking the nth power of both sides.

WARM-UP

Simplify these expressions:

1. $2(\frac{1}{2}x - 2) + 4$ **2.** $\frac{1}{3}(3x - 3) + 1$

3. $(x - 1) + 1$ **4.** $(x + 2) - 2$

5. $2(\frac{1}{2}x)$ **6.** $\frac{1}{4}(4x)$

7. $(\sqrt{x})^2$ for $x \geq 0$ **8.** $\sqrt{x^2}$ for $x \geq 0$

9. $\sqrt[3]{x^3}$ **10.** $(\sqrt[3]{x})^3$

IN THIS SECTION, we return to solving equations. We work with a more formal way of thinking about solving equations (inverse functions) and apply it to power and root functions. We then use powers and roots to solve equations and formulas in a variety of settings.

Inverse Functions

The answer to each simplification in the Warm-up is x, because the two operations in each set are inverse (or opposite) operations. Each operation undoes the other operation in the expression.

Like operations, functions have inverses. Two functions are **inverse functions** if *they undo each other.* For example, an "add two" function, $y = x + 2$, would be undone by a "subtract two" function, $y = x - 2$:

$$5 + 2 - 2 = 5$$

Some inverse functions are easy to find and name; others (as we will see in Chapter 7) must be specially defined.

We will now find inverse functions in three different ways: from a table, from ordered pairs on a graph, and from an equation.

INVERSE FUNCTIONS FROM A TABLE   To find the inverse function from the table for a function, swap the input and output (x and y) columns and find the rule for the new table.

EXAMPLE   Finding an inverse function from a table   The function $y = 4x$ is shown in Table 3. Its inverse is in Table 4. What equation describes the second table? What is the inverse function to $y = 4x$?

Input, x	Output, $f(x)$ $y = 4x$
1	4
2	8
3	12
4	16

Table 3

Input, x	Output: Inverse Function
4	1
8	2
12	3
16	4

Table 4

Solution   See the Answer Box.   ●

INVERSE FUNCTIONS FROM A GRAPH   *To find the inverse function from the graph of a function, swap the numbers (x, y) in each ordered pair and plot the resulting graph.*

EXAMPLE 2   Finding an inverse function from a graph   The function $y = \frac{1}{2}x - 2$ is shown in Figure 7. Identify four ordered pairs on the function and graph the inverse function. What equation describes the new graph? What is the inverse function to $y = \frac{1}{2}x - 2$?

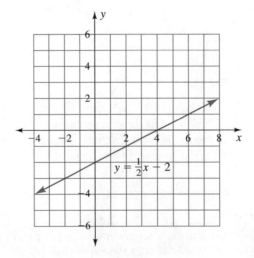

Figure 7

Solution   Ordered pairs on $y = \frac{1}{2}x - 2$ are $(-4, -4)$, $(-3, -2)$, $(0, -2)$, and $(4, 0)$. Ordered pairs on the inverse function are $(-4, -4)$, $(-2, -3)$, $(-2, 0)$, and $(0, 4)$. The graph is shown in Figure 8. For the equation of the inverse, see the Answer Box.

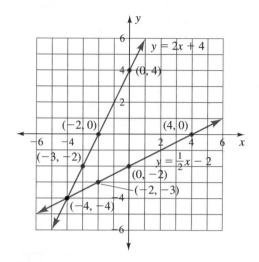

Figure 8   ●

EXAMPLE **3**   Finding an inverse function from a graph   The right side of the graph of $y = x^2$ is shown in Figure 9. Identify four ordered pairs on the function and graph the inverse function. What equation describes the new graph? What is the inverse function to $y = x^2$, $x \geq 0$, $y \geq 0$?

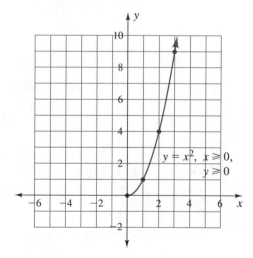

Figure 9

Solution   Ordered pairs on the function are $(0, 0)$, $(1, 1)$, $(2, 4)$, and $(3, 9)$. For ordered pairs and a graph of the inverse function, see the Answer Box.   ●

Think about it 1: If we draw the left side of the parabola in Figure 9 and then list the ordered pairs with x and y swapped, we obtain points below the x-axis. Why are these not part of the inverse function to $y = x^2$, $x \geq 0$, $y \geq 0$?

INVERSE FUNCTIONS FROM AN EQUATION   *To find the inverse function from the equation of a function, swap x and y and solve (if possible) for y. For example, to find the inverse function to $y = 4x$, we reverse x and y as follows:*

$$x = 4y \qquad \text{Multiply both sides by } \tfrac{1}{4} \text{ to solve for } y.$$
$$\tfrac{1}{4}x = y$$

The finding that $y = \frac{1}{4}x$ is the inverse function to $y = 4x$ agrees with Example 1. To find the inverse function to $y = \frac{1}{2}x - 2$, we again reverse x and y:

$$x = \tfrac{1}{2}y - 2 \qquad \text{To solve for } y, \text{ first add 2.}$$
$$x + 2 = \tfrac{1}{2}y \qquad \text{Then multiply by 2.}$$
$$2x + 4 = y$$

The finding that $y = 2x + 4$ is the inverse function to $y = \frac{1}{2}x - 2$ agrees with Example 2.

EXAMPLE **4**   Finding inverse functions from an equation   Find the inverse function to each of the following equations.

a. $y = 3x - 3$ **b.** $y = x - 1$

c. $y = \frac{1}{2}x$ **d.** $y = \sqrt{x}, x \geq 0, y \geq 0$

Solution   **a.**

$$y = 3x - 3 \qquad \text{Swap } x \text{ and } y \text{ to find the inverse function.}$$
$$x = 3y - 3 \qquad \text{Add 3 to each side.}$$
$$x + 3 = 3y \qquad \text{Multiply by } \tfrac{1}{3} \text{ on each side.}$$
$$\tfrac{1}{3}(x + 3) = y \qquad \begin{array}{l}\text{Apply the distributive property to change to}\\ y = mx + b \text{ form.}\end{array}$$
$$\tfrac{1}{3}x + 1 = y$$

The inverse function to $y = 3x - 3$ is $y = \frac{1}{3}x + 1$.

b.

$$y = x - 1 \qquad \text{Swap } x \text{ and } y \text{ to find the inverse function.}$$
$$x = y - 1 \qquad \text{Add 1 to each side.}$$
$$x + 1 = y$$

The inverse function to $y = x - 1$ is $y = x + 1$.

c.

$$y = \tfrac{1}{2}x \qquad \text{Swap } x \text{ and } y \text{ to find the inverse function.}$$
$$x = \tfrac{1}{2}y \qquad \text{Multiply both sides by 2.}$$
$$2x = y$$

The inverse function to $y = \frac{1}{2}x$ is $y = 2x$.

d. $y = \sqrt{x}, x \geq 0, y \geq 0$ Swap x and y (in the inequality conditions, too).

$$x = \sqrt{y}, y \geq 0, x \geq 0 \qquad \text{Square both sides.}$$
$$x^2 = y, y \geq 0, x \geq 0$$

The inverse function to $y = \sqrt{x}, x \geq 0, y \geq 0$ is $y = x^2, x \geq 0, y \geq 0$.   ●

Example 3 and part d of Example 4 suggest that powers and roots are inverse functions. We now formally define power and root functions.

Power and Root Functions

For the purpose of working with inverse functions, we will define the power and root functions only in terms of positive integer powers and integer roots. However, some rational exponents and related roots will be used in selected exercises, projects, and examples of calculator regression.

The **power function** is defined by $y = x^n$, *where n is a positive integer.* The identity function ($y = x$), squaring function ($y = x^2$), cubing function ($y = x^3$), and other positive integer powers of x are all power functions.

The **root function** is defined by $y = \sqrt[n]{x}$ *with* $x \geq 0$ *for all functions with an even index* n. The square root function ($y = \sqrt{x}$, $x \geq 0$), cube root function ($y = \sqrt[3]{x}$), and fourth root function ($y = \sqrt[4]{x}$, $x \geq 0$) are all root functions.

Think about it 2: Write the functions $y = \sqrt{x}$, $y = \sqrt[3]{x}$, and $y = \sqrt[4]{x}$, $x \geq 0$ with rational exponents.

The graphs of the root functions show limitations on the even roots. As shown in (a) and (c) of Figure 10, the square root and the fourth root have graphs only in the first quadrant. The limitations on the even roots explain why the even power functions have limitations when we are finding inverse functions. The odd roots, in (b) and (d) of Figure 10, have no limitations on inputs or outputs. Thus, the inverses to the odd roots have no limitations either.

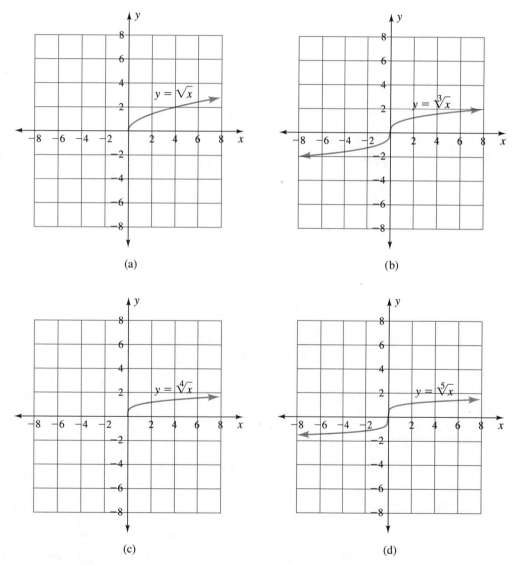

(a)

(b)

(c)

(d)

Figure 10

EXAMPLE **5**   Finding an inverse function to an odd root

a. Use ordered pairs to find the inverse function to $y = x^3$.

b. Check by finding the inverse function from the equation.

Solution **a.** Several ordered pairs on $y = x^3$ are $(-2, -8)$, $(-1, -1)$, $(0, 0)$, $(1, 1)$, and $(2, 8)$. Ordered pairs on the inverse are $(-8, 2)$, $(-1, -1)$, $(0, 0)$, $(1, 1)$, and $(8, 2)$. The pairs $(-8, -2)$ and $(8, 2)$ suggest that the inverse function is $y = \sqrt[3]{x}$. The ordered pairs are graphed in Figure 11.

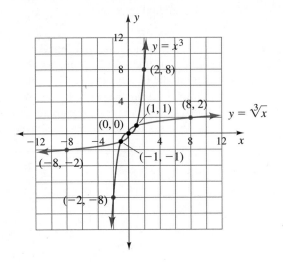

Figure 11

The graph suggests that the cube function and cube root are inverses, and so in part b we take the cube root to solve the equation.

b. To find the inverse function to $y = x^3$, we swap x and y.

$$x = y^3 \qquad \text{Take the cube root of both sides.}$$
$$\sqrt[3]{x} = y$$

The inverse function to $y = x^3$ is $y = \sqrt[3]{x}$. ●

Inverse Functions for Powers and Roots

> When n is odd, the inverse function to $y = x^n$ is $y = \sqrt[n]{x}$.
>
> When n is even, the inverse function to $y = x^n$, $x \geq 0$, $y \geq 0$ is $y = \sqrt[n]{x}$, $x \geq 0$, $y \geq 0$.

Solving nth Power Equations

In earlier sections, we squared both sides of an equation and took the square root of both sides of an equation. Taking the nth root of both sides of an equation is the technique for solving power equations of the form $y = ax^n$.

Solving nth Power Equations

> To solve equations containing nth powers, take the nth root of both sides. For even nth roots, write any limitations on the inputs and outputs.

EXAMPLE **6** Solving a power equation Solve $12 = 6(1 + r)^{18}$ for r.

Solution

$$12 = 6(1 + r)^{18}$$ Divide by 6.

$$2 = (1 + r)^{18}$$ Take the 18th root of both sides, either by raising each side to the 1/18 power or by showing an 18th root, $\sqrt[18]{}$.

$$2^{1/18} = (1 + r)^{18 \cdot \frac{1}{18}}$$ Simplify.

$$2^{1/18} = (1 + r)$$ Subtract 1 from each side.

$$2^{1/18} - 1 = r$$ Evaluate with a calculator.

$$r \approx 0.039, \text{ or } 3.9\%$$ ●

The variable r in the equation in Example 6 describes the annual rate of increase needed for a $6 per hour wage to double in 18 years. The solution says that an annual increase of 3.9% will cause the wage rate to double in 18 years.

ANNUAL RATE OF GROWTH AND INFLATION   The compound interest formula, with $n = 1$, describes annual growth in the value of an investment such as a house. When $n = 1$, the compound interest formula simplifies:

$$S = P\left(1 + \frac{r}{n}\right)^{nt} = P\left(1 + \frac{r}{1}\right)^{1 \cdot t} = P(1 + r)^{t}$$

A term often used to describe the annual growth rate of prices of houses and other goods is *inflation*. **Inflation** is *a measure of the change in prices over time.*

EXAMPLE   Solving a power equation: car prices and inflation   A 1968 four-door automatic Volvo sedan cost $3600. A similar style 1995 Volvo sedan cost $23,820. Suppose that the price increase was due entirely to inflation—what was the annual rate of inflation?

Solution   Because we want the annual rate of inflation, we let $n = 1$ in the compound interest formula. The price changes over a time period of $t = 27$ yr.

$$S = P(1 + r)^{t}$$ Let S = 23,820, P = 3600, t = 27.

$$23,820 = 3,600(1 + r)^{27}$$

In solving for r, we reverse the order of operations on r, as in Example 6.

$$23,820 = 3,600(1 + r)^{27}$$ Divide by 3600.

$$\frac{23,820}{3,600} = (1 + r)^{27}$$ Take the 27th root.

$$\left(\frac{23,820}{3,600}\right)^{\frac{1}{27}} = (1 + r)^{27 \cdot \frac{1}{27}}$$ Simplify exponents.

$$\left(\frac{23,820}{3,600}\right)^{\frac{1}{27}} = 1 + r$$ Subtract 1.

$$\left(\frac{23,820}{3,600}\right)^{\frac{1}{27}} - 1 = r$$

Evaluating, we obtain $r \approx 0.072$, or about 7% inflation per year. The Volvo increased in price at a rate of 7% per year.   ●

DOWNWARD CHANGE IN VALUE AND DEPRECIATION   **Depreciation** occurs when *a piece of equipment loses value with use.* The ability to calculate various forms of depreciation is a skill needed by a tax preparer or a CPA (certified public accountant). In Example 8, we use the compound interest formula with $n = 1$ to find an annual rate of depreciation for a truck.

EXAMPLE 8   Solving a power equation: value and depreciation   Suppose that in 5 years the value of a delivery truck drops from \$40,000 to \$13,107.20. What is the annual rate of depreciation?

Solution   We let the starting value, P, be \$40,000 and the ending value, S, be \$13,107.20. Replacing t with 5 years, we solve for r.

$$S = P(1 + r)^t$$

$$13{,}107.20 = 40{,}000(1 + r)^5$$

$$\frac{13{,}107.20}{40{,}000} = (1 + r)^5 \qquad \text{Take the 5th root.}$$

$$\left(\frac{13{,}107.20}{40{,}000}\right)^{1/5} = 1 + r \qquad \text{Subtract 1.}$$

$$\left(\frac{13{,}107.20}{40{,}000}\right)^{1/5} - 1 = r \qquad \text{Evaluate with a calculator.}$$

$$r = -0.20$$

The rate is negative because the value decreases over time: $r = 20\%$ depreciation. The truck loses 20% of its remaining value each year. ●

Think about it 3: What is 20% of \$40,000? If the truck lost this amount each year for 5 years, what value would remain? When might this method be a good way to calculate the truck's value? When might the method in Example 8 be more appropriate?

SOLVING FORMULAS   In Examples 9 and 10, we solve formulas by taking the nth root.

EXAMPLE 9   Solving a formula   The formula $V = x^3$ relates the length x of the side of a cube to the volume V. Solve for x.

Solution   The variable is cubed, so we take the cube root.

$$V = x^3$$
$$\sqrt[3]{V} = \sqrt[3]{x^3}$$
$$x = \sqrt[3]{V}$$ ●

EXAMPLE 10   Solving a formula   The formula $H = M/r^3$ is related to magnets. Solve for r, applying a proportion as appropriate.

Solution   The formula can be made into a proportion:

$$H = \frac{M}{r^3} \qquad \text{Visualize } \frac{H}{1}, \text{ and cross multiply.}$$

$$H \cdot r^3 = M \qquad \text{Divide by } H.$$

$$r^3 = \frac{M}{H} \qquad \text{Take the cube root.}$$

$$r = \sqrt[3]{\frac{M}{H}}$$ ●

Solving nth Root and Square Root Equations

We close the section by solving equations containing roots.

Solving nth Root Equations

> To solve equations containing nth roots, take the nth power of both sides.
>
> If $a = b$, then $a^n = b^n$ for any positive integer n.
>
> Even powers may introduce extra answers, so always check answers.

Because equations containing roots require careful checking, only our first example will involve a root other than a square root.

EXAMPLE **11**   Solving a 5th root equation   Solve $\sqrt[5]{x} = 4$ for x.

Solution
$$\sqrt[5]{x} = 4 \qquad \text{Take the 5th power of each side.}$$
$$\left(\sqrt[5]{x}\right)^5 = 4^5 \qquad \text{Simplify.}$$
$$x = 1024$$

●

Think about it 4: Why are the following statements different?

If $a = b$, then $a^n = b^n$.
If $a^n = b^n$, then $a = b$.

I n the remaining examples, we focus on square roots. Solutions include graphs and algebraic notation. In many algebraic equations such as that in Example 12, squaring gives a single answer that satisfies the equation.

EXAMPLE **12**   Solving equations containing square roots

a. Find inputs for which the equation $y = \sqrt{x - 5}$ has real-number solutions.
b. Solve $\sqrt{x - 5} = 3$ using algebraic notation.
c. Solve the equation from a graph.

Solution
a. The radicand, $x - 5$, is zero or positive if $x \geq 5$.
b.
$$\sqrt{x - 5} = 3 \qquad \text{Square both sides.}$$
$$(\sqrt{x - 5})^2 = 3^2 \qquad \text{Simplify.}$$
$$x - 5 = 9 \qquad \text{From part a, } x - 5 \text{ is positive.}$$
$$x = 14$$
Check: $\sqrt{14 - 5} \overset{?}{=} 3$ ✔

c. The intersection of $y = \sqrt{x - 5}$ with $y = 3$ in Figure 12 gives the solution to $\sqrt{x - 5} = 3$. This point of intersection is $(14, 3)$, so $x = 14$. The graph also shows the condition on $y = \sqrt{x - 5}$, $x \geq 5$, in that there are no points on the square root graph to the left of $x = 5$.

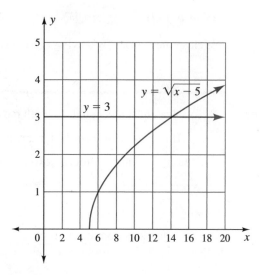

Figure 12

EXTRANEOUS ROOTS Example 13 illustrates the importance of checking solutions.

EXAMPLE **13** Solving equations containing square roots

a. Find inputs for which the equation $y = \sqrt{2 - x}$ has real-number solutions.

b. Solve $\sqrt{2 - x} = x + 1$ with algebraic notation.

c. Solve the equation by graphing.

Solution

a. The radicand, $2 - x$, must be zero or positive. Thus, $2 - x \geq 0$ and $2 \geq x$. The solutions are real for all $x \leq 2$.

b.

$\sqrt{2 - x} = x + 1$	Square both sides.
$(\sqrt{2 - x})^2 = (x + 1)^2$	Replace $(\sqrt{2 - x})^2$ with $(2 - x)$.
$2 - x = x^2 + 2x + 1$	From part a, $2 - x$ is positive.
$0 = x^2 + 3x - 1$	Apply the quadratic formula.

Either $x \approx 0.303$ or $x \approx -3.303$

The results are possible solutions. They need to be checked in the original equation.

Check:

$$\sqrt{2 - 0.303} \overset{?}{=} 0.303 + 1 \quad ✔$$

The result $x \approx 0.303$ gives a true statement.

$$\sqrt{2 - (-3.303)} \overset{?}{=} -3.303 + 1$$

The result $x \approx -3.303$ gives $2.303 = -2.303$, which is false. The solution $x \approx -3.303$ must be discarded.

c. The graph in Figure 13 shows a single point of intersection, at $x \approx 0.303$, and confirms our symbolic solution.

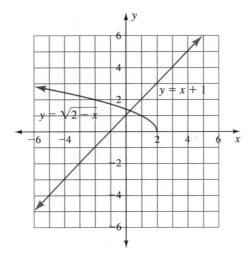

Figure 13 ●

The extra result in Example 13 is given a special name: an extraneous root. *Possible solutions that give false statements* are called **extraneous roots.** Extraneous roots are introduced in the squaring step, as we will see in Example 14.

EXAMPLE **14**   Solving equations containing square roots   Solve $\sqrt{x} + 3 = 1$ both symbolically and graphically.

Solution

$$\sqrt{x} + 3 = 1 \qquad \text{Subtract 3 from both sides.}$$
$$\sqrt{x} = -2 \qquad \text{This equation has no real-number solution.}$$

If we do not notice that $\sqrt{x} = -2$ has no real-number solution and square both sides, we discover our oversight when we check our answer.

$$(\sqrt{x})^2 = (-2)^2$$
$$x = 4$$

Check: $\sqrt{4} + 3 \overset{?}{=} 1$ gives a false statement, so $x = 4$ is an extraneous root. There is no real-number solution to this equation. In Figure 14, we see that the graph of $y = \sqrt{x} + 3$ has the y-intercept (0, 3) as its lowest point, so it cannot pass through the line $y = 1$. Hence, there are no solutions to $\sqrt{x} + 3 = 1$.

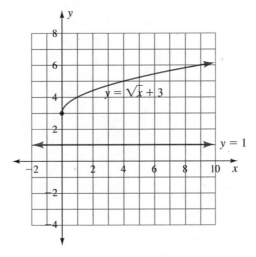

Figure 14   ●

SQUARING TWICE   Example 15 illustrates that it may be necessary to square twice in order to remove all radicals from an equation and solve for x.

EXAMPLE **15**   Solving equations containing square roots

a. Find inputs for which the equations $y = \sqrt{2x + 14}$ and $y = \sqrt{x} + 3$ have real-number solutions.

b. Solve $\sqrt{2x + 14} = \sqrt{x} + 3$ using symbols.

c. Solve the equations by graphing.

Solution a. The radicand $2x + 14$ is non-negative if $2x + 14 \geq 0$. Solving for x gives $x \geq -7$. The radicand x is non-negative if $x \geq 0$. To satisfy both requirements, we limit our choices for x to $x \geq 0$.

b. In solving for x, we will need to square both sides twice.

$$\sqrt{2x + 14} = \sqrt{x} + 3 \qquad \text{Square both sides.}$$
$$(\sqrt{2x + 14})^2 = (\sqrt{x} + 3)^2$$
$$2x + 14 = x + 3\sqrt{x} + 3\sqrt{x} + 9 \qquad \text{Add like terms.}$$
$$2x + 14 = x + 6\sqrt{x} + 9 \qquad \text{Isolate the radical term.}$$
$$x + 5 = 6\sqrt{x} \qquad \text{Square both sides.}$$
$$x^2 + 10x + 25 = 36x \qquad \text{Subtract the right side from the left side.}$$
$$x^2 - 26x + 25 = 0 \qquad \text{Apply the quadratic formula (or factor).}$$

Either $x = 1$ or $x = 25$

Check:

$$\sqrt{2(1) + 14} \stackrel{?}{=} \sqrt{1} + 3 \quad \text{✔}$$
$$\sqrt{2(25) + 14} \stackrel{?}{=} \sqrt{25} + 3 \quad \text{✔}$$

Both solutions check, so there should be two points of intersection on the graph.

c. The graphs of $y = \sqrt{2x + 14}$ and $y = \sqrt{x} + 3$ are shown in Figure 15. The graphs intersect at $(1, 4)$ and $(25, 8)$, so $x = 1$ and $x = 25$ are solutions.

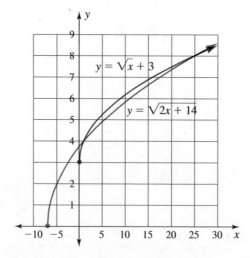

Figure 15

Checklist for Solving
Radical Equations

1. Look for extraneous roots.
2. Look for solutions that give negative expressions under the square root sign.
3. Look for problems that cannot be done because of the principal square root definition—for example, problems such as $\sqrt{x + 4} = -4$.
4. Discard possible solutions that are contrary to the geometry or problem setting.

ANSWER BOX

Warm-up: Each expression simplifies to x. **Example 1:** The inverse function to $y = 4x$ is $y = \frac{1}{4}x$. **Example 2:** The inverse function to $y = \frac{1}{2}x - 2$ is $y = 2x + 4$. **Example 3:** The graph through $(0, 0)$, $(1, 1)$, $(4, 2)$, and $(9, 3)$ is $y = \sqrt{x}$, $x \geq 0$, $y \geq 0$. (See left-hand figure below.)

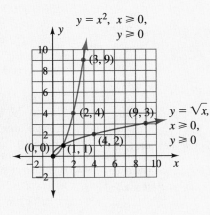

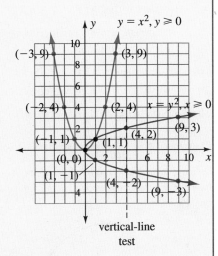

vertical-line
test

Think about it 1: A graph that has points above and below the x-axis fails the vertical-line test and is not a function. The equation is $x = y^2$, $x \geq 0$. (See right-hand figure above.) **Think about it 2:** $y = x^{1/2}$, $y = x^{1/3}$, $y = x^{1/4}$. The root functions all can be written as $y = x^n$.
Think about it 3: 20% of $40,000 is $8000. If the truck lost $8000 in value each year for 5 years, 0 value would remain. This method is practical when the truck really has no value at the end of 5 years. The method in Example 8 is appropriate when the truck retains some value—as scrap metal, as a source of parts, or as a used truck.
Think about it 4: "If $a = b$, then $a^n = b^n$" refers to taking the nth power of both sides and is true for all $a = b$. "If $a^n = b^n$, then $a = b$" is false if n is an even number and if $a = 2$ and $b = -2$.

EXERCISES 6.4

In Exercises 1 to 4, name the ordered pairs that are in the inverse to each set.

1. $\left(1, \frac{1}{2}\right), \left(2, \frac{1}{4}\right), \left(3, \frac{1}{8}\right), \left(4, \frac{1}{16}\right)$

2. $(1, 2), (2, 4), (3, 6), (4, 8)$

3. $(1, 2), (2, 5), (3, 8), (4, 11)$

4. $\left(1, \frac{1}{3}\right), \left(2, \frac{1}{9}\right), \left(3, \frac{1}{27}\right), \left(4, \frac{1}{81}\right)$

In Exercises 5 to 8, graph the equation, show four ordered pairs on the graph and on the inverse function, and graph the inverse.

5. $y = 2^x$

6. $y = 3^x$

7. $y = 3^x - 3$

8. $y = 2^x - 4$

In Exercises 9 to 26, use algebra to find the inverse function.

9. $y = 3x$

10. $y = 2x - 2$

11. $y = 4x - 2$

12. $y = x - 2$

13. $y = x + 3$

14. $y = \frac{1}{2}x$

15. $y = \sqrt{x}, x \geq 0, y \geq 0$

16. $y = \sqrt[4]{x}, x \geq 0, y \geq 0$

17. $y = \sqrt[5]{x}$

18. $y = x^3$

19. $y = x^4, x \geq 0, y \geq 0$

20. $y = x^5$

21. $y = x$

22. $y = \frac{1}{x}$

23. $y = -\frac{1}{x}$

24. $y = -x$

25. $y = 1 - x$

26. $y = 3 - x$

27. What is surprising about the inverse functions to the equations in Exercises 21 to 25 odd?

28. What do the graphs of the functions in Exercises 21 to 26 have in common?

Complete the sentences in Exercises 29 and 30.

29. To solve for x in a power equation $y = x^n$, ...

30. To solve for x in a root equation $y = \sqrt[n]{x}$, ...

The formula $A = \dfrac{\pi r^2 \theta}{360}$ relates the radius, r, and angle measure, θ, of a sector of a circle to the area of the sector, A. Apply this formula to answer the questions in Exercises 31 to 34. Round to the nearest tenth of an inch.

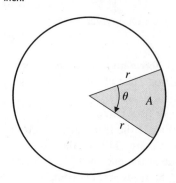

31. A pizza has a 5-inch radius. Find the area of a slice of pizza cut with a central angle of 44°.

32. A pizza slice has an area of 11 square inches. The central angle formed by the slice is 20°. What is the radius of the circle?

33. A pizza slice has an area of 22 square inches. The central angle formed by the slice is 20°. What is the radius of the circle?

34. A pizza has a 10-inch radius. Find the area of a slice of pizza cut with a central angle of 22°.

35. Solve for the radius r of a sphere given the volume, $V = \frac{4}{3}\pi r^3$.

36. Solve for the temperature T in the Stefan-Boltzmann law for the rate of total radiant energy emission, $Q_b = \sigma A T^4$.

37. Solve for the radius r of a tube in the Poiseuille formula for the flow of liquids, $\nu = \pi p r^4 / 8Ln$.

38. Solve for the velocity V in the kinetic energy equation, $K = mV^2/2$.

Exercises 39 to 46 use the compound interest formula as applied to annual growth or depreciation.

39. In 1985, a house was valued at $55,000. In 1998, the same house was valued at $119,000. What annual rate of growth describes the change in value? At this rate of growth, what will the value be in 2010, to the nearest thousand?

40. In 1937, the maximum yearly earnings to which Social Security taxes applied was $3000. In 1998, the maximum was $68,400. What annual rate of growth describes the change in maximum earnings? At this rate of growth, what will the maximum earnings be in 2010, to the nearest thousand?

41. Suppose Microsoft stock was purchased in 1994 for $55. At the same time in 1995, it was worth $87. What is the annual rate of gain? If this same rate continues, what will the stock be worth in the year 2010?

42. A house cost $14,000 in 1970. In 1995, the same house was worth $75,000. What is the annual rate of increase in value? What will the house be worth in 2010?

43. In 1974, a Texas Instruments SR11 calculator cost $96. The calculator could do only addition, subtraction, multiplication, division, and square root. In 1995, a similar calculator cost $5. What annual rate of interest describes the change in price?

44. Suppose a wheat harvester cost $150,000 in 1990 and wore out in 8 years. What annual depreciation rate was incurred if the value of the machine for scrap metal was $5000 in 1998?

45. Suppose a car was purchased in 1990 for $40,000. What annual depreciation rate was incurred if the car was worth $3000 in the year 2000?

46. A Panasonic home fax machine cost $470 in 1994. A better Panasonic fax cost $290 in 1998. What is the annual rate of price change?

In Exercises 47 to 49, solve the equations by taking the appropriate root. Round to the nearest thousandth.

47. $x^3 = 10, x^4 = 10, x^5 = 10, x^6 = 10$

48. $x^3 = 200, x^4 = 200, x^5 = 200, x^6 = 200$

49. What do you observe about the answers in Exercises 47 and 48?

Exercises 50 to 54 refer to the graphs and definitions of $y = \sqrt{x}$, $y = \sqrt[4]{x}$, $y = \sqrt[3]{x}$, *and* $y = \sqrt[5]{x}$.

50. Why do the graphs of the even roots have no points in the second and third quadrants?

51. In the definition of the even nth root of a, what phrase indicates why the even root graphs start at the origin?

52. What do we know about the outputs to even root graphs that explains why there are no points on the graphs in the fourth quadrant?

53. The even nth root of a is real for non-negative inputs. For what set of inputs is the odd principal nth root a real number?

54. How do the principal odd root graphs show that the principal odd nth root of a is real?

55. Why are there limitations to the set of inputs (domain) for graphs of equations containing square roots?

56. Why are there limitations to the set of outputs (range) for equations containing square roots?

57. The graph of $y = \sqrt{3x - 5}$ intersects the graph of $y = \sqrt{2x + 1}$ once. We obtain two numbers when solving $\sqrt{3x - 5} = \sqrt{2x + 1}$ symbolically. What can you conclude about one of the two numbers?

58. Why is there a difference in the graphs of $y = \sqrt{x} - 7$ and $y = \sqrt{x - 7}$?

In Exercises 59 to 62, solve with the graph in the figure.

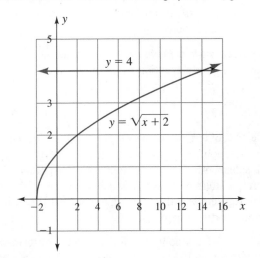

59. $\sqrt{x + 2} = 4$
60. $\sqrt{x + 2} = 2$

61. $\sqrt{x + 2} = -1$
62. $\sqrt{x + 2} = -4$

In Exercises 63 to 84, for what x are the radical equations defined? Solve using symbols, and then check your solution by substitution and by using a graph.

63. $\sqrt{2 - x} = 2$
64. $\sqrt{2 - x} = 3$

65. $\sqrt{x - 5} = 2$
66. $\sqrt{x - 5} = 1$

67. $\sqrt{x - 5} - 6 = -2$
68. $\sqrt{x - 5} - 4 = -1$

69. $\sqrt{2x + 14} = 0$
70. $\sqrt{2x + 14} = 2$

71. $\sqrt{2x + 14} = 6$
72. $\sqrt{2x + 14} = 4$

73. $\sqrt{3x - 5} = 2$
74. $\sqrt{3x - 5} = 4$

75. $\sqrt{3x - 5} = -1$
76. $\sqrt{5x - 1} = -1$

77. $\sqrt{5x - 1} = 3$
78. $\sqrt{5x - 1} = 2$

79. $\sqrt{2x + 7} = \sqrt{x + 2}$

80. $\sqrt{3x + 1} = \sqrt{2x + 1}$

81. $\sqrt{3x - 2} = \sqrt{x + 4}$

82. $\sqrt{7x + 4} = \sqrt{4x + 4}$

83. $\sqrt{3x - 3} = \sqrt{4x - 1}$

84. $\sqrt{3x - 1} = \sqrt{5x - 3}$

85. Solve $r = \sqrt{24L}$ for L. L is the length of skid marks on dry concrete at r mph.

86. Solve $r = \sqrt{12L}$ for L. L is the length of skid marks on wet concrete at r mph.

87. Evaluate the formulas in Exercises 85 and 86 for a 50-foot skid mark. Make a ratio of rates in mph for wet to dry pavement. What conclusion can you draw that is relevant to driving?

88. Solve $v = \sqrt{2gs}$ for s, where v is the final velocity of a falling object, g is the acceleration due to gravity, and s is the total distance fallen.

89. The distance seen in miles from a height of h feet on Earth is $d = \sqrt{3h/2}$.

  a. If we can see approximately 29 miles from the top of the Washington Monument in Washington, D.C., how tall is it?

  b. If we can see approximately 35.8 miles from the top of the Transamerica Pyramid in San Francisco, estimate its height.

  c. The distance seen in miles from a height of h feet on the moon is $d = \sqrt{3h/8}$. Repeat parts a and b for the moon's formula. Why are the answers so different?

Projects

90. *Finding the Area under the Square Root Curve with Graphing Calculator Power Regression.* This exploration will introduce you to finding the formula for the area between the x-axis and a square root curve.

  a. On small-grid graph paper, graph $y = \sqrt{x}$ for x in the interval [0, 25].

b. Make a table with x as the input and the area under the graph of $y = \sqrt{x}$ as the output. Your table should start with 1 as the first input. For the output, estimate the number of squares under the graph up to each input. One way to estimate the area is to count the square if half or more is under the curve and to ignore the square otherwise. Count any unmatched half squares.

c. Enter the data from your table into calculator lists. Calculate a power regression for the lists. Write your equation as $y = ax^b$. Record the coefficient of regression r. The coefficient of regression indicates how well your data fit the calculated equation. An r of 1 or -1 indicates a perfect fit.

91. *Water Supply.* The table shows the population served by a water supply pipe of the indicated inner diameter. The estimates are based on a supply of 60 gallons of water per day per person. An assumption is made that there is a drop of 50 feet per mile between the water reservoir and the city pipes. The data are from a 1909 engineer's handbook.

Diameter of Pipe (inches)	Population Served
6	1647
10	5908
14	13,706
18	25,677
22	42,433
26	64,447
30	91,580
34	125,840
40	188,320
48	297,600
60	511,200
72	800,000
80	1,064,000

From John C. Trautwine, *Trautwine's Engineer's Pocket-Book*, John Wiley & Sons, 1909, p. 653.

a. Graph the data. What shape does the graph have?

b. Use the statistical features on a graphing calculator to fit a curve with the power regression option.

c. How does the consumption of water per person per day in this 1909 table compare with your current water consumption? Find your water usage from your household utility bill, if possible.

92. *Planetary Motion.* The speed of a planet in orbit about the sun varies directly with its distance from the sun. The astronomical unit AU represents the average distance from Earth to the sun—150 million kilometers.

a. Enter the paired data from the table below into a graphing calculator. *Note:* All orbital period data must be in the same units for statistical calculation and graphing.

Planet	Average Distance from Sun	Orbital Period
Mercury	0.387 AU	
Venus	0.723 AU	225 days
Earth	1.000 AU	365 days
Mars	1.52 AU	
Jupiter	5.2 AU	11.9 years
Saturn	9.54 AU	29.5 years
Uranus	19.18 AU	84 years
Neptune	30.07 AU	
Pluto	39.44 AU	

b. Use power regression to fit a curve to the data. Record the regression equation in the form $y = ax^b$.

c. Use your equation to predict the orbital period in days or in years, as appropriate, for Mercury, Mars, Neptune, and Pluto.

93. *Exploration with a Formula.* An almanac for theater set design and construction gives the formula

$$A = \frac{1}{2}ns\sqrt{r^2 - \frac{s^2}{4}}$$

for the area of a regular n-sided polygon (such as a pentagon, hexagon, or octagon). A *regular polygon* has n equal sides with an equal angle at each corner. Although the following activity can be done with any regular figure, we will use a square so that the results are more readily apparent.

a. Let $n = 4$, and solve the formula for r in terms of the area A and side s. The positions of r and s are shown in the figure.

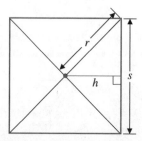

b. Let $A = 64$ in^2 in your result in part a. Graph r in terms of s.

c. Using the table feature on your calculator and the results from part b, fill in r in the table.

s	r	h
4		
8		
12		
16		

d. The variable r describes the hypotenuse of a right triangle within the square. The expression $s/2$ describes one leg of the right triangle. Let h be the other leg, or height, of the right triangle. (See the figure.) Use the Pythagorean theorem to write a formula for h. Then calculate h and complete the table.

e. Using the results in the table, draw each square (side s) to scale. Within each square, draw the triangle on each side formed with h as a height. Using the formula for the area of a triangle $\left(A = \frac{1}{2}bh\right)$, calculate the area of each triangle (even if the triangles overlap) and the sum of the areas. The designs on the chapter opening page show what your squares should look like.

f. You started with a formula describing the area of a square. You let that area be 64 square units. What created all the other squares? What do they have in common?

CHAPTER ⑥ SUMMARY _____

Vocabulary

For definitions and page references, see the Glossary/Index.

conjugate products

cube root

depreciation

digits

exponential equation

extraneous roots

index of a radical

inflation

inverse function

−1 exponents

−2 exponents

multiplication property of radicals

nth root

placeholders

power

power function

power properties

principal nth root

principal square root

product property of like bases

quotient property of like bases

radical sign

rational exponent

rationalizing

real-number conjugates

recursive formula

root function

scientific notation

significant digits

similar radicals

zero exponents

Concepts

6.0 Exponents and Scientific Notation

The definition of an exponent includes $a^0 = 1$, $a \neq 0$ and $a^{-n} = \dfrac{1}{a^n}$, $a \neq 0$.

Scientific notation has one nonzero digit before the decimal point. This number is multiplied by a power of 10.

To change scientific notation to regular decimal notation and the reverse, remember that numbers larger than 1 have a positive exponent on the 10 and small numbers between 0 and 1 have a negative exponent on the 10.

Significant digits include zeros between nonzero digits, zeros following a nonzero digit after a decimal point, and zeros that are placeholders and are marked with an overbar.

6.1 Rational Exponents

Rational exponent expressions may be simplified with this property:

$$b^{m/n} = (b^m)^{1/n} = (b^{1/n})^m, \quad \text{where } b^{1/n} \text{ is defined}$$

Rational exponents may appear in two types of equations:

1. Power equations have the variable in the base and a constant exponent: $y = x^n$.

2. Exponential equations have a constant base and the variable in the exponent: $y = b^x$. Exponential equations appear in Sections 6.0 and 6.1 and again in Chapter 7.

For m and n as rational numbers and p and q as integers, the following real number properties hold.

Product property: $b^m \cdot b^n = b^{m+n}$

Quotient property: $\dfrac{b^m}{b^n} = b^{m-n}, \quad b \neq 0$

Power properties: $(b^m)^n = b^{m \cdot n}$

$$(a \cdot b)^m = a^m \cdot b^m$$

$$\left(\frac{a}{b}\right)^m = \frac{a^m}{b^m}, \quad b \neq 0$$

The number of compounding periods per year in the compound interest formula $A = P(1 + r/n)^{nt}$ is n. The frequency of compounding may be annual ($n = 1$), semi-annual ($n = 2$), quarterly ($n = 4$), monthly ($n = 12$), weekly ($n = 52$), or daily ($n = 365$).

When rounding money amounts, round down if you are receiving the money; round up if you are paying it.

6.2 Roots and Rational Exponents

For all numbers b for which $\sqrt[n]{b}$ is defined,

$$\sqrt[n]{b} = b^{1/n}$$

If n is odd, $b^{n/n} = b$. If n is even, $b^{n/n} = |b|$.

If m/n is a rational number in lowest terms and if $\sqrt[n]{b}$ is defined, then

$$b^{m/n} = (b^{1/n})^m = (\sqrt[n]{b})^m = \sqrt[n]{b^m}$$

The following power properties of radical expressions correspond to the power properties of rational exponents summarized above:

$$\sqrt[p]{\sqrt[q]{b}} = \sqrt[pq]{b}, \quad b > 0$$
$$\sqrt[p]{a \cdot b} = \sqrt[p]{a} \cdot \sqrt[p]{b}, \quad a > 0, \quad b > 0$$
$$\sqrt[p]{\frac{a}{b}} = \frac{\sqrt[p]{a}}{\sqrt[p]{b}}, \quad a > 0, \quad b > 0$$

6.3 Operations with Radicals

Radical expressions can be added or subtracted if they can be changed to similar radicals.

To multiply multiple-term radical expressions, apply the distributive property.

The product of real-number conjugates is a rational number.

To rationalize a one-term expression, multiply by a number whose product with the radical makes an exact root.

To rationalize a two-term expression, multiply by the conjugate.

Checklist for simplifying a radical expression:

1. For a radical with index 2, all perfect square factors have been removed.

2. For a radical with index 3, all perfect cube factors have been removed.

3. The index and exponent of factors in the radicand have no common factors.

4. Exponents in the radicand are smaller than the index.

5. Radicals have been eliminated from the denominator.

6. All possible operations have been performed.

6.4 Inverse Functions; Solving Power and Root Equations

To find the inverse from the table for a function, swap the input and output (x and y) columns and find the rule for the new table. Check whether the inverse is a function.

To find the inverse from the graph of a function, swap the numbers (x, y) in each ordered pair and plot the resulting graph. Check whether the inverse is a function.

To find the inverse from the equation of a function, swap x and y in the equation and in the restrictions and solve (if possible) for y.

To solve equations containing nth powers, take the nth root of both sides. For even roots, limit inputs to $x \geq 0$, as needed, and outputs to $y \geq 0$ or use absolute value.

To solve equations containing nth roots, take the nth power of both sides. It may be necessary to first isolate the radical expression or to square the equations twice. If $a = b$, then $a^n = b^n$ for any positive integer n. Even powers may introduce extraneous roots that do not solve the equation, so always check answers.

Checklist for solving radical equations:

1. Look for extraneous roots.

2. Look for solutions that give negative expressions under the square root sign.

3. Look for problems that cannot be done because of the principal square root definition—for example, problems such as $\sqrt{x + 4} = -4$.

4. Discard possible solutions that are contrary to the geometry or problem setting.

To use the compound interest formula for annual rate of growth, inflation, or depreciation, let $n = 1$ in $A = P(1 + r/n)^{nt}$.

CHAPTER ⑥ REVIEW EXERCISES

In Exercises 1 to 6, simplify.

1. $\dfrac{a^0 b^{-1}}{c^2}$

2. $\dfrac{a^3 b^{-2}}{c^{-1}}$

3. $\left(\dfrac{2x}{y^2}\right)^4$

4. $\dfrac{2}{x^{-1}}$

5. $\left(\dfrac{2x^2}{y}\right)^3$

6. $\left(\dfrac{x^{-1}}{2y^2}\right)^{-3}$

7. Change to decimal notation.

    a. 3.45×10^{-15} b. 6.400×10^{-3}

    c. 4.005×10^5 d. 4.7800×10^3

8. Simplify mentally and write the answer in scientific notation.

    a. $\dfrac{4.5 \times 10^{-2}}{0.9 \times 10^5}$

    b. $\dfrac{3.6 \times 10^4}{0.9 \times 10^{-3}}$

9. According to a 1995 publication of the Hanford Health Information Network, radioactivity released by the Hanford Nuclear Site into the Columbia River between World War II and 1970 included the amounts listed below. Calculate the total disintegrations per second by multiplying the amount released by the curie number.

A curie is 37×10^9 disintegrations per second. Write the answers in both scientific notation and decimal notation.

a. Arsenic-76: 2,500,000 curies

b. Neptunium-239: 6,300,000 curies

c. Phosphorus-32: 230,000 curies

d. Sodium-24: 12,000,000 curies

e. Zinc-65: 490,000 curies

A disintegration is the release of an alpha, beta, or gamma particle.

10. Suppose you start at $7 per hour and are given a 4% raise at the end of each year. Find your hourly wage after 10 years. Find your hourly wage if you get a 2% raise every six months for the 10 years.

11. Find the amount of money in a savings account if $1200 is deposited at 6% interest compounded daily and the money is left for 7 years.

12. Find the amount of money in a savings account if $1600 is deposited at 5% interest compounded monthly and the money is left for 4 years.

13. Repeat Exercise 11 using $2\frac{1}{2}$ years.

14. Repeat Exercise 12 using $3\frac{3}{4}$ years.

In Exercises 15 to 18, simplify without a calculator.

15. a. $64^{1/2}$ **b.** $64^{1/6}$ **c.** $64^{5/6}$

16. a. $32^{1/5}$ **b.** $16^{3/4}$ **c.** $32^{4/5}$

17. a. $16^{0.75}$ **b.** $25^{3/2}$ **c.** $27^{4/3}$

18. a. $9^{1.5}$ **b.** $8^{4/3}$ **c.** $4^{5/2}$

Simplify the expressions in Exercises 19 and 22. Assume the variables represent positive numbers.

19. a. $(b^4)^{3/4}$ **b.** $a^{2/3} \cdot a^{1/3}$ **c.** $x^{3/4} \cdot x^{1/2}$

 d. $\dfrac{b^{3/4}}{b^{1/2}}$ **e.** $b \cdot b^x$

20. a. $(a^2)^{3/2}$ **b.** $x^{1/3} \cdot x^{2/3}$ **c.** $a^{1/4}a^{3/2}$

 d. $\dfrac{b^{2/3}}{b^{1/2}}$ **e.** $x \cdot x^n$

21. a. $x^{1/2} \cdot x^{1/4}$ **b.** $\dfrac{x^n}{x}$ **c.** $(x^{1/4})^{2/3}$

 d. $\dfrac{a^{2/3}}{a^{4/3}}$ **e.** $\dfrac{a}{a^x}$

22. a. $b^{3/4} \cdot b^{1/4}$ **b.** $(a^{\frac{4}{3}})^{\frac{3}{4}}$ **c.** $\dfrac{x^n}{x^{n-1}}$

 d. $x \cdot x^{n-1}$ **e.** $b^{1/2} \cdot b$

Change each expression in Exercises 23 and 24 to exponential notation, and simplify.

23. a. $\sqrt[3]{-8}$ **b.** $\sqrt[4]{\frac{625}{16}}$ **c.** $\sqrt[3]{8^2}$

24. a. $\sqrt[3]{\frac{27}{8}}$ **b.** $\sqrt[5]{-32}$ **c.** $\sqrt{4^3}$

Change each expression in Exercises 25 and 30 to radical notation. Simplify. Radical notation may vary.

25. a. $125^{2/3}$ **b.** $64^{1/3}$ **c.** $32^{2/5}$

26. a. $64^{3/2}$ **b.** $32^{3/5}$ **c.** $27^{2/3}$

27. a. $(-64)^{1/3}$ **b.** $-64^{1/2}$ **c.** $(-16)^{1/4}$

28. a. $-16^{0.75}$ **b.** $(-64)^{1/2}$ **c.** $-27^{1/3}$

29. a. $x^{2/3}$ **b.** $x^{1.5}$ **c.** $a^{0.75}$

30. a. $a^{3/2}$ **b.** $a^{1.25}$ **c.** $x^{0.8}$

Simplify Exercises 31 and 32. Write your observations about patterns in the answers. Explain why the patterns hold. Show four significant digits.

31. a. $\sqrt[3]{8000}$ **b.** $\sqrt[3]{800}$ **c.** $\sqrt[3]{80}$

 d. $\sqrt[3]{8}$ **e.** $\sqrt[3]{0.8}$ **f.** $\sqrt[3]{0.08}$

 g. $\sqrt[3]{0.008}$

32. a. $\sqrt[3]{27000}$ **b.** $\sqrt[3]{2700}$ **c.** $\sqrt[3]{270}$

 d. $\sqrt[3]{27}$ **e.** $\sqrt[3]{2.7}$ **f.** $\sqrt[3]{0.27}$

 g. $\sqrt[3]{0.027}$

In Exercises 33 and 34, simplify. Identify restrictions necessary to have real-number outputs.

33. a. $\sqrt{9n}$ **b.** $\sqrt{90n^2}$

 c. $\sqrt{0.09x}$ **d.** $\sqrt{900x^4}$

34. a. $\sqrt{0.9n^2}$ **b.** $\sqrt{9000x^2}$

 c. $\sqrt{0.009x}$ **d.** $\sqrt{90x^3}$

Simplify in Exercises 35 to 38. Assume the variables can be any real number.

35. a. $\sqrt{z}\sqrt{z^5}$ **b.** $\sqrt[3]{x^2} \cdot \sqrt[3]{x}$ **c.** $\sqrt[3]{x^5} \cdot \sqrt[3]{x^1}$

36. a. $\sqrt[4]{32}$ **b.** $\sqrt{72b^3}$ **c.** $\sqrt[4]{32x^6y^5z^4}$

37. a. $\sqrt[3]{x} \cdot \sqrt[2]{x}$ **b.** $\sqrt{x^3} \cdot \sqrt{x^2}$ **c.** $\sqrt[2]{\sqrt[3]{x}}$

38. a. $\sqrt{x^3} \cdot \sqrt{x^4}$ **b.** $\sqrt[3]{x} \cdot \sqrt[4]{x}$ **c.** $\sqrt[3]{\sqrt[4]{x}}$

Simplify the expressions in Exercises 39 to 42 by performing the indicated operations.

39. a. $\sqrt{3} + 2\sqrt{12}$

 b. $\sqrt{9x} + \sqrt{x}$

40. a. $\sqrt{x} + \sqrt{x^3}$

 b. $\sqrt{16x} - \sqrt{x}$

41. a. $(3 + \sqrt{6})(3 - \sqrt{6})$

   b. $(3 - \sqrt{x})(3 - \sqrt{x})$

42. a. $(2 - \sqrt{2})(2 - \sqrt{2})$

   b. $(5 - \sqrt{2x})(5 + \sqrt{2x})$

43. Rationalize the denominators.

   a. $\dfrac{1}{b - \sqrt{a}}$

   b. $\dfrac{1}{\sqrt{a} + \sqrt{b}}$

   c. $\dfrac{1}{\sqrt[3]{9x}}$

44. Use $(a + b)^2$ as a model to multiply out these problems.

   a. $(2^x + 2^{-x})^2$ **b.** $(c^x + c^{-x})^2$

   c. $(b^{1/2} + b^{1/3})^2$ **d.** $(a^{1/3} + a^{1/2})^2$

   e. $(\sqrt{7} + \sqrt{5})^2$ **f.** $(\sqrt{a} + \sqrt{b})^2$

In Exercises 45 and 46, substitute the numbers into the quadratic equation to show whether the numbers are solutions.

45. $x^2 - 2x - 1 = 0$, where $x = 1 - \sqrt{2}$. What is the other solution?

46. $x^2 - 4x - 2 = 0$, where $x = 2 - \sqrt{6}$. What is the other solution?

47. What is the inverse to each set of ordered pairs? Is it a function?

   a. $(3, 3), (4, 5), (5, 7)$

   b. $(-2, -1), (0, -2), (2, -3)$

48. Make a table with four entries, and graph $y = 2^x + 1$. Make a table for the inverse function, and sketch it on the graph.

49. Find the inverse function to each equation.

   a. $y = 2x - 1$ **b.** $y = 3x$

   c. $y = x^3$ **d.** $y = \sqrt{x}, x \geq 0, y \geq 0$

   e. $y = -x$ **f.** $y = \dfrac{1}{x}$

In Exercises 50 to 53, set up an equation based on $A = P(1 + r)^t$, and solve it.

50. Suppose that a starting wage of $5.50 has increased to $8.00 over a period of 10 years. What annual rate of increase does this change represent?

51. Suppose that a starting wage of $5.50 has increased to $9.00 over a period of 10 years. What annual rate of increase does this change represent?

52. A car cost $7000 new in 1983. A comparable car cost $17,000 in 1995. What annual rate of increase in cost (inflation) does this change represent?

53. A car cost $10,000 new in 1990. A comparable car cost $15,000 in 1995. What annual rate of increase in cost (inflation) does this change represent?

54. Solve for V: $r = \sqrt[3]{\dfrac{3V}{4\pi}}$.

55. Solve the formula for the period of a pendulum, given by $T = 2\pi\sqrt{L/g}$, for g, the acceleration due to gravity.

56. Solve each of the following for y.

   a. $x^3 + y^3 = a^3$

   b. $x^2 = \dfrac{8}{27k}(y - k)^3$

Solve the equations in Exercises 57 to 60 for the indicated variable.

57. Force of attraction between two charges separated by a distance r: $F = \dfrac{Q_1 Q_2}{kr^2}$ for r

58. Gravitational attraction between two masses with r as the distance between their centers: $F = k\dfrac{M_1 M_2}{r^2}$ for M_1

59. Lemniscate curve: $r^3 = a^2 p$ for r

60. Gerono's lemniscate: $x^4 = a^2(x^2 - y^2)$ for y

61. a. Solve $\sqrt{8x - 1} = \sqrt{5x - 1}$ from the graph below.

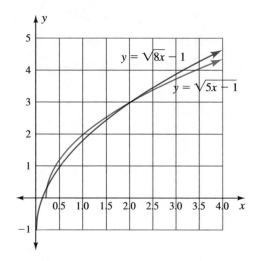

b. Solve $2 = \sqrt{5x - 1}$ with symbols and from the graph.

c. Solve $\sqrt{8x - 1} = 1$ with symbols and from the graph.

d. Solve $\sqrt{8x - 1} = \sqrt{5x - 1}$ with symbols.

62. **a.** Solve $\sqrt{2x} + 1 = \sqrt{3x - 5}$ from the graph below.

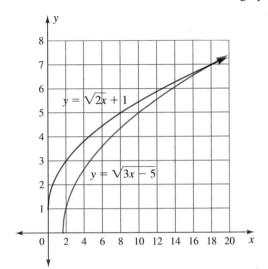

b. Solve $\sqrt{2x} + 1 = 5$ with symbols and from the graph.

c. Solve $\sqrt{3x - 5} = 0$ with symbols and from the graph.

d. Solve $\sqrt{2x} + 1 = \sqrt{3x - 5}$ with symbols.

63. **a.** Solve $\sqrt{3x} + 2 = \sqrt{6x - 8}$ from the graph below.

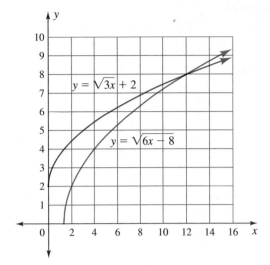

b. Solve $\sqrt{6x - 8} = 4$ with symbols and from the graph.

c. Solve $\sqrt{3x} + 2 = 2$ with symbols and from the graph.

d. Solve $\sqrt{3x} + 2 = \sqrt{6x - 8}$ with symbols.

CHAPTER **6** TEST

Simplify the expressions in Exercises 1 to 3.

1. $\dfrac{a^0 b^2}{b^{-1}}$ **2.** $\dfrac{a^{-2}b^3}{(bc)^2}$ **3.** $\left(\dfrac{x^2 y}{3x}\right)^{-2}$

4. Explain why $2x^2 \neq (2x)^2$.

5. Explain why $\sqrt{4x^2} \neq 2x$.

6. **a.** Write in decimal notation: 3.450×10^{-5}.

b. Simplify $\dfrac{6.3 \times 10^{-2}}{7 \times 10^4}$.

7. Simplify; note restrictions so that the outputs are real numbers.

a. $\sqrt{36x}$ **b.** $\sqrt{0.36x^2}$ **c.** $\sqrt{360x^4}$

8. Show whether $a\left(\dfrac{b}{a}\right)^4 = b\left(\dfrac{b}{a}\right)^3$.

9. Find the amount of money in an account if $2000 is deposited at 8% interest compounded quarterly and the money is left for 3 years.

10. A car cost $12,000 new in 1990. A comparable car cost $18,000 in 1995. What annual rate of increase in cost (inflation) does this change represent?

11. Solve the formula $M = \dfrac{mgl}{\pi r^2 s}$ for r.

12. Solve the formula $F = \dfrac{mv^2}{r}$ for v.

In Exercises 13 to 15, change each expression to radical notation. Simplify.

13. $64^{2/3}$ **14.** $32^{4/5}$ **15.** $81^{3/4}$

16. Simplify

a. $\sqrt[4]{\frac{16}{81}}$ **b.** $\sqrt[3]{-27}$ **c.** $-\sqrt[3]{-64}$

17. Simplify:

a. $(b^2)^{3/2}$, $b \geq 0$ **b.** $b^{2/3}b^{1/3}$, $b \geq 0$ **c.** $3 \cdot 3^n$

18. Simplify by performing the indicated operation.

a. $(2 - \sqrt{x})(2 + \sqrt{x})$

b. $(8 - \sqrt{3})(3 + \sqrt{8})$

19. Rationalize the denominators.

a. $\dfrac{1}{x + \sqrt{y}}$ **b.** $\dfrac{1}{\sqrt[3]{x}}$

20. Substitute $x = 4 - \sqrt{10}$ into $x^2 - 8x + 6 = 0$ to show whether it is a solution.

21. What is the inverse to the function described by the ordered pairs $(-1, -1)$, $(-2, -3)$, and $(-3, -5)$? Is it a function?

22. Make a table containing four ordered pairs, and sketch a graph of $y = \frac{1}{2}x$. Make a table for the inverse function to $y = \frac{1}{2}x$, and sketch its graph.

23. Find the inverse function to $3x - 2y = 6$.

24. **a.** Solve $\sqrt{x + 1} = \sqrt{2x - 1}$ from the graph below.

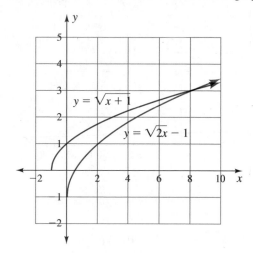

b. Solve $\sqrt{x + 1} = 2$ with symbols. Confirm with the graph.

c. Solve $\sqrt{2x - 1} = 1$ with symbols and from the graph.

d. Solve $\sqrt{x + 1} = \sqrt{2x - 1}$ with symbols.

CUMULATIVE REVIEW OF CHAPTERS I TO 6

1. Solve $5 - 3(x + 2) = 32$.

2. Simplify $\dfrac{6 + 9x}{15}$.

3. Solve for b: $A = \dfrac{a + b + c}{3}$.

4. Solve $3 - x < 16 < 10 - x$ for $a < x < b$ form.

5. In January, the County Electric bill was $68.46 for 1120 kilowatt hours of electricity. In February, the bill was $83.33 for 1410 kilowatt hours. Define variables and find the equation County Electric uses for billing. What assumptions did you make?

6. If $f(x) = 3 - x$, what is $f(3)$? $f(-2)$?

7. **a.** What is the slope of the line AB connecting $A(4, -2)$ and $B(-1, 1)$?

   b. What is the slope of a line perpendicular to AB?

8. Find a linear function through the origin that is parallel to $3y - 6x = 3$.

9. What are the set of inputs (domain) and set of outputs (range) for the constant function $y = 4$?

10. Show whether the lengths 3.5, 12, and 12.5 are the sides of a right triangle.

11. Let $f(x) = 2x^2 - 5x + 6$.

    a. Find $f(-2)$ and $f(0)$.

    b. What is the meaning of $f(0)$ on the graph of $f(x)$?

c. What is the vertex of $f(x)$?

d. Show the quadratic formula and use it to find the x-intercepts.

12. Solve each equation using the graph in the figure.

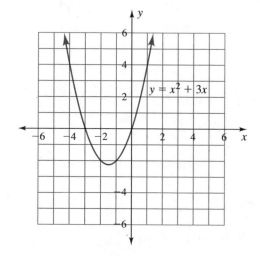

a. $x^2 + 3x = 4$ **b.** $x^2 + 3x = 0$

c. $x^2 + 3x = -5$

13. Solve the equations in parts a and b of Exercise 12 by factoring.

14. Solve $x^2 + 3x > 0$ from the figure in Exercise 12.

15. What needs to be added to $x^2 + 7x$ to make the square of a binomial?

16. Multiply $(x^3 - 3x^2 + 3x - 1)(x - 1)$.

17. Give the next number in each pattern. Identify whether the sequence is from a linear function, a quadratic function, or neither. Fit an equation to the linear and quadratic sequences.

  a. 2, 5, 10, 17, 26   **b.** −5, −2, 1, 4, 7

  c. −3, 1, 7, 15, 25   **d.** −2, −1, 1, 5, 13

18. Multiply and simplify using the properties of imaginary numbers.

  a. $(8 - 2i)(8 + 2i)$ **b.** $(5 - 4i)(3 + 2i)$

  c. $\dfrac{1}{2 + i} \cdot \dfrac{2 - i}{2 - i}$

19. Find all the solutions, real and imaginary, to $0 = x^3 - 1$.

20. Describe how to get the graph of $y = (x - a)^2$ from that of $y = x^2$.

21. Suppose the dive path of the clavadistas (cliff divers) is approximated by the equation $y = 0.09x - 0.16x^2$, where x and y are in feet. Let the diver start at the origin and let the surface of the water be 115 feet below the diver.

  a. What is the vertex of the dive?

  b. What is the horizontal distance the diver has traveled when he reaches the water?

22. Tender Touch Diapers cost $8.99 per package of 56 diapers. Suppose a baby uses 10 diapers each day. Set up a unit analysis to find the cost of a 365-day supply of diapers.

Match each expression in Exercises 23 to 28 with the equivalent expression below.

a. $\dfrac{x}{y + x}$ **b.** $1 + \dfrac{b}{a - b}$ **c.** $\dfrac{x}{y} + 1$

d. $\dfrac{x}{y - x} + \dfrac{y}{x - y}$ **e.** $\dfrac{1}{b} + \dfrac{1}{a}$ **f.** $\dfrac{-a}{a - 1}$

23. $\dfrac{x + y}{y}$ **24.** $\dfrac{a + b}{ab}$ **25.** $\dfrac{x}{x + y}$

26. $\dfrac{a}{1 - a}$ **27.** $\dfrac{a}{a - b}$ **28.** $\dfrac{x - y}{y - x}$

29. Multiply and simplify: $\dfrac{x^2 + 3x}{x - 1} \cdot \dfrac{x^2 - 1}{3x^2}$.

30. Divide $x^3 - 1$ by $x - 1$.

31. The ideal Body Mass Index (BMI) relates height and weight. The formula

$$19 \le \frac{(150)(704.5)}{x^2} \le 24$$

shows the BMIs for people weighing 150 pounds. The variable x represents height in inches. For what range of heights in inches should people weigh 150 pounds?

32. Change to decimal notation:

  a. 3.80×10^{-16}

  b. 4.23×10^{15}

33. A capsule of three types of bacteria contains at least 0.825 billion each of Lactobacillus acidophilus, Lactobacillus bifidus, and Streptococcus faecium. What is the total bacterial content of the capsule? Write the total in both decimal notation and scientific notation.

34. Simplify these expressions. Assume the variables represent any real number.

  a. $32^{4/5}$ **b.** $\sqrt[3]{125^2}$ **c.** $\sqrt[4]{-81}$

  d. $-\sqrt{2}\sqrt{8}$ **e.** $x^{3/5} \cdot x^{1/3}$ **f.** $\dfrac{x^{1/2}}{x^{-3/2}}$

  g. $\sqrt{18a^3}$ **h.** $\sqrt[3]{16x^3y^4z^5}$ **i.** $(x^{\frac{2}{3}})^{\frac{1}{2}}$

  j. $\sqrt{8x} + \sqrt{2x} + \sqrt{50x}$

  k. $(2 - \sqrt{2})(2 + \sqrt{2})$

35. Solve for x in the following equations. Use algebraic notation, and then graph the left and right sides to check.

  a. $x + 2 = 3\sqrt{x}$ **b.** $x = \sqrt{6 - x}$

36. Find the annual rate of interest if $1500 grows to $2500 when interest is compounded annually for 6 years. Repeat for interest compounded monthly.

37. Solve for B in Steinmetz's equation, related to the magnetization of iron or steel: $W = nB^{1.6}$. (*Hint:* Change the decimal exponent to a fraction.)

38. The length of the line between two ordered pairs is

$$d = \sqrt{(x_2 - x_1)^2 + (y_2 - y_1)^2}$$

Find the length of the three segments connecting $(-2, 2)$, $(4, 1)$, and $(3, -5)$. Identify the resulting figure as completely as possible.

7

Exponential and Logarithmic Functions

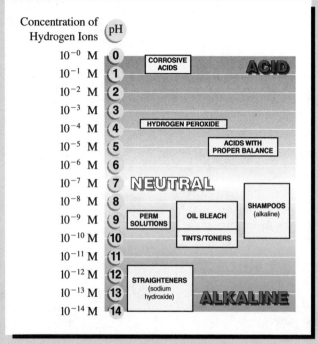

Figure 1

Chemists rate common hair products according to their pH, a measure of acidity or basicity. Acid-balanced shampoos, bleaches, and hair straighteners all have a pH number; see Figure 1. The pH scale is based on the exponential and logarithmic functions of this chapter. We will work with pH in Section 7.3.

7.0   Patterns, Geometric Sequences, and Exponential Equations

OBJECTIVES

- Create tables and graphs, given activity settings or problem situations.
- Find the nth term, a_n, for a geometric sequence.
- Identify sequences created by linear, quadratic, or exponential functions that are restricted to positive integer inputs.
- Find the function of an exponential sequence with exponential regression.
- Show that the function and the nth term expression for a given sequence of numbers are equivalent.

WARM-UP

Give the next number in each sequence.

1. 3, 6, 12, 24, ____   **2.** 5, 8, 11, 14, ____

3. 1, 4, 9, 16, ____   **4.** 9, 5, 1, −3, ____

5. 3, 10, 21, 36, ____   **6.** 81, 27, 9, 3, ____

7. 4, 12, 36, 108, ____

I N THIS SECTION, we explore patterns created by exponential expressions and write exponential expressions as geometric sequences. We distinguish geometric sequences from arithmetic and quadratic sequences. We find equations from geometric sequences using patterns and calculator regression.

Patterns

In Examples 1 and 2, we consider two applications that generate patterns.

EXAMPLE

Exploring patterns: steel in Japanese swords   The traditional Japanese sword is made from steel heated to a temperature higher than 1200°C (2200°F), hammered, folded, and hammered and folded again many times. This process removes impurities and creates a blade that does not break or bend.

How many layers of steel would there be in a sword blade created by heating and folding the steel in half ten times?

a. Model the folding process with a piece of paper (three folds are shown in Figure 2). Then make a table with number of folds as input and number of layers of paper in the stack as output. When you get to the point where the paper becomes too thick to fold, predict the number of layers of paper that would be in the stack.

b. Graph the data from your table.

c. Describe the number of layers of paper in the stack in terms of the number of folds.

d. Is the pattern in the table and graph an increasing or a decreasing function?

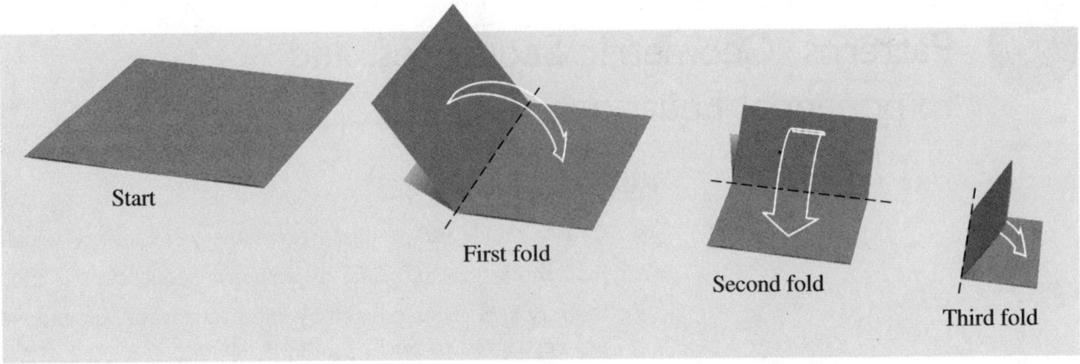

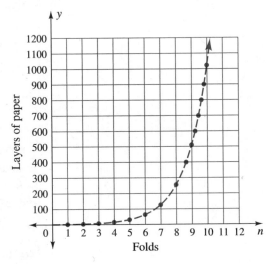

Figure 2

Solution   **a.** The number of layers of paper for each number of folds is given in Table 1.

b. The graph of Table 1 is shown in Figure 3. The points are connected with a dashed line to show the shape of the curve.

Fold	Layers of Paper
1	2
2	4
3	8
4	16
5	32
6	64
7	128
8	256
9	512
10	1024

Table I

Figure 3

c. The number of layers of paper in the stack is always a power of 2. Thus, for n folds, the number of layers of paper in the stack is 2^n.

d. The function is increasing.

EXAMPLE **2**   Exploring patterns: superball bounces   Suppose a superball bounces to $\frac{2}{3}$ of its previous height with each successive bounce. We start by dropping the ball from a height of 108 inches (see Figure 4).

a. Make a table for the first five heights of the ball. Use 1 to 5 as inputs and the heights of the ball as outputs.

b. Graph the data from your table.

c. Describe the nth height of the ball.

d. Is the pattern in the table and graph an increasing or a decreasing function?

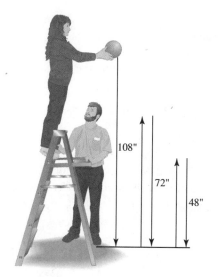

Figure 4

Solution   **a.** Table 2 shows the height to be $\frac{2}{3}$ of the prior height.

b. The graph of Table 2 is in Figure 5. The points are connected with a dashed line to show the shape of the curve.

Input, n	Output: Height, y
1	108
2	$108\left(\frac{2}{3}\right) = 72$
3	$108\left(\frac{2}{3}\right)\left(\frac{2}{3}\right) = 48$
4	$108\left(\frac{2}{3}\right)\left(\frac{2}{3}\right)\left(\frac{2}{3}\right) = 32$
5	$108\left(\frac{2}{3}\right)\left(\frac{2}{3}\right)\left(\frac{2}{3}\right)\left(\frac{2}{3}\right) = 21\frac{1}{3}$

Table 2

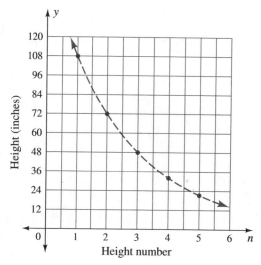

Figure 5

c. The table suggests that the nth height will be $108\left(\frac{2}{3}\right)^{n-1}$.

d. The function is decreasing.

Geometric Sequences

Recall that a sequence is a function with inputs from the set of positive integers. Because the inputs in Examples 1 and 2 are positive integers—the number of folds and the number of heights—we can think of the outputs as sequences.

FINDING DIFFERENCES   We identify arithmetic (linear) and quadratic sequences by finding differences. Arithmetic sequences are characterized by a constant common first difference between terms. When the second differences in a sequence are constant, we have a **quadratic sequence,** which *can be described by a quadratic function.*

In Example 3, we find whether any common differences occur for the sequences described by 2^n, $108\left(\frac{2}{3}\right)^{n-1}$, and 3^n.

EXAMPLE **3**   Finding differences   Find the first and second differences for the sequences given by

a. 2^n **b.** $108\left(\frac{2}{3}\right)^{n-1}$ **c.** 3^n

Solution   **a.** 2^n: 2, 4, 8, 16, 32
∨ ∨ ∨ ∨
2, 4, 8, 16   The first differences are the original sequence.
(Check that the second differences are also the original sequence.)

b. $108\left(\frac{2}{3}\right)^{n-1}$:   108   72   48   32 $21\frac{1}{3}$
∨   ∨   ∨   ∨
36   24   16 $10\frac{2}{3}$ The first differences are half the numbers in the original sequence.
∨   ∨   ∨
12   8 $5\frac{1}{3}$ The second differences are half the numbers in the first differences.

c. 3^n: 3, 9, 27, 81, 243
∨ ∨ ∨ ∨
6, 18, 54, 162   The first differences are twice the original sequence.
∨ ∨ ∨
12, 36, 108   The second differences are twice the first differences. ●

In Example 3, *the differences are either the same as or a multiple of the numbers in the original sequence.* This distinctive pattern suggests a new type of sequence. However, it does not tell us the slope or coefficients in the equation, as do the constant differences in linear and quadratic sequences. And the repetition of the sequence in the differences occurs in various types of sequences (see Exercise 48). We need a more informative tool to describe these sequences.

FINDING RATIOS   Considering how the sequences in Exercises 1 and 2 are formed—by multiplying by a constant—gives us another way to describe the sequences. Instead of finding differences between terms of the sequence, we *divide* the terms.

EXAMPLE **4**   Finding ratios   For each sequence, find the ratio of consecutive terms by dividing each term by the preceding term, a_n/a_{n-1}.

a. 2^n: 2, 4, 8, 16, 32   **b.** $108\left(\frac{2}{3}\right)^{n-1}$: 108, 72, 48, 32, $21\frac{1}{3}$

c. 3^n: 3, 9, 27, 81, 243

Solution   We divide the second term by the first, the third by the second, and so forth.

a. $\dfrac{4}{2} = 2, \dfrac{8}{4} = 2, \dfrac{16}{8} = 2$, and $\dfrac{32}{16} = 2$. Every ratio is 2.

b. $\dfrac{72}{108} = \dfrac{2}{3}, \dfrac{48}{72} = \dfrac{2}{3}, \dfrac{32}{48} = \dfrac{2}{3}$, and $\dfrac{21\frac{1}{3}}{32} = \dfrac{2}{3}$. Every ratio is $\dfrac{2}{3}$.

c. $\dfrac{9}{3} = 3, \dfrac{27}{9} = 3, \dfrac{81}{27} = 3$, and $\dfrac{243}{81} = 3$. Every ratio is 3. ●

Think about it: Show, without a calculator, that $\dfrac{21\frac{1}{3}}{32} = \dfrac{2}{3}$.

Example 4 suggests that instead of a common difference, these sequences have a common ratio. This finding leads to the definition of a geometric sequence:

Definition of Geometric Sequence

A **geometric sequence** is a sequence of terms in which the ratio of each term to the preceding term is the same for all terms.

The ratio obtained from the consecutive terms is the **common ratio, *r*.**

THE *n*TH TERM OF A GEOMETRIC SEQUENCE   The common ratio is essential for finding the *n*th term, a_n, of a geometric sequence, just as the common difference was important in finding the *n*th term of an arithmetic sequence, $a_n = a_1 + (n - 1)d$.

The ***n*th term of a geometric sequence** is
$$a_n = a_1 \cdot r^{n-1}$$
where a_1 is the first term of the sequence and r is the common ratio of terms.

EXAMPLE **5**

Exploring the *n*th term   Consider the sequence 4, 12, 36, 108, 324.

a. Show whether the sequence is geometric.

b. How many times do we multiply by the common ratio, $r = 3$, to get each term in the sequence?

c. What is a_n for the sequence?

d. Why does the *n*th term formula, $a_n = a_1 \cdot r^{n-1}$, make sense?

Solution **a.** $\frac{12}{4} = 3$, $\frac{36}{12} = 3$, $\frac{108}{36} = 3$, and $\frac{324}{108} = 3$. The common ratio is 3. Thus, the sequence is geometric.

b. $a_1 = 4$
$a_2 = 4 \cdot 3 = 12$ We multiply once by the common ratio to get a_2.
$a_3 = 4 \cdot 3 \cdot 3 = 36$ We multiply twice by the common ratio to get a_3.
$a_4 = 4 \cdot 3 \cdot 3 \cdot 3 = 108$ We multiply three times by the common ratio to get a_4.
$a_5 = 4 \cdot 3 \cdot 3 \cdot 3 \cdot 3 = 324$ We multiply four times by the common ratio to get a_5.

c. $a_n = 4 \cdot 3^{n-1}$

d. For each term a_n in the geometric sequence, we multiply by $n - 1$ common ratios. The common ratio is the base in the exponential expression $a_1 \cdot r^{n-1}$. Because we start each multiplication with the first term, a_1, the first term is also needed in finding the *n*th term, a_n.   ●

Exponential Functions

In Sections 6.0 and 6.1, we worked with exponential expressions. We now define the exponential function and relate it to geometric sequences.

Definition of Exponential Function

An **exponential function** is written as
$$f(x) = a \cdot b^x$$
where the coefficient a is a constant, the base b is positive but not equal to 1, and the exponent x is a real number.

Any exponential function with the set of inputs (domain) limited to the positive integers can be written as a geometric sequence. Thus, if x is limited to positive integers, the function $f(x) = 4 \cdot 3^{x-1}$ has the same set of outputs as the sequence in Example 5, $a_n = 4 \cdot 3^{n-1}$. Going from sequences to functions requires more limitations (more on that after Example 6).

EXAMPLE **6**

Finding *n*th terms and functions   Identify each sequence as either arithmetic, quadratic, or geometric. Write a linear, quadratic, or exponential function for each.

a. 3, 6, 12, 24, 48   **b.** 5, 8, 11, 14, 17

c. 1, 4, 9, 16, 25   **d.** 9, 5, 1, −3, −7

e. 3, 10, 21, 36, 55   **f.** 81, 27, 9, 3, 1

Solution   **a.** *Sequence:*   3, 6, 12, 24, 48
                            ∨ ∨ ∨ ∨
First differences: 3, 6, 12, 24
The differences are a repeat of the sequence itself. This suggests a geometric sequence. We check for a common ratio and find $r = 2$. The sequence is geometric with $a_1 = 3$ and $r = 2$ in $a_n = ar^{n-1}$, so $a_n = 3 \cdot 2^{n-1}$. The function is exponential, $f(x) = 3 \cdot 2^{x-1}$.

b. *Sequence:*   5, 8, 11, 14, 17
                            ∨ ∨ ∨ ∨
First differences: 3, 3, 3, 3
The first differences are constant, so the function is linear with slope $m = 3$. The y-intercept, where $x = 0$, is 3 less than the first term: $b = 5 - 3 = 2$. The linear function is $f(x) = 3x + 2$, or $y = 3x + 2$. If we use $a_n = a_1 + (n - 1)d$, we get

$$a_n = 5 + (n - 1)3$$
$$= 5 + 3n - 3$$
$$= 3n + 2 \qquad \text{The same as the linear equation}$$

We check these results with linear regression.

c. *Sequence:*   1, 4, 9, 16, 25
                            ∨ ∨ ∨ ∨
First differences: 3, 5, 7, 9
                            ∨ ∨ ∨
Second differences: 2, 2, 2   The second differences are constant.
The sequence is made up of the squares of the integers, so $f(x) = x^2$. We check this result with quadratic regression.

d. *Sequence:*   9, 5, 1, −3, −7
                            ∨ ∨ ∨ ∨
First differences: −4, −4, −4, −4   The first differences are constant.
The function is linear with slope $m = -4$. Because the slope is −4, the y-intercept will be 4 larger than the first term: $b = 9 + 4 = 13$. The linear function is $f(x) = -4x + 13$, or $y = -4x + 13$. If we use $a_n = a_1 + (n - 1)d$, we get

$$a_n = 9 + (n - 1)(-4)$$
$$= 9 + (-4)n - 1(-4)$$
$$= -4n + 13 \qquad \text{The same as the linear equation}$$

We check these results with linear regression.

e. *Sequence:*   3, 10, 21, 36, 55
                               ∨   ∨   ∨   ∨
First differences: 7, 11, 15, 19
                                 ∨   ∨   ∨
Second differences: 4,   4,   4   The second differences are constant.

The sequence is quadratic. With quadratic regression, we find the sequence to be $a_n = 2n^2 + n$, from the function $f(x) = 2x^2 + x$.

f. *Sequence:*   81, 27,   9,   3,   1
                                ∨   ∨   ∨   ∨
First differences: −54, −18, −6, −2
                                   ∨   ∨   ∨
Second differences: −36, −12, −4

Neither the first nor the second differences are constant. The first differences are negatives of numbers twice the size of those in the sequence. We divide consecutive terms and find a common ratio, $\frac{1}{3}$. The sequence is geometric with $a_1 = 81$ and $a_n = a_1 r^{n-1}$. Thus $a_n = 81 \cdot \left(\frac{1}{3}\right)^{n-1}$. The function is exponential, $f(x) = 81 \cdot \left(\frac{1}{3}\right)^{x-1}$.   ●

COMPARING EXPONENTIAL FUNCTIONS AND GEOMETRIC SEQUENCES   In both the exponential function and the geometric sequence, the variable is in the exponent, but there are two important differences between the exponential function and the geometric sequence.

1. The domains are different. The exponential function is defined for any real-number input, whereas the geometric sequence is limited to positive integer inputs. This means that 3^π and $3^{\sqrt{2}}$ are meaningful exponential expressions, but the irrational numbers π and $\sqrt{2}$ would not be meaningful inputs to n in $a_n = ar^{n-1}$ in a geometric sequence.

2. The base is more restricted for the exponential function than for the geometric sequence. A geometric sequence may have a positive or a negative base (the common ratio), whereas the base of the exponential function must be positive. The base $b = 1$ is excluded from the exponential function because it represents a *constant* function (see Section 2.4).

The differences between a geometric sequence and an exponential function are summarized in Table 3.

Function	**Inputs (Domain)**	**Base**
Geometric sequence $a_n = a_1 r^{n-1}$	n is a positive integer	$r < 0$ or $r > 0$
Exponential function $y = ab^x$.	x is a real number	$b > 0$, $b \neq 1$

Table 3   Differences between geometric sequences and exponential functions

EXPONENTIAL REGRESSION   In addition to the linear and quadratic regression used in Example 6, graphing calculators also have exponential regression. Because the base of an exponential function must be positive and not equal to 1, we can use calculator exponential regression to find the rule for any geometric sequence with a common ratio that is positive and not equal to 1.

Graphing Calculator Technique:
Exponential Regression

Clear lists.

Place in list 1, L_1, the sequence inputs or term numbers, such as 1, 2, 3, 4, 5.

Place in list 2, L_2, the terms of the sequence, such as 3, 6, 12, 24, 48.

Return to the menu and calculate an exponential regression for L_1 and L_2.

The equation data are given as a and b for $y = ab^x$, followed by the correlation coefficient r. As before, if $r = 1$ or $r = -1$, the equation is a perfect fit for the data. The closer r is to 1 or -1, the better the fit. Zero indicates the worst fit.

In this example, $a = 1.5$ and $b = 2$, so the exponential function is $y = 1.5 \cdot 2^x$.

EXAMPLE **7**

Using exponential regression   Explain why exponential regression can be applied to each sequence, find the function, and compare the function with the earlier result.

a. 2, 4, 8, 16, 32 (from Example 1)

b. 108, 72, 48, 32, $21\frac{1}{3}$ (from Example 2)

c. 4, 12, 36, 108, 324 (from Example 5)

d. 3, 6, 12, 24, 48 (from part a, Example 6)

e. 81, 27, 9, 3, 1 (from part f, Example 6)

Solution   We first check to see that the common ratio in each sequence is positive and not equal to 1. If this condition is satisfied, then we use calculator regression to find the exponential function. Assume that $x = n$.

a. The common ratio is $\frac{4}{2} = 2$. Exponential regression gives $y = 1 \cdot 2^x$. This agrees with 2^n from Example 1.

b. The common ratio is $\frac{72}{108} = \frac{2}{3}$. Exponential regression gives $y = 162\left(\frac{2}{3}\right)^x$. This does not look like $108\left(\frac{2}{3}\right)^{n-1}$ from Example 2.

c. The common ratio is $\frac{12}{4} = 3$. Exponential regression gives $a = 1\frac{1}{3}$, or $\frac{4}{3}$; thus, $y = \frac{4}{3} \cdot 3^x$. This does not look like $a_n = 4 \cdot 3^{n-1}$ from Example 5.

d. The common ratio is $\frac{6}{3} = 2$. Exponential regression gives $1.5 \cdot 2^x$. This does not look like $f(x) = 3 \cdot 2^{x-1}$ from part a of Example 6.

e. The common ratio is $\frac{27}{81} = \frac{1}{3}$. Exponential regression gives $y = 243 \cdot \left(\frac{1}{3}\right)^x$. This does not look like $f(x) = 81 \cdot \left(\frac{1}{3}\right)^{n-1}$ from part f of Example 6. ●

Exponential regression gives expressions in $y = ab^x$ notation, rather than the $a_n = ar^{n-1}$ form obtained from the nth term function. In Example 8, we show that the two expressions in parts b, c, d, and e of Example 7 are equivalent. Try the example yourself before looking up the properties or reading the solution.

EXAMPLE **8**

Showing exponential expressions to be equivalent   Apply the definitions and properties of exponents (pages 408–410) to show that these pairs of exponential expressions are equivalent.

a. $162\left(\frac{2}{3}\right)^x$ and $108\left(\frac{2}{3}\right)^{x-1}$

b. $\frac{4}{3} \cdot 3^x$ and $4 \cdot 3^{x-1}$

c. $1.5 \cdot 2^x$ and $3 \cdot 2^{x-1}$

d. $243 \cdot \left(\frac{1}{3}\right)^x$ and $81 \cdot \left(\frac{1}{3}\right)^{x-1}$

Solution   The steps are described in part a. Parts b, c, and d follow similar steps.

a. $108\left(\frac{2}{3}\right)^{x-1}$

$= 108\left(\frac{2}{3}\right)^{x+(-1)}$ Change subtraction to adding the opposite.

$= 108\left(\frac{2}{3}\right)^{x} \cdot \left(\frac{2}{3}\right)^{-1}$ Apply the product property: $b^{m+n} = b^m \cdot b^n$.

$= 108\left(\frac{2}{3}\right)^{x} \cdot \left(\frac{3}{2}\right)$ Apply the reciprocal property: $b^{-1} = \frac{1}{b}$.

$= 108 \cdot \left(\frac{3}{2}\right) \cdot \left(\frac{2}{3}\right)^{x}$ Use the commutative property of multiplication.

$= 162\left(\frac{2}{3}\right)^{x}$ Multiply the first two factors.

b. $4 \cdot 3^{x-1} = 4 \cdot 3^x \cdot 3^{-1} = 4 \cdot 3^x \cdot \frac{1}{3} = 4 \cdot \frac{1}{3} \cdot 3^x = \frac{4}{3} \cdot 3^x$

c. $3 \cdot 2^{x-1} = 3 \cdot 2^x \cdot 2^{-1} = 3 \cdot 2^x \cdot \frac{1}{2} = 3 \cdot \frac{1}{2} \cdot 2^x = 1.5 \cdot 2^x$

d. $81 \cdot \left(\frac{1}{3}\right)^{x-1} = 81 \cdot \left(\frac{1}{3}\right)^x \cdot \left(\frac{1}{3}\right)^{-1} = 81 \cdot \left(\frac{1}{3}\right)^x \cdot 3 = 81 \cdot 3 \cdot \left(\frac{1}{3}\right)^x = 243 \cdot \left(\frac{1}{3}\right)^x$ ●

In Example 9, we apply exponential regression to a sequence with a negative base and observe that the regression does not work.

EXAMPLE **9**   Show whether the sequence 5, -10, 20, -40, 80 is geometric, and apply both the nth term formula and exponential regression to find an equation.

Solution   We divide consecutive terms:

$$\frac{-10}{5} = -2, \qquad \frac{20}{-10} = -2, \qquad \frac{-40}{20} = -2, \qquad \frac{80}{-40} = -2$$

The common ratio is $r = -2$. The sequence is geometric with $a_1 = 5$ and $a_n = 5 \cdot (-2)^{n-1}$. We observe that the base of this expression is (-2), so it will not be an exponential function. We enter 1, 2, 3, 4, 5 in L_1 and 5, -10, 20, -40, 80 in L_2. We obtain a "Domain Error," because the calculator detects the negative base, in violation of the definition of an exponential function. We cannot fit an exponential function to the sequence.   ●

ANSWER BOX

Warm-up: 1. 48 **2.** 17 **3.** 25 **4.** -7 **5.** 55 **6.** 1 **7.** 324 **Think about it:** One way to simplify this complex fraction is to multiply both numerator and denominator by the least common denominator, 3:

$$\frac{21\frac{1}{3}}{32} \cdot \frac{3}{3} = \frac{64}{96} = \frac{2 \cdot 32}{3 \cdot 32} = \frac{2}{3}$$

EXERCISES **7.0**

In Exercises 1 to 12, do the following:

a. *Give the common ratio for each sequence.*

b. *Find the expression for the nth term, a_n.*

c. *Find the sequence's equation with exponential regression.*

d. *Use the properties of exponents to show that your description of a_n is the same as that given by $y = ab^x$ from exponential regression.*

1. 0.5, 1.5, 4.5, 13.5

2. 0.6, 1.8, 5.4, 16.2

3. 10, 20, 40, 80

4. $\frac{1}{16}, \frac{1}{8}, \frac{1}{4}, \frac{1}{2}, 1, 2$

5. $\frac{1}{4}, \frac{1}{2}, 1, 2, 4$

6. 8, 16, 32, 64, 128

7. 1, 2, 4, 8, 16

8. 4, 2, 1, $\frac{1}{2}, \frac{1}{4}$

9. 32, 16, 8, 4, 2

10. 243, 81, 27, 9

11. 3, 1, $\frac{1}{3}, \frac{1}{9}, \frac{1}{27}$

12. $\frac{1}{9}, \frac{1}{3}, 1, 3, 9$

State whether each sequence in Exercises 13 to 22 is related to a linear function (arithmetic sequence), quadratic function, or exponential function (geometric sequence). Find the equation for each.

13. 2, 8, 18, 32

14. $-3, -7, -11, -15$

15. 0.5, 1.5, 4.5, 13.5

16. 7.5, 11, 14.5, 18, 21.5

17. 3, 7, 12, 18

18. 4, 9, 15, 22

19. 30, 38, 46, 54, 62

20. 1.2, 3.6, 10.8, 32.4

21. 5, 10, 20, 40, 80

22. 12, 24, 40, 60

In Exercises 23 to 28, change the expression into a product or a quotient (such as $a \cdot 2^x$, $2^x/a$, $a \cdot 3^x$, or $3^x/a$), using the properties of exponents. (Hint: In Example 8, we changed $a \cdot b^{x-1}$ expressions to $c \cdot b^x$.)

23. 2^{x+2}

24. 3^{x-1}

25. 2^{x-1}

26. 3^{x+2}

27. $\left(\frac{1}{2}\right)^{-x}$

28. $\left(\frac{1}{3}\right)^{-x}$

In Exercises 29 to 36, change each expression into the form $2^{n\pm a}$ or $3^{n\pm a}$, using the properties of exponents. (Hint: Write the coefficients as powers of the given base.)

29. $2^n/2$

30. $2 \cdot 2^n$

31. $4 \cdot 2^n$

32. $2^n/4$

33. $3^n/3$

34. $3 \cdot 3^n$

35. $9 \cdot 3^n$

36. $3^n/9$

In Exercises 37 and 38, write the rule for each sequence using $f(x) = 2^{x\pm a}$.

37. a. $\frac{1}{8}, \frac{1}{4}, \frac{1}{2}, 1, 2$ **b.** 4, 8, 16, 32, 64

38. a. $\frac{1}{2}, 1, 2, 4, 8$ **b.** 1, 2, 4, 8, 16

In Exercises 39 and 40, write the rule for each sequence using $f(x) = \left(\frac{1}{2}\right)^{x\pm a}$.

39. a. $4, 2, 1, \frac{1}{2}, \frac{1}{4}$ **b.** $8, 4, 2, 1, \frac{1}{2}$

40. a. 64, 32, 16, 8, 4 **b.** $\frac{1}{16}, \frac{1}{32}, \frac{1}{64}, \frac{1}{128}$

Show that the numbers in Exercises 41 to 44 represent geometric sequences whose equations cannot be determined with exponential regression. Explain why.

41. $2, -6, 18, -54, 162$

42. $6, -12, 24, -48, 96$

43. $64, -32, 16, -8, 4$

44. $81, -27, 9, -3, 1$

45. The 88 keys on a piano cover a range of notes from lowest A, called A_0, to high C, called C_8. The sound frequencies in cycles per second associated with the notes form a sequence of numbers called the chromatic scale. The sequence of frequencies is most clearly seen for the A notes (see the table). What equation describes the sequence?

Note	Cycles per second
A_0	27.50
A_1	55.00
A_2	110.0
A_3	220.0
A_4	440.0
A_5	880.0
A_6	1760
A_7	3520

46. a. Suppose the superball in Example 2 bounces to $\frac{1}{2}$ its previous height each time. Using both a_n and exponential regression, find the equation of the nth height. Find the tenth height. Describe how the change from $\frac{2}{3}$ to $\frac{1}{2}$ alters the equation.

b. Using both a_n and exponential regression, write the equation of the nth height for the superball in Example 2 if it starts at a height of 135 inches.

Projects

47. *Stacking Paper.* Refer to Example 1, as needed.

a. If there are 500 sheets of paper in a 2-inch ream of paper, what is the thickness of one sheet?

b. How thick a stack would result from folding a piece of paper ten times? (It may be easier to think in terms of the paper's being cut in half rather than folded.)

c. How many folds (cuts) would be needed to make a stack of paper 1 mile high?

d. How many folds (cuts) would be needed to make a stack that reached from Earth to the sun, 93 million miles away?

48. *Sequences.* It is tempting to say that any sequence whose differences are terms of the same sequence is geometric. Find the first and second differences for the sequences in parts a to d. Are the sequences geometric? Explain why or why not.

a. 1, 1, 2, 3, 5, 8

b. 2, 4, 6, 10, 16, 26

c. 4, 5, 9, 14, 23, 37

d. 2, 5, 7, 12, 19, 31

e. Return to the project in Exercise 72 in Section 3.4 (page 219), and complete it.

49. *Japanese Swords.* Research Japanese sword making. Knowing the following terms may assist in your research: The hard steel made by folding is called *kawagane*. The manner of folding and hammering creates different blade patterns, called *kitae-hada*. (This vocabulary is from the exhibition catalog "Living National Treasures of Japan," Museum of Fine Arts, Boston, November 3, 1982–January 2, 1983.)

7.1 Exponential Functions and Graphs

OBJECTIVES

- Identify the coefficient, the base, and the exponent in an exponential expression.
- Graph exponential functions.
- Examine how the base and coefficient of an exponential expression affect the graph of an exponential function.
- Identify the *y*-intercept for an exponential function.

WARM-UP

Complete the table for these equations.

Input, x	$y = 2^x$	$y = 2^{x-1}$	$y = 2^{x+2}$
1			
2			
3			
4			
5			
6			
7			
8			
9			
10			

IN THIS SECTION, we continue our work with exponential functions by exploring two applications, identifying bases and exponents, and examining graphs. If you do not do the Exploration in Example 1 in class, you might want to try it at home. You need 48 coins and about 10 minutes. You may be surprised by the outcome and better appreciate the nature of exponential functions.

EXAMPLE **1**   Exploring patterns in coin tosses

a. *Understand the problem:* Before you carry out this activity, read the following instructions and describe, in writing, what you think will happen:

Place 48 coins (such as pennies) in a sandwich bag or similar plastic bag. Shake the bag, and then toss the coins into a low-sided box (such as a box for overhead projector transparencies). Remove from the box the coins that are "heads up." Count the remaining coins. Repeat this procedure with the remaining coins until all the coins have been removed.

b. *Plan:* Make some predictions. In how many tosses will half the coins be gone? In how many tosses will all the coins be gone?

c. *Carry out the plan:* Make a table, with the first row showing toss 0 as input and 48 coins as output. Start the activity, and record the results after each toss. Make a graph from your table.

d. *Check:* Are your results reasonable? Compare your experimental results with the theoretical or expected results. Make a table reflecting what would happen if the coin toss behaved as we would expect. Decide if and how you will round numbers. State your rule for rounding. Predict a theoretical equation describing the number of coins remaining (output) after each toss (input). Draw the theoretical graph on the same axes as the graph in part c.

e. *Extension:* Carry out an exponential regression on the tabular data in part c. Enter the toss number in list 1 and the number of coins remaining in list 2. Begin with toss 0 and 48 coins. Do not enter the final data point (with 0 coins remaining), because the calculator function will indicate an error. Record the exponential regression equation in your summary. Compare the regression equation and its graph with the theoretical equation and its graph.

Solution   The results from the penny toss are exponential and have a theoretical equation of the form $y = ab^x$. A possible table, the theoretical equation, and its graph are in the Answer Box.   ●

EXAMPLE **2**   Exploring patterns: a reward for the inventor of chess   The story is told that the game of chess was invented by a member of the court of a maharajah (prince) in India. The prince was so delighted with the game that he offered the inventor any reward. The reward requested by the inventor seemed simple enough: a grain of rice on the first square, two grains on the second square, four grains on the third square, and so forth, with the number of grains of rice on each successive square doubled for the entire board. Supposedly, the prince eagerly agreed to the request.

A chessboard is shown in Figure 6. A grain of rice has been placed on the first square, two grains of rice on the second square, four grains on the third square.

Figure 6

a. Make a table with the numbers of the squares as input and the numbers of grains of rice on the squares as output. Then graph the data from your table. How many grains of rice will be on the 10th square?

b. Describe a pattern relating the number of the square to the number of grains of rice on the square.

c. How many grains of rice will be on the 20th square?

d. How many grains of rice will be on the last, or 64th, square?

Solution   **a.** The input-output table appears in Table 4; the graph is in Figure 7. A dashed line connects the points to show the shape of the graph. On the 10th square, there will be 512 grains of rice. If we write 512 as a power of 2, we get 2^9— the exponent is 1 less than the number of the square holding the rice.

Square	Grains of Rice
1	1
2	2
3	4
4	8
5	16
6	32
7	64
8	128
9	256
10	512

Table 4

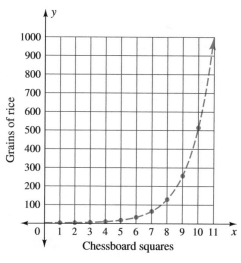

Figure 7

b. The number of grains of rice on each square is a power of 2. The exponent on each is 1 less than the number of the square holding the rice. Thus, the number of grains of rice on square n is 2^{n-1}.

c. On the 20th square, the number of grains of rice will be $2^{19} = 524,288$.

d. On the last square, the 64th square, the number of grains of rice will be 2^{63}. This number is considerably larger than the calculator display can accommodate and shows in scientific notation as approximately 9.22×10^{18}, which is equal to

9,220,000,000,000,000,000, or 9.22 quintillion grains of rice   ●

Exponential Functions

The equations for the coin toss and the grains of rice illustrate exponential functions. We restate the definition of an exponential function here.

Definition of Exponential Function

> An **exponential function** is written as
> $$f(x) = a \cdot b^x$$
> where the coefficient a is a constant, the base a is positive but not equal to 1, and the exponent x is a real number.

EXAMPLE **3**   *Naming the parts in an exponential equation*   Identify the coefficient, the base, and the exponent in each of these exponential functions.

a. $y = 3 \cdot 2^x$

b. $y = 1000(1 + 0.065)^t$

c. $y = 19,000(1 - r)^t$, r constant

d. $y = 2000\left(1 + \dfrac{0.08}{12}\right)^{12t}$

Solution   **a.** Coefficient is 3, base is 2, exponent is x.

b. Coefficient is 1000, base is $(1 + 0.065)$, exponent is t.

c. Coefficient is 19,000, base is $(1 - r)$, exponent is t.

d. Coefficient is 2000, base is $\left(1 + \dfrac{0.08}{12}\right)$, exponent is $12t$.   ●

We now turn to the graphs of exponential functions.

Exponential Graphs

As the next few examples show, the exponential graph has several distinctive properties and features.

BASES AND GRAPHS   In Example 1, the number of coins remaining decreased with each toss. The theoretical equation for the coin toss (see the Answer Box) was $y = 48\left(\frac{1}{2}\right)^x$. The graph declined rapidly from left to right.

In Example 2, the number of grains of rice doubled with each chessboard square. The equation was $y = 2^{x-1}$. The graph rose rapidly from left to right.

In Example 4, we compare two somewhat simpler graphs.

EXAMPLE    Comparing graphs and bases   Compare the tables and graphs for $y = 2^x$ and $y = \left(\frac{1}{2}\right)^x$.

Solution   The tables for $y = 2^x$ and $y = \left(\frac{1}{2}\right)^x$ contain the same output numbers, but for different inputs. In Table 5, the outputs for the two functions are the same numbers in opposite order. The change in base from 2 to $\frac{1}{2}$ changes the graph from increasing to decreasing; see Figure 8.

x	$y = 2^x$	$y = \left(\frac{1}{2}\right)^x$
-2	$\frac{1}{4}$	4
-1	$\frac{1}{2}$	2
0	1	1
1	2	$\frac{1}{2}$
2	4	$\frac{1}{4}$

Table 5

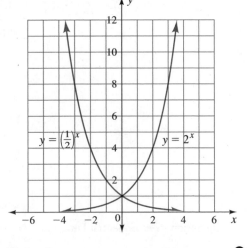

Figure 8   ●

The examples suggest that when the base is a positive fraction less than 1, the graph falls from left to right. When the base is larger than 1, the graph rises from left to right.

Graphs of Exponential Functions

> The graph of $y = b^x$ decreases from left to right for $0 < b < 1$.
>
> The graph of $y = b^x$ increases from left to right for $b > 1$.

An exponential function is defined only for positive bases, $b \neq 1$. The above summary does not mention a negative base, a base of 0, or a base of 1. You will explore these bases in Exercises 41 to 44.

Think about it 1: Write $\frac{1}{2}$ as a power of 2. Write $y = \left(\frac{1}{2}\right)^x$ in another way. Does this second way of writing $y = \left(\frac{1}{2}\right)^x$ fit our graphing rules?

FEATURES OF THE EXPONENTIAL GRAPH   Example 5 explores how the graph changes as the base of the exponential function changes.

EXAMPLE 5

Finding key features on an exponential graph   Making a table and a graph, compare $y = 2^x$, $y = 3^x$, and $y = 4^x$ for inputs on the interval -2 to 2. Discuss the y-intercept and the relative positions of the graphs. What is y for $x = 1$? What is the x-intercept for each graph?

Solution

x	$y = 2^x$	$y = 3^x$	$y = 4^x$
-2	$\frac{1}{4}$	$\frac{1}{9}$	$\frac{1}{16}$
-1	$\frac{1}{2}$	$\frac{1}{3}$	$\frac{1}{4}$
0	1	1	1
1	2	3	4
2	4	9	16

Table 6

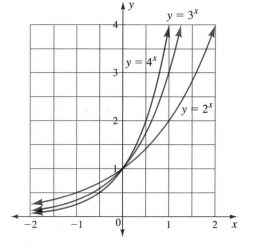

Figure 9

Table 6 shows a common point at $(0, 1)$. The graphs in Figure 9 all have the same y-intercept, $(0, 1)$, as we can see from the table. The y-intercept acts as a pivot point, with $y = 4^x$ highest on the right and lowest on the left, $y = 2^x$ lowest on the right and highest on the left, and $y = 3^x$ between the other two graphs on both sides.

At $x = 1$, $2^1 = 2$, $3^1 = 3$, and $4^1 = 4$. Thus, the point $(1, b)$ is on each graph.

If we trace along the graphs near the x-axis, none of the graphs touch the x-axis. The calculator rounds the output to zero if we trace far enough, but the graph never actually touches the x-axis. There are no x-intercepts. ●

Example 5 suggests three conclusions about exponential graphs:

Features of Exponential Graphs

The graph of $y = b^x$ passes through $(0, 1)$ and $(1, b)$.

For $b > 1$, the larger the base in $y = b^x$, the steeper the graph.

The graphs approach but do not touch the x-axis.

When a *graph approaches without touching the x-axis,* we say that the x-axis is a **horizontal asymptote.**

The exponential graphs we have seen thus far also help clarify the domain and range of the function $f(x) = b^x$. The graphs in Example 5 extend to the left and right without end and are entirely above the x-axis, implying the following:

Domain and Range

> The domain for $f(x) = b^x$ is all real numbers.
>
> The range for $f(x) = b^x$ is $f(x) > 0$.

HORIZONTAL SHIFTS   The Japanese swords in Example 1 of Section 7.0 and the grains of rice in Example 2 gave sequences related to the functions $f(x) = 2^x$ and $f(x) = 2^{x-1}$. These functions both represent powers of 2. We compare their graphs in Example 6.

EXAMPLE **6**   Comparing graphs and exponents   **Graph and compare** $f(x) = 2^x$ and $f(x) = 2^{x-1}$.

Solution   When the functions $f(x) = 2^x$ and $f(x) = 2^{x-1}$ are graphed on the same axes (see Figure 10), the graphs are identical in shape, but $f(x) = 2^{x-1}$ is shifted 1 unit to the right of $f(x) = 2^x$.

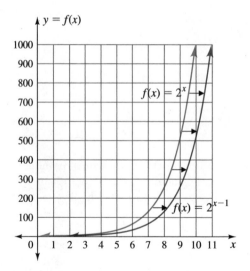

Figure 10   ●

Think about it 2: How does the graph of $f(x) = (x - 1)^2$ compare with the graph of $f(x) = x^2$? How does this result compare with the one in Example 6? How does the graph of $f(x) = (x + 2)^2$ compare with the graph of $f(x) = x^2$?

EXAMPLE **7**   Comparing graphs and exponents   How does the graph of $f(x) = 2^{x+2}$ compare with the graph of $f(x) = 2^x$?

Solution   The graphs of $f(x) = 2^x$ and $f(x) = 2^{x+2}$ (see Figure 11) are identical in shape, but the graph of $f(x) = 2^{x+2}$ is shifted 2 units to the left of that of $f(x) = 2^x$.

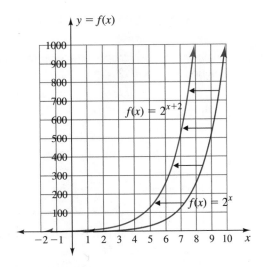

Figure II

In summary, Examples 6 and 7 show the following:

Horizontal Shifts

For $h > 0$, the graph of $y = b^{x-h}$ is shifted h units to the right of that of $y = b^x$.

For $h > 0$, the graph of $y = b^{x+h}$ is shifted h units to the left of that of $y = b^x$.

y-INTERCEPTS   In Examples 8 and 9, we explore the role of the coefficient a in $f(x) = ab^x$.

EXAMPLE **8**   Comparing y-intercepts   Graph $f(x) = 1 \cdot 2^x$, $f(x) = 5 \cdot 2^x$, and $f(x) = 10 \cdot 2^x$. Identify the y-intercept for each graph.

Solution   The graphs, in Figure 12, show three distinct y-intercepts:

The y-intercept for $f(x) = 1 \cdot 2^x$ is 1.

The y-intercept for $f(x) = 5 \cdot 2^x$ is 5.

The y-intercept for $f(x) = 10 \cdot 2^x$ is 10.

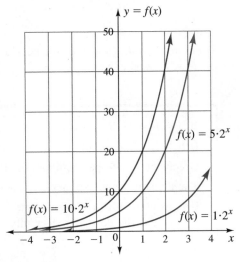

Figure 12

Think about it 3: Do the graphs in Figure 12 intersect?

EXAMPLE **9**   Finding the y-intercept   How can we find the y-intercept for any exponential function $f(x) = ab^x$?

Solution   At the y-intercept, $x = 0$.

$y = ab^x$ Let $x = 0$.

$y = ab^0$ Substitute $b^0 = 1$.

$y = a$ ●

The y-intercept for an Exponential Function

| Because $x = 0$ at the y-intercept, the coefficient a is the y-intercept for the graph of $y = ab^x$. |

EXAMPLE **10**   Finding y-intercepts   Predict the y-intercept for each function and then find the intercept.

a. $y = 2b^x$ **b.** $y = 1000(1 + r)^x$ **c.** $y = 3^{x+1}$ **d.** $y = 2^{x-3}$

Solution   **a.** Because $a = 2$, the y-intercept is 2. When $x = 0$,

$y = 2b^0 = 2 \cdot 1 = 2$

b. Because $a = 1000$, the y-intercept is 1000. When $x = 0$,

$y = 1000(1 + r)^0 = 1000 \cdot 1 = 1000$

c. To find the coefficient, we change to ab^x form.

$y = 3^{x+1} = 3^x \cdot 3^1 = 3 \cdot 3^x$

We predict that the y-intercept is 3. When $x = 0$,

$y = 3^{0+1} = 3^1 = 3$

d. To find the coefficient, we change to ab^x form.

$y = 2^{x-3} = 2^x \cdot 2^{-3} = 2^x \cdot \frac{1}{8} = \frac{1}{8} \cdot 2^x$

We predict that the y-intercept is $\frac{1}{8}$. When $x = 0$,

$y = 2^{0-3} = 2^{-3} = \frac{1}{8}$ ●

Think about it 4: How does the coefficient a on $y = ab^x$ help us find an appropriate window for the graph?

Summary: Properties of the Exponential Graph

| If $0 < b < 1$, the graph of $y = b^x$ decreases from left to right. |
| If $b > 1$, the graph of $y = b^x$ increases from left to right. |
| Graphs of $y = b^x$ pass through $(0, 1)$ and $(1, b)$. |
| The domain of the exponential function is all real numbers. |
| Graphs of $y = b^x$ are in only the first and second quadrants. Thus, the range is the positive real numbers. |

Graphs of $y = b^x$ approach the x-axis without touching or crossing it. The x-axis is a horizontal asymptote.

For $h > 0$, the graph of $y = b^{x-h}$ is shifted h units to the right of the graph of $y = b^x$.

For $h > 0$, the graph of $y = b^{x+h}$ is shifted h units to the left of the graph of $y = b^x$.

Let $x = 0$ to find the y-intercept. The coefficient a is the y-intercept for the graph of $y = ab^x$.

ANSWER BOX

Warm-up: For $y = 2^x$: 2, 4, 8, 16, 32, 64, 128, 256, 512, 1024; for $y = 2^{x-1}$: 1, 2, 4, 8, 16, 32, 64, 128, 256, 512; for $y = 2^{x+2}$: 8, 16, 32, 64, 128, 256, 512, 1024, 2048, 4096 **Example 1:** According to the table, if we start with 48 coins and the coins are equally likely to land heads up, the coins will be gone after 6 tosses. In the table, fractional coins have been rounded down to next lowest whole coin. The theoretical equation is $y = 48\left(\frac{1}{2}\right)^x$; the theoretical graph is shown in the figure.

Toss	Coins Remaining
0	48
1	24
2	12
3	6
4	3
5	1
6	0

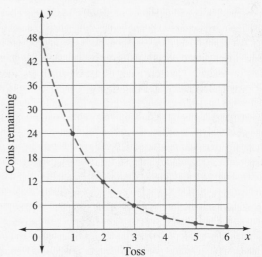

Think about it 1: $2^{-1} = \frac{1}{2}$; $y = \left(\frac{1}{2}\right)^x$ is equivalent to $y = 2^{-x}$. The rules are for $y = b^x$. Equations of the form $y = b^{-x}$ need to be changed to $y = \left(\dfrac{1}{b}\right)^x$ form before the rules are applied. **Think about it 2:** The graph of $f(x) = (x - 1)^2$ is shifted 1 unit to the right of the graph of $f(x) = x^2$. This is the same relationship found between the graphs in Example 6. The graph of $f(x) = (x + 2)^2$ is shifted 2 units to the left of the graph of $f(x) = x^2$. **Think about it 3:** The graphs do not intersect because $10 \cdot 2^x > 5 \cdot 2^x > 1 \cdot 2^x$ for each x. Try tracing the graphs to the left and using the up and down cursor keys to compare the functions at $x = -4$. **Think about it 4:** The coefficient a indicates the y-intercept and suggests that the minimum y should be below the intercept and the maximum y should be above it.

EXERCISES 7.1

In Exercises 1 to 4, identify the base and the exponent.

1. a. a^x **b.** $4x^3$

   c. 2^{-x} **d.** x^a

2. a. $3x^5$ **b.** 3^x

   c. $(x-2)^3$ **d.** 3^{x-2}

3. a. π^x **b.** $2x^4$

   c. $a_1 r^{n-1}$, r constant   **d.** $100(1.06)^t$

4. a. x^π **b.** $3e^x$

   c. Pe^{rt} **d.** $4 \cdot 2^n$

In Exercises 5 to 10, predict without graphing whether the graph will be increasing or decreasing from left to right.

5. a. $y = 5^x$ **b.** $y = \left(\frac{3}{4}\right)^x$

6. a. $y = \left(\frac{2}{3}\right)^x$ **b.** $y = 4^x$

7. a. $y = \left(\frac{3}{2}\right)^x$ **b.** $y = 0.5^x$

8. a. $y = 1.5^x$ **b.** $y = 0.75^x$

9. a. $y = (1 + 0.05)^x$ **b.** $y = \left(\frac{4}{3}\right)^x$

10. a. $y = 2.5^x$ **b.** $y = (1 - 0.05)^x$

11. The figure contains graphs of $y = 2^x$, $y = 10^x$, and $y = 2.72^x$. Match each equation with its graph—a, b, or c. Do not use a calculator.

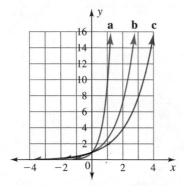

12. The figure contains graphs of $y = \pi^x$, $y = 1.5^x$, and $y = 2^x$. Match each equation with its graph—a, b, or c. Do not use a calculator.

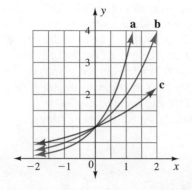

13. For what set of inputs is the graph of $y = 2^x$ below the x-axis?

14. For what set of inputs x is the output to $y = 3^x$ negative?

15. For what set of inputs x is the output to $y = 2^x$ negative?

16. For what set of inputs is the graph of $y = 3^x$ below the x-axis?

17. Evaluate $3 \cdot 2^0$, $4 \cdot 2^0$, and $\frac{1}{2} \cdot 2^0$. What do these results tell us about the y-intercept of $y = a \cdot 2^x$?

18. Explain why the letter a in $y = ab^x$ is not part of the base.

Name the y-intercept in Exercises 19 to 21. Change each equation to the form $y = 2^{x \pm a}$ or $y = 3^{x \pm a}$.

19. a. $y = 16 \cdot 2^x$

   b. $y = 8 \cdot 2^x$

20. a. $y = \frac{1}{2} \cdot 2^x$

   b. $y = 3^x/9$

21. a. $y = 27 \cdot 3^x$

   b. $y = 3^x/27$

Change each equation in Exercises 22 to 24 to $y = ab^x$ form. Name the y-intercept.

22. a. $y = 3^{x-2}$

   b. $y = 3^{x+3}$

23. a. $y = 2^{x-4}$

   b. $y = 2^{x+1}$

24. a. $y = \left(\frac{1}{4}\right)^{x+1}$

   b. $y = 2^{x-2}$

25. Describe the role of the coefficient a in the graph of the exponential equation $y = ab^x$.

26. Evaluate 2^1, 3^1, and 5^1. What do these results tell us about the y-coordinate of the point where $x = 1$ in $y = b^x$?

27. Why does the graph of $y = b^x$ pass through $(1, b)$?

28. Why does the graph of $y = b^x$ pass through $(0, 1)$?

29. Graphs of exponential functions such as $y = 2^x$ have the x-axis as a horizontal asymptote. Name a non-exponential function from a different chapter whose graph has a horizontal asymptote (the graph approaches the x-axis without touching or crossing it).

30. Compare the effects of the exponents on the graphs of $y = 2^{-x}$ and $y = 2^x$.

In Exercises 31 and 32, complete the table for the indicated functions. Match each function with one of the graphs in the figure. Describe the effect of adding or subtracting 1 in the exponent.

31.

x	$y = 2^{x-1}$	$y = 2^x$	$y = 2^{x+1}$
-2			
-1			
0			
1			
2			

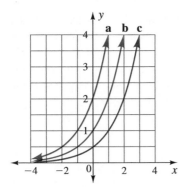

32.

x	$y = 3^{x-1}$	$y = 3^x$	$y = 3^{x+1}$
-2			
-1			
0			
1			
2			

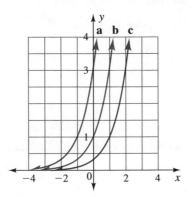

Using a graph of $y = 3^x$ as a reference, sketch a graph of each function in Exercises 33 to 36. Label the y-intercept and $f(1)$.

33. $f(x) = 3^{x+3}$ **34.** $f(x) = 3^{x-2}$

35. $f(x) = 3^{x-3}$ **36.** $f(x) = 3^{x+2}$

37. Where do the graphs of $y = 3^x$ and $y = 3^{x+1}$ intersect? Why?

38. Where do the graphs of $y = 2^x$ and $y = 3^x$ intersect? Why?

39. Refer to the figure below. The base a is not given, so answer the questions from the graph.

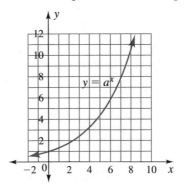

  a. Estimate x for $a^x = 2$.

  b. Estimate x for $a^x = 4$.

  c. Estimate y corresponding to a^2.

  d. Estimate y corresponding to a^5.

  e. Use a graphing guess-and-check process to find a graph similar to the figure and thereby estimate the base a.

40. Refer to the figure below. The base d is not given, so answer the questions from the graph.

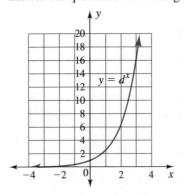

  a. Estimate x for $d^x = 2$.

  b. Estimate x for $d^x = 4$.

  c. Estimate y corresponding to d^2.

  d. Estimate y corresponding to d^3.

  e. Use a graphing guess-and-check process to find a graph similar to the figure and thereby estimate the base d.

41. a. Predict the graph of $y = 1^x$.

  b. Graph $y = 1^x$.

  c. What do you observe?

d. Is the function defined for all inputs?

e. What is the range of the function?

42. a. Predict the graph of $y = 0^x$.

    b. Graph $y = 0^x$.

    c. What do you observe?

    d. Is the function defined for all inputs?

    e. What is the range of the function?

43. a. Graph $y = (-2)^x$. Use a decimal zoom to assure decimal inputs.

    b. What do you observe?

    c. Is this equation defined for all inputs?

    d. Is this equation an exponential function?

44. a. Graph $y = (-1)^x$. Use a decimal zoom to assure decimal inputs.

    b. What do you observe?

    c. Is this equation defined for all inputs?

    d. Is this equation an exponential function?

Projects

45. *Total Rice.* Add a column to Table 4 in Example 2. In this column, enter, for each square, the total number of grains of rice needed for it plus all the squares preceding it.

    a. How does the cumulative total for each square compare with the number of grains of rice needed for that square?

    b. How many grains of rice will be needed on the 11th square?

    c. How many grains of rice will be needed in total for the first 11 squares?

    d. How many grains of rice will be needed in total for the whole chessboard?

46. *Pounds of Rice.* See Example 2 and Exercise 45. Explain how you would estimate the number of grains in a pound of rice, and then do it. How much of the chessboard would a pound of rice fill? Estimate the total weight of the rice needed by the maharajah to satisfy the inventor's request. How does this quantity of rice compare with the world's annual production of rice in 1991, which was 520 million metric tons?

47. *A Line of Rice Grains.* See Example 2 and Exercise 45. How many grains of rice, placed end to end, would it take to reach 1 inch (or to reach 5 centimeters)? Specify the type of rice you choose. Estimate the length of a line created by placing end to end the total number of grains of rice needed to satisfy the chess inventor's request. Give your answer in suitable units.

48. *Recognizing Expressions.* Tell whether each expression in the given exercises is a power (of a variable) expression, an exponential expression, or neither. How do we limit a power expression to create a power function? How do we limit an exponential expression to create an exponential function?

    a. Exercise 1

    b. Exercise 2

    c. Exercise 3

    d. Exercise 4

49. *Comparing Power and Exponential Functions.* Compare the graphs of $y = x^2$ and $y = 2^x$.

    a. What kind of function is each?

    b. What are the domain and range of each?

    c. What are the x-coordinates of their points of intersection?

    d. Discuss the differences in the graphs as x goes to the left, at 0, and as x goes to the right.

    e. Which graph rises faster for $x > 4$?

    f. Repeat parts a to e to compare the graphs of $y = x^3$ and $y = 3^x$.

7.2 Solving Exponential Equations: Like Bases and Logarithms _____

OBJECTIVES

- Use like bases to solve exponential equations.
- Find the inverse to an exponential function by table, graph, and equation.
- Change between exponential equations and logarithmic equations.
- Evaluate common logarithms with a calculator.

WARM-UP

Solve each equation for x. Guess and check with a calculator, as needed, to find the exponents to the nearest thousandth.

1. $10^x = 1$ **2.** $10^x = 10$ **3.** $10^x = 20$

4. $10^x = 100$ **5.** $2^x = 2$ **6.** $2^x = 4$

7. $2^x = 6$ **8.** $2^x = 8$

THIS SECTION INTRODUCES two tools for solving exponential equations. The first uses bases in exponential expressions, and the second uses the inverse to the exponential function. We graph the inverse to the exponential function and evaluate it with a calculator.

Like Bases Property

In Example 1, we write both sides of the equations from the Warm-up as exponential expressions, if possible, and observe how the exponents help us solve the equations.

EXAMPLE   Changing to like bases   If possible, change the right-hand side of each equation in the Warm-up to an exponential expression, with the base given on the left-hand side. What is the value of the variable in the exponent?

Solution   **a.** $10^x = 10^0$, $x = 0$ **b.** $10^x = 10^1$, $x = 1$

c. 20 is not an integral power of 10.   **d.** $10^x = 10^2$, $x = 2$

e. $2^x = 2^1$, $x = 1$ **f.** $2^x = 2^2$, $x = 2$

g. 6 is not an integral power of 2.   **h.** $2^x = 2^3$, $x = 3$ ●

Example 1 suggests that equal exponential expressions with the same bases have equal exponents. This property is called the like bases property.

Like Bases Property

> Suppose $a > 0$ and $a \neq 1$.
>
> $a^x = a^y$ if and only if $x = y$

The **like bases property** says that *if two exponential expressions are equal and have a common positive base, not 1, the exponents must be equal.*

Think about it 1: Name a pair of numbers that shows why we cannot say that if $x^2 = y^2$ then $x = y$.

> To use the like bases property:
>
> **1.** Change the exponential expressions on both sides of an equation to the same base, if possible.
>
> **2.** Set the exponents equal.
>
> **3.** Solve for the variable.

EXAMPLE **2**   Solving equations with the like bases property   Use the like bases property to solve the equations for x.

a. $8^x = 1024$ b. $4^x = 512$ c. $25^x = 125$

d. $\left(\frac{1}{2}\right)^x = 32$ e. $\left(\frac{1}{3}\right)^x = 81$ f. $3^{x+2} = 9^{2x-5}$

Solution   a.

$8^x = 1024$	Change to a common base.
$(2^3)^x = 2^{10}$	Apply the power property of exponents.
$2^{3x} = 2^{10}$	Set the exponents equal.
$3x = 10$	Divide by 3.
$x = \frac{10}{3}$	

Check: $8^{10/3} \overset{?}{=} 1024$ ✔

b.

$4^x = 512$	Change to a common base.
$(2^2)^x = 2^9$	Apply the power property of exponents.
$2^{2x} = 2^9$	Set the exponents equal.
$2x = 9$	Divide by 2.
$x = \frac{9}{2}$	

Check: $4^{9/2} \overset{?}{=} 512$ ✔

c.

$25^x = 125$	Change to a common base.
$(5^2)^x = 5^3$	Apply the power property of exponents.
$5^{2x} = 5^3$	Set the exponents equal.
$2x = 3$	Divide by 2.
$x = \frac{3}{2}$	

Check: $25^{3/2} \overset{?}{=} 125$ ✔

d.

$\left(\frac{1}{2}\right)^x = 32$	Change to a common base.
$(2^{-1})^x = 2^5$	Apply the power property of exponents.
$2^{-1x} = 2^5$	Set the exponents equal.
$-1x = 5$	Divide by -1.
$x = -5$	

Check: $\left(\frac{1}{2}\right)^{-5} \overset{?}{=} 32$ ✔

e.

$\left(\frac{1}{3}\right)^x = 81$	Change to a common base.
$(3^{-1})^x = 3^4$	Apply the power property of exponents.
$3^{-1x} = 3^4$	Set the exponents equal.
$-1x = 4$	Divide by -1.
$x = -4$	

Check: $\left(\frac{1}{3}\right)^{-4} \overset{?}{=} 81$ ✔

f.

$3^{x+2} = 9^{2x-5}$	Change to a common base, placing the exponent on the right in parentheses.
$3^{x+2} = 3^{2(2x-5)}$	Multiply with the distributive property on the right side.
$3^{x+2} = 3^{4x-10}$	Set the exponents equal.
$x + 2 = 4x - 10$	Add 10 to both sides, and subtract x from both sides.
$12 = 3x$	Divide both sides by 3.
$4 = x$	

Check: $3^{4+2} \overset{?}{=} 9^{2(4)-5}$ ✔   ●

Think about it 2: If $2^{x+1} = 2^1 \cdot 2^x$, how could we rewrite 3^{x+2} to eliminate the addition sign?

The like bases property is useful, even though its application is limited to equations in which bases are easily made alike. Solving equations with it is good practice and reminds us of important numerical relationships between certain bases and exponents.

Inverse Functions and Exponential Equations

The equations $10^x = 20$ and $2^x = 6$ in the Warm-up could not be solved with like bases. There is no integer or rational-number power of 10 that makes 20, nor is there any integer or rational-number power of 2 that makes 6. We now consider a solution method that is more general than guess and check or applying the like bases property. (We will solve $y = 2^x$ in Section 7.3.)

To solve an equation such as $y = 10^x$ for x, we need an inverse function. From Section 6.4, we have three ways to find an inverse function: from a table, from a graph, and with symbols. We will solve $y = 10^x$ for x using each of these three methods.

INVERSE FUNCTIONS FROM A TABLE   *To find the inverse from a function listed in a table, swap the input and output (x and y) columns and find the rule for the new table.*

EXAMPLE 3   Finding the inverse   Given Table 7 for $y = 10^x$, find the table for the inverse.

Input, x	Output, $y = 10^x$
-2	$\frac{1}{100}$
-1	$\frac{1}{10}$
0	1
1	10
2	100
3	1000

Table 7

Solution   By swapping the numbers in the columns of Table 7, we can see that an inverse exists.

Input to Inverse, x	Output to Inverse, y
$\frac{1}{100}$	-2
$\frac{1}{10}$	-1
1	0
10	1
100	2
1000	3

Table 8   ●

INVERSE FUNCTIONS FROM A GRAPH   *To find the inverse from the graph of a function, swap the numbers (x, y) in each ordered pair and plot the resulting graph.*

EXAMPLE 4   Finding the inverse   Use Tables 7 and 8 to graph $y = 10^x$ for $-2 \le x \le 1$ and to sketch the inverse. Label both axes from -2 to 10. Is the inverse a function?

Solution   The graphs of $y = 10^x$ and its inverse are in Figure 13. A vertical line placed anywhere on the graph of the inverse passes through the graph exactly once. Thus, the inverse of the exponential function $y = 10^x$ passes the vertical-line test and is also a function.

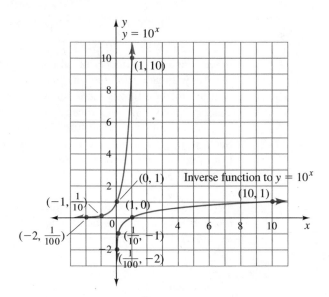

Figure 13

INVERSE FUNCTIONS FROM AN EQUATION   *To find the inverse from the equation of a function, swap x and y and solve (if possible) for y.*

EXAMPLE **5**   Finding the inverse   Given the equation $y = 10^x$, find the inverse.

Solution   We swap x and y and then solve (if possible) for y.

$$y = 10^x$$
$$x = 10^y$$

There are no algebraic steps to use to solve for y.

Think about it 3: Show the result from doing each of the following operations to both sides of $x = 10^y$: adding y, subtracting y, multiplying by y, dividing by y, and taking the $1/y$ power (yth root).

Logarithmic Functions

Because algebra fails to solve $y = b^x$ for x, mathematicians had to invent a solution. The invention of what led to the inverse to the exponential function took place shortly before 1620, when the Pilgrims landed at Plymouth, Massachusetts. In 1614, John Napier (1550–1617) called his invention *logarithms*, roughly translated from the Greek as *number words* or *number ratios*. Note the similarity between arithmetic and logarithm.

Logarithmic Function

> The logarithmic function is the inverse to the exponential function, $f(x) = b^x$.
>
> The logarithmic function is $f(x) = \log_b x$, where x is a positive real number and b is a positive real number not equal to 1. Read $\log_b x$ as "log base b of x."

The graphs of exponential functions, $f(x) = b^x$, $b > 1$, and logarithmic functions, $f(x) = \log_b x$, look similar to those of $f(x) = 10^x$ and its inverse function,

$f(x) = \log_{10} x$ (Figure 13). We will do more with graphs of logarithmic functions in Section 7.3.

Changing Exponential Equations to Logarithmic Equations

In 1728, Leonhard Euler modified the definition of a logarithm. We use his version of the definition today to solve exponential equations.

Solving Exponential Equations, $y = b^x$, for x

> The **logarithm** is the exponent to which we raise a base to obtain a given number. If $y = b^x$, where b is a fixed positive number not equal to 1, then x is the logarithm base b of y.
>
> $y = b^x$ is equivalent to $x = \log_b y$.

The base b is the same for both the exponential equation and the logarithmic equation. This fact will help you keep track of where the different numbers and letters go when you change between exponential and logarithmic equations. In the next example and the related exercises, the equivalent equations are placed in tables to emphasize their relationship.

EXAMPLE **6**

Changing between exponential and logarithmic equations Find the missing entries in Table 9, changing the logarithmic equations to exponential equations and the exponential equations to logarithmic equations.

	Exponential Equation, $y = b^x$	Logarithmic Equation, $x = \log_b y$
a.	$2 = 10^x$	
b.		$x = \log_{10} 20$
c.	$10^x = 200$	
d.		$1 = \log_{10} 10$
e.	$10^2 = 100$	
f.		$0 = \log_{10} 1$

Table 9

Solution The first three equations contain the variable x.
a. If $2 = 10^x$, the base is 10 and

$$x = \log_{10} 2$$

b. If $x = \log_{10} 20$, the base is 10 and

$$10^x = 20$$

c. If $10^x = 200$, the base is 10 and

$$x = \log_{10} 200$$

The second three equations are common facts about powers of 10.
d. If $1 = \log_{10} 10$, the base is 10 and

$$10^1 = 10$$

This fact reminds us that $b^1 = b$; any number to the first power is itself.
e. If $10^2 = 100$, the base is 10 and

$$2 = \log_{10} 100$$

This fact states that the square of 10 is 100.

f. If $0 = \log_{10} 1$, the base is 10 and

$$10^0 = 1$$

This fact reminds us that $b^0 = 1$, $b \neq 0$.

The completed version of Table 9 appears in Table 10.

Exponential Equation, $y = b^x$	Logarithmic Equation, $x = \log_b y$
$2 = 10^x$	$x = \log_{10} 2$
$10^x = 20$	$x = \log_{10} 20$
$10^x = 200$	$x = \log_{10} 200$
$10^1 = 10$	$1 = \log_{10} 10$
$10^2 = 100$	$2 = \log_{10} 100$
$10^0 = 1$	$0 = \log_{10} 1$

Table 10

A common error is to place the base in an incorrect position in the logarithmic expression. *The base b is written in subscript position, small and below the main line of type containing the rest of the logarithmic expression.*

Common Logarithms

Now that we have solved $y = 10^x$ for x and obtained $x = \log_{10} y$, it would be reasonable to expect to find some numbers for x. We will work first with logarithms having base 10, because base 10 is one of the two most commonly occurring bases. The other is base e, and we will consider it in Section 7.5.

Logarithms with base 10 are called **common logarithms.** The common logarithm of a number is evaluated with the $\boxed{\text{LOG}}$ key on a calculator.

Evaluating Logarithms by Calculator

To find the common logarithm, $\log_{10} x$, on a graphing calculator, press $\boxed{\text{LOG}}$ followed by the number x.

On some scientific calculators, enter the number x followed by $\boxed{\text{LOG}}$.

EXAMPLE Finding common logarithms with a calculator Complete this table.

Input, x	Output, $y = \log_{10} x$
$\frac{1}{100}$	
$\frac{1}{10}$	
1	
10	
100	
1000	

Table 11

Solution The solution table is same as Table 8 in Example 3. Remember: *y, the logarithm base 10 of x, is the exponent to which we raise 10 to obtain x.*

In Example 8, we look for patterns while finding the common logarithms with a calculator.

EXAMPLE **8**   Finding common logarithms with a calculator   What patterns do you observe? Round your answers to five decimal places.

a. $\log_{10} 2$ b. $\log_{10} 20$ c. $\log_{10} 200$

d. $\log_{10} 3$ e. $\log_{10} 30$ f. $\log_{10} 300$

g. $\log_{10} 0.2$ h. $\log_{10} 0.02$ i. $\log_{10} 0.002$

Solution   a. 0.30103   b. 1.30103   c. 2.30103

d. 0.47712   e. 1.47712   f. 2.47712

g. -0.69897 h. -1.69897 i. -2.69897

The decimal portions are the same for log base 10 of 2, 20, and 200. Similarly, the decimal portions are the same for log base 10 of 3, 30, and 300. For log base 10 of 0.2, 0.02, and 0.002, the decimal portions are the same and the numbers are negative. The decimal portion is not the same for 2 as for 0.2. Other observations are possible.   ●

EXAMPLE **9**   Evaluating common logarithms   Evaluate these logarithms on the calculator. Use the ⬚LOG⬚ key for log base 10. Change each to an exponential equation to verify the calculator result.

a. $\log_{10} 0 = x$ b. $\log_{10}(-2) = x$

Solution   Graphing calculators may give an error message for both logarithms. The message indicates that the logarithm of zero is undefined and that the logarithm of a negative number is not a real number.

a. In $\log_{10} 0 = x$, the base is 10, so we are seeking the solution to $10^x = 0$. Since an exponential function is always positive, there is no real-number solution to this equation.

b. In $\log_{10}(-2) = x$, the base is 10, so $10^x = -2$ is an equivalent equation. There is no real-number power of 10 that results in a negative number.   ●

In Example 10, we add to our table a third column in which we solve the equation.

EXAMPLE **10**   Solving equations   Complete Table 12, changing the exponential equations to logarithmic equations and the logarithmic equations to exponential equations. Use

facts about powers of numbers to find x before using a calculator.

Exponential Equation, $y = b^x$	Logarithmic Equation, $x = \log_b y$	Solve for x
	$\log_2 128 = x$	
	$\log_{10} x = 2$	
$x^5 = 32$		
$10^x = 240$		
	$\log_{10} 40 = x$	
$10^{-3} = x$		

Table 12

Solution The variable x is placed in several different positions, so the definition of a logarithm must be used carefully. The completed version of Table 12 appears as Table 13.

Exponential Equation, $y = b^x$	Logarithmic Equation, $x = \log_b y$	Solve for x
$2^x = 128$	$\log_2 128 = x$	$2^x = 2^7$ $x = 7$
$10^2 = x$	$\log_{10} x = 2$	$10^2 = 100$ $x = 100$
$x^5 = 32$	$\log_x 32 = 5$	$2^5 = 32$ $x = 2$
$10^x = 20$	$\log_{10} 20 = x$	$x \approx 1.30103$
$10^x = 40$	$\log_{10} 40 = x$	$x \approx 1.60206$
$10^{-3} = x$	$\log_{10} x = -3$	$x = 0.001$

Student Note: You solved $10^x = 20$ in the Warm-up for this section.

Table 13

●

ANSWER BOX

Warm-up: 1. $x = 0$ **2.** $x = 1$ **3.** $x = 1.301$ **4.** $x = 2$ **5.** $x = 1$
6. $x = 2$ **7.** $x \approx 2.585$ **8.** $x = 3$ **Think about it 1:** For example,
$x = 2$ and $y = -2$. **Think about it 2:** $3^{x+2} = 3^x \cdot 3^2 = 9 \cdot 3^x$ **Think about it 3:** $x + y = 10^y + y$; $x - y = 10^y - y$; $xy = y \cdot 10^y$;
$x/y = 10^y/y$; $x^{1/y} = 10$. None of the expressions simplify. Thus, none of the operations both remove y from the exponent and permit us to solve for y.

EXERCISES

In Exercises 1 to 16, solve the equation for n or x. Recall that $2^{-1} = \frac{1}{2}$ and $3^{-2} = \frac{1}{9}$.

1. a. $2^x = 256$ **b.** $16^x = 256$

2. a. $4^n = 256$ **b.** $8^n = 256$

3. a. $4^n = 8$ **b.** $8^n = 16$

4. a. $8^n = 4$ **b.** $4^n = 128$

5. a. $16^n = 4$ **b.** $100^x = 1000$

6. a. $8^n = 2$ **b.** $100^x = 10$

7. a. $64^x = \frac{1}{8}$ **b.** $125^n = \frac{1}{5}$

8. a. $64^n = \frac{1}{4}$ **b.** $25^x = \frac{1}{125}$

9. a. $8^n = \frac{1}{2}$ **b.** $81^x = \frac{1}{3}$

10. a. $4^x = \frac{1}{64}$ **b.** $27^x = \frac{1}{9}$

11. a. $\left(\frac{1}{2}\right)^{x+5} = 32$ **b.** $\left(\frac{1}{8}\right)^{x+3} = 8$

12. a. $\left(\frac{1}{3}\right)^{x-5} = 81$ **b.** $\left(\frac{1}{9}\right)^{x-2} = 9$

13. a. $2^{x-1} = 2^{2x+5}$ **b.** $3^{2x-1} = 3^{x+3}$

14. a. $3^{x-2} = 3^{2x+3}$ **b.** $2^{2x+3} = 2^{x-5}$

15. a. $2^{x+5} = 4^{x-1}$ **b.** $5^{x+1} = 125^{x-3}$

16. a. $3^{x+4} = 9^{x-5}$ **b.** $5^{2x-1} = 625^x$

17. a. Make a table for $f(x) = 3^x$. Use integers on the interval $[-2, 4]$.

     b. Show the table for the inverse.

     c. Graph $f(x) = 3^x$ and show the graph of the inverse on the same axes. Is the inverse a function?

     d. Start with $y = 3^x$ and find the inverse with symbols.

18. Repeat Exercise 17 for $f(x) = 2^x$.

Complete the statements listed in Exercises 19 to 24. In Exercises 19 to 22, choose from domain or range.

19. The set of outputs from a function is its _____ .

20. The set of inputs of a function is its _____ .

21. Because we swap x and y, the domain of a function is the _____ of its inverse.

22. Because we swap x and y, the range of a function is the _____ of its inverse.

23. If the ordered pair (a, b) belongs to a function, then the ordered pair _____ belongs to its inverse.

24. If the graph of an inverse to a function passes the vertical-line test, then the inverse is a _____ .

Complete the tables in Exercises 25 and 26.

25.

Exponential Equation	Logarithmic Equation
	$\log_3 27 = 3$
	$\log_2 8 = 3$
$10^1 = 10$	
$5^3 = 125$	
	$\log_3\left(\frac{1}{9}\right) = -2$

26.

Exponential Equation	Logarithmic Equation
$6^3 = 216$	
$5^{-1} = \frac{1}{5}$	
	$\log_2 16 = 4$
	$\log_4\left(\frac{1}{16}\right) = -2$
	$\log_6 6 = 1$

Write the equations in Exercises 27 to 30 as exponential equations.

27. a. $\log_{10} 1000 = 3$ **b.** $\log_{10} 1 = 0$

c. $\log_3 81 = 4$ **d.** $\log_{10} 100 = 2$

28. a. $\log_{10} 10 = 1$ **b.** $\log_{10} 0.01 = -2$

c. $\log_2 32 = 5$ **d.** $\log_5 125 = 3$

29. a. $\log_{10} 0.001 = -3$ **b.** $\log_4 1 = 0$

c. $\log_m n = k$ **d.** $\log_5\left(\frac{1}{25}\right) = -2$

30. a. $\log_3 3 = 1$ **b.** $\log_{10} 10000 = 4$

c. $\log_3\left(\frac{1}{27}\right) = -3$ **d.** $\log_p r = m$

Write the equations in Exercises 31 to 34 as logarithmic equations.

31. a. $2^5 = 32$ **b.** $2^1 = 2$

c. $2^0 = 1$ **d.** $10^2 = 100$

32. a. $3^2 = 9$ **b.** $3^5 = 243$

c. $3^0 = 1$ **d.** $10^{-2} = 0.01$

33. a. $10^{-3} = 0.001$ **b.** $f^d = g$

c. $3^{-2} = \frac{1}{9}$ **d.** $4^0 = 1$

34. a. $2^{10} = 1024$ **b.** $2^{-1} = \frac{1}{2}$

c. $10^0 = 1$ **d.** $d^a = c$

In Exercises 35 to 38, start by finding the logs using $\boxed{\text{LOG}}$ on a calculator. Switch to finding the logs mentally as soon as possible. Round to five decimal places.

35. a. $\log_{10} 1$ **b.** $\log_{10} 6$

c. $\log_{10} 10$ **d.** $\log_{10} 60$

e. $\log_{10} 100$ **f.** $\log_{10} 600$

36. a. $\log_{10} 1000$ **b.** $\log_{10} 7000$

c. $\log_{10} 100{,}000$ **d.** $\log_{10} 700{,}000$

e. $\log_{10} 7{,}000{,}000$ **f.** $\log_{10} 1{,}000{,}000$

37. a. $\log_{10} 0.6$ **b.** $\log_{10} 0.1$

c. $\log_{10} 0.01$ **d.** $\log_{10} 0.06$

e. $\log_{10} 0.006$ **f.** $\log_{10} 0.000006$

38. a. $\log_{10} 0.7$ **b.** $\log_{10} 0.07$

c. $\log_{10} 0.001$ **d.** $\log_{10} 0.0001$

e. $\log_{10} 0.0007$ **f.** $\log_{10} 0.00007$

39. a. How are the logs of 6, 60, and 6000 related?

b. How are the logs of 0.6, 0.06, and 0.006 related?

40. a. How are the logs of 7000, 700, 70, and 7 related?

b. How are the logs of 0.00007, 0.0007, and 0.07 related?

Change the equations in Exercises 41 and 42 to logarithmic equations, and find x. Round to five decimal places.

41. a. $17 = 10^x$

b. $125 = 10^x$

c. $10^x = 400$

d. $10^x = 0.05$

42. a. $25 = 10^x$

b. $100 = 10^x$

c. $10^x = 300$

d. $0.28 = 10^x$

Complete the tables in Exercises 43 and 44.

43.

Exponential Equation	Logarithmic Equation	Solve for x
	$\log_5 x = 0$	
	$\log_x 64 = 3$	
$10^x = -10$		

44.

Exponential Equation	Logarithmic Equation	Solve for x
$2^x = 256$		
	$\log_x 125 = 3$	
	$\log_3 (-4) = x$	

Write the equations in Exercises 45 to 48 as exponential equations, and find x. Do not use a calculator.

45. a. $\log_7 x = 2$ **b.** $\log_3 3 = x$

    c. $\log_{10} 0.01 = x$ **d.** $\log_{10} x = 0$

    e. $\log_x 64 = 3$ **f.** $\log_a x = 1$

46. a. $\log_2 512 = x$ **b.** $\log_4 x = 4$

    c. $\log_2 x = 6$ **d.** $\log_{10} x = -1$

    e. $\log_x 256 = 4$ **f.** $\log_a x = 0$

47. a. $\log_2 x = 8$ **b.** $\log_{10} x = -2$

    c. $\log_2 \left(\frac{1}{4}\right) = x$ **d.** $\log_{10} 1 = x$

    e. $\log_x 100 = 2$ **f.** $\log_a a = x$

48. a. $\log_3 x = 1$ **b.** $\log_x 128 = 7$

    c. $\log_8 x = 2$ **d.** $\log_3 1 = x$

    e. $\log_3 27 = x$ **f.** $\log_a a^2 = x$

49. What is the line of symmetry for the graph of a function and its inverse?

50. Write how we say $\log_2 x$ in words.

Complete the statements in Exercises 51 to 57.

51. The logarithm is the _____ to which we raise a base to obtain a given number.

52. The _____ of an exponential equation must be a positive number not equal to 1.

53. The _____ of a logarithmic equation must be a positive number not equal to 1.

54. The set of inputs (domain) to an exponential equation, $y = b^x$, is _____ .

55. The set of outputs (range) to a logarithmic equation, $y = \log_b x$, is _____ .

56. The set of inputs (domain) to a logarithmic equation, $y = \log_b x$, is _____ .

57. The set of outputs (range) to an exponential equation, $y = b^x$, is _____ .

58. Solve for n: $\log_8 \sqrt[3]{2} = n$.

59. Error analysis: What is wrong with this equation?

$$\frac{\log_{10} 2}{\log_{10}} = 2$$

60. Can $\frac{5 \text{ feet}}{\log} \cdot \log_{10} 5$ be simplified?

61. If the exponents in an equation are equal, are the bases equal also? Consider $x^2 = 3^2$ in your explanation.

62. Why is it difficult to solve $2^n = 10$ with the like bases property?

63. Suppose two competing research groups examine bacterial growth data. The first group concludes that the population of bacteria is growing according to the rule $P = 4^{6t}$. The second group concludes that the population is growing according to the rule $P = 8^{4t}$. You have been asked to compare their results. What do you suggest?

64. Johanna solves $3^{x+2} = 243$ by first substituting $9 \cdot 3^x$ for 3^{x+2} and then dividing both sides by 9 to obtain $3^x = 27$. Explain whether her process is correct.

MID–CHAPTER **7** TEST

1. Identify the following sequences as either arithmetic (linear), quadratic, or geometric (exponential). Fit an equation to the linear and quadratic sequences. Find the nth term, a_n, for the geometric sequences. Then use exponential regression on a calculator to find $f(x)$. Show, using properties of exponents, that your results are the same.

    a. 3, 9, 27, 81, 243

    b. −1, 5, 11, 17, 23

    c. 3, 6, 12, 24, 48

    d. 81, 27, 9, 3, 1

    e. 1, 6, 15, 28, 45

    f. $\frac{1}{8}, \frac{1}{4}, \frac{1}{2}, 1, 2$

2. Sketch an exponential curve. Mark the points on the graph illustrating these facts:

    a. With the exception of $b = 0$, any number b raised to the zeroth power is 1.

b. Any number raised to the first power is itself: $b^1 = b$.

3. a. Describe the general shape of the graphs of the exponential functions $f(x) = 2^x$ and $f(x) = 2^{x-1}$ as we move from left to right on the coordinate axes.

b. Explain how the two graphs are related.

c. Where do the graphs of $y = 2^x$ and $y = 2^{x-1}$ intersect? Why?

4. Refer to a graph of $y = 2^x$.

a. Estimate the exponent on 2 that gives 10 as an output.

b. What equation did you solve in part a?

5. Compare the graphs of $y = \frac{1}{2} \cdot 3^x$ and $y = 2 \cdot \left(\frac{1}{3}\right)^x$.

a. Explain what part of the equation shows the y-intercept.

b. Explain how you know whether the equation indicates an increasing or decreasing function.

Solve Exercises 6 and 7 without a calculator. Show your steps.

6. a. $2^x = 32$ **b.** $\left(\frac{1}{2}\right)^x = 4$

  c. $\left(\frac{1}{2}\right)^x = \frac{1}{16}$ **d.** $5^x = \frac{1}{125}$

e. $10^x = -10$ **f.** $10^x = 0.01$

7. a. $\log_4 x = -1$ **b.** $\log_{10} 0.0001 = x$

  c. $\log_3 x = 3$ **d.** $\log_2 x = 0$

  e. $\log_4 2 = x$ **f.** $\log_4 -2 = x$

8. a. Make a table for $y = 4^x$ for integer inputs $[-2, 3]$.

b. Show the table for the inverse function.

c. Graph $y = 4^x$ and its inverse on one set of axes.

d. Use $y = 4^x$ to solve for the inverse.

9. Complete the table:

Exponential Equation	Logarithmic Equation	Solve for x
$10^x = 15$		
	$\log_{10} 13 = x$	
$3^x = 81$		
	$\log_3 x = 9$	
$4^{x+1} = 64$		

7.3

Applications of Exponential and Logarithmic Functions _____

OBJECTIVES

- Identify the base for a common logarithm.
- Change equations between exponential and logarithmic form to solve pH and Richter scale problems.
- Apply the change of base formula to evaluate logarithms.
- Graph logarithmic functions on a calculator.
- Solve applications with compound interest and exponential growth and decay.
- Estimate doubling time with the rule of 72.

WARM-UP

Change to an exponential equation and find the value of x or y. Write the answers to Exercises 1 and 2 in scientific notation.

1. $-9.2 = \log_{10} x$ **2.** $6.9 = \log_{10} y$

3. $x = \log_{10} 1$ **4.** $x = \log_{10} 10$

5. $y = \log_2 \left(\frac{1}{2}\right)$ **6.** $y = \log_2 1$

7. $y = \log_2 8$

Solve by guess and check.

8. $2 = (1.08)^t$ **9.** $2000 = 1000(1 + 0.07)^t$

IN THE PRIOR SECTION, we used common logarithms to solve exponential equations with base 10. We now continue with common logarithms but change to a more standard notation. We use the change of base formula to graph functions and solve equations containing bases other than 10. We apply our exponential and logarithmic techniques to compound interest and growth applications.

Common Logarithms and Their Applications

As mentioned earlier, logarithms with base 10 are called *common logarithms*. *Base 10 logarithms are usually written without a base:*

$$y = \log x \quad \text{is} \quad \log_{10} x.$$

EXAMPLE **1**

Identifying bases   Name the bases for these logarithms. If the base is a variable, state any restrictions.

a. $\log_2 8 = x$ **b.** $\log_4 x = 3$ **c.** $\log 100 = x$ **d.** $y = \log_5 x$
e. $\log_x 16 = 4$ **f.** $f = \log_c d$ **g.** $\log d = c$

Solution   **a.** The base is 2.

b. The base is 4.

c. The base is missing, so it is 10: $\log_{10} 100 = x$.

d. The base is 5.

e. The base is x, $x > 0$, $x \neq 1$.

f. The base is c, $c > 0$, $c \neq 1$.

g. The base is missing, so it is 10: $\log_{10} d = c$. ●

Until you are thoroughly familiar with logarithms, however, it is recommended that you write the base 10 as you work examples, exercises, and test questions.

pH SCALE   The pH scale describes the relative acidity or basicity of a substance. To the chemist, pH is a measure of the concentration of positive hydrogen ions, H^+, in a solution. The pH scale generally ranges from 1 to 14, but pH can be zero, negative, or greater than 14. Pure distilled water at 25°C (about room temperature) is said to be *neutral* and has a pH of 7. Solutions with pH greater than 7 are *basic*. Solutions with pH less than 7 are *acidic*. The pH ratings for common beauticians' supplies are shown in Figure 1 on the chapter opening (page 472).

The formula for the pH scale depends on logarithms:

$$\text{pH} = -\log [H^+]$$

where $[H^+]$ is the symbol for hydrogen ion concentration. The pH formula is included here to provide additional practice with common logarithms and scientific notation.

EXAMPLE **2**

Evaluating a common logarithm: the pH of water   Find the pH of water if water has an H^+ concentration of 1.0×10^{-7} M. (The M stands for *moles* and may be ignored for this application.)

Solution   As mentioned above,

$$\text{pH} = -\log(1.0 \times 10^{-7}) = 7$$

No base shows in the logarithm, so it is assumed to be a base 10, or common, logarithm. Suggested keystrokes are

●

In some chemical applications, we are given the pH of a number and need to find the hydrogen ion concentration, [H$^+$]. This involves substituting the pH number into the formula pH $= -\log[$H$^+]$ and solving for [H$^+$].

To solve pH $= -\log[$H$^+]$ for [H$^+$], remember to do two things:

1. Write the base, 10, in the equation.
2. Multiply both sides by -1 to get the negative sign on the other side of the equation.

EXAMPLE 3   Solving a common logarithm: the pH of borax   Suppose borax, a cleaning agent, has a pH of 9.2. Is borax acidic or basic? What is the hydrogen ion concentration?

Solution   Borax is basic because its pH is larger than the neutral value of 7 for distilled water. To find the hydrogen ion concentration in moles (M), we substitute the pH of 9.2 into the pH formula and solve for [H$^+$].

pH $= -\log[$H$^+]$	Substitute 9.2 for pH and write in the base, 10.
$9.2 = -\log_{10}[$H$^+]$	Multiply both sides by -1.
$-9.2 = \log_{10}[$H$^+]$	Change to an exponential equation.
$10^{-9.2} = [$H$^+]$	Simplify the expression by evaluating $10^{-9.2}$.
$[$H$^+] \approx 6.31 \times 10^{-10}$ M	

The p in pH represents a function. The p function is *negative log of.* In function notation, pH would be written $p(H)$. The H could be replaced by the appropriate concentration of any ion, but most commonly the concentration of the hydrogen ion, H$^+$, is used.

RICHTER SCALE   According to an article called "Abandoning Richter" (R. Monastersky, *Science News,* October 15, 1994), seismologists have pretty much given up use of the Richter scale to measure the relative magnitude of earthquakes but have neglected to inform the general public. Of the dozen or more available measurements of earthquakes, two have been generally adopted by scientists as replacements for the Richter number: the *preliminary magnitude* (somewhat like the Richter magnitude and available immediately after the earthquake is detected) and the *moment magnitude* (a measure of the total seismic wave energy released in an earthquake but not available until an hour or two after the earthquake because of the need to gather extensive data and perform calculations). But because the better known Richter scale provides a good illustration of use of a logarithmic scale, it will be used in the following explanation, examples, and related exercises.

Two earthquakes are compared by finding the ratio of their energy intensities, I. The Richter number, R, is related to the intensity of the earthquake by the formula

$$R = \log I$$

Note that this is a base 10, or common, logarithm.

EXAMPLE 4   Evaluating a common logarithm: Richter measurements   Calculate the Richter number for each of these earthquakes.

a. San Francisco, 1906 (503 deaths): $I = 199,526,000$

b. Iran, 1990 (40,000 deaths): $I = 50,120,000$

c. Alaska, 1964 (131 deaths): $I = 251,190,000$

Solution   **a.** $R = \log I = \log 199{,}526{,}000 \approx 8.3$

   b. $R = \log I = \log 50{,}120{,}000 \approx 7.7$

   c. $R = \log I = \log 251{,}190{,}000 \approx 8.4$ ●

To compare relative sizes, we divide the intensities, not the logarithms. In Example 5, we solve for the intensity by changing the logarithmic equation to an exponential equation, and then we compare intensities.

EXAMPLE **5**   Solving a common logarithm: San Francisco earthquake, 1989   For the San Francisco earthquake in 1989, $R = 6.9$.

a. Calculate the intensity, I, of this earthquake.

b. Divide the intensity of the 1906 quake by that of the 1989 quake to estimate the ratio of intensities.

Solution   **a.** $R = \log I$ Substitute 6.9 for R and write in the base 10.

   $6.9 = \log_{10} I$ Change to an exponential equation.

   $10^{6.9} = I$ Calculate the intensity.

   $I \approx 7{,}943{,}000$

b. The ratio of the intensity of the 1906 quake to that of the 1989 quake is

$$\frac{199{,}526{,}000}{7{,}943{,}000} \approx 25$$

The 1906 earthquake was rated about 25 times as intense as the 1989 earthquake.   ●

When working with common logarithms in applications, remember:

Common Logarithm in Applications

> **1.** Write the base, 10, before doing any operations or simplifications.
>
> **2.** For $y = -\log_{10} x$ or $y = a\log_{10} x$, move the negative or constant a to the other side:
>
> $$-y = \log_{10} x \qquad \text{or} \qquad \frac{y}{a} = \log_{10} x$$
>
> **3.** Use the equivalence of $y = \log_b x$ and $x = b^y$ to solve equations.

Change of Base Formula, Base 10

We used guess and check in Section 7.2 to solve the equation $2^x = 6$ and again in Warm-up Exercises 8 and 9 to solve $2 = (1.08)^t$ and $2000 = 1000(1 + 0.07)^t$. We can solve these equations much more easily with logarithms.

The change of base formula shown below changes a logarithm with any base to logarithms with base 10. The formula can be evaluated with a calculator.

Change of Base Formula, Base 10

> $$\log_b a = \frac{\log a}{\log b}$$

The change of base formula can be written with any base:

$$\log_b a = \frac{\log_c a}{\log_c b}$$

The formula applies to any real numbers a, b, and c for which the logarithm is defined. The proof of the change of base formula is in Section 7.4.

EXAMPLE **6**   Applying the change of base formula   Change these equations to logarithmic form and solve. In part a, round to three decimal places; in parts b and c, round to the nearest tenth.

a. $2^x = 6$ **b.** $2 = (1.08)^t$ **c.** $2000 = 1000(1 + 0.07)^t$

Solution   **a.** $2^x = 6$ Change to a logarithmic equation.

$\quad x = \log_2 6$ Apply the change of base formula.

$\quad x = \dfrac{\log 6}{\log 2}$ Enter the expression into a calculator. If your calculator automatically writes a parenthesis after log, close the parentheses on both the 6 and the 2.

$\quad x \approx 2.585$

Check: $2^{2.585} \stackrel{?}{=} 6$ ✔

b. $2 = (1.08)^t$ Change to a logarithmic equation.

$\quad t = \log_{1.08} 2$ Apply the change of base formula.

$\quad t = \dfrac{\log 2}{\log 1.08}$ Enter the expression into a calculator.

$\quad t \approx 9.0$ Remember that there will be some rounding error when you check.

Check: $(1.08)^9 \stackrel{?}{=} 2$ ✔

c. $2000 = 1000(1 + 0.07)^t$ Divide both sides by 1000.

$\quad 2 = (1.07)^t$ Change to a logarithmic equation.

$\quad t = \log_{1.07} 2$ Apply the change of base formula.

$\quad t = \dfrac{\log 2}{\log 1.07}$ Enter the expression into a calculator.

$\quad t \approx 10.2$ Remember that there will be some rounding error when you check.

Check: $1000(1 + 0.07)^{10.2} \stackrel{?}{=} 2000$ ✔   ●

Graphing Logarithmic Functions on a Calculator

In Example 7, we use the change of base formula to complete a table and a graph for $y = \log_2 x$.

EXAMPLE **7**   Applying the change of base formula to calculator graphing   Use the change of base formula to rewrite $y = \log_2 x$ in base 10. Make a table and a graph for $y = \log_2 x$.

Solution   With the change of base formula,

$$y = \log_2 x = \frac{\log x}{\log 2}$$

One possible table appears in Table 14, and the graph in Figure 14.

Input, x	Output, $y = \log_2 x$
-1	No real number
0	No real number
$\frac{1}{2}$	-1
1	0
2	1
4	2
8	3

Table 14

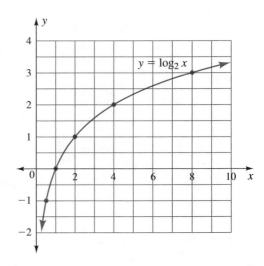

Figure 14   ●

Example 8 poses questions to guide a closer examination of the graph of $y = \log_2 x$.

EXAMPLE **8**   Investigating the properties of the logarithmic graph

   a. Describe any x- or y-intercepts of the graph of $y = \log_2 x$ in Figure 14.

   b. Describe any other key points.

   c. Set the table feature on a graphing calculator to a minimum of 1 and a change in input of -0.1. What happens as x approaches zero? Change the minimum to 0.1 and the change in input to -0.01. What happens as x approaches zero? What happens at zero? What happens below zero?

Solution   **a.** The graph of $y = \log_2 x$ crosses the x-axis at $(1, 0)$. There is no y-intercept.

   b. The graph of $y = \log_2 x$ passes through the point $(2, 1)$ because $2^1 = 2$.

   c. The output becomes more negative as x passes through smaller decimal numbers between 0 and 1. We say that the graph approaches the y-axis without touching or crossing it as x approaches zero. The y-axis is a vertical asymptote to the function.

   The output is not real when x is zero or negative, so the graph does not cross the y-axis. The function is not defined in the real numbers for $x \leq 0$.   ●

All graphs of $y = \log_b x$ have similar shapes. As with exponential graphs, there are certain points and characteristics that make it possible to recognize the graphs or to sketch them with a minimum number of points.

Summary: Properties of the Logarithmic Graph	• Graphs of $y = \log_b x$ pass through the x-intercept $(1, 0)$ because $0 = \log_b 1$ implies $b^0 = 1$. • Graphs of $y = \log_b x$ pass through $(b, 1)$ because $1 = \log_b b$ implies $b^1 = b$. • Graphs of $y = \log_b x$ have no y-intercept because $y = \log_b 0$ is undefined in the real numbers.

- Graphs of $y = \log_b x$ have no points in the second or third quadrants because, for negative x, $y = \log_b x$ is undefined in the real numbers. Hence, *the domain is the set of positive real numbers.*
- Graphs of $y = \log_b x$ have negative outputs for inputs between 0 and 1.
- Graphs of $y = \log_b x$ approach the y-axis as x approaches zero and rise slowly to the right as x increases. Hence, *the range is the set of all real numbers.*

Compound Interest

In Section 6.1 (page 421), we increased wages P by a given percent r each year according to the compound interest formula, $S = P(1 + r)^t$. When wages increased twice a year (page 422), it was necessary to use the compound interest formula for compounding n times a year,

$$S = P\left(1 + \frac{r}{n}\right)^{nt}$$

After using the compound interest formulas in Section 6.1 to find S, future wages or account balances, we used them in Section 6.4 (page 455) to find r, the annual rate of growth, inflation, or depreciation.

Returning to these settings, we now use logarithms and the change of base formula as we solve for the time, t.

EXAMPLE **9**

Finding the number of years to reach a target value   If Myrna starts at $5.50 per hour and receives a 6% annual wage increase, in how many years will her wage be $8.27 per hour?

Solution   To prevent rounding error in the solution, we will not simplify any expression that results in rounding; we will enter the unsimplified expression into the calculator. The initial amount, P, is $5.50 per hour. The annual rate of growth, r, is 6%. The future amount, S, is $8.27 per hour.

$$S = P(1 + r)^t$$

$$8.27 = 5.50(1 + 0.06)^t \qquad \text{Divide each side by \$5.50.}$$

$$\frac{8.27}{5.50} = (1.06)^t \qquad \text{Change to a logarithmic equation.}$$

$$\log_{1.06}\left(\frac{8.27}{5.50}\right) = t \qquad \text{Apply the change of base formula.}$$

$$t = \frac{\log\left(\dfrac{8.27}{5.50}\right)}{\log 1.06} \qquad \text{Enter the expression into a calculator.}$$

$$t \approx 7 \text{ yr}$$

It will take 7 years for her hourly wage to grow from $5.50 to $8.27 at 6% per year. ●

Exponential Growth and Decay

DOUBLING TIME   In Example 11 of Section 6.1, we used guess and check and a graph to solve for the number of years needed to double a $1000 investment at 7% interest compounded weekly. We now solve the same formula with logarithms and the change of base formula.

EXAMPLE **10** Solving with logarithms and change of base Solve this compound interest formula for t:

$$2000 = 1000\left(1 + \frac{0.07}{52}\right)^{52t}$$

Solution

$$2000 = 1000\left(1 + \frac{0.07}{52}\right)^{52t} \qquad \text{Divide each side by 1000.}$$

$$2 = \left(1 + \frac{0.07}{52}\right)^{52t} \qquad \text{Change to a logarithmic equation.}$$

$$\log_{(1+0.07/52)} 2 = 52t \qquad \text{Apply the change of base formula.}$$

$$\frac{\log 2}{\log(1 + 0.07/52)} = 52t \qquad \text{Multiply each side by } \frac{1}{52}.$$

$$\frac{1}{52} \cdot \frac{\log 2}{\log(1 + 0.07/52)} = t$$

$$t \approx 9.9$$

It will take approximately 9.9 years for the investment to double at 7% interest, compounded weekly. ●

*T*he *time required for money to double* in Example 10 is called the **doubling time.** This is an important concept in studying **exponential growth,** *the behavior described by an increasing exponential function.*

EXAMPLE **11**

Finding doubling time: garbage production Suppose garbage production in a large city increases by 8% per year. In how many years will garbage production double? From a graph, find when it will double a second time.

Solution The situation is described by compound interest calculated once a year, $n = 1$.

$$S = P(1 + r)^t \qquad \text{Doubling means } S = 2P, \text{ so substitute } r = 0.08 \text{ and } S = 2P.$$

$$2P = P(1 + 0.08)^t \qquad \text{Divide both sides by } P. \text{ Simplify the base.}$$

$$2 = (1.08)^t \qquad \text{Change to a logarithmic equation.}$$

$$\log_{1.08} 2 = t \qquad \text{Apply the change of base formula.}$$

$$t = \frac{\log 2}{\log 1.08} \approx 9 \text{ yr}$$

It will take about 9 years for the amount of garbage produced annually to double. This assumes that the 8% change in garbage production includes changes in both population and rate of throwing things away. A sudden increase in population or in recycling could alter the doubling time.

The graph in Figure 15 shows the ratio of future garbage production (S) to the current year's production (P); that is, $y = S/P = (1.08)^t$. The graph intersects $y = 2$ at 9 years. The graph intersects $y = 4$ at 18 years—the second doubling time.

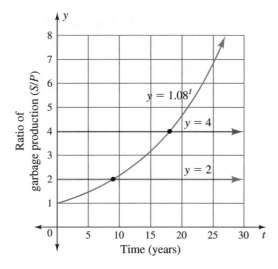

Figure 15

RULE OF 72   In business, the rule of 72 provides an important shortcut in estimating doubling time, given the annual rate of interest.

Rule of 72	Dividing 72 by the annual interest rate (expressed as a whole number) estimates the number of years required for the initial amount P to double. $72 \div$ interest rate $=$ doubling time For 8% interest, $72 \div 8 \approx 9$ yr doubling time.

HALF-LIFE   In Example 12, we return to the truck depreciation setting from Example 8 of Section 6.4 (page 456).

EXAMPLE 12

Solving with logarithms and change of base   Suppose a truck depreciates 20% each year. The initial value is $40,000. Solve the equation below to find out in how many years the truck will be worth half the original value:

$$20,000 = 40,000(1 + (-0.20))^t$$

Solution

$$20,000 = 40,000(1 + (-0.20))^t \qquad \text{Divide each side by 40,000.}$$

$$\frac{20,000}{40,000} = (1 - 0.20)^t \qquad \text{Simplify.}$$

$$0.5 = (0.80)^t \qquad \text{Change to a logarithmic equation.}$$

$$\log_{0.80} 0.5 = t \qquad \text{Apply the change of base formula.}$$

$$\frac{\log 0.5}{\log 0.80} = t$$

$$t \approx 3.1 \text{ yr}$$

In about 3.1 years, the truck will be worth half its original value.

Think about it 1: Why does the 0.80 in the solution to Example 12 make sense?

The time required for the value to drop to half its original size in Example 12 is called the **half-life.** This is an important concept in studying **exponential decay,** *the behavior described by a decreasing exponential function.*

EXAMPLE **13**   Finding the half-life   Suppose a 10-microgram mass of radioactive iodine-131 decreases by 8.3% each day. In how many days will half the mass be gone?

Solution   Half the mass is 5 micrograms (5×10^{-6} gram).

$$5 = 10(1 - 0.083)^t \qquad \text{Divide by 10 on each side.}$$
$$0.5 = (1 - 0.083)^t \qquad \text{Simplify within the parentheses.}$$
$$0.5 = (0.917)^t \qquad \text{Change to a logarithmic equation.}$$
$$\log_{0.917} 0.5 = t \qquad \text{Apply the change of base formula.}$$
$$t = \frac{\log 0.5}{\log 0.917}$$
$$t \approx 8 \text{ days}$$

The half-life of iodine-131 is 8 days.

USING REGRESSION EQUATIONS   An exponential regression equation can be the source of r, the annual rate of growth, as demonstrated in Example 14.

EXAMPLE **14**   Finding r, the annual rate of growth (or decline): population decline   Use the population facts in Table 15 for Sharon, Pennsylvania, to examine a case of population decline.

Year	Years since 1930, x	Population, y
1930	0	25,908
1940	10	25,622
1980	50	19,057
1990	60	17,533

Table 15   Population of Sharon, Pennsylvania

a. Plot the data in Table 15.

b. Fit an exponential regression equation to the data. Graph the equation.

c. What is the annual population change r?

d. Predict the population in the year 2000.

e. If the population continues to fall, in what year will the population be half that of 1930?

Solution   **a.** We let x be the number of years after 1930. The data are plotted in Figure 16.

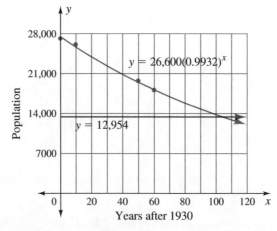

Figure 16

b. The exponential regression equation is $y \approx 26{,}600(0.9932)^x$. Because the exponential regression equation is fit to all four pieces of data, it does not have the 1930 population as its initial, or starting, value. Had we used only the first and last years' data, the equation would have been $y \approx 25{,}908(0.9935)^x$.

c. To estimate the annual population change, we compare the base of the exponential equation $y = 26{,}600(0.9932)^x$ with that of $S = P(1 + r)^t$. We find that $0.9932 = 1 + r$ and $-0.0068 = r$. The average annual population change is $r \approx -0.68\%$ per year.

d. For the year 2000, $x = 2000 - 1930 = 70$. Substituting $x = 70$ into the regression equation gives

$$f(70) = 26{,}600(0.9932)^{70} \approx 16{,}500$$

e. Half the original population is $25{,}908/2 = 12{,}954$. We let $y = 12{,}954$ and solve for x from the graph, finding $x \approx 105$.

  We get a similar result from an equation. Using the annual population change of 0.68% with a starting population of 25,908, we find the time required for the population to drop to 12,954.

$12{,}954 = 25{,}908(1 - 0.0068)^t$	Divide each side by 25,908.
$0.5 = (0.9932)^t$	Change to a logarithmic equation.
$\log_{0.9932} 0.5 = t$	Apply the change of base formula.
$t = \dfrac{\log 0.5}{\log 0.9932}$	
$t \approx 102 \text{ yr}$	

Assuming the population trend continues, the population will be half the original size about 100 years after 1930, or in 2030. ●

Think about it 2: Why might an exponential function represent population decline more realistically than a linear function?

Growth Rates

In exponential growth or increasing exponential functions, the base is greater than 1. The annual rate, r, is positive.

In exponential decay or decreasing exponential functions, the base is between 0 and 1. Thus, the annual rate, r, is negative.

ANSWER BOX

Warm-up 1. $x \approx 6.310 \times 10^{-10}$ **2.** $y \approx 7.943 \times 10^6$ **3.** $x = 0$
4. $x = 1$ **5.** $y = -1$ **6.** $y = 0$ **7.** $y = 3$ **8.** $t \approx 9.0$ **9.** $t \approx 10.2$
Think about it 1: When 20% is lost, 80% remains. So the value remaining each year is 80% of the prior year's value. **Think about it 2:** An exponential function never goes through the x-axis. It is likely that even a ghost town will have one or two residents or a caretaker for property owners. A linear function would pass through the x-axis and into the fourth quadrant, representing a zero and then a negative population.

EXERCISES 7.3

In Exercises 1 and 2, identify the base.

1. **a.** $\log_3 9 = 2$ **b.** $\log 100 = 2$

  c. $4^3 = 64$ **d.** $25 = 5^2$

  e. $\log x = y$ **f.** $\log_m h = c$

2. a. $\log_2 32 = 5$ **b.** $81 = 3^4$

  c. $\log 10 = 1$ **d.** $2^4 = 16$

  e. $\log_c m = h$ **f.** $\log y = x$

3. Find the pH of the following substances.

  a. NaOH (sodium hydroxide), with an H^+ concentration of 1.0×10^{-14} M

  b. HCl (hydrochloric acid), with an H^+ concentration of 0.20 M

  c. Lemon juice, with $[H^+] = 5.01 \times 10^{-3}$ M

  d. Milk of magnesia, with $[H^+] = 1.00 \times 10^{-10}$ M

  e. Apple cider, with $[H^+] = 1.26 \times 10^{-3}$ M

4. Find the pH of the following substances.

  a. Baking soda, with $[H^+] = 3.16 \times 10^{-9}$ M

  b. Soft drink, with $[H^+] = 0.01$ M

  c. Banana, with $[H^+] = 2.512 \times 10^{-5}$ M

  d. Milk, with $[H^+] = 2.512 \times 10^{-7}$ M

  e. Hominy, with $[H^+] = 1 \times 10^{-8}$ M

5. Write the word *base* or *acid* beside each answer in Exercise 3 to identify the nature of the substance.

6. Write the word *base* or *acid* beside each answer in Exercise 4 to identify the nature of the substance.

In Exercises 7 and 8, change the pH values back to hydrogen ion concentrations using the pH formula.

7. a. pH = 4.1 (tomatoes)

  b. pH = −0.5

  c. pH = 12.0 (washing soda)

  d. pH = 3.5 (acid rain)

8. a. pH = 7.4 (blood plasma)

  b. pH = −0.8

  c. pH = 5.8 (potatoes)

  d. pH = 3.3 (strawberries)

9. For what range of $[H^+]$ will the pH be negative? (*Hint:* Examine the graph of $y = -\log x$. For what set of inputs is the output of $y = -\log x$ negative?) For what range of $[H^+]$ will the pH be positive?

10. Describe the two steps that make it difficult to solve the equation $y = -\log x$ for x.

In Exercises 11 and 12, write the intensity in scientific notation. Calculate the Richter number for each earthquake.

11. a. Guatemala, 1976 (23,000 deaths): $I = 31,623,000$

  b. Mexico City, 1985 (4200 deaths): $I = 125,890,000$

12. a. Armenia, 1988 (55,000 deaths): $I = 6,310,000$

  b. Agaña, Guam, 1993: $I = 100,000,000$

In Exercises 13 and 14, calculate the intensity for each earthquake. Round to the nearest thousand.

13. a. Chile, Valparaiso, 1906: $R = 8.6$

  b. Northridge, California, 1994 (61 deaths): $R = 6.8$

14. Southern California, 1992 (1 death in two earthquakes on June 28):
  a. $R = 7.5$ **b.** $R = 6.6$

15. Compare the 1989 San Francisco earthquake, at $R = 6.9$, and the 1994 Northridge earthquake, at $R = 6.8$. First calculate the intensities, and then find how many times as strong the earlier earthquake was.

16. Compare the 1976 Tangshan, China, earthquake, at $R = 8.2$, and the 1993 Hokkaido, Japan, earthquake, at $R = 7.7$. First calculate the intensities, and then find how many times as strong the earlier earthquake was.

In Exercises 17 to 20, estimate x. Then solve, using logarithms and the change of base formula as needed. Round to two decimal places. Comment on any patterns you observe in the answers.

17. a. $2^x = 1$ **b.** $2^x = 3$ **c.** $2^x = 4$

  d. $2^x = 5$ **e.** $2^x = 6$ **f.** $2^x = 10$

  g. $2^x = 12$ **h.** $2^x = 20$ **i.** $2^x = 24$

18. a. $3^x = 2$ **b.** $3^x = 4$ **c.** $3^x = 6$

  d. $3^x = 9$ **e.** $3^x = 12$ **f.** $3^x = 18$

19. a. $5^x = 2$ **b.** $5^x = 5$ **c.** $5^x = 6$

  d. $5^x = 10$ **e.** $5^x = 25$ **f.** $5^x = 30$

20. a. $4^x = 3$ **b.** $4^x = 4$ **c.** $4^x = 5$

  d. $4^x = 12$ **e.** $4^x = 16$ **f.** $4^x = 20$

In Exercises 21 and 22, use the change of base formula, as needed, to solve the equations. Round to five decimal places.

21. a. $3^x = 10$

  b. $x = \log_4 6$

  c. $20 = 10(1.07)^x$

  d. $15 = 30(0.95)^x$

  e. $30 = 60(0.5)^x$

22. a. $4^x = 20$

   b. $x = \log_5 4$

   c. $30 = 15(1.04)^x$

   d. $10 = 20(0.93)^x$

   e. $16 = 32(0.25)^x$

23. Use a calculator to graph $y = \log_3 x$. Identify the x-intercept and $f(1)$.

24. Repeat Exercise 23 for $y = \log_5 x$.

25. On a calculator, graph $y = \log_3 x$ together with $y = \log_6 x$. Identify the x-intercept. Describe the relative positions of the two graphs to the right of the x-intercept and to the left of the x-intercept.

26. True or false: For $x > 1$ and b larger than 10, the graph of $y = \log_b x$ is above the graph of $y = \log_{10} x$.

27. True or false: For $0 < x < 1$ and b smaller than 10, the graph of $y = \log_b x$ is above the graph of $y = \log_{10} x$.

28. True or false: The point $(1, 0)$ is on every graph of $y = \log_b x$.

29. True or false: The point $(1, b)$ is on every graph of $y = \log_b x$.

30. True or false: For $0 < x < 1$ and $y = \log_b x$, y is negative.

31. On a calculator, graph $y_1 = x$, $y_2 = 2^x$, and $y_3 = \log_2 x$. Select a square window. Describe the positions of y_2 and y_3 relative to the line $y = x$. Explain why $(0, 1)$ lies on y_2 and $(1, 0)$ lies on y_3. Explain why $(1, 2)$ lies on y_2 and $(2, 1)$ lies on y_3. Name another pair of points (a, b) and (b, a) that lie on y_2 and y_3, respectively.

32. On a calculator, graph together $y = x$, $y = \sqrt{x}$, and $y = x^2$ for $x \geq 0$. Select a square window. Where do the graphs cross? Why?

In Exercises 33 to 36, find how long (to the nearest year) it will take for $1000 to grow to $1 million at the given interest rate.

33. 2% annual interest   **34.** 8% annual interest

35. 10% annual interest   **36.** 14% annual interest

In Exercises 37 to 40, identify the annual growth rate r.

37. a. $y = 1000(1 + 0.08)^t$

   b. $y = 50(1.06)^x$

38. a. $y = 400(1 + 0.05)^t$

   b. $y = 30(1.04)^x$

39. a. $y = 150(1 - 0.03)^t$

   b. $y = 200(0.95)^x$

40. a. $y = 40(1 - 0.10)^t$

   b. $y = 100(0.93)^x$

41. Suppose water consumption in a large city increases by 6% per year. In how many years will water consumption double?

42. Suppose electricity usage in a large city increases by 4% per year. In how many years will electricity usage double?

43. Use the rule of 72 to give the doubling times for 1%, 2%, 5%, 8%, 10%, and 20% interest.

44. Does the rule of 72 describe a direct variation or an inverse variation between interest rate and doubling time? Explain why.

The article "A Population Exploding," in the December 1988 National Geographic Magazine, gives the following doubling times for the populations of the given countries if the populations continue to rise at their current rates. In Exercises 45 and 46, use the rule of 72 to estimate to the nearest tenth the percent annual growth rate being experienced in each country.

45. a. Kenya, 17 years   **b.** Brazil, 34 years

46. a. India, 35 years   **b.** China, 49 years

47. Suppose a $30,000 car loses 10% of its value each year. Write an exponential equation and use it to find in how many years (to the nearest hundredth) the value will be half of the original.

48. Suppose $5000 software loses 40% of its value each year. Write an exponential equation and use it to find in how many years (to the nearest hundredth) the value will be less than $100.

49. A bouncing ball's height is given by $h = 108\left(\frac{2}{3}\right)^{n-1}$, where $n = 1$ is the starting height of 108 inches.

   a. What is n when the height is half the original height?

   b. What is n when the height is 32 inches?

   c. What is n when the height is 1 inch?

50. In Exercise 49, suppose we start the ball at 144 inches and it bounces to $\frac{3}{4}$ of its previous height each time. The starting height is where $n = 1$.

   a. Write a formula for the ball's height in terms of n.

   b. Find what n must be for the height to be 60.75 inches.

   c. Find what n must be for the height to be below 20 inches.

51. When an initial amount P is doubled, the ending amount is $2P$. Using this fact and $S = P(1 + r)^t$, solve for t to find a formula for doubling time in terms of the rate, r. Does P appear in your formula? What does this say about doubling time and the initial amount of money?

52. When an initial amount P is halved, the ending amount is $\frac{1}{2}P$. Using this fact and $S = P(1 + r)^t$, solve for t to find a formula for half-life in terms of the rate, r. Does P appear in your formula? What does this say about half-life and the initial amount?

Exercises 53 to 58 list the five radioactive substances released in greatest quantity by the Hanford Nuclear Site into the Columbia River between World War II and 1970. The release data are in curies* and represent the initial value S_0.

a. *Estimate the number of half-lives in 1 year.*

b. *Fit an equation using exponential regression. Use $(0, S_0)$ and $(half\text{-}life, \frac{1}{2}S_0)$ for data points.*

53. Arsenic-76: 2,500,000 curies, half-life of 26.3 hours

54. Neptunium-239: 6,300,000 curies, half-life of 2.4 days

55. Phosphorus-32: 230,000 curies, half-life of 14.3 days

56. Neptunium decays into plutonium-239, with a half-life of 24,000 years. The amount of neptunium-239 released decayed into 1.7 curies of plutonium-239.

57. Sodium-24: 12,000,000 curies, half-life of 15 hours

58. Zinc-65: 490,000 curies, half-life of 245 days

59. In 1986, it was possible to find Glue Stics® on sale at 2 for $1.00. By 1995, Glue Stics cost $0.99 each.

    a. Estimate the cost of Glue Stics in 2005.

    b. Fit a linear regression equation to the cost data. Interpret the slope of the line. Graph the line.

    c. Fit an exponential regression equation to the cost data. Graph the equation. What is the y-intercept?

    d. From the exponential equation, find the rate of inflation (growth rate r).

    e. What factors are assumed with the linear model and with the exponential model?

    f. Use each equation to predict the price of Glue Stics in the year 2005.

60. Suppose a house that cost $28,000 in 1973 doubles in value every 10 years.

    a. Estimate the value of the house in the year 2000.

    b. Use exponential regression to fit an equation to the house values calculated in order to complete part a.

    c. From the exponential regression equation, find the annual growth rate r.

In Exercises 61 to 64, find an exponential regression equation using the data from 1930, 1940, 1980, and 1990. Let 1930 be $x = 0$, and express the other dates in terms of the number of years after 1930. Indicate the annual rate of population growth (or decline). Use your equation to predict the population in the year 2000. Round to the nearest thousand. Research and compare with the year 2000 census data.

61. Albuquerque, New Mexico: 1930, 26,570; 1940, 35,449; 1980, 332,920; 1990, 384,619

62. Longview, Texas: 1930, 5,036; 1940, 13,758; 1980, 65,762; 1990, 70,311

63. Salem, Massachusetts: 1930, 43,353; 1940, 41,213; 1980, 38,276; 1990, 38,091

64. Oshkosh, Wisconsin: 1930, 40,108; 1940, 39,089; 1980, 49,620; 1990, 55,006

65. The following data sets have the same outputs and different inputs. Find the regression equation for each set. Compare the results.

    a. (45, 50), (50, 25), (55, 12.5), (60, 6.25)

    b. (0, 50), (5, 25), (10, 12.5), (15, 6.25)

    c. (0, 50), (1, 25), (2, 12.5), (3, 6.25)

Each output is half the output from the prior data point, yet the regression equations for parts a and b contain base 0.87. Why is the base not 0.5?

Projects

66. *Wage Options (spreadsheet optional)*. Suppose that there are three wage options for food servers.

    Option 1: The base salary is $5.00 per hour. An annual raise of 25 cents per hour is given at the end of the year.

    Option 2: The base salary is $4.00 per hour. A raise is given annually. The raise is 25 cents per hour at the end of the first year, 30 cents per hour at the end of the second year, 35 cents per hour at the end of the third year, 40 cents per hour at the end of the fourth year, and so on.

    Option 3: The base salary is $4.00 per hour. At the end of each year, a 5% increase is made in the hourly wage.

    a. Evaluate each wage option for the first five years.

    b. Identify the type of function for each.

    c. Fit an equation to each.

    d. Discuss the circumstances under which each option would be the best.

    e. In how many years will the wage double under each option?

---

*A curie is 37×10^9 disintegrations per second. A disintegration is the release of an alpha, beta, or gamma particle. The data are from *Radionuclides in the Columbia River: Possible Health Problems in Humans and Effects on Fish,* a publication of the Hanford Health Information Network, 1995.

67. *Birthday Gift Options (spreadsheet optional).* When each of her grandchildren is born, a grandmother offers three options for birthday gifts.

> Option 1: A dollar for each year of the child's age.
>
> Option 2: A dollar at age 1, with a 10% increase on each subsequent birthday.
>
> Option 3: A dollar at age 1, with a $0.10 increase at age 2, a $0.20 increase at age 3, a $0.30 increase at age 4, a $0.40 increase at age 5, and so forth.

a. Evaluate each option for the first five years.

b. Identify the type of function for each.

c. Describe each option with an equation.

d. In how many years will the birthday gift double under each option?

e. Discuss the circumstances under which each option would be the best.

f. Spreadsheet extension: Show the cumulative totals for each year in order to compare total gifts received under each option.

68. *Decibel Ratings.* Research the subject of sound intensity and decibels. Find a chart showing the relative decibel ratings for various sounds, such as those of a jet aircraft and a gasoline-powered lawn mower. Create five exercises based on your research, and solve them.

69. *Garbage Research.* What size garbage can was your household using 10 years ago? Have you changed from a 30-gallon metal or plastic can to a 60- (or even 90-) gallon cart? Find out what the population of your city is now and what it was 10 years ago. Based on the change in garbage production in your household and the change in population in your city, estimate the change in garbage production over the past 10 years. Finally, contact the solid waste management agency for your city and ask what its records show as the change over the past 10 years.

7.4 Properties of Logarithms and the Logarithmic Scale _____

OBJECTIVES

- Prove the properties of logarithms.
- Apply properties of logarithms in writing expressions.
- Solve equations by taking the logarithm of both sides.
- Prove the change of base formula.
- Read a logarithmic scale.
- Graph data on semilog graph paper.

WARM-UP

Use a calculator to find the logarithms in Exercises 1 to 3. Round to three decimal places. Explain how the expressions in parts a and b might be related.

1. a. $\log_{10} 5 + \log_{10} 8$
   b. $\log_{10} 40$

2. a. $3 \log_{10} 8$
   b. $\log_{10} 8^3$

3. a. $\log_{10} 60 - \log_{10} 6$
   b. $\log_{10} 10$

4. Solve $3^x = \frac{1}{27}$ with the like bases property.

5. Solve $3^x = 15$ with the definition of logarithms and the change of base formula.

I N THIS SECTION, we prove properties of logarithms, solve equations containing logarithms, and solve equations by taking the logarithm of both sides. We examine exponential functions plotted on special (semilogarithmic) graphs.

Properties of Logarithms

The Warm-up suggests three properties of logarithms involving multiplication, exponents, and division.

Properties of Logarithms

If m, n, and b are positive numbers and $b \neq 1$, then

1. $\log_b (m \cdot n) = \log_b m + \log_b n$
2. $\log_b m^n = n \cdot \log_b m$
3. $\log_b \left(\dfrac{m}{n} \right) = \log_b m - \log_b n$

We have seen how the properties $b^m \cdot b^n = b^{m+n}$ and $b^{-1n} = (1/b)^n$ permit us to transform expressions resulting from the nth term of a geometric sequence into expressions resulting from exponential regression. These properties are important in proving the properties of logarithms.

In Examples 1 and 2, we prove the first two of the three properties of logarithms. To prove the third property, we return to the definitions and follow a line of thinking similar to that used in the proof of property 1. This proof is left as an exercise.

EXAMPLE **1**   Proving property 1   Prove $\log_b (m \cdot n) = \log_b m + \log_b n$.

Solution   We start by writing equations containing logarithmic expressions like those on the right side of property 1.

$x = \log_b m$ and $y = \log_b n$	Change to exponential equations.
$m = b^x$ and $n = b^y$	Multiply the left sides and the right sides of the exponential equations.
$mn = b^x \cdot b^y$	Apply the property of like bases: $a^m \cdot a^n = a^{m+n}$.
$= b^{x+y}$	Change to a logarithmic equation.
$\log_b mn = x + y$	Substitute for x and y.
$= \log_b m + \log_b n$	●

Think about it 1: What changes in Example 1 if we change the multiplication of m and n to a division?

To prove the power property of logarithms, we use the fact that a positive integer exponent means repeated multiplication of the base and then apply property 1.

EXAMPLE **2**   Proving property 2   Prove $\log_b m^n = n \cdot \log_b m$.

Solution We start by rewriting $\log_b m^n$ as n factors of m:

$$\log_b m^n = \log_b (m \cdot m \cdot m \cdot \cdots \cdot m) \qquad \text{Apply property I, to obtain } n \text{ terms.}$$

$$= \log_b m + \log_b m + \log_b m + \cdots + \log_b m \qquad \text{Change } n \text{ terms to } n \text{ times the term.}$$

$$= n \cdot \log_b m \qquad \bullet$$

J ohn Napier invented logarithms to aid in arithmetic calculation. In the next three examples, we apply the properties to the expressions in the Warm-up.

EXAMPLE **3** Investigating multiplication Combine these logarithmic expressions into an equation: $\log_{10} 5$, $\log_{10} 8$, and $\log_{10} 40$. Check by evaluating the logarithms, and show with exponents why the equation is true.

Solution From property 1,

$$\log_{10} 5 + \log_{10} 8 = \log_{10} (5 \cdot 8)$$
$$= \log_{10} 40$$

Check: $0.69897 + 0.90309 = 1.60206$ ✔

When we use the logarithms as the exponents on 10, we obtain

$$5 \cdot 8 \approx 10^{0.69897} \cdot 10^{0.90309} = 10^{0.69897+0.90309} = 10^{1.60206} \approx 40 \qquad \bullet$$

EXAMPLE **4** Investigating powers Combine these logarithmic expressions into an equation: $\log_{10} 8^3$ and $3 \log_{10} 8$. Check by evaluating the equation, and show with exponents why the equation is true.

Solution From property 2,

$$\log_{10} 8^3 = 3 \log_{10} 8$$

Check: $2.70927 = 3(0.90309)$ ✔

When we use the logarithms as the exponents on 10, we obtain

$$8^3 \approx 10^{0.90309} \cdot 10^{0.90309} \cdot 10^{0.90309}$$
$$= 10^{0.90309+0.90309+0.90309}$$
$$= 10^{3(0.90309)}$$
$$= 10^{2.70927} \approx 512 \qquad \bullet$$

EXAMPLE **5** Investigating division Combine these logarithmic expressions into an equation: $\log_{10} 60$, $\log_{10} 6$, and $\log_{10} 10$. Check by evaluating the equation, and show with exponents why the equation is true.

Solution From property 3,

$$\log_{10} 60 - \log_{10} 6 = \log_{10} \left(\tfrac{60}{6}\right)$$
$$= \log_{10} 10$$

Check: $1.77815 - 0.77815 = 1.00000$ ✔

When we use the logarithms as the exponents on 10, we obtain

$$\frac{60}{6} = \frac{10^{1.77815}}{10^{0.77815}} = 10^{1.77815-0.77815} = 10^{1.00000} = 10 \qquad \bullet$$

Solving Equations by Taking the Logarithm of Both Sides

We have solved exponential equations such as $3^x = \frac{1}{27}$ and $3^x = 15$ with a variety of methods: guess and check, graphing, the like bases property (if $a^x = a^y$, then $x = y$), changing to a logarithm (if $y = b^x$, then $x = \log_b y$), and the change of base formula. In Example 6, we review the like bases property and the change of base formula ($\log_b a = \log_{10} a / \log_{10} b$).

EXAMPLE **6**   Solving equations

a. Solve $3^x = \frac{1}{27}$ with the like bases property.

b. Solve $3^x = 15$ with the definition of the logarithm and the change of base formula.

Solution   **a.** $3^x = \frac{1}{27}$ Change $\frac{1}{27}$ to a power of 3.

$3^x = 3^{-3}$ Apply the like bases property.

$x = -3$

b. $3^x = 15$ Change to a logarithmic equation.

$\log_3 15 = x$ Apply the change of base formula.

$$x = \frac{\log 15}{\log 3} \approx 2.46497$$ ●

In addition to using the equation-solving methods reviewed in Example 6, we may also take the logarithm, to any base, of both sides of an equation.

Solving Exponential Equations	We may take the logarithm base b of both sides of an equation.

In Example 7, we solve the equations from Example 6 by taking the log base 10 of both sides.

EXAMPLE **7**   Taking the logarithm base b of both sides   Solve these equations by taking the log base 10 of both sides.

a. $3^x = \frac{1}{27}$ **b.** $3^x = 15$

Solution   **a.** $3^x = \frac{1}{27}$ Take the log base 10 of both sides.

$\log_{10} 3^x = \log_{10} \left(\frac{1}{27}\right)$ Use property 2 to move the exponent x.

$x \log_{10} 3 = \log_{10} \left(\frac{1}{27}\right)$ Divide both sides by $\log_{10} 3$.

$$x = \frac{\log_{10} \left(\frac{1}{27}\right)}{\log_{10} 3}$$ This is the change of base formula, applied to $x = \log_3 \left(\frac{1}{27}\right)$.

≈ -3

Check: $3^{-3} = \frac{1}{27}$ ✔

b. $3^x = 15$ Take the log base 10 of both sides.

$\log_{10} 3^x = \log_{10} 15$ Use property 2 to move the exponent x.

$x \log_{10} 3 = \log_{10} 15$ Divide both sides by $\log_{10} 3$.

$$x = \frac{\log_{10} 15}{\log_{10} 3}$$ This is the change of base formula, applied to $x = \log_3 15$.

≈ 2.46497

Check: $3^{2.46497} \approx 15$ ✔   ●

Think about it 2: Could we solve the equations in Example 7 by taking the logarithm base 3 of both sides?

In parts a and b of Example 7, our last step was evaluating an expression that was in fact the change of base formula. This suggests that we can prove the change of base formula by taking the log of both sides of an equation.

Proof of the Change of Base Formula

By taking the log of both sides of an equation and applying the properties of logarithms, we prove the change of base formula in Example 8.

EXAMPLE

Proving change of base   Prove the change of base formula,

$$\log_b a = \frac{\log_c a}{\log_c b}$$

Solution   Our plan is to work with one side of the equation and show that it equals the other side of the equation. We start by setting the left side of the formula equal to the variable.

$x = \log_b a$	Change to an exponential equation.
$b^x = a$	Take the log base *c* of both sides.
$\log_c b^x = \log_c a$	Apply property 2 to move the exponent *x*.
$x \log_c b = \log_c a$	Divide both sides by $\log_c b$.
$x = \dfrac{\log_c a}{\log_c b}$	This equation contains the right side of the formula. Substitute $\log_b a$ for *x*.
$\log_b a = \dfrac{\log_c a}{\log_c b}$	The change of base formula is true. ●

As you will find in the exercises, by replacing the log base *c* with the log base 10, we can prove the change of base formula that we have used on the calculator.

$$y = \log_b a = \frac{\log a}{\log b}$$

Application: Logarithmic Scale

We now turn to a visual representation of logarithms—the logarithmic scale. Have you ever wondered how illustrators fit small numbers (such as the growth rate of a child in miles per hour) and large numbers (such as the speed of light in miles per hour) on the same graph? Example 9 illustrates the problems encountered in graphing small and large numbers on the same scale.

EXAMPLE

Exploring the scale on the axes   In a table, list the outputs for $y = 2^x$ if the inputs are the integers from 1 to 10. Graph the table values, and discuss the limitations of the graph.

Solution The input-output table appears in Table 16, and the graph in Figure 17.

x	y = 2^x
1	2
2	4
3	8
4	16
5	32
6	64
7	128
8	256
9	512
10	1024

Table 16

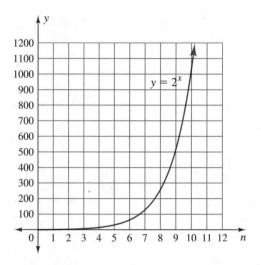

Figure 17

Finding a scale for the vertical axis in Figure 17 is a challenge. When we count by 100 on the vertical axis, the outputs between 0 and 100 are hard to read because they are so near the *x*-axis. If we enlarged the distance between 0 and 100 to, say, an inch, the vertical axis would need to be 12 inches tall and the graph would not fit on the page. ●

The solution to this scaling problem is to use a logarithmic scale. *When the spacing on an axis of a line graph is proportional to the logarithms of numbers,* the scale is called a **logarithmic scale.** Figure 18 shows a logarithmic scale with the positive powers of 10 in a horizontal line. The pH scale for hair products in Figure 1 on the chapter opener shows the negative powers of 10 forming a logarithmic scale on the vertical axis.

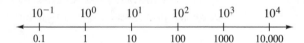

Figure 18

Think about it 3: What are the values of log 1, log 10, log 100, log 1000, and log 10,000?

When we place a logarithmic scale on one axis of a coordinate graph, we create a semilog graph. On a **semilog graph,** *one scale is uniform and the other scale is based on logarithms.* The logarithmic scale can be on either the horizontal or the vertical axis.

In Example 10, we create a semilog graph of the equation $y = 2^x$.

EXAMPLE **10**

Graphing on semilog axes Construct a semilog graph of $y = 2^x$ from the values shown in Table 16. Use special semilog paper, with the logarithmic scale on the vertical axis.

Solution Figure 19 shows a semilog graph of $y = 2^x$. The inputs are where we expect them to be, but the outputs, 2^x, are not in the same positions as in Figure 17. The numbers on the y-axis, although unusually spaced, make it possible to locate outputs in the same way as in other graphs.

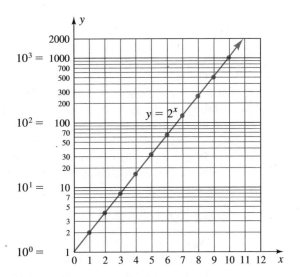

Figure 19

The graph in Figure 19 has a number of unique features. In Example 11, we compare it with the original graph of $y = 2^x$ in Figure 17.

EXAMPLE **11**   Identifying patterns   What distinguishes the semilog graph in Figure 19 from the graph in Figure 17?

Solution On the vertical axis, the distance between 2 and 20 is the same as that between 20 and 200. The distance between 10 and 100 is the same as that between 100 and 1000.

The spacing between the horizontal lines varies, but in a regular pattern.

The exponential curve in Figure 17 has been replaced by a straight line in Figure 19.

The smaller outputs, 2, 4, 8, and 16, are clearly shown in Figure 19, whereas in Figure 17 they are all near the x-axis and hard to read. The larger outputs, 512 and 1024, are harder to read in Figure 19.

So far, the labels on all our number lines (and, hence, on axes on coordinate graphs) have been equally spaced—say, counting by 10. On the logarithmic scale, we multiply by 10 between numbers on the number line. Thus, the equally spaced numbers are powers of 10, and the logarithms of the powers of 10 are the integers, $-3, -2, -1, 0, 1, 2, 3$.

Y ou may wonder about the spacing in the logarithmic scale. Figure 20 is a line with 0.1 spacing between the numbers 0 to 1.0. In Figure 20, dots are placed at the logarithms of the numbers 1 to 10. Observe that the spacing between the logarithms of the numbers 1 to 10 is similar to the spacing between the lines on the vertical axis in Figure 19. If we changed the number line to 1.0 to 2.0, the spacing between dots for the logarithms of 10 to 100 would be identical to that for the logarithms of 1 to 10 (see Exercise 52).

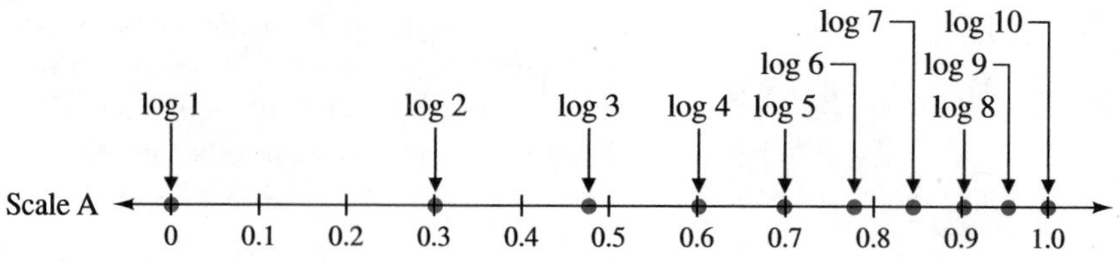

Figure 20

The purpose of this section is simply to ensure that you understand the logarithmic scale, how to plot data on a semilog graph, and how to read data from a semilog graph. In science classes, especially chemistry and physics, semilog graphs have traditionally been used to find equations from data. This process is being replaced with the use of regression analysis on graphing calculators. Project 59 gives you the opportunity to ask a physics or chemistry instructor about finding an exponential equation from a semilog graph.

ANSWER BOX

Warm-up: 1. a. 1.602 **b.** 1.602; $5 \cdot 8 = 40$ **2. a.** 2.709 **b.** 2.709; $3 \log 8 = \log 8^3$ **3. a.** 1 **b.** 1; $60/6 = 10$ **4.** $x = -3$ **5.** $x \approx 2.465$
Think about it 1: The exponents in b^x and b^y will be subtracted instead of added. **Think about it 2:** Yes, any base could be used.
Think about it 3: $\log 1 = 0$; $\log 10 = 1$; $\log 100 = 2$; $\log 1000 = 3$; $\log 10{,}000 = 4$.

EXERCISES 7.4

In Exercises 1 and 2, match an expression from the list of choices to each logarithmic expression. Write the original expression as well as the chosen expression.

1. Choose from
$\log (x - 2)$, $\log (x/2)$, $\log (2 - x)$, $\log (2 + x)$, $\log 2 - \log x$, $2 \log x$, $x \log 2$, $\log 2 + x$, $\log 2x$

  a. $\log 2 + \log x$

  b. $\log x^2$

  c. $\log (2/x)$

  d. $\log 2^x$

  e. $\log x - \log 2$

2. Choose from
$\log 3 + \log x$, $\log x^3$, $\log (3 - x)$, $\log (3/x)$, $\log 3^x$, $\log x - \log 3$, $\log 3 + x$, $\log (x - 3)$, $\log (3 + x)$

  a. $\log (x/3)$

  b. $\log 3x$

  c. $\log 3 - \log x$

  d. $x \log 3$

  e. $3 \log x$

Identify the property of logarithms that explains each fact in Exercises 3 to 6.

3. If $\log 2 \approx 0.30103$ and $\log 3 \approx 0.47712$, then $\log 6 \approx 0.77815$.

4. If $\log 2 \approx 0.30103$, then $\log (10 \cdot 2) \approx 1 + 0.30103$.

5. If $\log 3 \approx 0.47712$, then $\log 9 \approx 0.95424$.

6. If $\log 2 \approx 0.30103$, then $\log 8 \approx 0.90309$.

7. Show that $\log \sqrt{x} = \frac{1}{2} \log x$.

8. Show that $\log (1/x) = -\log x$.

9. Prove that $\log_b (m/n) = \log_b m - \log_b n$. Model your proof after Example 1.

10. In Example 8, replace the log base c with the log base 10, and prove the change of base formula $y = \log_b a = \log a/\log b$.

In Exercises 11 to 14, write the expression as a single logarithm and simplify where possible. Assume that all logarithms are defined. (Hint: Factoring may be helpful.)

11. a. $\log (x + 1) + \log (x - 1)$

  b. $\log (x^2 - 1) - \log (x - 1)$

12. a. $\log (x + 1) - \log (x^2 - 1)$

b. $\log (x - 2) + \log (x + 4)$

13. a. $\log (x^2 + 3x + 2) - \log (x + 1)$

b. $\log (x + 3) + \log (x - 2)$

14. a. $\log (x - 3) + \log (x + 3)$

b. $\log (x^2 - 9) - \log (x^2 + 6x + 9)$

Simplify the left side of each equation in Exercises 15 to 22, and then solve.

15. $\log_2 x^2 + \log_2 x = 6$

16. $\log (x^2 - x) - \log (x - 1) = 2$

17. $\log x + \log x = 2$

18. $\log_5 x^3 - \log_5 x^2 = 1$

19. $\log (x^2 + x) - \log (x + 1) = 1$

20. $\log_4 x + \log_4 1 = 2$

21. $\log_3 x^3 - \log_3 x = 4$

22. $\log_6 x + \log_6 x = 4$

Solve the equations in Exercises 23 to 38 in two ways:

a. Take the log of both sides.

b. Change to logarithmic equations and apply the change of base formula.

Round to four decimal places.

23. $5^x = 20$

24. $6^x = 24$

25. $3^{x+2} = 48$

26. $8^{x+1} = 36$

27. $4^{x-1} = 28$

28. $7^{x-1} = 35$

29. $9^{x+1} = 42$

30. $10^{x-1} = 25$

31. $1000(1.08)^t = 2000$

32. $1000(1.09)^t = 3000$

33. $15(1.07)^t = 45$

34. $10(1.06)^t = 20$

35. $1000(1 - 0.08)^t = 500$

36. $900(1 - 0.09)^t = 300$

37. $45(1 - 0.07)^t = 15$

38. $20(1 - 0.06)^t = 10$

39. One reference book gives the pH formula as

$$pH = \log \frac{1}{[H^+]}$$

Another reference has

$$pH = -\log [H^+]$$

Are these formulas the same or different? Why?

40. Why is $\log 3 = \frac{1}{2} \cdot \log 9$?

For Exercises 41 and 42, make a table for integer inputs -1 to 6. Graph the data from the tables on both regular graph paper and semilog graph paper. Photocopy the semilog grid below if needed.

41. $y = 3^x$

42. $y = 4^x$

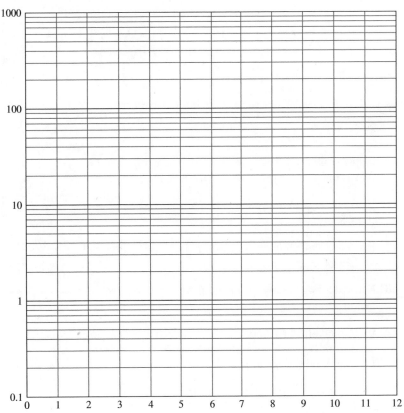

43. What shape is formed by exponential data on a semilog graph?

44. The spacing between the powers of 10 on the vertical axis of the semilog graphs in Exercises 41 and 42 is equal. The spacing represents the integers formed by the _____ of the numbers shown on the axis.

In Exercises 45 to 50, refer to the graph of the snag "recruitment" process. The graph shows the decay of a dead fir tree that has been left standing to serve as a wildlife habitat.

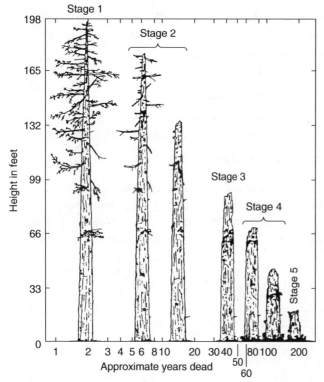

Snag "recruitment" process
Source: Rising from the Ashes, Shady Beach—A New Perspective on Recovery, U.S. Forest Service (Willamette National Forest), Department of Agriculture, 1990.

45. Approximately how many feet of height are lost between the second year and the sixth year?

46. Estimate how many years it takes for the tree to lose its top 66 feet.

47. Estimate how many years until the tree loses its second 66 feet of height.

48. Estimate how many years until the last 66 feet of height are lost.

49. Is the height loss more rapid between year 2 and year 12 or between year 100 and year 200? What natural events might explain this?

50. Estimate the height of the stump in year 200.

51. The graph below is a plot of $y = \log_{10} x$. Trace or photocopy the graph, including axes. Draw a horizontal line through each point on the graph of $y = \log_{10} x$ associated with one of the inputs x from 1 to 10. (You can trace the horizontal lines for the first three outputs, $f(1) = \log 1$, $f(2) = \log 2$, and $f(3) = \log 3$. You will need to draw the remaining horizontal lines, for $f(4)$, $f(5), f(6), \ldots$, and $f(10)$.) Compare the spacing of these horizontal lines with that of the lines in the second graph for Exercises 41 and 42.

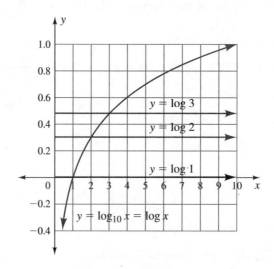

52. a. Trace or photocopy scales A, B, and C. Mark the results from scale A in Figure 20 onto the copy of scale A below.

b. Use the table feature of your calculator to find log x for $x = 10, 20, 30, \ldots, 100$. Place a dot on scale B below for each log value.

c. Use the table feature of your calculator to find log x for $x = 100, 200, 300, \ldots, 1000$. Place a dot on scale C below for each log value.

d. Describe the pattern formed by the dots on scales A, B, and C. Why does this occur?

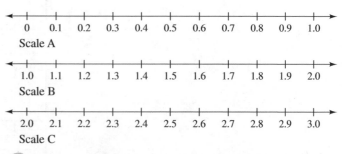

53. Place the depth data from the table on page 529 into calculator list L_1 and the percents of light penetration into calculator list L_2. Let a third calculator list be the log of the second list.

Depth (in meters)	Light Penetration (in percents)
0	100
1	45
2	39
10	22
50	5
100	0.5

Penetration of light through the clearest ocean water

Source: D. Ingmanson and W. J. Wallace, *Oceanography: An Introduction,* 4th ed. (Belmont, CA: Wadsworth Publishing, 1989), p. 114.

a. Fit an exponential regression equation to the data in L_1 and L_2.

b. Write a sentence that describes the relationship between depth of water and light penetration.

c. Fit a linear regression equation to L_1 and L_3. The resulting equation will be in the form $\log_{10} y = ax + b$.

d. Change the equation in part c into exponential form, and show that it is the same as the equation in part a.

Use the table below for Exercises 54 and 55.

Altitude (in kilometers)	Pressure (in millimeters of mercury)	Dry Air Density (in g/cm³)
20	44.1	0.000092
15	95.3	0.000199
10	205.1	0.000419
8	274.3	0.000524
4	466.6	0.000803
2	598.0	0.000990
1	674.8	0.001100
0	760.0	0.001223

Data from "Variation of Temperature, Pressure, and Density of the Atmosphere with Altitude," in *Handbook of Chemistry and Physics,* 36th ed. (Cleveland: Chemical Rubber Publishing Co., 1954), p. 3093.

54. Place altitude in calculator list L_1 and pressure in list L_2, and let $L_3 = \log L_2$.

a. Fit an exponential regression equation to the data in L_1 and L_2.

b. Write a sentence that describes the relationship between altitude and pressure.

c. Fit a linear regression equation, $\log y = ax + b$, to L_1 and L_3.

d. Change the equation in part c into exponential form.

55. Place altitude in calculator list L_1 and density in list L_2, and let $L_3 = \log L_2$.

a. Fit an exponential regression equation to L_1 and L_2.

b. Write a sentence that describes the relationship between altitude and air density.

c. Fit a linear regression equation, $\log y = ax + b$, to L_1 and L_3.

d. Change the equation in part c into exponential form.

Projects

56. *Scientific Notation and Logarithms.* Properties of exponents and $3 \approx 10^{0.47712}$ permit us to write these numbers in scientific notation and then as powers of 10.

$$30,000 = 3 \times 10^4 \approx 10^{0.47712} \times 10^4$$
$$= 10^{0.47712+4} = 10^{4.47712}$$
$$0.3 = 3 \times 10^{-1} \approx 10^{0.47712} \times 10^{-1}$$
$$= 10^{0.47712-1} = 10^{-0.52288}$$

In parts a to l, write the given number in scientific notation. Then use the following facts to find the exponent n, the logarithm of the given number:

$$4 \approx 10^{0.60206}$$
$$5 \approx 10^{0.69897}$$

a. 400, $10^n = 400$

b. 5000, $10^n = 5000$

c. 5,000,000, $10^n = 5,000,000$

d. 500,000, $10^n = 500,000$

e. 4000, $10^n = 4000$

f. 500, $10^n = 500$

g. 40,000, $10^n = 40,000$

h. 4,000,000, $10^n = 4,000,000$

i. 0.4, $10^n = 0.4$

j. 0.005, $10^n = 0.005$

k. 0.0004, $10^n = 0.0004$

l. 0.5, $10^n = 0.5$

m. What property of exponents allows us to add or subtract the exponents in parts a to l?

n. Explain why each n in parts a to l is a logarithm.

o. How can scientific notation tell us the first number of a logarithm of a number greater than 1?

p. When a number is written in scientific notation, $a \times 10^n$, the number a is always between what numbers? Write those numbers as powers of 10.

q. What numbers have negative logarithms? Why?

57. *Antilogs.* Changing back to the original number—finding $x = 900$ when $\log x = 2.95424$—is called *finding the antilog*. Antilog 2.95424 means $10^{2.95424}$ and is approximately 900. Evaluate the expressions or solve the equations in parts a to p.

a. antilog 3.29907

b. antilog 0.9132

c. $10^{2.5}$

d. $\log x = 1.69897$

e. antilog 3.28892

f. $\log x = 0.49715$

g. $10^{1.5}$

h. antilog 2.000

i. 1.30103 is $\log x$.

j. 2.30103 is $\log x$.

k. antilog $-3 = x$

l. antilog 1

m. antilog -2.000

n. antilog -2.5

o. antilog -1

p. antilog 0

58. *Arithmetic with Logarithms.* This project is designed to give you the flavor of how calculations with logarithms change multiplication into addition. Suppose we wish to use logarithms to find the area of a golf hole—a circle with a diameter of 4.25 inches. The radius will be $\frac{1}{2}(4.25)$, or 2.125, inches.

$$A = \pi r^2 = 3.14159265 \cdot (2.125)^2 \text{ in}^2$$

To calculate the product, we find the logarithm of each number (to five decimal places) with the calculator, write each number as a power of 10, add the exponents, and calculate the resulting power of 10.

$3.14159265 \cdot 2.125 \cdot 2.125$

$$= 10^{0.49715} \cdot 10^{0.32736} \cdot 10^{0.32736}$$
$$= 10^{0.49715+0.32736+0.32736}$$
$$= 10^{1.15187}$$
$$= 14.2 \text{ in}^2$$

Because our least accurate number, 4.25, has only three significant digits, we round our final answer to three significant digits: 14.2 in^2.

In the past, the steps in finding the logarithm and then the antilog were done with a table of logarithms. Today, working directly with a calculator and the $\boxed{\pi}$ key, we obtain $\pi r^2 \approx 14.18625433$, or 14.2 in^2.

Use logarithms to do the following computations.

a. Find the volume of a sphere of diameter 4.25 inches: $V = \frac{4}{3}\pi r^3$

b. Find the surface area of a sphere of diameter 4.25 inches: $A = 4\pi r^2$

c. Find the volume of a cone with a base diameter of 4.25 inches and height of 2 inches: $V = \frac{1}{3}\pi r^2 h$

59. *Science Graphs.* Talk with a physics or chemistry instructor about how you can find the exponential equation from a linear graph on semilog paper. Obtain some chemistry data, and give two examples of graphs and equations for those graphs.

7.5 The Natural Number e in Exponential and Logarithmic Functions

OBJECTIVES

- Calculate powers and roots of e.
- Apply the formula for continuously compounded interest.
- Find logarithms for base e.
- Change natural logarithmic equations to exponential equations and solve.
- Change exponential equations to natural logarithmic equations and solve.

WARM-UP

The exclamation point, !, is used to denote "factorial." The **factorial** of a number is *the number itself multiplied by every lower positive integer.* Thus,

$$3! = 3 \cdot 2 \cdot 1 = 6$$

and

$$7! = 7 \cdot 6 \cdot 5 \cdot 4 \cdot 3 \cdot 2 \cdot 1 = 5040$$

Look in your calculator manual to see how to find the factorial, !.
Evaluate these factorial expressions by hand.

1. $2!$

2. $5!$

3. $8!$

Use a calculator to do the next two problems.

4. Evaluate $52!$, the number of ways a deck of 52 cards can be arranged.

5. Suppose $0!$ and $1!$ are defined as 1. Evaluate the following expression by adding one term at a time to the prior answer. Record your sum term by term.

$$\frac{1}{0!} + \frac{1}{1!} + \frac{1}{2!} + \frac{1}{3!} + \frac{1}{4!} + \frac{1}{5!} + \frac{1}{6!} + \frac{1}{7!} + \frac{1}{8!}$$

THIS SECTION INTRODUCES the constant e—the natural number—and its role in exponential and logarithmic functions. Of particular importance is the role of e in calculating compound interest and as a base for many exponential and logarithmic functions.

Compound Interest

EXAMPLE Exploring compound interest calculations   An investor has $1000 to deposit for a year. Shopping around, she discovers that the current interest rate is 5%, but it is being calculated in four different ways: annually, monthly, daily, and hourly. Calculate the amount of money she would have with each method after 1 year.

Solution   The compound interest formula is

$$S = P\left(1 + \frac{r}{n}\right)^{nt}$$

If interest is compounded annually, then $n = 1$ and

$$S = P\left(1 + \frac{r}{1}\right)^{1t}$$

$$= P(1 + r)^t$$

$$= 1000(1 + 0.05)^1 = 1000(1.05) = \$1050$$

If interest is compounded monthly, then $n = 12$ and

$$S = P\left(1 + \frac{r}{12}\right)^{12t}$$

$$= 1000\left(1 + \frac{0.05}{12}\right)^{12} \approx 1051.161\,898 \approx \$1,051.16$$

If interest is compounded daily, then $n = 365$ and

$$S = P\left(1 + \frac{r}{365}\right)^{365t}$$

$$= 1000\left(1 + \frac{0.05}{365}\right)^{365} \approx 1051.267\,496 \approx \$1,051.26$$

Student Note: Continue to round down if interest is received, and up if interest is paid.

If interest is compounded hourly, then $n = 365 \cdot 24 = 8760$ and

$$S = P\left(1 + \frac{r}{8760}\right)^{8760t}$$

$$= 1000\left(1 + \frac{0.05}{8760}\right)^{8760} \approx 1051.270\,947 \approx \$1,051.27$$

Rounded to the nearest cent, there is a one-cent difference between daily interest and hourly interest. This may surprise you, because there were differences between annual and monthly interest and between monthly and daily interest.

The results in Example 1 are even more apparent in a graph. Figure 21 shows the graph of $y = 1000(1 + 0.05/x)^{x \cdot 1}$ for x between 0 and 365 and y between 1050 and 1052. As x gets large, the curve flattens out, so there is little change in y as we trace.

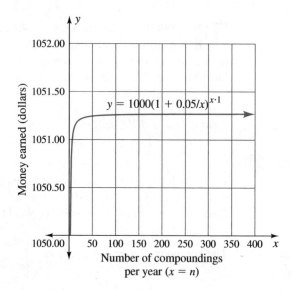

Figure 21

Think about it 1: Is there a y-intercept in Figure 21?

The Natural Number e

Mathematicians looked for an upper limit to the compound interest formula as the number of compoundings increased toward infinity. They held the interest rate constant ($r = 1$) and studied $(1 + 1/n)^n$.

EXAMPLE **2**   Looking for an upper limit to $(1 + 1/n)^n$ Evaluate $(1 + 1/n)^n$ for $n = 1, 2, 12, 52, 365$, and 8760.

Solution   Table 17 contains values of $(1 + 1/n)^n$ for several inputs. As we found in Example 1, there is considerable change in the output for lower inputs and little change in the output for higher inputs. This suggests that the expression is approaching a limit near 2.718.

n	$(1 + 1/n)^n$
1	2.0000
2	2.2500
12	2.613035
52	2.692597
365	2.714567
8760	2.718127

Table 17   ●

Although the process of finding the upper limit requires calculus, Example 2 suggests that we can observe the limit by looking at the behavior of $(1 + 1/n)^n$ for large n. For $n = 8750$, we are within one ten-thousandth of the limit.

The limiting number is approximately

2.718 281 828 459 045

The limit is irrational (it cannot be written in the form a/b, where a and b are integers and $b \neq 0$). This limit is called the **natural number** and is given the special letter e.

The letter e was chosen by Leonhard Euler, a noted Swiss mathematician (1707–1783) whose image appears on the Swiss 10-franc note (Figure 22). It is believed unlikely that Euler named the number for himself; however, considering his work with the number, it would have been appropriate.

Figure 22

The factorial expression in Warm-up Exercise 5 adds up to approximately 2.718 278 8. The sum is approaching e. We can get as close as we like to e by adding more terms: $1/9!$, $1/10!$, and so forth.

Think about it 2: How are pi and the golden ratio (phi) like e?

Calculating with e

The number e is so important that it appears on every scientific and business calculator. It usually is obtained in a shifted position with a key labeled ⌈LN⌉ (more about ⌈LN⌉ later).

EXAMPLE **3**   Evaluating expressions containing e   Evaluate these expressions with a calculator.

a. e^2 **b.** $e^{0.5 \cdot 3}$ **c.** e **d.** 2^e **e.** $e^{-\pi/2}$ **f.** $\sqrt{e}$

Solution   **a.** To evaluate e^2, we enter e to the exponent 2. Using ⌈2nd⌉ ⌈e^x⌉ 2 ⌈ENTER⌉, we obtain 7.389 056 099. Because e is almost always used with an exponent, the ⌈e^x⌉ key has the exponent sign, ⌈^⌉, built in to save keystrokes.

b. To evaluate $e^{0.5 \cdot 3}$, we enter e to the exponent (0.5 · 3). The parentheses must enclose the exponent. Using ⌈2nd⌉ ⌈e^x⌉ (0.5 × 3) ⌈ENTER⌉, we obtain 4.481 689 07.

c. To evaluate e, we enter e to the exponent 1. Using ⌈2nd⌉ ⌈e^x⌉ 1 ⌈ENTER⌉, we obtain 2.718 281 828.

d. To evaluate 2^e, we enter 2 to the exponent e to the exponent 1. Using 2 ⌈^⌉ ⌈2nd⌉ ⌈e^x⌉ 1 ⌈ENTER⌉, we obtain the calculator display 2^e^1, followed by the answer 6.580 885 991.

e. To evaluate $e^{-\pi/2}$, we use parentheses to enclose the exponent. Using ⌈2nd⌉ ⌈e^x⌉ $(-\pi/2)$ ⌈ENTER⌉, we obtain 0.207 879 576 4.

f. To evaluate $\sqrt{e}$, we enter the square root followed by e to the exponent 1. The square root of e is approximately 1.648 721 271.   ●

Practice doing the problems in Example 3 to make sure you are using the calculator correctly.

Continuously Compounded Interest

Because $(1 + 1/n)^n$ has a limiting value, e, we are able to calculate interest continuously. With **continuously compounded interest,** *the number of compounding periods approaches infinity* (and the length of the compounding period approaches zero). The limiting value, e, is the base in the formula for continuously compounded interest.

Continuously Compounded Interest

> The formula for continuously compounded interest is
>
> $$S = Pe^{rt}$$
>
> where S is the future value, P is the present value (principal), r is the annual rate of interest, and t is the time in years.

EXAMPLE **4**   Finding the future value for continuously compounded interest   Calculate the value of $1000 after 1 year, with interest compounded continuously at 5%, and compare the results with those of Example 1.

Solution $S = Pe^{rt} = 1000 \cdot e^{0.05 \cdot 1} \approx 1051.271\ 096 \approx \$1,051.27$

With daily interest, the value was $1051.267 496; with hourly interest, it was $1051.270 947. The continuously compounded interest yields only a slightly higher value.   ●

As you work through Example 5, practice enclosing the exponent in parentheses when evaluating on a calculator.

EXAMPLE **5**   Finding the future value for continuously compounded interest   Find the value of $1000 if interest is compounded continuously at the given rate for the given time.

a. 5% for 3 years   **b.** 3% for 5 years

c. 6.93% for 10 years   **d.** 6% for 11.55 years

Solution **a.** $S = Pe^{rt} = 1000 \cdot e^{0.05 \cdot 3} = \$1161.83$

b. $S = Pe^{rt} = 1000 \cdot e^{0.03 \cdot 5} = \$1161.83$

Parts a and b give the same answer.

c. $S = Pe^{rt} = 1000 \cdot e^{0.0693 \cdot 10} = \$1999.71$

d. $S = Pe^{rt} = 1000 \cdot e^{0.06 \cdot 11.55} = \$1999.71$

Parts c and d give the same answer.   ●

Example 6 illustrates a typical use of $S = Pe^{rt}$.

EXAMPLE **6**   Finding the initial or present value with continuously compounded interest   What amount of money would you need to invest at 8% interest compounded continuously to have $1,000,000 in 20 years?

Solution We are missing the initial amount of money P (the present value or principal). *Because interest is compounded continuously, we use the $S = Pe^{rt}$ formula.*

$$S = Pe^{rt}$$ Substitute the facts.

$$1,000,000 = Pe^{0.08 \cdot 20}$$ Divide by $e^{0.08 \cdot 20}$ to solve for P.

$$\frac{1,000,000}{e^{0.08 \cdot 20}} = P$$

$$P \approx \$201,896.52$$

In 20 years, at 8% interest compounded continuously, $202,000 will grow to approximately $1,000,000.   ●

Logarithms with Base e

In Example 6, we solved for the present value or principal. Because both time and rate of interest are in the exponent in the $S = Pe^{rt}$ formula, we need logarithms to solve for t or r. In Example 7, we solve for the length of time needed for money to double at 8% interest compounded continuously.

EXAMPLE **7**   Finding doubling time with continuous compounding   What length of time is needed for money to double at 8% interest compounded continuously?

Solution When money doubles, the future value is double the starting value, and $S = 2P$. Thus, we can proceed as follows:

$$S = Pe^{rt}$$ Substitute the facts.

$$2P = Pe^{0.08t}$$ Divide by P.

$$2 = e^{0.08t}$$ Change to a logarithmic equation.

$$\log_e 2 = 0.08t$$ Divide by 0.08.

$$\frac{\log_e 2}{0.08} = t$$ Apply the change of base formula.

$$t = \frac{\frac{\log 2}{\log e^1}}{0.08}$$

$$t \approx 8.7 \text{ yr}$$ ●

Caution: On many graphing calculators, correct use of parentheses in the change of base formula and with other logarithmic expressions is essential.

We will return to using logarithms to solve for the rate in Example 9, after an introduction to the special notation for logarithms with base *e*. It is not necessary to use the change of base formula on $\log_e 2$, as we did in Example 7, because log base *e* has its own calculator key. The key for log base *e* is LN (the same key as is used in shifted position for e^x). LN stands for **natural logarithm,** *the logarithm with the natural number as its base.*

Special Notation for Logarithms with Base *e*

The natural logarithm of *x* has a special notation:

$$\log_e x = \ln x$$

EXAMPLE **8**   Using the natural logarithm   Write $t = \log_e 2/0.08$ with a natural logarithm, and evaluate it with the natural logarithm key, LN .

Solution
$$t = \frac{\log_e 2}{0.08}$$

$$= \frac{\ln 2}{0.08}$$

$$\approx 8.7 \text{ yr}$$ ●

Caution: Until you have considerable experience with logarithms, write *every* logarithm with its base. For example, in solving equations, use $\log_{10} x$ and $\log_e x$ when doing steps with algebraic notation and change to log *x* and ln *x* just before evaluating with a calculator.

EXAMPLE **9**   Solving for the rate, *r*   A house cost $55,000 in 1985. Its value 14 years later is $119,000. Assuming the value of the house has grown continuously, what is the rate?

Solution   The word *continuously* points to the continuous compounding formula.

$$S = Pe^{rt}$$ Let $S = 119,000$, $P = 55,000$, and $t = 14$.

$$119,000 = 55,000e^{r \cdot 14}$$ Divide by 55,000.

$$\frac{119,000}{55,000} = e^{r \cdot 14}$$ Change to a logarithmic equation and simplify the fraction.

$$\log_e \frac{119}{55} = r \cdot 14$$ Change to natural log notation and divide by 14.

$$\frac{\ln \frac{119}{55}}{14} = r$$ Evaluate.

$$r \approx 0.055$$

The rate is approximately 5.5% growth.   ●

Summary: Calculators and Logarithms

- The common logarithm is abbreviated log x and means $\log_{10} x$. Use $\boxed{\text{LOG}}$ on the calculator.
- The natural logarithm is abbreviated ln x and means $\log_e x$. Use $\boxed{\text{LN}}$ on the calculator.
- All other logarithms are written $\log_b x$ and are entered on the calculator with the change of base formula, log x/log b.

Changing between Exponential and Logarithmic Form

Example 10 illustrates the importance of including the bases when we change a logarithmic equation to its corresponding exponential equation.

EXAMPLE 10 Changing forms of equations   Complete Table 18, changing the exponential equations to logarithmic equations and the logarithmic equations to exponential equations. Where appropriate, solve for x.

Exponential Equation	Logarithmic Equation	Solve for x (as needed)
	ln $e = 1$	
	log $100 = 2$	
$e^0 = 1$		
$10^x = 1000$		
	ln $2 = x$	

Table 18

Solution   We write the base on the logarithm before changing to an exponential equation. The completed table appears in Table 19.

Exponential Equation	Logarithmic Equation	Solve for x (as needed)
$e^1 = e$	ln $e = 1$ $\log_e e = 1$	
$10^2 = 100$	log $100 = 2$ $\log_{10} 100 = 2$	
$e^0 = 1$	$\log_e 1 = 0$ ln $1 = 0$	
$10^x = 1000$	$\log_{10} 1000 = x$ log $1000 = x$	$x = 3$
$e^x = 2$	ln $2 = x$ $\log_e 2 = x$	$x \approx 0.693$

Table 19

In Section 7.3, we used the rule of 72 to estimate doubling time in years for an annual rate of interest r. There is a comparable rule for estimating the doubling time for an investment calculated with the continuously compounded interest formula: the rule of 69.

Rule of 69

Dividing 69 by the annual interest rate (expressed as a whole number) estimates the number of years required for the initial amount, P, to double with continuously compounded interest.

$$69 \div \text{interest rate (as a whole number)} = \text{doubling time}$$

For 10% interest, $69 \div 10 = 6.9$ yr doubling time.

Because rate and time are both in the exponent of $S = Pe^{rt}$, the rule of 69 can also be used to find the rate at which we need to invest money to double it in a given time period:

$$69 \div \text{time to double} = \text{interest rate needed}$$

We now prove the rule of 69. Because the rule requires the interest rate as a whole number, we use $100r$ to change a percent to a whole number in the last step.

EXAMPLE 11   Proving the rule of 69   Start with $S = Pe^{rt}$. Show why the rule of 69 works by showing that the doubling time at an annual rate r is approximately $69/100r$.

Solution   For money to double, we must have $S = 2P$.

$$S = Pe^{rt} \qquad \text{Substitute } 2P \text{ for } S.$$
$$2P = Pe^{rt} \qquad \text{Divide by } P.$$
$$2 = e^{rt} \qquad \text{Change to a logarithmic equation.}$$
$$\log_e 2 = rt \qquad \text{Divide by } r.$$
$$\frac{\log_e 2}{r} = t \qquad \text{Calculate } \log_e 2 \text{ with ln 2.}$$
$$t = \frac{\ln 2}{r}$$
$$\approx \frac{0.693}{r} \qquad \text{Multiply numerator and denominator by 100.}$$
$$= \frac{69.3}{100r}, \text{ or about } \frac{69}{100r}$$

Thus, the rule of 69 works because the doubling time is $(\ln 2)/r$.   ●

Applications: Natural Logarithms in Chemistry

As we saw with pH, many disciplines adopt special notation, so it is important to look at definitions carefully in any applied situation. In the next example, we apply properties of exponents and logarithms to solving chemical equations commonly found in first-year chemistry courses. As before, remember to write the bases in all logarithmic expressions.

EXAMPLE 12   Solving logarithmic equations: Boltzmann equation   Solve for Ω in the Boltzmann equation, $S = k \ln \Omega$.

Solution

$$S = k \ln \Omega \qquad \text{Divide both sides by } k.$$

$$\frac{S}{k} = \ln \Omega \qquad \text{Write ln as log base } e.$$

$$\frac{S}{k} = \log_e \Omega \qquad \text{Change to an exponential equation.}$$

$$e^{S/k} = \Omega \qquad \bullet$$

Some logarithmic equations in applications may seem intimidating because they contain many variables or unfamiliar symbols (such as the Greek letter omega, Ω, in Example 12). Often, difficulties stem from forgetting that the base is e.

EXAMPLE **13**   Solving logarithmic equations   Solve $E = E^0 - \dfrac{RT}{nF} \ln Q$ for the reactant quotient Q. (*Note:* Many chemistry texts use a zero superscript to represent standard conditions such as 1 atm and 25°C. In this formula, the E^0 should not be interpreted as the zeroth power of E.)

Solution

$$E = E^0 - \frac{RT}{nF} \ln Q \qquad \text{Subtract } E^0.$$

$$E - E^0 = -\frac{RT}{nF} \ln Q \qquad \text{Multiply by } -\frac{nF}{RT}.$$

$$-\frac{nF}{RT}(E - E^0) = \ln Q \qquad \text{Write ln as log base } e.$$

$$-\frac{nF}{RT}(E - E^0) = \log_e Q \qquad \text{Change to exponential form.}$$

$$e^{-\frac{nF}{RT}(E - E^0)} = Q \qquad \bullet$$

ANSWER BOX

Warm-up: 1. 2 **2.** 120 **3.** 40,320 **4.** 8.066×10^{67} **5.** 1, 2, 2.5, 2.666..., 2.708 333..., 2.716 66..., 2.718 055 5..., ≈2.718 253 96, ≈2.718 278 76 **Think about it 1:** There is no y-intercept because the expression $0.05/x$ is undefined if $x = 0$. **Think about it 2:** All three numbers are irrational.

EXERCISES **7.5**

In Exercises 1 and 2, assume a deposit of $1000 is made to a savings account at the rate of interest shown. Find the amount of money in the account after 3 years if interest is compounded annually, monthly, and daily (assume 365 days). How much is gained by compounding monthly and daily rather than annually?

1. 4%   **2.** 7%

Evaluate the expressions in Exercises 3 to 6 with a calculator. Round to the nearest hundredth.

3. a. e^2 **b.** e^π **c.** $2\sqrt{e}$

4. a. $e^{1.5}$ **b.** π^e **c.** $1/2\sqrt{e}$

5. a. 3^e **b.** $e^{0.4 \cdot 2}$ **c.** $e^{\pi/2}$

6. a. 4^e **b.** $e^{2/\pi}$ **c.** $e^{0.6 \cdot 3}$

7. Describe how to get e^e on your calculator.

8. Between which two of the numbers 1^2, 2^2, 3^2, and 4^2 does the expression e^2 fall?

In Exercises 9 to 14, find the value of the principal after one year, with interest compounded continuously. Round to the nearest cent.

9. $P = \$1000$, 8% annual interest

10. $P = \$1000$, 6% annual interest

11. $P = \$1000$, 10% annual interest

12. $P = \$1000$, 12% annual interest

13. $P = \$1000$, $4\frac{1}{2}\%$ annual interest

14. $P = \$1000$, $21\frac{1}{2}\%$ annual interest

15. What amount of money would need to be invested at 10% compounded continuously to have $500,000 in 12 years?

16. What amount of money would need to be invested at 10% compounded continuously to have $500,000 in 10 years?

17. What amount of money would need to be invested at 10% compounded continuously to have $500,000 in 20 years?

18. Would the answer to Exercise 17 change if the money were invested at 20% compounded continuously for 10 years? Why?

Complete the tables in Exercises 19 and 20.

19.

Logarithmic Equation	Show base e in logarithm	Exponential Equation	$y = ?$
		$y = e^1$	
		$y = e^0$	
$y = \ln(-1)$			
$y = \ln e^2$			
$y = \ln e^e$			

20.

Logarithmic Equation	Show base e in logarithm	Exponential Equation	$y = ?$
$y = \ln e$			
$y = \ln 1$			
$y = \ln 0$			
		$y = e^{-1}$	
		$y = e^e$	

In Exercises 21 to 24, solve for x. Round to four decimal places.

21. a. $e^x = 4$ b. $\ln x = -2$

22. a. $\ln x = -1$ b. $e^x = 1.5$

23. a. $\ln x = 1.5$ b. $e^x = 2$

24. a. $e^x = 0.5$ b. $\ln x = 0.5$

25. Estimate each rate of interest, using the rule of 69.

   a. The rate of interest required for money to double in 8 years

   b. The rate of interest required for money to double in 11 years

   c. An expression for the rate of interest required for money to double in t years

26. Using $S = Pe^{rt}$, find the rate of interest for part a of Exercise 25.

27. Using $S = Pe^{rt}$, find the rate of interest for part b of Exercise 25.

28. Using $S = Pe^{rt}$, find an expression for the rate of interest required for money to double in t years.

29. Using $S = Pe^{rt}$, find the amount of time required for money to triple at 8% interest.

30. Using $S = Pe^{rt}$, find the amount of time required for money to triple at 6% interest.

31. Using $S = Pe^{rt}$, find an expression for the amount of time required for money to triple at a rate r of interest.

32. Is the relation of interest rate and years to doubling time, as expressed in the rule of 69, inversely proportional? If so, what is the constant of variation?

33. Find the amount of time required for a city's population to halve when its rate of decrease is 8% per year. Use $S = Pe^{rt}$.

34. Find the amount of time required for a city's population to halve when its rate of decrease is 4% per year. Use $S = Pe^{rt}$.

35. a. To investigate the change of base formula using $\ln x$ instead of $\log x$, evaluate $\log_6 8$ using the ratios $\ln a / \ln b$ and $\log a / \log b$.

   b. Evaluate $\log 8$, $\log 6$, $\ln 8$, and $\ln 6$ separately.

   c. What do you observe?

36. The change of base formula may contain the natural logarithm instead of the common logarithm:

$$\log_b a = \frac{\ln a}{\ln b}$$

Using as a model the proof shown in Example 8 of Section 7.4, write a proof for the natural logarithm change of base formula.

Solve each chemistry formula in Exercises 37 to 40 for the variable indicated. The zero superscript represents standard conditions, not a zero exponent.

37. Solve $\ln K = -\dfrac{Ea}{RT} + C$ for K, the rate constant.

38. Solve $\Delta G^0 = -RT \ln K$ for K, the equilibrium constant.

39. Solve $E = -\dfrac{RT}{nF} \ln [\text{H}^+]$ for $[\text{H}^+]$, the hydrogen ion concentration.

40. Solve $\ln P = \dfrac{-\Delta H_{vap}}{RT} + B$ for P, the pressure.

41. The normal probability curve in statistics is described by

$$y = \frac{1}{\sqrt{2\pi}} e^{-x^2/2}$$

The curve describes the distribution of a set of numbers with a mean of 0 and a standard deviation of 1.

a. Graph the normal probability equation for $x = -3$ to $x = 3$ and $y = -1$ to $y = 1$.

b. Estimate the highest point on the normal curve.

c. What part of the equation shows the y-intercept?

d. If the total area under the curve equals 1 square unit, estimate the area under the curve between $x = -1$ and $x = 1$.

42. Velocity is a function of time, t. The velocity in feet per second of a sky diver in free fall is given by

$$v = \frac{g}{a}(1 - e^{-at})$$

where g is the acceleration due to gravity, 32.2 ft/sec^2. The factor a contains information on the mass of the skydiver and the resisting force of the air. The a values in parts a to c have been calculated for a 135-pound female sky diver. Substitute a into the equation for velocity. Use a calculator graph to find the terminal (maximum) velocity and the time t required to be within 1 foot per second of terminal velocity.

a. Horizontal position (arms and legs spread from body): $a = 0.183$

b. Fetal position: $a = 0.129$

c. Nose dive (arms along body): $a = 0.1586$

d. Change the terminal velocity in part c to miles per hour.

43. Suppose the population of Detroit, Michigan, in t years after 1950 can be estimated by the function $f(t) = 1,849,568e^{-0.0147t}$. Round answers to three significant digits.

a. What was the population in 1950?

b. What was the population 20 years later, in 1970?

c. What was the population 40 years later, in 1990?

d. Which part of the function indicates whether the population is increasing or decreasing?

e. When would the population be half the population in 1950?

f. Predict the population in the year 2000.

44. Suppose the population of El Paso, Texas, in t years after 1950 can be estimated by the function $f(t) = 130,485e^{0.0343t}$. Round answers to three significant digits.

a. What was the population in 1950?

b. What was the population 20 years later, in 1970?

c. What was the population 40 years later, in 1990?

d. Which part of the function indicates whether the population is increasing or decreasing?

e. When would the population be twice the population in 1950?

f. Predict the population in the year 2000.

45. Find the ordered pair (x, y) where the maximum y value is attained by the function $y = x^{1/x}$, $x > 0$. *Hint:* Trace on a graph. (This problem, called the Jakob Steiner problem, is from Heinrich Dörrie, *100 Great Problems of Elementary Mathematics,* Dover, New York, 1965.)

46. When we solve $S = Pe^{rt}$ for P, we find two answers that appear to be different: $P = Se^{-rt}$ and $P = S/e^{rt}$. What property of exponents explains why they are the same?

47. Evaluate the factorials, as defined in the Warm-up.

a. 4!   **b.** 10!

c. 6!   **d.** 9!

e. Evaluate the expression for the 9 terms shown, and compare the result with e^2.

$$2^0 + \frac{2^1}{1!} + \frac{2^2}{2!} + \frac{2^3}{3!} + \frac{2^4}{4!} + \frac{2^5}{5!} + \frac{2^6}{6!} + \frac{2^7}{7!} + \frac{2^8}{8!} + \cdots$$

f. Evaluate

$$1 - \tfrac{1}{2} + \tfrac{1}{3} - \tfrac{1}{4} + \tfrac{1}{5} - \tfrac{1}{6} + \tfrac{1}{7} - \tfrac{1}{8} + \cdots - \cdots$$

for 15 terms. Compare your result with $\ln 2$.

Projects

48. *Graphing with e^x and $\ln x$.*   Graph the equations with a calculator, and record sketches on paper. Label the curves with their equations, and label intercepts and horizontal or vertical lines that the graphs approach. Mention any graphing shortcuts that you recall from earlier graphing exercises.

a. Graph $y = e^{-x}$ and $y = -e^{-x}$.

b. Graph $y = -e^x$ and $y = e^x$.

c. Graph $y = e^x + 1$ and $y = e^{x+1}$.

d. Graph $y = e^{x-1}$ and $y = e^x - 1$.

e. Graph $y = \ln x$ and $y = \ln x - 1$.

f. Graph $y = \ln x + 1$ and $y = \ln (x + 1)$.

g. Graph $y = e^x$ and $y = \ln x$, together with $y = x$. Change to a square window. Describe the relative positions of the graphs.

49. *Exceeding the Accuracy of the Calculator.* In evaluating compound interest for n larger than 8760 (hourly), it is possible to exceed the capacity of a graphing calculator and obtain results that are wrong.

a. Calculate the amount in an account at 5% interest compounded every hour for 1 year. Start with $1000.

b. Calculate the amount in an account at 5% interest compounded every minute for 1 year. Start with $1000.

c. Calculate the maximum amount that should be in the account. Use $S = Pe^{rt}$.

d. Verify that your graphing calculator gives a number above the limit in part b. How far above the limit is the number?

e. Set $Y_1 = 1000(1 + 0.05)/x)^{x \cdot 1}$. Start a table with 525,600, and set the table change at 1. Move the cursor down the Y_1 column, and comment on what happens to the outputs.

f. Set Xmin = 525600, Xmax = 525605, Ymin = 1051.271, and Ymax = 1051.2712, and graph. Com-

pare the result with the smooth curve that should result.

50. *The Black Death.* Read about a plague called the Black Death, which occurred in the fourteenth century. One good source is Barbara Tuchman's *A Distant Mirror: The Calamitous 14th Century* (New York: Ballantine Books, 1978), Chapter 5. This 700-page book provides an excellent, well-documented account, vividly descriptive of the disease and human (inhumane?) behaviors. No encyclopedia will give you as true a feeling for what went on during this period.

a. Describe the course of the plague with a graph. Place time (the years are available) on the horizontal axis and total number of deaths (the numbers will not be available, so your curve will be somewhat generic) on the vertical axis.

b. Describe the causes of the plague and how it was spread.

c. Describe what people of the time thought caused the plague.

d. What inhumane behavior shocked you the most?

51. *Plague of 1664–1666.* Another plague occurred between 1664 and 1666. What impact did this plague have on Isaac Newton's work? What did he discover during this period? See E. T. Bell, *Men of Mathematics* (New York: Simon and Schuster, 1965), Chapter 6.

CHAPTER 7 SUMMARY

Vocabulary

For definitions and page references, see the Glossary/Index.

change of base formula

common logarithm

common ratio

continuously compounded interest

doubling time

e, the natural number

exponential decay

exponential function

exponential growth

factorial

geometric sequence

half-life

horizontal asymptote

like bases property

logarithm

logarithmic function

logarithmic scale

natural logarithm

nth term of a geometric sequence

quadratic sequence

rule of 72

rule of 69

semilog graph

Concepts

This chart summarizes compound interest, where $S =$ future value, $P =$ present value (or principal), $r =$ annual rate of interest, $t =$ time in years (or whatever other unit matches that of r), n is the number of times a year the interest is calculated (compounded), and e, the natural number, is an irrational constant.

Compound Interest	$S = P\left(1 + \dfrac{r}{n}\right)^{nt}$
	$n = 1$ (annual) $n = 2$ (semiannual) $n = 4$ (quarterly) $n = 12$ (monthly) $n = 52$ (weekly) $n = 365$ (daily)
Annual Compounding $n = 1$	$S = P(1 + r)^t$ Inflation Depreciation
Continuously Compounded Interest	$S = Pe^{rt}$

This chart summarizes important ideas about linear, quadratic, and exponential functions, sequences, and variation. See the related chart in Chapter 5 for inverse variation.

Linear Equations $y = mx + b$ and $x = c$ Linear Functions $f(x) = mx + b$ Domain: Real numbers	Arithmetic Sequences $a_n = a_1 + (n-1)d$ Domain: Positive integers Constant first differences	Direct Variation Direct Proportion $y = kx$ Linear Variation $y = mx + b$ Domain: Real numbers
Quadratic Equations $y = ax^2 + bx + c$ Quadratic Functions $f(x) = ax^2 + bx + c$ Domain: Real numbers	Quadratic Sequences Domain: Positive integers Constant second differences	Quadratic Variation $y = kx^2$ Domain: Real numbers
Exponential Equations $y = ab^x$ $b > 0, b \neq 1$ Exponential Functions $f(x) = ab^x$ Domain: Real numbers Range: Positive real numbers	Geometric Sequences $a_n = a_1 r^{n-1}$ $r < 0$ or $r > 0$ Domain: Positive integers Differences are the same as or a multiple of the original sequence. Constant ratio of terms	

7.0 Patterns, Geometric Sequences, and Exponential Equations

Geometric sequences have differences that are either the same as or a multiple of the original sequence. However, they are not the only sequence with this pattern. A common ratio uniquely defines a geometric sequence.

Any exponential function with the domain limited to the positive integers can be written as a geometric sequence.

Use properties of exponents to change exponential expressions b^{x+c} to ab^x, where $a = b^c$.

7.1 Exponential Functions and Graphs

The coefficient a is the y-intercept of the graph of $y = ab^x$.

If $0 < b < 1$, the graph of $y = b^x$ decreases from left to right.

If $b > 1$, the graph of $y = b^x$ increases from left to right.

The domain for $f(x) = b^x$ is all real numbers.

The range for $f(x) = b^x$ is $f(x) > 0$.

The x-axis is a horizontal asymptote for the graph of $y = b^x$.

The graph of $y = b^x$ passes through $(0, 1)$ and $(1, b)$.

For $b > 1$, the larger the base in $y = b^x$, the steeper the graph.

For $h > 0$, the graph of $y = b^{x-h}$ is shifted h units to the right of the graph of $y = b^x$.

For $h > 0$, the graph of $y = b^{x+h}$ is shifted h units to the left of the graph of $y = b^x$.

7.2 Solving Exponential Equations: Like Bases and Logarithms

To use the like bases property, change the exponential expressions on both sides of an equation to the same base, set the exponents equal, and solve for the variable.

The logarithmic function, $y = \log_b x$, is the inverse function to the exponential function, $y = b^x$, $b > 0$, $b \neq 1$, $x \geq 0$.

The equations $y = b^x$ and $x = \log_b y$ are equivalent.

The base b for an equivalent pair of exponential and logarithmic equations is the same.

7.3 Applications of Exponential and Logarithmic Functions

Common, or base 10, logarithms are written without the base: $\log_{10} x = \log x$.

Use the $\boxed{\text{LOG}}$ key on a calculator to obtain the common logarithm of a number.

In order to change $y = ab^x$ to an equivalent logarithmic equation, divide both sides by a.

The change of base formula may be written with any base. Base 10 is usually convenient:

$$\log_b a = \frac{\log a}{\log b}$$

The logarithmic function has the set of positive real numbers as its domain and all real numbers as its range.

Graphs of logarithmic functions $y = \log_b x$ or $b^y = x$ pass through the x-intercept $(1, 0)$ because $b^0 = 1$ and pass through $(b, 1)$ because $b^1 = b$.

Graphs of logarithmic functions have no y-intercept because $y = \log_b 0$ is undefined.

The rule of 72 gives a way to estimate doubling time for interest compounded annually. The doubling time formula for rate r is $t = 72/100r$.

When data involve years, let the first year be zero and the other dates be in terms of the number of years after the first year.

7.4 Properties of Logarithms and the Logarithmic Scale

Properties of logarithms: If m, n, and b are positive numbers and $b \neq 1$, then

1. $\log_b (m \cdot n) = \log_b m + \log_b n$
2. $\log_b m^n = n \cdot \log_b m$
3. $\log_b \dfrac{m}{n} = \log_b m - \log_b n$

The graph of an exponential function becomes a line on semilog paper.

7.5 The Natural Number e

When the number of compoundings per year, n, is increased without bound, the compound interest formula $S = P(1 + r/n)^{nt}$ becomes $S = Pe^{rt}$.

The formula $S = Pe^{rt}$ is used whenever interest is compounded continuously.

The number e is the natural number, $e \approx 2.718\ 281\ 828\ 459\ 045$. To get e on a calculator, find e^1 with $\boxed{e^x}$ 1.

The logarithms with base e, $\log_e x$, are usually written $\ln x$. The calculator key for natural logarithms is $\boxed{\text{LN}}$.

The rule of 69 applies to continuously compounded growth. The formula $t = 69/100r$ permits us to estimate the time required for a quantity to double at rate r or the rate needed for a quantity to double in a specified time.

When working with logarithmic equations involving either base 10 or base e, rewrite the expressions to show the base.

Various techniques can be used to solve exponential and logarithmic equations:

1. Guess and check (Sections 6.0, 6.1, 7.0, and 7.1)
2. Graphing (Section 7.1)
3. Changing to like bases and applying the like bases property (Section 7.2)
4. Changing exponential equations, $y = b^x$, to logarithmic equations (Sections 7.2 and 7.3)
5. Dividing $y = ab^x$ on both sides by a and changing to a logarithmic equation (Sections 7.2 and 7.3)
6. Changing logarithmic equations to exponential equations (Section 7.2)
7. Taking the logarithm of both sides (Section 7.4)

CHAPTER **7** REVIEW EXERCISES

In Exercises 1 to 6, find the next number in each sequence. Find whether the sequence is arithmetic, quadratic, or geometric, and then fit a linear, quadratic, or exponential equation.

1. 9, 27, 81, 243
2. 10, 13, 16, 19, 22
3. 1, 3, 7, 13, 21
4. $-7, -4, 1, 8, 17$
5. 5, 8, 11, 14, 17
6. $\frac{1}{8}, \frac{1}{4}, \frac{1}{2}, 1, 2$

Find the nth term a_n for the sequences in Exercises 7 to 10. Then use exponential regression to find $f(x)$. Using properties of exponents, show that your results are the same.

7. $\frac{1}{4}, \frac{1}{2}, 1, 2, 4$
8. 8, 16, 32, 64, 128
9. $\frac{1}{16}, \frac{1}{8}, \frac{1}{4}, \frac{1}{2}, 1$
10. $6, 2, \frac{2}{3}, \frac{2}{9}, \frac{2}{27}$

In Exercises 11 and 12, change the equations to $y = ab^x$ form and identify the y-intercept of the graph of each equation.

11. **a.** $y = 2^{x-3}$ **b.** $y = 3^{x+1}$ **c.** $y = 3^{x-3}$
12. **a.** $y = 2^{x+4}$ **b.** $y = 2^{x-2}$ **c.** $y = 3^{x+2}$

In Exercises 13 and 14, name the y-intercept of the graph of each equation. Change the equation to $y = b^{x \pm n}$ form.

13. **a.** $y = 2 \cdot 2^x$ **b.** $y = \frac{1}{9} \cdot 3^x$
14. **a.** $y = 81 \cdot 3^x$ **b.** $y = \frac{1}{16} \cdot 2^x$

Solve each equation in Exercises 15 to 20 for n or for x.

15. **a.** $4^x = 64$ **b.** $2^x = 2$
     c. $a^x = \dfrac{1}{a}, a \neq 0$ **d.** $b^n = 1, b \neq 0$

16. **a.** $\left(\frac{1}{2}\right)^x = \frac{1}{8}$ **b.** $16^x = 8$
     c. $a^x = 1, a \neq 0$ **d.** $\pi^x = \dfrac{1}{\pi^2}$

17. **a.** $25^n = 125$ **b.** $27^n = \frac{1}{3}$
     c. $\left(\frac{1}{25}\right)^n = 125$ **d.** $\left(\frac{1}{16}\right)^n = 16$

18. **a.** $49^n = 343$ **b.** $64^n = \frac{1}{4}$
     c. $\left(\frac{1}{4}\right)^n = 64$ **d.** $(0.1)^n = 100$

19. **a.** $\left(\frac{1}{100}\right)^n = 10$ **b.** $100^n = 10$
     c. $4^{x-4} = 64$ **d.** $27^{x+1} = 81$

20. **a.** $\left(\frac{1}{10}\right)^n = 100$ **b.** $\left(\frac{1}{100}\right)^n = 100$
     c. $4^{0.5x+2} = 1$ **d.** $25^{x-2} = \frac{1}{5}$

21. Explain why $f(x) = \left(\frac{1}{2}\right)^x$ and $f(x) = 2^{-x}$ have the same graph.

22. Explain why $y = -3^x$ and $y = (-3)^x$ are not the same.

23. Graph the equations, and label the graphs with their rules:

$$y = 2^x, \quad y = 2^{-x}, \quad y = -2^x, \quad y = -2^{-x}$$

24. Graph the equations, and label the graphs with their rules:

$$y = 3^x, \quad y = 3^{-x}, \quad y = -3^x, \quad y = -3^{-x}$$

25. Graph $y = 2^x$, and explain how to obtain the graph of these equations:

a. $y = 2^{x+1}$ **b.** $y = 2^{x-2}$

26. Graph $y = 3^x$, and explain how to obtain the graph of these equations:

a. $y = 3^{x-1}$ **b.** $y = 3^{x+3}$

For Exercises 27 and 28, complete the table.

27.

Exponential Equation	Logarithmic Equation	Solve for x
	$\log_2 16 = x$	
	$\log_x 25 = 2$	
	$\log_3 81 = x$	
	$\log_{10} x = \frac{1}{2}$	
$10^x = 19$		
	$\log_4 x = 0$	

28.

Exponential Equation	Logarithmic Equation	Solve for x
$10^3 = x$		
	$\log_x 16 = 4$	
$5^x = 1$		
	$\log_{10} 10 = x$	
$4^x = 9$		
	$\log_4 4 = x$	

Solve the equations in Exercises 29 to 52. Round to three decimal places.

29. $10^x = 0.1$

30. $10^x = 1$

31. $36 = 10^x$

32. $15 = 10^x$

33. $10^x = 0.75$

34. $10^{1.5} = x$

35. $4^{x+1} = 32$

36. $4^{x+2} = 1$

37. $3^{2x} = 6$

38. $4^{2x} = 0.2$

39. $\log_{10} x = -1$

40. $\log_{10} 0.0001 = x$

41. $\log_2 x = 1$

42. $\log_3 x = 0$

43. $\log_2 2 = x$

44. $\log_3 x = -2$

45. $\log 10000 = x$

46. $\log 0.01 = x$

47. $\log x = 2$

48. $\log_2 x = 3$

49. $\log_3 9 = x$

50. $\log_7 7 = x$

51. $\log_{27} 9 = x$

52. $\log_2 x = 0$

53. Show, using symbols, that $y = \log_3 x$ is the inverse function to $y = 3^x$.

54. Show, using symbols, that $y = \log_2 x$ is the inverse function to $y = 2^x$.

55. Graph $y = 3^x$ and $y = \log_3 x$ on a calculator with a square window.

a. Explain what facts make the points (0, 1) and (1, 0) appear on the two graphs.

b. Explain what facts make the points (1, 3) and (3, 1) appear on the two graphs.

c. Graph $y = x$. What do you observe about the graphs now?

56. Graph $y = 2^x$ and $y = \log_2 x$ on a calculator with a square window.

a. Explain what facts make the points (0, 1) and (1, 0) appear on the two graphs.

b. Explain what facts make the points (1, 2) and (2, 1) appear on the two graphs.

c. Graph $y = x$. What do you observe about the graphs now?

In Exercises 57 to 59, calculate the amount in a savings account if $P = \$1000$, $r = 6\%$ interest, and $t = 2$ years. Then calculate how long it will take for the money to double at 6% interest.

57. $n = 4$

58. $n = 12$

59. $n = 365$

60. Use the rule of 72 to estimate how long it will take money to double at 6% interest, 8% interest, and 12% interest, compounded annually.

Write each expression in Exercises 61 to 64 as a single logarithm. Simplify, where possible.

61. $\log(x - 1) + \log(x - 2)$

62. $\log(x^2 + 4x + 4) - \log(x + 2)$

63. $\log(x^2 + x) - \log x$

64. $\log(x^2 + x + 2) + \log x$

65. Show that $\log(x^2 + 2x + 1) = 2\log(x + 1)$.

66. Show that $\log (8x^3) = 3 \log (2x)$.

67. Show that $\log \sqrt{x} = \frac{1}{2} \log x$.

68. Show that $\frac{1}{3} \log x = \log \sqrt[3]{x}$.

In Exercises 69 and 70, simplify the left side and then solve the equation.

69. $\log_3 x^2 - \log_3 x = 2$

70. $\log (x^2 - 4) - \log (x + 2) = 2$

71. If we know $\log 5 = 0.69897$, how do we get $\log 500$ without a calculator?

72. If we know $\log 3000 = 3.47712$, how do we get $\log 30$ without a calculator?

73. What do $\log 300$, $\log 400$, $\log 500$, and $\log 600$ have in common?

74. Explain how $\log x$ and $\log (x/10)$ are related.

75. Explain how $\log x$ and $\log (1000x)$ are related.

76. What is the shape of the graph obtained when we plot exponential data on a semilog graph?

77. Sketch a logarithmic scale with labels 10^0 to 10^4. Locate these points on the scale: 3, 8, 30, 300, 800, and 3000.

78. Evaluate, rounding to three significant digits.

    a. e^e **b.** 3^e **c.** e^3

79. Evaluate, rounding to three significant digits.

    a. $\ln 3$ **b.** $\ln \pi$ **c.** $\ln e$

80. If you start with $1000, find the amount of money in a savings account after 5 years with continuous compounding at these interest rates.

    a. 6%   **b.** 8%   **c.** 12%

81. A savings account holds $1000 after 5 years. Find the starting amount of money under continuous compounding at these interest rates.

    a. 6%   **b.** 8%   **c.** 12%

82. Find the number of years required for $1000 to double at these interest rates.

    a. 6%   **b.** 8%   **c.** 12%

83. Using the rule of 69, estimate the interest rate required for money to double every 5 years with continuous compounding. Confirm your answer with an exponential equation.

In Exercises 84 to 91, solve for x.

84. $e^x = \dfrac{1}{e^2}$ **85.** $e^x = 1$

86. $(1/e)^x = e$ **87.** $e^x = e$

88. $\ln x = 3$ **89.** $\ln x = -2$

90. $\ln x = 1/e$ **91.** $\ln x = e$

92. A 200-sheet, 5-subject notebook cost $1.99 in 1986. In 1995, the same notebook cost $2.99. Let the input be x, the number of years since 1986.

    a. Fit an exponential regression equation to the data.

    b. Use your equation to predict the cost of the notebook in the year 2005.

93. The yearly economic impact of salmon fishing on one West Coast state has been estimated as follows: 1989, $46.5 million; 1990, $29.6 million; 1991, $18.8 million; 1992, $15.4 million; 1993, $8.0 million; 1994, $4.0 million. Fit an exponential equation to the data for the years 1989 to 1994. Predict the economic impact in the year 2005.

94. The cost of a movie ticket in 1960 was $1.00. If the cost doubles every 12 years, estimate the cost in the year 2000. Fit an exponential equation.

In Exercises 95 and 96, find a population equation with exponential regression. Let the input be the number of years after 1930. State the rate of growth or decay. Predict the population in the year 2000 with your equation. State whether your estimate seems reasonable.

95. Euclid, Ohio: 1930, 12,751; 1940, 17,866; 1980, 59,999; 1990, 54,875

96. Hoboken, New Jersey: 1930, 59,261; 1940, 50,115; 1980, 42,460; 1990, 33,397.

97. The interest on the national debt of the United States consumes a sizable portion of the annual federal budget. Historical national debt data follow: 1910, $1.1 billion; 1920, $24 billion; 1930, $16.1 billion; 1940, $43 billion; 1945, $258 billion; 1950, $256 billion; 1960, $284 billion; 1970, $370 billion; 1980, $908 billion; 1985, $1,823 billion; 1990, $3,233 billion; 1993, $4,351 billion.

    a. Graph the data. Start with 1910 as year zero, and mark the graph in decades. Use $200 billion as the scale on the vertical axes.

    b. Fit an appropriate regression equation to the data.

    c. History question: What events caused the large jumps in the debt?

    d. Use the graph and the equation to predict the current debt. Compare with the actual current debt.

    e. Create a semilog-style graph by graphing the log of the debt on the vertical axis. How does this change the graph?

98. The cost of first-class postage for a one-ounce letter in the United States has been as follows: 1932, 3¢; 1958, 4¢; 1963, 5¢; 1968, 6¢; 1971, 8¢; 1975, 13¢; 1978, 15¢; 1981, 20¢; 1987, 22¢; 1989, 25¢; 1991, 29¢; 1995, 32¢; 1999, 33¢.

  a. Graph the data on a calculator.

  b. Fit an appropriate regression equation to the data.

  c. Compare the data with the graph of the regression equation.

  d. State the average annual rate of growth r.

  e. Use your equation to predict the current rate. Compare your prediction with the current rate for first-class postage.

CHAPTER **7** TEST

In Exercises 1 to 4, do the following:

a. Identify the type of sequence (arithmetic, quadratic, or geometric).

b. For arithmetic and geometric sequences, write the equation using a_n and the appropriate calculator regression. Show that the results are the same with both methods.

c. For quadratic sequences, use quadratic regression to find the equation.

1. 3, 6, 9, 12, 15

2. $\frac{1}{8}, \frac{1}{4}, \frac{1}{2}, 1, 2$

3. $\frac{1}{4}, \frac{1}{8}, \frac{1}{16}, \frac{1}{32}$

4. 3, 12, 27, 48, 75

5. The population of Klamath Falls, Oregon, in 1930 was 16,093 and in 1940 was 16,497. Fit both a linear and an exponential equation to the data. Use your equations to predict the population of this city in 1990. The actual population in 1990 was 17,737; comment on the accuracy of your predictions.

6. Historical gasoline prices for a West Coast city are as follows: 1920, $0.27; 1930, $0.21; 1940, $0.21; 1950, $0.28; 1960, $0.34; 1970, $0.35; 1980, $1.19; 1990, $1.38.

  a. Graph the data.

  b. Would you use a linear, quadratic, or exponential equation to model these data? Explain your choice.

  c. Fit the chosen type of equation to the data with calculator regression, and use the graph and the equation to predict the current price.

7. Solve the equations for x or for n.

  a. $3^x = 27$

  b. $\left(\frac{1}{16}\right)^n = 64$

  c. $\frac{1}{1000} = 0.1^x$

  d. $4^{0.5x} = 16$

  e. $4^x = \frac{1}{64}$

  f. $100^n = 100$

  g. $27^x = 9$

  h. $4^{x+1} = 8$

8. Solve the equations for x.

  a. $10^x = 3$

  b. $5^{2x} = 0.25$

  c. $10^x = -1$

  d. $e^x = 1/e$

  e. $5^{x-2} = \frac{1}{125}$

  f. $\log_2 x = -2$

  g. $\log_5 x = 3$

  h. $\log_x 32 = 5$

  i. $\log x = 1$

  j. $\ln x = 1.5$

9. Evaluate or simplify these expressions.

  a. $\log_7 1$

  b. $\log 1$

  c. $\log_2 x + \log_2 (x + 1)$

  d. $\log_3 (x^2 - 9) - \log_3 (x + 3)$

10. Suppose $y = b^{x+2}$. Change the equation to the form $y = ab^x$. Find the y-intercept.

11. Sketch a graph of $y = b^x$ for $x < 0$.

12. Explain how to find r from $S = 100(0.93)^x$.

13. Explain how $\log x$ and $\log (100x)$ are related.

14. Give two pairs of points such that (a, b) is on the graph of $y = 10^x$ and (b, a) is on the graph of $y = \log x$. Explain why this is true.

15. The formula for pH is $\text{pH} = -\log [\text{H}^+]$.

  a. Find the pH for sea water with $[\text{H}^+] = 5.01 \times 10^{-9}$ M.

  b. Find $[\text{H}^+]$, given that the pH of household ammonia is 11.9.

16. Plot (1, 2), (5, 32), (8, 256), and (9, 512) on a semilog graph with positive integers 1 to 10 on the horizontal axis and powers of 10 (10^0 to 10^3) in a logarithmic scale on the vertical axis.

17. What is the equation of the exponential graph in the figure? (*Hint:* What is y when $x = 0$? What is it when $x = 1$?) Sketch the inverse function.

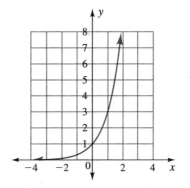

18. What is the equation of the logarithmic graph in the figure? (*Hint:* What is x when $y = 0$? What is x when $y = 1$?) Sketch the inverse function.

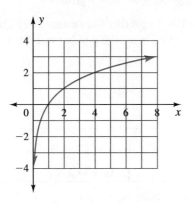

19. How long will it take $1000 to increase to $1360 at 7.5% interest compounded annually?

20. How long will it take $1000 to increase to $1906 at 7.5% interest compounded continuously?

8

Systems of Equations

Figure 1

The three chocolate rabbits shown in Figure 1 are selections at Janelle's candy store. Although pricing usually depends on weight, some smaller styles of rabbits require more labor and are priced accordingly. The number made of each style depends on historic demand and anticipated sales. We will return to the chocolate rabbits in Section 8.3.

The chapter begins with a review of guess and check and the substitution and elimination methods of solving systems of equations. We review graphical solutions and obtain equations from quantity-value tables in Section 8.1. Matrices are introduced as a method for solving systems of two (Section 8.2) and three or more (Section 8.3) linear equations. The chapter closes with an overview of conic sections (Section 8.4), nonlinear systems of equations (Section 8.5), and systems of inequalities (Section 8.6).

8.0 Solving Systems of Equations by Substitution and Elimination ___

OBJECTIVES

* Solve problems by guess and check.
* Write a system of equations using information obtained by guess and check.
* Find the solutions to a system of linear equations by substitution and elimination.
* Solve application problems.

WARM-UP

1. Add 3 to both sides of the equation $x - 3 = 8$. Do you obtain an equation with the same solution as the original equation?

2. Add x to both sides of the equation $2x - 3 = 9 - x$. Do you obtain an equation with the same solution as the original equation?

3. Suppose $x = 3$. Add x to the left side of $y - x = 4$ and add 3 to the right side. Have you added equal values to both sides of the equation?

IN THIS SECTION, we review guess and check as a method of building linear equations that describe problem settings. We solve systems of equations by substitution and elimination, and then we apply substitution to the derivation of the roadbed transition curve from Section 4.5.

Systems of Equations

A **system of equations** is *a set of two or more equations to be solved for values of the variables that satisfy all of the equations, if such values exist. The set of numbers that make all equations true* is the **solution to the system.**

Systems of equations are commonly used in geometry, science, business, and economics to describe relationships and conditions. The systems we study in this chapter will suggest some of these applications.

Computer and calculator technology now allows us to solve systems of linear equations with ease. Our purpose here is to examine basic equation solving so as to better understand the work done by the technology.

Solution Methods for
Systems of Equations

We can solve systems of equations in many ways:

1. Guess and check (Section 8.0)
2. Substitution (Section 8.0)
3. Elimination (Section 8.0)
4. Graphing and calculator graphing (Section 8.1)
5. Calculator matrices (Sections 8.2 and 8.3)

Guess and Check

Organizing information about a problem and guessing the solution helps us in several ways—for instance, guessing may improve our understanding. If we don't solve the problem by guessing, we are frequently able to set up equations that we can use to solve the problem. Recall the four problem-solving steps in the guess-and-check process, as introduced in Section 1.1:

Move toward *understanding* by reading carefully.

Plan by considering what might be reasonable inputs and preparing a table in which to record guesses.

Carry out the plan by working through the problem with the chosen input.

Check by comparing the result with the conditions or requirements of the original problem.

In Example 1, we will use a table to organize our steps and record our guesses so that we learn from each guess. *Each row in the table represents a guess.*

EXAMPLE **1** Solving problems by guess and check: time allocation   Jian's academic time includes the time spent attending class, plus 2 hours of study time for each hour spent in class. Each week, he spends twice as many hours on academics as he does at his job. He spends 28 more hours each week sleeping than at his job. He spends the remaining 44 hours of the week on other tasks. How many hours does Jian devote to sleep, work, and academics?

a. Make a table to determine the hours spent at each activity.

b. Write equations that describe the problem.

Solution   **a.** *Understand and plan:* The three activities—sleep, work, and academics—make natural column headings for a guess-and-check table. The hours spent on other tasks and total hours form the two other columns. The total hours column should equal 168:

$$\frac{24 \text{ hours}}{1 \text{ day}} \cdot \frac{7 \text{ days}}{1 \text{ week}} = \frac{168 \text{ hours}}{1 \text{ week}}$$

Carry out the plan: A reasonable starting guess is 8 hours of sleep per night, which, times 7 nights per week, gives 56 hours of sleep per week. This guess results in 184 total hours for the week and so is too high. For our next guess, we drop the hours of sleep to 50. This guess results in 160 total hours and so is too low. Our third guess for sleep is 52 hours, a number between 50 and 56.

Sleep	**Work**	**Academics**	**Other**	**Total Hours**
56	28	56	44	184 (too high)
50	22	44	44	160 (too low)
52	24	48	44	168 (correct)

Table 1

As Table 1 shows, Jian sleeps 52 hours a week, works 24 hours, and spends 48 hours attending class and studying.

Check: The results satisfy the conditions in the problem. ✔

b. We will let s = hours of sleep, w = hours of work, and a = hours on academics. Using these variables, we can describe the table setup with equations.

$s - 28 = w$ From the first two columns

$a = 2w$ From the second and third columns

$s + w + a + 44 = 168$ From each row

We will solve these equations later in this section.

Think about it 1: If x equals the number of hours spent sleeping, what one equation, in terms of x, describes Example 1?

The problem in Example 2 does not have whole-number answers, so continued guessing is not productive. We must use the patterns formed within the table to find equations.

EXAMPLE **2**

Setting up equations by guess and check: another time allocation   Tana's priority is to get through school. She plans to spend twice as many hours in class as she does at work. She studies 3 hours for every hour in class. To maintain her health and sanity, she wants to limit the total time allocated to work and school to 60 hours per week. How many hours of classes, study, and work should she plan?

a. Make a guess-and-check table with four entries.

b. Write equations that describe the problem.

Solution

a. ***Understand and plan:*** The hours allocated to work, class, and study make natural column headings in a table. The total time should be 60 hours.

Carry out the plan: We start Table 2 with a guess of 10 hours of work per week. This guess leads us to 90 total hours, which is too high. We cut the number of work hours in half and obtain a total of 45 hours, which is too low. Subsequent guesses of 8 and 7 work hours get us closer to the 60 hours.

Work	Class	Study	Total
10	20	60	90 (too high)
5	10	30	45 (too low)
8	16	48	72 (too high)
7	14	42	63 (too high)

Table 2

b. Letting w = hours of work, c = hours in class, and s = hours of study, we can describe the table with equations.

$$c = 2w \qquad \text{From the first and second columns}$$
$$s = 3c \qquad \text{From the second and third columns}$$
$$w + c + s = 60 \qquad \text{From each row}$$

We will solve these equations later in this section. ●

Think about it 2: If x = the number of hours spent working, what one equation, in terms of x, describes Example 2?

Substitution Method

The substitution method for solving equations is particularly valuable because it applies to both linear and nonlinear equations. (We will work with nonlinear equations in later sections.) The **substitution method** of solving equations obtains its name from the replacement process: *we replace variables with equivalent expressions or numbers.*

SOLVING FOR ONE VARIABLE   The first step in solving a system of equations by substitution is to solve an equation for one of its variables.

EXAMPLE **3** Solving for one variable   Solve each equation for the indicated variable.

    a. $x - 2y = 6$ for y

    b. $2y - 3x = -6 - y$ for y

    c. $y = -2x^2 + 2$ for x^2

    d. $x^2 + y^2 = 4$ for y

Solution   **a.** We solve $x - 2y = 6$ for y as follows:

$$x - 2y = 6 \qquad \text{Subtract } x \text{ from both sides.}$$
$$-2y = 6 - x \qquad \text{Divide both sides by } -2.$$
$$y = \frac{6 - x}{-2} \qquad \text{Multiply both numerator and denominator by } -1.$$
$$y = \frac{x - 6}{2}$$

b. We solve $2y - 3x = -6 - y$ for y as follows:

$$2y - 3x = -6 - y \qquad \text{Add } y \text{ to both sides.}$$
$$3y - 3x = -6 \qquad \text{Add } 3x \text{ to both sides.}$$
$$3y = 3x - 6 \qquad \text{Divide both sides by 3.}$$
$$y = x - 2$$

c. We solve $y = -2x^2 + 2$ for x^2 as follows:

$$y = -2x^2 + 2 \qquad \text{Subtract 2 from both sides.}$$
$$y - 2 = -2x^2 \qquad \text{Divide both sides by } -2.$$
$$\frac{y - 2}{-2} = x^2 \qquad \text{Multiply both numerator and denominator by } -1.$$
$$\frac{2 - y}{2} = x^2$$

d. We solve $x^2 + y^2 = 4$ for y as follows:

$$x^2 + y^2 = 4 \qquad \text{Subtract } x^2 \text{ from both sides.}$$
$$y^2 = 4 - x^2 \qquad \text{Take the square root of both sides.}$$
$$y = \pm\sqrt{4 - x^2}$$

We obtain two equations for y: $y = \sqrt{4 - x^2}$ and $y = -\sqrt{4 - x^2}$.   ●

Caution: In part b of Example 3, it would not have been correct to leave y terms on both sides.

Definition of "In Terms of"
> To solve for the variable y **in terms of** x means to isolate y on one side of the equation and obtain an expression on the other side of the equal sign that does not contain y.

MAKING THE SUBSTITUTION   The second step in the substitution method for a system of equations is to substitute the expression from the first step into the second equation, as shown in Example 4.

EXAMPLE **4**

Solving by substitution   Solve this system of linear equations by substitution:

$$2y + x = -9 \qquad \text{(1)}$$
$$4y - 3x = 12 \qquad \text{(2)}$$

Solution   We solve the first equation for x:

$$2y + x = -9 \qquad\qquad \text{Subtract 2y from both sides.}$$
$$x = -9 - 2y$$

Then we substitute $-9 - 2y$ for x in the second equation:

$$4y - 3x = 12$$
$$4y - 3(-9 - 2y) = 12 \qquad\qquad \text{Apply the distributive property.}$$
$$4y + 27 + 6y = 12 \qquad\qquad \text{Subtract 27 from both sides and add y terms.}$$
$$10y = -15 \qquad\qquad \text{Divide by 10.}$$
$$y = -1.5$$
$$2(-1.5) + x = -9 \qquad\qquad \text{Substitute } -1.5 \text{ for y in the first equation.}$$
$$-3 + x = -9 \qquad\qquad \text{Add 3 to both sides.}$$
$$x = -6$$

The solution to the system is $x = -6$ and $y = -1.5$.

Check (in both equations): $2(-1.5) + (-6) \stackrel{?}{=} -9$ ✔

$$4(-1.5) - 3(-6) \stackrel{?}{=} 12 \text{ ✔}$$

●

Think about it 3: In Example 4, why do we choose to solve the first equation for x?

Solving Two Equations by Substitution

---

1. Solve one equation for a first variable.
2. Substitute the resulting expression into the second equation. The second equation now has one variable. Solve for the second variable.
3. Substitute the value of the second variable into the first equation, and find the value of the first variable.
4. Check the solutions.

SOLVING SYSTEMS OF THREE EQUATIONS   In Examples 5 and 6, we solve the systems of three equations from Examples 1 and 2. Substitution permits us to replace two of the three variables in one equation.

EXAMPLE **5**

Solving by substitution   Solve the system of equations in Example 1 by substitution:

$$s - 28 = w \qquad\qquad \text{(1)}$$
$$a = 2w \qquad\qquad \text{(2)}$$
$$a + s + w + 44 = 168 \qquad\qquad \text{(3)}$$

Solution   Each equation contains w.

$$s - 28 = w \qquad\qquad \text{Solve for s in terms of w.}$$
$$s = w + 28 \qquad\qquad \text{Substitute this equation and } a = 2w$$
$$\text{into the third equation.}$$

$$a + s + w + 44 = 168$$
$$(2w) + (w + 28) + w + 44 = 168$$
$$4w + 72 = 168$$
$$4w = 96$$
$$w = 24$$

We now find s and a, given $w = 24$.

$$s = 24 + 28 = 52$$
$$a = 2(24) = 48$$

The solution is 24 hours of work, 52 hours of sleep, and 48 hours of academics.

●

EXAMPLE **6**   Solving by substitution   Solve the system of equations in Example 2 by substitution:

$$c = 2w$$
$$s = 3c$$
$$c + s + w = 60$$

Solution   Each equation contains c.

$c = 2w$ Solve for w in terms of c.

$\dfrac{c}{2} = w$ Substitute this equation and $s = 3c$ into the third equation.

$$c + s + w = 60$$
$$c + 3c + \frac{c}{2} = 60$$
$$4.5c = 60$$
$$c = 13\tfrac{1}{3} \text{ hr}$$

We now find w and s, given $c = 13\tfrac{1}{3}$ hr.

$$w = \frac{13\tfrac{1}{3}}{2} = 6\tfrac{2}{3} \text{ hr}$$
$$s = 3\left(13\tfrac{1}{3}\right) = 40 \text{ hr}$$

●

$\mathbf{S}$ ubstitution may be used to change a system of three equations to a system of two equations, as shown in Example 7.

EXAMPLE **7**   Solving by substitution   Solve the following system of equations by substitution:

$$a + b + c = 3 \qquad (1)$$
$$2a + 3b + c = 13 \qquad (2)$$
$$2a - b = 0 \qquad (3)$$

Solution   Because the third equation is missing a variable, we can solve that equation for one variable and substitute into the other two equations, creating two equations in two unknowns.

$2a - b = 0$ Solve for b.

$2a = b$

We replace b with $2a$ in the first and second equations.

$$a + 2a + c = 3 \qquad \text{gives} \quad 3a + c = 3 \qquad (4)$$
$$2a + 3(2a) + c = 13 \quad \text{gives} \quad 8a + c = 13 \qquad (5)$$

We solve the fourth equation for c, getting $c = 3 - 3a$, and then substitute for c in the second equation.

$$8a + (3 - 3a) = 13 \qquad \text{Simplify.}$$
$$5a + 3 = 13 \qquad \text{Subtract 3 from each side.}$$
$$5a = 10 \qquad \text{Divide by 5.}$$
$$a = 2$$

We substitute $a = 2$ into the third equation, $2a - b = 0$, to find $b = 4$. We substitute $a = 2$ and $b = 4$ into the second equation, $a + b + c = 3$, to find $c = -3$. The solution to the system is $a = 2$, $b = 4$, and $c = -3$.

Check: $2 + 4 + (-3) \overset{?}{=} 3$ ✔

$\qquad\quad 2(2) + 3(4) + (-3) \overset{?}{=} 13$ ✔

$\qquad\quad 2(2) - (4) \overset{?}{=} 0$ ✔   ●

Examples 5, 6, and 7 suggest the following strategies for solving three equations by substitution:

Solving Three Equations by Substitution	• Write two equations in terms of the same variable and substitute them into the third equation. Solve the remaining one equation in one unknown. • Solve one equation in terms of a single variable and substitute into each of the other two equations. Solve the remaining two equations in two unknowns.

Elimination Method

In solving equations by the elimination method, we must pay close attention to the form of the equation. The standard form of a linear equation is the most general form of a linear equation in two variables.

Standard Form of a Linear Equation	All linear equations in two variables may be written as $\qquad ax + by = c$ where a, b, and c are any real numbers.

We will use the standard form in this section and in Sections 8.2 and 8.3. We will return to the slope-intercept form, $y = mx + b$, when graphing in Section 8.1.

The **elimination method** of solving a system of equations is *a process in which one variable is removed from the system of equations by adding (or subtracting) two equations.* Elimination depends on the equality of the left and right sides of an equation. Because the sides are equal, we may use the addition property of equations to add two equations together.

EXAMPLE **8**   Solving by elimination   Solve this system of linear equations by elimination:

$$y = 2x + 4 \qquad (I)$$
$$x = 10 + y \qquad (2)$$

Solution   We arrange each equation in standard form.
Equation 1:

$$y = 2x + 4 \qquad \text{Subtract } 2x \text{ from both sides.}$$
$$-2x + y = 4$$

Equation 2:

$$x = 10 + y \qquad \text{Subtract } y \text{ from both sides.}$$
$$x - y = 10$$

Then we write the equations together, one above the other:

$$
\begin{array}{rl}
-2x + y = 4 & (I) \\
\underline{x - y = 10} & (2) \qquad \text{Add the equations.} \\
-x = 14 & \\
x = -14 &
\end{array}
$$

Substituting $x = -14$ into the first equation, we can solve for y:

$$y = 2(-14) + 4$$
$$y = -24$$

The solution to the system is $x = -14$ and $y = -24$.

Check: $(-24) \stackrel{?}{=} 2(-14) + 4$ ✔
$(-14) \stackrel{?}{=} 10 + (-24)$ ✔   ●

Adding the equations together is productive only if the terms for one variable have opposite coefficients. Then adding the terms yields zero, eliminating that variable.

Eliminating Like Terms

> If two equations in standard form contain like terms with opposite coefficients, then adding the equations will eliminate those terms.

In Example 9, we multiply the second equation by -3, the opposite of the numerical coefficient on x in the first equation. This creates equations with opposite coefficients on the x-terms.

EXAMPLE **9**   Solving by elimination   Solve these equations:

$$3x + 2y = 8 \qquad (I)$$
$$x + 5y = -6 \qquad (2)$$

Solution   To obtain the opposite to the x-term in the first equation, we multiply the second equation by -3:

$$
\begin{array}{rl}
-3(x + 5y) = -3(-6) & \\
-3x - 15y = 18 & \\
\underline{3x + 2y = 8} & \text{Add the first equation.} \\
0 - 13y = 26 & \text{Divide by } -13. \\
y = -2 &
\end{array}
$$

$$3x + 2(-2) = 8 \quad \text{Substitute } y = -2 \text{ in the first equation.}$$
$$x = 4$$

The solution to the system is $x = 4$ and $y = -2$.

Check: $3(4) + 2(-2) \overset{?}{=} 8$ ✔

 $4 + 5(-2) \overset{?}{=} -6$ ✔

●

I n Example 10, both equations must be multiplied to produce like terms with opposite coefficients.

EXAMPLE **10**

Solving by elimination   Solve these equations:

$$4x + 2y = -35$$
$$-3x + 5y = 81.5$$

Solution   To obtain 12 and -12 as the respective coefficients on x, we multiply the first equation by 3 and the second equation by 4:

$$3(4x + 2y) = 3(-35)$$
$$4(-3x + 5y) = 4(81.5)$$

$$\begin{array}{rl} 12x + 6y = & -105 \\ -12x + 20y = & \underline{326} \\ 26y = & 221 \\ y = & 8.5 \end{array} \quad \begin{array}{l} \\ \text{Add the equations.} \\ \text{Divide by 26.} \\ \ \end{array}$$

$$4x + 2(8.5) = -35 \quad \text{Substitute } y = 8.5 \text{ in the first equation.}$$
$$x = -13$$

The solution to the system is $x = -13$ and $y = 8.5$.

Check: $4(-13) + 2(8.5) \overset{?}{=} -35$ ✔

 $-3(-13) + 5(8.5) \overset{?}{=} 81.5$ ✔

●

Think about it 4: Which variable would have been eliminated by adding the equations in Example 10 if we had multiplied the first equation by -5 and the second by 2?

Solving Two Equations by Elimination

1. Arrange the equations in standard form.
2. Multiply one or both equations by numbers that create opposite values (additive inverses) of the coefficients on one pair of like terms.
3. Add the equations to eliminate one variable, and solve for the variable that remains.
4. Use substitution to find the other variable.
5. Check the solutions.

Application: Roadbed Transition Curves

The number of unknowns in a system determines the number of equations needed to solve the system. When we have three unknowns in a system, we need three equations.

In Example 11, we return to the *transition curve* parabolic model for a highway roadbed (Section 4.5). To find the equation of the parabolic transition curve, we need to find the three unknowns a, b, and c in $y = ax^2 + bx + c$. We therefore need three equations. One of the equations is $y = ax^2 + bx + c$. The other two equations come from relationships within the problem setting. Those relationships are based on the formula for the slope m of the parabolic roadbed at any point (x, y):

$$m = 2ax + b \qquad \text{The slope of a parabola at } (x, y)$$

The parameters a and b are the same as in $y = ax^2 + bx + c$, because this slope formula is from a calculus operation known as the derivative of $y = ax^2 + bx + c$.

EXAMPLE **11** Finding an equation: roadbed transition curve A roadbed is to pass through point G on one hill and point H on the next hill. A side view of the required roadbed is shown in Figure 2. There is to be a 3% downgrade slope at G and a 2% upgrade slope at H. Suppose point G has coordinates $(0, 1000)$ with units in feet. Point H has an x-coordinate of 1500 feet because it is 1500 feet away from point G.

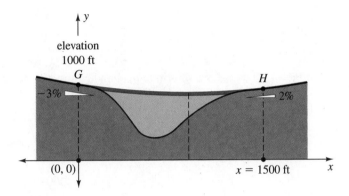

Figure 2

a. Use the slope formula and the x-coordinate at point G to write one equation.

b. Use the slope formula and the x-coordinate at point H to write a second equation.

c. Use $y = ax^2 + bx + c$ and the ordered pair at point G to write a third equation.

d. Use the information from parts a, b, and c to write the quadratic equation that meets the roadbed requirements.

e. Find the elevation (y-coordinate) at point H.

Solution In $y = ax^2 + bx + c$, we have three unknowns: a, b, and c. The facts give us information for three equations:

$$m = 2ax + b \qquad \text{The slope at } G$$
$$m = 2ax + b \qquad \text{The slope at } H$$
$$y = ax^2 + bx + c \qquad \text{The parabola through } (0, 1000)$$

a. The slope formula permits us to enter the x-coordinate and the slope at that point and solve for a or b. Point G at $(0, 1000)$ has $x = 0$ and a 3% downgrade slope; thus, $m = -\frac{3}{100}$.

$$m = 2ax + b \qquad \text{Substitute } m = -\tfrac{3}{100} \text{ and } x = 0.$$
$$-\tfrac{3}{100} = 2a(0) + b$$
$$b = -\tfrac{3}{100}$$

b. We use the slope formula at point H with $x = 1500$ and a 2% upgrade slope, $m = \frac{2}{100}$.

$$m = 2ax + b \qquad \text{Substitute } m = \tfrac{2}{100} \text{ and } x = 1500.$$

$$\tfrac{2}{100} = 2a(1500) + b$$

$$\tfrac{2}{100} = 3000a + b$$

c. Because the road passes through point G, we may obtain another equation by substituting the ordered pair at G, $(0, 1000)$, into the parabola's equation:

$$y = ax^2 + bx + c \qquad \text{Substitute } x = 0 \text{ and } y = 1000.$$

$$1000 = a(0)^2 + b(0) + c$$

$$c = 1000$$

d. From part a, $b = -\frac{3}{100}$. Substituting this value into the equation from part b, we have

$$\tfrac{2}{100} = 3000a + b$$

$$\tfrac{2}{100} = 3000a - \tfrac{3}{100} \qquad \text{Add } \tfrac{3}{100} \text{ to both sides.}$$

$$\tfrac{5}{100} = 3000a \qquad \text{Divide by 3000.}$$

$$\tfrac{5}{300,000} = a \qquad \text{Simplify.}$$

$$a = \tfrac{1}{60,000}$$

We substitute a, b, and c into $y = ax^2 + bx + c$ to obtain the quadratic equation describing the roadbed:

$$y = \tfrac{1}{60,000}x^2 - \tfrac{3}{100}x + 1000$$

e. To find the elevation at H, we place $x = 1500$ into the transition curve equation:

$$y = \tfrac{1}{60,000}x^2 - \tfrac{3}{100}x + 1000$$

$$= \tfrac{1}{60,000}(1500)^2 - \tfrac{3}{100}(1500) + 1000$$

$$= 992.5 \text{ ft} \qquad \bullet$$

The placement of the origin in Example 11 helped us find the coefficients a, b, and c.

Ordered Pairs and Equations

1. The location of the origin, $(0, 0)$, in an application is arbitrary. Place the origin at any convenient, reasonable position. Because $f(0)$ is the y-intercept, a convenient origin makes it easier to find equations or solve for the constant term in an equation.

2. If a point lies on a graph, its ordered pair makes the equation true for that graph. Thus, the ordered pair for the point may be substituted into the equation to find an equation or to solve for an unknown.

ANSWER BOX

Warm-up: 1. $x = 11$; yes **2.** $3x - 3 = 9$; yes **3.** $y = 7$; yes **Think about it 1:** $x + (x - 28) + 2(x - 28) + 44 = 168$ **Think about it 2:** $x + 2x + 3(2x) = 60$ **Think about it 3:** The coefficient on x is 1, so solving for x saves doing a division and avoids fractions. **Think about it 4:** The y would be eliminated.

EXERCISES 8.0

Set up guess-and-check tables for Exercises 1 to 6. Make three to four guesses, and then write equations. Solve your equations using any method.

1. Celine works twice as many hours as she spends in class. She studies 3 hours for every hour in class. To maintain her health and sanity, she limits her total time commitment to work and school to 60 hours per week. How many hours does she plan to spend on each activity: attending class, studying, and working?

2. Vaughn works 35 hours per week. He studies 3 hours for every hour in class. To maintain his health and sanity, he limits his total time commitment to work and school to 60 hours per week. How many hours does he spend attending class and studying?

3. An ounce of macadamia nuts (salted and roasted in oil) contains 222 calories. There are twice as many grams of carbohydrates as protein. The total weight of the protein, fat, and carbohydrates is 28 grams. There are 4 calories per gram of protein and carbohydrates and 9 calories per gram of fat. How many grams are there each of carbohydrates, fat, and protein?

4. A loosely packed cup of raisins contains 489 calories. It contains 4 more grams of protein than of fat. There are 121 grams of carbohydrates, fat, and protein altogether. There are 4 calories per gram of protein and carbohydrates and 9 calories per gram of fat. How many grams are there each of carbohydrates, fat, and protein?

5. A banana contains 121 calories. It contains equal amounts of protein and fat. There are 29 grams of carbohydrates, fat, and protein altogether. There are 4 calories per gram of protein and carbohydrates and 9 calories per gram of fat. How many grams are there each of carbohydrates, fat, and protein?

6. An ounce of sunflower seeds contains 170 calories. There is 1 gram more of protein than of carbohydrates. The total weight of the protein, fat, and carbohydrates is 25 grams. There are 4 calories per gram of protein and carbohydrates and 9 calories per gram of fat. How many grams are there each of carbohydrates, fat, and protein?

In Exercises 7 to 22, solve the equations for the indicated variable.

7. $3x + y = 4$ for y

8. $5x - y = 7$ for y

9. $x - 3y = 7$ for x

10. $2y - x = 6$ for x

11. $4y - 2x = 5$ for x

12. $3y - 2x = 11$ for x

13. $4x - 2y = -3$ for y

14. $2x - 3y = -5$ for y

15. $y = x^2 + 5$ for x^2

16. $x = 3y^2 - 4$ for y^2

17. $x - 3 = -2y^2 + x^2$ for y^2

18. $y + 5 = -x^2 - y^2$ for x^2

19. $x^2 + y^2 = 8$ for y

20. $x^2 - y^2 = 4$ for y

21. $x^2 - y^2 = 10$ for x

22. $x^2 + y^2 = 5$ for x

Solve the systems of equations in Exercises 23 to 36. Choose either substitution or elimination, but use each method at least twice.

23. $y = 3x - 4$
    $3y - 2x = 9$

24. $2y = 3x - 10$
    $2.5x + y = 3$

25. $0.6x + y = 3$
    $1.4x + y + 3 = 0$

26. $5y + 2x - 9 = 0$
    $y + 1.2x = 3$

27. $x + 2y = 9.4$
    $3x - 5y = 4$

28. $x - 3y = -1.4$
    $2x - 7y + 1.6 = 0$

29. $y = 1.8x - 6.6$
    $y + 1.4x = 3$

30. $2x + 3y = 10.9$
    $2x + 16.1 = 3y$

31. $5x = 14.7 + 6y$
    $2x + 6y = -4.2$

32. $3x + 4y = 4$
    $2x + y = 5$

33. $2x + 4y = 0$
    $3x - 2y = 24$

34. $3x + 2y = 2$
    $-4x + 3y = 37$

35. $5x - 2y = -9$
    $6x + 7y = 8$

36. $4x - 5y = 7$
    $3x - 2y = 0$

Solve the systems of equations in Exercises 37 to 44.

37. $2a + 3b + 2c = 1$
    $3a + 2c = 10$
    $a + b = 1$

38. $3a + 4b + c = -2$
    $2a - c = -10$
    $4b - 4a = 15$

39. $2a + b - 4c = 17$
    $5a - 2b = 27$
    $b + 3c + 7 = 0$

40. $4a + b - c = -12$
    $b + c = 2$
    $2a + c = 5$

41. $x + y + z = 2$
    $x + 3z = 0$
    $2z = 7 + x + y$

42. $2x + y + z = 9$
    $x = 3 + y + z$
    $3x = 2z$

43. $2x + 3y - z = 11$
    $3x + y + 2z = 13$
    $2y = z + 7$

44. $3x = 2y + z$
    $4x + z + 17 = 3y$
    $2z + 5y + 4 = 0$

45. Can all equations whose graphs are straight lines be written in the slope-intercept form of a linear equation, $y = mx + b$?

46. Can all equations whose graphs are straight lines be written in the standard form, $ax + by = c$?

47. Write the equation of the vertical line $x = 2$ in standard form, and identify a, b, and c.

48. Write the equation of the horizontal line $y = 3$ in standard form, and identify a, b, and c.

49. Solve for y only:

$$acx + bcy = ce$$
$$acx + ady = af$$

50. Solve for x only:

$$adx + bdy = de$$
$$bcx + bdy = bf$$

51. Explain when using substitution may be more convenient than using elimination.

52. Explain when using elimination may be more convenient than using substitution.

In Exercises 53 to 56, the road grades connecting two hills are to be designed with a parabolic transition curve. The slope of the parabolic transition curve at any point is given by $m = 2ax + b$. The curve itself is $y = ax^2 + bx + c$.

53. In the figure, point K has a 4% downgrade slope and point L has a 3% upgrade slope. The horizontal distance between points K and L is 2000 feet. The elevation of point K is 700 feet, and its coordinates are $(0, 700)$. Find the equation for the road grade.

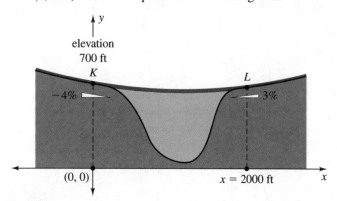

54. In the figure for Exercise 53, suppose that point K has a 3% downgrade slope and point L has a 5% upgrade slope. As before, the horizontal distance between points K and L is 2000 feet. The elevation of point K is 700 feet, and its coordinates are $(0, 700)$. Find the equation for the road grade.

55. Point M in the figure has a 5% downgrade slope, and point N has a 6% upgrade slope. The horizontal distance between points M and N is 2400 feet. The elevation of point N is 1200 feet, and its coordinates are $(2400, 1200)$. Find the equation for the road grade.

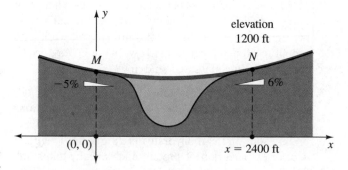

56. In the figure for Exercise 55, suppose that point M has a 4% downgrade slope and point N has a 2% upgrade slope. As before, the horizontal distance between points M and N is 2400 feet. The elevation of point N is 1200 feet, and its coordinates are $(2400, 1200)$. Find the equation for the road grade.

Project

57. *Roadbed Spreadsheet.* Design a spreadsheet to do Example 11 and the related exercises. When given slopes m_1 and m_2 at points $(0, y_1)$ and (x_2, y_2), respectively, the spreadsheet should calculate the coefficients a, b, and c for the transition curve $y = ax^2 + bx + c$ between the two hills. Design the spreadsheet to also calculate the height of the roadbed every 50 feet along the horizontal between the two reference points (between $x = 0$ and $x = x_2$).

8.1   Solving Systems of Two Linear Equations by Graphing _____

OBJECTIVES

- Solve systems of linear equations by graphing.
- Identify systems of two linear equations having a single solution, an infinite number of solutions, or no solution.
- Explain the algebraic results from solving systems containing equations whose graphs are parallel or coincident.
- Solve quantity and value problems by writing and solving systems of equations.

> **WARM-UP**
>
> These exercises review material from Section 2.2.
>
> 1. Find the equation of a line parallel to $y = -\frac{3}{2}x + 2$ that passes through the point (2, 5).
>
> 2. Find the equation of a line perpendicular to $y = -\frac{3}{2}x + 2$ that passes through the point (3, 0).

IN THIS SECTION, we use graphs of linear equations to understand why some systems of linear equations have a unique solution while others have no solution. We examine quantity-value tables as a way of obtaining linear equations from word problems.

Graphing

Graphing gives a method of finding the numerical solution to a system of linear equations and lends a visual meaning to the solution. *Because every point on the graph of an equation makes the equation true, the point of intersection of two graphs is the ordered pair that makes both equations true.* Thus, the point of intersection of two graphs identifies an ordered pair as the solution to the two equations.

An advantage of the graphical solution is that we are reminded by the ordered pair itself that a system of two linear equations has two variables (unknowns) and must be solved for both variables. A common error is to solve a system for only one of the two variables.

In a graph drawn by hand, the point of intersection can be located with only limited accuracy. Intersection features of graphing calculators and built-in solvers in computer programs permit us to locate coordinates of the point of intersection that are correct to many decimal places.

EXAMPLE

Solving with a graph   Solve the following system of linear equations from a calculator graph:

$$3y + 6 = 2x \qquad (1)$$
$$2y + 3x = 16 \qquad (2)$$

Solution   First, we solve the equations for y.
Equation 1:

$$3y + 6 = 2x \qquad \text{Subtract 6.}$$
$$3y = 2x - 6 \qquad \text{Divide by 3.}$$
$$y = \tfrac{2}{3}x - 2$$

Equation 2:

$$2y + 3x = 16 \qquad \text{Subtract } 3x.$$
$$2y = -3x + 16 \qquad \text{Divide by 2.}$$
$$y = -\tfrac{3}{2}x + 8$$

We enter the equations in the calculator under $\boxed{\text{Y=}}$ and trace to estimate the intersection on the graph. We can then confirm the intersection with the INTERSECT option. To the nearest thousandth, the intersection is (4.615, 1.077); see Figure 3.

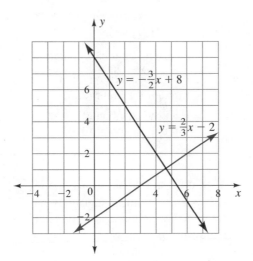

Figure 3

Graphing Calculator Technique:
Solving Two Equations Graphically

> Solve each equation for y as a function of x. If the equations are not in x and y, solve each for one variable in terms of the second variable. Enter the two functions of x as Y_1 and Y_2 under $\boxed{Y=}$. Set the viewing window. Graph, and verify that the point of intersection, if any, shows in the viewing window.
>
> Use the INTERSECT or SOLVE option to find the point of intersection of the two functions.

Special Cases

We now examine two special geometric and algebraic results for a system of two linear equations.

COINCIDENT LINES   *Two lines that are described by the same equation* are **coincident lines.** Because coincident lines have all points in common, we say that *a system of equations whose graphs form coincident lines has an infinite number of solutions.* Nonvertical coincident lines have the same slope and the same y-intercept. They are easy to identify because we obtain the same equation when we transform the equations to $y = mx + b$ form.

EXAMPLE

Identifying coincident lines   Solve each equation in this system for y. What can you conclude about the graphs of the equations?

$$2x - 3y = 6$$
$$y + 2 = \tfrac{2}{3}x$$

Solution   Equation 1:

$$2x - 3y = 6 \qquad \text{Subtract 6 and add 3y on both sides.}$$
$$2x - 6 = 3y \qquad \text{Divide by 3.}$$
$$\tfrac{2}{3}x - 2 = y$$

Equation 2:

$$y + 2 = \tfrac{2}{3}x \qquad \text{Subtract 2.}$$
$$y = \tfrac{2}{3}x - 2$$

The equations are both $y = \frac{2}{3}x - 2$, so they describe the same line (see Figure 4).

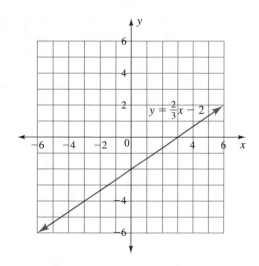

Figure 4 ●

I n Example 3, we examine the algebraic results from solving a system containing equations with coincident graphs.

EXAMPLE **3**   Finding the algebraic results for coincident lines   Solve the system of equations in Example 2 by substitution.

$$2x - 3y = 6$$
$$y + 2 = \tfrac{2}{3}x$$

Solution   Equation 2:

$$y + 2 = \tfrac{2}{3}x$$ Solve for *y*.
$$y = \tfrac{2}{3}x - 2$$ Substitute into Equation I.

Equation 1:

$$2x - 3y = 6$$ Let $y = \tfrac{2}{3}x - 2$.
$$2x - 3\left(\tfrac{2}{3}x - 2\right) = 6$$ Simplify.
$$2x - 2x + 6 = 6$$ Add like terms.
$$6 = 6$$

The variable drops out, and the remaining statement, $6 = 6$, is always true. This result indicates that all coordinate pairs (x, y) that make the first equation true also make the other equation true.   ●

Coincident Lines

A true statement such as $0 = 0$ in the solution to a system of equations implies that there are an infinite number of solutions. In the case of two equations in two variables, the graphs are coincident lines.

PARALLEL LINES   Parallel lines provide a second special case in working with a system of two linear equations. *Two lines in the coordinate plane that have no intersection* are **parallel lines.** Because parallel lines have no point of intersection, we say that *a system of equations whose graphs are parallel lines has no* *real-number solution.* Nonvertical parallel lines have the same slope and different y-intercepts.

EXAMPLE **4** Identifying parallel lines   Solve each equation in this system for y. What can you conclude about the graphs of the equations?

$$3x + 2y = 4 \qquad \text{(I)}$$
$$8 - \tfrac{3}{2}x = y \qquad \text{(2)}$$

Solution   Equation 1:

$$3x + 2y = 4 \qquad \text{Subtract } 3x.$$
$$2y = -3x + 4 \qquad \text{Divide by 2.}$$
$$y = -\tfrac{3}{2}x + 2$$

Equation 2:

$$y = -\tfrac{3}{2}x + 8$$

The equations have the same slope, $-\tfrac{3}{2}$, but different y-intercepts. The lines are parallel (see Figure 5).

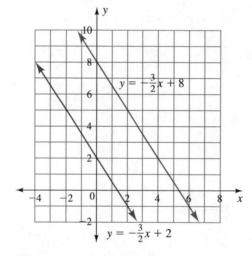

Figure 5   ●

In Example 5, we examine the algebraic results from solving a system of linear equations whose graphs are parallel lines.

EXAMPLE **5** Finding the algebraic results for parallel lines   Solve the system of equations in Example 4 by substitution.

$$3x + 2y = 4 \qquad \text{(I)}$$
$$8 - \tfrac{3}{2}x = y \qquad \text{(2)}$$

Solution   We substitute the expression for y from the second equation into the first equation:

$$3x + 2y = 4 \qquad \text{Let } y = 8 - \tfrac{3}{2}x.$$
$$3x + 2\left(8 - \tfrac{3}{2}x\right) = 4 \qquad \text{Simplify.}$$
$$3x + 16 - 3x = 4 \qquad \text{Add like terms.}$$
$$16 = 4$$

The variable drops out and the remaining statement is false, implying that there are no solutions to the system of equations. Because the equations describe parallel lines (Example 3), it is reasonable that there is no solution. ●

Parallel Lines	A false statement such as $0 = 1$ in the solution to a system of equations implies that there are no real-number solutions. In the case of two equations in two variables, the graphs are parallel.

Graphing a System of Two Linear Equations

In graphing a system of two linear equations, we have three possible outcomes (see the figures below):

1. The equations describe lines intersecting in exactly one point (point A). The system of equations has exactly one solution—the ordered pair at the point of intersection.

2. The equations describe the same line. The graphs are coincident (line BC). The system has an infinite number of solutions because the coordinates of every point on the graph of the first equation satisfy the second equation.

3. The equations describe parallel lines (lines DE and FG). The graphs have no point of intersection; hence, the system has no real-number solution.

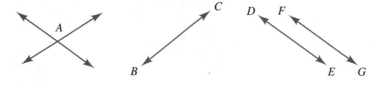

Table 3 summarizes the geometry and algebra associated with solving a system of two linear equations.

Geometry	Intersecting lines	Coincident lines	Parallel lines
Algebra	The equations can be solved for x and y.	The variables drop out, and the remaining statement is true.	The variables drop out, and the remaining statement is false.
Solution	One ordered pair is the solution.	An infinite number of ordered pairs satisfy the system.	There is no real-number solution to the system.

Table 3

Quantity-Value Tables

In order to describe word problems with equations, we need to identify the facts provided. In the guess-and-check tables in Section 8.0, there was only one fact about each item: the number of hours worked, slept, or spent in school. Many problems describe two or more facts about each item. In the highway transition curve problems, points on the roadbed had a coordinate position and a slope. In Exercises 8.0, the food items had a mass in grams and an energy value in calories.

The facts are frequently related to quantity and value or quantity and rate. **Quantity** *answers the question How many? or How much?* **Value** may be *monetary worth.* Value is most commonly given as a rate. The word **per** identifies *a value or rate,* such as percent, cents per quarter, kilometers per hour, or cost per item.

We summarize the quantity and value (or rate) in a quantity-value table. The **quantity-value table** is *characterized by multiplying the quantity Q of each item by the value V (or rate) of each item to obtain the item's total value, Q · V.*

We use quantity-value tables to organize data in Examples 6 and 7 and to build a system of equations in Example 8.

EXAMPLE   Setting up the quantity-value table: interest on debt   Suppose Heidi has a $12,500 student loan at 7.5%, $1200 in credit card debt at 21%, and an $8000 car loan at 8.9%. How much in total has she borrowed? What is the total annual interest paid? What is the average annual rate of interest for the loans? Organize the information in a quantity-value table.

Solution   The information is organized in Table 4.

Type of Loan	Quantity: Amount of Loan (in dollars)	Value: Rate of Interest	Q · V: Interest Paid (in dollars)
Student loan	12,500	0.075	937.50
Credit card balance	1200	0.21	252.00
Car loan	8000	0.089	712.00
Total	21,700		1901.50

Table 4

Heidi has borrowed $21,700. The total annual interest paid is $1901.50. The average annual interest rate is calculated by dividing total interest paid by the total amount of the loans:

$$\text{Average rate} = \frac{\text{interest paid}}{\text{total loans}} = \frac{1901.50}{21,700.00} \approx 0.088$$

The average annual interest rate paid on the loans is about 8.8%. The 8.8% could be placed in the last row of the value column in Table 4.   ●

The value or rate may also be the number assigned to a specific outcome or activity. For example, the grade point for a specific letter grade is an example of a rate based on an outcome: 4 points per credit for an A, 3 points per credit for a B, 2 points per credit for a C, and so forth.

EXAMPLE   Setting up a quantity-value table: GPA calculations   What is a student's grade point average (GPA) if she received an A in a 3-credit psychology course, a B in a 4-credit trigonometry course, and a C in a 4-credit computer science course? Show the information in a quantity-value table.

Solution   In Table 5, we set up a table with the quantity, the value, and their product in separate columns.

Subject and Grade	Quantity (credits)	Value (points per credit)	$Q \cdot V$ (points)
Psychology, A	3	4	12
Trigonometry, B	4	3	12
Computer science, C	4	2	8
Total	11		32

Table 5

To calculate the grade point average, we add the items in the $Q \cdot V$ column in Table 5 and divide by the total credits:

$$\text{GPA} = \frac{\text{total } Q \cdot V}{\text{total } Q} = \frac{32 \text{ points}}{11 \text{ credits}} = 2.91 \text{ points per credit}$$

If we were to place the GPA into the table, it would go into the last row of the value column.   ●

In many settings, we use a quantity-value table to find a given blended value instead of an average. In Example 8, we set up a quantity-value table and use it to write equations to determine quantities needed for a specified blend.

EXAMPLE **8**

Setting up equations with a quantity-value table: dog food blend   How many pounds of dog food A, containing 16% protein, need to be blended with dog food B, containing 5% protein, to obtain a 500-pound blend with 7% protein? Set up equations and solve with substitution. Then check with a graph.

Solution   We will let x be the number of pounds of dog food A and y be the number of pounds of dog food B.

Dog Food	Quantity (pounds)	Value (percent protein, converted to decimal)	$Q \cdot V$ (pounds of protein)
A	x	0.16	$0.16x$
B	y	0.05	$0.05y$
Total	500	0.07	$(500)(0.07) = 35$

Table 6

The average, or the blended value, is 7%. We multiply 500 pounds by 7% to obtain the sum of the $Q \cdot V$ column in Table 6: 35 pounds.

There are two equations within the table. From the quantity column, we have

$$x + y = 500$$

From the $Q \cdot V$ column, we have

$$0.16x + 0.05y = 35$$

We obtain $y = 500 - x$ from the first equation, and we substitute into the second equation:

$$0.16x + 0.05y = 35 \qquad \text{Substitute } y = 500 - x.$$

$$0.16x + 0.05(500 - x) = 35 \qquad \text{Apply the distributive property.}$$

$$0.16x + 25 - 0.05x = 35 \qquad \text{Add like terms, and subtract 25 from both sides.}$$

$$0.11x = 10 \qquad \text{Divide by 0.11.}$$

$$x \approx 90.9 \text{ lb of dog food A}$$

$$y = 500 - x$$

$$y \approx 409.1 \text{ lb of dog food B}$$

Because the blend has a slightly higher protein level than dog food B, a relatively smaller amount of dog food A is needed.

The graphs of the two equations are shown in Figure 6. The graphs intersect near (100, 400), in agreement with the symbolic work.

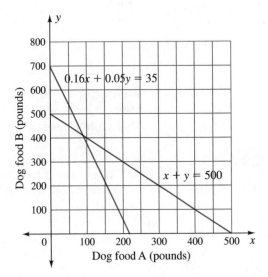

Figure 6

Summary: Quantity-Value Tables

- Adding the items in the quantity column gives the total quantity.
- Adding the items in the $Q \cdot V$ column gives the same number as multiplying the total quantity by the average value (across the last row).
- The total of the $Q \cdot V$ column divided by the total of the quantity column gives an average value. The last number in the value column is an average value and is not found by adding the items in the value column.

ANSWER BOX

Warm-up: 1. $y = -\frac{3}{2}x + 8$ **2.** $y = \frac{2}{3}x - 2$

EXERCISES

Solve the systems of equations in Exercises 1 to 12 by graphing. For each system that describes coincident or parallel lines, describe the result of solving the system by substitution or elimination.

1. $x + y = 5$
$x - y = -3$

2. $x - y = 6$
$y - 3 = x$

3. $2x - 3 = 2y$
$x - y = 1.5$

4. $y - x = 4$
$y + x = 6$

5. $2y = 1.5x + 1$
$4y - 3x + 8 = 0$

6. $y - 2x = 3$
$y + x = -2$

7. $x + 2y = 4$
$x - y = 3$

8. $6y - 4x = 3$
$y = \dfrac{2x}{3} + \dfrac{1}{2}$

9. $2x - y = 4$
$4x + y = -1$

10. $y = \dfrac{2x}{3} - 2$
$3y - 2x = 4$

11. $y = \dfrac{x}{3} + 2$
$3y - 6 = x$

12. $3x + 2y = -2$
$3y + x = \frac{1}{2}$

Identify the quantity and value words and numbers in Exercises 13 to 18.

13. Invest $1600 at 5% annual interest and $12,000 at 8% annual interest.

14. Blend 200 kilograms of Salvadoran coffee at $15.40 per kilogram with 300 kilograms of Guatemalan coffee at $18.70 per kilogram.

15. Earn $6.50 per hour for 20 hours at the first job and $7.25 per hour for 15 hours at the second job.

16. Sell 200 five-pound bags of onions for $2.98 each and 100 ten-pound bags for $4.49 each.

17. Blend 100 milliliters of 8% solution with 1000 milliliters of water at 0% solution.

18. Earn 10 points for first place, 5 points for second place, and 1 point for third place. Team A had 3 first-place finishes and 2 third-place finishes. Team B had 1 first-place finish and 5 second-place finishes.

In Exercises 19 to 22, what question might be asked about the sentence? What facts must be added by the reader because they are not stated in the sentence?

19. Samantha has 15 dimes and 12 quarters.

20. Jessika has 25 pennies and 12 nickels.

21. An isosceles triangle has one angle 10° larger than the two equal angles.

22. A parallelogram has one interior angle three times another interior angle.

Use quantity-value tables as needed to set up equations in Exercises 23 to 34. Use either substitution or graphing to answer the questions.

23. What is the GPA of a student who earns an A in computer science (4 credit hours), a B in psychology (3 credit hours), a C in English composition (3 credit hours), and a B in physical education (1 credit hour)?

24. What is the GPA of a student with 22 credit hours at a grade of A, 37 credit hours at a grade of B, and 16 credit hours at a grade of C?

25. Suppose that a student has a 3.08 GPA with 75 credits. He wants to raise his GPA to a 3.25. How many credit hours at a grade of A does he need?

26. A student has 3 hours at a grade of A, 30 hours at a grade of B, and 5 hours at a grade of C. How many more hours at a grade of B does she need to raise her GPA to a 3.00?

27. How many pounds of dog food A, containing 15% protein, need to be blended with dog food B, containing 10% protein, to obtain a 1000-pound blend with 12% protein?

28. How many pounds of cat food A, containing 8% protein, need to be blended with cat food B, containing 13% protein, to obtain a 500-pound blend with 9% protein?

29. An inheritance of $75,000 is to be split and invested at 6% and 9% interest to produce an average 8% interest rate. How much money is to be invested at each rate?

30. A personal bank loan costs 12% interest, and a credit card loan costs 20% interest. How should $10,000 be borrowed to keep the interest rate at 15%?

31. A small business averages 13% interest on loans totaling $10,000. Credit card debt is 21%, and bank interest is 11%. How much has been borrowed from each source?

32. A money market account pays 5% interest, and a savings account pays 3% interest. How much is saved in each account if the average interest on the $2400 total savings is 4.5%?

33. A doctor prescribes 0.5 liter of a 5% glucose solution. How much distilled water (0% glucose) and 50% glucose solution must be blended to obtain the proper solution?

34. A nurse's aide must prepare 4000 milliliters of a 0.5% potassium permanganate solution for an astringent. How much distilled water and 4% potassium permanganate solution must be blended?

Salespeople are often paid a small salary plus a percent commission on sales. In Exercises 35 to 40, find the level of sales for which the two compensation packages are equal.

Package 1 is $250 plus 10% of sales.
Package 2 is $350 plus 8% of sales.
Package 3 is $550 plus 4% of sales.
Package 4 is $750 plus 4% of sales.
Package 5 is $550 plus 6% of sales.

35. Packages 1 and 2   **36.** Packages 2 and 3

37. Packages 3 and 4   **38.** Packages 3 and 5

39. Packages 1 and 5   **40.** Packages 4 and 5

41. If we increase the percent of sales in a compensation package, how does its graph change?

42. If we increase the fixed amount in a compensation package, how does its graph change?

43. Describe how you might write a pair of equations with no solution.

44. Describe how you might write a pair of equations with an infinite number of solutions.

Projects

45. *Linear Equations.* Conditions on the constants a, b, and c in $ax + by = c$ are given below. Match each set of conditions with one of the linear graphs described in parts a to f.

Set 1: $a = 0, b > 0, c > 0$
Set 2: $b = 0, a > 0, c > 0$
Set 3: $c = 0, a > 0, b > 0$
Set 4: $b = 1, a > 0, c > 0$
Set 5: $a = 1, b > 0, c > 0$
Set 6: $b = -1, a > 0, c > 0$

a. The graph has a y-intercept at $(0, c)$ and slope $-a$.

b. The graph is a horizontal line.

c. The graph passes through the origin with slope $-a/b$.

d. The graph has a y-intercept at $(0, c/b)$ and slope $-1/b$.

e. The graph has a y-intercept at $(0, -c)$ and slope a.

f. The graph is a vertical line.

46. *Hiker's Dilemma.* One beautiful morning, Kristin starts at the rim of the Grand Canyon and hikes down to Phantom Ranch on the Colorado River, arriving in the afternoon. The next morning, she leaves the ranch and returns to the rim, on the same trail.

a. Explain why there is a point on the trail at which she arrives at the same time on both days. Although it does not matter, you may assume a uniform rate of travel walking down the trail and another uniform rate walking up the trail.

b. Describe times of departure and arrival that would make it impossible for her to arrive at one point at the same time on both days.

8.2 Solving Systems of Two Linear Equations with Matrices _____

OBJECTIVES

• Multiply matrices by hand and with a calculator.

• Find the determinant of a matrix by hand and with a calculator.

• Translate a system of linear equations into a matrix equation.

• Use the determinant to identify systems of linear equations with a unique solution.

• Use a calculator to solve a matrix equation, if a solution exists.

• Use algebra and/or graphing to find whether a system without a unique solution represents parallel or coincident lines.

IN THIS SECTION, we explore matrices and the application of calculator matrix operations to solving a system of two linear equations. These examples use systems of two linear equations in two variables so that the properties of the systems can be verified graphically. In Section 8.3, matrix operations are extended to systems of three or more linear equations.

Matrices

Matrices are a powerful tool in mathematics. In this text, we will use them to provide a general procedure for solving a system of linear equations, just as the quadratic formula

$$x = \frac{-b \pm \sqrt{b^2 - 4ac}}{2a}$$

provides a general procedure for solving equations of the form $ax^2 + bx + c = 0$. After a brief introduction to matrices, we will use matrix operations on a calculator to solve linear equations.

A **matrix** is *an array of numbers in rows and columns placed between brackets. The number of rows and the number of columns* determine the **shape of a matrix,** $r \times c$. The matrix below has two rows and three columns.

	column 1	column 2	column 3
row 1	a	b	c
row 2	d	e	f

The matrix is said to be a 2 by 3, or 2×3, matrix. When we refer to a matrix without the word *matrix*, we place its name in brackets. Thus, the array of numbers for matrix A will have the symbol $[A]$.

ADDITION AND SUBTRACTION OF MATRICES   As with other sets of numbers, operations such as addition, subtraction, and multiplication are defined for matrices.

In order *to be added (or subtracted), matrices must have the same shape*—that is, the same number of rows and columns. We add by adding numbers in the same position inside the matrix. We subtract by subtracting numbers in the same position inside the matrix.

Here is how we add 2×1 matrices:

$$\begin{bmatrix} a \\ b \end{bmatrix} + \begin{bmatrix} c \\ d \end{bmatrix} = \begin{bmatrix} a + c \\ b + d \end{bmatrix}$$

Here is how we subtract 1×3 matrices:

$$[a \quad b \quad c] - [d \quad e \quad f] = [a - d \quad b - e \quad c - f]$$

We will not need addition and subtraction in order to solve a system of equations. The operations are included here only for completeness of presentation.

MULTIPLICATION OF MATRICES   The expression $[A][X]$ implies multiplication of $[A]$ and $[X]$. The product $[A][X]$, where

$$[A] = \begin{bmatrix} a & b \\ c & d \end{bmatrix} \quad \text{and} \quad [X] = \begin{bmatrix} x \\ y \end{bmatrix}$$

is precisely the left side of the system of equations

$$ax + by = e$$
$$cx + dy = f$$

Thus,

$$[A][X] = \begin{bmatrix} a & b \\ c & d \end{bmatrix}\begin{bmatrix} x \\ y \end{bmatrix} = \begin{bmatrix} ax + by \\ cx + dy \end{bmatrix}$$

EXAMPLE **1**   Multiplying matrices   Multiply these matrices.

a. $\begin{bmatrix} 1 & 1 \\ 1 & -2 \end{bmatrix}\begin{bmatrix} x \\ y \end{bmatrix}$ **b.** $\begin{bmatrix} -2 & 1 \\ -0.5 & 1 \end{bmatrix}\begin{bmatrix} x \\ y \end{bmatrix}$

Solution   **a.** $\begin{bmatrix} 1 & 1 \\ 1 & -2 \end{bmatrix}\begin{bmatrix} x \\ y \end{bmatrix} = \begin{bmatrix} 1x + 1y \\ 1x - 2y \end{bmatrix}$

b. $\begin{bmatrix} -2 & 1 \\ -0.5 & 1 \end{bmatrix}\begin{bmatrix} x \\ y \end{bmatrix} = \begin{bmatrix} -2x + 1y \\ -0.5x + 1y \end{bmatrix}$ ●

To multiply a 2×2 matrix by another 2×2 matrix, we use the following rule:

$$\begin{bmatrix} a & b \\ c & d \end{bmatrix}\begin{bmatrix} e & f \\ g & h \end{bmatrix} = \begin{bmatrix} ae + bg & af + bh \\ ce + dg & cf + dh \end{bmatrix}$$

EXAMPLE **2**   Multiplying matrices   Multiply:

$$\begin{bmatrix} 1 & 2 \\ 3 & 4 \end{bmatrix}\begin{bmatrix} p & q \\ r & s \end{bmatrix}$$

Solution $\begin{bmatrix} 1 & 2 \\ 3 & 4 \end{bmatrix}\begin{bmatrix} p & q \\ r & s \end{bmatrix} = \begin{bmatrix} 1p + 2r & 1q + 2s \\ 3p + 4r & 3q + 4s \end{bmatrix}$ ●

Note that we multiply a row in the first matrix times a column in the second matrix. The product of the first numbers in the row and column is added to the product of the second numbers.

IDENTITIES AND INVERSES   In arithmetic, when we multiply any number by 1 we get the same number back. When we multiply a number by its multiplicative inverse, or reciprocal, we get 1. That is,

$$a \cdot 1 = a \qquad 4 \cdot 1 = 4 \qquad \text{I is the identity.}$$

$$a \cdot \frac{1}{a} = 1 \qquad 4 \cdot \frac{1}{4} = 1 \qquad \frac{\text{I}}{n} \text{ is the inverse.}$$

In matrices, when we multiply a matrix by an **identity matrix,** *I,* we *get the same matrix back.* When we multiply a matrix by its **matrix inverse,** we *get an identity matrix, I.* For our work in this section, the 2×2 identity matrix will be written

$$I = \begin{bmatrix} 1 & 0 \\ 0 & 1 \end{bmatrix}$$

The numbers in a matrix inverse will depend on the original matrix, but the symbol for the matrix will be labeled with a superscript -1. Thus, *the matrix inverse of* $[A]$ *is* $[A]^{-1}$, if the inverse exists. Although the -1 superscript is in the same position as an exponent, it is not an exponent. In most cases, we will use a calculator to find the matrix inverse. The instructions to Exercises 29 to 32 show how to find the inverse of a 2×2 matrix by hand.

In Example 3, we check that multiplying a matrix by an identity matrix I yields the original matrix and that multiplying a matrix by its inverse gives an identity matrix I.

EXAMPLE 3   Multiplying matrices   For the matrices $[A] = \begin{bmatrix} 1 & -1 \\ -3 & 2 \end{bmatrix}$, $[A]^{-1} = \begin{bmatrix} -2 & -1 \\ -3 & -1 \end{bmatrix}$, and $[I] = \begin{bmatrix} 1 & 0 \\ 0 & 1 \end{bmatrix}$, calculate

a. $[A][I]$ **b.** $[A][A]^{-1}$ **c.** $[A]^{-1}[A]$

Solution   **a.** $[A][I] = \begin{bmatrix} 1 & -1 \\ -3 & 2 \end{bmatrix}\begin{bmatrix} 1 & 0 \\ 0 & 1 \end{bmatrix}$

$$= \begin{bmatrix} 1 \cdot 1 + -1 \cdot 0 & 1 \cdot 0 + -1 \cdot 1 \\ -3 \cdot 1 + 2 \cdot 0 & -3 \cdot 0 + 2 \cdot 1 \end{bmatrix}$$

$$= \begin{bmatrix} 1 & -1 \\ -3 & 2 \end{bmatrix} = [A]$$

b. $[A][A]^{-1} = \begin{bmatrix} 1 & -1 \\ -3 & 2 \end{bmatrix}\begin{bmatrix} -2 & -1 \\ -3 & -1 \end{bmatrix}$

$$= \begin{bmatrix} 1 \cdot (-2) + (-1) \cdot (-3) & 1 \cdot (-1) + (-1) \cdot (-1) \\ -3 \cdot (-2) + 2 \cdot (-3) & -3 \cdot (-1) + 2 \cdot (-1) \end{bmatrix}$$

$$= \begin{bmatrix} 1 & 0 \\ 0 & 1 \end{bmatrix} = [I]$$

c. $[A]^{-1}[A] = \begin{bmatrix} -2 & -1 \\ -3 & -1 \end{bmatrix}\begin{bmatrix} 1 & -1 \\ -3 & 2 \end{bmatrix}$

$$= \begin{bmatrix} -2 \cdot 1 + (-1) \cdot (-3) & -2 \cdot (-1) + (-1) \cdot 2 \\ -3 \cdot 1 + (-1) \cdot (-3) & -3 \cdot (-1) + (-1) \cdot 2 \end{bmatrix}$$

$$= \begin{bmatrix} 1 & 0 \\ 0 & 1 \end{bmatrix} = [I]$$

The results in Example 3 are consistent with our definitions:

• The product of a matrix and an identity matrix is the original matrix:

$$[A][I] = [A]$$

• The product of a matrix and its inverse is an identity matrix:

$$[A][A]^{-1} = [A]^{-1}[A] = I$$

CALCULATING THE DETERMINANT   There is no reciprocal (or multiplicative inverse) for zero because $\frac{1}{0}$ is undefined. We have a similar situation with matrices. In the introduction to the matrix inverse, note the phrase *if the inverse exists.*

There are times when the matrix inverse has the matrix equivalent of a zero in the denominator. Thus, we say that

If [A] has a determinant of zero, then $[A]^{-1}$ is undefined.

The **determinant** is a *value*, defined only for square matrices, *that determines whether the corresponding system of equations has a unique solution.* How we calculate the determinant, det [A], depends on the size of the matrix. For a 2 × 2 matrix,

$$\text{If} \quad [A] = \begin{bmatrix} a & b \\ c & d \end{bmatrix}, \quad \text{then det } [A] = ad - bc.$$

A nonzero value for the determinant of matrix *A* tells us that the inverse, $[A]^{-1}$, exists.

EXAMPLE **4**   Finding the determinant   Calculate the determinant for these matrices:

a. $\begin{bmatrix} 1 & 1 \\ 1 & -2 \end{bmatrix}$ **b.** $\begin{bmatrix} -2 & 1 \\ -0.5 & 1 \end{bmatrix}$ **c.** $\begin{bmatrix} 2 & 1 \\ 2 & 1 \end{bmatrix}$

Solution   **a.** The value of the determinant is $1(-2) - 1(1) = -3$.

b. The value of the determinant is $-2(1) - (-0.5)(1) = -1.5$.

c. The value of the determinant is $2(1) - 2(1) = 0$.   ●

Solving a System of Equations

As in solving quadratic equations, the first step in the matrix solution of linear equations is to place the equations in standard form.

Standard Form of a Linear Equation

All linear equations in two variables may be written as

$$ax + by = e$$

where *a*, *b*, and *e* are real numbers.

(The letter *c* has been changed here to the letter *e* so that we can more clearly distinguish the left and right sides of equations in subsequent steps.)

We arrange the equations in standard form, with the variable terms on the left and the constant terms on the right:

$$ax + by = e$$
$$cx + dy = f$$

We are now ready to change the system of equations in standard form to matrix form.

SETTING UP THE MATRIX EQUATION   A **matrix equation** *contains matrices instead of coefficients, variables, and constants.* We use the symbols $[A][X] = [B]$ to describe the matrix equation in which [A] is the matrix of coefficients on the variables, [X] is the matrix of variables, and [B] is the matrix of constants.

Matrix *A* is the **matrix of coefficients** and *contains the coefficients of the variable terms from the left side of each equation.* If one variable is missing, we write zero as its coefficient. The shape of matrix *A* is 2 by 2, meaning two rows and two columns. We write matrix A by placing the coefficients of the variables in the form

$$[A] = \begin{bmatrix} a & b \\ c & d \end{bmatrix}$$

Matrix X is the **matrix of variables** and *contains the two variables x and y*:

$$[X] = \begin{bmatrix} x \\ y \end{bmatrix}$$

The shape of matrix X is 2 by 1, meaning two rows and one column.

Matrix B is the **matrix of constants** and *contains the constant term from the right side of each equation*. The shape of matrix B is 2 by 1. We write matrix B by placing the constant terms in the form

$$[B] = \begin{bmatrix} e \\ f \end{bmatrix}$$

We can combine the matrices above into the matrix equation

$$\begin{bmatrix} a & b \\ c & d \end{bmatrix} \begin{bmatrix} x \\ y \end{bmatrix} = \begin{bmatrix} e \\ f \end{bmatrix}$$

A shorter way to write the matrix equation is $[A][X] = [B]$.

EXAMPLE **5** Writing matrix equations   Write these systems of equations as matrix equations.

a. $x + y = 3$ **b.** $y = 2x + 4$ **c.** $y = -2x + 3$
   $x - 2y = 6$ $y = 0.5x - 2$ $y + 2x = 4$

Solution **a.** The equations are already in standard form:

$$x + y = 3$$
$$x - 2y = 6$$

The matrix equation is

$$\begin{bmatrix} 1 & 1 \\ 1 & -2 \end{bmatrix} \begin{bmatrix} x \\ y \end{bmatrix} = \begin{bmatrix} 3 \\ 6 \end{bmatrix}$$

b. The equations must first be arranged in standard form. It is convenient to move the x terms to the left side:

$$-2x + y = 4$$
$$-0.5x + y = -2$$

The matrix equation is

$$\begin{bmatrix} -2 & 1 \\ -0.5 & 1 \end{bmatrix} \begin{bmatrix} x \\ y \end{bmatrix} = \begin{bmatrix} 4 \\ -2 \end{bmatrix}$$

c. The equations may be written in standard form as

$$2x + y = 3$$
$$2x + y = 4$$

The matrix equation is

$$\begin{bmatrix} 2 & 1 \\ 2 & 1 \end{bmatrix} \begin{bmatrix} x \\ y \end{bmatrix} = \begin{bmatrix} 3 \\ 4 \end{bmatrix}$$

SOLVING THE MATRIX EQUATION   To solve a matrix equation, we must do the matrix equivalent of dividing both sides of $[A][X] = [B]$ by the matrix A. It is nearly that simple on a calculator, but we do need to look at the steps leading to the calculator solution. The calculator solution relies on *inverses* and *identities*.

As an introduction to the matrix process, look carefully at the role played by the reciprocal, or multiplicative inverse, in Example 6.

EXAMPLE **6**

Solving with inverses   Solve $\frac{2}{3}x = 50$.

Solution

$$\frac{2}{3}x = 50 \qquad \text{Multiply both sides by the reciprocal, } \tfrac{3}{2}.$$
$$\frac{3}{2} \cdot \frac{2}{3}x = \frac{3}{2} \cdot 50 \qquad \text{Simplify.}$$
$$x = 75$$

●

In Example 6, we multiplied by the multiplicative inverse of $\frac{2}{3}$ to solve the equation. To solve a matrix system $[A][X] = [B]$, we multiply both sides by the inverse of matrix A, $[A]^{-1}$, if the inverse exists.

$$[A][X] = [B]$$
$$[A]^{-1}[A][X] = [A]^{-1}[B]$$
$$[I][X] = [A]^{-1}[B]$$
$$[X] = [A]^{-1}[B]$$

Student Note: The multiplication by $[A]^{-1}$ is on the left on both sides—more about this in the exercises.

Thus, the solution to $[A][X] = [B]$ is $[X] = [A]^{-1}[B]$.

Using a Calculator to Solve $[A][X] = [B]$

To solve a system of linear equations with matrices, we take the numerical coefficients and constants from the equations, place them in matrices, and then solve on the calculator. Here is the process:

Matrix Solution of a System of Linear Equations

> 1. Set up the matrix equation $[A][X] = [B]$.
> 2. Enter $[A]$ into the calculator.
> 3. Enter $[B]$ into the calculator.
> 4. Calculate det $[A]$. If det $[A] = 0$, stop; if det $[A] \neq 0$, proceed to step 5.
> 5. Calculate $[A]^{-1}[B]$, and record the answer.

EXAMPLE **7**

Solving with matrices   Solve the system of equations with matrices.

$$x - y = 3$$
$$2y - 3x = -6$$

Solution

To set up the matrix equation, we first arrange the equations in standard form. In this case, only the second equation must be rearranged.

$$x - y = 3$$
$$-3x + 2y = -6$$

We select the coefficients from the variables and place them in matrix A:

$$[A] = \begin{bmatrix} 1 & -1 \\ -3 & 2 \end{bmatrix} \qquad \text{$[A]$ is a 2-row by 2-column matrix.}$$

We write the variables in matrix X:

$$[X] = \begin{bmatrix} x \\ y \end{bmatrix} \qquad \text{$[X]$ is a 2-row by 1-column matrix.}$$

We take the constants from the right side and place them in matrix B:

$$[B] = \begin{bmatrix} 3 \\ -6 \end{bmatrix} \qquad \text{[B] is a 2-row by 1-column matrix.}$$

We then place $[A]$, $[X]$, and $[B]$ into the matrix equation $[A][X] = [B]$:

$$\begin{bmatrix} 1 & -1 \\ -3 & 2 \end{bmatrix} \begin{bmatrix} x \\ y \end{bmatrix} = \begin{bmatrix} 3 \\ -6 \end{bmatrix}$$

After entering $[A]$ and $[B]$ into the calculator, we calculate det $[A]$. The determinant of A, det $[A]$, is -1. Because it is not equal to zero, we can continue with the solution process.

The next step is to calculate $[A]^{-1}[B]$ and set the answer equal to the variable matrix $[X]$:

$$[X] = \begin{bmatrix} x \\ y \end{bmatrix} = \begin{bmatrix} 0 \\ -3 \end{bmatrix}$$

The solution to the system is $x = 0$ and $y = -3$. This means that the graphs of the equations intersect at $(0, -3)$. ●

The following box gives more details on the graphing calculator steps for the matrix solution.

Graphing Calculator Technique: Matrix Solution of a System of Linear Equations

To solve the system $x - y = 3$ and $2y - 3x = -6$ with matrices, follow this general procedure, modifying it as necessary to fit your calculator.

1. Set up the matrix equation $[A][X] = [B]$ by hand. (See Example 7.)

2. Enter matrix A into the calculator.

   a. Under the matrix menu, choose EDIT or NEW.

   b. Enter the shape of matrix A as 2×2.

   c. Enter the matrix A numbers in the order 1, -1, -3, 2. Usually ENTER is needed after each number.

   d. Leave with QUIT or EXIT.

3. Enter matrix B into the calculator.

   a. Under the matrix menu, choose EDIT or NEW.

   b. Enter the shape of matrix B as 2×1.

   c. Enter the matrix B numbers in the order 3, -6. Usually ENTER is needed after each number.

   d. Leave with QUIT or EXIT.

4. Calculate det $[A]$.

   a. Under the matrix menu or an option within it, find the determinant, det.

   b. Under the matrix menu, select $[A]$, and press ENTER to evaluate the determinant.

   c. If det $[A] = 0$, stop; if det $[A] \neq 0$, proceed to step 5.

5. Calculate $[A]^{-1}[B]$.

   a. Under the matrix menu, select $[A]$.

   b. Obtain $^{-1}$ with the $\boxed{x^{-1}}$ key. **Do not use $\boxed{\wedge}$ $\boxed{+/-}$ 1; it will not work.**

   c. Under the matrix menu, select $[B]$.

   d. Press ENTER to evaluate.

   e. Record the answer.

EXAMPLE   Solving with matrices   Solve the systems of equations in parts a and b of Example 5 with calculator matrix operations.

a. $x + y = 3$ **b.** $y = 2x + 4$
$\quad x - 2y = 6$ $\quad y = 0.5x - 2$

Solution   **a.** For $x + y = 3$ and $x - 2y = 6$, the matrix equation $[A][X] = [B]$ is

$$\begin{bmatrix} 1 & 1 \\ 1 & -2 \end{bmatrix} \begin{bmatrix} x \\ y \end{bmatrix} = \begin{bmatrix} 3 \\ 6 \end{bmatrix}$$

We enter matrix A, the 2×2 matrix

$$[A] = \begin{bmatrix} 1 & 1 \\ 1 & -2 \end{bmatrix}$$

Then we enter matrix B, the 2×1 matrix

$$[B] = \begin{bmatrix} 3 \\ 6 \end{bmatrix}$$

Calculation of det $[A]$ yields

det $[A] = -3$

Because det $[A] \neq 0$, we may continue with the solution process.
   Calculation of $[A]^{-1}[B]$ gives a 2×1 matrix equal to the matrix of variables:

$$[X] = \begin{bmatrix} x \\ y \end{bmatrix} = \begin{bmatrix} 4 \\ -1 \end{bmatrix}$$

Matrix X shows that $x = 4$ and $y = -1$. The graphs of the equations intersect at $(4, -1)$.

b. For $y = 2x + 4$ and $y = 0.5x - 2$, the matrix equation $[A][X] = [B]$ is

$$\begin{bmatrix} -2 & 1 \\ -0.5 & 1 \end{bmatrix} \begin{bmatrix} x \\ y \end{bmatrix} = \begin{bmatrix} 4 \\ -2 \end{bmatrix}$$

We enter matrix A, the 2×2 matrix

$$[A] = \begin{bmatrix} -2 & 1 \\ -0.5 & 1 \end{bmatrix}$$

Then we enter matrix B, the 2×1 matrix

$$[B] = \begin{bmatrix} 4 \\ -2 \end{bmatrix}$$

Calculation of det $[A]$ yields

det $[A] = -1.5$

Because det $[A] \neq 0$, we may continue with the solution process.
   Calculation of $[A]^{-1}[B]$ gives a 2×1 matrix equal to the matrix of variables:

$$[X] = \begin{bmatrix} x \\ y \end{bmatrix} = \begin{bmatrix} -4 \\ -4 \end{bmatrix}$$

Matrix X shows that $x = -4$ and $y = -4$. The graphs of the equations intersect at $(-4, -4)$.   ●

W hat can go wrong? To solve the matrix equation on a calculator, we need the expression $[A]^{-1}[B]$. For a matrix of reasonable size, the calculator will find $[A]^{-1}$, if it exists. Unfortunately, not all matrices have inverses.

In order to have an inverse, a matrix A must meet two conditions:

1. $[A]$ must be a **square matrix,** *having the same number of rows and columns.*

2. The determinant of $[A]$, det $[A]$, cannot be zero.

Thus, as indicated above, it is important to always find whether or not a system has a unique solution by checking for a zero determinant value.

Example 9 shows how one older model calculator gives a false result when we seek a matrix solution without verifying the existence or uniqueness of the solution. Although this problem may not occur on more recent calculators, always remember to think critically about the results you obtain from technology.

EXAMPLE **9**   Exploring a false calculator solution

a. Set up a matrix solution of

$$3x - y = 2$$
$$-6x + 2y = -4$$

b. Enter the matrices A and B into the calculator.

c. Evaluate the determinant, $ad - bc$, by hand and with det $[A]$.

d. Find the solution $[X] = [A]^{-1}[B]$, if possible, with the calculator.

e. If a calculator solution exists, substitute it into the original equations and comment on the results.

Solution   **a.** The matrix system is

$$\begin{bmatrix} 3 & -1 \\ -6 & 2 \end{bmatrix} \begin{bmatrix} x \\ y \end{bmatrix} = \begin{bmatrix} 2 \\ -4 \end{bmatrix}$$

b. Matrix A is 2×2, and matrix B is 2×1.

c. Calculated by hand, the determinant is

$$ad - bc = 3(2) - (-1)(-6) = 0$$

On the older model calculator,

$$\det [A] \approx -6\text{E}-13$$

which is scientific notation for -6×10^{-13}. Newer model calculators give 0.

d. The zero value of the hand-calculated determinant indicates that the two linear equations represent either parallel or coincident lines. We must not proceed further with the calculator. However, we might not recognize that $-6\text{E}-13$ represents a number near zero and continue with a calculator solution. On the older model, $[X] = [A]^{-1}[B]$ is [2, 0], or the coordinate (2, 0). Newer models indicate a "Singular Matrix" error.

e. When we substitute the "solution" (2, 0) into the original equations, we obtain

$$3(2) - 0 \neq 2 \quad \text{and} \quad -6(2) + 2(0) \neq -4$$

Thus, the coordinate (2, 0) does not make either equation true. It is not a solution.   ●

Whenever the calculator gives a determinant value near zero, such as $-6E-13$, consider the result to be a warning that the equations may represent either parallel or coincident lines, and do not proceed with a matrix solution.

EXAMPLE **10**

Identifying the type of system

a. Plot the graphs of the equations from Example 9,

$$3x - y = 2$$
$$-6x + 2y = -4$$

Find whether they are parallel or coincident.

b. Show algebraically why the graphs are either parallel or coincident.

Solution

a. The graphs are shown in Figure 7. The graphs are coincident, and the equations represent the same line.

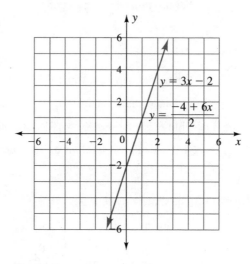

Figure 7

b. We solve the equations for y:

$$3x - y = 2$$
$$3x - 2 = y$$

and

$$-6x + 2y = -4$$
$$2y = 6x - 4 \qquad \text{Divide by 2.}$$
$$y = 3x - 2$$

The equations are the same. There should be an infinite number of solutions to the system; however, the result $(2, 0)$ from Example 9 is not one of them.

●

Summary: Solving a System of Linear Equations on the Calculator

1. Change the system of equations into a matrix equation.
2. Specify the shape of matrix A, and enter matrix A into the calculator.
3. Specify the shape of matrix B, and enter matrix B into the calculator.

4. Calculate the determinant of matrix A, det $[A]$. If det $[A] = 0$, stop; if det $[A] \neq 0$, go on to step 5.

5. Calculate $[A]^{-1}[B]$. The numbers are the values of the variables for the system of equations and appear in the same order as in the matrix of variables $[X]$.

ANSWER BOX

Warm-up: 1. $2x - 3y = -1$ **2.** $3x - 4y = 2$ **3.** $7x - 2y = 8$
4. $12x - 5y = -4$

EXERCISES 8.2

For Exercises 1 to 10, use the following matrices.

$$[A] = [2 \quad 3] \qquad [B] = [2 \quad 4] \qquad [C] = \begin{bmatrix} 1 & 2 & 3 \\ 4 & 5 & 6 \\ 7 & 8 & 9 \end{bmatrix}$$

$$[D] = \begin{bmatrix} 6 & 7 & 8 \\ 9 & 10 & 11 \\ 12 & 13 & 14 \end{bmatrix} \qquad [E] = \begin{bmatrix} 1 \\ 2 \\ 3 \end{bmatrix} \qquad [F] = \begin{bmatrix} 4 \\ 5 \\ 6 \end{bmatrix}$$

1. Describe the shape of matrix A.

2. Describe the shape of matrix C.

3. Describe the shape of matrix F.

4. How do we describe the shape of a matrix—row by column or column by row?

5. Add $[A]$ and $[B]$.

6. Add $[C]$ and $[D]$.

7. Add $[E]$ and $[F]$.

8. Find $[B] - [A]$.

9. Find $[D] - [C]$.

10. Find $[F] - [E]$.

For Exercises 11 to 34, use the following matrices.

$$[A] = \begin{bmatrix} 1 & 3 \\ -1 & -2 \end{bmatrix} \qquad [B] = \begin{bmatrix} 2 & 3 \\ 4 & 5 \end{bmatrix} \qquad [C] = \begin{bmatrix} 4 & -2 \\ 1 & 2 \end{bmatrix}$$

$$[D] = \begin{bmatrix} -2 & -1 \\ 1 & 2 \end{bmatrix} \qquad [E] = \begin{bmatrix} 2 \\ -1 \end{bmatrix} \qquad [X] = \begin{bmatrix} x \\ y \end{bmatrix}$$

Find the product of the matrices in Exercises 11 and 12 without using a calculator.

11. $[A][X]$

12. $[B][X]$

Find the product of the matrices in Exercises 13 to 16 by hand or using a calculator.

13. $[A][E]$

14. $[B][E]$

15. $[C][E]$

16. $[D][E]$

Find the products in Exercises 17 to 24. Do at least one without using a calculator.

17. $[A][B]$

18. $[A][C]$

19. $[B][A]$

20. $[C][A]$

21. $[C][D]$

22. $[B][C]$

23. $[D][C]$

24. $[C][B]$

In Exercises 25 to 28, find the determinant of each matrix by hand and then by calculator.

25. det $[A]$

26. det $[B]$

27. det $[C]$

28. det $[D]$

In Exercises 29 to 32, find the inverse by hand, and then check with a calculator. For a 2 × 2 matrix, the inverse of $[K]$, $[K]^{-1}$, may be calculated as follows:

if $[K] = \begin{bmatrix} a & b \\ c & d \end{bmatrix}$ and det $[K] = ad - bc$,

then $[K]^{-1} = \begin{bmatrix} \dfrac{d}{\det [K]} & \dfrac{-b}{\det [K]} \\ \dfrac{-c}{\det [K]} & \dfrac{a}{\det [K]} \end{bmatrix}$

29. $[A]^{-1}$

30. $[B]^{-1}$

31. $[C]^{-1}$

32. $[D]^{-1}$

33. When we multiply a matrix by its inverse, we obtain the identity matrix, *I*. Record the matrix obtained from

    a. $[A][A]^{-1}$ b. $[A]^{-1}[A]$

34. Why is there no inverse for matrix *E*?

Solve Exercises 35 to 42 with matrices. If no unique solution is possible, explain why.

35. $1x + 2y = 5$
    $4x + 4y = 6$

36. $1x + 2y = 5$
    $3x + 4y = 6$

37. $1x + 2y = 5$
    $2x + 4y = 6$

38. $1x + 2y = 5$
    $1x + 4y = 6$

39. $1x + 2y = 5$
    $0x + 4y = 6$

40. $1x + 2y = 5$
    $-1x + 4y = 6$

41. $1x + 2y = 5$
    $-2x + 4y = 6$

42. $1x + 2y = 5$
    $-3x + 4y = 6$

43. Describe what Exercises 35 to 42 have in common. For the line $ax + 4y = 6$, how will changing the coefficient *a* on *x* change the graph? What part of the graph of $ax + 4y = 6$ stays the same?

44. What will all the solutions (if they exist) to the systems in Exercises 35 to 42 have in common?

Solve Exercises 45 to 50 with matrices. If no unique solution is possible, explain why.

45. $5x = 28.2 + 2y$
    $3x + 5y = -5.4$

46. $4x + 3y = 25$
    $3x + 2.5 = 2y$

47. $3y = 2x + 12$
    $y + 2 = 2x/3$

48. $y = 0.25x + 2$
    $4y = x + 8$

49. $x - 1.5y = 3.5$
    $2x - 7 = 3y$

50. $4x + 2 = y$
    $x - 0.25y = 0.5$

In Exercises 51 to 60, set up equations. Quantity-value tables may be helpful. Solve by any method: substitution, elimination, graphing, or matrices.

51. How many kilograms of Honduran coffee at $18.70 per kilogram need to be blended with Indonesian coffee at $23.65 per kilogram to produce 200 kilograms of a blend selling for $19.80 per kilogram?

52. How many kilograms of Honduran coffee at $18.70 per kilogram need to be blended with Indonesian coffee at $23.65 per kilogram to produce 200 kilograms of a blend selling for $22.00 per kilogram?

53. How much hydrogen peroxide in a 3% solution must be blended with hydrogen peroxide in a 20% solution to obtain 1000 liters of a 6.4% solution?

54. How much 24% carbamide peroxide solution must be blended with 8% carbamide peroxide solution to obtain 500 liters of a 10% carbamide peroxide solution?

55. How much water and 20% hydrogen peroxide solution should be blended by a manufacturer to obtain 1000 gallons of a 3% hydrogen peroxide solution?

56. How much anhydrous glycerol containing no carbamide peroxide must be blended with a 24% carbamide peroxide solution to produce 200 gallons of a 10% carbamide peroxide solution suitable for treating mouth sores?

57. Suppose 9 dimes and 13 quarters weigh 3.5 ounces. On the same scale, 12 dimes and 9 quarters weigh 3 ounces. Estimate the weight of each coin.

58. Suppose 15 nickels and 7 copper-clad zinc pennies weigh 3.5 ounces. On the same scale, 5 nickels and 16 pennies weigh 2.5 ounces. Estimate the weight of each coin.

59. Farzaneh earns $1782 interest on a total of $23,600 placed in two investments. The investments earn 4.5% and 8.5%. How much money is in each investment?

60. Polly earns $2343 interest on a total of $29,800 placed in two investments. The investments earn 3.5% and 8.5%. How much money is in each investment?

61. In parts a to d, think about real numbers.

    a. What number *n* is the *additive identity*—that is, $a + n = a$?

    b. What number *n* is the *multiplicative identity*—that is, $a \cdot n = a$?

    c. What is the product of a number and its multiplicative inverse?

    d. What linear function is the *identity function*? (*Hint:* See Section 2.4.)

    e. Why do we call the matrix *I* in Exercise 33 the identity matrix?

62. a. Evaluate the determinant for $[E] = \begin{bmatrix} 2 & -1 \\ -4 & 2 \end{bmatrix}$.

    b. Evaluate $[E]^{-1}$ without a calculator. Describe the result.

    c. Evaluate $[E]^{-1}$ with a calculator. Explain the probable meaning of the phrase *singular matrix*.

63. What do we call the property of multiplication indicating that $4 \cdot 5 = 5 \cdot 4$? Examine the answers to Exercises 17 to 24. Is there a similar property for the multiplication of matrices? Explain.

Project

64. **Calculator Quirks.** As the solution to the system

    $$3x - y = 4 \quad \text{and} \quad -6x + 2y = -8$$

    one older calculator gives (10, 10), a false solution, while another model gives $(-2, -10)$, one of an infinite number of solutions. Set up the matrix equation, solve, and describe how your calculator reacts to the problem. Give three ordered pairs that do satisfy the system.

MID–CHAPTER TEST

In Exercises 1 and 2, solve the equation for y.

1. $5x - 2y = 4$

2. $2x - \frac{1}{2}y = 5$

In Exercises 3 and 4, solve the system of equations by either substitution or elimination. Show your work.

3. $2y - 3x = 8$
    $y = 1.5x - 3$

4. $3y = 2x - 6$
    $y = -\frac{5}{6}x + 1$

In Exercises 5 and 6, set up a matrix equation for the system of equations, enter [A] and [B], indicate the value of det [A], and then, if appropriate, calculate $[A]^{-1}[B]$. Explain whether the system is solvable by calculator.

5. $y = -0.8x - 1.8$
    $y = -1.2x + 3$

6. $y = 2.5x + 4$
    $2y - 5x = 8$

7. Write a set of equations whose graphs are two parallel lines, set up matrix A, and evaluate det $[A]$.

8. Wilt Chamberlain and Elgin Baylor scored a total of 54,568 points in their careers in the National Basketball Association. Wilt scored 8270 more points than Elgin. How many did each score?

9. The equation $y = ax^2 + bx + c$ is the equation of a transition curve between two hills. The slope of the curve at any point (x, y) is $m = 2ax + b$. At $(0, 600)$, there is a slope of -5%. The slope is to be 3% at $(2000, 580)$. What are the coefficients a, b, and c of the transition curve?

10. A nurse needs 1 pint of 2% hydrogen peroxide solution for a topical antiseptic. The over-the-counter bottle contains a 3% solution. How much water and 3% hydrogen peroxide solution need to be blended?

8.3 Matrix Solutions of Systems of Three or More Linear Equations

OBJECTIVES

- Identify a graphical system as containing graphs of consistent, inconsistent, or dependent equations.
- Solve a system of three or more linear equations with matrices on a calculator.
- Describe possible implications of a zero determinant.
- Set up and solve systems of equations for applications.

WARM-UP

Translate each sentence into an equation.

1. There are five times as many medium rabbits, m, as large rabbits, l.

2. There are three times as many small rabbits, s, as medium rabbits, m.

3. There are a third as many quarters, q, as dimes, d.

THIS SECTION INTRODUCES systems of linear equations containing three or more variables. We examine the formal names for systems of equations having no solution, one solution, and an infinite number of solutions. We look at possible graphs of these systems and find a matrix solution when a unique solution exists.

Linear Equations

The linear equation $y = mx + b$ or $ax + by = c$ is said to be of **degree 1** because *the highest exponent on the variables in any term is 1.* The graphs of these equations are lines formed by points described by ordered pairs. Although the graphs of these equations are straight lines, the word *linear* refers not to the straight line but to the exponent 1 on the variables.

If *a, b, c, d,* and *e* are constants, $ax + by + cz = d$ is a linear equation in three variables ($x, y,$ and z) and $aw + bx + cy + dz = e$ is a linear equation in four variables ($w, x, y,$ and z). A **linear equation** is *any equation that is of the first degree in its variable(s)*.

The graphs of linear equations in three variables are **planes,** or *infinite flat surfaces.* The *points on the planes are described by three numbers,* or **ordered triples.** When we graph three planes on a three-dimensional graph, there are many possible arrangements (see Figure 8). As a result, we need to use more general vocabulary than the words *intersecting, parallel,* and *coincident,* which summarized the graphs of two-variable linear equations.

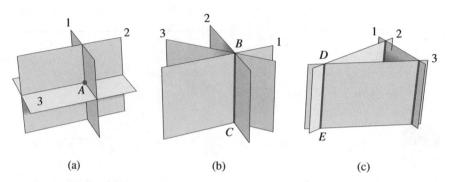

(a)   (b)   (c)

Figure 8

We describe the equations in terms of the number of solutions that the system contains.

1. If *a system of linear equations has exactly one solution,* the equations are **consistent.**

If the system of linear equations doesn't have exactly one solution, there are two possibilities:

2. If *a system of linear equations has no solution,* the equations are **inconsistent.**

3. If *a system of equations has an infinite number of solutions,* the equations are **dependent.**

EXAMPLE **1**   Identifying systems of linear equations   Identify as consistent, inconsistent, or dependent the two-variable linear equations whose graphs are

a. parallel lines

b. coincident lines

c. intersecting lines

Solution   **a.** Parallel lines have no intersection, and hence their equations have no solution. The equations describing parallel lines are inconsistent.

b. Coincident lines have all points in common, an infinite number of solutions, so their equations are dependent.

c. Intersecting lines have exactly one point of intersection, so their equations are consistent.   ●

EXAMPLE **2**   Identifying types of equations   Based on the intersections (or lack of intersections) of the three planes shown, identify the graphs in Figure 8 as representing consistent, inconsistent, or dependent linear equations.

Solution   In Figure 8(a), there is one point of intersection, point *A*. The equations describing these graphs are consistent.

In Figure 8(b), there is a line of intersection, *BC*, for all three planes. The line contains an infinite number of points and, hence, an infinite number of solutions. The equations describing these graphs are dependent.

In Figure 8(c), although there is an intersection line, *DE*, the line is not part of the plane labeled with a 1. There is no common intersection of all three planes and, hence, no solution to the system. The equations describing the graphs are inconsistent.   ●

Summary: Graphing Outcomes for a System of Three Linear Equations

In graphing a system of three linear equations, we have three possible outcomes:

1. The equations describe planes that intersect in exactly one point. The system of equations will have exactly one solution, described by the ordered triple specifying the point of intersection. The equations are consistent.

2. The equations describe planes that have no common intersection. The system will have no real-number solution. The equations are inconsistent.

3. The equations describe three planes that all intersect in a line. The system will have an infinite number of solutions because a line contains an infinite number of points. The equations are dependent.

Solving Systems of Three Equations

In Example 3, we return to the system of equations given in Example 7 of Section 8.0 and solve it with calculator matrices. The example reminds us that any letters can be variables.

EXAMPLE **3**   Solving a system of equations   Solve the system of equations

$$a + b + c = 3$$

$$2a + 3b + c = 13$$

$$2a - b = 0$$

a. Set up the matrix equation $[A][X] = [B]$.

b. Calculate the determinant of the matrix of coefficients, det $[A]$, and solve the system if possible.

Solution   **a.** The equations are already in standard form. We write the matrix equation directly from the equations:

$$\begin{bmatrix} 1 & 1 & 1 \\ 2 & 3 & 1 \\ 2 & -1 & 0 \end{bmatrix} \begin{bmatrix} a \\ b \\ c \end{bmatrix} = \begin{bmatrix} 3 \\ 13 \\ 0 \end{bmatrix}$$

$[A]$ is 3×3, while $[X]$ and $[B]$ are 3×1. We enter $[A]$ and $[B]$ into the calculator.

b. Because det $[A] = -5$, a nonzero number, it is possible to solve the system. We calculate $[A]^{-1}[B]$ to find the variable matrix, $[X]$.

$$[X] = \begin{bmatrix} a \\ b \\ c \end{bmatrix} = \begin{bmatrix} 2 \\ 4 \\ -3 \end{bmatrix}$$

Thus, $a = 2$, $b = 4$, and $c = -3$. There is a unique solution, so the equations in the system are consistent.   ●

When we solve a system of equations on a calculator, *we obtain a zero determinant, det $[A] = 0$, for both inconsistent and dependent equations. To find which type—inconsistent or dependent—the equations are, we will solve the system by hand.* Examples 4 and 5 explore an inconsistent system and a dependent system.

EXAMPLE   Exploring an inconsistent system   Solve the system of equations

$$x + y - 2z = 2 \qquad \text{(I)}$$
$$x - z = 4 \qquad \text{(2)}$$
$$y - z = 1 \qquad \text{(3)}$$

a. Set up a matrix equation $[A][X] = [B]$.

b. Calculate the determinant of the matrix of coefficients, det $[A]$, and solve the system if possible.

c. If the determinant is zero, solve the system algebraically to find whether the equations are inconsistent or dependent.

Solution   **a.** We arrange the equations in standard form, with zero coefficients on missing variables:

$$x + y - 2z = 2$$
$$x + 0y - z = 4$$
$$0x + y - z = 1$$

We set up the matrix equation:

$$\begin{bmatrix} 1 & 1 & -2 \\ 1 & 0 & -1 \\ 0 & 1 & -1 \end{bmatrix} \begin{bmatrix} x \\ y \\ z \end{bmatrix} = \begin{bmatrix} 2 \\ 4 \\ 1 \end{bmatrix}$$

$[A]$ is 3×3, while $[X]$ and $[B]$ are 3×1.

b. Because det $[A] = 0$, it is not possible to solve the system.

c. The variable z appears in all three equations. We solve the second and third equations in terms of z and substitute for x and y in the first equation.

$$x = 4 + z \qquad \text{(2)}$$
$$y = 1 + z \qquad \text{(3)}$$
$$ x + y - 2z = 2 \qquad \text{(I)}$$
$$(4 + z) + (1 + z) - 2z = 2$$
$$4 + z + 1 + z - 2z = 2 \qquad \text{The variable } z \text{ drops out.}$$
$$5 = 2$$

We obtain a false statement. There is no real-number solution to the system, so the equations are inconsistent. ●

As in Section 8.2, when the variable drops out, leaving a false statement, the system has no solutions and the equations are inconsistent. The graphs of the equations may be parallel planes, or they may be planes that intersect in such a way that they have no common point in all three planes, as in Figure 8(c).

EXAMPLE 5 Exploring a dependent system   Solve the system of equations

$$x - y - 2z = 3 \qquad (1)$$
$$x - z = 5 \qquad (2)$$
$$y + z = 2 \qquad (3)$$

a. Set up a matrix equation $[A][X] = [B]$.

b. Calculate the determinant of the matrix of coefficients, det $[A]$, and solve the system if possible.

c. If the determinant is zero, solve the system algebraically to find whether the equations are inconsistent or dependent.

Solution **a.** We arrange the equations in standard form:

$$x - y - 2z = 3$$
$$x + 0y - z = 5$$
$$0x + y + z = 2$$

We set the matrix equation:

$$\begin{bmatrix} 1 & -1 & -2 \\ 1 & 0 & -1 \\ 0 & 1 & 1 \end{bmatrix} \begin{bmatrix} x \\ y \\ z \end{bmatrix} = \begin{bmatrix} 3 \\ 5 \\ 2 \end{bmatrix}$$

$[A]$ is 3×3, while $[X]$ and $[B]$ are 3×1.

b. Because det $[A] = 0$, it is not possible to solve the system.

c. The variable z appears in all three equations. We solve the second and third equations in terms of z and substitute for x and y in the first equation.

$$x = 5 + z \qquad\qquad\qquad\qquad (2)$$
$$y = 2 - z \qquad\qquad\qquad\qquad (3)$$
$$x - y - 2z = 3 \qquad (1)$$
$$(5 + z) - (2 - z) - 2z = 3$$
$$5 + z - 2 + z - 2z = 3 \qquad \text{The variable } z \text{ drops out.}$$
$$3 = 3$$

We obtain a true statement. There are an infinite number of solutions to the system; hence, the equations are dependent. ●

When the variable drops out, leaving a true statement such as $3 = 3$, the system has an infinite number of solutions and the equations are dependent. In the three-variable case, the three equations may describe three planes intersecting in a line, as in Figure 8(b), or two or more equations may describe the same plane.

Table 7 summarizes the geometry and algebra associated with solving a system of three linear equations.

	Consistent Equations, det $[A] \neq 0$	**Inconsistent Equations, det $[A] = 0$**	**Dependent Equations, det $[A] = 0$**
Geometry	Planes intersect in one common point.	There is no common intersection.	There is linear or planar intersection.
Algebra	Equations can be solved to yield unique x, y, and z values.	The variables drop out, and the resulting statement is false.	The variables drop out, and the resulting statement is true.
Solution	There is one solution, an ordered triple.	There is no solution.	An infinite number of ordered triples satisfy the system.

Table 7

Applications

COIN PUZZLES   The coin puzzle in Example 6 illustrates that some information may not be given in a problem.

EXAMPLE **6**

Solving applications by matrices: counting coins   Alison has 61 coins altogether, for a total of $4.69. She has one-third as many quarters as dimes. She also has some pennies. How many of each coin does she have?

Solution

We let p = number of pennies, d = number of dimes, and q = number of quarters. The system of equations is

Student Note: In many problems, remember that there may be both quantity and value.

$p + d + q = 61$ This equation gives the quantity of coins.

$0.01p + 0.10d + 0.25q = \$4.69$ This equation gives the monetary value.

$3q = d$ If she has one-third as many quarters as dimes, then the number of quarters can be multiplied by 3 to get the number of dimes.

The matrix equation $[A][X] = [B]$ is

$$\begin{bmatrix} 1 & 1 & 1 \\ 0.01 & 0.10 & 0.25 \\ 0 & -1 & 3 \end{bmatrix} \begin{bmatrix} p \\ d \\ q \end{bmatrix} = \begin{bmatrix} 61 \\ 4.69 \\ 0 \end{bmatrix}$$

$[A]$ is 3×3, while $[X]$ and $[B]$ are 3×1. The value of the determinant, det $[A]$, is 0.51. The solution is $[X] = [A]^{-1}[B]$.

$$[X] = \begin{bmatrix} 29 \\ 24 \\ 8 \end{bmatrix} = \begin{bmatrix} p \\ d \\ q \end{bmatrix}$$

Alison has 29 pennies, 24 dimes, and 8 quarters.

Think about it 1: We used the equation $3q = d$ to describe dimes in terms of quarters. What equation describes quarters in terms of dimes?

CANDY PRODUCTION   In Examples 7 and 8, we return to the chocolate rabbits of the chapter opening.

EXAMPLE 7

Solving applications by matrices: chocolate rabbits   Fenton makes three sizes of chocolate rabbits (see Figure 1 on page 549). From prior sales records, he estimates that the store can sell five times as many small rabbits as medium rabbits and five times as many medium rabbits as large rabbits. If he wants to make 620 rabbits altogether, how many of each should he make? Identify variables, set up a system of equations, prepare a matrix equation, and solve.

Solution   We will let s = number of small rabbits, m = number of medium rabbits, and l = number of large rabbits. The equations are $s + m + l = 620$, $s = 5m$, and $m = 5l$. The system of equations is

$$
\begin{aligned}
s + m + l &= 620 \\
s - 5m + 0 &= 0 \\
0 + m - 5l &= 0
\end{aligned}
$$

The matrix equation $[A][X] = [B]$ is

$$
\begin{bmatrix} 1 & 1 & 1 \\ 1 & -5 & 0 \\ 0 & 1 & -5 \end{bmatrix}
\begin{bmatrix} s \\ m \\ l \end{bmatrix} =
\begin{bmatrix} 620 \\ 0 \\ 0 \end{bmatrix}
$$

$[A]$ is 3×3, while $[X]$ and $[B]$ are 3×1.

The value of the determinant, det $[A]$, is 31. It is nonzero, so we may proceed with a matrix solution. The solution is

$$
[X] = [A]^{-1}[B] = \begin{bmatrix} 500 \\ 100 \\ 20 \end{bmatrix} = \begin{bmatrix} s \\ m \\ l \end{bmatrix}
$$

Thus, Fenton should make 500 small rabbits, 100 medium rabbits, and 20 large rabbits. ●

EXAMPLE 8

Solving applications by matrices: more chocolate rabbits   As he begins production, Fenton finds that the small rabbits with their hand-painted eggs take longer to make than he had planned. He consults with his brother Lee, who suggests making the same total number of rabbits, but dropping to three times as many small rabbits as medium rabbits and limiting the large rabbits to 20. Set up a system of equations, prepare a matrix equation, and solve.

Solution   Again, we let s = number of small rabbits, m = number of medium rabbits, and l = number of large rabbits. The equations are $s + m + l = 620$, $s = 3m$, and $l = 20$. The system of equations is

$$
\begin{aligned}
s + m + l &= 620 \\
s - 3m + 0 &= 0 \\
0 + 0 + l &= 20
\end{aligned}
$$

The matrix equation $[A][X] = [B]$ is

$$
\begin{bmatrix} 1 & 1 & 1 \\ 1 & -3 & 0 \\ 0 & 0 & 1 \end{bmatrix}
\begin{bmatrix} s \\ m \\ l \end{bmatrix} =
\begin{bmatrix} 620 \\ 0 \\ 20 \end{bmatrix}
$$

[A] is 3×3, while [X] and [B] are 3×1. The value of the determinant, det [A], is -4. The solution is

$$[X] = [A]^{-1}[B] = \begin{bmatrix} 450 \\ 150 \\ 20 \end{bmatrix} = \begin{bmatrix} s \\ m \\ l \end{bmatrix}$$

This plan calls for making 450 small rabbits, 150 medium rabbits, and 20 large rabbits. ●

We will return to the candy-making scenario in the exercises, as we explore the meaning of multiplication of matrices.

FITTING POLYNOMIAL EQUATIONS TO DATA   Matrices have traditionally been used to fit polynomial equations to sets of data. Most graphing calculators can fit up to a fourth-degree, or quartic, equation with their regression options. The method in Example 9 can be employed to fit equations of higher degree by using $n + 1$ points for an nth-degree equation. In Example 9, we find a fourth-degree equation, so we can check our results with the calculator regression option. Observe that the first step in Example 9 is essentially the same as substituting an ordered pair into the roadbed transition curve equation.

EXAMPLE **9**

Solving applications by matrices   Fit a quartic (fourth-degree polynomial) equation to the points $(-2, 1)$, $(0, 2)$, $(1, -1)$, $(2, 4)$, and $(3, 3)$. Use $y = ax^4 + bx^3 + cx^2 + dx + e$ and each point (x, y) to generate a system of five linear equations. Compare your results with those from quartic regression.

Solution   We substitute each x-coordinate and y-coordinate for x and y:

$$
\begin{array}{llll}
a(16) + b(-8) + c(4) + d(-2) + e = & 1 & \text{For } (-2, 1) \\
a(0)\ + b(0)\ \ + c(0) + d(0)\ \ + e = & 2 & \text{For } (0, 2) \\
a(1)\ \ + b(1)\ \ + c(1) + d(1)\ \ + e = & -1 & \text{For } (1, -1) \\
a(16) + b(8)\ \ + c(4) + d(2)\ \ + e = & 4 & \text{For } (2, 4) \\
a(81) + b(27)\ + c(9) + d(3)\ \ + e = & 3 & \text{For } (3, 3)
\end{array}
$$

Then we set up a matrix equation $[A][X] = [B]$:

$$
\begin{bmatrix}
16 & -8 & 4 & -2 & 1 \\
0 & 0 & 0 & 0 & 1 \\
1 & 1 & 1 & 1 & 1 \\
16 & 8 & 4 & 2 & 1 \\
81 & 27 & 9 & 3 & 1
\end{bmatrix}
\begin{bmatrix} a \\ b \\ c \\ d \\ e \end{bmatrix}
=
\begin{bmatrix} 1 \\ 2 \\ -1 \\ 4 \\ 3 \end{bmatrix}
$$

[A] is 5×5, while [X] and [B] are 5×1. The value of the determinant of the matrix of coefficients, det [A], is 1440. Evaluating $[X] = [A]^{-1}[B]$ gives

$$
[X] = \begin{bmatrix} -0.725 \\ 2.0167 \\ 3.025 \\ -7.3167 \\ 2 \end{bmatrix} = \begin{bmatrix} a \\ b \\ c \\ d \\ e \end{bmatrix}
$$

The equation is

$$y = -0.725x^4 + 2.0167x^3 + 3.025x^2 - 7.3167x + 2$$

With the calculator, we place the x-coordinates in calculator list L_1 and the y-coordinates in calculator list L_2. Applying quartic regression to L_1 and L_2 yields the same equation. ●

Think about it 2: What are the parameters a, b, c, d, and e in fraction notation? (*Hint:* Obtain $[X] = [A]^{-1}[B]$ on the display, press ENTER to obtain the matrix containing the parameters, and then convert to fraction notation with the INTO FRACTIONS option.)

ANSWER BOX

Warm-up: 1. $m = 5l$ **2.** $s = 3m$ **3.** $3q = d$ or $q = \frac{1}{3}d$ **Think about it 1:** $q = \frac{1}{3}d$ **Think about it 2:** The fractions are $-\frac{29}{40}$, $\frac{121}{60}$, $\frac{121}{40}$, $-\frac{439}{60}$, and $\frac{2}{1}$.

EXERCISES 8.3

1. Tell whether the three planes in each figure have a single point of intersection, no common point of intersection, or a line or plane of intersection. Identify each figure as being the graph of consistent, inconsistent, or dependent equations. Assume that planes 1, 2, and 3 in part c and planes 1 and 2 in part d are parallel.

   a.
   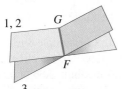

   b.

   c.
   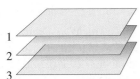

   d.
   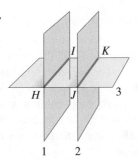

2. Identify each of the following as describing consistent, inconsistent, and/or dependent equations.

   a. For $[A][X] = [B]$, the value of the determinant of $[A]$ is zero.

   b. For $[A][X] = [B]$, the value of the determinant of $[A]$ is nonzero.

   c. The three equations have an infinite number of solutions.

   d. The three equations have no solution.

   e. The three equations have exactly one solution.

   f. An algebraic solution yields a false statement.

   g. An algebraic solution yields a true statement.

   h. The planes do not intersect.

   i. The planes intersect in a common point.

   j. The planes intersect in a common line.

   k. The planes intersect in a common plane.

In Exercises 3 to 14, solve the systems of three equations. Use algebra as needed to distinguish between inconsistent and dependent equations.

3. $\begin{aligned} x + y - z &= 2 \\ x - y + z &= 3 \\ -x + y + z &= 4 \end{aligned}$

4. $\begin{aligned} x + y - z &= 2 \\ x - z &= 4 \\ x + y &= 3 \end{aligned}$

5. $\begin{aligned} 3y + 5z &= 0 \\ 3x + 7z &= 0 \\ x + y + 4z &= 3 \end{aligned}$

6. $\begin{aligned} 2x - y + z &= 1 \\ x + z &= -2 \\ x - y &= 3 \end{aligned}$

7. $\begin{aligned} 2x - y &= 3 \\ x + 2z &= 4 \\ -2x + y + z &= 1 \end{aligned}$

8. $\begin{aligned} 2x + y - z &= 3 \\ 5x - 4z &= 0 \\ 5y + 3z &= 0 \end{aligned}$

9. $\begin{aligned} -2x + 3z &= 0 \\ -x + y + 3z &= -1 \\ x + y &= -1 \end{aligned}$

10. $\begin{aligned} x - z &= 3 \\ x + 3y + 2z &= 6 \\ -2x + y &= 3 \end{aligned}$

11. $a + 2b + 3c = 2$
$4a + 5b + 6c = 3$
$a + b + c = 1$

12. $a + 2b + 3c = 2$
$4a + 5b + 6c = 5$
$a + b + c = 1$

13. $a + b + c = 4$
$a - b - c = 0$
$-a + b - c = 2$

14. $a - b + c = 4$
$a - b - c = 0$
$-a + b - c = 2$

Set up systems of equations in Exercises 15 to 22, and solve.

15. An 8-ounce container of lowfat yogurt contains 148 calories. The cup contains a total of 32 grams of protein, fat, and carbohydrates. There are 12 more grams of carbohydrates than of fat. Protein and carbohydrates have 4 calories per gram; fat has 9 calories per gram. How many grams each of protein, fat, and carbohydrates are there?

16. An ounce of dry-roasted cashew nuts (salted) contains 169 calories. There are 5 more grams of carbohydrates than of protein. The total weight of protein, fat, and carbohydrates is 26 grams. Protein and carbohydrates have 4 calories per gram; fat has 9 calories per gram. How many grams each of protein, fat, and carbohydrates are there?

17. Shelby has $5.80 in nickels, dimes, and quarters in her piggy bank. There are 44 coins altogether, with 3 more quarters than dimes. How many of each type of coin does she have?

18. Zachary has nickels, dimes, and quarters in his piggy bank. He has 58 coins with a total value of $6.50. There are twice as many dimes as nickels. How many of each coin does he have?

19. In 1987, Ayako Okamoto and Curtis Strange were the leading money winners in their respective professional golf associations. Together, they won $412,545 more than Corey Pavin won in 1991. Curtis won $459,907 more than Ayako. Ayako and Corey together won $1,445,464. How much did each win?

20. In 1978, Nancy Lopez and Tom Watson were the leading money winners in their respective professional golf associations. Together, they won $43,750 less than Betsy King won in 1993. Betsy won $406,179 more than Nancy. Tom won $233,563 less than Betsy. How much did each win?

21. Vienna blend coffee is a mixture of Ethiopian ($10.95 per pound), French roast ($8.95), Guatemalan ($8.50), and light Colombian ($8.25). The blend calls for twice as much light Colombian as Guatemalan and four times as much French roast as Ethiopian. The cost of the blended coffee is $8.59 per pound. The person blending the coffee wants to use 50-pound bags, so round answers to the nearest 50 pounds. How many pounds of each are used to make 1000 pounds of Vienna blend?

22. "Four ingredients," a stir-fry favorite, contains shrimp, barbequed pork, beef, and chicken sauteed with vegetables. The weight of the uncooked ingredients is 20 ounces. The weight of the pork and beef together is the same as the weight of the vegetables. The chicken weighs twice as much as the beef. The shrimp, beef, and vegetables together weigh 12 ounces. The shrimp and chicken together weigh 10 ounces. How many ounces are there of each ingredient?

In Exercises 23 and 24, set up and solve a matrix equation that fits these points to a fourth-degree polynomial equation. Check with quartic regression.

23. $(-2, -7), (-1, -9), (0, -5), (1, -1), (2, 45)$

24. $(-2, -32), (-1, -3), (1, 1), (3, 13), (4, -38)$

25. Lee is surprised that his plan in Example 8 reduced the number of small rabbits by only 50. He changes to an equal number of small and medium rabbits and keeps total rabbits at 620, with 20 large rabbits. Write the new matrix equation.

26. To solve his problem in Example 8, Fenton consults with Janelle, who suggests making twice as many small rabbits as medium rabbits, 20 large rabbits, and a total of 620 rabbits. Write the new matrix equation.

Exercises 27 to 29 examine applications of matrix multiplication.

27. Janelle's accountant, Amelie, wants to find the total income from various combinations of products. Small rabbits sell for $6, medium for $12, and large for $30. The prices are shown in matrix A. Four production options are listed in matrix B; each column gives the numbers of small, medium, and large rabbits for one option. The product $[A][B]$ gives a matrix of total revenue for the various options.

$$[A][B] = [6 \quad 12 \quad 30] \begin{bmatrix} 500 & 450 & 400 & 300 \\ 100 & 150 & 200 & 300 \\ 20 & 20 & 20 & 20 \end{bmatrix}$$

a. Identify the shapes of matrices A and B, and predict the shape of the product $[A][B]$.

b. Enter the matrices into the calculator in $[A]$ and $[B]$, use the multiplication key to multiply, and store the result in matrix C.

28. Oliko, Janelle's production assistant, wants to calculate the manufacturing time for each production option.

She estimates 6 minutes to make a small rabbit and 4 minutes to make a medium or large rabbit.

a. Express the time required to make one of each rabbit as a fraction of an hour.

b. Change the elements in matrix A in Exercise 27 so that they represent the time in hours required to make each rabbit instead of the sale price of each rabbit. Enter the times as fractions (the calculator will change the entries to decimals). Find the total production time $[A][B]$.

c. Find the cost of each production option, assuming that workers cost $9 per hour (minimum wage plus benefits). *Hint:* Replay the prior operation, multiply by 9, and store the result in matrix D.

29. Janelle's purchasing agent, Juan, reports that chocolate costs $3.50 per pound. The small rabbit contains 4.5 ounces of chocolate; the medium, 12 ounces; and the large, 3 pounds.

a. Write an expression for the weight, in pounds, of each rabbit.

b. Calculate the weight of the chocolate used for each production option. Round to the nearest tenth of a pound.

c. Calculate the cost of chocolate used for each production option. Round to the nearest dollar. *Hint:* Replay the prior operation, multiply by $3.50 (the cost per pound), and store the result in matrix E.

In Exercises 30 and 31, use addition or subtraction of matrices found as answers to prior exercises. Hint: D = [522 507 492 462].

30. To obtain the total cost for each production option, add the costs for labor (from Exercise 28) and for chocolate (from Exercise 29): $[D] + [E]$.

31. To obtain the profit for each production option, subtract the costs from the revenue, using the matrices from prior exercises: $[C] - [D] - [E]$. From this profit, Janelle must pay sales clerks, rent, utilities, taxes, and so forth.

32. Explain why a system of equations cannot be both inconsistent and dependent.

Projects

33. *Cramer's Rule.* Research Cramer's rule, and explain how determinants are used to find the solutions to a system of equations.

34. *Computer (and Calculator) Testing.* It may come as a surprise to you that the computation of pi to millions of digits is useful in checking the computational accuracy of a computer.

During the 1970s, the calculation of the inverse, $[A]^{-1}$, of a so-called Hilbert matrix was used as a test of the inversion program. Test your calculator inversion package by calculating $[A]^{-1}$ for successively larger Hilbert matrices and multiplying $[A][A]^{-1}$. The product $[A][A]^{-1}$ must be an identity matrix with ones on the main diagonal (upper left to lower right) and zeros elsewhere. Summarize your results by listing the Hilbert matrix and its inverse for a 2×2, 3×3, 4×4, and 5×5 matrix. The $n \times n$ Hilbert matrix is

$$
\begin{bmatrix}
1 & \dfrac{1}{2} & \dfrac{1}{3} & \dfrac{1}{4} & \cdots & \dfrac{1}{n} \\[2mm]
\dfrac{1}{2} & \dfrac{1}{3} & \dfrac{1}{4} & \dfrac{1}{5} & \cdots & \dfrac{1}{n+1} \\[2mm]
\dfrac{1}{3} & \dfrac{1}{4} & \dfrac{1}{5} & \dfrac{1}{6} & \cdots & \dfrac{1}{n+2} \\[2mm]
\dfrac{1}{4} & \dfrac{1}{5} & \dfrac{1}{6} & \dfrac{1}{7} & \cdots & \dfrac{1}{n+3} \\[2mm]
\vdots & \vdots & \vdots & \vdots & & \vdots \\[2mm]
\dfrac{1}{n} & \dfrac{1}{n+1} & \dfrac{1}{n+2} & \dfrac{1}{n+3} & \cdots & \dfrac{1}{2n-1}
\end{bmatrix}
$$

8.4 Conic Sections

OBJECTIVES

- Use specific parameters and formulas to find equations for conic sections.
- Identify the standard forms of conic sections centered at the origin.
- Graph conic sections on a calculator, using two functions when required.
- Explore, with a graphing calculator, how changing parameters changes the shape or orientation of the curve.

I N THIS SECTION, we explore a set of curves called conic sections. The discussion starts with an overview of conic sections and then introduces the equations of circles and ellipses. Two of the conic section curves were presented earlier: parabolas in Chapters 3 and 4 and hyperbolas in Chapter 5. We revisit the parabola and hyperbola and examine standard forms of their equations.

Conical Surfaces and Conic Sections

A **conical surface** is *a double cone.* We generate a conical surface with a line. Model the formation of a conical surface with a pencil as a representation of a line. Hold the pencil gently in the middle (lengthwise). With the other hand, move the lower end of the pencil in a circular motion. The upper half of the pencil will also move in a circular motion. The surface modeled by the motion of the two halves of the pencil is the conical surface shown in each part of Figure 9.

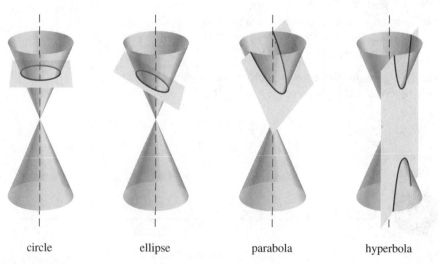

circle   ellipse   parabola   hyperbola

Figure 9

Conic sections are *the set of curves obtained by slicing the conical surface with a plane. The vertical line through the center of the conical surface* is the **axis.** The orientation of the slice relative to the axis or to the surface itself determines the type of curve obtained; see Figure 9.

When we slice the conical surface perpendicular to the axis, we obtain a *circle* as the intersection. A slice tilted slightly relative to the axis gives an *ellipse* as the intersection. A slice parallel to the conical surface gives a *parabola*. A slice parallel to the axis intersects the conical surface in two places and forms the two branches of a *hyperbola*. The hyperbola is the only curved conic section in two parts.

A slice along the axis itself forms two straight lines (a "degenerate" conic section); see Figure 10. A slice that just touches the conical surface forms a straight line. A slice through the center of the surface—where the points of the cones meet—forms a point (another degenerate conic section).

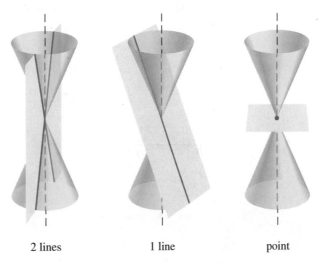

| 2 lines | 1 line | point |

Figure 10

The four conic sections are shown together in the sketch of paths of a spacecraft in Figure 11. If the orbital velocity is less than the escape velocity, the spacecraft will follow an elliptical or circular orbit. If the orbital velocity is greater than or equal to the escape velocity, the spacecraft will follow a parabolic or hyperbolic path and leave Earth's orbit.

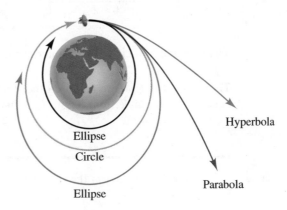

Figure 11

We begin a more detailed look at conic sections by examining the circle. We will then consider the ellipse, hyperbola, and parabola.

Circle

Definition of a Circle	A **circle** is the set of all points $P(x, y)$ equidistant from a point called the center C. The constant distance between the point P and the center C is the **radius** r.

We obtain the equation of a circle directly from its definition by applying the distance formula to two points: P with coordinates (x, y) on the circle and C with coordinates $(0, 0)$ at the origin (Figure 12).

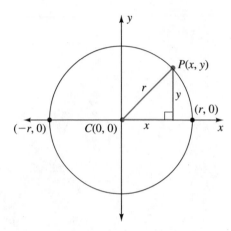

Figure 12

EXAMPLE **1**   Finding the equation of a circle   Suppose a circle has radius r and center at the origin; find its equation.

Solution   By definition, the point P located anywhere on the circle must be a constant distance r from the center C. We substitute the coordinates of P and C and the distance r into the distance formula:

$$d = \sqrt{(x_2 - x_1)^2 + (y_2 - y_1)^2}$$
$$r = \sqrt{(x - 0)^2 + (y - 0)^2} \qquad \text{Square both sides to eliminate the radical.}$$
$$r^2 = x^2 + y^2$$

The equation of a circle with center at the origin is $x^2 + y^2 = r^2$.   ●

Think about it 1: If we had used the Pythagorean theorem directly to find the equation of the circle from Figure 12, what are the sides of the right triangle that we would have used?

Equation of a Circle	A circle with center at the origin and radius r has the equation $$x^2 + y^2 = r^2, \quad r > 0$$

In Example 2, we graph first with a sketch and then with a calculator. Sketches are important because they prevent many errors.

EXAMPLE **2**   Sketching and graphing circles   Make a sketch of a circle with center at the origin and radius 4. Write the equation of the circle, and then show the graph on a calculator.

Solution   The circle with center at the origin and radius 4 will pass through points exactly 4 units from the origin in each horizontal and vertical direction. Four such points on the axes are $(4, 0)$, $(-4, 0)$, $(0, 4)$, and $(0, -4)$. These four points provide the basis of a good sketch of the circle, shown in Figure 13.

The equation will have $r = 4$ in $x^2 + y^2 = r^2$:

$$x^2 + y^2 = 4^2$$
$$x^2 + y^2 = 16$$

We solve for y:

$$y^2 = 16 - x^2$$
$$y = \pm\sqrt{16 - x^2}$$

It takes two equations to graph the circle, as shown in Figure 14.   ●

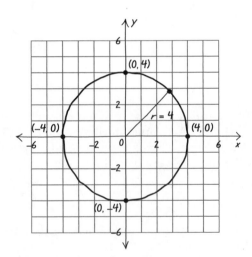

Figure 13

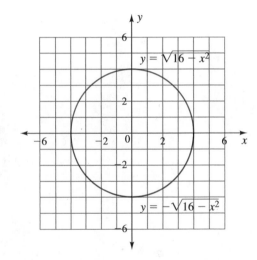

Figure 14

Graphing Calculator Technique:
Graphing a Circle

Two problems may arise when you enter the two equations needed to graph a circle. First, the two pieces of the circle may not meet at or near the *x*-axis if the points for which the calculator evaluates the equations are not "close enough" to the *x*-intercepts to complete the circle. Second, the circle may appear oval on the calculator because the horizontal scale spreads the numbers over a longer distance than the vertical scale does. To correct this problem, choose the square window option. Unfortunately, fixing the appearance of the circle may reopen the gap between the upper and lower halves.

Think about it 2: Why does it take two equations to graph a circle? *Hint:* Is the equation of a circle a function? (See Section 2.1.)

Ellipse

In the following example, we observe what happens when we place a numerical coefficient on one of the variable terms in the equation of a circle.

EXAMPLE **3**   Comparing graphs

a. Graph and compare these equations:

$$x^2 + y^2 = 1$$

$$\frac{x^2}{4} + y^2 = 1$$

$$\frac{x^2}{8} + y^2 = 1$$

Hint: Write the equation as $x^2/L_1 + y^2 = 1$, solve for y, and enter 1, 4, and 8 into calculator list L_1.

b. Compare the results in part a with those for $x^2 + y^2/L_1 = 1$.

Solution   a. $\dfrac{x^2}{L_1} + y^2 = 1$

$$y^2 = 1 - \frac{x^2}{L_1}$$ We need two equations to solve for y.

$$y_1 = \sqrt{1 - \frac{x^2}{L_1}}$$

$$y_2 = -\sqrt{1 - \frac{x^2}{L_1}}$$

The first equation, $x^2 + y^2 = 1$, forms a circle of radius 1. The second equation, $x^2/4 + y^2 = 1$, forms an ellipse; and the third, $x^2/8 + y^2 = 1$, forms a still wider ellipse. Placing a number under the x^2 term seems to stretch the circle horizontally (see Figure 15).

b. To graph $x^2 + y^2/L_1 = 1$, we solve for y, obtaining

$$y_1 = \sqrt{L_1}\sqrt{1 - x^2} \quad \text{and} \quad y_2 = -\sqrt{L_1}\sqrt{1 - x^2}$$

This time, we have a vertical stretching as a result of dividing the y^2 term by 4 and 8 (see Figure 16).

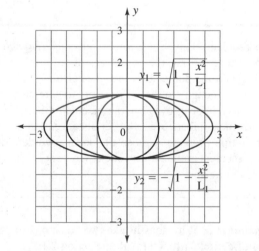

Figure 15   **Figure 16**   ●

Example 3 suggests that slight variations in the equation of a circle will produce an oval shape called an ellipse.

Equation of an Ellipse

> An equation written in the form $\dfrac{x^2}{a^2} + \dfrac{y^2}{b^2} = 1$, where a and b are nonzero real numbers, has a graph that is an ellipse.

The similarity in structure between the equation for a circle and that for an ellipse is more apparent when we divide the circle equation by r^2 on both sides:

$$x^2 + y^2 = r^2$$
$$\frac{x^2}{r^2} + \frac{y^2}{r^2} = 1$$

If a and b are equal, then the equation of an ellipse is that of a circle with radius $a = b = r$. Thus, *the key distinction between the circle and the ellipse is that the coefficients on x^2 and y^2 are equal for a circle and different for an ellipse.*

Although Example 3 suggests how we might predict whether the ellipse stretches horizontally or vertically, it is usually easier to calculate the x- and y-intercepts of the equation and sketch the graph.

EXAMPLE 4

Sketching ellipses Sketch these ellipses by calculating the x- and y-intercepts.

a. $\dfrac{x^2}{4} + \dfrac{y^2}{9} = 1$ **b.** $\dfrac{x^2}{25} + \dfrac{y^2}{1} = 1$

Solution **a.** Substituting $y = 0$, we find the x-intercepts to be $(\pm 2, 0)$. Similarly, for $x = 0$, the y-intercepts are $(0, \pm 3)$. The graph is in Figure 17.

b. Substituting $y = 0$, we find the x-intercepts to be $(\pm 5, 0)$. Similarly, for $x = 0$, the y-intercepts are $(0, \pm 1)$. The graph is in Figure 18.

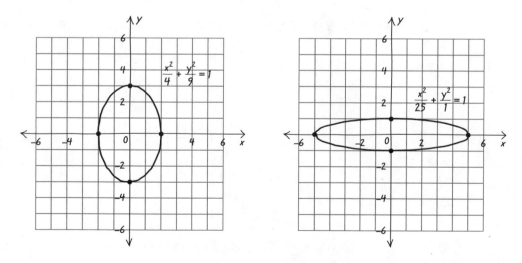

Figure 17　　　　　　　　　　　　**Figure 18**　　　　　　　●

For the circle, the x- and y-intercepts are $\pm r$. For the ellipse centered at the origin, the x-intercepts are $\pm a$ and the y-intercepts are $\pm b$. We can graph the ellipse by marking the x- and y-intercepts and drawing a smooth curve that passes through each of the intercepts.

EXAMPLE 5

Finding equations What is the equation of an ellipse with intercepts $(-5, 0)$, $(5, 0)$, $(0, -4)$, and $(0, 4)$?

Solution   The x-intercepts are ± 5, so $a = \pm 5$ and $a^2 = 25$. The y-intercepts are ± 4, so $b = \pm 4$ and $b^2 = 16$. We substitute into

$$\frac{x^2}{a^2} + \frac{y^2}{b^2} = 1$$

obtaining

$$\frac{x^2}{25} + \frac{y^2}{16} = 1$$

●

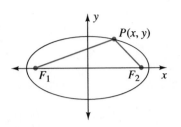

Figure 19

Before reading the formal definition of an ellipse, imagine splitting the center of a circle into two points and attaching a radius from a point $P(x, y)$ on the circle to each of the "centers," F_1 and F_2, in Figure 19.

Definition of an Ellipse

> An **ellipse** is the set of all points $P(x, y)$ such that the sum of the distances from P to two fixed points F_1 and F_2 is constant.

The points F_1 and F_2 are foci of the ellipse (see Figure 19). In a circle, the radius is a constant distance. In an ellipse, the two distances P to F_1 and P to F_2 add to a constant.

Hyperbola

The hyperbola has an equation similar to that of an ellipse, but its graph is an entirely different shape. Hyperbolas are used in the design of cooling towers for many nuclear power plants. See Figure 20 in Example 6 for a graph of the "cooling tower" style hyperbola.

In Example 6, we explore the effect of changing the addition operation in the ellipse formula to a subtraction.

EXAMPLE **6**   Graphing hyperbolas   Solve for y, and graph the resulting equations.

a. $\dfrac{x^2}{4} - \dfrac{y^2}{9} = 1$ **b.** $\dfrac{y^2}{16} - \dfrac{x^2}{9} = 1$

Solution   **a.**
$$\frac{x^2}{4} - \frac{y^2}{9} = 1$$

$$\frac{x^2}{4} - 1 = \frac{y^2}{9}$$

$$9\left(\frac{x^2}{4} - 1\right) = y^2$$

$$\pm 3\sqrt{\frac{x^2}{4} - 1} = y$$

The graph is shown in Figure 20.

b. $\dfrac{y^2}{16} - \dfrac{x^2}{9} = 1$

$$\frac{y^2}{16} = 1 + \frac{x^2}{9}$$

$$y^2 = 16\left(1 + \frac{x^2}{9}\right)$$

$$y = \pm 4\sqrt{1 + \frac{x^2}{9}}$$

The graph is shown in Figure 21.

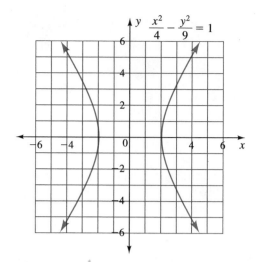

Figure 20

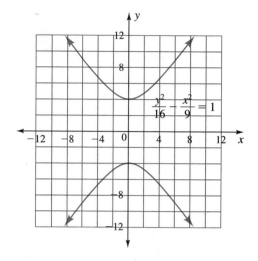

Figure 2I

Both the graphs are hyperbolas centered at the origin.

●

Equations of Hyperbolas

An equation written in the form $\dfrac{x^2}{a^2} - \dfrac{y^2}{b^2} = 1$ or $\dfrac{y^2}{b^2} - \dfrac{x^2}{a^2} = 1$, where a and b are nonzero real numbers, has a graph that is a hyperbola.

The hyperbola may cross either the x- or the y-axis, depending on its equation.

EXAMPLE **7**   Finding intercepts   Determine the x- and y-intercepts of these hyperbolas.

    a. $\dfrac{x^2}{4} - \dfrac{y^2}{9} = 1$ **b.** $\dfrac{y^2}{16} - \dfrac{x^2}{9} = 1$

Solution   **a.** $\dfrac{x^2}{4} - \dfrac{y^2}{9} = 1$ Let $y = 0$ for x-intercepts.

$\dfrac{x^2}{4} - \dfrac{0}{9} = 1$ Solve for x.

$x = \pm 2$

$\dfrac{x^2}{4} - \dfrac{y^2}{9} = 1$ Let $x = 0$ for y-intercepts.

$\dfrac{0}{4} - \dfrac{y^2}{9} = 1$ Solve for y^2.

$y^2 = -9$

There is no real-number solution to $y^2 = -9$ and no y-intercept (see Figure 20).

b. $\dfrac{y^2}{16} - \dfrac{x^2}{9} = 1$ Let $y = 0$.

$\dfrac{0}{16} - \dfrac{x^2}{9} = 1$ Solve for x^2.

$x^2 = -9$

There is no real-number solution to $x^2 = -9$ and no x-intercept (see Figure 21).

$$\frac{y^2}{16} - \frac{x^2}{9} = 1 \qquad \text{Let } x = 0.$$

$$\frac{y^2}{16} - \frac{0}{9} = 1 \qquad \text{Solve for } y.$$

$$y = \pm 4$$

●

For completeness, we include a formal definition of the hyperbola. Like the ellipse, it has two foci, F_1 and F_2 (see Figure 22), but its definition refers to a difference in distances rather than the sum of distances found in the definition of the ellipse.

Definition of a Hyperbola

A **hyperbola** is the set of all points $P(x, y)$ such that the absolute value of the difference of the distances from P to two fixed points F_1 and F_2 is a constant positive number.

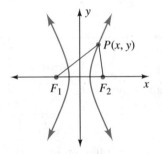

Figure 22

Section 5.2 introduced a different type of hyperbola. Such a hyperbola, which *approaches but does not intersect either of the rectangular coordinate axes,* is called a **rectangular hyperbola.** The graph of the reciprocal function $f(x) = 1/x$, shown in Figure 23, is a rectangular hyperbola. In Section 5.2, we changed the equations of several hyperbolas from the form $y = k/x$ to the form $xy = k$ in order to obtain the inverse variation equations.

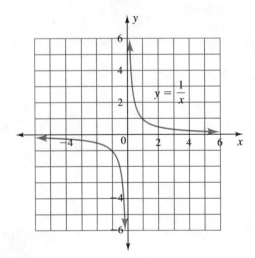

Figure 23

The orientation of a rectangular hyperbola is different from that of a conic section hyperbola—as shown in Figure 23, the graph is turned, or rotated, from a conic section hyperbola position. The equation of a rectangular hyperbola, $y = k/x$, can be written in conic section form, but doing the mathematics requires trigonometry.

Think about it 3: Is the rectangular hyperbola a function? Is the conic section hyperbola, as shown in this section, a function?

Parabola

axis of symmetry

vertex

Figure 24

In Chapters 3 and 4, we worked extensively with quadratic functions $f(x) = ax^2 + bx + c$. The graphs of these quadratic equations were parabolas with a vertical axis of symmetry passing through the vertex (see Figure 24). We saw parabolas in the paths of objects in motion, roadbed transition curves, museum arches, and support cables for suspension bridges.

Although the graphs of quadratic functions are parabolas, not all parabolas are quadratic functions. The conic sections expand our set of parabolas to include those with a horizontal axis of symmetry. The simplest of these curves is given by $x = y^2$.

EXAMPLE **8**   Graphing parabolas   Make a table and a graph of $y = x^2$ and $x = y^2$.

Solution   The tables for the two equations appear as Table 8 and Table 9.

x	$y = x^2$
-2	4
-1	1
0	0
1	1
2	4

Table 8

$x = y^2$	y
4	-2
1	-1
0	0
1	1
4	2

Table 9

The graphs of the parabolas $y = x^2$ and $x = y^2$ are shown in Figure 25.

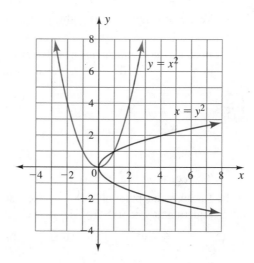

Figure 25

Think about it 4: If we solve $x = y^2$ for y, what two functions together describe this parabola with a horizontal axis?

Section 4.4 introduced the vertex form of a parabola, $y = a(x - h)^2 + k$, with $a \neq 0$ and vertex (h, k).

EXAMPLE **9** Comparing conic section and vertex forms of a parabola

a. What effect does locating the vertex at the origin have on the vertex form of a parabola, $y = a(x - h)^2 + k$?

b. Describe the parabola if $a > 0$.

c. Describe the parabola if $a < 0$.

Solution   a. $y = a(x - h)^2 + k$
$$= a(x - 0)^2 + 0$$
$$= ax^2$$

b. If $a > 0$, the parabola opens up.

c. If $a < 0$, the parabola opens down.   ●

In our work with conic sections, we have focused on vertices at the origin. Examples 8 and 9 suggest the conclusions in Table 10.

Way Parabola Opens	Equation with Vertex at the Origin	Function?
Upward	$y = ax^2$, a is positive	Yes
Downward	$y = ax^2$, a is negative	Yes
Right	$x = ay^2$, a is positive	No
Left	$x = ay^2$, a is negative	No

Table 10

EXAMPLE **10** Finding equations   What is the equation for each of these parabolas?

a. Vertex at origin, opens left, passes through $(-4, 2)$

b. Vertex at origin, opens right, passes through $(2, -3)$

Solution   a. Because the parabola has a vertex at the origin and opens to the left, the equation is $x = ay^2$. To find a, we substitute $(-4, 2)$ and solve for a.

$$x = ay^2 \qquad \text{Substitute } (-4, 2).$$
$$-4 = a(2)^2 \qquad \text{Simplify.}$$
$$-4 = a \cdot 4 \qquad \text{Divide by 4.}$$
$$a = -1$$

The equation is $x = -1y^2$.

b. Because the parabola has a vertex at the origin and opens downward, the equation is $y = ax^2$. To find a, we substitute $(2, -3)$ and solve for a.

$$y = ax^2 \qquad \text{Substitute } (2, -3).$$
$$-3 = a(2)^2 \qquad \text{Simplify.}$$
$$-3 = a \cdot 4 \qquad \text{Divide by 4.}$$
$$a = -\tfrac{3}{4}$$

The equation is $y = -\tfrac{3}{4}x^2$.   ●

For completeness, we include the formal definition of a parabola.

Definition of a Parabola

A **parabola** is the set of all points $P(x, y)$ such that the distance from P to a fixed point F is equal to the distance from P to a fixed line l.

The point F is the focus, and the line l is the directrix (see Figure 26).

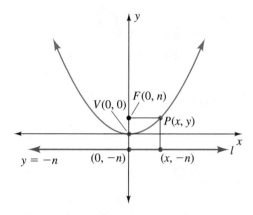

Figure 26

The distance from a point to a line is always the perpendicular distance. Consequently, in Figure 26, the coordinates of the point directly below $P(x, y)$ on the directrix are $(x, -n)$. We return to this definition and illustration in Exercise 48.

Identifying a Conic Section from Its Equation

First, arrange the equation in standard form, and then look carefully at the squared terms.

The equation of a *circle* centered at the origin with radius r is $x^2 + y^2 = r^2$, or

$$\frac{x^2}{r^2} + \frac{y^2}{r^2} = 1$$

The x^2 and y^2 terms have the same coefficients.

The equation of an *ellipse* centered at the origin is

$$\frac{x^2}{a^2} + \frac{y^2}{b^2} = 1$$

The coefficients on the x^2 and y^2 terms are different but have the same sign.

The equation of a *hyperbola* centered at the origin is

$$\frac{x^2}{a^2} - \frac{y^2}{b^2} = 1 \qquad \text{or} \qquad \frac{y^2}{b^2} - \frac{x^2}{a^2} = 1$$

The x^2 and y^2 terms have opposite signs.

The equation of a *parabola* with vertex at the origin has one of the following forms:

$$y = ax^2 \qquad \text{or} \qquad x = ay^2$$

The parabola has either x or y squared but not both.

Work done in earlier chapters allows us to recognize other parabolic and hyperbolic forms:

- The vertex form of a *parabola* is $y = a(x - h)^2 + k$, and the quadratic form is $y = ax^2 + bx + c$. In either case, the parabola still has a square on only one of the two variables x or y.

- A *rectangular hyperbola* has equation $y = k/x$ or $xy = k$. The x and y are multiplied, and no squared term is present.

ANSWER BOX

Warm-up: 1. right triangle; $8^2 + 15^2 = 17^2$ **2.** right triangle; $1.5^2 + 2^2 = 2.5^2$ **3.** right triangle; $10^2 + \sqrt{21}^2 = 11^2$ **4.** $9^2 + 16^2 \neq 25^2$; does not even make a triangle. **5.** $\sqrt{20} \approx 4.472$ **6.** 10 **7.** $\sqrt{x^2 + y^2}$ **8.** $\sqrt{(x - h)^2 + (y - k)^2}$ **Think about it 1:** x, y, and r **Think about it 2:** When we solve for y, we get two equations. The graphs of the equations create the upper and lower halves of the circle. The equation of a circle, $x^2 + y^2 = r^2$, is not a function because the circle fails the vertical-line test. However, each of the two equations whose graphs form the circle is a function. **Think about it 3:** The rectangular hyperbola is a function; the conic section hyperbolas shown here are not functions. **Think about it 4:** $x = \sqrt{y}$ and $x = -\sqrt{y}$.

EXERCISES 8.4

In Exercises 1 and 2, making a sketch on graph paper may be helpful.

1. List four points at a distance of 3 units from $(0, 0)$.

2. List four points at a distance of 4 units from $(0, 0)$.

Write an equation fitting each of the descriptions in Exercises 3 to 20.

3. Circle, center at origin, radius $= 5$

4. Circle, center at origin, radius $= 1.5$

5. Ellipse, center at origin, x-intercepts $(\pm 5, 0)$, y-intercepts $(0, \pm 2)$

6. Ellipse, center at origin, x-intercepts $(\pm 4, 0)$, y-intercepts $(0, \pm 1)$

7. Hyperbola, center at origin, passes through $(0, \pm 4)$, no x-intercept, $a = 3$

8. Hyperbola, center at origin, passes through $(\pm 5, 0)$, no y-intercept, $b = 4$

9. Parabola, vertex at origin, opens to the right, passes through $(3, -1)$

10. Parabola, vertex at origin, opens to the left, passes through $(-2, -2)$

11. Parabola, vertex at origin, opens downward, passes through $(3, -1)$

12. Parabola, vertex at origin, opens downward, passes through $(-2, -2)$

13. Circle, center at origin, radius $= 4$

14. Ellipse, center at origin, x-intercepts $(\pm 3, 0)$, y-intercepts $(0, \pm 4)$

15. Ellipse, center at origin, x-intercepts $(\pm 1, 0)$, y-intercepts $(0, \pm 5)$

16. Parabola, vertex at origin, opens to the left, passes through $(-2, 1)$

17. Hyperbola, center at origin, passes through $(0, \pm 3)$, no x-intercept, $a = 5$

18. Hyperbola, center at origin, passes through $(\pm 4, 0)$, no y-intercepts, $b = 2$

19. Parabola, vertex at origin, opens upward, passes through $(-2, 1)$

20. Circle, center at origin, radius $= 25$

21. What happens to the equation of a circle if $r < 0$? if $r = 0$?

22. What happens to the equation of an ellipse if a and b are equal?

In Exercises 23 to 44, identify the conic section (circle, ellipse, hyperbola, parabola, or straight line). Find any x- or y-intercepts. Write the equations needed to graph the conic section on a calculator. Some equations may relate to material from earlier chapters.

23. $y = x^2 + 2$

24. $y = -\dfrac{1}{x}$

25. $x^2 + y^2 = 9$

26. $x = y^2 + 2$

27. $x^2 - y^2 = 4$

28. $y = x^2 - 4x + 4$

29. $\dfrac{x^2}{4} + y^2 = 1$

30. $x^2 + \dfrac{y^2}{4} = 1$

31. $4x^2 + y^2 = 100$

32. $-2y - 10 + 5x = 0$

33. $2x^2 + 2y^2 - 8 = 0$

34. $xy = 5$

35. $y = x$

36. $y = 2x^2$

37. $x^2 - 4y^2 = 1$

38. $y = 2x - 1$

39. $x = 4y^2$

40. $25y^2 - 4x^2 = 100$

41. $\dfrac{y^2}{1} - \dfrac{x^2}{4} = 1$

42. $x^2 + y^2 = 2$

43. $x = -y^2$

44. $\dfrac{x^2}{4} + \dfrac{y^2}{25} = 1$

45. How does the shape of the hyperbola $\dfrac{x^2}{1} - \dfrac{y^2}{b^2} = 1$ change as b^2 gets large? (*Hint:* Use $b^2 = 1$, $b^2 = 5$, and $b^2 = 10$, and graph.)

46. How does the shape of the hyperbola $\dfrac{x^2}{a^2} - \dfrac{y^2}{1} = 1$ change as a^2 gets large? (*Hint:* Use $a^2 = 1$, $a^2 = 5$, and $a^2 = 10$, and graph.)

47. a. Make a table and graph for $x = -y^2$. Let x be $\{-9, -4, -1, 0, 1, 4, 9\}$.

b. Why might a student think that the equation $y = \sqrt{-x}$ has imaginary outputs and yet produces a graph?

c. Under what circumstances is $y = -\sqrt{-x}$ meaningful?

48. a. Referring to Figure 26, write an expression for the distance between the focus (point F) and the arbitrary point P on the parabola.

b. Referring to Figure 26, write an expression for the distance between the arbitrary point P on the parabola and the point $(x, -n)$ on the directrix.

c. Explain how the origin, lying between points $F(0, n)$ and $(0, -n)$, satisfies the definition of a parabola. Why is the origin labeled V?

d. Simplify the expressions in the equation

$$\sqrt{(x - 0)^2 + (y - n)^2} = \sqrt{(x - x)^2 + (y - (-n))^2}$$

and derive the formula

$$y = \dfrac{1}{4n}x^2$$

for a parabola with vertex at the origin. The parameter n is the distance between the vertex and the focus of a parabola.

What are the coordinates of the focus for each of the parabolas in parts e to h?

e. $y = x^2$

f. $y = -2x^2$

g. $y = 0.5x^2$

h. $y = -0.25x^2$

We may use matrices to fit any three noncollinear points to a circle. A general equation of a circle, not necessarily centered at the origin, is

$$x^2 + y^2 + Dx + Ey + F = 0$$

In Exercises 49 to 52, use the coordinates given to write three equations containing D, E, and F; solve the system; and state the resulting equation of the circle.

49. $(-12, 5)$, $(12, 5)$, $(5, -12)$

50. $(6, -8)$, $(6, 8)$, $(-8, 6)$

51. $(-4, -1)$, $(-5, 6)$, $(3, 6)$

52. $(-1, -5)$, $(6, 2)$, $(-1, 3)$

Projects

53. *Happy Face Calculating*

a. Given the shifted equation of a circle $(x - h)^2 + (y - k)^2 = r^2$, with center at (h, k), create a set of equations that can be used to produce the happy face shown in the figure on a calculator.

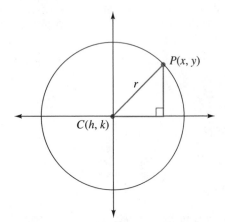

b. By applying the distance formula to the coordinates for *C* and *P* on the circle in the figure, derive the equation of a circle given in part a.

54. *Waxed Paper Parabola Construction*

a. On a sheet of waxed paper about the size of a notebook page, draw a horizontal line segment 20 centimeters long, positioned about 5 centimeters from the bottom edge of the paper.

b. Mark the line segment into centimeters.

c. Place a point *A* about 15 centimeters above the midpoint of the segment.

d. Fold the paper so that the first centimeter mark on the line segment lies on point *A*. Crease the paper along the fold. Reopen the paper.

e. Fold the paper so that the second centimeter mark lies on point *A*. Crease the paper again. Reopen the paper.

f. Repeat the folding and creasing for each of the centimeter marks on the line segment.

The resulting shape is called an *envelope curve*. Explain how the envelope curve you have created satisfies the definition of a conic section.

55. *More Waxed Paper Construction*s

a. Start by drawing a large circle (about 6 inches in diameter), with center *C*, on a piece of waxed paper. Mark points evenly around the circle (a protractor can be used to get 36 equally spaced points).

b. Place a point *P* somewhere outside the circle. Fold the paper so that *P* touches one point on the circumference of the circle. Crease the paper and unfold. Fold again so that *P* touches the next point on the circumference of the circle. Crease and unfold. Repeat for each point on the circumference. What is the conic section?

c. Repeat part a for a point *P* somewhere inside the circle. What is the conic section now?

56. *String Conic Section.* Tie a piece of string in a loop so that the circumference of the loop is about 10 inches. Place a piece of notebook paper on a sheet of cardboard, and insert two thumbtacks about 2 inches apart near the center of the paper. Lay the string so that it loops around the tacks. Place a pencil point inside the loop, and pull the string tight. Keeping the string tight, move the pencil along the paper, tracing out a curve. What conic section is formed? Explain how the string satisfies the definition of that conic section.

8.5 Solving Nonlinear Systems of Equations by Substitution and Graphing

OBJECTIVES

- Predict the number of solutions to a system including nonlinear equations by identifying the conic sections within the system.
- Solve systems including nonlinear equations with a calculator.
- Solve systems algebraically by substitution.

I N THIS SECTION, we identify conic sections within systems of equations, predict the number of solutions to systems, and solve systems graphically and algebraically. The systems of equations include exponential and logarithmic equations as well as conic sections.

Nonlinear Systems Including Conic Sections

We have seen several applications of systems of equations in earlier chapters. In Chapter 3, we worked with parabolas and lines in relating area and perimeter formulas and in finding the time at which an Olympic diver would be at a specified height. In Chapter 5, we solved systems involving linear equations and rectangular hyperbolas. This section extends the solution of systems to other combinations of conic sections, called **nonlinear systems** because *at least one equation is not linear*.

Applications of systems of conic sections are common. In trapshooting, for a shot to be a "hit," the parabolic path of the shot must intersect the parabolic path of the clay target. In tennis, the circular motion of the tennis racket must intersect the parabolic path of the tennis ball. When a dog catches a Frisbee or a baseball player makes a running catch, the intersection is accomplished intuitively. In contrast, ensuring that the flight of a space vehicle on a path around the sun intersects the elliptical orbit of a planet requires calculation by computer.

In the examples in this section, we will solve systems by starting with substitution and then employing a variety of algebraic techniques to complete the solution. Suggestions for graphical solutions to the examples are included in the exercises. Matrices, based on linear systems, will not work with nonlinear systems.

EXAMPLE **1**

Solving a nonlinear system   Identify the conic equations, predict the number of solutions, and solve the system by substitution:

$$y = 2x^2 - 3x - 5$$
$$2x - y = 2$$

Solution   The system contains a parabola and a line. There may be up to two points of intersection (see Figure 27).

For the substitution, we solve the linear equation for y:

$$2x - y = 2 \qquad \text{Add } y \text{ to both sides.}$$
$$2x = 2 + y \qquad \text{Subtract 2.}$$
$$2x - 2 = y$$

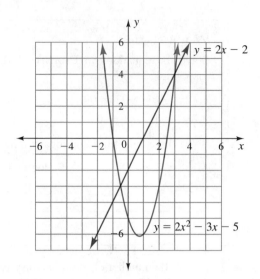

Figure 27

We then substitute $y = 2x - 2$ into the quadratic equation:

$$y = 2x^2 - 3x - 5$$
$$2x - 2 = 2x^2 - 3x - 5 \qquad \text{Add 2 and subtract } 2x \text{ on both sides.}$$
$$0 = 2x^2 - 5x - 3 \qquad \text{Solve with the quadratic formula.}$$

With $a = 2$, $b = -5$, and $c = -3$, we have

$$x = 3 \quad \text{and} \quad x = -0.5$$

Now we substitute each value of x into the linear equation:

$$y = 2(3) - 2$$
$$= 4 \qquad \text{At } x = 3, y = 4.$$
$$y = 2(-0.5) - 2$$
$$= -3 \qquad \text{At } x = -0.5, y = -3.$$

Check: We must check both coordinates in both equations:

$$4 \overset{?}{=} 2(3)^2 - 3(3) - 5 \; \checkmark \qquad\qquad\qquad 2(3) - 4 \overset{?}{=} 2 \; \checkmark$$
$$-3 \overset{?}{=} 2(-0.5)^2 - 3(-0.5) - 5 \; \checkmark \qquad 2(-0.5) - (-3) \overset{?}{=} 2 \; \checkmark$$

●

Think about it 1: What is the significance of the points $(3, 4)$ and $(-0.5, -3)$ in the graphs of the equations in Example 1?

In Example 1, we used the quadratic formula to solve for x. Example 2 will also require use of the quadratic formula within its solution.

EXAMPLE **2** Solving a nonlinear system   Identify the conic sections, predict the number of solutions, and solve the system:

$$y = 2x^2 + 3x - 6$$
$$x^2 - y + 4 = 0$$

Solution   Because only one variable is squared in each equation, the system contains two parabolas. The x^2 indicates vertical axes of symmetry. There are two intersections (see Figure 28).

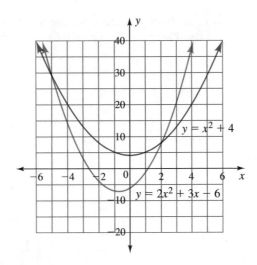

Figure 28

We solve $x^2 - y + 4 = 0$ for y:

$$x^2 + 4 = y$$

Then **we** substitute $y = x^2 + 4$ into the quadratic equation for y:

$$y = 2x^2 + 3x - 6$$
$$x^2 + 4 = 2x^2 + 3x - 6 \qquad \text{Subtract } x^2 + 4 \text{ from both sides.}$$
$$0 = x^2 + 3x - 10 \qquad \text{Solve with the quadratic formula.}$$

With $a = 1$, $b = 3$, and $c = -10$, we have

$$x = 2 \qquad \text{and} \qquad x = -5$$

Now we find the corresponding y for each x. At $x = 2$,

$$y = (2)^2 + 4$$
$$y = 8$$

At $x = -5$,

$$y = (-5)^2 + 4$$
$$y = 29$$

Check: We must check both coordinates in both equations:

$$8 \stackrel{?}{=} 2(2)^2 + 3(2) - 6 \ \vee \qquad (2)^2 - 8 + 4 \stackrel{?}{=} 0 \ \vee$$
$$29 \stackrel{?}{=} 2(-5)^2 + 3(-5) - 6 \ \vee \qquad (-5)^2 - 29 + 4 \stackrel{?}{=} 0 \ \vee$$

The coordinates of the points of intersection are $(2, 8)$ and $(-5, 29)$. ●

Think about it 2: What is the greatest possible number of intersections for a system of two quadratic equations?

In Example 3, we apply reasoning and long division to find the solutions to a fourth-degree equation. Repeat the problem with a graphing calculator.

EXAMPLE **3**   Solving a nonlinear system   Identify the conic sections, predict the number of solutions, and solve the system:

$$y = x^2 - 1$$
$$x = y^2 - 1$$

Solution   These are both parabolas, but one has a vertical axis and the other has a horizontal axis. There may be up to four points of intersection (see Figure 29).

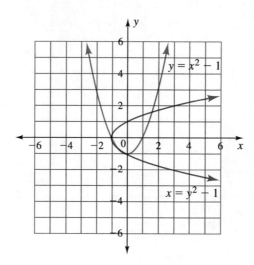

Figure 29

If we substitute $y = x^2 - 1$ into $x = y^2 - 1$, we obtain

$$x = (x^2 - 1)^2 - 1 \qquad \text{Square the expression } x^2 - 1.$$
$$x = x^4 - 2x^2 + 1 - 1 \qquad \text{Change to "= 0" form.}$$
$$x^4 - 2x^2 - x = 0 \qquad x \text{ is a common factor.}$$
$$x(x^3 - 2x - 1) = 0$$

We must guess and check to factor the expression on the left. Because the last term in $x^3 - 2x - 1$ is -1, we try $(x - 1)$ and $(x + 1)$. By polynomial long division (Section 5.5), we find $(x + 1)$ is a factor, and

$$\frac{x^3 - 2x - 1}{x + 1} = x^2 - x - 1$$

Thus, our fourth-degree equation becomes

$$x(x + 1)(x^2 - x - 1) = 0$$

Two solutions are $x = 0$ and $x = -1$. The quadratic formula applied to $x^2 - x - 1 = 0$ gives $x \approx 1.618$ and $x \approx -0.618$ as the other two solutions. We substitute each value of x into $y = x^2 - 1$ to find the corresponding y.

$$y = (-1)^2 - 1 \qquad \text{For } x = -1$$
$$= 0$$
$$y = (-0.618)^2 - 1 \qquad \text{For } x \approx -0.618$$
$$\approx -0.618$$
$$y = 0^2 - 1 \qquad \text{For } x = 0$$
$$= -1$$
$$y = (1.618)^2 - 1 \qquad \text{For } x \approx 1.618$$
$$\approx 1.618$$

There are four points of intersection: $(-1, 0)$, $(-0.618, -0.618)$, $(0, -1)$, and $(1.618, 1.618)$.

Check: The coordinates agree with the graph in Figure 29, so we will not check them in the equations. ✔   ●

I n solving systems where one expression must be squared, we substitute so as to square the simplest possible expression.

EXAMPLE **4**

Solving a nonlinear system   Identify the conic sections, estimate the number of solutions, and solve the system:

$$y = 1/x$$
$$x^2 + y^2 = 4$$

Solution   The equations describe a rectangular hyperbola and a circle. There could be up to four points of intersection and, hence, four solutions (see Figure 30).

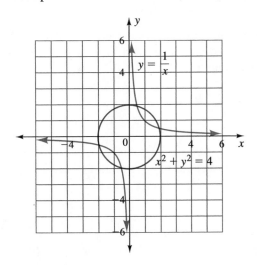

Figure 30

We substitute $y = 1/x$ into the equation of a circle:

$$x^2 + y^2 = 4 \qquad \text{Substitute } y = 1/x.$$

$$x^2 + \left(\frac{1}{x}\right)^2 = 4 \qquad \text{Simplify.}$$

$$x^2 + \frac{1}{x^2} = 4 \qquad \text{Multiply both sides by } x^2.$$

$$x^4 + 1 = 4x^2 \qquad \text{Subtract } 4x^2 \text{ from both sides.}$$

$$x^4 - 4x^2 + 1 = 0$$

We can solve $x^4 - 4x^2 + 1 = 0$ with the quadratic formula. We let $n = x^2$ and $n^2 = x^4$:

$$n^2 - 4n + 1 = 0$$

Evaluating the quadratic equation for $a = 1$, $b = -4$, and $c = 1$ gives

$$n \approx 3.73205 \qquad \text{and} \qquad n \approx 0.26795$$

We replace n with x^2:

$$x^2 \approx 3.73205 \qquad \text{and} \qquad x^2 \approx 0.26795$$
$$x \approx \pm 1.932 \qquad \text{and} \qquad x \approx \pm 0.518$$

We substitute each x into $y = 1/x$ and obtain four ordered pairs representing the four points of intersection of the graphs:

$$(1.932, 0.518), (-1.932, -0.518), (0.518, 1.932), \text{ and } (-0.518, -1.932).$$

Check: The coordinates agree with the graph in Figure 30, so we will not check them in the equations. ✔   ●

Think about it 3: What is the reciprocal of 1.932? Why are the x and y in Example 4 thus related?

EXAMPLE **5** Solving a nonlinear system   Identify the conic sections, predict the number of solutions, and solve the system:

$$\frac{x^2}{9} + \frac{y^2}{4} = 1$$

$$\frac{x^2}{1} - \frac{y^2}{4} = 1$$

Solution By comparing the x^2 and y^2 terms, we observe that the first equation is an ellipse and the second is a hyperbola. They could intersect in at most four places (see Figure 31).

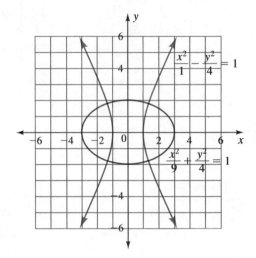

Figure 31

Because $y^2/4$ is in both equations, we solve the hyperbola equation for $y^2/4$:

$$\frac{x^2}{1} - \frac{y^2}{4} = 1$$

$$x^2 - 1 = \frac{y^2}{4}$$

We next substitute $x^2 - 1$ for $y^2/4$ in the ellipse equation:

$$\frac{x^2}{9} + \frac{y^2}{4} = 1$$

$$\frac{x^2}{9} + x^2 - 1 = 1 \qquad \text{Using } x^2 = \frac{9x^2}{9}, \text{ add the } x^2 \text{ terms.}$$

$$\frac{10x^2}{9} = 2$$

$$x^2 = 1.8$$

$$x = \pm\sqrt{1.8}$$

$$x \approx \pm 1.342$$

Then we substitute $x^2 = 1.8$ in the hyperbola equation:

$$\frac{y^2}{4} = 1.8 - 1$$

$$y^2 = 3.2$$

$$y = \pm\sqrt{3.2}$$

$$y \approx \pm 1.789$$

The solutions, described as ordered pairs, are $(-1.342, -1.789)$, $(-1.342, 1.789)$, $(1.342, -1.789)$, and $(1.342, 1.789)$.

Check: The coordinates agree with the intersections in Figure 31. ✔ ●

Nonlinear Systems Including Exponential and Logarithmic Equations

In Examples 6 and 7, we return to exponential and logarithmic equations.

EXAMPLE **6**

Solving a nonlinear system   Solve this system of equations by substitution.

$$y = 2^x$$
$$y = 5 - 2^x$$

Solution   The equations are graphed in Figure 32. There is one point of intersection.

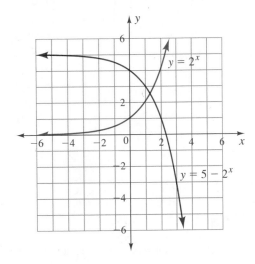

Figure 32

The equations are both solved for y. Substitution for y gives

$2^x = 5 - 2^x$	Add 2^x to both sides.
$2^x + 2^x = 5$	Add like terms.
$2 \cdot 2^x = 5$	Divide by 2.
$2^x = 2.5$	Change to a logarithmic equation.
$\log_2 2.5 = x$	Apply the change of base formula.
$x = \dfrac{\log 2.5}{\log 2}$	
≈ 1.322	

Then we solve for y:

$$y \approx 2^{1.322}$$

$$\approx 2.5$$

The solution to the system of equations is $x \approx 1.322$, $y \approx 2.5$.

Check: $2.5 \overset{?}{=} 2^{1.322}$ ✔

$2.5 \overset{?}{=} 5 - 2^{1.322}$ ✔ ●

Think about it 4: Which step in the solution to Example 6 indicates that y is exactly 2.5 and not approximately 2.5?

The system in Example 7 contains logarithms. Look for similarities between the solution of the exponential system in Example 6 and the solution of the logarithmic system in Example 7.

EXAMPLE **7** Solving a nonlinear system Graph this system of equations and then solve by substitution.

$$y = \log_2 x$$

$$y = 1 - \log_2 x$$

Solution The system is graphed in Figure 33. There is one point of intersection.

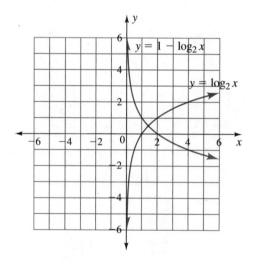

Figure 33

We use substitution to replace y in one of the equations:

$\log_2 x = 1 - \log_2 x$	Add $\log_2 x$ to both sides.
$\log_2 x + \log_2 x = 1$	Add like terms.
$2 \log_2 x = 1$	Divide both sides by 2.
$\log_2 x = \tfrac{1}{2}$	Change to an exponential equation.
$2^{1/2} = x$	
$x \approx 1.414$	

Because $y = \log_2 x$ and the solution above indicates that $\log_2 x = \tfrac{1}{2}$, we have $y = \tfrac{1}{2}$. Thus, the solution to the system is $x \approx 1.414$ and $y = \tfrac{1}{2}$.

Check: $\tfrac{1}{2} \overset{?}{=} \log_2 1.414$ ✔

$\tfrac{1}{2} \overset{?}{=} 1 - \log_2 1.414$ ✔ ●

Think about it 5: What is another way of writing $2^{1/2}$?

The closing example is based on the observation that the values of y in Examples 6 and 7 were rational. Sometimes we can evaluate expressions without knowing x.

EXAMPLE **8** Evaluating expressions without knowing x

a. If $2^x = \frac{4}{3}$, what is 2^{x+2}?

b. If $2^x = \frac{1}{3}$, what is 2^{x+1}?

c. If $3^x = \frac{1}{10}$, what is 3^{x+2}?

Solution **a.** $2^{x+2} = 2^x \cdot 2^2 = \frac{4}{3} \cdot 4 = \frac{16}{3}$

b. $2^{x+1} = 2^x \cdot 2^1 = \frac{1}{3} \cdot 2 = \frac{2}{3}$

c. $3^{x+2} = 3^x \cdot 3^2 = \frac{1}{10} \cdot 9 = \frac{9}{10}$

ANSWER BOX

Warm-up: 1. $x = 3, -0.5$ **2.** $x = 2, -5$ **3.** $n \approx 3.732, 0.268$
4. $x \approx 1.322$ **5.** $x \approx 1.414$ **6.** $x^2 - x - 1$ **Think about it 1:** $(3, 4)$
and $(-0.5, -3)$ are the points of intersection of the graphs of $y = 2x - 2$ and $y = 2x^2 - 3x - 5$. **Think about it 2:** Two parabolas (one with a horizontal axis and one with a vertical axis) could have up to four points of intersection, as shown in Example 3. **Think about it 3:** The reciprocal is 0.518. The hyperbola $y = 1/x$ is the reciprocal function, so each point (x, y) must satisfy that condition. **Think about it 4:** When we divided by 2 in $2 \cdot 2^x = 5$, we obtained $2^x = 2.5$. Because one of the original equations was $y = 2^x$, we know at this point that $y = 2.5$. **Think about it 5:** $2^{1/2} = \sqrt{2}$

EXERCISES 8.5

In Exercises 1 to 20, identify the conic sections (straight line, parabola, circle, ellipse, or hyperbola), and solve the system. Write the solution as coordinate points. Leave answers in radical form or round to three decimal places.

1. $y = x^2 - 3x + 2$
$y = -x + 5$

2. $y = 2x^2 - 3x + 2$
$y = x + 8$

3. $\dfrac{x^2}{4} - \dfrac{y^2}{9} = 1$
$y = x^2 - 4$

4. $y = x^2 - 2$
$x^2 + y^2 = 4$

5. $y = 5x^2 + 2x$
$y = -x^2 + x + 2$

6. $y = 2x^2 + 4x$
$y = x^2 + 2x + 3$

7. $x^2 + \dfrac{y^2}{4} = 1$
$y = x^2 - 1$

8. $y = x^2 - 2$
$y = \dfrac{1}{x}$

9. $x^2 + y^2 = 16$
$y = x^2 - 4$

10. $y = 15x^2 + 5x - 1$
$y = -5x^2 + 4x$

11. $y = -x^2 + 2$
$y = x^2 - 6$

12. $x^2 + y^2 = 16$
$y = x^2 - 2$

13. $\dfrac{x^2}{4} + y^2 = 1$
$y = x^2 - 1$

14. $y = \dfrac{2}{x}$
$x^2 + y^2 = 4$

15. $x^2 + y^2 = 4$
$x^2 - y^2 = 4$

16. $x^2 + y^2 = 4$
$x^2 - y^2 = 2$

17. $y = \dfrac{1}{x}$
$x + y = 4$

18. $x^2 + y^2 = 4$
$y = 2x + 1$

19. $x^2 + y^2 = 9$
$\dfrac{x^2}{16} + \dfrac{y^2}{4} = 1$

20. $\dfrac{x^2}{4} - \dfrac{y^2}{1} = 1$
$\dfrac{x^2}{9} + \dfrac{y^2}{4} = 1$

21. A graph of the system in Example 3 requires three equations:

$$y = x^2 - 1$$
$$y = \sqrt{x + 1}$$
$$y = -\sqrt{x + 1}$$

Find the dimensions of a viewing window that clearly shows these three (of the four) points of intersection: $(-1, 0)$, $(-0.618, -0.618)$, and $(0, -1)$.

22. Solve $x^4 - 2x^2 - x = 0$ by graphing, and show on a sketch that the solutions also are the x-coordinates of the points of intersection for the system in Example 4:

$$y = \dfrac{1}{x}$$
$$x^2 + y^2 = 4$$

23. Are $(-1, 0)$ and $(1, 0)$ solutions to the following system? Why or why not?

$$y = -2x^2 + 2$$
$$x^2 + (y - 2)^2 = 4$$

24. Are $(\pm 2, 2)$ and $(\pm\sqrt{3}, 1)$ solutions to the following system? Why or why not?

$$y = x^2 - 2$$
$$x^2 + (y - 2)^2 = 4$$

In Exercises 25 to 32, solve the system of equations, and write the solution as an ordered pair.

25. $y = 3^x$
$y = 2 - 3^x$

26. $y = 3^{x+2}$
$y = 1 - 3^x$

27. $y = 2^{x+2}$
$y = 4 + 2^x$

28. $y = 2^{x+1}$
$y = 1 - 2^x$

29. $y = \log x$
$y = 3 - \log x$

30. $y = \log_3 x$
$y = 1 - 2 \log_3 x$

31. $y = \log_4 x$
$y = 2 - \log_4 x$

32. $y = \frac{1}{2} \log_2 x$
$y = 6 - \log_2 x$

In Exercises 33 to 40, answer the question without solving for x.

33. If $2^x = \frac{1}{3}$, what is 2^{x+2}?

34. If $3^x = 4$, what is 3^{x+2}?

35. If $3^x = 2$, what is 3^{x+1}?

36. If $2^x = \frac{3}{4}$, what is 2^{x+1}?

37. If $2^x = 5$, what is 2^{x-1}?

38. If $3^x = 6$, what is 3^{x-2}?

39. If $4^x = 3$, what is 4^{x-2}?

40. If $4^x = 2$, what is 4^{x-1}?

41. In the solution to Example 7, we divided both sides of the equation $2 \log_2 x = 1$ by 2 and obtained $\log_2 x = \frac{1}{2}$, with the single solution $x = 2^{1/2}$. If we had applied properties of logarithms, we would have obtained $\log_2 x^2 = 1$. Change this new equation to exponential notation, and solve for x. How is the new solution the same? How is it different? What must we conclude?

42. Think back to the meaning of the degree of an equation. Explain why the conic sections (including the rectangular hyperbola, $xy = k$) are second-degree equations.

Projects

43. *Intersections of Circle and Parabola.* What are the possible numbers of intersections of a parabola and the circle $x^2 + y^2 = 4$? Make sketches and give equations.

44. *Intersections of Hyperbola and Straight Line.* What are the possible numbers of intersections of the hyperbola $x^2/4 - y^2/4 = 1$ with a straight line? Make sketches and give equations.

8.6 Solving Systems of Inequalities

OBJECTIVES

- Identify quadrants described by inequalities.
- Describe quadrants with inequalities.
- Solve systems of linear inequalities.
- Find inequalities for conic sections.
- Solve systems of nonlinear inequalities.

WARM-UP

Sketch the graphs of these equations.

1. $y = x + 4$
   $y = 4 - x$

2. $x^2 + y^2 = 4$

3. $\dfrac{x^2}{4} - \dfrac{y^2}{9} = 1$

4. $x = 2y^2$
   $x + y = 3$

5. $y = \frac{1}{3}x^2$
   $x^2 + y^2 = 9$

I N THIS SECTION, we extend our work with linear and nonlinear equations to inequalities.

Inequalities and Quadrants

In many applications, we make assumptions about the inputs or outputs that limit the graphs to the first quadrant or to the first and second quadrants. We usually describe these limitations with inequalities. In Examples 1 and 2, we examine the inequalities needed to describe quadrants, as we distinguish between the logical statements "*a* or *b*" and "*a* and *b*."

In Section 1.0, we defined "*a* or *b*" as meaning that either *a is true or b is true or both are true.*

EXAMPLE 1 Identifying quadrants Which quadrants are described by the statement $x > 0$ or $y > 0$?

Solution The first and fourth quadrants contain $x > 0$; see Figure 34(a). The first and second quadrants contain $y > 0$; see Figure 34(b). Because either statement may be true, the quadrants described by $x > 0$ or $y > 0$ are the first, second, and fourth quadrants; see Figure 34(c).

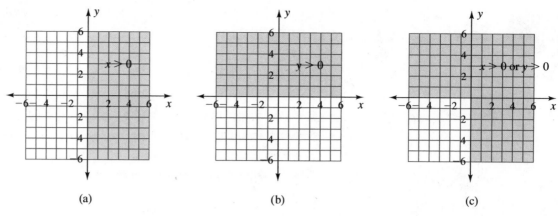

Figure 34

In Section 1.0, we defined "*a and b*" as indicating that *both a and b must be true*. Example 2 gives two applications of the logical *and*.

EXAMPLE **2** Describing quadrants Describe these quadrants with inequalities and a graph.

a. First quadrant

b. Third quadrant

Solution **a.** Both x and y are positive in the first quadrant, so we write $x > 0$ and $y > 0$. The first quadrant is shaded in Figure 35 to indicate all points where $x > 0$ and $y > 0$.

b. Both x and y are negative in the third quadrant, so we write $x < 0$ and $y < 0$. The third quadrant is shaded in Figure 36.

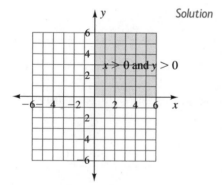

Figure 35

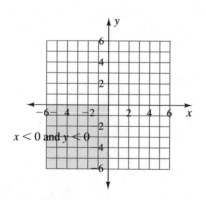

Figure 36

Systems of Linear Inequalities

As indicated in Section 1.4, we solve an inequality in two variables by drawing the graph of the equation (boundary) and then shading the half-plane that makes the inequality true. For a system of inequalities, we graph equations and shade the appropriate regions. The region where the half-planes (shadings) overlap is the solution to the system.

The shaded region in Figure 37 highlights the upper and lower limits on the antibiotic level required in the bloodstream to effectively combat a bacterial infection. The vertical axis is labeled with names for the limits rather than with numbers. The horizontal axis is time in hours. In this model of the antibiotic level, the patient should have taken a pill every 8 hours after an initial injection but did not. The antibiotic level is dropping close to the point where the antibiotic becomes ineffective.

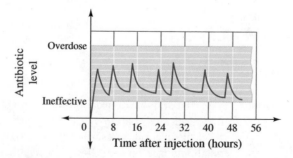

Figure 37

EXAMPLE **3**

Student Note: See Section 1.4 for other examples.

Solving a system of linear inequalities   Graph each inequality and find the region that satisfies both inequalities.

$$y \geq x + 4$$
$$y \leq 4 - x$$

Solution   The graphs are shown in Figure 38. The area where the two half-planes overlap is the solution. The test point $(-4, 4)$ makes each inequality true and confirms the solution.

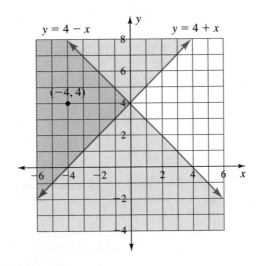

Figure 38

Second-Degree Inequalities

Linear inequalities are characterized by a boundary line and a half-plane. When inequalities have boundary lines that are curves (such as parabolas or circles), the shape of the solution set varies with the shape of the curve.

Graphing Second-Degree Inequalities

> To graph a second-degree inequality, graph the equation as a boundary line. Then use a test point to find which side of the boundary to shade.

As before, a dashed line excludes the boundary and a solid line includes the boundary in the solution set.

EXAMPLE **4**

Graphing second-degree inequalities   Graph each inequality.

a. $x^2 + y^2 > 4$

b. $\dfrac{x^2}{4} - \dfrac{y^2}{9} \leq 1$

Solution   **a.** The equation $x^2 + y^2 = 4$ is the equation of a circle with radius 2. The inequality symbol $>$ indicates that the graph of the boundary should be dashed. The origin makes the inequality false. The graph is shown in Figure 39.

b. The equation $\dfrac{x^2}{4} - \dfrac{y^2}{9} = 1$ is the equation of a hyperbola with x-intercepts $(\pm 2, 0)$. The inequality symbol $\leq$ indicates that the graph of the boundary should be solid. The origin makes the inequality true. The graph is shown in Figure 40.

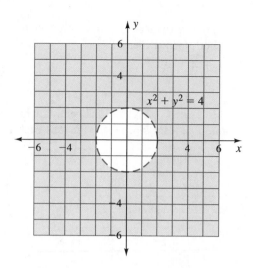

Figure 39

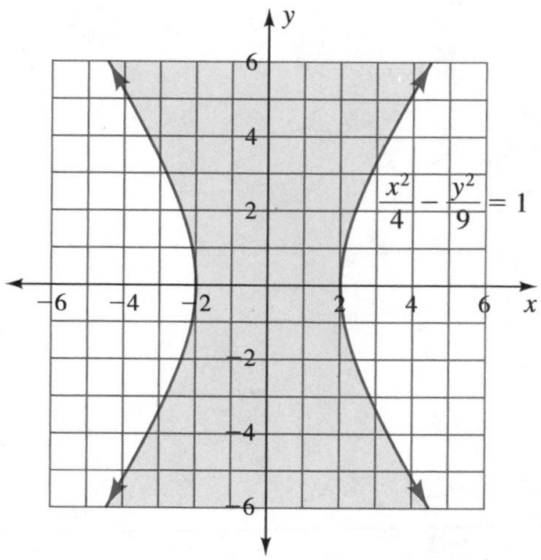

Figure 40   ●

Solving Systems of Nonlinear Inequalities

In solving a system of inequalities, graph each inequality separately and then look for the overlapping region.

EXAMPLE **5**

Solving a system of nonlinear inequalities   Find the solution set for each inequality, and then identify the solution to the system.

$$x \geq 2y^2$$
$$x + y \leq 3$$

Solution   This system has a parabola (vertex at the origin) opening to the right and a line with intercepts at $(3, 0)$ and $(0, 3)$. Using $(1, 0)$ as a test point, we shade to the left of the line and inside the parabola. The overlap of regions, shown in Figure 41, is the solution to the system.

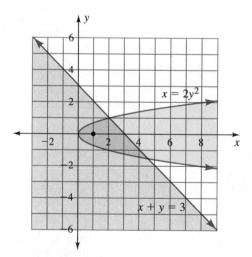

Figure 41   ●

EXAMPLE **6**   Solving a system of nonlinear inequalities   Find the solution set for each inequality, and then identify the solution to the system.

$$x^2 + y^2 \leq 9$$
$$y \geq \tfrac{1}{3}x^2$$

Solution   This system has a circle of radius 3 and a parabola (vertex at the origin) opening upward. Using (0, 1) as a test point, we shade inside the circle and inside the parabola. The overlap of regions, shown in Figure 42, is the solution to the system.

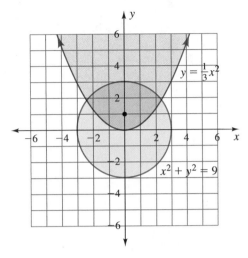

Figure 42

ANSWER BOX

Warm-up: 1. See the lines graphed in Figure 38. **2.** See the circle graphed in Figure 39. **3.** See the hyperbola graphed in Figure 40.
4. See the parabola and line graphed in Figure 41. **5.** See the parabola and circle graphed in Figure 42.

EXERCISES **8.6** _____

Name the quadrants that satisfy the inequalities in Exercises 1 to 8.

1. $x < 0$ and $y > 0$ **2.** $x > 0$ and $y < 0$

3. $x < 0$ and $y < 0$ **4.** $y > 0$

5. $y < 0$ **6.** $x > 0$

7. $x < 0$

8. $x < 0$ and $y > 0$ or $x > 0$ and $y < 0$

Write inequalities that describe the quadrants listed in Exercises 9 to 12.

9. Quadrant 2

10. Quadrant 4

11. Quadrants 2 and 4

12. Quadrants 1 and 4

In Exercises 13 to 16, name the inequalities shown in the graph.

13.

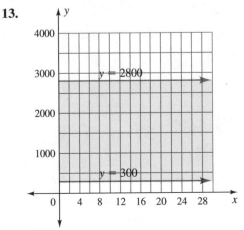

14.

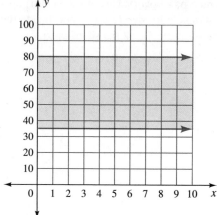

15.

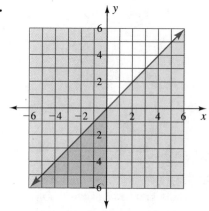

16.

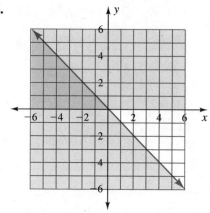

In Exercises 17 to 20, graph each inequality, and find the region that satisfies both inequalities.

17. $y \geq x - 3$
$2x - y \leq 5$

18. $x + y \leq 5$
$\dfrac{x}{2} + 2 \geq y$

19. $2x + 2y < 5$
$3x - 2y > 6$

20. $y - 6 < 3x$
$y > -3x$

In Exercises 21 to 28, graph each inequality.

21. $y \geq \dfrac{1}{x}$

22. $y \leq x^2$

23. $x^2 + \dfrac{y^2}{4} \leq 1$

24. $x^2 + y^2 \leq 4$

25. $\dfrac{x^2}{4} + y^2 \geq 1$

26. $\dfrac{x^2}{9} - \dfrac{y^2}{4} \geq 1$

27. $x > y^2$

28. $x < -y^2$

In Exercises 29 to 34, graph each inequality and indicate the solution to the system.

29. $y \leq \dfrac{x^2}{2}$
$x^2 + 4y^2 \geq 4$

30. $x^2 + y^2 \geq 9$
$4x^2 + y^2 \leq 4$

31. $y \geq \dfrac{1}{x}$
$y \leq x$

32. $y \geq -\dfrac{1}{x}$
$x^2 + y^2 \leq 4$

33. $y \leq -\dfrac{1}{x}$
$x^2 + 4y^2 \leq 4$

34. $y \geq -\dfrac{x^2}{2}$
$x^2 + y^2 \leq 9$

CHAPTER SUMMARY

Vocabulary

For definitions and page references, see the Glossary/Index.

axis	conic sections	degree 1	elimination method
circle	conical surface	dependent equations	ellipse
coincident lines	consistent equations	determinant	hyperbola

identity matrix
inconsistent equations
in terms of
linear equations
matrix
matrix equation
matrix inverse
matrix of coefficients
matrix of constants
matrix of variables
nonlinear system
ordered triples
parabola
parallel lines

per
planes
quantity
quantity-value table
radius
rectangular hyperbola
shape of a matrix
solution to the system
square matrix
standard form of a linear
  equation
substitution method
system of equations
value

Concepts

8.0 Solving Systems by Substitution and Elimination

To solve equations by substitution:

1. Solve one equation for a first variable.

2. Substitute the resulting expression into the second equation. The second equation now has one variable. Solve for the second variable.

3. Substitute the value of the second variable into the first equation, and find the value of the first variable.

4. Check the solutions.

To solve equations by elimination:

1. Arrange the equations in standard form.

2. Multiply one or both equations by numbers that create opposite values (additive inverses) of the coefficients on one pair of like terms.

3. Add the equations to eliminate one variable, and solve for the variable that remains.

4. Use substitution to find the other variable.

5. Check the solutions.

8.1 Solving Systems by Graphing

A true statement such as $0 = 0$ in the solution to a system of equations implies that there are an infinite number of solutions. In the case of two equations in two variables, the lines are coincident.

A false statement such as $0 = 1$ in the solution to a system of equations implies that there are no real-number solutions. In the case of two equations in two variables, the lines are parallel.

In quantity-value tables:

- Adding the items in the quantity column gives the total quantity.

- Adding the items in the $Q \cdot V$ column gives the same number as multiplying the total quantity by the average value (across the last row).

- The total of the $Q \cdot V$ column divided by the total of the quantity column gives an average value. The last number in the value column is not found by adding the items in the value column.

8.2 Solving Systems with Matrices

Matrices must have the same shape in order to be added or subtracted.

The product of a matrix and an identity matrix is the original matrix:

$$[A][I] = [A]$$

The product of a matrix and its inverse is an identity matrix:

$$[A][A]^{-1} = [I]$$

Matrices must be square and have a nonzero determinant in order to have an inverse.

Use the value of the determinant to verify that a unique matrix solution exists before actually solving the system. For a 2×2 matrix:

$$\text{If}\quad [A] = \begin{bmatrix} a & b \\ c & d \end{bmatrix}, \quad \text{then}\quad \det [A] = ad - bc.$$

To solve a system of linear equations on the calculator:

1. Change the system of equations into a matrix equation.

2. Specify the shape of matrix A, and enter matrix A into the calculator.

3. Specify the shape of matrix B, and enter matrix B into the calculator.

4. Calculate the value of the determinant of matrix A, det $[A]$. If det $[A] = 0$, stop; if det $[A] \neq 0$, go on to step 5.

5. Calculate $[A]^{-1}[B]$. The numbers are the values of the variables for the system of equations and appear in the same order as in the matrix of variables $[X]$.

8.3 Matrix Solutions of Linear Systems

If a system of equations has exactly one solution, the system is consistent. The determinant of a matrix for a consistent system is nonzero.

If the determinant is zero, the system is either inconsistent or dependent. A system of equations is inconsistent if it has no solution. A system of equations is dependent if it has an infinite number of solutions.

To find what is causing a zero determinant value in a system of three or more equations, solve the system algebraically.

8.4 Conic Sections

To identify a conic section from its equation, arrange the equation in standard form.

For a circle, the x^2 and y^2 terms have the same sign and the same coefficients.

For an ellipse, the coefficients on the x^2 and y^2 terms are different but have the same sign.

For a hyperbola, the x^2 and y^2 terms have opposite signs.

A rectangular hyperbola has equation $y = k/x$ or $xy = k$. The x and y are multiplied, and no squared term is present.

The parabola has either x or y squared, but not both.

8.5 Solving Nonlinear Systems

Systems of nonlinear equations may be solved by substitution and by graphing.

8.6 Solving Systems of Inequalities

Systems of linear and nonlinear inequalities may be solved by graphing the inequalities separately and identifying the overlapping region.

CHAPTER ⑧ REVIEW EXERCISES

In Exercises 1 to 4, using guess and check and set up equations as necessary. Find the three dimensions of the boxes. It may not be necessary to specify which number is length, width, or height in all problems.

1. A box has rectangular sides. The sum of the length, width, and height is 12 inches. The areas of two surface rectangles are 15 square inches and 12 square inches. Two solutions are possible.

2. The three different-sized rectangles forming the sides of a box have areas of 48 square inches, 40 square inches, and 30 square inches.

3. A box with rectangular sides has a length equal to the sum of the width and the height. The width is 2 more than the height. The volume of the box is 240 cubic inches. Specify the length, width, and height.

4. A box has rectangular sides. The sum of the length, width, and height is 21 inches. The areas of two surface rectangles are 30 square inches and 24 square inches. Two solutions are possible.

In Exercises 5 to 14, first solve the system of equations by substitution or elimination. Then set up a matrix equation for each, and solve with a calculator.

5. $3x + 5y = 5$
   $2x - 4y = 18$

6. $5x - 102 = 9y$
   $3.5x + 1.5y = 9$

7. $1.5x = 3.5y + 19$
   $5.5x + 1.5y = 41$

8. $y = -x/3 - 1/3$
   $2.5x - 4y = 20.5$

9. $-3x + 4y = 5$
   $y = \frac{3}{4}x + 2$

10. $x + 2y - z = -1$
    $-3x - y + z = -4$
    $x + 3z = -3$

11. $3x - y + z = 4$
    $-x + z = 2$
    $-y + 4z = 10$

12. $2x - 4z = 20$
    $5y + 3z = -19$
    $-2x + 3y = -14$

13. $2x + y - z = 3$
    $x + y + 2z = 4$
    $x - y - 3z = 6$

14. $x + y + z = 180$
    $x = 2y$
    $x = z$

Use these matrices for Exercises 15 to 20.

$$[A] = \begin{bmatrix} 2 & 4 \\ 1 & 3 \end{bmatrix} \quad [B] = \begin{bmatrix} 1 & 2 \\ 4 & 3 \end{bmatrix} \quad [C] = \begin{bmatrix} 1.5 & -2 \\ -0.5 & 1 \end{bmatrix}$$

15. $[A] + [B]$

16. $\det [A]$

17. $[A][B]$

18. $[A][C]$

19. $[C][A]$

20. $[A][I]$

Write equations for the conic sections described in Exercises 21 to 24.

21. A parabola with vertex at the origin, opening to the left, and passing through $(-3, 1)$

22. A circle with center at the origin and radius $= 16$

23. An ellipse centered at the origin with x-intercepts $(\pm 3, 0)$ and y-intercepts $(0, \pm 2)$

24. A hyperbola centered at the origin with y-intercepts $(0, \pm 4)$, no x-intercepts, and $a = 3$

In Exercises 25 to 34, identify the straight lines, parabolas, hyperbolas, circles, and ellipses in each system. Solve the system of equations.

25. $y = x^2 + x - 2$
    $y = -x - 3$

26. $y = -x^2 - x - 2$
    $y = x + 2$

27. $y = x^2 - 4x - 1$
    $y = -x + 3$

28. $y = x^2 + 2x - 5$
    $y = -x - 1$

29. $x^2 + y^2 = 16$
    $y = x^2 + 1$

30. $y = \dfrac{1}{x}$
    $x^2 + y^2 = 2$

31. $y = x^2 - x$
    $y = -x^2 + 3$

32. $y = -x^2 + 1$
    $y = x^2 - 5$

33. $y = x^2 + 1$
    $y = x^2 - 5$

34. $y = x^2 - 1$
    $\dfrac{x^2}{4} + y^2 = 1$

Solve the systems of equations in Exercises 35 to 38.

35. $y = 2^x$
$y = 4 - 2^x$

36. $y = 3^x$
$y = 7 - 3^{x+1}$

37. $y = \log x$
$y = 3 - 2 \log x$

38. $y = 5 - \log x$
$y = \log x$

In Exercises 39 to 46, solve with systems of equations.

39. Jeanne Maiden-Naccarato and Tish Johnson lead the Women's International Bowling Congress with the most 300 games (a perfect score in bowling). Jeanne has 2 more perfect games than Tish. Together they have 40 perfect games. How many does each have?

40. An order for disinfectant requires 2000 milliliters of a 3% Lysol solution from a 25% stock solution. How much water and stock solution need to be blended?

41. A cup of raisins has 215 total milligrams in calcium, phosphorus, and iron. The amount of phosphorus is 1 milligram less than twice the amount of calcium. There is 47 times as much phosphorus as iron. How many milligrams are there of each substance?

42. A cup of skim milk contains 201 calories. The cup contains a total of 49 grams of protein, fat, and carbohydrates. There are 10 more grams of carbohydrates than of protein. Protein and carbohydrates have 4 calories per gram; fat has 9 calories per gram. How many grams each of protein, fat, and carbohydrates are there?

43. A 1-ounce snack bag of walnuts contains 184 calories. Protein and carbohydrates have 4 calories per gram; fat has 9 calories per gram. The total weight of protein, fat, and carbohydrates is 26 grams. There are 9 more grams of fat than of protein. How many grams each of protein, fat, and carbohydrates are there?

44. Audrey has $12.40 in her piggy bank. She has 74 coins altogether, in nickels, dimes, and quarters. The number of quarters is 5 less than double the number of nickels. How many of each coin does she have?

45. Sandra has $86 in her billfold. The bills consist of $10, $5, and $1 denominations. The number of $5 bills is 2 less than the total number of $10 and $1 bills. If she has 18 bills altogether, how many of each does she have?

46. Anders purchased textbooks, paperback books, and notebooks at the bookstore. Individual prices of the items are $26.00, $3.50, and $1.50, respectively. Individual weights of the items are 2 pounds, 1 pound, and 0.5 pound, respectively. The number of notebooks is the same as the total number of textbooks and paperback books. The total cost of the purchase was $152.50. The total weight was 17 pounds. How many of each item were purchased?

In Exercises 47 and 48, $y = ax^2 + bx + c$ is the equation of a transition curve between two hills. The slope of the curve at any point (x, y) is $m = 2ax + b$.

47. At (2800, 300), there is to be a slope of 4%. The slope is to be -3% at a horizontal distance of 2800 feet to the left. What are the coefficients a, b, and c of the transition curve?

48. At (3000, 500), there is to be a slope of 2%. The slope is to be -6% at a horizontal distance of 3000 feet to the left. What are the coefficients a, b, and c of the transition curve?

Use a graph to solve the inequalities or systems of inequalities given in Exercises 49 to 54.

49. $x^2 + y^2 \geq 4$

50. $\dfrac{x^2}{4} \geq y$

51. $y \geq x$
$\dfrac{x^2}{4} + \dfrac{y^2}{9} \leq 1$

52. $y \leq -\dfrac{1}{x}$
$x \geq y$

53. $x^2 + y^2 \leq 9$
$y \geq 2x^2$

54. $y \leq x^2 + 2x + 1$
$y \geq x + 2$

CHAPTER **8** TEST _____

1. Solve by guess and check or reasoning: The volume of a box with rectangular sides is 240 cubic inches. The area of one rectangular surface is 20 square inches, and the sum of the length, width, and height is 24 inches. Find the length, width, and height.

In Exercises 2 to 11, solve the systems by any method. Show sketches if you do the problem graphically, or show the algebra if you do the problem symbolically. Identify the conic sections in Exercises 6 to 9.

2. $2x - 3y = 6$
$4x + 5y = -32$

3. $x + 2y = -8$
$0.5x + y = -4$

4. $2x + y + 2z = 5$
$x + 3z = 4$
$4x + 3y = 0$

5. $3x + y + z = 8$
$2x - y - 2z = 17$
$-x - y + 3z = -9$

6. $y = 3x^2 - 4x - 5$
$y = x - 3$

7. $y = x^2 - 3x + 2$
$y = \frac{1}{2}x - 2$

8. $x^2 + y^2 = 4$
$-x^2 + y^2 = 4$

9. $y = x^2 + 1$
$x^2 + y^2 = 4$

10. $y = 3^x$
$y = 5 - 3^x$

11. $y = \log_3 x$
$y = 4 - \log_3 x$

12. Carol Heiss won the World Championship in figure skating for 5 straight years. Peggy Fleming won the championship for 3 straight years. Peggy's first championship came 10 years after Carol's. The sum of their first winning years is 3922. In what years did they first win?

13. An ounce of pecan halves contains 199 calories. There are 3 fewer grams of protein than of carbohydrates. The total weight of protein, fat, and carbohydrates is 26 grams. Protein and carbohydrates have 4 calories per gram; fat has 9 calories per gram. How many grams each of protein, fat, and carbohydrates are there?

14. An eye lubricant is to contain 1.4% polyvinyl alcohol. If 400 gallons are to be made, how much distilled water and 16% polyvinyl solution need to be blended by the manufacturer?

15. Riley has a total of 59 pennies, nickels, and quarters. He has 10 more quarters than nickels and a total of $6.35. How many of each coin does he have?

16. Marcilio buys cans of juice at $0.89 each, uniform apples at $0.59 each, and packages of Oreos® at $3.98 each. Individual weights are 17 ounces each for the juice, 8 ounces each for the apples, and 16 ounces per package of Oreos. The total weight of his purchase is 158 ounces. The total cost of the purchase is $15.07. He buys 11 items altogether. How many of each item does he purchase?

17. The equation $y = ax^2 + bx + c$ is the equation of a transition curve between two hills. The slope of the curve at any point (x, y) is given by $m = 2ax + b$. At $(0, 1500)$, there is to be a slope of -3%. The slope is to be 6% of $x = 3200$ ft. What are the coefficients a, b, and c of the transition curve?

18. If $[A] = \begin{bmatrix} 3 & -5 \\ -1 & 2 \end{bmatrix}$ and $[B] = \begin{bmatrix} 2 & 5 \\ 1 & 3 \end{bmatrix}$, find

   a. $[A][B]$

   b. $[A]^{-1}$

   c. $\det [B]$

19. Set up and solve a matrix equation for Exercise 2.

20. Graph each inequality and indicate the region that solves the system.

$$x^2 + y^2 \geq 16$$
$$6x \leq -y^2$$

FINAL EXAM REVIEW _____

PART I: SHORT ANSWERS

In exercises where words are enclosed in brackets, [ ], choose the correct word to complete the sentence.

Chapter 1

1. Evaluate $-x^2$ for $x = -2$.

2. Simplify $5 - 3(2 - x)$.

3. Simplify $\dfrac{xy^2}{x^2y}$.

4. Evaluate $\sqrt{(x + 2)^2 - 9}$ for $x = -2$.

5. Solve for x: $x - 6 + 3(2 - x) = 8$.

6. Solve for x: $2x + 3 = -4x$.

7. Solve and sketch a line graph of the solution: $3 - x \leq 8$

8. Solve with algebraic notation and with a graph: $x + 7 \leq 4x - 8$.

9. The -4 in $-4xyz$ is the [coordinate, coefficient] of xyz.

Chapter 2

10. If $f(x) = 3x + 4$, what is $f(-2)$?

11. A line parallel to the x-axis may be described by [a constant, an identity] function.

12. The graphs of $y = 2x - 3$ and $y = -2x + 3$ are [parallel, perpendicular, intersecting] lines.

13. Which equation in Exercise 12 is a decreasing function?

14. What is the equation of a line perpendicular to $y = 1.5x$ passing through the origin?

15. The range of the absolute value function is _____.

16. What is the slope of the line passing through (x_1, y_1) and (x_2, y_2)?

17. What is the equation of a line passing through the points $(-40, -40)$, $(0, 32)$, and $(100, 212)$? Show steps, and check with linear regression.

Chapter 3

18. If $f(x) = 2x^2 + x$, what is $f(-2)$?

19. If $f(x) = 2x^2$, what is $f(a + b)$?

20. Factor $4a^2 - b^2$.

21. What name is given to the graph of a quadratic equation, $y = ax^2 + bx + c$?

22. Complete the statement: The solutions of a quadratic equation, $ax^2 + bx + c = 0$, are _____.

23. State the Pythagorean theorem formula.

24. Solve $ax^2 + b = 0$ for x.

25. $(x + \underline{\quad})^2 = x^2 + 5x + \underline{\quad}$

26. Simplify $\sqrt{0.36x^2}$ for any real number x.

27. Give an example of an irrational number.

Chapter 4

28. What is the vertex of $y = 2x^2 - 8x - 5$?

29. Multiply $(2 - i)(3 + i)$.

30. What is the conjugate of $-3 - 9i$?

31. Simplify $\sqrt{-16}$.

32. The vertex tells us the _____ or _____ value of a quadratic function.

33. What are the solutions to $x^2 + 2x + 5 = 0$?

34. The expression $b^2 - 4ac$ is called the [determinant, discriminant, discriminate].

35. What is the maximum number of possible solutions to the equation $ax^4 + bx^3 + cx^2 + dx + e = 0$?

Chapter 5

36. What is the additive inverse, or opposite, of $x - 2$?

37. $x^3 - x$ is the least common denominator for fractions with denominators of $x^2 + x$, x, and what other expression?

38. $y = k/x$ is an equation for [direct, inverse] variation.

39. Solve for the missing sides of the similar triangles in the figure.

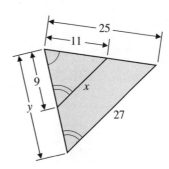

40. Without graphing, finding its equation, or using linear regression, how can we use two ordered pairs to determine whether a linear function passes through the origin?

41. The intensity of illumination from a light source is given by $I = P/r^2$, where P is in watts and r is distance from the light source in meters. Solve for r.

In Exercises 42 to 50, simplify or perform the indicated operations.

42. $\dfrac{4x^2}{28x}$

43. $12x\left(\dfrac{1}{12} + \dfrac{2}{3x}\right)$

44. $\dfrac{5x^2 + 15x}{6x + 18}$

45. $\dfrac{a^2 + 2a}{2a - 6} \cdot \dfrac{a^2 + a - 12}{a + 2}$

46. $\dfrac{x + 1}{x - 3} + \dfrac{x}{x^2 - 9}$

47. $\dfrac{x - 2}{x^2 - 3x + 2} - \dfrac{2x}{x - 2}$

48. $\dfrac{x^2 - 8x + 12}{x^2 - 1} \div \dfrac{2x - x^2}{x^2 - 5x - 6}$

49. Divide $x^3 + 2x^2 - 3x + 4$ by $x + 1$.

50. Divide $x^3 - 2x^2 + 3x + 4$ by $x - 1$.

In Exercises 51 to 55, indicate for what values the expressions in each equation are defined, and solve the equation.

51. $\dfrac{x - 5}{10} = \dfrac{x + 3}{12}$

52. $\dfrac{x + 1}{8} = \dfrac{2}{x + 1}$

53. $\dfrac{x + 3}{x - 2} = \dfrac{2}{1}$

54. $\dfrac{x}{2} = \dfrac{3}{5 - x}$

55. $\dfrac{2}{x - 4} - \dfrac{1}{x} = \dfrac{x + 4}{3x}$

Chapter 6

56. The principal square root of a number is always _____.

57. Simplify $\sqrt[3]{8x^6y^{12}}$.

58. Change 34×10^{-2} to correct scientific notation.

59. Write $0.000\,000\,567$ in correct scientific notation.

60. Simplify $64^{1/3}$.

61. What is the distance between (x_1, y_1) and (x_2, y_2)?

62. We use a ____ exponent to indicate the reciprocal of a number.

63. If $x = 3^y$ and $x = 1$, what is y?

Simplify the expressions in Exercises 64 to 68.

64. $\sqrt{(3x + 2)^2}$

65. $(3x + 2)^2$

66. $(3 + 2i)^2$

67. $(\sqrt{3} + 2)^2$

68. $(3 + 2i)(3 - 2i)$

69. Change these expressions to radical form:

    a. $x^{1.5}$ **b.** $x^{n/m}$ **c.** $x^{1/2}(y^2)^{1/4}$

70. Change these expressions to exponent form:

    a. $\sqrt[3]{x^2}$ **b.** $\sqrt[4]{a^3 b^4}$, $b > 0$

71. Simplify and leave in radical form:

    a. $\sqrt[3]{a^3 b^4 c^5}$ **b.** $\sqrt[4]{w^4 x^5 y^8 z^{12}}$ for any real numbers

Chapter 7

72. Describe how to distinguish among arithmetic sequences (linear functions), quadratic sequences, and geometric sequences (exponential functions).

73. Complete this statement of the like bases property: If $a^x = a^y$ and $a > 0$, $a \neq 1$, then _____.

Solve the equations in Exercises 74 to 80 for n or for x.

74. $2^x = 2$ **75.** $16^x = 8$

76. $\left(\dfrac{1}{2}\right)^x = 8$ **77.** $0.01^x = 100$

78. $25^n = 0.04$ **79.** $9^{x+1} = \dfrac{1}{27}$

80. $4^{n-2} = 2$

81. Write $y = \log x$ in exponential form.

82. What is log base e of 30.21?

83. Write $\log_b a = c$ in exponential form.

84. In geometric sequences, consecutive terms have a common _____.

85. Apply the change of base formula to $y = \log_b a$ using log base 10.

86. Write as a logarithmic equation $y = 5^{x-2}$.

87. Name the base in $2^6 = 64$.

88. Name the base in $\log_4 1024 = 5$.

89. When we write a logarithm without a base, we mean log base _____.

90. The natural logarithm has base _____.

91. Calculating $y = $ antilog 2.3 means raising _____ to the power 2.3.

In Exercises 92 to 106, solve for x. Round decimals to the nearest thousandth.

92. $3^x = 19$ **93.** $2^{x+1} = 19$

94. $17^x = 10$ **95.** $10^x = 17$

96. $\log_8 1 = x$ **97.** $\log_2 x = 1$

98. $\log_4 x = 0$ **99.** $\log_{10} 2 = x$

100. $\log_{10} x = -2$ **101.** $\log_{10} 1000 = x$

102. $\log_{10} 0.001 = x$ **103.** $\log_9 27 = x$

104. $\ln x = -1$ **105.** $\ln x = 2$

106. $\ln x = e$

In Exercises 107 and 108, find the time needed for $1000 to grow to $2500 under the given conditions.

107. Interest is compounded semiannually at 9.4%.

108. Interest is continuously compounded at 8%.

Chapter 8

Solve the systems of equations in Exercises 109 and 110.

109. $y = 79 - x$ **110.** $5x + 3y = 17.8$
$2x - 3y = 28$ $4x - 2y = 5$

111. Set up a matrix equation for the following system and solve with a calculator:

$$2x + 3y - z = 10.7$$
$$3x - 4y + z = 4.2$$
$$4x - y + 2z = 19.6$$

112. In solving a system of two equations whose graphs are parallel, we may expect a result that is [false, such as $0 = 4$; true, such as $0 = 0$].

113. A system of two equations that has an infinite number of solutions might contain [parallel lines, perpendicular lines, coincident lines].

114. In the first Little League World Series, in 1947, Williamsport, PA beat Lock Haven, PA by 9 points. The sum of their scores was 23 points. Write and solve a system of equations that gives the scores.

In Exercises 115 and 116, identify the conic section represented by each equation and then solve the system of nonlinear equations. Write the solutions as coordinate points.

115. $x^2 + y^2 = 11$ **116.** $2x^2 + y^2 = 31$
$x^2 - y^2 = 7$ $y = x^2 + 2$

117. Sketch the solution set for this system of inequalities:

$$x^2 + y^2 \le 4$$
$$x \ge \tfrac{1}{4}y^2$$

118. What is the "determinant value" of this matrix?

$$\begin{bmatrix} \text{CON} & \text{TIONS!} \\ \text{ULA} & \text{GRAT} \end{bmatrix}$$

PART 2: EXTENDED PROBLEMS

Chapter 1

1. a. A league bowling handicap h is calculated in terms of the bowler's average a, where $h = 0.8(200 - a)$. Bowling scores (and hence averages) vary from 0 to 300. People with averages over 200 are considered excellent bowlers. A handicap is never negative. What is the handicap for an average of 100? 150? 200? 250?

b. Write and graph an equation for the scores if the average and the handicap are added. Then write and graph an equation for the average scores without the handicap.

Chapter 2

2. Any person in Sweden finding an ancient gold "treasure" must sell the object to the Swedish government at the current price of gold by weight plus 10%.

a. Give the equation for the finder's sale price.

b. In building the equation, does it matter whether we add the 10% to the weight or to the price per ounce?

c. In July 1995, a Swede announced he had found a gold necklace weighing about 17 ounces. The necklace was estimated to be 2000 years old. At the time, gold was selling for about $384 per ounce. What should he receive from the sale?

d. The newspaper article estimated that the man would get about $7450. Does this match the calculation in part c?

Chapter 3 (and Parts of Chapter 4)

3. For the equation $y = -4x^2 + 2x - 5$, give the domain, range, axis of symmetry, and vertex. Explain why some of the answers are related.

In Exercises 4 and 5, solve the equation by at least three different methods.

4. $0 = 5x^2 + x - 4$

5. $0 = 5x^2 + 8x - 4$

In Exercises 6 to 8, find the original quadratic equation and give the solutions, real or complex. State what the solutions indicate about the graph of the quadratic function related to the original equation.

6. $t = \dfrac{-6 \pm \sqrt{36 + 64}}{32}$

7. $x = \dfrac{-2 \pm \sqrt{4 + 60}}{2}$

8. $x = \dfrac{8 \pm \sqrt{64 - 80}}{10}$

9. Find a quadratic equation describing a graph that passes through $(-1, -6)$, $(1, -2)$, and $(5, 30)$.

10. For what x is $5x^2 + 8x - 4 \le 0$?

Chapter 4

11. Give two different quadratic equations whose graphs pass through the points $(-2, 0)$ and $(3, 0)$.

12. Use the graph of $f(x) = x^2 - 2x - 3$ in the figure to answer the following questions.

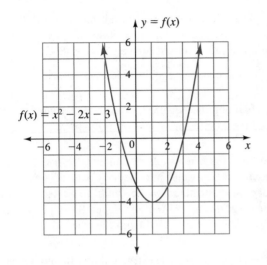

a. For what inputs x, if any, is $f(x)$ less than zero? Use both an inequality and an interval to answer.

b. As we trace the graph from left to right, for what inputs x is the graph increasing (rising)? Use both an inequality and an interval to answer.

13. Explain what is the same and what is different about each of these algebraic statements. Be as specific as possible. Tables or graphs may help.

$$x^2 - 3x - 10$$
$$y = x^2 - 3x - 10$$
$$0 = x^2 - 3x - 10$$

14. What is the significance of c in the graph of $y = ax^2 + bx + c$?

15. What is the significance of a in the graph of $y = ax^2 + bx + c$?

16. The graph of a quadratic function $f(x) = ax^2 + bx + c$ does not cross the x-axis. What does this tell us about the solutions to $f(x) = 0$?

17. Show by substitution whether $-3 - 2i$ is a solution to $x^2 + 6x + 13 = 0$. If so, what is the other solution?

Chapter 5

In Exercises 18 to 20, simplify.

18. $\dfrac{\dfrac{55 \text{ mi}}{1 \text{ hr}} \cdot \dfrac{10 \text{ hr}}{1 \text{ day}}}{\dfrac{7 \text{ mi}}{1 \text{ gal}}}$

19. $\dfrac{y + \dfrac{1}{y}}{y - \dfrac{1}{y}}$

20. $\dfrac{3 - \dfrac{x}{2}}{x + \dfrac{6}{x}}$

21. What is the reciprocal of $\dfrac{x}{2} + \dfrac{3}{x}$?

22. In 1993, the record for the fastest women's tennis service was 115 miles per hour. This record, set at Wimbledon, was held jointly by Brenda Shultz of the Netherlands and Jana Novotna of Czechoslovakia. How long does it take a 115-mile-per-hour service to travel the 78-foot length of the court?

23. Play the teacher; look at the solutions below, presented by four different students attempting to solve the formula $A = \frac{1}{2}h(a + b)$ for b. Show whether each student has or has not used correct algebra to get to the given answer. Which might be the most correct answer? Why?

a. $b = \dfrac{2A}{h} - a$ **b.** $b = \dfrac{A}{\frac{1}{2}h} - a$

c. $b = \dfrac{A - \frac{1}{2}ah}{\frac{1}{2}h}$ **d.** $b = \dfrac{2A - ah}{h}$

Chapter 6

24. Solve $\sqrt{3x - 11} = \sqrt{5x} - 3$ from the graph in the figure below.

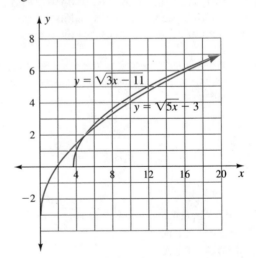

25. Solve $\sqrt{3x - 11} = 1$ with symbols and from the graph in the figure above.

26. Solve $\sqrt{5x} - 3 = 5$ with symbols.

27. Solve $\sqrt{x - 9} = 9 - \sqrt{x}$ for x.

28. How much will be in a savings account if it starts with $1000 and receives 5% interest compounded annually for 10 years? compounded monthly?

29. At what rate of interest, compounded annually, will $1000 double in 10 years? in 8 years? in $5\frac{3}{8}$ years?

Chapter 7

In Exercises 30 to 34,

a. *Identify the type of sequence: arithmetic, quadratic, or geometric.*

b. *For arithmetic and geometric sequences, write the equation using a_n and using linear or exponential regression on a calculator. Show that the results are the same with both methods.*

c. *For quadratic sequences, use quadratic regression to find the equation.*

d. *Find the sum to 10 terms for each geometric sequence. (See Appendix 2.)*

e. *Find the infinite sum if it is defined for the sequence. (See Appendix 2.)*

30. $\frac{1}{16}, \frac{1}{32}, \frac{1}{64}, \frac{1}{128}$

31. 27, 9, 3, 1

32. 4, 8, 16, 32, 64

33. 1, 6, 15, 28, 45

34. 2, 6, 10, 14, 18

35. The yearly sales figures for Rich Products Corporation (Associated Press, June 25, 1995) are as follows: 1945, $28,000; 1946, $474,000; 1952, $1.4 million; 1960, $3.86 million; 1976, $100 million; 1994, $1 billion. Graph the data. Fit an exponential equation to the data. Discuss your results, and include annual growth rate.

Chapter 8

36. A student is attempting to solve the system of equations

$$2x - 3y = 18$$
$$5x + 2y = 26$$

The student correctly solves the first equation for x. The student then substitutes for x in the first equation and from the results concludes that the two original equations have graphs that are coincident lines. Work through each step of the student's solution. Explain why the student thought the lines were coincident, and then solve the system correctly.

37. Jesse Owen won an Olympic gold medal by completing a 100-meter run in 10.3 seconds. Florence Griffith Joyner set a world record of 10.49 seconds in the 100-meter run 52 years later. Joyner's time would have won the 1956 men's 100-meter run. The sum of the years for Owen's and Joyner's races is 3924. Write and solve a system of equations that gives the year for each race.

38. Ward bought 11 grocery items, with a total weight of 13 pounds 15 ounces and a total cost of $13.58. A can of tomatoes weighs 17 ounces, a package of pizza sauce weighs 16 ounces, and a jar of salsa weighs 2 pounds 4.5 ounces. A can of tomatoes costs $0.50, a package of pizza sauce costs $1.62, and a jar of salsa costs $2.86. Set up and solve the system of equations needed to find how many of each product Ward bought. (*Note:* 16 ounces = 1 pound.)

39. The equation of a transition curve between two hills is $y = ax^2 + bx + c$. The slope of the curve at any point (x, y) is $m = 2ax + b$. At $(0, 800)$, there is to be a slope of -2%. At a horizontal distance of 2500 feet to the right, the slope is to be 4%. What are the coefficients a, b, and c of the transition curve?

40. The equation of a transition curve between two hills is $y = ax^2 + bx + c$. The slope of the curve at any point (x, y) is $m = 2ax + b$. At $(3200, 1000)$, there is to be a slope of 4%. At a horizontal distance of 3200 feet to the left, the slope is to be -5%. What are the coefficients a, b, and c of the transition curve?

APPENDIX I

Selected Formulas

Distance, rate, and time $d = rt$

Simple interest $I = prt$, where I is the interest earned on p dollars at an annual interest rate r for t years

Compound interest $S = P(1 + r)^t$, where S is the future value, P is the dollars put in or principal or present value, r is the annual interest rate, and t is the time in years

Triangle*   Area $= \frac{1}{2}$ base $\cdot$ height $= \frac{1}{2}bh$

Square   Area $=$ side $\cdot$ side $= s^2$
Perimeter $= 4 \cdot$ side $= 4s$

Rectangle   Area $=$ length $\cdot$ width $= lw$
Perimeter $= 2l + 2w$

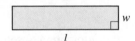

Parallelogram   Area $=$ base $\cdot$ height $= bh$

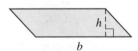

Trapezoid   Area $= \frac{1}{2}$ height $\cdot$ (sum of parallel sides)
$= \frac{1}{2}h(a + b)$

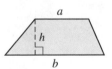

Circle   Area $= \pi r^2$, $r =$ radius
Circumference $= 2\pi r = \pi d$
Diameter $= d = 2r$

Rectangular prism (box)   Surface area $= 2lw + 2hl + 2hw$
Volume $= lwh$

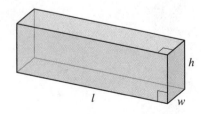

*In all formulas, the base and height (or length and width) refer to dimensions that are perpendicular.

Cylinder	Surface area $= 2\pi r^2 + 2\pi rh$ Volume $= \pi r^2 h$, $r =$ radius, $h =$ height

Sphere	Surface area $= 4\pi r^2$ Volume $= \left(\frac{4}{3}\right)\pi r^3$

Cone	Surface area $= \pi r l + \pi r^2$ Volume $= \frac{1}{3}\pi r^2 h$

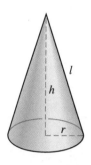

Distance formula (length of a line segment on a graph)

$$d = \sqrt{(x_2 - x_1)^2 + (y_2 - y_1)^2}$$

Exponential equation

$$y = ab^x, \ b > 0, \ b \neq 1$$

Linear equation

$$ax + by = c \ \text{(standard)}$$
$$y = mx + b \ \text{(slope-intercept)}$$

Logarithmic equation

$$y = \log_b x, \ b > 0, \ b \neq 1$$

Pythagorean theorem (where c is always the hypotenuse)

$$a^2 + b^2 = c^2$$

Quadratic equation

$$y = ax^2 + bx + c$$

Quadratic formula

If $ax^2 + bx + c = 0$, then
$$x = \frac{-b + \sqrt{b^2 - 4ac}}{2a} \ \text{or}$$
$$x = \frac{-b - \sqrt{b^2 - 4ac}}{2a}$$

Slope formula

$$m = \frac{y_2 - y_1}{x_2 - x_1}$$

y-intercept

$$b = y - mx$$

Sequences

Arithmetic nth term:
$$a_n = a_1 + (n - 1)d$$

Arithmetic sum:
$$S_n = \frac{n}{2}(a_1 + a_n)$$

Geometric nth term:
$$a_n = a_1 r^{n-1}$$

Geometric sum:
$$S_n = \frac{a_1(1 - r^n)}{1 - r} \text{ for } n \text{ terms}$$

$$S_n = \frac{a_1}{1 - r} \text{ for an infinite number of terms}$$

APPENDIX 2

Sequences and Their Sums

Arithmetic Sequences

In Section 2.4, we examined arithmetic sequences. Arithmetic sequences are characterized by a constant first difference, d. We use the first term, a_1, of the sequence and the constant difference to build the nth term, a_n.

nth Term of an Arithmetic Sequence

$$a_n = a_1 + (n - 1)d$$

ARITHMETIC SEQUENCES AND LINEAR FUNCTIONS An arithmetic sequence is associated with a linear function, where the constant difference of the arithmetic sequence is the slope of the linear function. Subtracting the common difference from the first term gives the y-intercept for the linear function. The domain for arithmetic sequences is the positive integers, whereas the domain for linear functions is all real numbers.

SUMS OF ARITHMETIC SEQUENCES The story is told of an eighteenth-century teacher in Germany who wanted to keep his young students busy for a considerable period. He asked them to add the numbers from 1 to 100. Within a few minutes, before the teacher could get comfortable doing something else, one student raised his hand and said that he had the answer. The teacher was amazed at the 10-year-old boy's technique. The teacher supposedly took the boy, Johann Friederich Carl Gauss (1777–1855), to a private tutor who was able to help Gauss get into college at age 15. Gauss became one of history's most highly regarded mathematicians.

Gauss was a superb problem solver. While his classmates assumed that they should start adding numbers one at a time, Gauss reconsidered the assumptions and came up with a pairing of numbers, as shown in Example 1.

EXAMPLE **1** Solving Gauss's problem Add the sequence of numbers from 1 to 100,

$$1 + 2 + 3 + 4 + 5 + \cdots + 97 + 98 + 99 + 100$$

Solution Gauss's technique was to add pairs of numbers:

$$1 + 2 + 3 + 4 + 5 + \cdots + 97 + 98 + 99 + 100$$

Because $1 + 100 = 101$, $2 + 99 = 101$, $3 + 98 = 101$, and so forth, all Gauss needed to finish the problem was the number of such pairs, which is 50. The sum is $50 \cdot 101 = 5050$. ●

639

Gauss's technique for addition was based on adding the first and last terms and multiplying by half the number of terms. The process works for adding any arithmetic sequence.

EXAMPLE ❷   Adding arithmetic sequences

a. To identify the arithmetic sequence shown below, give the first term, a_1, the common difference, d, and the number of terms, n.

$$8, 12, 16, 20, 24, 28, 32, 36, 40, 44, 48, 52$$

b. Use Gauss's pairing method to add the numbers.

Solution   **a.** *Sequence:*   8, 12, 16, 20, 24, ...   The common difference is 4.

Differences: 4,   4,   4,   4, ...   Arithmetic
$a_1 = 8$, $d = 4$, and $n = 12$ terms.

b. 8, 12, 16, 20, 24, 28, 32, 36, 40, 44, 48, 52

There are 12 numbers, and 12/2 = 6 pairs, each adding to 60. The sum is 6(60) = 360.   ●

Example 3 shows that Gauss's method also works with an odd number of numbers.

EXAMPLE ❸   Adding arithmetic sequences   Suppose you earn $20,000 during the first year and receive a $1000 pay increase at the end of each subsequent year.

a. Write the arithmetic sequence for your earnings over the first 5 years. Give the first term, a_1, the common difference, d, and the number of terms, n.

b. Use Gauss's pairing method to add the numbers.

Solution   **a.** $a_i = \$20,000$, $d = \$1000$, and $n = 5$. The sequence is

$$\$20,000, \$21,000, \$22,000, \$23,000, \$24,000$$

b. $20,000 + $21,000 + $22,000 + $23,000 + $24,000

There are $2\frac{1}{2}$ pairs, each adding to $44,000. The sum is

$$2\tfrac{1}{2}(\$44,000) = \$110,000$$ ●

Gauss's pairing method leads us to a general formula for the sum of n terms in an arithmetic sequence.

Sum of n Terms of an Arithmetic Sequence

To add an arithmetic sequence, add the first and last terms and multiply by half the number of terms. The sum, S_n, of n terms is

$$S_n = \frac{n}{2}(a_1 + a_n)$$

EXAMPLE ❹   Finding the number of terms   Add 5, 8, 11, 14, 17, ..., 77.

Solution   In the sequence 5, 8, 11, 14, 17, ..., 77, $a_1 = 5$, $d = 3$, and $a_n = 77$. To find the number of terms n, we substitute into

$$a_n = a_1 + (n - 1)d$$

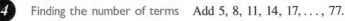

and solve for n.

$$77 = 5 + (n - 1) \cdot 3 \qquad \text{Subtract 5 from both sides.}$$
$$72 = (n - 1) \cdot 3 \qquad \text{Divide by 3.}$$
$$24 = n - 1 \qquad \text{Add 1 to both sides.}$$
$$25 = n$$

Thus, the sequence contains 25 terms. Substituting a_1, d, a_n, and n into the sum formula

$$S_n = \frac{n}{2}(a_1 + a_n)$$

we obtain

$$\tfrac{25}{2}(5 + 77) = 1025 \qquad\qquad \bullet$$

The sum of a sequence is called a **series**. This is an awkward word choice, because in ordinary English *series* means one event right after another or an order of a number of things. This ordinary English meaning is very close to the sequence concept, so it is easy to confuse the two words. We will emphasize the word *sum* and the adding concept rather than the word *series*.

Geometric Sequences

In Section 7.0, we examined geometric sequences. Geometric sequences are characterized by a constant ratio, r, of consecutive terms. We use the first term, a_1, of the sequence and the constant ratio to build the nth term, a_n.

nth Term of a Geometric Sequence

$$a_n = a_1 r^{n-1}$$

GEOMETRIC SEQUENCES AND EXPONENTIAL FUNCTIONS   A geometric sequence is associated with an exponential function, where the constant ratio of the geometric sequence is the base of the exponential function. Dividing the common ratio into the first term of the sequence gives the y-intercept for the exponential function. The domain for geometric sequences is the positive integers, whereas the domain for exponential functions is all real numbers. (See also page 479.)

SUMS OF GEOMETRIC SEQUENCES

EXAMPLE **5**

Finding sums: doubling sequence   One at a time, add the terms of the sequence $a_n = 2^{n-1}$, giving the grains of rice on a chessboard in Example 2 of Section 7.1.

Solution   The sequence $a_n = 2^{n-1}$ is 1, 2, 4, 8, 16, 32, 64, ... The sums of the terms, one at a time, are

$$S_1 = 1$$
$$S_2 = 1 + 2 = 3$$
$$S_3 = 1 + 2 + 4 = 7$$
$$S_4 = 1 + 2 + 4 + 8 = 15$$
$$S_5 = 1 + 2 + 4 + 8 + 16 = 31$$

The sums are 1 less than 2^n, or $S_n = 2^n - 1$. $\qquad\qquad \bullet$

In Example 5, both the sequence and the sum were based on powers of 2. This suggests that there is a formula for finding the sum of the terms.

Sum of n Terms of a
Geometric Sequence

The **sum of** n terms of **a geometric sequence** is given by

$$S_n = \frac{a_1(1 - r^n)}{1 - r}$$

In Example 6, we apply the formula to the doubling sequence in Example 5.

EXAMPLE **6** Finding sums Use the sum formula $S_n = \dfrac{a_1(1 - r^n)}{1 - r}$ on the sequence in Example 5.

Solution The first term, a_1, is 1. The common ratio, r, is 2.

$$S_n = \frac{a_1(1 - r^n)}{1 - r}$$

$$= \frac{1(1 - 2^n)}{1 - 2} = \frac{1 - 2^n}{-1} = 2^n - 1$$

The equation is the same as the one in Example 5. For 64 chessboard squares, the total number of grains of rice is $2^{64} - 1$. (See the projects in Exercises 45, 46, and 47 of Section 7.1 for investigations into the quantity of rice needed to fill the request.) ●

In Example 7, we return to the superball bounce (Example 2 of Section 7.0).

EXAMPLE **7** Finding sums: superball bounce, continued The first 10 heights reached by the superball form the sequence 108, 72, 48, 32, 21.1, ..., 2.8 (see Figure 1). What is the total of these 10 heights?

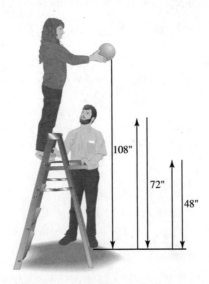

108"

72"

48"

Figure 1

Solution In order to find the sum of the first 10 heights, we add the numbers in the sequence 108, 72, 48, 32, 21.1,..., 2.8 using the sum formula, with $a_1 = 108$, $r = \frac{2}{3}$, and $n = 10$:

$$S_n = \frac{a_1(1 - r^n)}{1 - r}$$

$$S_{10} = \frac{108\left[1 - \left(\frac{2}{3}\right)^{10}\right]}{1 - \frac{2}{3}}$$

$$\approx 318.38 \text{ in.}$$

(*Note:* When entering the sum expression in the calculator, remember to put parentheses around the denominator.) ●

There is an important difference between the geometric sequences in Examples 6 and 7. The common ratio for the rice is 2. The sum grows larger faster and faster with each term added. The common ratio for the superball heights is $\frac{2}{3}$. The sum grows, but more slowly with each bounce.

When a geometric sequence has a common ratio between 0 and 1, the exponential expression describes exponential decay. The diminishing value of terms in the geometric sequence makes it possible to *add all the terms in an infinite sequence,* for an **infinite sum.**

Figure 2 shows an area model of the sum of the first five terms of $a_n = \left(\frac{1}{2}\right)^n$. Observe that the sum of the first five terms is less than 1. Because we are adding half the remaining area each time, the sum will stay less than 1.

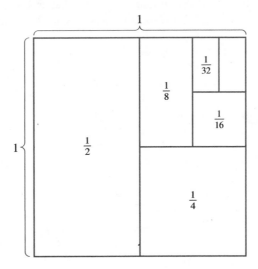

Figure 2

In Example 8, we examine a graphical model of the sum of the sequence described by $a_n = \left(\frac{1}{2}\right)^n$. The sum to n terms is $S_n = 1 - \left(\frac{1}{2}\right)^n$.

EXAMPLE

Exploring sums with a graph Graph the sum formula $S_n = 1 - \left(\frac{1}{2}\right)^n$ with x replacing n (see Figure 2). Trace or use a table to find the sum at 10 and 15 terms. Does there seem to be an upper limit?

Solution We change the sum formula to a formula in x and y: $y = 1 - \left(\frac{1}{2}\right)^x$. Figure 3 is the graph. The sum for 10 terms is at (10, 0.99902); for 15 terms, it is at (15, 0.99997). The graph rises toward but does not cross $y = 1$. The number 1 appears to be an upper limit for the graph and for the sum of the terms of the sequence. We say that the line $y = 1$ is a horizontal asymptote for the graph of $y = 1 - \left(\frac{1}{2}\right)^x$.

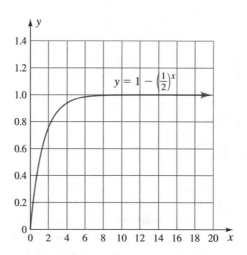

Figure 3

To add all the terms of a geometric sequence, we use the sum to infinity formula.

Sum to Infinity Formula

If the common ratio, r, is between 0 and 1, the sum to an infinite number of terms for a geometric sequence with first term a_1 is

$$S_\infty = \frac{a_1}{1 - r}$$

EXAMPLE

Finding sums   What is the sum to an infinite number of terms for the sequence $a_n = \left(\frac{1}{2}\right)^n$ from Example 8?

Solution   With $a_1 = \frac{1}{2}$ and $r = \frac{1}{2}$,

$$S_\infty = \frac{\frac{1}{2}}{1 - \frac{1}{2}} = 1$$

The formula is consistent with the graphical and tabular results.

APPENDIX ❷ EXERCISES

ARITHMETIC SEQUENCES

In Exercises 1 to 6, what is the sum of the first 20 terms of each arithmetic sequence?

1. 18, 16, 14, 12, 10, . . .

2. 31, 27, 23, 19, 15, . . .

3. 10, 18, 26, 34, 42, . . .

4. 4, 10, 18, 28, 40, . . .

5. 3, 9, 27, 81, . . .

6. −16, −9, −2, 5, 12, . . .

7. Explain why Gauss's system of adding pairs of numbers would not be helpful in adding a sequence such as 1, 3, 6, 10, 15, 21, 28, 36, 45, 55, 66.

In Exercises 8 to 11, assume the setting is a sequence, write a_n, and then answer the question. The questions do not ask for sums.

8. A population census of the United States is required by the Constitution. The census is taken every 10 years. How many times will the census be taken between 1790 and 2000?

9. Dominique bought a block of basketball tickets numbered consecutively from 47 to 113. How many seats does she have?

10. Ravi Tej bought a block of tickets to a play. The tickets are numbered consecutively from 53 to 99. How many tickets did he receive?

11. The Summer Olympic Games are held every 4 years. If the first modern Olympics was held in 1896, what should be the total number of games held between 1896 and 2000? World events bonus: Why were the games canceled in 1916, 1940, and 1944?

12. Shareen changes her car's oil every 3000 miles. She bought the car at 22,000 miles and immediately changed the oil. At 109,000 miles, she has just changed the oil again. How many oil changes has she made altogether?

13. At the end of the first hour of a treatment crisis, Lari tests a sample of the water leaving the city waste water system and directs that the water be tested every 3 hours thereafter. At the 145th hour, a test is made. How many tests have been made altogether?

14. Solve the formula $a_n = a_1 + (n - 1)d$ for n. If you were to solve Exercises 12 and 13 by reasoning, how would that reasoning explain the form of the answer in this exercise?

Projects

15. *Sum of Odd Numbers*

   a. Find a_n for the sequence of odd numbers, 1, 3, 5, 7, 9, ... using the formula for the nth term of an arithmetic sequence.

   b. Use linear regression on the points (1, 1), (2, 3), (3, 5), (4, 7), ... to find an expression for the nth term.

   c. What pattern is formed by adding odd numbers as follows?

   $$1$$
   $$1 + 3$$
   $$1 + 3 + 5$$
   $$1 + 3 + 5 + 7$$

   and so on.

   d. How can we relate the pattern observed in part a to the odd numbers pictured in the figure?

   e. Find the sum of n odd numbers using the formula for the sum of an arithmetic sequence.

16. *Trapezoid Area.* Build a model or draw a figure to suggest why the formula for the sum of an arithmetic sequence,

$$\text{sum} = \tfrac{1}{2}n(a_i + a_n)$$

and the formula for the area of a trapezoid,

$$A = \tfrac{1}{2}h(a + b)$$

have the same form. Blocks or grid paper may be helpful.

GEOMETRIC SEQUENCES

In Exercises 17 to 22, find the sum to 10 terms, and find the infinite sum, if appropriate.

17. 9, 27, 81, 243, ...

18. 3.5, 7, 14, 28, ...

19. 64, 32, 16, 8, 4, ...

20. 8, 4, 2, 1, $\frac{1}{2}$, ...

21. 3, 9, 27, 81, ...

22. $\frac{1}{2}, \frac{1}{4}, \frac{1}{8}, \frac{1}{16}, \ldots$

23. Find the sum of the first 20 heights of the superball in Example 7.

24. Find the sum in Exercise 23 to an infinite number of bounces.

25. Find the sum of the first 20 heights of the superball if the initial height is 135 inches and each rebound is $\frac{2}{3}$ of the previous height.

26. Find the sum in Exercise 25 to an infinite number of bounces.

27. Find the infinite sum of the sequence $1, \frac{1}{2}, \frac{1}{4}, \frac{1}{8}, \frac{1}{16}, \ldots$.

28. Find the infinite sum of the sequence $1, \frac{1}{3}, \frac{1}{9}, \frac{1}{27}, \ldots$.

29. What is the sum to infinity of the distances traveled by a child on a swing if the swing starts with an initial distance of 18 feet and goes $\frac{5}{6}$ of the prior distance on each subsequent swing?

30. Explain why

$$S_n = \frac{a_1(1 - r^n)}{1 - r}$$

yields the same value as

$$S_n = \frac{a_1(r^n - 1)}{r - 1}$$

Projects

31. *Repeating Decimals.* The repeating decimal 0.3333333... may be written as the sum of a geometric sequence:

$$0.3 + 0.03 + 0.003 + 0.0003 + \cdots$$

a. Write this sequence in fractions. Determine the first term and the common ratio, and find the sum to an infinite number of terms. Write the sum as a fraction.

b. Convert 0.4444444... to a fraction, as you did the repeating decimal in part a.

c. Convert 0.12121212... to a fraction, as you did the repeating decimal in part a.

d. Convert 0.42424242... to a fraction, as you did the repeating decimal in part a.

e. Convert 0.99999999... to a fraction, as you did the repeating decimal in part a.

32. You normally give your teenage daughter $10 per week. On January 1, she asks that her weekly allowance for the coming year be 1¢ the first week, 2¢ the second week, 4¢ the third week, 8¢ the fourth week, and so on, doubling each week as the year progresses.

a. How much will she be paid in the 10th week?

b. How much will she be paid in the 20th week?

c. How much will she be paid altogether in the first 20 weeks?

d. Make a graph of her allowance for the first 12 weeks.

e. Is this plan a good deal for her?

f. What is the total allowance paid for the entire year?

APPENDIX 3

The Binomial Theorem

This section introduces an arrangement of numbers that appears in many applications of mathematics including genetics, probability, and transportation. This symmetrical arrangement of numbers will give us a shortcut for finding powers of binomials.

Exploring Patterns

We begin this section with three questions, followed by three examples that suggest how we might answer the questions.

QUESTION 1   The lines in Figure 1 represent city streets. If one can travel *only south or east* on the streets, how many different ways are there to get from point A to point B?

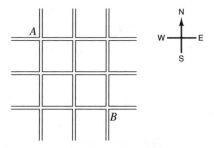

Figure 1

QUESTION 2   What are all the possible birth orders for a family with 4 children? Assume there are no twins.

QUESTION 3   What is $(a + b)^4$?

To solve each question, we might *consider an easier problem.* Solving a simpler but related problem is a good strategy for building understanding of a problem and designing a plan to solve it. The bonus in this set of three questions is that all have related solutions.

EXAMPLE ❶   Exploring street grids   For Question 1, consider the number of ways to get to each street intersection, one at a time. Use Figure 2.

a. Find the number of ways from A to B_1.

b. Find the number of ways from A to B_2.

c. Find the number of ways from A to B.

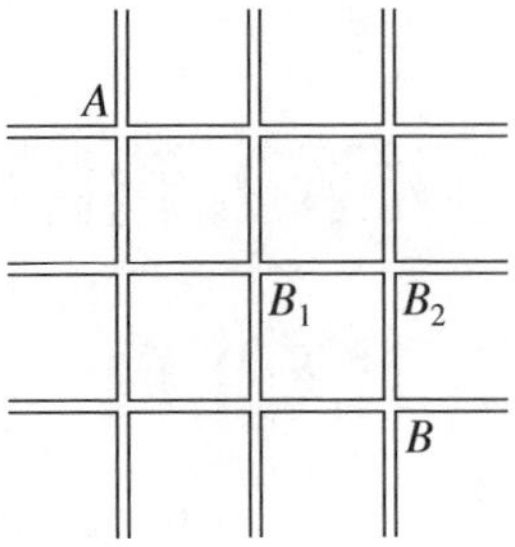

Figure 2

Solution   **a.** As shown in Figure 3, there are 2 ways to leave position *A*: 1 to the east and 1 to the south. We record a 1 at each adjacent intersection in Figure 3(a). Because we are limited to moving east or south, each of the ways gives 1 route to B_1. Thus, from *A* to B_1, there are 2 routes. We record a 2 at B_1 to indicate the 2 routes.

b. If we count routes to B_2, we find there is 1 additional route to B_2, beyond our results recorded earlier. Thus, there are 3 ways to get from *A* to B_2. We record a 3 at B_2 in Figure 3(b).

c. Continuing to record numbers on the intersections confirms that there are 6 routes to *B* from *A*, as shown in Figure 3(c).

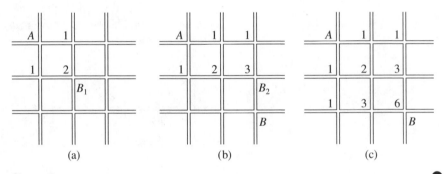

(a)   (b)   (c)

Figure 3

EXAMPLE **2**   Exploring birth order   Simplify Question 2 in the following way.

a. What are the possible children in a single-child family?

b. What are the possible birth orders for a 2-child family?

c. What are the possible birth orders for a 3-child family?

d. Return to Question 2.

Solution   **a.** The one child is either a boy or a girl:

Girl   Boy

b. With two children, we have these birth orders:

Girl Girl   Girl Boy   Boy Boy
                 Boy Girl

c. With three children, we have these birth orders:

GGG   GGB   GBB   BBB
         GBG   BGB
         BGG   BBG

d. With four children, we have these birth orders:

GGGG	GGGB	GGBB	BBBG	BBBB
	GGBG	GBGB	BBGB	
	GBGG	GBBG	BGBB	
	BGGG	BBGG	GBBB	
		BGBG		
		BGGB		

Thus, we have

1 birth order for 4 girls and 0 boys,

4 birth orders for 3 girls and 1 boy,

6 birth orders for 2 girls and 2 boys,

4 birth orders for 1 girl and 3 boys, and

1 birth order for 0 girls and 4 boys. ●

EXAMPLE

Exploring binomial powers   To answer Question 3, take the products, one power at a time.

a. Find $(a + b)^2$, using a table or mentally.

b. Find $(a + b)^3$ by multiplying the answer in part a by $(a + b)$.

c. Find $(a + b)^4$ by multiplying the answer in part b by $(a + b)$.

Solution   **a.** $(a + b)^2$ means $(a + b)(a + b)$. We can multiply these expressions with a table, as we did in Section 3.4.

Multiply	a	$+b$
a	a^2	$+ab$
$+b$	$+ab$	$+b^2$

Thus,

$$(a + b)^2 = a^2 + 2ab + b^2$$

b. $(a + b)^3 = (a + b)(a + b)(a + b)$
$$= (a + b)(a + b)^2$$
$$= (a + b)(a^2 + 2ab + b^2)$$

We can take the answer from the table in part a and multiply it by $(a + b)$ in a new table.

Multiply	a^2	$+2ab$	$+b^2$
a	a^3	$+2a^2b$	$+ab^2$
$+b$	$+a^2b$	$+2ab^2$	$+b^3$

Thus,

$$(a + b)^3 = a^3 + 3a^2b + 3ab^2 + b^3$$

c. Similarly,

$$(a + b)^4 = (a + b)(a + b)^3$$
$$= (a + b)(a^3 + 3a^2b + 3ab^2 + b^3)$$

Multiply	a^3	$+3a^2b$	$+3ab^2$	$+b^3$
a	a^4	$+3a^3b$	$+3a^2b^2$	$+ab^3$
$+b$	a^3b	$+3a^2b^2$	$+3ab^3$	$+b^4$

$$(a + b)^4 = a^4 + 4a^3b + 6a^2b^2 + 4ab^3 + b^4$$

We now examine the three questions together.

EXAMPLE **4** What do the answers to Questions 1, 2, and 3 have in common?

Solution The number 6 appears in each answer. This is not coincidental. Look at the pattern of numbers in this summary of each step in Example 3.

$$(a + b)^0 = \mathbf{1}$$
$$(a + b)^1 = \mathbf{1}a + \mathbf{1}b$$
$$(a + b)^2 = \mathbf{1}a^2 + \mathbf{2}ab + \mathbf{1}b^2$$
$$(a + b)^3 = \mathbf{1}a^3 + \mathbf{3}a^2b + \mathbf{3}ab^2 + \mathbf{1}b^3$$
$$(a + b)^4 = \mathbf{1}a^4 + \mathbf{4}a^3b + \mathbf{6}a^2b^2 + \mathbf{4}ab^3 + \mathbf{1}b^4$$

If we compare the numerical coefficients in the expressions with the numbers in Figure 4, an expanded street grid from Question 1, we see the same set of numbers. The coefficients of $(a + b)^4$ are the same as the diagonal numbers at the intersections four blocks from A. These numbers are part of an array called **Pascal's triangle.**

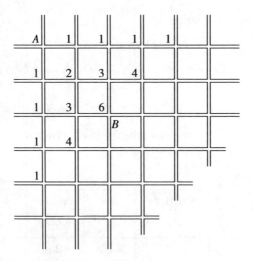

Figure 4

Here are the first few rows of Pascal's triangle. This array of numbers is filled with patterns, and it could be a course of study all to itself.

$$1$$
$$1 \quad 1$$
$$1 \quad 2 \quad 1$$
$$1 \quad 3 \quad 3 \quad 1$$
$$1 \quad 4 \quad 6 \quad 4 \quad 1$$
$$1 \quad 5 \quad 10 \quad 10 \quad 5 \quad 1$$
$$1 \quad 6 \quad 15 \quad 20 \quad 15 \quad 6 \quad 1$$
$$1 \quad 7 \quad 21 \quad 35 \quad 35 \quad 21 \quad 7 \quad 1$$

Compare the numbers of birth orders listed in parts a, b, and c of Example 2 with Pascal's triangle. Each set of birth orders matches a row of the triangle. ●

Think about it: What do you observe about the exponents on a? What do you observe about the exponents on b? (The answer is in the Binomial Theorem box.)

Using the Binomial Theorem

Through Pascal's triangle, each question is related to a concept called the **binomial theorem.** This theorem (derived fact), dealing with the powers of binomials, permits us to write out $(a + b)$ to any power we want.

Binomial Theorem

> The binomial theorem for $(a + b)^n$ includes these properties:
>
> • The numerical coefficients are the row starting $1 \ n \ldots$ in Pascal's triangle.
> • The exponents on a decrease, term by term.
> • The exponents on b increase, term by term.

EXAMPLE 5   Building powers of binomials   Write out the indicated power of the given binomial. Obtain the numbers from the appropriate row of Pascal's triangle. Use descending exponents on the first term of the binomial and ascending exponents on the second. Note how subtraction creates alternating signs within the answers to parts b and c.

a. $(x + 1)^4$ **b.** $(y - 1)^3$ **c.** $(1 - 3a)^4$

Multiplying out these expressions with the table method as a check is left to the exercises.

Solution   **a.** $(x + 1)^4 = 1x^4 + 4x^3 \cdot 1^1 + 6x^2 \cdot 1^2 + 4x \cdot 1^3 + 1 \cdot 1^4$
$$= 1x^4 + 4x^3 + 6x^2 + 4x + 1$$

The number coefficients (1, 4, 6, 4, and 1) come from Pascal's triangle. The exponents on x descend—the first term contains x^4, and the last term contains no x. The powers of 1 all equal 1.

b. $(y - 1)^3 = 1y^3 + 3y^2 \cdot (-1)^1 + 3y^1 \cdot (-1)^2 + 1 \cdot (-1)^3$
$$= 1y^3 - 3y^2 + 3y - 1$$

The initial number coefficients (1, 3, 3, and 1) come from Pascal's triangle. The y exponents descend. The expression $(y - 1)$ is the same as $[y + (-1)]$. The powers of (-1) alternate between positive and negative, so the signs in the answer alternate.

c. $(1 - 3a)^4 = 1 + 4(-3a)^1 + 6(-3a)^2 + 4(-3a)^3 + 1(-3a)^4$
$$= 1 + 4(-3a) + 6(9a^2) + 4(-27a^3) + 1(81a^4)$$
$$= 1 - 12a + 54a^2 - 108a^3 + 81a^4$$

The initial number coefficients (1, 4, 6, 4, and 1) come from Pascal's triangle. The exponents on a are in ascending order, and $(1 - 3a) = [1 + (-3a)]$. Observe that if the exponent on $(-3a)$ is even, the term is positive. If the exponent on $(-3a)$ is odd, the term is negative. ●

APPENDIX ③ EXERCISES

In Example 5, the binomial expressions were expanded using the binomial theorem. Check the expansions in Exercises 1 to 3 by doing the multiplications.

1. $(x + 1)^4 = (x + 1)(x + 1)(x + 1)(x + 1)$
$$= (x + 1)(x + 1)(x^2 + 2x + 1)$$
$$=$$
$$=$$

2. $(y - 1)^3 = (y - 1)(y - 1)(y - 1)$
$$= (y - 1)(y^2 - 2y + 1)$$
$$=$$

3. $(1 - 3a)^4 = (1 - 3a)(1 - 3a)(1 - 3a)(1 - 3a)$
$$= (1 - 3a)(1 - 3a)(1 - 6a + 9a^2)$$
$$= (1 - 3a)(1 - 9a + 27a^2 - 27a^3)$$
$$=$$

4. Describe how to get from one row to the next in Pascal's triangle. How is a row of Pascal's triangle related to the table method of multiplying polynomials?

5. Complete Pascal's triangle to the row starting 1 10.... Note that each row begins and ends with a 1 and that the other terms in the row are related to the row above by addition.

In Exercises 6 to 9, use the binomial theorem and Pascal's triangle to finish the expressions.

6. $(x + y)^8 = 1x^8 + 8x^7y^1 + 28x^6y^2 + 56x^5y^3 +$

7. $(a + b)^9 = 1a^9 + 9a^8b^1 + 36a^7b^2 + 84a^6b^3 +$

8. $(x - y)^9 = 1x^9 - 9x^8y^1 + 36x^7y^2 - 84x^6y^3 +$

9. $(a - b)^8 = 1a^8 - 8a^7b^1 + 28a^6b^2 - 56a^5b^3 +$

10. Why do the signs alternate in Exercises 8 and 9?

11. It is a common error to say $(a + b)^2 = a^2 + b^2$. Explain what is wrong with that statement. Which row of Pascal's triangle shows the coefficients of $(a + b)^2$? How does a table method of multiplying $(a + b)^2$ show what is wrong with the statement?

Use Pascal's triangle and the binomial theorem to write out the binomial powers in Exercises 12 to 24.

12. $(x + 1)^3$

13. $(x - y)^3$

14. $(1 + z)^4$

15. $(x + y)^6$

16. $(x + 1)^6$

17. $(b - 1)^4$

18. $(b - 1)^5$

19. $(1 + z)^5$

20. $(x + 2)^4$

21. $(x + 3)^4$

22. $(x - 2y)^3$

23. $(x - 3y)^3$

24. $(2a - b)^3$

25. Evaluate the first ten powers of 2: $2^1 = 2$, $2^2 = 4$, $2^3 = 8$, etc. Add the numbers in each row of Pascal's triangle. What do you observe?

26. List the 32 possible birth orders for 5 children. They fall into 6 different groups: all girls, 4 girls, 3 girls, 2 girls, 1 girl, and 0 girls. How many birth orders are in each group?

27. Use a calculator to make a list of the powers of 11 from 11^0 to 11^6. Compare your list with Pascal's triangle. What do you observe? How is the list similar to Pascal's triangle? How is it different?

28. Use the patterns started in Example 1 to find the number of ways from A to B on the street grids in parts a to d.

a.

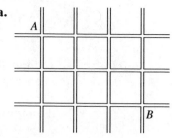

b.

d.

c.

Calculator Objectives

Here is a checklist of important calculator objectives, to complement the mathematical objectives at the beginning of each section of the text. Use of these scientific calculator keys is assumed: change of sign, reciprocal, parentheses, exponent, pi, store, and recall.

Use grouping symbols appropriately when entering equations and expressions.

Use the absolute value key.

Build a table, evaluate from a table, and solve equations.

Enter an equation, set a viewing window, graph, trace, and zoom.

Solve equations from a table and a graph.

Graph an inequality in two variables.

Plot data and use graphs to predict the type of equation.

Fit linear, quadratic, power, and exponential equations with regression.

Use the coefficient of correlation to find whether a slope is positive or negative.

Use a table to list terms in an arithmetic sequence.

Explore the effects of changing parameters on the graph of a linear, quadratic, absolute value, or exponential equation.

Use programs for evaluating formulas.

Select and deselect scientific notation mode.

Enter numbers in scientific notation with the $\boxed{\text{EXP}}$, $\boxed{\text{EE}}$, or $\boxed{\text{EEX}}$ key.

Correctly read a calculator display showing numbers in scientific notation.

Use the square root key, $\boxed{\sqrt{}}$, and recognize the calculator's display for expressions undefined in the real numbers.

Evaluate and graph logarithmic functions using the change of base.

Evaluate and graph expressions containing the natural number, e.

Solve systems of two or three linear equations.

Graph conic sections centered at the origin.

See "Calculators, graphing" in the Glossary/Index for page references to discussions of the various calculator techniques.

Answers to Selected Odd-Numbered Exercises and Tests

As you compare your answers to those listed here, keep these hints in mind:

- Don't give up too quickly.
- Have confidence that you worked the exercise correctly.
- Check that you looked up the right answer.
- Check that you copied the exercise correctly.
- See if you can use algebraic notation or simplification to change your answer to match the text's answer.

If the answers still don't match, try working the exercise again:

- Make sure you thoroughly understand the exercise. Read the exercise aloud. Shut the book and say it in your own words.
- On a separate piece of paper, copy the exercise from the text.
- Work the exercise without looking at your first attempt.
- Let the problem rest for an hour or two or overnight. Sometimes the solutions to problems become clear when you step away from them.
- Compare your work with that of another student. (Do this only after you have done the problem twice.)
- Review the text material and related examples.
- Go on to another problem. You can continue doing homework without having completed each and every exercise.
- At the next opportunity, ask your instructor to review your work. Note that this does not mean that you should ask him or her to show you how to do the exercise.

Only after trying the above steps should you assume either that your work is wrong or that the four to six human beings who worked every exercise for this book made an error. The latter is possible, and the author and publisher would appreciate corrections.

EXERCISES 1.0

1. **a.** natural numbers **b.** opposites **c.** reciprocals
   d. whole numbers **e.** rational numbers
3. **a.** $<$ **b.** simplify **c.** $\geq$ **d.** set **e.** $\sqrt{-1}$
5. **a.** commutative property for addition **b.** inequality
   c. factored form of the distributive property
   d. associative property for multiplication
   e. logic phrase "a or b"
7. **a.** positive **b.** negative **c.** positive **d.** zero
   e. positive
9. integer 11. irrational
13. **a.** $-18, -12, 45, 5$ **b.** $-4, -8, -12, -3$
   c. $-\frac{1}{4}, \frac{3}{4}, -\frac{1}{8}, -\frac{1}{2}$ **d.** $-\frac{3}{4}, -\frac{1}{12}, \frac{5}{36}, 1\frac{1}{4}$
   e. $2.25, 2.75, -0.625, -10$ **f.** $\frac{1}{12}, -1\frac{7}{12}, -\frac{5}{8}, -\frac{9}{10}$

15. **a.** -9 **b.** 14 **c.** 4.1 **d.** 4.1 **e.** 3 **f.** $-6\frac{1}{2}$
17. **a.** 9 **b.** -8 **c.** -9 **d.** $9x^2$ **e.** $3x^2$ **f.** $-3x^2$
19. **a.** $A \approx 27.8 \text{ m}^2$ **b.** $V \approx 3.5 \text{ in}^3$ **c.** $d = 132$
   d. $x = 13$
21. **a.** $x = -3$ **b.** $d = 10$ **c.** $m = \frac{3}{4}$ **d.** $x = 1\frac{2}{3}$
   e. $d = 2.04\overline{6}$
23. **a.** $3x^2 - 6x$ **b.** $2a(a - b)$
25. **a.** $6x + 12$ **b.** $5(3y + 2)$ 27. **a.** $x^2 + 5x$ **b.** $b(a + c)$
29. **a.** $-6 + 8x$ **b.** $-ab + ac$ **c.** $x^3 - 2x^2 - 3x$
31. **a.** $6, 6(x - 9)$ **b.** $15, 15(x - 15)$ **c.** $2a, 2a(3a - 4)$
33. **a.** 0 **b.** 210 **c.** 30 **d.** 190
35. **a.** $>$ **b.** $=$ **c.** $>$ **d.** $<$ **e.** $<$ **f.** $>$
37. $x \leq 2$ 39. $-1 < x < 5$
41. [number line with arrows, marks at -1 and 0] 43. $-2 < x < 4$ 45. $x < 2$ or $x > 4$
47. $x < -3$ or $x > 2$ 49. $x \leq -4$ or $x \geq 1$
51. Possible answer: A subtraction can be restated as an addition problem.

EXERCISES 1.1

1. **a.**

Input x	Output y
1	3
2	6
3	9
4	12
10	30

   b. 3 **c.** $y = 3x$
   d. for $x = 50, y = 150$;
   for $x = 100, y = 300$

3. **a.**

Input x	Output y
1	2
2	5
3	8
4	11
5	14
10	29

   b. 3 **c.** $y = 3x - 1$
   d. for $x = 50, y = 149$;
   for $x = 100, y = 299$

5. $10 - 2x = y$ 7. $y = \frac{1}{2}x + 10$ 9. $\frac{x}{15} + 2 = y$
11. $y = 7x - 14$ 21. $y = 1.29x - 0.45$ 23. $y = 1.25 - 0.05x$
25. $y = 125x + 85$
27. **a.** evaluate **b.** even numbers **c.** input numbers
   d. quotient **e.** constant

29. a. product   **b.** variable   **c.** expression   **d.** square of x
   e. sum

31. a. $x, 2, 3$ **b.** r, π, none   **c.** $x, -1, 4$ **d.** x, 1 and 1, -1

33. $3x^2 + 5x + 6$ **35.** $4a^2 + 3ab + 2b^2$

37. a. $a + 2b$ **b.** $2x^2 + 2x - 3$

39. a. $8ac - 3ac + 7bc$ **b.** $x^2 + 8x$

41. a. $15, \frac{1}{3}$ **b.** $16, \frac{3}{10}$ **c.** $240, \frac{22}{15}$

43. a. $be, \dfrac{a}{n}$ **b.** $b, a + c$ **c.** $1, \dfrac{x - 2}{2}$

45. a. $x, \dfrac{y + z}{y}$ **b.** $x, \dfrac{x + 2}{2}$ **c.** $xy, \dfrac{y - 2x}{2}$

47. a. 64   **b.** 64   **49. a.** 32   **b.** 4   **51. a.** -9 **b.** 36

53. a. $C = 0$ **b.** $C = -40$ **c.** $C = 37$

55. a. $L = 199$ **b.** $L = 200$ **c.** $L = 500$

57. a. $V = 14.1$ cm^3 **b.** $V = 113.1$ cm^3 **c.** $V = 904.8$ cm^3

59. a. 10   **b.** 13   **c.** 17

EXERCISES 1.2

1.

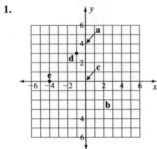

3. a. y-axis   **b.** origin   **c.** (x, y) **d.** x-axis
   e. perpendicular

5. a. $\{0\}$ **b.** $\{\ \}$ **c.** scale   **d.** counterclockwise
   e. coordinate plane

7. a. J **b.** E **c.** G **d.** K

9. a. $(-18, 0)$ **b.** $A(-40, -40)$

11.

x	y
-3	-3
-2	-1
-1	1
0	3
1	5
2	7
3	9

13.

x	y
-3	6
-2	5
-1	4
0	3
1	2
2	1
3	0

15.

x	y
-3	9
-2	7
-1	5
0	3
1	1
2	-1
3	-3

17.

x	y
-3	12
-2	6
-1	2
0	0
1	0
2	2
3	6

19.

x	y
-3	-7
-2	-2
-1	1
0	2
1	1
2	-2
3	-7

21.

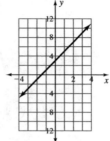

23.

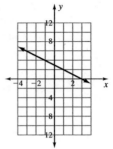

25.

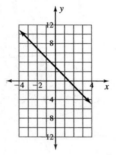

27.

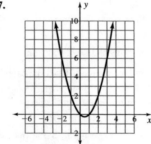

29.

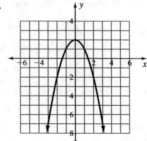

31. a. Exercises 21, 23, 25   **b.** Exercises 27, 29   **c.** $y = ax + b$

33.

Weight (lb)	Index
100	18.3
110	20.2
120	22.0
130	23.8
140	25.7

35.

Weight (lb)	Index
130	18.7
140	20.1
150	21.6
160	23.0
170	24.4

37. a. $W = 130$ lb   **b.** $W = 110$ lb

39. a. $W = 160$ lb   **b.** $W = 170$ lb

41. a. $x = -3$ **b.** $x = 2$

43. a. $x = -1$ **b.** $x = 3$

45. a. $x = 2$ or $x = -1$ **b.** $x = -3$ or $x = 4$
   c. no real-number solution   **d.** $x = 0.5$

47. a. $x = 0$ **b.** $x = -2$ or $x = 2$ **c.** $x = -4$ or $x = 4$
   d. no real-number solution

49. Find the output for an input of 2.

51. Find the *x*-coordinate for the point of intersection of the graphs for $y = 3x + 4$ and $y = 7$.

53. independent: *s*; dependent: *A*

55. independent: *s*; dependent: *A*

57. independent: *A*; dependent: *E*

59. independent: weight of package; dependent: cost of shipping

61. finding the independent variable

63. finding the dependent variable

65. finding the independent variable

MID–CHAPTER I TEST

1. a. -2 **b.** $-\frac{3}{4}$ **c.** 7   **d.** -0.6

2. a. 1700   **b.** 21   **c.** -160 **d.** $\frac{3}{22}$

3. a. -16 **b.** 16   **c.** -16 **d.** $20\frac{1}{2}$ **e.** $-\frac{1}{3}$ **f.** $-\frac{3}{5}$
   g. $3\sqrt{10} \approx 9.5$ **h.** $\frac{37}{30} \approx 1.2$

4. a. $3x + 6$ **b.** $-3x + 6$ **c.** $-4x - 13$ **d.** $2x$

5. a. $2, 2(2x - 9)$ **b.** $x, x(x + 5)$ **c.** $n, n(mn - p^2)$
   d. $7xy, 7xy(9x - 7y)$

6. a. 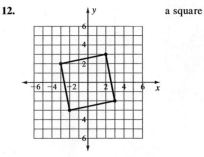 (number line: open circle at −4, 0)

   b. (number line: 0, open circle at 6)

   c. (number line: −3, 0, 2)

   d. (number line: 3, 6)

   e. (number line: open circle −2, 0, open circle 3)

   f. (number line: 0) all real numbers

7. a. variable: *x*; constant term: -4; coefficient: -1
   b. variable: *x*; constants: none; coefficients: $2, -1$

8. a. $-x^3 + 13x - 12$ **b.** $a^3 - 3a^2b - ab^2 + b^3$

9. a. $\frac{x}{y}$ **b.** $\frac{1}{y}$ **c.** $\frac{x - z}{x}$

10. a. independent: *s*; dependent: *h*; $h = 2\sqrt{3}$ yd
   b. independent: *d*; dependent: *A*; $A = 6.25\pi$ cm^2

11. 2, 3

12. 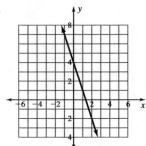   a square

13. a. Tables will vary.

(graph)

b. Tables will vary.

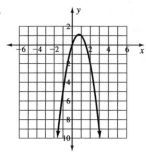

14. a. $x = -1$ or $x = 5$ **b.** $x = 1$ or $x = 3$ **c.** $x = 2$

15. a. $y > 0$ **b.** $x < 0$ **c.** $y < 0$ **d.** $x \geq 0$ **e.** $x > 0$
   f. $y \leq 0$

16. a.

Number of Pairs of Pens	Total Number of Panels
1	7
2	12
3	17
4	22
10	52
50	252
100	502

   b. $y = 5x + 2$

17. j   **18.** a   **19.** g   **20.** h   **21.** i   **22.** b   **23.** e

24. c

EXERCISES I.3

1. a. -10 **b.** -2 **c.** 1   **d.** 25   **e.** 1   **f.** -18

3. a. 3   **b.** $\frac{1}{5}$ **c.** 3   **d.** 0   **e.** $\frac{8}{7}$ **f.** $-\frac{3}{8}$

5. a. $x = 4$ **b.** $x = -9$ **c.** $x = -2$ **d.** $x = -12$
   e. $x = 28$ **f.** $x = 32$ **g.** $x = -36$ **h.** $x = -32$
   i. $x = 40$

7. take off jacket, take off vest, take off shirt; dressing and undressing   **9.** inverse not meaningful; taking pictures

11. order not important

13. a. $x = 6$ **b.** $x = 5$ **c.** $x = -1$ **d.** $x = 2$

15. a. $x = -13$ **b.** $x = 50$ **c.** $x = 41$ **d.** $x = -22$

17. $x = -1$ **19.** $x = 15$ **21.** $x = 24$ **23.** $x = 32$

25. $x = -3$ **27.** $x = -9$ **29.** $x = 12$ **31.** $x = 1.5$

33. $x = -5.5$ **35.** $x = 35$ **37.** $x = 38$ **39.** $t = \frac{D}{r}$

41. $b = y - mx$ **43.** $r^3 = \frac{3V}{4\pi}$ **45.** $p = \frac{1}{N}$ **47.** $b = \frac{2A}{h} - a$

49. $a_1 = a_n - (n - 1)d$ **51.** $a = 3A - b - c$

53. $a_1 = \frac{2S}{n} - a_n$ **55.** $T_c = -ET_h + T_h$

57. $T = \frac{V - 344}{0.6} + 20$ **67.** $d = \dfrac{\frac{2S}{n} - 2a}{n - 1}$

EXERCISES I.4

1. $y = -2$ is below $y = -2x + 3$ for $x < 2.5$.

3. $y = -2x + 3$ is above $y = 0$ for $x < 1.5$.

5. $y = x + 4$ is above $y = -\frac{1}{2}x + 1$ for $x > -2$.

7. $y = 0$ is below $y = x + 4$ for $x > -4$.

9. $y = -\frac{1}{2}x + 1$ is below or same as $y = 0$ for $x \geq 2$.

11. $y = -2x + 3$ is above or same as $y = 3$ for $x \leq 0$.

13. $y = x + 4$ is below or same as $y = 4$ for $x \leq 0$.

15. $x > -1$

17. $x < \frac{3}{2}$

19. $x < \frac{1}{3}$

21. $x \geq \frac{1}{3}$

23. $x \leq -\frac{1}{3}$

25. $y = x - 2$ is below $y = x$ for all real numbers.

27. For example, $x - 2 > x$ **29.** $x > \frac{1}{3}$ **31.** $x < 6$

33. The negative changes the relative value of the sides; for example,
$20 > 10 \rightarrow (-1)(20) > (-1)(10) \rightarrow -20 < -10$.

35.

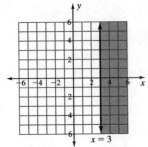

37.

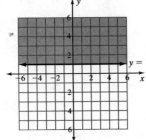

39.

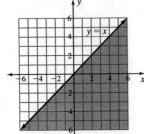

41.

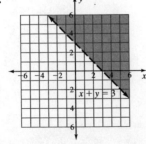

43.

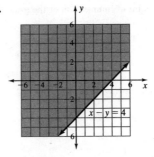

45.

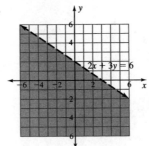

47.

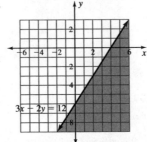

49. $15x + 50y \geq 500,000$. If only student tickets are sold, sales of 33,334 tickets are needed. If only regular tickets are sold, sales of 10,000 tickets are needed. Other ticket sales on the boundary line represent possible combinations exactly meeting the goal.

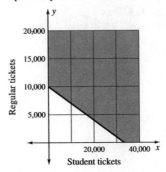

51. $500x + 250y \leq 2500$

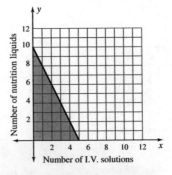

53. A dot indicates that the value is a solution, whereas a small circle indicates that the value is excluded from the solution.

55. Select a test point. If the test point makes the inequality true, then shade the half-plane that contains the point. Otherwise, shade the other half-plane.

CHAPTER 1 REVIEW EXERCISES

1. Associative properties for addition and multiplication, commutative properties for addition and multiplication, distributive property for multiplication over addition

3. Absolute value, braces, brackets, fraction bar, parentheses, square root

5. boundary line, half-plane

7. sum, difference, product, quotient

9. input-output relationships, dependent variable, independent variable

11. multiplicative inverses, reciprocals

	Input x	Input y	Output $x + y$	Output $x - y$
13. a.	−7	3	−4	−10
b.	3	−7	−4	10
c.	7	3	10	4

	Input x	Input y	Output xy	Output $x + y$
15. a.	3	4	12	7
b.	−2	3	−6	1
c.	−7	−3	21	−10
d.	−3	5	−15	2

17. $2\frac{2}{3}$; associative property for addition

19. 1300; commutative property for multiplication

21. 4   **23.** 16.5π in^2 **25.** \$7.75   **29.** $\frac{x}{15} + 8$

31. $-26x + 27y + 9$ **33.** -21 **35.** $A = 12.25\pi$ in^2

37.

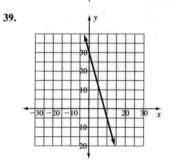

39.

41. $x = 4$ **43.** $F = 104$ **45.** $n = 12$ **47.** $r = \dfrac{I}{Pt}$

49. $b = \dfrac{c - a}{Y}$ **51.** $3x, 3x(2x + 5)$

---

53. $x > 1$

55. $x \geq 3$

57. a. $x = 0$ **b.** $x > 0$

59. a. $x = -1$ **b.** $x \leq -1$

63.

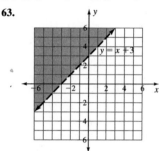

CHAPTER 1 TEST

1. opposite   **2.** ordered pair   **3.** independent

5. a. $-30, -24, 81, 9$ **b.** $-1, -2, -0.75, -3$
   c. $1\frac{1}{4}, 1\frac{3}{4}, -\frac{3}{8}, -6$

6. a. 748; associative property for addition
   b. 989; commutative property for addition

7. a. 160   **b.** -9 **c.** 9   **d.** 500   **e.** $6x - 8$

8. Four times the difference between x and 3

9. $b = 3A - a - c$

10. $b = \dfrac{2A}{h} - a$ **11.** $3xy$

12. a. $x = -5$ **b.** $x = 2$ **c.** $x = -2$
   d. $x > 2$ **e.** $x < -2$
   f. $x < 1$

14.

$y = \frac{1}{3}x + \frac{5}{3}$

EXERCISES 2.0

1. 6.5, 42, 42.5   **3.** \$28.00   **5.** 1073 bars

7. a. $y = 4 + 0.5(x - 2)$ for $x > 2$, $y = 4$ for $1 \leq x \leq 2$
   b. dot graph
   c.

Cost (dollars) / Number of faxes

9. a. $y = 20 + 5(x - 3)$ for $x > 3$, x rounded up to next integer; $y = 20$ for $0 < x \leq 3$ **b.** step graph

c.

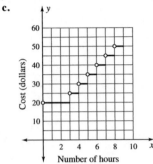

11. a. $y = 65 + 0.15(x - 100)$ for $x > 100$, $y = 65$ for $0 < x \leq 100$ **b.** dot graph (dots too close together to be seen on graph)   **c.**

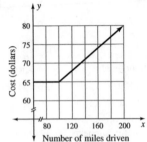

13. a. $y = 85 + 4.75(x - 10)$ for $x > 10$, $y = 85$ for $0 < x \leq 10$
b. dot graph (dots too close together to be seen on graph)   **c.**

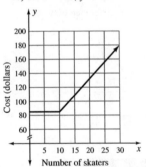

15. a. $y = 19.95 + 0.25(x - 100)$ for $x > 100$, x rounded up to next integer; $y = 19.95$ for $0 \leq x \leq 100$ **b.** step graph

c.

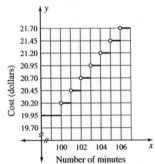

17. 11 pages   **19.** 10 hr   **21.** 1300 mi   **23.** 13 skaters
25. 140 min

EXERCISES 2.1

1. a.

x	y
1	11
2	26
3	41
4	56

b. $f(x) = 15x - 4$

3. a.

x	y
18	−4
6	−2
2	0
6	2
18	4

b. not a function

5. a.

x	y
−5	0
−2	21
0	25
2	21
5	0

b. $f(x) = 25 - x^2$

7. function   **9.** not a function   **11.** not a function
13. not a function   **15.** function   **17.** not a function
19. $y = x + 2$ **21.** $y = \frac{1}{3}x$
23. the first letter in the spelling of a state
25. a. 6:00 A.M.   **b.** undefined   **c.** 6:00 A.M.   **d.** 7:00 A.M.

27.

x	$f(x)$
−2	−5
−1	−3
0	−1
1	1
2	3
3	5
4	7

29.

x	$f(x)$
−2	5
−1	0
0	−3
1	−4
2	−3
3	0
4	5

31.

x	$f(x)$
−2	4
−1	7
0	8
1	7
2	4
3	−1
4	−8

33. a. 3   **b.** 11   **c.** $2n + 1$ **d.** $2n + 2m + 1$
35. a. 10   **b.** 0   **c.** $\square^2 + \square - 2$ **d.** $n^2 + n - 2$
e. $n^2 - 2nm + m^2 + n - m - 2$
37. 10   **39.** 9
41. Exercise 14   **43.** Exercise 15
45. x is any real number; $f(x) \geq -2$
47. x is any real number; $f(x) \leq 6$
49. a. domain   **b.** negative numbers plus zero
51. a. range   **b.** positive numbers
53. a. domain   **b.** positive numbers
55. a. range   **b.** negative numbers plus zero
57. a. $C(r) = 2\pi r$ **b.** $A(r) = \pi r^2$
59. a. $V(l, w, h) = lwh$ **b.** $V(r, h) = \pi r^2 h$
61. $I(p, r) = pr$
63. a. −1   **b.** 3   **c.** −5   **d.** 7
65. a. $x = 3$ or $x = -1$ **b.** $x = 6$ or $x = -4$
c. $x = 4$ or $x = -2$ **d.** $x = 9$ or $x = -7$
67. a. $x = -2$ or $x = 2$ **b.** $x = -1$ or $x = 1$
c. $x = -4$ or $x = 4$ **d.** $x = -6$ or $x = 6$

EXERCISES 2.2

1. a. x-intercept: −1; y-intercept: 2
b. x-intercept: 3; y-intercept: −2
3. x-intercept: −1 and 1; y-intercept: −1   **5.** linear   **7.** linear
9. not linear   **11.** 2b   **13.** 1a   **15.** Exercise 4   **23.** $-\frac{1}{3}$
25. $\frac{1}{2}$ **27.** $\frac{3}{5}$ **29.** $-\frac{3}{2}$ **31.** undefined   **33.** 0
35. a. 0   **b.** positive   **c.** 0   **d.** day 8 to 9   **e.** 2
37. a. \$0.06 tax/\$ sales
b. x-intercept: (0, 0); for zero sales, the tax is zero dollars.
c. y-intercept: (0, 0); there is zero sales tax for zero sales.
39. a. −\$0.75/trip
b. x-intercept: $\left(26\frac{2}{3}, 0\right)$; the maximum number of trips is 26.
c. y-intercept: (0, 20); the original value of the ticket is \$20.
41. nonlinear
43. $\frac{1}{2}, -2, \frac{1}{2}, -2$; opposite lines are parallel; adjacent lines are perpendicular.

45.

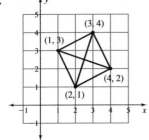

Diagonals are perpendicular. **a.** 3 **b.** $-\frac{1}{3}$

47. The slope is $-\dfrac{b}{a}$.

MID–CHAPTER 2 TEST

1. true   **2.** false; it has a zero in the denominator.

3. false; the change in the output becomes smaller for a given change in the input.

4. true   **5.** true   **6.** true   **7.** false; they multiply to -1.

8. false; $x = c$ is a vertical line.   **9.** non-negative   **10.** domain

11. undefined   **12.** $(0, y)$

13. a. -2 **b.** 4   **c.** -20 **d.** $3a - 5$
   e. $3(a + b) - 5$ or $3a + 3b - 5$

14. a. 0   **b.** 6   **c.** 30   **d.** $a^2 - a$
   e. $(a + b)^2 - (a + b)$ or $a^2 + 2ab + b^2 - a - b$

15. a. all real numbers   **b.** $y \geq 0$ **c.** yes

16. a. all real numbers   **b.** all real numbers   **c.** yes

17. a. $-5 \leq x \leq 1$ **b.** $-3 \leq y \leq 3$ **c.** no

18. a. 4   **b.** $-\frac{1}{4}$ **c.** $-\frac{1}{2}$ **d.** 2   **e.** 2   **f.** undefined
   d and e are parallel; a and b, c and d, c and e are perpendicular.

19. a.

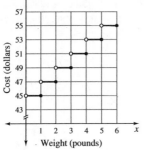

Months

b. 0.15   **c.** -0.01 **d.** -0.17 **e.** Jan. to Feb.
f. amount of price change per month
g. Iraq's invasion of Kuwait and the Gulf War

20. ≈ 59 mi

21. $y = 16.45 + 0.29(x - 30)$ for $x > 30$, x rounded up to next integer; $y = 16.45$ for $0 < x \leq 30$

22.

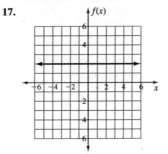

The segment of the graph with a y value of \$53 has an x value of over 4 pounds, up to and including 5 pounds.

EXERCISES 2.3

1. slope = \$0.055 per dollar; y-intercept = 0

3. slope = \$3.00 per person; y-intercept = \$10

5. slope = 2π; vertical axis intercept = 0

7. slope = μ; vertical axis intercept = 0

9. slope = b; vertical axis intercept = a **11.** $y = 8x - 4$

13. $y = \frac{1}{2}x - 8$ **15.** $y = -2x$ **17.** $y = \frac{8}{3}x - 2$

19. $y = -x + 2$

21. a. Pulse rate is a function of age.   **c.** $P = 0.5(220 - x)$
   d. $P = 0.7(220 - x)$ **e.** 85 to 119   **f.** 30

23. \$300; \$0.025; $C = 0.025x + 300$

25. $y = -\frac{1}{3}x + 5\frac{2}{3}$ **27.** $y = 3x - 11$ **29.** $y = -2x + 9$

31. $C = 43.70x + 75$; \$75; \$43.70

33. $F = \frac{9}{5}C + 32$ **35.** $y \approx 19.2x - 296$; lines are nearly the same

37. $y = -\frac{2}{3}x$ **39.** $y = 2x$ **41.** $y = -\frac{8}{5}x + 6\frac{1}{5}$

43. $y = \frac{4}{3}x + 3\frac{2}{3}$ **45.** $y = -\frac{2}{5}x - \frac{1}{5}$ **47.** linear; $y = -x + 17$

49. $y \approx 0.135x + 1.29$

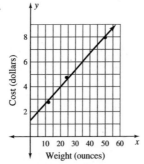

Weight (ounces)

51. Equation will approximate $y \approx 1051.3x - 32.9$.

EXERCISES 2.4

1. a. $-4, -2, -1, \ldots$; next term: $\frac{1}{2}$
   b. $-4, -4, -4, \ldots$; next term: -8
   c. $8, 8, 8, 8, \ldots$; next term: 27
   d. $-6, 10, -14, 18, \ldots$; next term: -12

3. 1b: $a_n = -4n + 12$ matches $y = -4x + 12$;
   1c: $a_n = 8n - 21$ matches $y = 8x - 21$

5. a. $a_1 = -1, d = 4$ **b.** 35   **c.** $a_n = 4n - 5$
   d. $y = 4x - 5$

7. a. $a_1 = 11, d = 6$ **b.** 65   **c.** $a_n = 6n + 5$
   d. $y = 6x + 5$

9. a. $a_1 = 9, d = 2$ **b.** 27   **c.** $a_n = 2n + 7$
   d. $y = 2x + 7$

11. a. $a_1 = 7, d = -3$ **b.** -20 **c.** $a_n = -3n + 10$
   d. $y = -3x + 10$

13. if the common difference is constant

15. 400, 320, 240, 160, 80, 0; 5 months; \$100 interest;
   25% of total borrowed

17.   domain: all real numbers;
                              range: 2

19.

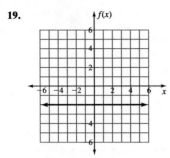

domain: all real numbers;
range: −2

21. constant function   **23.** constant function
25. constant function   **27.** identity function

29.

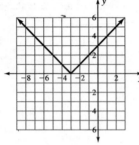

domain: all real numbers;
range: $y \geq 0$

31.

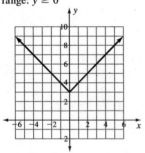

domain: all real numbers;
range: $y \geq 3$

33.

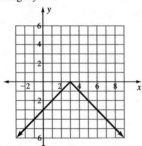

domain: all real numbers;
range: $y \leq 0$

35.

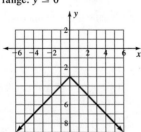

domain: all real numbers;
range: $y \leq -3$

37. a. $x = -1$ or $x = 4$ **b.** $x = 0$ or $x = 3$ **c.** $x = \frac{3}{2}$
d. no solution

39. a. −2   **b.** 2   **c.** 3, 0   **d.** $y = -2x + 3, x \leq 1.5$ **e.** 0
f. $y = 2x - 3, x \geq 1.5$ **g.** 0

41. a. $x = -10$ or $x = 2$ **b.** no solution   **c.** $x = -6$ or $x = -2$
d. $x = 4$ or $x = -10$

43. a. $x = -2$ **b.** $x =$ all real numbers   **c.** $-6 < x < 2$
d. $x =$ all real numbers   **e.** no solution
f. $x = 1$ or $x = -5$ **g.** $x < -5$ or $x > 1$

CHAPTER 2 REVIEW EXERCISES

1. vertical-line test   **3.** dot graph   **5.** relevant domain

7. common difference   **9.** identity function

11. absolute value function (the squaring function is also correct but
not in the list.)   **13.** subscripts

15. line of best fit, linear regression, point-slope equation, rule for
the nth term of a sequence, slope-intercept equation

17. ≈114 lb, ≈144 lb

19. $x =$ amount of purchase, $y =$ total cost; $y = 1.07x$; straight line

21. $x =$ number of children, $y =$ total cost; $y = 40 + 4.5(x - 10)$
for $x > 10$, $y = 40$ for $0 < x \leq 10$; dot graph

23. $x =$ number of minutes for a call, $y =$ total cost;
$y = 1.08 + 0.63(x - 2)$ for $x > 2$, x rounded up to next integer;
$y = 1.08$ for $0 < x \leq 2$; step graph

25. a. $-6.2 \leq x \leq 2.2$ **b.** $-3 \leq y \leq 3$ **c.** no

27. a. all real numbers   **b.** $y \geq 0$ **c.** yes

29. a. 1   **b.** 4   **c.** $\frac{1}{2}$ **d.** 8   **e.** 2   **f.** $x \geq 1$
g. none   **h.** $x \leq 2$ **i.** 1   **j.** $\mathbb{R}$ **k.** $y > 0$

31. a. $\{-1\}$ **b.** $\{-4, 2\}$ **c.** $\{-5, 3\}$ **d.** $\{-3, 1\}$
e. $\{\ \}$ or $\varnothing$

33. a. −8.9°C   **b.** 2.4°C   **c.** 4 km   **d.** 1 km   **e.** 15.7°C

35. 0   **37.** 0   **39.** 15   **41.** $2\square^2 - 3\square + 1$

43. a. $y = -\frac{1}{3}x$ **b.** $y = -\frac{1}{3}x + 5$

45. −1; $y = -x + 4$; neither; 4; 4   **47.** −1; $y = -x$; neither; 0; 0

49. 2; $y = 2x + 1$; neither; $-\frac{1}{2}$; 1   **51.** 0; $y = 3$; horizontal; none; 3

53. The lines formed by Exercises 45 and 47 are parallel.

55. nearly linear

57. a. $y = 0.065x$
b. slope = $0.065 tax/$ purchase; y-intercept = 0; no tax if
nothing is purchased

59. a. $y = 500 + 45x$
b. slope = $45/hour of repair; y-intercept = $500; basic
inspection cost

61. a. using first and last, $y \approx 6.67x + 0.33$.   **b.** $7.00
c. A linear equation is not appropriate. The amount of plastic
needed to make the larger combs is larger, relatively speaking.

63. $y \approx 11,528 - 42x$ **65.** 43   **67.** 1   **69.** 1, 5, 9, 13

71. $a_n = -2n + 20$ **73.** $a_n = 3n - 8$ **75.** $C = 350; constant

77. $C = 5x + 350$; increasing   **79.** $V = 35 - 1.50x$; decreasing

81.

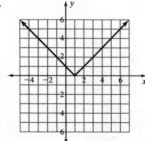

83. domain: all real numbers; range: all real numbers

85. domain: all real numbers; range: $y \geq -1$

87. domain: all real numbers; range: $y \geq 0$

89. domain: all real numbers; range: 1

CHAPTER 2 TEST

1.

Number of Copies	Total Cost (dollars)
1	5
2	5
3	7
4	9
5	11
6	13

Points should not be connected, since the numbers of copies are integer values.

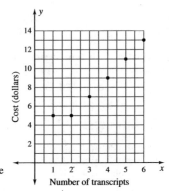

2. a. not a function **b.** function **c.** function **d.** not a function **3. a.** 12 **b.** -4 **c.** 4

4. a. $-\frac{2}{7}$ **b.** $y = -\frac{2}{7}x + \frac{24}{7}$ **c.** $-\frac{2}{7}$ **d.** $y = \frac{7}{2}x - 8$

5. a. zero **b.** negative; decreasing **c.** linear **d.** constant **e.** constant **f.** positive integers

6. a. $y = 7x + 2.5$ **b.** \$7 per mile

7. a. input = time in minutes = x; output = cost in dollars = y **b.** (1, 0.13), (19, 2.11) **c.** $y = 0.11x + 0.02$

8. $y \approx 10.1x - 13.8$

9. a. 50; arithmetic **b.** 54 **c.** 243 **d.** 19; arithmetic

10.

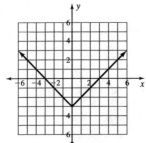

11.

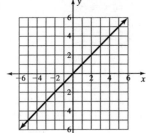

12.

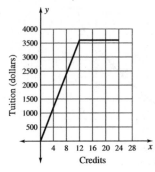

13.

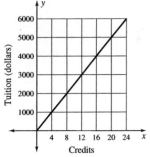

14. Times can vary.

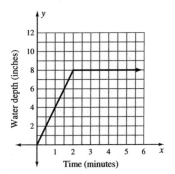

15.

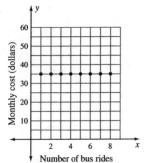

CUMULATIVE REVIEW OF CHAPTERS 1 AND 2

1.

Input x	Input y	Output xy	Output $x + y$	Output $x - y$
-2	4	-8	2	-6
-3	7	-21	4	-10
2	-3	-6	-1	5
-3	-2	6	-5	-1
-1	-6	6	-7	5
-5	-2	10	-7	-3
3	-2	-6	1	5
2	-9	-18	-7	11

3. a. opposites **b.** factors **c.** reciprocals **d.** factoring **e.** sets

5. $2ac - 2bc$ **7.** 56.25π ft^2 **9.** does not simplify

11. $x = 15$ **13.** $3x = x + 15; x = 7.5$

15. a product to a sum **17.** $x \geq 0$

19. subtraction to addition of the opposite number

21. ≈ 124.7 lb $\leq x \leq 157.5$ lb **23.** 1, 2, 3, 4

25. 1, 0, 1, 4 **27.** slope $= -\frac{2}{5}; y = -\frac{2}{5}x - \frac{7}{5}$

29. $y = -\frac{1}{3}x$ **31.** (4, 4); $f(x) = x$ **33.** (4, 4); $f(x) = 4$

35. $y = 65x + 500$

37. linear function

EXERCISES 3.0

1. a. $1 \cdot 48, 2 \cdot 24, 3 \cdot 16, 4 \cdot 12, 6 \cdot 8$
   b. $1 \cdot 36, 2 \cdot 18, 3 \cdot 12, 4 \cdot 9, 6 \cdot 6$
   c. $1 \cdot 72, 2 \cdot 36, 3 \cdot 24, 4 \cdot 18, 6 \cdot 12, 8 \cdot 9$

3. a. not a polynomial   **b.** binomial

5. a. monomial   **b.** trinomial

7.

m	n	$m + n$	$m \cdot n$
-3	-5	-8	15
3	4	7	12
2	6	8	12
3	5	8	15
-4	-6	-10	24
-2	-12	-14	24
-3	-8	-11	24
2	-6	-4	-12

9. a. $x^3 - 8$ **b.** $x^3 - 9x^2 + 27x - 27$

11. a. $x^3 + 3x^2 + 3x + 1$ **b.** $a^3 + b^3$

13. $x^3 - 3x^2y + 3xy^2 - y^3$ **15.** $6x^2 + 7x - 3$

17. $9x^2 + 6x + 1$ **19.** $x^3 + 8$ **21.** $x^2 + 3x - 18$

23. $x^2 - 3x - 18$ **25.** $x^2 - 11x + 18$ **27.** $x^2 - 16$

29. $x^2 - 10x + 25$ **31.** $2x^2 + 5x - 12$

33. $2x^2 - 5x - 12$ **35.** $9x^2 + 3x - 2$

37. $x^2 + 4x + 4$ **39.** $4x^2 - 1$ **41.** $x^3 - 8$

43. $x^4 - 13x^2 + 4$ **45.** $(2x - 3)(6x + 1)$

47. $(3x - 4)(x - 3)$ **49.** $(x + 4)(x + 3)$

51. $(x - 3)(x + 4)$ **53.** $(x - 3)(x - 4)$

55. $(x - 12)(x + 1)$ **57.** $(x + 7)(x - 4)$

59. $(2x + 5)(3x + 2)$ **61.** $(2x + 5)(3x - 2)$

63. $(6x - 5)(x - 2)$ **65.** $(3x - 2)(5x + 3)$

67. $(2x + 1)(3x + 5)$ **69.** $(10x + 1)(x + 6)$

71. $2(3x - 1)(x - 5)$ **73.** $x(x^2 + 2x + 4)$ **75.** $x(x - 3)^2$

77. $a(a^2 - ab + b^2)$ **79.** $a(a^2 + ab + b^2)$

81. They are all squares of binomials; Exercises 28 and 29

83. $b = \pm13, \pm8, \pm7$ **85.** $b = \pm19, \pm8, \pm1$

87. $3(x - 4)(x + 4)$ **89.** $3(2x - 3)(2x + 3)$

91. $5(2x - 3)(2x + 3)$ **93.** $7(2 - 3x)(2 + 3x)$

95. a. factors   **b.** monomial   **c.** binomial   **d.** trinomial
   e. square   **f.** perfect cube

EXERCISES 3.1

1. $x = \pm15$ **3.** $x = \pm1.1$ **5.** $x = \pm100$ **7.** no real number

9. 225   **11.** 7   **13.** 121   **15.** 10   **17.** 1.4   **19.** ≈ 8.660

21. ≈ 6.325 **23.** ≈ 14.142 **25.** ≈ 4.243, Exercise 20

27. ≈ 14.142, Exercise 23   **29.** ≈ 9.487, Exercise 24   **31.** $5\sqrt{6}$

33. $4\sqrt{5}$ **35.** $4\sqrt{2}$ **37.** ≈ 2.414 **39.** ≈ -0.366

41. ≈ 2.414 **43.** ≈ -0.366 **45.** $\dfrac{2 + \sqrt{2}}{2} \approx 1.707$

47. ≈ 1.707 **49.** $2\sqrt{2} \approx 2.828$

51. a. Both graphs are identical for $y \geq 0$. However, $y\sqrt{x}$ is a function, whereas $x = y^2$ is not.
   b. $4 = 2^2, 2 = \sqrt{4}; 9 = 3^2, 3 = \sqrt{9}; 16 = 4^2, 4 = \sqrt{16}$
   c. $y = \sqrt{x}$ has restriction $y \geq 0$; $x = y^2$ does not.
   d. x cannot be a negative number if y is a real number.

53. a. $x = 4\sqrt{5} \approx 8.944$ **b.** $x = \sqrt{14} \approx 3.742$
   c. $x = 2\sqrt{14} \approx 7.483$

55. a. $x = 8\sqrt{2} \approx 11.314$ **b.** $x = 12\sqrt{2} \approx 16.971$
   c. $x = 4\sqrt{2} \approx 5.657$ **57.** a right triangle

59. a right triangle   **61.** not a right triangle   **63.** a right triangle

65. a right triangle   **67.** ≈ 16 ft 11.6 in.   **69.** ≈ 23 ft 0.3 in.

71. $\sqrt{881}$ ft ≈ 29.7 ft   **73.** $4\sqrt{17}$ ft ≈ 16.5 ft   **75.** $20\sqrt{2}$ m

77. 2 ft   **79.** 6 in.   **81.** Both ratios are $\sqrt{2}$.

83. $4\sqrt{3}$ in. ≈ 6.9 in.   **85.** $\dfrac{5\sqrt{3}}{2}$ in. ≈ 4.3 in.   **87.** $A = \dfrac{x^2\sqrt{3}}{4}$

EXERCISES 3.2

1. a. 729; neither   **b.** 46; linear   **c.** 11; linear
   d. 49; quadratic

3. a. 3   **b.** 5   **c.** 2   **d.** 1

5. a. input variable: r; $a = \pi, b = 2\pi(3), c = 0$
   b. input variable: T; $a = \dfrac{g}{4\pi^2}, b = 0, c = 0$
   c. input variable: x; $a = \frac{1}{2}, b = \frac{1}{2}, c = 0$
   d. input variable: x; $a = 1, b = -2, c = 1$
   e. input variable: x; $a = 2, b = 4, c = 4$

7. x-intercepts: (1, 0), (3, 0); y-intercept: (0, 3); vertex: (2, −1)

9.

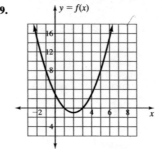

11. $(x - 3)(x - 1)$; setting factors equal to 0 and solving for x gives x-intercepts.

13. The symmetry of the table gives $f(6) = 15$.

15. a. $\{-1, 5\}$ **b.** $\{1, 3\}$ **c.** $\{-2, 6\}$ **d.** { } or $\varnothing$

17. a. $(-2, 0), (4, 0)$ **b.** $(0, -8)$ **c.** $x = 1$ **d.** $(1, -9)$
   e. $\{1\}$ **f.** $\{-1, 3\}$ **g.** $(-2, 4)$ **h.** $y \geq -9$
   i. $(x + 2)(x - 4)$; set the factors equal to 0 and solve for x.

19.

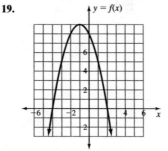

   a. $(-4, 0), (2, 0)$
   b. $(0, 8)$
   c. $x = -1$
   d. $(-1, 9)$
   e. $\{-4, 2\}$
   f. $\{-2, 0\}$
   g. { } or $\varnothing$
   h. $\{-3, 1\}$
   i. $y \leq 9$
   j. $-(x + 4)(x - 2)$; set the factors equal to 0 and solve for x.

21.

r	A
0	0
1	44.0
2	100.5
3	169.6
4	251.3
5	345.6
6	452.4

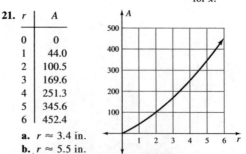

   a. $r \approx 3.4$ in.
   b. $r \approx 5.5$ in.
   c. The top and bottom surface area varies with the square of the radius.

23. a. initial or starting height of 32.8 ft   **b.** no
   c. ≈ -0.6 sec; not logical   **d.** 0 and ≈ 0.4 sec
25. ≈ 1.55 sec
27. $0 = -\frac{1}{2}(9.81)t^2 + 35$; $t \approx 2.7$ sec
29. $0 = -\frac{1}{2}(9.81)t^2 + 1.5t + 35$; $t \approx 2.8$ sec
33. Set each of the factors equal to 0 and solve for x.
35. $x = -4$

MID–CHAPTER 3 TEST

1. a. $x^3 - 7x^2 + 27x - 27$ **b.** $x^3 - 27$ **c.** $7a + b$
   d. $5a + b$
2. a. $x^2 - 25$ **b.** $x^2 - 2x + 1$ **c.** $-x^2 - 2x + 3$
   d. $-4x^2 + 9$
3. a. $(3x - 4)(x + 3)$ **b.** $x(x - 1)$ **c.** $(2x - 1)(3x + 4)$
   d. $3x(x + 1)^2$
4. $\sqrt{8}$ **5.** $\sqrt{50}$ **6.** $\sqrt{48}$ **7.** $\sqrt{72}$ **8.** $\sqrt{27}$ **9.** $\sqrt{12}$
10. $x^2 = 16$ has two solutions, $\{-4, 4\}$; whereas $x = \sqrt{16}$ has one, $\{4\}$
11. a. ≈ 1.816 **b.** ≈ 0.048 **12. a.** yes   **b.** yes   **c.** no
13. a. $x = 12$ **b.** $x = 8\sqrt{6}$ **c.** $x = 6.5$
14. a. 24; quadratic   **b.** 28; quadratic   **c.** 23; linear
   d. 29; other
15. $a = \frac{1}{2}$, $b = \frac{1}{2}$, $c = 1$
16. $a = 1000$, $b = 2000$, $c = 1000$
17. a.

x	$A(x)$
0	0
1	6
2	24
3	54
4	96
5	150
6	216
7	294
8	384
9	486
10	600

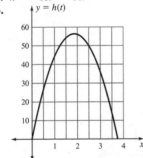

   b. It is 4 times the original.
18. a. $h = -16t^2 + 60t$
   b.

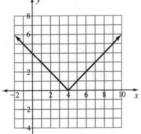

   c. (0, 0), (3.75, 0); the first is at the instant the ball is thrown.
   d. $\{1.75, 2\}$; on the way up and down
   e. 56.25 ft

EXERCISES 3.3

1. $\{-2, 3\}$ **3.** $\left\{-1, \frac{1}{2}\right\}$ **5.** $\{\pm 4\}$ **7.** $\{-5, 1\}$
9. $\{-1, 11\}$ **11.** $\{-7, -1\}$ **13.** $\{-2, 6\}$
15. $\left\{\pm \frac{1}{2}\right\}$ **17.** $\left\{-\frac{1}{2}\right\}$ **19.** $\{\pm 2\}$ **21.** $\{\pm 15\}$
23. $\{\pm 11\}$ **25.** $\{\pm 4\}$ **27.** $\{\pm 3\}$ **29.** $\{\pm\sqrt{3}\} \approx \{\pm 1.7\}$
31. $\{\pm\sqrt{5}\} \approx \{\pm 2.2\}$

33. a.

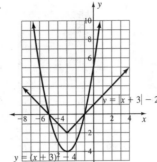

$y = |x + 3| - 2$

$y = (x + 3)^2 - 4$

   b. same
   c. x-intercepts are the solutions.

35. $4, -3$ **37.** $\frac{5}{2}, -3$ **39.** $\frac{4}{3}, -\frac{1}{2}$
41. a. $d = 2\sqrt{\dfrac{A}{\pi}}$
   b. 0.7854; the keys are next to each other, in a square.
43. ≈ 80 in.; ≈ 6.7 ft   **45.** $r \approx 2.8$ in.
47. a.

x	$f(x)$
-3	5
-1	3
0	2
1	1
3	1
5	3

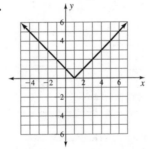

   b. $f(0) = f(4) = 2$
   c. $x = 2$

49.

51.

53. $\{\pm 4\}$ **55.** $\{-5, 1\}$ **57.** $\{3, 7\}$ **59.** $\{2, 6\}$
61. The vertex of the graph is now at $(0, -2)$ instead of $(2, 0)$.
63. $b\sqrt{a}$ **65.** $a\sqrt{b}$ **67.** $a^2\sqrt{b}$ **71.** ≈ 9.2 sec
73. a. $L = \dfrac{r^2}{12}$ **b.** ≈ 252 ft   **c.** ≈ 35 mph

EXERCISES 3.4

1. $x^2 + 6x + 9$ **3.** $x^2 - 12x + 36$ **5.** $x^2 - x + \frac{1}{4}$
7. $4x^2 - 12x + 9$ **9.** $(x - 2)^2$ **11.** $(x + 4)^2$ **13.** $(x + 9)^2$
15. $(x - 8)^2$

17.

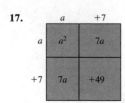

$(a + 7)^2 = a^2 + 14a + 49$

19.

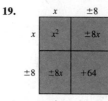

$(x + 8)^2 = x^2 + 16x + 64$ or
$(x - 8)^2 = x^2 - 16x + 64$

21. $x^2 + 4x + 4$ **23.** $x^2 - 18x + 81$ **25.** $x^2 - 7x + \frac{49}{4}$

27. $x^2 + 12x + 36 = 49$; $(x + 6)^2 = 49$

29. $x^2 - 11x + \frac{121}{4} = \frac{49}{4}$; $\left(x - \frac{11}{2}\right)^2 = \frac{49}{4}$

31. $a_2 = 4$, $a_1 = -3$, $a_0 = 5$ **33.** $a_2 = 1$, $a_1 = 0$, $a_0 = -1$

35. $a_2 = 3$, $a_1 = 0$; $a_0 = -1$

37. $A = \pi r^2 + 4\pi r$; input $= r$; $a = \pi$, $b = 4\pi$, $c = 0$

39. $r = n^2 + n$; input $= n$; $a = 1$, $b = 1$, $c = 0$

41. $A = Pr^2 + 2Pr + P$; input $= r$; $a = P$, $b = 2P$, $c = P$

43. $x = -1$; $a = 1$, $b = -4$, $c = -5$; $x^2 - 4x - 5 = 0$

45. $x = \frac{3}{2}$; $a = 2$, $b = 5$, $c = -12$; $2x^2 + 5x - 12 = 0$

47. $x = 1$; $a = 3$, $b = 1$, $c = -4$; $3x^2 + x - 4 = 0$

49. $\{-3 \pm \sqrt{17}\}$; irrational   **51.** $\{-4, -2\}$; rational

53. $\left\{-1, -\frac{1}{2}\right\}$; rational

55. $\left\{\dfrac{-3 \pm \sqrt{-7}}{4}\right\}$; no real-number solution

57. $\left\{\dfrac{-5 \pm \sqrt{61}}{6}\right\}$; irrational   **59.** $\approx\{4.226, 15.774\}$

61. $\approx\{5.918, 14.082\}$

63. a. $-16.1t^2 + 6t + 32.8 = 0$; $t \approx 1.6$ sec
   b. $-16.1t^2 + 6t + 32.8 = 23$; $t \approx 1.0$ sec
   c. $-16.1t^2 + 6t + 32.8 = 33$; $t < 0.1$ sec, $t \approx 0.3$ sec
   d. $-16.1t^2 + 6t + 32.8 = 35$; no real-number solution
   e. $-16.1t^2 + 4t + 32.8 = 0$; $t \approx 1.6$ sec
   f. $-4.905t^2 + 3t + 10 = 0$; $t \approx 1.8$ sec
   g. $-4.905t^2 + 2t + 10 = 5$; $t \approx 1.2$ sec

65. cannot simplify; radicand does not contain a perfect square factor.

67. $|2x - 1|$

EXERCISES 3.5

1. $-3 < x < 5$; $(-3, 5)$;

3. $-4 < x \le 2$; $(-4, 2]$;

5. $x > 5$; $(5, +\infty)$; set of numbers greater than 5

7. $(-\infty, -2)$; set of numbers less than -2;

9. $x \le -3$; set of numbers less than -3, including -3;

11. $x \ge 4$; set of numbers greater than 4, including 4;

13. $x < -1$ or $x > 2.5$ **15.** $x < 2$ or $x > 3$

17. $-2 \le x \le 7$ **19.** $x \le -1$ or $x \ge 6$ **21.** $\mathbb{R}$, $x \ne -3$

23. $\{ \ \}$ or $\varnothing$ **25.** $-5 < x < 3$ **27.** $x \le -12$ or $x \ge 2$

29. $-3 \le x \le 6$ **31.** $x < -18$ or $x > 1$ **33.** $\mathbb{R}$

35. $\{ \ \}$ or $\varnothing$ **37.** $\approx 12 < x < 100$ **39.** ≈ 28 ft   **41.** 50 ft

43. 25 ft on either side of bridge center: $-25 < x < 25$, where $x = 0$ is at center of bridge

45. seasonal products such as ski equipment and heating oil

47. Dec. 29   **49. a.** $\approx\$6,600,000$ **b.** $\$4,250,000$ **c.** yes

51. $\approx 1,680,000$ metric tons   **53.** c   **55.** f   **57.** a

59. x^2 is always ≥ 0, so if $y \ge x^2$, $y \ge 0$.

CHAPTER 3 REVIEW EXERCISES

1. a. binomial   **b.** not a polynomial   **c.** not a polynomial
   d. binomial   **e.** trinomial   **f.** not a polynomial

3. a. $x^2 - 9$ **b.** $4x^2 - 20x + 25$ **c.** $x^3 - 1$
   d. $n^2 + 8n + 16$ **e.** $6x^2 - x - 12$ **f.** $x^4 - 3x^2 + 1$

5. a. $(x + 4)(x - 1)$ **b.** $x(2x - 3)$ **c.** $(2x + 3)(x - 1)$
   d. $(3x + 2)^2$ **e.** $x(x - 1)$ **f.** $3(x + 1)^2$

7. a. $5\sqrt{3}$ **b.** $2\sqrt{2}$ **c.** $4\sqrt{2}$

9. a. $1 - \sqrt{6}$ **b.** $1 + \sqrt{6}$

11. a. $12.5^2 = 7.5^2 + 10^2$ **b.** $30^2 = 18^2 + 24^2$
   c. $(\sqrt{13})^2 = (\sqrt{5})^2 + (\sqrt{8})^2$ **d.** $6^2 = 4^2 + (\sqrt{20})^2$

13. $z = x\sqrt{2}$; $\dfrac{z}{x} = \sqrt{2}$ **15.** false; $y \ge 0$ **17.** 42; quadratic

19. 14; linear

21. a.

x	$f(x)$
-2	0
-1.5	-2.25
-1	-4
-0.5	-5.25
0	-6
0.5	-6.25
1	-6
1.5	-5.25
2	-4

b. $(-2, 0)$, $(3, 0)$ **c.** $x = 0.5$ **d.** $(0.5, -6.25)$

e.

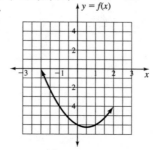

f. $\{-3, 4\}$ **g.** $\{-2, 3\}$ **h.** $\{ \ \}$ or $\varnothing$

23.

r	A
0.00	20,000
0.01	19,602
0.02	19,208
0.03	18,818
0.04	18,432
0.05	18,050
0.06	17,672
0.07	17,298
0.08	16,928

25. $a = \sqrt{\dfrac{C^2 - 2\pi^2 b^2}{2\pi^2}}$ **27.** $\{2, 3\}$ **29.** $\{3\}$ **31.** $\left\{-1\frac{1}{3}, 1\right\}$

33. $\left\{\pm\frac{2}{3}\right\}$

35. a. $h = -16t^2 + 72t$
   b. $(0, 0)$, $(4.5, 0)$; the first is at the instant the ball is thrown.
   c. $\{2, 2.5\}$; on the way up and down   **d.** 81 ft

37. $x = -6 + 2\sqrt{15} \approx 1.746$ **39.** $x = \dfrac{9 - \sqrt{21}}{5} \approx 0.883$

43. $x \le -7$ or $x \ge 1$ **45.** $x < -\frac{1}{2}$ or $x > 3$

47. $x < -1$ or $x > 2\frac{1}{4}$ **49.** true   **51.** true

53. a. ≈ 504.8 ft   **b.** ≈ 504.7585 ft   **c.** ≈ 504.7594 ft

55. a. Graph 3   **b.** Graph 1   **c.** Graph 4   **d.** Graph 2

CHAPTER 3 TEST

1. $x = \dfrac{-b \pm \sqrt{b^2 - 4ac}}{2a}$ **2.** $12x^2 - 25x + 12$ **3.** $x^3 - 27$

4. $(2x - 3)(x - 2)$ **5.** $x = \pm 65$ **6.** $7\sqrt{2}$ **7.** $2 - \sqrt{6}$

8. a. $x = 3\sqrt{3}$ **b.** $x = \dfrac{7\sqrt{2}}{2}$

9. a. 54; quadratic   **b.** -17; linear   **c.** 32; linear
d. -5; quadratic

10. $\{1, 3\}$

11. a. $\{-4, -1\}$ **b.** $\{-7, 2\}$ **c.** $\{-6, 1\}$ **d.** $\{\ \}$ or $\varnothing$
e. $x = -2.5$

12. a. $x \le -6$ or $x \ge 1$ **b.** $-6 \le x \le 1$ **c.** $x < -6$ or $x > 1$

13. $\{\frac{3}{2}, 2\}$

15. a.

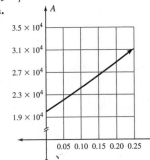

n (percent)	A (dollars)
0	20,000
1	20,402
2	20,808
3	21,218
4	21,632
5	22,050
6	22,472
7	22,898
8	23,328
9	23,762
10	24,200

b. 7%   **c.** $\approx 22.5\%$

16. $s = \dfrac{v^2}{2g}$ **17.** ≈ 300 steps

18. $x = \sqrt{4}, x = 2; \sqrt{x^2} = 2, x = \pm 2$

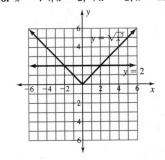

EXERCISES 4.0

1. $x^2 + x - 12$ **3.** $2x^2 - 5x - 12$ **5.** $x^2 + 6x + 9$

7. $x^2 - 36$ **9.** $4x^2 - 20x + 25$ **11.** Exercises 4, 5, 8, 9, 10

13. shift right 4 units   **15.** $x^2 + 8x + 16$

17. $x^2 - 14x + 49 = (x - 7)^2$ **19.** $x^2 - 5x + 6.25$

21. $x^2 + 7x + 12.25 = (x + 3.5)^2$

23. $x^2 - 24x + 144 = (x - 12)^2$ **25.** $x^2 - 16$ **27.** $4x^2 - 9$

29. shift down 16 units: $\{-4, 4\}$ **31.** $(x + 12)(x - 12)$

33. $(0.5x + 0.1)(0.5x - 0.1)$ **35. b.** $(-1, 0)$ **c.** $(4, 0)$

37. b. shift left 2 units   **c.** shift right 1 unit

39. $x^3 + 1$ **41.** $x^3 - 9x^2 + 27x - 27$

43. $x^3 - x^2 - x + 1$ **45.** Exercises 39, 44

47. $x^2 + 5x + 25$

49. a. $(x + 1)(x^2 - x + 1)$ **b.** $(x + 5)(x^2 - 5x + 25)$
c. $(x - 10)(x^2 + 10x + 100)$

51. a. perfect square trinomial   **b.** $(x + 1)^2$ **c.** $x = -1$
d. crosses once at $(-1, 0)$

53. a. difference of squares   **b.** $(x + 1)(x - 1)$ **c.** $\{-1, 1\}$
d. once at $(1, 0)$ and touches at $(-1, 0)$

55. no; $(x + 1)^3$ is x^3 shifted left 1 unit; $x^3 + 1$ is x^3 shifted up
1 unit.

57. 5   **59.** 3

61. Possible solution:
a. not possible   **b.** $y = -6$ **c.** $y = -1$ **d.** $y = 0$
e. not possible

63. a. $y = 6$ **b.** $y \approx 5.5$ **c.** $y = -6$ **d.** $y \approx -0.8$
e. $y = -1$

65. $\{0.382, 1, 2.618\}$

67. $y = x^3 + 1$ crosses x-axis; $y = x^2 + 1$ does not.

69. Possible solutions: $f(x) = 2, \{-2.7, -1.4\}; f(x) = 6$, no
solution; $f(x) = -1, \{-2.9, -0.7, 0, 0.5\}; f(x) = 0,$
$\{-2.831, -1\}$

EXERCISES 4.1

1. $a = -1$ **3.** $a = 1.5$ **5.** $f(x) = 2x^2 - 2x - 12$

7. $f(x) = -2x^2 + 2x + 12$ **9.** $f(x) = -0.5x^2 + x + 7.5$

11. $f(x) = 2x^2 - 6x - 20$ **13.** $f(x) = -0.12x^2 + 4.8x - 36$

15. $f(x) = 0.032x^2$

17. $y = -\frac{3}{40}(x - 40)(x - 10)$; the vertex is on the axis of symmetry.
The x-component of the vertex can be determined by averaging
the x-intercept values: $x = \dfrac{10 + 40}{2}$, or $x = 25$.
vertex: $(25, 16.875)$

19. quadratic; $y = x^2 + 6x$

21. linear; $y = 4x$

23. quadratic; $y = 2x^2 + 13x - 7$

25. quadratic; $y = x^2 + 4x$

27. linear; $y = 12x + 1$

29. neither

31. quadratic; $y = -x^2 + 10x$

33. Graphed points seem to follow a parabolic shape;
$y \approx 3.14x^2 - 0.007x + 0.002$

EXERCISES 4.2

1. a is negative.

3. $y = \frac{4}{125}x^2$; positive coefficient on x^2 for parabola that opens
upward

5.

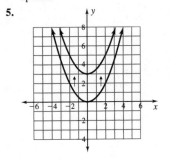

7.

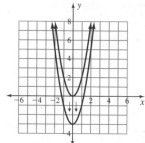

9. $h(x), m(x), q(x)$ **11.** $f(x), g(x), h(x), j(x), m(x), q(x)$ **13.** $j(x)$

15. $g(x)$ **17.** $j(x)$ **19.** $h(x)$ **21.** $p(x)$ **23.** steeper

25. m

27. $y = x^2$; $(0, 0)$; $y = x^2 - 1x$: $\left(\frac{1}{2}, -\frac{1}{4}\right)$; $y = x^2 - 2x$: $(1, -1)$

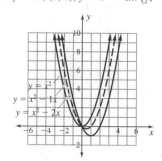

29.

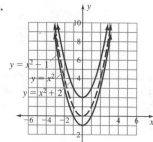

31. $y = x^2 - 3$ or $y = -x^2 - 3$

33. a. $(0, 20), (200, 170), (400, 20)$
   b. $y = -0.00375x^2 + 1.5x + 20$

EXERCISES 4.3

1. 2 real solutions   **3.** 2 complex solutions   **5.** 1 real solution

7. 2 complex solutions   **9.** 2 real solutions

11. a. $2i\sqrt{5}$ **b.** $2i\sqrt{10}$ **c.** $6i\sqrt{2}$

13. a. $i\sqrt{3}$ **b.** $3i\sqrt{6}$ **c.** $3i\sqrt{3}$

15. a. $16 + 0i$ **b.** $0 + 4i$ **c.** $6 + i\sqrt{6}$

17. a. $3 + i\sqrt{3}$ **b.** $\frac{1}{2} - \frac{1}{2}i\sqrt{6}$ **19.** $2 - 3i$ **21.** $1 - i$

23. $-3 - 2i$ **25.** $4 + 2i$ **27.** $-8i$ **29.** $11 - 15i$

31. $-2a$ **33.** $-17 + 4i$ **35.** $-15 + 7i$

37. a. 25 **b.** 25 **39. a.** 3 **b.** 3 **41.** $x^2 + 1$

43. $x^2 + 9$ **45.** $\{\pm 2\sqrt{2}\} = \{\pm 2.8\}$

47. $\{-1 \pm i\sqrt{3}\} \approx \{-1 \pm 1.7i\}$ **49.** $\{2 \pm 2i\}$ **51.** $\{-4, 1\}$

53. $\{2\}$ **55.** $\{-1 \pm i\}$ **57.** $\{-2.5 \pm 0.5i\}$

59. $\{-2, 1 \pm i\sqrt{3}\} \approx \{-2, 1 \pm 1.7i\}$ **61.** $\{\pm 3, \pm 3i\}$

63. $\{-2, -1, 0\}$ **65.** true, use the discriminant.

67. true; multiply a number and its conjugate (see Exercises 37 and 40).

69. c   **71.** b   **73.** a, e

MID–CHAPTER 4 TEST

1. a. $4x^2 - 9$; difference of squares
   b. $4x^2 - 12x + 9$; perfect square trinomial
   c. $4x^2 + 12x + 9$; perfect square trinomial
   d. $6x^2 + 5x - 6$

2. a. $x^3 - 8$ **b.** $(x + 3)(x^2 - 3x + 9)$

3. The graph is shifted 8 units below $f(x) = x^3$. Both graphs have same shape.

4. $y = (x + 2)^2$ is 2 units to the left of $y = x^2$; vertex and x-intercept are $(-2, 0)$ for $y = (x + 2)^2$, $(0, 0)$ for $y = x^2$; y-intercept is $(0, 4)$ for $y = (x + 2)^2$, $(0, 0)$ for $y = x^2$.

5. $y = -\frac{1}{2}(x + 3)(x - 4)$ or $y = -\frac{1}{2}x^2 + \frac{1}{2}x + 6$

6. a. quadratic; $y = x^2 - 10x + 24$
   b. linear; $y = 5x - 1$
   c. linear; $y = 7x - 10$
   d. neither; $(-1)^n$ or $y = 1$ for odd x, $y = -1$ for even x

7. The second differences are a constant 9.81. A quadratic regression is most appropriate; $f(x) \approx 4.9x^2$.

8. Both graphs share the same vertex; $y = -2x^2$ opens down and is steeper than $y = x^2$.

9. $y = x^2 - 4$ is 4 units below $y = x^2$. The vertex and y-intercept are shifted from $(0, 0)$ for $y = x^2$ to $(0, -4)$ for $y = x^2 - 4$. The x-intercepts are $(-2, 0)$ and $(2, 0)$ for $y = x^2 - 4$ and $(0, 0)$ for $y = x^2$.

10. a. $0 + 4i$ **b.** $4 + 0i$ **c.** $3 + 2i$ **d.** $0 + 2i\sqrt{14}$

11. a. 13 **b.** 5 **c.** $x^2 + 1$ **d.** 13

12. a. both 13; the conjugates give real-number products.
   b. both 5

13. $16; 2; \{-1, 3\}$ **14.** $\{2, -1 \pm i\sqrt{3}\} \approx \{2, -1 \pm 1.7i\}$

EXERCISES 4.4

1. a. $y = (x + 2)^2$ **b.** $y = (x - 3)^2$

3. a. $y = x^2 + 2$ **b.** $y = x^2 - 1$

5. a. $y = x^2 - 1$ **b.** $y = -x^2 - 3$

7. a. $y = (x + 1)^2 - 2$ **b.** $y = (x - 2)^2 + 1$

9. shift left 2 units   **11.** shift right 4 units

13. shift right 3 units, up 4 units

15. shift left 4 units, up 3 units

17. $(1, 0)$ **19.** $(-3, 4)$ **21.** $(2, -3)$

23. $y = \frac{1}{8}(x - 3)^2 + 1$

25. $y = 2(x + 1)^2 - 1$ **27.** $y = -2(x - 3)^2 - 2$

29. a. Possible answer: $(10, 0)$; $y = -0.24x^2 + 2.4x$
   b. $y = -0.24x^2 + 2.4x$

31. a. $(0, -25), (20, -15)$
   b. $y = 0.0375x^2 - 1.5x$ or $y = 0.0375(x - 20)^2 - 15$
   c. Vertex is now at $(20, -15)$.

33. $y = -\frac{1}{9}(x - 30)^2 + 100$ **35.** $y = (x + 5)^2$

37. $y = (x - 3)^2$ **39.** $y = (x + 5)^2 + 5$

41. $y = (x - 3)^2 - 1$

EXERCISES 4.5

1. x-intercepts, $(0, 0), (2, 0)$; vertex $(1, -1)$

3. x-intercepts, $(-7, 0), (3, 0)$; vertex $(-2, 25)$

5. $(-4, -1)$

7. $(2, 1)$ **9.** $\left(\frac{1}{4}, -\frac{25}{8}\right)$

11. maximum height ≈ 62.9 ft, distance ≈ 251.6 ft

13. no; ≈ 310.6 ft

15.

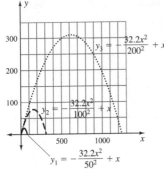

$$y_3 = -\frac{32.2x^2}{200^2} + x$$

$$y_2 = -\frac{32.2x^2}{100^2} + x$$

$$y_1 = -\frac{32.2x^2}{50^2} + x$$

    a. quadrupled   **b.** quadrupled

19. $l = 30$ ft, $w = 15$ ft; $A = 450$ ft^2 **21.** $\approx$30.3 ft

23. 690 ft; drain at (1143, 677)   **25.** 1188 ft; drain at (1091, 1161)

27. $\approx$155 ft

29. The other intercept is the same distance $|h - x|$ from the vertex on the opposite side.

CHAPTER 4 REVIEW EXERCISES

1. $x^2 - 8x + 12$; other   **3.** $x^2 + 13x + 12$; other

5. $x^2 - 4x + 4$; perfect square trinomial

7. $x^2 - 30x + 225 = (x - 15)^2$ **9.** $x^2 + 20x + 100 = (x + 10)^2$

11. $(x + 7)(x - 7)$ **13.** $(2x - 1)^2$ **15.** cannot be factored

17. $(2x + 3)^2$ **19.** $a^3 - 3a^2b + 3ab^2 - b^3$ **21.** $a^3 + b^3$

23. Exercise 19   **25.** $(a - 2)(a^2 + 2a + 4)$

27. $(x + 3)(x^2 - 3x + 9)$ **29.** $(x - 10)(x^2 + 10x + 100)$

31. a. $y = -5$ **b.** $y = -4$ **c.** $y = 2$ **d.** not possible
    e. not possible

33. a. not possible   **b.** $y = 5$ **c.** $y = 4$ **d.** $y = 2$
    e. not possible

35. a. $\approx\{0.5, 3.5\}$ **b.** $\approx\{0, 4\}$

37. a. $\approx\{-3, -1, 2\}$ **b.** $\approx\{-3.5, 1.5, 0\}$

39. $y = x^2 - 2x - 8$ **41.** $y = -\frac{1}{3}x^2 + \frac{2}{3}x + \frac{8}{3}$

43. $y = 2x^2 + 3x + 1$ **45.** $y = 3x - 4$ **47.** $y = 16 - x^2$

49. neither

51. a. $y = 99.93 - 0.000989x$, where $x =$ elevation
    b. $y = 0.00000000276x^2 - 0.00107x + 100.2$, where
    $x =$ elevation
    c.

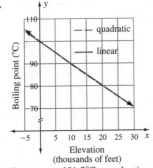

    d. linear: $y \approx 101.2$°C; quadratic: $y \approx 101.6$°C

53. steeper   **55.** shifted down 9 units

57. $y = x^2 - 9$ crosses x-axis twice; $y = x^2 + 9$ does not cross
    x-axis.

59. Both factors are the same; double root.

61. 0; 1 double root; 1 x-intercept

63. -8; 2 complex; no x-intercepts   **65.** c   **67.** b   **69.** b

71. a. $4i$ **b.** $5i\sqrt{2} \approx 7.07i$

73. $4 + 3i$

75. a. $7 + 24i$ **b.** 25   **c.** 25

77. $\left\{3, -\frac{3}{2} \pm \frac{3i\sqrt{3}}{2}\right\}$

79. $\{-1, 2\}$; vertex $(0.5, -2.25)$

81. no x-intercepts; vertex $(1, 3)$

83.

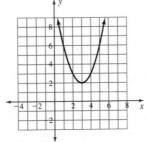

85.

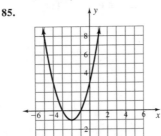

87. $y = (x + 2)^2 + 5$

89. a. $\{-2, 4\}$ **b.** $x = 1$ **c.** $(1, -27)$

91. $y = -0.111x^2 + 6.667x$

93. a. $\approx(271, 135)$ **b.** $\approx(752, 376)$ **c.** $\approx(2484, 1242)$
    d. $\approx(3882, 1941)$

95. $A = 40w - w^2$; vertex (20, 400); maximum area is square with
    $w = l = 20$; $20^2 = 400$.

97. $x < 8.8$ or $x > 51.2$

CHAPTER 4 TEST

1. $x^3 - 9x^2 + 27x - 27$

2. a. $(x + 7)(x - 7)$ **b.** $(x - 3)(x^2 + 3x + 9)$

3. Graph is shifted up 2 units; no x-intercepts; vertex is at $(0, 2)$.

4. Graph is turned upside down; vertex remains at $(0, 0)$.

5. The first parabola, because its graph opens down and thus could
    describe the path of a ball; the other opens up.

6. a. 54; quadratic; $y = x^2 + 5x - 12$
    b. -17; linear; $y = -8x + 31$ **c.** 32; linear; $y = 6x - 4$
    d. -5; quadratic; $y = \frac{1}{2}x^2 - \frac{19}{2}x + 34$

7. 16; 2 real; 2 x-intercepts

8. a. $6i$ **b.** $5i\sqrt{3} \approx 8.7i$

9. a. $1 + 2i$ **b.** $4 + 2i$

10. a. 26   **b.** 26   **c.** $-24 - 10i$

11. $\left\{1, -\frac{1}{2} + i\frac{\sqrt{3}}{2}\right\}$

12. a. $\{-2, 1.5\}$ **b.** $x = -\frac{1}{4}$ **c.** $\left(-\frac{1}{4}, -6\frac{1}{8}\right)$
    d. $x < -2$ or $x > 1.5$ **e.** possible answer: $y = -10$
    f. $-6\frac{1}{8} < y < -6$

13. $\{-1, 2\}$; vertex $\left(\frac{1}{2}, -2\frac{1}{4}\right)$ **15.** $y = (x + 2)^2 - 5$

16. shift right 3 units, down 1 unit   **17.** $y = 3x^2 + 6x - 5$

18.

x	y
1	4
2	12
3	24
4	40
5	60

quadratic; $y = 2x^2 + 2x$

19. a. $y = lw$ **b.** $y = 15w - 2w^2$
   c. vertex $\left(\frac{15}{4}, 28\frac{1}{8}\right)$; graph shows maximum area of 28.125 ft^2 at $w = 3.75$ ft.

CUMULATIVE REVIEW OF CHAPTERS 1 TO 4

1. $-4x + 17$ **3.** $y = \frac{3}{2}x$ **5. a.** $n = \dfrac{c}{t^2}$ **b.** $n = \dfrac{s - 2a + d}{d}$
7. yes; $20.5^2 = 20^2 + 4.5^2$
9. a. 60; quadratic; $y = x^2 + 4x$ **b.** 43; linear; $y = 7x + 1$
   c. 36; linear; $y = 6x$ **d.** 99; quadratic; $y = 2x^2 + 5x - 3$
11. a. $\{-1, 3\}$ **b.** $-1 \le x \le 3$ **c.** $x \le 0$ or $x \ge 2$
   d. $-2 < x < 4$
13. a. $\{-3 \pm i\}$ **b.** $\left\{-1, \frac{4}{5}\right\}$ **c.** $\left\{-1, \frac{3}{7}\right\}$ **d.** $\left\{\frac{1}{3}, 1\frac{1}{2}\right\}$
15. a. $(4 - x)(4 + x)$ **b.** $(x - 4)(x^2 + 4x + 16)$

EXERCISES 5.0

1. a. 21   **b.** 30   **c.** 128   **3. a.** 15   **b.** 45   **c.** 40
5. a. 30   **b.** 30   **c.** 96   **7.** $\frac{7}{20}; \frac{3}{20}; \frac{1}{40}; \frac{5}{2}$
9. $\frac{19}{12}; -\frac{1}{12}; \frac{5}{8}; \frac{9}{10}$ **11.** $3\frac{5}{6}; -1\frac{1}{6}; 3\frac{1}{3}; \frac{8}{15}$ **13.** $3\frac{9}{20}; \frac{19}{20}; 2\frac{3}{4}; 1\frac{19}{25}$
15. 31 cm   **17.** $\dfrac{1}{48 \text{ in}^2}$ **19.** 60 lb   **21.** 50 kw
23. a. 16.1 ft   **b.** 144.9 ft   **c.** 1304.1 ft
25. a. 9.81 m/sec   **b.** 19.62 m/sec   **c.** 39.24 m/sec
27. 0.3 L   **29.** 7.62 m   **31.** 5.6 yd^3 **33.** 0.058 ft^3
35. 200 g   **37.** $\approx$59 ft/sec   **39.** 20 microdrops/min
41. 193 total calories   **a.** 9 fat calories
   b. 164 carbohydrate calories   **c.** 20 protein calories
43. 196 total calories   **a.** 72 fat calories
   b. 108 carbohydrate calories   **c.** 16 protein calories
45. a. multiply   **b.** divide   **47. a.** add   **b.** subtract
49. Wrong; find the common denominator before adding
51. Right; method always works; may result in fractional expression in numerator and/or denominator.
53. Wrong; find a common denominator before adding.

EXERCISES 5.1

1. a. $\frac{3}{1}$ **b.** $\frac{1}{2}$ **c.** $\frac{1}{3}$ **3. a.** $\frac{1}{100}$ **b.** $\frac{2}{15}$ **c.** $\frac{8}{3}$
5. a. $\frac{75}{28} \approx 2.679$ to 1   **b.** $\frac{5625}{784} \approx 7.175$ to 1
7. a. $\approx$2.011 mm^2 **b.** 49.736 to 1   **9.** 32
11. a. $x = 8.4$ **b.** $x = 2.4$ **c.** $x \approx 54.9$ **13.** 15 ft
15. $a = 6\frac{2}{3}, b = 6$ **17.** $f = 12, g = 6\sqrt{3}$
19. $e = \dfrac{10\sqrt{3}}{3}, f = \dfrac{20\sqrt{3}}{3}$ **21.** $f = 2\sqrt{5}, g = \sqrt{15}$
23. $n = 2\sqrt{2}, x = 4\sqrt{2}, y = 4\sqrt{2}$ **25.** 4.5 ft   **27.** 5.9 ft
29. a. $x =$ weight in lb, $y =$ distance stretched in in., $y = \frac{2}{5}x$
   b. direct
31. a. $x =$ credit hr, $y =$ cost in \$, $y = 80x + 25$
   b. not proportional   **c.** y-intercept = student fees
33. a. $x =$ purchase price, $y =$ tax, $y = 0.075x$ **b.** direct

35. $k = 55$ mi/hr; 440 mi   **37.** $k = \$14.49$/CD; \$144.90
39. $k \approx 0.367$ oz/in^2; $\approx$0.827 oz
41. $y = 0.49x$; $y = 0.26x + 0.8$; connecting charge   **43.** are
45. $y = mx$ **47. a.** $k = \frac{1}{2}$ **b.** $k = 1$ **c.** $k = \frac{1}{3}\pi$
49. $\approx$2,594,576 m^3 **51.** $C = kr; k = 2\pi$ **53.** $A = ks^2; k = 1$
55. $D = krt; k = 1$ **57.** $A = ks^2; k = 6$
61. unit conversion upside down
63. wrote cross multiplication results as a fraction instead of an equation.

EXERCISES 5.2

1. $y = 1,000,000/x$; $y =$ yr, $x =$ yd^3/yr; $k = 1,000,000$ yd^3
3. $y = 5.74 \times 10^8/x$; $y =$ yr, $x =$ metric tons/yr;
   $k = 5.74 \times 10^8$ metric tons
5. $y = 100/x$; $y =$ points/problem, $x =$ no. of problems;
   $k = 100$ points
7. inverse variation; $k = 5600\dfrac{\text{tons}}{\text{day}} \cdot$years
9. direct variation; $k = \$0.21875$/item
11. inverse variation; $k = 90$ mi
13. inverse variation; $k = 12$ worker·days
15. direct variation; $k = \$1100$/semester
17. $x = 52.5$ lb; $k = 210$ lb·ft
19. $x = 60$ servings; $k = 60$ cups
21. $x \approx 92$ yr; $k = 119,700,000$ metric tons
23. $x = 1200$ ft/gal; $k = 300$ ft^2/gal   **25.** 4 times greater
27. $\dfrac{x_1}{x_2} = \dfrac{y_2}{y_1}; \dfrac{y_1}{y_2}$ has been inverted.

29.

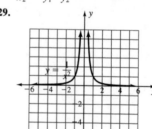

   a. undefined
   b. $x = 0$
   c. Graph approaches x-axis from above x-axis.

31. a.

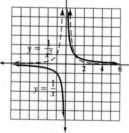

   b. $x > 1$ **c.** $x = 1$
   d. $1/x^2$ is positive for $x < 0$; $1/x$ is negative for $x < 0$.

33. a.

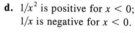
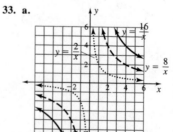

b. in first and third quadrants

c. $\dfrac{2}{x}, \dfrac{8}{x}, \dfrac{16}{x}$ progressively farther from origin; $\dfrac{16}{x}$ above $\dfrac{8}{x}$ and $\dfrac{2}{x}$ in first quadrant, below in third

d. the numerator

e. increases distance from x-axis

f. closer to x-axis

g. above in first quadrant, below in third

h. get close to x-axis

35.

Rate (mph)	Time (hours)
1	6
2	3
3	2
4	1.5
5	1.2
6	1
7	0.857
8	0.75
9	$\frac{2}{3}$
10	0.6

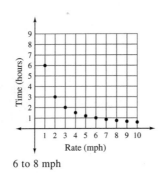

Rate (mph)

6 to 8 mph

37. a. ≈ 18.5 min

b.

Speed (mph)	Time (minutes)
60	20
65	18.462
70	17.143
75	16
80	15

c. between 89 and 90 mph

39. a. 10 mA   **b.** inversely proportional   **c.** $k = 1.2$ volt·amp

EXERCISES 5.3

1. $x = -3$ **3.** $a = 4$ **5.** defined for all $\mathbb{R}$

7. a. $-x - y$ **b.** $x - y$ **c.** $x - y$

9. a. $x - 3, x \neq 3$ **b.** $-2 - x, x \neq -2$ **c.** $b - a, b \neq a$
   d. $x - 3, x \neq 3$

11. a. $2a; 8a/5, a \neq 0$ **b.** $2xy; x/5y, x \neq 0, y \neq 0$

13. a. already simplified; $y \neq -2x$ **b.** $x; \dfrac{y}{2+y}, x \neq 0, y \neq -2$

15. a. $(x - 3); -\dfrac{1}{x+3}, x \neq \pm 3$ **b.** $(x + 3); \dfrac{x-3}{x+2}, x \neq -3, -2$

17. a. $2(a + 2b); c/2d, a \neq -2b, d \neq 0$
   b. $3x; \dfrac{2x+1}{2(2x-1)}, x \neq 0, \frac{1}{2}$

19. a. $(x - 2); -(x + 3), x \neq 2$ **b.** $(x - 3); -\frac{1}{2}, x \neq 3$

21. a. x **b.** $\dfrac{1}{a^3 b^2}$ **c.** $\dfrac{b^3}{a^3}$ **d.** $\dfrac{y}{x^2}$

23. a. $\dfrac{a+3}{a(a+2)}$ **b.** $x + 1$ **25. a.** $x + 2$ **b.** $\dfrac{-2a-1}{2a(2+3a)}$

27. a. $\dfrac{x}{x-2}$ **b.** $\dfrac{a}{(a+5)(a-5)}$

29. a. 1   **b.** $\dfrac{-(x-y)}{x+y}$ or $\dfrac{y-x}{x+y}$

31. a. $a^2 + ab + b^2$ **b.** $\dfrac{(x+y)(x-y)}{x^2 - xy + y^2}$ or $\dfrac{x^2 - y^2}{x^2 - xy + y^2}$

33. a. $\dfrac{\frac{1}{a}}{\frac{1}{b}} = \dfrac{b}{a}$ **b.** $\dfrac{\frac{1}{b}}{a} = \dfrac{1}{ab}$ **c.** $\dfrac{\frac{1}{b}}{b} = \dfrac{1}{b^2}$ **d.** $\dfrac{\frac{a}{b}}{\frac{1}{b}} = a$

35. $\dfrac{x(x-2)}{x+1}$ or $\dfrac{x^2 - 2x}{x+1}$ **37.** $\dfrac{-1}{x+4}$ **39.** 60 sec

41. $4\frac{1}{3}$ gal/hr   **43.** ≈ 6 cans/\$ **45.** 96 stitches/ft

47. ≈ 0.3 page/min

49. must factor first, but $x^2 + x + 2$ does not factor, so x is not a common factor.

51. must factor first, but numerator factors to $(x - 1)(x - 1)$, so $(x - 2)$ is not a common factor.

53. Multiplication is not distributive over multiplication.

MID–CHAPTER 5 TEST

1. $\frac{4}{3}$ **2.** $\frac{26}{45}$ **3.** $\dfrac{b}{es}$ **4.** $1\frac{2}{25}$ **5.** $\dfrac{3d}{2c}$ **6.** $x - 2$ **7.** $\dfrac{1}{a+3}$

8. $1 + c$ **9.** $\dfrac{ac}{b}$ **10.** $\dfrac{b^3}{c^3}$ **11.** $\dfrac{x-4}{x-2}$

12. $x(x - 1)$ or $x^2 - x$ **13.** -1 **14.** $\frac{8}{15}$ **15.** $\frac{5}{6}$

16. $x = -4, x = -2, x = 2$ **17.** 3906.25 gal

18. 6.048×10^{-9} mph   **19.** $a \approx 5.14$ **20.** $x = 11$

21. $n = 8, p = 13.2$ **22.** $y = \frac{3}{2}x$; proportion

23. $y = 0.45x + 0.15$; not a proportion

24. a. 400 lb·ft   **b.** 5 ft from pivot point

25. $y = \dfrac{100}{x}$; $y = $ no. of cookies/child, $x = $ no. of children; $k = 100$ cookies

26. \$15.15   **27.** 84

EXERCISES 5.4

1. a. $\dfrac{3-x}{4}$ **b.** $-\dfrac{1}{x}, x \neq 0$ **c.** $\dfrac{4-x^2}{x+3}, x \neq -3$
   d. $\dfrac{3-x}{x^2+1}$

3. a. 12   **b.** $2a$ **c.** $a^2 b^2$ **d.** $x(x-2)(x+2)$
   e. $(x+3)(x-3)^2$

5. a. $\frac{5}{6}$ **b.** $-\dfrac{1}{2a}, a \neq 0$ **c.** $\dfrac{b^3 - ac}{a^2 b^2}, a \neq 0, b \neq 0$
   d. $\dfrac{8x+10}{x(x-2)(x+2)}, x \neq \pm 2, 0$ **e.** $\dfrac{2x^2 + 24x}{(x+3)(x-3)^2}, x \neq \pm 3$

7. $\dfrac{2a+5b}{4ab}, a \neq 0, b \neq 0$ **9.** $\dfrac{x^2 + x - 3}{x(x-3)}, x \neq 0, 3$

11. $\dfrac{4x^2 - 7x + 8}{x(x-1)}, x \neq 0, 1$ **13.** $\dfrac{x^2 - 3x + 3}{(x-3)^2}, x \neq 3$

15. $\dfrac{5b+3}{(b+1)(b-1)}, b \neq \pm 1$ **17.** $\dfrac{2b-3a}{ab(2+b)}, a \neq 0, b \neq -2, 0$

19. $\dfrac{x^2 + 3x - 9}{x(x-3)^2}, x \neq 0, 3$ **21.** $\dfrac{t_1 + t_2}{t_1 t_2}$ **23.** $\dfrac{m_1 + m_2}{m_1 m_2}$

25. yes   **27.** ≈ 14.3 hr; ≈ 51.4 mph   **29.** ≈ 59.6 mph

31. ≈ 3.2 min   **33.** $\dfrac{x^4 + 4x^3 + 12x^2 + 24x + 24}{24}$

35. $\dfrac{RTv^2 - av + ab}{v^2(v-b)}$ **37.** $t = \dfrac{t_1 t_2 t_3}{t_2 t_3 + t_1 t_3 + t_1 t_2}$; no   **39.** 7

41. $a = \dfrac{3v}{4\pi b^2}$ **43.** $\dfrac{3x}{2(x+6)}$ **45.** $\dfrac{2E}{2R+r}$

47. a. multiply   **b.** divide   **49. a.** add   **b.** multiply

51. $\frac{21}{26}$; $\frac{3}{2} + \frac{7}{4} = \frac{13}{4}$ **53.** $\dfrac{bd}{ad+bc}$

EXERCISES 5.5

1. $\mathbb{R}$ except $x = 4$; $x + 1$ **3.** $\mathbb{R}$ except $x = 3$; $x - \dfrac{4}{x - 3}$

5. $\mathbb{R}$ except $x = 2$; $x - 1 - \dfrac{6}{x - 2}$

7. $\mathbb{R}$ except $x = 1$; $x - 2 - \dfrac{6}{x - 1}$

9. $\mathbb{R}$ except $x = 0$; $x - 3 - \dfrac{4}{x}$

11. $\mathbb{R}$ except $x = -1$; $x - 4$

13. $(x + 1)(x - 4)$; quotient has a remainder of zero.

15.

x	$f(x)$
-1	0
0	-4
1	-6
2	-6
3	-4
4	0

They are the same.

17.

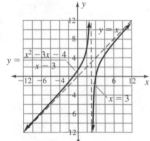

a. Graph is approaching the restriction at $x = 3$.
b. Graph is approaching the function $y = x$.
c.

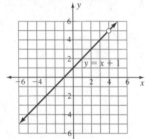

Graph is linear with a hole at $x = 4$.
d.

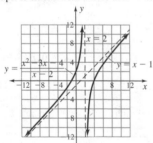

Graph is nearly vertical at its restriction $x = 2$.

19. $x^2 + 2x + 4 + \dfrac{7}{x - 2}$

21. $x^2 + x + 1$; denominator is a factor.   **23.** $x^2 - \dfrac{1}{x}$

25. $x^2 - x + 1 - \dfrac{2}{x + 1}$

27.

x	$f(x)$
2	7
1	0
0	-1
-1	-2

same

29. $x^2 - x + 1$; denominator is a factor.

31. $x^3 - x^2 + x - 1 + \dfrac{2}{x + 1}$

33. $x^3 - 3x^2 + 3x - 1$; denominator is a factor.

35. $3x^2 - 2x + 2 - \dfrac{3}{x + 1}$

37. Subtract the degree of the divisor from the degree of the dividend.

39. $(-1, 3)$ **41.** $(1, 0)$ **45.** true

47. $f(-2) = f(5) = 6$; remainders equal 6.   **49.** remainder

51. same

EXERCISES 5.6

1. $x = 24$ **3.** $r = \dfrac{180\,L}{\pi\theta}$, $\theta \neq 0$ **5.** $x = 5\frac{5}{11}$ **7.** $n = 5$

9. no variables in denominators   **17.** $x = 7$

19. $\{-2, 5\}$; $x \neq -5$ **21.** $\{-1.5, 5\}$; $x \neq \frac{1}{2}$ **23.** $x = 64$

25. $3 - 2x$ **27.** $6x + 1$ **29.** $-x - 2$ **31.** $3x + 5$

33. $x \neq 0$; $x = \frac{15}{19}$ **35.** $x \neq 0$; $x = 6$ **37.** $x \neq 0$; $\{\ \}$ or $\varnothing$

39. $x \neq 0$; $x = 3$ **41.** $x \neq \frac{1}{2}, 1$; $x = \frac{3}{2}$ **43.** $x \neq -2$; $\{-7, 4\}$

45. $x \neq 1$; $\{-7, 6\}$ **47.** $x \neq 0$; $\{-1.5, 6\}$ **49.** $x \neq 2$; $\{\ \}$ or $\varnothing$

51. $x \neq \pm 1$; $\left\{-\frac{1}{3}, 4\right\}$ **53.** $x \neq 3, 6$; $\{-3, 4\}$

55. $x \neq 0, 2$; $x = 6$ **57.** $x \neq -3, 0$; $\left\{-\frac{3}{2}, 5\right\}$

59. $x \neq \pm 1$; $\{-3, 4\}$ **61.** $x \neq -3$; $\{-7, 11\}$ **63.** $x \neq 0, 1$; $\left\{\frac{1}{2}, 2\right\}$

65. $\dfrac{1}{6} + \dfrac{1}{x} = \dfrac{1}{2}$; $x = 3$ min   **67.** $\dfrac{1}{20} + \dfrac{1}{x} = \dfrac{1}{12}$; $x = 30$ min

69. $\dfrac{1}{10{,}000} + \dfrac{1}{4000} = \dfrac{1}{R}$; $R \approx 2857$ ohms

71. $\dfrac{1}{5} + \dfrac{1}{3} + \dfrac{1}{3} = \dfrac{1}{x}$; $x \approx 1.15$ min   **73.** 14.3 cm, 5.7 cm

75. 15.5 cm, 6.7 cm   **77.** $a = \dfrac{bc}{b - c}$ **79.** $c = \dfrac{ab}{a + b}$

81. $R = \dfrac{E}{I} - r$

83. did not distribute $15x(x + 2)$ over $\dfrac{2}{x}$;

$15x + 3(x + 2) = 11x(x + 2)$

85. forgot to multiply by 2 in the second term;

$15x + 3(x + 2) = 11x(x + 2)$

CHAPTER 5 REVIEW EXERCISES

1. $2, -\frac{2}{3}, \frac{8}{9}, \frac{1}{2}$ **3.** $\dfrac{ac + ab}{bc}, \dfrac{ac - ab}{bc}, \dfrac{a^2}{bc}, \dfrac{c}{b}$

5. **a.** $\dfrac{9y}{5x}$ **b.** -1 **c.** $1 - c$

7. **a.** $3(x - 2)$ **b.** $\dfrac{1}{a + b}$ **c.** $2 + 4x$

9. **a.** $\dfrac{x - 6}{x - 5}$ **b.** $\dfrac{2x - 1}{x + 2}$ **c.** $\dfrac{x - 1}{x - 4}$ **11.** $x - 6$

13. 3 to 16   **15.** $x = -3, 1$ **17.** $\dfrac{y}{6x}$ **19.** 11   **21.** $\dfrac{2}{1 - x}$

23. $\dfrac{x+1}{x-3}$ **25.** $\dfrac{x+1}{x-1}$ **27.** $-\dfrac{x-4}{x^2-1}$ or $\dfrac{x+4}{1-x^2}$

29. y approaches $-\infty$. **31.** y approaches $-\infty$.

33. ≈ 114.2 yr (365 days)

35. a. $\approx 5.115 \times 10^{-8}$ mph **b.** ≈ 525 in.; ≈ 43.75 ft

37. $y = 5.6, x = 4.8$ **39.** $r = 10, s = 10, t = 12.5$

41. $y = 0.25x + 6.00$; $\$6.00 = $ cost of developing

43. $y = 1.87x$; direct

45. $D = k\sqrt{h}$; $D = $ distance, $h = $ height

47. $V = klwh$, $V = $ volume, $l = $ length, $w = $ width, $h = $ height

49. $d = ks^2$ **51.** $s = k\sqrt{df}$ **53.** $y = \dfrac{90}{x}$, $k = 90$ credits

55. $y = \dfrac{10,000}{x}$; $k = \$10,000$

57. a. total hours minus 3 hours not traveling

b.

Total Trip Time (hours)	Rate (mph)
4	300
5	150
6	100
7	75
8	60

 c. Replacing t by $t - 3$ causes a shift of 3 units to the right.

59. a. 600 mph **b.** 3000 mph **c.** 30,000 mph

 d. not possible

61. $h = \dfrac{3V}{\pi r^2}$ **63.** 0.00178 grain **65.** $x + 2$ **67.** no

69. $x \neq 0$; $x^2 - 3x + 3 - \dfrac{1}{x}$ **71.** $x \neq 1$; $x^2 - 2x + 1$

73. $x \neq 2$; $x^2 - x + 1 + \dfrac{1}{x-2}$ **75.** $x \neq 3$; $x^2 + 3 + \dfrac{8}{x-3}$

77.

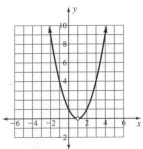

Hole is created when a factor of the denominator is also a factor of the numerator.

79. $x = 12.8$; $x \neq 0$ **81.** $\{-7, 2\}$; $x \neq -5$ **83.** $x = 26$

85. $x = 4$; $x \neq -2, 0$ **87.** $\left\{\frac{4}{7}, 3\right\}$; $x \neq 0, 1$

89. a. $y = -\frac{4}{5}x + 12$ **b.** $y = 40/x$

 c.

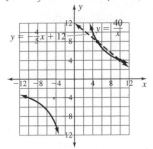

 d. Both contain $(5, 8)$ and $(10, 4)$; both are decreasing. The function in part a is straight; the one in part b is a curve.

CHAPTER 5 TEST

1. $\dfrac{b}{ac}$, $a \neq 0, b \neq 0, c \neq 0$ **2.** already simplified, $b \neq 3$

3. $\dfrac{1}{5c}$, $a \neq 0, c \neq 0$ **4.** $\dfrac{3c^2}{2b}$, $b \neq 0, c \neq 0$

5. $-\dfrac{x+5}{x+4}$, $x \neq -4, 5$ **6.** $\dfrac{2}{a+3}$, $a \neq -3, 6$

7. $\dfrac{3x-1}{2x-1}$, $x \neq \frac{1}{2}, 2$ **8.** $\dfrac{2x+3}{x+2}$, $x \neq -2$

9. $\dfrac{2y}{x}$, $x \neq 0, y \neq 0, 7$ **10.** $\dfrac{(x-3)^2}{(x+1)^2}$, $x \neq -3, -1$

11. $(x+4)(x+2)$, $x \neq -4, 2$ **12.** $5x + 3$, $x \neq -1, 0$

13. $\dfrac{8x}{(x-2)(x+2)}$, $x \neq \pm 2$ **14.** $\dfrac{x-3}{x+2}$, $x \neq -2, 0$

15. 1000 cm^2 **16.** $\frac{1}{2}$ hr/gal **17.** $\dfrac{3-x^2}{x^2+2}$, $x \neq 0$

18. $\dfrac{-x^2}{3(3+x)}$, $x \neq -3, 0, 3$ **19.** ≈ 0.0173 oz/hr

20. ≈ 0.0135 oz/hr **21.** 112.5 ft **22.** ≈ 50 mph

23. Volume varies directly with the cube of the radius; $k = \frac{4}{3}\pi$

24. $EF = 10$; $AB = 3\sqrt{29}$; $DE = 2\sqrt{29}$

25. a. "cover the same distance" **b.** 360 mi

 c. $60 \cdot 6 = t \cdot 55$, $t \approx 6.5$ hr

26. $\{-3, 6\}$; $x \neq 1$ **27.** $\frac{1}{4} \pm i(\sqrt{23}/4)$; $x \neq 0$

28. $x = 7$; $x \neq 0, 4$ **29.** $x = 37$ **30.** $\left\{-\frac{3}{4}, 1\right\}$, $x \neq -1, 0$

31. $\mathbb{R}$, $x \neq -2$; $x^2 + 4x + 4$

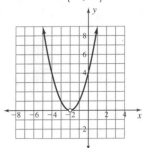

32. $\mathbb{R}$, $x \neq -4$; $x^2 + 2x + 4 - \dfrac{8}{x+4}$

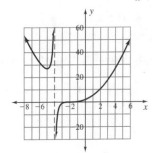

EXERCISES 6.0

1. a. $\dfrac{2}{x}$ **b.** $\dfrac{x}{y}$ **c.** 1 **d.** 4 **e.** 1

3. a. $\dfrac{n}{m}$ **b.** 1 **c.** $\dfrac{ab}{c}$ **d.** 1 **e.** 0.4

5. a. $\dfrac{1}{4y^2}$ **b.** $\dfrac{2}{x^4}$ **c.** $\dfrac{y^2}{4x^2}$ **d.** $\dfrac{8c^3}{a^3}$ **e.** $\dfrac{x^4}{3y^4}$ **f.** b^{n+1}

7. a. $\dfrac{c^3}{27a^6}$ **b.** $\dfrac{c^4}{a^2}$ **c.** c^3 **d.** $2a^2$ **e.** $3a^4b^3$ **f.** 3^{x+1}

9. Exponent does not affect the 3. Answer is $\dfrac{3}{x^2}$.

15. $9^3 + 10^3$, $1^3 + 12^3$

17. a. $\dfrac{1}{t} = \dfrac{1}{t_1} + \dfrac{1}{t_2}$ **b.** $g = 9.81 \text{ m/sec}^2$ **c.** $P = \dfrac{A}{e^{rt}}$

   d. $d = \dfrac{1 \text{ g}}{\text{cm}^3}$ **e.** $p = \dfrac{31 \text{ lb}}{\text{in}^2}$

19. $\dfrac{-1}{x(x + h)}$ **21.** $\dfrac{x^2 + 2x}{(x + 1)^2}$

23. a. 10^{-3} **b.** 10^{-5} **c.** 10^4 **d.** 10^{-2} **e.** 10^6 **f.** 10^{-6}

25. a. 3×10^3 **b.** 3.5×10^2 **c.** 3.500×10^5
   d. 3.50×10^{-3}

27. 1; 2; 4; 3

29. a. 29,979,000,000 cm/sec
   b. 0.000 000 000 000 000 000 000 001 672 6 g
   c. $-0.000\ 000\ 000\ 000\ 000\ 000\ 160\ 22$ C
   d. 0.000 000 000 000 000 000 160 22 J

31. 6×10^4 **33.** 8×10^8

35. a. 1.25×10^{-3}; 0.001 25   **b.** 2.06×10^2; 206
   c. 2.13×10^{10}; 21,300,000,000
   d. 4.23×10^{-15}; 0.000 000 000 000 004 23

37. a. 6.70×10^8 mph   **b.** 5.36×10^9 mph   **c.** 1.81×10^{10} mph
   d. 1.88×10^{12} mph

EXERCISES 6.1

1. a. $x = \frac{2}{3}$ **b.** $x = \frac{3}{2}$ **c.** $x = \frac{3}{4}$

3. $3 \cdot 3 \cdot 3 \cdot 3 = (3 \cdot 3) \cdot (3 \cdot 3)$

5. a. $x = 5$ **b.** $x = 3\frac{1}{3} = \frac{10}{3}$ **c.** $x = 2\frac{1}{2} = \frac{5}{2}$ **d.** $x = 2$
   e. $x = 1\frac{2}{3} = \frac{5}{3}$ **f.** $x = 1\frac{3}{7} = \frac{10}{7}$ **g.** $x = 1\frac{1}{4} = \frac{5}{4}$

7. a. 8   **b.** 2   **c.** 2   **9. a.** 4   **b.** 64   **c.** 216   **d.** 64

11. a. 32   **b.** 512   **c.** 125   **d.** 27

17. a. $x \approx 4.5$ **b.** $x \approx 4.7$ **c.** $x \approx 5$

19. a. 125   **b.** 32   **c.** 1024   **d.** 0.3

21. a. annual   **b.** semiannual   **c.** quarterly   **d.** monthly
   e. weekly   **f.** daily

23. $1 + \dfrac{r}{h}$ **25. a.** 11,272.72   **b.** 11,268.25

27. a. 12,702.37   **b.** 12,682.42

29. $10,000 earnings, 4% compounded monthly/quarterly for 3 years

31. $9.77   **33.** $1414.77

35. $\approx$\$6.55; $\approx$\$6.94; $\approx$\$7.36; rounding error   **37.** smaller value

39. $\approx$\$1149.73   **41.** $\approx$\$1196.50   **43.** $\approx$\$1447.88

45. $\approx$\$1390.44   **49.** 23.04 to 31.56   **51.** 15.21 to 20.10

53. 17.99 to 23.80

EXERCISES 6.2

1. a. $0.2\sqrt{10} \approx 0.632$ **b.** 2   **c.** $2\sqrt{10} \approx 6.325$ **d.** 20
   e. $20\sqrt{10} \approx 63.246$

3. a. $200\sqrt{10} \approx 632.456$ **b.** 2000   **c.** 0.2   **d.** 0.002

5. $10\sqrt{n}$ **7.** $0.1\sqrt{n}$ **9.** $\dfrac{\sqrt{n}}{10}$ **11. a.** 4   **b.** 2

13. a. -3 **b.** 5

15. a. not a real number   **b.** not a real number

17. a. -10 **b.** -10 **19. a.** $\sqrt[6]{x}$ **b.** $x^{1/3}$ **c.** $\sqrt{x}$

21. a. $(\sqrt{x})^3$ or $\sqrt{x^3}$ **b.** $x^{3/4}$ **c.** $(\sqrt[5]{x})^4$ or $\sqrt[5]{x^4}$

23. a. $(-8)^{5/3} = -32$ **b.** $27^{2/3} = 9$

25. a. $16^{1/2} = 4$ **b.** not a real number

27. a. $|y|$ **b.** z^4 **c.** $|x^3|$ **29. a.** x^2 **b.** x^2 **c.** $|y|$

31. a. 6   **b.** -9 **c.** 4   **d.** 2

33. a. 3   **b.** 3   **c.** 4   **d.** 2   **35. a.** $a^2, a \geq 0$ **b.** b

37. a. $x^{13/12}$ **b.** $x^{17/12}$ **c.** $\dfrac{1}{x^{1/3}}$ **d.** $x^{11/12}$

39. a. $x^{11/6}$ **b.** $x^{5/12}$ **c.** $x^{1/2}$ **41. a.** $x^{1/3}$ **b.** $\dfrac{8x^3}{y^6}$ **c.** $\dfrac{1}{x^2}$

43. a. $2\sqrt[3]{2}$ **b.** $5a^2\sqrt{3a}$ **c.** $2x^2\sqrt[4]{2x}$ **d.** $-2bc\sqrt[5]{a^4c}$
   e. $xy^2z^3\sqrt[3]{xy^2}$

45. a. $\approx 2.61 \times 10^3$ m/sec   **b.** $\approx 3.95 \times 10^4$ m/sec
   c. $\approx 1.05 \times 10^4$ m/sec   **d.** $\approx 4.53 \times 10^3$ m/sec

MID–CHAPTER 6 TEST

1. $\sqrt{x}, x \geq 0$ **2.** $x^{-1}, x \neq 0$ **3.** $\dfrac{b^2}{a}$ **4.** $\dfrac{a^2c^2}{b}$ **5.** a^{x+1}

6. a. $\dfrac{9}{4x^2}$ **b.** xy^2

7. a. 0.0043   **b.** 1.23×10^{-4} **c.** $3 \times 10^{-5} = 0.00003$
   d. two

8. a. $1418.73   **b.** $1212.11

9. a. $x = 3$ **b.** $x = 3$ **c.** $x = 3$ **d.** $x = 0$

10. A 2% wage increase twice a year is the same as a 4.04% annual increase.

11. a. $2\sqrt{n}; n \geq 0$ **b.** $2|n|\sqrt{10}$ **c.** $20\sqrt{n}; n \geq 0$
   d. $20\sqrt{10n}$

12. a. $(\sqrt[3]{8})^2 = 4$ **b.** $\sqrt[3]{125} = 5$

13. a. $(\sqrt[4]{16})^3 = 8$ **b.** $(\sqrt[3]{a^3})^2 = a^2$ **14. a.** -4 **b.** $2|y|$

15. a. $3x^2y\sqrt{x}$ **b.** $2xy^2\sqrt[3]{3x^2}$

EXERCISES 6.3

1. $6\sqrt{2}$ **3.** $5\sqrt{5}$ **5.** $8\sqrt{3}$ **7.** $6\sqrt{x}$ **9.** $3.1\sqrt{x}$

11. not like terms   **13.** $3\sqrt{ab}$ **15.** $2\sqrt[3]{x}$ **17.** not like terms

19. $5\sqrt[3]{x}$ **21.** $2a\sqrt[4]{ab}$ **23.** $3x\sqrt[4]{y}$

25. a. 1   **b.** $27 - 10\sqrt{2}$ **27. a.** $x^2 - 5$ **b.** $x^2 - 2x\sqrt{7} + 7$

29. a. $x + 6\sqrt{x} + 9$ **b.** $a - 6\sqrt{a} + 9$

31. a. $1 - 2\sqrt{a} + a$ **b.** $a - 2\sqrt{ab} + b$

37. a. The product is -1.   **b.** ≈ 1.6180; ≈ 0.6180
   c. $\dfrac{2}{1 + \sqrt{5}} = -\dfrac{1 - \sqrt{5}}{2}$; $\dfrac{2}{1 - \sqrt{5}} = -\dfrac{1 + \sqrt{5}}{2}$; negative reciprocals

39. $x^3 + 2$

41. a. $3 - \sqrt{2}$ **b.** $3 + \sqrt{a}$ **c.** $a + \sqrt{b}$ **d.** $\sqrt{2} + \sqrt{3}$

43. a. $\dfrac{4\sqrt{5}}{5}$ **b.** $\dfrac{4\sqrt{6}}{3}$ **45. a.** $\dfrac{a\sqrt{c}}{c}$ **b.** $\sqrt{a}$ **47.** $\dfrac{7 + \sqrt{5}}{11}$

49. $\dfrac{x^2 + x\sqrt{y}}{x^2 - y}$ **51. a.** $\dfrac{\sqrt[3]{4}}{2}$ **b.** $\sqrt[3]{2}$ **c.** $\sqrt[3]{9}$

57. slope $= -\frac{3}{5}$; $d = 2\sqrt{34} \approx 11.662$

59. slope $= \frac{11}{9}$; $d = \sqrt{202} \approx 14.213$

61. slope $= \frac{b}{a}$; $d = \frac{\sqrt{a^2 + b^2}}{2}$

EXERCISES 6.4

1. $\left(\frac{1}{2}, 1\right), \left(\frac{1}{4}, 2\right), \left(\frac{1}{8}, 3\right), \left(\frac{1}{16}, 4\right)$ **3.** $(2, 1), (5, 2), (8, 3), (11, 4)$

5.

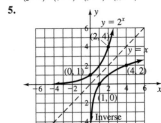

7.

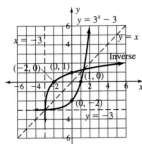

9. $y = \dfrac{x}{3}$ **11.** $y = \frac{1}{4}x + \frac{1}{2}$ **13.** $y = x - 3$

15. $y = x^2, x \geq 0, y \geq 0$ **17.** $y = x^5$

19. $y = \sqrt[4]{x}, x \geq 0, y \geq 0$ **21.** $y = x$ **23.** $y = -\dfrac{1}{x}$

25. $y = 1 - x$ **27.** same as original function

29. Take the nth root of both sides of the equation.

31. $\approx 9.6 \text{ in}^2$ **33.** $\approx 11.2 \text{ in.}$ **35.** $r = \sqrt[3]{\dfrac{3V}{4\pi}}$

37. $r = \sqrt[4]{\dfrac{8vLn}{\pi p}}$ **39.** $\approx 6.1\%, \approx \$243,000$

41. $\approx 58.18\%$; $\approx \$84,486.60$ **43.** $\approx -13\%$ **45.** $\approx 22.8\%$

47. 2.154; 1.778; 1.585; 1.468

49. The base approaches 1 as the exponent gets larger.

51. if a is non-negative   **53.** all real numbers

55. The square root of a negative number is not a real number.

57. extraneous root   **59.** $x = 14$ **61.** { } or $\varnothing$

63. $x \leq 2$; $x = -2$ **65.** $x \geq 5$; $x = 9$ **67.** $x \geq 5$; $x = 21$

69. $x \geq -7$; $x = -7$ **71.** $x \geq -7$; $x = 11$

73. $x \geq \frac{5}{3}$; $x = 3$ **75.** $x \geq \frac{5}{3}$; { } or $\varnothing$ **77.** $x \geq \frac{1}{5}$; $x = 2$

79. $x \geq 0$; {1, 9}   **81.** $x \geq 0$; $x = 12$ **83.** $x \geq 1$; $x = 4$

85. $L = \dfrac{r^2}{24}$

87. $\approx 34.64 \text{ mph}$; $\approx 24.49 \text{ mph}$; $\dfrac{r_{\text{wet}}}{r_{\text{dry}}} = \dfrac{\sqrt{2}}{2}$; slow down when pavement is wet.

89. a. 560 ft   **b.** ≈ 850 ft

   c. $2242\frac{2}{3}$ ft; ≈ 3420 ft; The formula $d = \sqrt{\dfrac{3h}{8}}$ involves multiplying the output by a factor of 8, whereas the formula $d = \sqrt{\dfrac{3h}{2}}$ involves multiplying the output by a factor of 2.

CHAPTER 6 REVIEW EXERCISES

1. $\dfrac{1}{bc^2}$ **3.** $\dfrac{16x^4}{y^8}$ **5.** $\dfrac{8x^6}{y^3}$

7. a. 0.000 000 000 000 000 003 45   **b.** 0.006 400   **c.** 400,500

   d. 4780.0

9. a. $9.25 \times 10^{16} = 92,500,000,000,000,000$

   b. $2.331 \times 10^{17} = 233,100,000,000,000,000$

   c. $8.51 \times 10^{15} = 8,510,000,000,000,000$

   d. $4.44 \times 10^{17} = 444,000,000,000,000,000$

   e. $1.813 \times 10^{16} = 18,130,000,000,000,000$

11. \$1826.29   **13.** \$1394.18   **15. a.** 8   **b.** 2   **c.** 32

17. a. 8   **b.** 125   **c.** 81

19. a. b^3 **b.** a **c.** $x^{5/4}$ **d.** $b^{1/4}$ **e.** b^{x+1}

21. a. $x^{3/4}$ **b.** x^{n-1} **c.** $x^{1/6}$ **d.** $a^{-2/3}$ or $\dfrac{1}{a^{2/3}}$ **e.** a^{1-x}

23. a. $(-8)^{1/3} = -2$ **b.** $\left(\frac{625}{16}\right)^{1/4} = \frac{5}{2}$ **c.** $8^{2/3} = 4$

25. a. $(\sqrt[3]{125})^2 = 25$ **b.** $\sqrt[3]{64} = 4$ **c.** $(\sqrt[5]{32})^2 = 4$

27. a. $\sqrt[3]{-64} = -4$ **b.** $-\sqrt{64} = -8$

   c. not a real number

29. a. $(\sqrt[3]{x})^2$ **b.** $(\sqrt{x})^3$ **c.** $(\sqrt[4]{a})^3$

31. a. 20   **b.** ≈ 9.283 **c.** ≈ 4.309 **d.** 2   **e.** ≈ 0.9283

   f. ≈ 0.4309 **g.** 0.2

33. a. $3\sqrt{n}, n \geq 0$ **b.** $3|n|\sqrt{10}$ **c.** $0.3\sqrt{x}, x \geq 0$ **d.** $30x^2$

35. a. $|z^3|$ **b.** x **c.** x^2

37. a. $x^{5/6}, x \geq 0$ **b.** $x^{5/2}, x \geq 0$ **c.** $x^{1/6}, x \geq 0$

39. a. $5\sqrt{3}$ **b.** $4\sqrt{x}, x \geq 0$ **41. a.** 3   **b.** $9 - 6\sqrt{x} + x$

43. a. $\dfrac{b + \sqrt{a}}{b^2 - a}$ **b.** $\dfrac{\sqrt{a} - \sqrt{b}}{a - b}$ **c.** $\dfrac{(\sqrt[3]{9x})^2}{9x}$ **45.** $1 + \sqrt{2}$

47. a. $(3, 3), (5, 4), (7, 5)$; yes   **b.** $(-1, -2), (-2, 0), (-3, 2)$; yes

49. a. $y = \frac{1}{2}x + \frac{1}{2}$ **b.** $y = \dfrac{x}{3}$ **c.** $y = \sqrt[3]{x}$

   d. $y = x^2, x \geq 0, y \geq 0$ **e.** $y = -x$ **f.** $y = \dfrac{1}{x}$

51. $\approx 5.05\%$ **53.** $\approx 8.45\%$ **55.** $g = \dfrac{4L\pi^2}{T^2}$ **57.** $r = \sqrt{\dfrac{Q_1 Q_2}{kF}}$

59. $r = \sqrt[3]{a^2 p}$

61. a. $\approx \{0.25, 2\}$ **b.** $x = 1$ **c.** $x = \frac{1}{2}$ **d.** $\left\{\frac{2}{9}, 2\right\}$

63. a. $x = 12$ **b.** $x = 4$ **c.** $x = 0$ **d.** $x = 12$

CHAPTER 6 TEST

1. b^3 **2.** $\dfrac{b}{a^2 c^2}$ **3.** $\dfrac{9}{x^2 y^2}$ **4.** $(2x)^2 = 2^2 \cdot x^2 = 4x^2$

5. $\sqrt{4x^2} = \sqrt{4} \cdot \sqrt{x^2} = 2|x|$

6. a. 0.000 034 50   **b.** 9×10^{-7} or 0.000 000 9

7. a. $6\sqrt{x}, x \geq 0$ **b.** $0.6|x|$ **c.** $6x^2\sqrt{10}$

8. yes; both equal $\dfrac{b^4}{a^3}$.   **9.** \$2536.48   **10.** $\approx 8.45\%$

11. $r = \sqrt{\dfrac{mgl}{\pi sM}}$ **12.** $v = \sqrt{\dfrac{Fr}{m}}$ **13.** $(\sqrt[3]{64})^2 = 16$

14. $(\sqrt[5]{32})^4 = 16$ **15.** $(\sqrt[4]{81})^3 = 27$ **16. a.** $\frac{2}{3}$ **b.** -3 **c.** 4

17. a. b^3 **b.** b **c.** 3^{n+1}

18. a. $4 - x$ **b.** $24 - 3\sqrt{3} + 16\sqrt{2} - 2\sqrt{6}$

19. a. $\dfrac{x - \sqrt{y}}{x^2 - y}$ **b.** $\dfrac{(\sqrt[3]{x})^2}{x}$

20. yes; $(4 - \sqrt{10})^2 - 8(4 - \sqrt{10}) + 6 = 0$

21. $(-1, -1), (-3, -2), (-5, -3)$; yes

22.

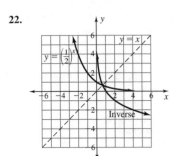

23. $3y - 2x = 6$ or $y = \frac{2}{3}x + 2$

24. a. $x = 8$ **b.** $x = 3$ **c.** $x = 2$ **d.** $x = 8$

CUMULATIVE REVIEW OF CHAPTERS 1 TO 6

1. $x = -11$ **3.** $b = 3A - a - c$ **5.** $y \approx 11.03 + 0.051x$

7. a. $-\frac{3}{5}$ **b.** $\frac{5}{3}$ **9.** domain: $\mathbb{R}$; range: 4

11. a. 24; 6   **b.** It is the y-intercept or $f(x)$-intercept.

    c. $\left(1\frac{1}{4}, 2\frac{7}{8}\right)$ **d.** $x = \dfrac{-b + \sqrt{b^2 - 4ac}}{2a}$; no real x-intercepts

15. $\frac{49}{4}$

17. a. 37; quadratic; $y = x^2 + 1$ **b.** 10; linear; $y = 3x - 8$

    c. 37; quadratic; $y = x^2 + x - 5$ **d.** 29; neither; $y = 2^{x-1} - 3$

19. $\left\{ 1, -\frac{1}{2} \pm i\dfrac{\sqrt{3}}{2} \right\}$ **21. a.** $\approx(0.28, 0.01)$ **b.** $\approx$27.1 ft

23. c   **25.** a   **27.** b   **29.** $\dfrac{(x + 3)(x + 1)}{3x}$

31. $\approx$66 in. $\leq x \leq$ 75 in.   **33.** $2.475 \times 10^9 = 2{,}475{,}000{,}000$

35. a. $\{1, 4\}$ **b.** $x = 2$ **37.** $B = \left(\dfrac{W}{n}\right)^{5/8}$

EXERCISES 7.0

1. a. 3   **b.** $a_n = 0.5 \cdot 3^{n-1}$ **c.** $y = 0.167 \cdot 3^x$

3. a. 2   **b.** $a_n = 10 \cdot 2^{n-1}$ **c.** $y = 5 \cdot 2^x$

5. a. 2   **b.** $a_n = \left(\frac{1}{4}\right) \cdot 2^{n-1}$ **c.** $y = 0.125 \cdot 2^x$

7. a. 2   **b.** $a_n = 2^{n-1}$ **c.** $y = 0.5 \cdot 2^x$

9. a. $\frac{1}{2}$ **b.** $a_n = 32\left(\frac{1}{2}\right)^{n-1}$ **c.** $y = 64 \cdot 0.5^x$

11. a. $\frac{1}{3}$ **b.** $a_n = 3 \cdot \left(\frac{1}{3}\right)^{n-1}$ **c.** $y = 9 \cdot 0.333^x$

13. quadratic; $y = 2x^2$ **15.** exponential; $y = 0.167 \cdot 3^x$

17. quadratic; $y = 0.5x^2 + 2.5x$ **19.** linear; $y = 8x + 22$

21. exponential; $y = 2.5 \cdot 2^x$ **23.** $4 \cdot 2^x$ **25.** $2^x/2$ **27.** 2^x

29. 2^{n-1} **31.** 2^{n+2} **33.** 3^{n-1} **35.** 3^{n+2}

37. a. $f(x) = 2^{x-4}$ **b.** $f(x) = 2^{x+1}$ **39. a.** $\left(\frac{1}{2}\right)^{x-3}$ **b.** $\left(\frac{1}{2}\right)^{x-4}$

41. $r = -3$; $a_n = 2(-3)^{n-1}$; base is negative, not an exponential function

43. $r = -\frac{1}{2}$; $a_n = 64\left(-\frac{1}{2}\right)^{n-1}$; base is negative, not an exponential function

45. $a_n = 27.5(2)^{n-1}$

EXERCISES 7.1

1. a. base $= a$; exponent $= x$ **b.** base $= x$; exponent $= 3$

    c. base $= 2$; exponent $= -x$ **d.** base $= x$; exponent $= a$

3. a. base $= \pi$; exponent $= x$ **b.** base $= x$; exponent $= 4$

    c. base $= r$; exponent $= n - 1$

    d. base $= 1.06$; exponent $= t$

5. a. increasing   **b.** decreasing

7. a. increasing   **b.** decreasing

9. a. increasing   **b.** increasing

11. $a = 10^x$; $b = 2.72^x$; $c = 2^x$ **13.** { } or $\emptyset$ **15.** { } or $\emptyset$

17. $3, 4, \frac{1}{2}$; y-intercept $= a$

19. a. y-intercept $= 16$; $y = 2^{x+4}$ **b.** y-intercept $= 8$; $y = 2^{x+3}$

21. a. y-intercept $= 27$; $y = 3^{x+3}$ **b.** y-intercept $= \frac{1}{27}$; $y = 3^{x-3}$

23. a. $y = \frac{1}{16} \cdot 2^x$; y-intercept $= \frac{1}{16}$ **b.** $y = 2 \cdot 2^x$; y-intercept $= 2$

25. Represents y-intercept value   **27.** when $x = 1$, $b^x = b$

31. $a = 2^{x+1}$; $b = 2^x$; $c = 2^{x-1}$

33.

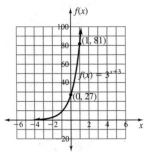

35.

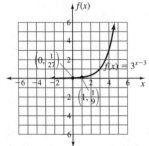

37. no intersection; $3^x \neq 3^{x+1}$ for all x

39. a. $x \approx 2.5$ **b.** $x \approx 4.8$ **c.** $y \approx 1.8$

    d. $y \approx 4.3$ **e.** $a \approx 1.34$

41. c. horizontal line at $y = 1$ **d.** yes   **e.** $y = 1$

43. b. points only   **c.** no   **d.** no; negative base

EXERCISES 7.2

1. a. $x = 8$ **b.** $x = 2$ **3. a.** $n = \frac{3}{2}$ **b.** $n = \frac{4}{3}$

5. a. $n = \frac{1}{2}$ **b.** $x = \frac{3}{2}$ **7. a.** $x = -\frac{1}{2}$ **b.** $n = -\frac{1}{3}$

9. a. $n = -\frac{1}{3}$ **b.** $x = -\frac{1}{4}$ **11. a.** $x = -10$ **b.** $x = -4$

13. a. $x = -6$ **b.** $x = 4$ **15. a.** $x = 7$ **b.** $x = 5$

17. a.

x	y
-2	$\frac{1}{9}$
-1	$\frac{1}{3}$
0	1
1	3
2	9
3	27
4	81

b.

x input to inverse	y output to inverse
$\frac{1}{9}$	-2
$\frac{1}{3}$	-1
1	0
3	1
9	2
27	3
81	4

c.

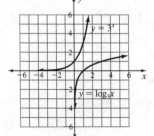

d. $y = \log_3 x$

19. range   **21.** range   **23.** (b, a)

25.

Exponential Equation	Logarithmic Equation
$3^3 = 27$	$\log_3 27 = 3$
$2^3 = 8$	$\log_2 8 = 3$
$10^1 = 10$	$\log_{10} 10 = 1$
$5^3 = 125$	$\log_5 125 = 3$
$3^{-2} = \frac{1}{9}$	$\log_3 \left(\frac{1}{9}\right) = -2$

27. a. $10^3 = 1000$ **b.** $10^0 = 1$
c. $3^4 = 81$ **d.** $10^2 = 100$

29. a. $10^{-3} = 0.001$ **b.** $4^0 = 1$
c. $m^k = n$ **d.** $5^{-2} = \frac{1}{25}$

31. a. $\log_2 32 = 5$ **b.** $\log_2 2 = 1$ **c.** $\log_2 1 = 0$
d. $\log_{10} 100 = 2$

33. a. $\log_{10} 0.001 = -3$ **b.** $\log_f g = d$ **c.** $\log_3 \left(\frac{1}{9}\right) = -2$
d. $\log_4 1 = 0$

35. a. 0   **b.** 0.77815   **c.** 1   **d.** 1.77815   **e.** 2
f. 2.77815

37. a. -0.22185 **b.** -1 **c.** -2 **d.** -1.22185
e. -2.22185 **f.** -5.22185

39. a. Decimal portion is the same; all are positive.
b. Decimal portion is the same; all are negative.

41. a. $\log_{10} 17 = x; x \approx 1.23045$ **b.** $\log_{10} 125 = x; x \approx 2.09691$
c. $\log_{10} 400 = x; x \approx 2.60206$
d. $\log_{10} 0.05 = x; x \approx -1.30103$

43. $5^0 = x, x = 1; x^3 = 64, x = 4; \log_{10} (-10) = x, \{\ \}$ or $\varnothing$

45. a. $7^2 = x, x = 49$ **b.** $3^x = 3, x = 1$
c. $10^x = 0.01, x = -2$ **d.** $10^0 = x, x = 1$
e. $x^3 = 64, x = 4$ **f.** $a^1 = x, x = a$

47. a. $2^8 = x, x = 256$ **b.** $10^{-2} = x, x = \frac{1}{100}$
c. $2^x = \frac{1}{4}, x = -2$ **d.** $10^x = 1, x = 0$
e. $x^2 = 100, x = 10$ **f.** $a^x = a, x = 1$

49. $y = x$ **51.** exponent   **53.** base   **55.** $\mathbb{R}$ **57.** $y > 0$

59. The denominator is missing an input value for the logarithmic expression.

61. No; $x = -3$ is another solution.

63. Both groups are growing at the same rate.

MID–CHAPTER 7 TEST

1. a. geometric; $a_n = 3 \cdot 3^{n-1}; f(x) = 3^x$
b. arithmetic; $y = 6x - 7$
c. geometric; $a_n = 3 \cdot 2^{n-1}; f(x) = 1.5 \cdot 2^x$
d. geometric; $a_n = 81 \cdot \left(\frac{1}{3}\right)^{n-1}; f(x) = 243 \cdot \left(\frac{1}{3}\right)^x$
e. quadratic; $y = 2x^2 - x$
f. geometric; $a_n = \left(\frac{1}{8}\right) \cdot 2^{n-1}; f(x) = 0.0625 \cdot 2^x$

2.

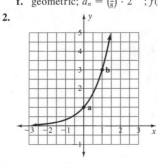

3. a. approach x-axis to left and rise rapidly to right
b. Graph of 2^{x-1} is that of 2^x shifted right 1 unit.
c. no intersection; $2^x \neq 2^{x-1}$ for all x

4. a. ≈ 3.3
b. $\log_2 10 = x$ or $2^x = 10$

5. Graph of $y = \frac{1}{2} \cdot 3^x$ grows rapidly as x gets large; that of $y = 2 \cdot 3^{-x}$ approaches zero.
a. constants $\frac{1}{2}$ and 2
b. $0 < b < 1$ for a decreasing function

6. a. $x = 5$ **b.** $x = -2$ **c.** $x = 4$ **d.** $x = -3$
e. $\{\ \}$ or $\varnothing$ **f.** $x = -2$

7. a. $x = \frac{1}{4}$ **b.** $x = -4$ **c.** $x = 27$ **d.** $x = 1$
e. $x = \frac{1}{2}$ **f.** $\{\ \}$ or $\varnothing$

8. a.

x	y
-2	$\frac{1}{16}$
-1	$\frac{1}{4}$
0	1
1	4
2	16
3	64

b.

x	y
$\frac{1}{16}$	-2
$\frac{1}{4}$	-1
1	0
4	1
16	2
64	3

c.

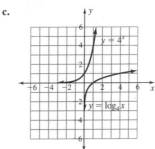

d. $y = \log_4 x$

9.

Exponential Equation	Logarithmic Equation	Solve for x
$10^x = 15$	$\log_{10} 15 = x$	$x \approx 1.17609$
$10^x = 13$	$\log_{10} 13 = x$	$x \approx 1.11394$
$3^x = 81$	$\log_3 81 = x$	$x = 4$
$3^9 = x$	$\log_3 x = 9$	$x = 19683$
$4^{x+1} = 64$	$\log_4 64 = x + 1$	$x = 2$

EXERCISES 7.3

1. a. 3   **b.** 10   **c.** 4   **d.** 5   **e.** 10   **f.** m

3. a. 14   **b.** 0.7   **c.** 2.3   **d.** 10   **e.** 2.9

5. a. base   **b.** acid   **c.** acid   **d.** base   **e.** acid

7. a. 7.94×10^{-5} M   **b.** 3.16 M   **c.** 1.0×10^{-12} M
d. 3.16×10^{-4} M

9. $[H^+] > 1; [H^+] < 1$ **11. a.** $10^7; 7.5$ **b.** $10^8; 8.1$

13. a. 398,107,000   **b.** 6,310,000

15. $\approx 7,943,000; \approx 6,310,000; \approx 1.3$ times as strong

17. a. $x = 0$ **b.** $x \approx 1.58$ **c.** $x = 2$ **d.** $x \approx 2.32$
e. $x \approx 2.58$ **f.** $x \approx 3.32$ **g.** $x \approx 3.58$ **h.** $x \approx 4.32$
i. $x \approx 4.58$

19. a. $x \approx 0.43$ **b.** $x = 1$ **c.** $x \approx 1.11$ **d.** $x \approx 1.43$
e. $x = 2$ **f.** $x \approx 2.11$

21. a. $x \approx 2.09590$ **b.** $x \approx 1.29248$ **c.** $x \approx 10.24477$
d. $x \approx 13.51341$ **e.** $x = 1$

23. x-intercept $= 1; f(1) = 0$

25. x-intercept $= 1$ for both graphs; the graph of $\log_6 x$ is above that of $\log_3 x$ to the left of the x-intercept and is below it to the right of the x-intercept.

27. false   **29.** false

31. Graphs mirror each other across $y = x$. $2^0 = 1$ is the same fact as $\log_2 1 = 0$; $2^1 = 2$ is the same fact as $\log_2 2 = 1$. Answers may vary: $(2, 4)$ is on y_2, $(4, 2)$ is on y_3.

33. ≈349 yr   **35.** ≈73 yr   **37. a.** 8%   **b.** 6%

39. a. −3%   **b.** −5%   **41.** ≈11.9 yr

43. 72 yr, 36 yr, 14.4 yr, 9 yr, 7.2 yr, 3.6 yr

45. a. 4.2%   **b.** 2.1%

47. $(0.9)^t = 0.5$, $t ≈ 6.58$ yr

49. a. $n ≈ 2.7$ **b.** $n = 4$ **c.** $n ≈ 12.5$

51. $t = \dfrac{\log 2}{\log (1 + r)}$; P does not appear in formula; doubling time is not dependent on P.

53. a. ≈333   **b.** $y = 2,500,000(0.974)^x$, x in hours

55. a. ≈26   **b.** $y = 230,000(0.953)^x$, x in days

57. a. ≈584   **b.** $y = 12,000,000(0.955)^x$, x in hours

59. b. $y = 0.0544x + 0.5$
   c. $y = 0.5 \cdot 1.079^x$; y-intercept is \$0.50 cost in 1986.

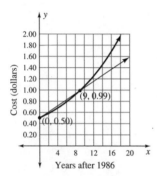
Years after 1986

   d. $r = 7.9\%$
   e. The linear model assumes that the price changes a constant amount each year. The exponential model assumes a percent increase each year.
   f. linear: ≈\$1.53; exponential: ≈\$2.12

61. $y = 24,780(1.049)^x$; 4.9%; 705,000

63. $y ≈ 42,738(0.998)^x$; −0.2%; 37,000

65. a. $y ≈ 25,600(0.87)^x$ **b.** $y = 50(0.87)^x$
   c. $y = 50(0.5)^x$

EXERCISES 7.4

1. a. $\log 2x$ **b.** $2 \log x$ **c.** $\log 2 - \log x$ **d.** $x \log 2$
   e. $\log \left(\dfrac{x}{2}\right)$

3. property 1   **5.** property 2

11. a. $\log (x^2 - 1)$ **b.** $\log (x + 1)$

13. a. $\log (x + 2)$ **b.** $\log (x^2 + x - 6)$ **15.** $x = 4$

17. $x = 10$ **19.** $x = 10$ **21.** $x = 9$ **23.** $x ≈ 1.8614$

25. $x ≈ 1.5237$ **27.** $x ≈ 3.4037$ **29.** $x ≈ 0.7011$

31. $t ≈ 9.0065$ **33.** $t ≈ 16.2376$ **35.** $t ≈ 8.3130$

37. $t ≈ 15.1385$

41.

x	y
−1	$\frac{1}{3}$
0	1
1	3
2	9
3	27
4	81
5	243
6	729

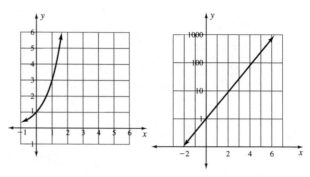

43. a straight line   **45.** ≈22 ft   **47.** 60 to 70 yr

49. years 2 and 12; wind breakage

51.

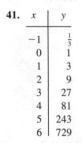

Spacing is the same (on a larger scale).

53. a. $y = 52.2(0.95)^x$
   b. The deeper the water, the less the amount of light penetration.
   c. $\log_{10} y = -0.02x + 1.72$
   d. $y = 10^{-0.02x + 1.72}$; error due to rounding

55. a. $y = 0.0013(0.881)^x$
   b. Density decreases as altitude increases.
   c. $\log_{10} y = -0.0552x - 2.882$
   d. $y ≈ 0.0013(0.881)^x$

EXERCISES 7.5

1. ≈\$1124.86, ≈\$1127.27, ≈\$1127.48; ≈\$2.62, ≈\$2.85

3. a. ≈7.39   **b.** ≈23.14   **c.** ≈3.3

5. a. ≈19.81   **b.** ≈2.23   **c.** ≈4.81   **7.** $e^{(e^1)}$

9. ≈\$1083.29   **11.** ≈\$1105.17   **13.** ≈\$1046.03

15. ≈\$150,597.11   **17.** ≈\$67,667.64

19.

Logarithmic Equation	Show Base e in Logarithm	Exponential Equation	$y = ?$
$\ln y = 1$	$\log_e y = 1$	$y = e^1$	$y = e$
$\ln y = 0$	$\log_e y = 0$	$y = e^0$	$y = 1$
$y = \ln (-1)$	$y = \log_e (-1)$	$e^y = -1$	{ } or $\varnothing$
$y = \ln e^2$	$y = \log_e e^2$	$e^y = e^2$	$y = 2$
$y = \ln e^e$	$y = \log_e e^e$	$e^y = e^e$	$y = e$

21. a. $x ≈ 1.3863$ **b.** $x ≈ 0.1353$

23. a. $x ≈ 4.4817$ **b.** $x ≈ 0.6931$

25. a. 8.625%   **b.** ≈6.27%   **c.** ≈69/t

27. ≈6.30%

29. ≈13.7%   **31.** $t = \dfrac{\ln 3}{r}$ **33.** $t ≈ 8.7$ yr

35. a. ≈1.1606; 1.1606
   b. ≈0.903; ≈0.778; ≈2.079; ≈1.792
   c. Ratios are equal; logarithms are not.

37. $K = e^{\left(-\frac{Ea}{RT} + C\right)}$ **39.** $[H^+] = e^{\left(-\frac{EnF}{RT}\right)}$

41. a.

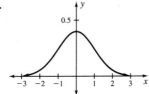

b. $y \approx 0.4$ at $x = 0$ **c.** y-intercept $= \dfrac{1}{\sqrt{2\pi}}$

d. ≈ 0.65

43. a. 1,850,000   **b.** 1,380,000   **c.** 1,030,000
   d. the sign of the exponent   **e.** ≈ 1997 **f.** 887,000

45. $(e, e^{1/e})$

47. a. 24   **b.** 3,628,800   **c.** 720   **d.** 362,880
   e. ≈ 7.3873; $e^2 \approx 7.3891$ **f.** ≈ 0.72537; $\ln 2 \approx 0.69315$

CHAPTER 7 REVIEW EXERCISES

1. 729; geometric; $y = 9 \cdot 3^{x-1}$ or $y = 3 \cdot 3^x$

3. 31; quadratic; $y = x^2 - x + 1$ **5.** 20; arithmetic; $y = 3x + 2$

7. $a_n = \frac{1}{4}(2)^{n-1}$; $f(x) = 0.125 \cdot 2^x$

9. $a_n = \frac{1}{16}(2)^{n-1}$; $f(x) = 0.03125 \cdot 2^x$

11. a. $y = \frac{1}{8} \cdot 2^x$; $\frac{1}{8}$ **b.** $y = 3 \cdot 3^x$; 3   **c.** $y = \frac{1}{27} \cdot 3^x$; $\frac{1}{27}$

13. a. 2; $y = 2^{x+1}$ **b.** $\frac{1}{9}$; $y = 3^{x-2}$

15. a. $x = 3$ **b.** $x = 1$ **c.** $x = -1$ **d.** $n = 0$

17. a. $n = \frac{3}{2}$ **b.** $n = -\frac{1}{3}$ **c.** $n = -\frac{3}{2}$ **d.** $n = -1$

19. a. $n = -\frac{1}{2}$ **b.** $n = \frac{1}{2}$ **c.** $x = 7$ **d.** $x = \frac{1}{3}$

21. $\left(\frac{1}{2}\right)^x = (2^{-1})^x = 2^{-x}$

23.

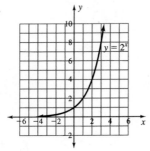

25.

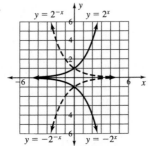

   a. shift 1 unit to the left of $y = 2^x$
   b. shift 2 units to the right of $y = 2^x$

27.

Exponential Equation	Logarithmic Equation	Solve for x
$2^x = 16$	$\log_2 16 = x$	$x = 4$
$x^2 = 25$	$\log_x 25 = 2$	$x = 5$
$3^x = 81$	$\log_3 81 = x$	$x = 4$
$10^{1/2} = x$	$\log_{10} x = \frac{1}{2}$	$x = \sqrt{10} \approx 3.162$
$10^x = 19$	$\log_{10} 19 = x$	$x \approx 1.2788$
$4^0 = x$	$\log_4 x = 0$	$x = 1$

29. $x = -1$ **31.** $x \approx 1.556$ **33.** $x \approx -0.125$ **35.** $x = 1.5$

37. $x \approx 0.815$ **39.** $x = 0.1$ **41.** $x = 2$ **43.** $x = 1$

45. $x = 4$ **47.** $x = 100$ **49.** $x = 2$ **51.** $x = \frac{2}{3}$

55. a. $3^0 = 1$, $\log_3 1 = 0$
   b. $3^1 = 3$, $\log_3 3 = 1$
   c. mirror each other across $y = x$

57. $\approx \$1126.49$, ≈ 11.64 yr   **59.** $\approx \$1127.48$, ≈ 11.55 yr

61. $\log(x^2 - 3x + 2)$ **63.** $\log(x + 1)$

65. $\log(x^2 + 2x + 1) = \log(x + 1)^2 = 2\log(x + 1)$

67. $\log \sqrt{x} = \log x^{1/2} = \frac{1}{2}\log x$ **69.** $x = 9$

71. $\log 500 = \log(5 \cdot 100) = \log 5 + \log 100 =$
   $\log 5 + 2 \approx 2.69897$

73. $2 +$ a decimal

75. $\log(1000x)$ is 3 greater than $\log x$.

77.

79. a. ≈ 1.10 **b.** ≈ 1.14 **c.** 1

81. a. $\approx \$740.82$ **b.** $\approx \$670.32$ **c.** $\approx \$548.81$

83. $\approx 13.8\%$; $2 = e^{5r}$, $r \approx 0.138$

85. $x = 0$ **87.** $x = 1$

89. $x = e^{-2}$ **91.** $x = e^e$

93. $y \approx 49.3(0.626)^x$ with $x = 0$ for 1989; ≈ 0.027 million

95. $P \approx 13,426(1.0265)^x$; $\approx 2.65\%$; $\approx 84,000$; not reasonable

97. a.

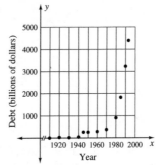

   b. $y \approx (4.1986 \times 10^9)(1.0868)^x$
   c. wars, spending in 1980s
   e.

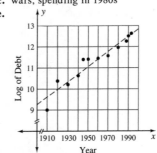

   Graph becomes somewhat linear.

CHAPTER 7 TEST

1. a. arithmetic   **b.** $a = 3n$, $f(x) = 3x$

2. a. geometric   **b.** $a_n = \left(\frac{1}{8}\right)2^{(n-1)}$, $f(x) = 0.0625 \cdot 2^x$

3. a. geometric   **b.** $a_n = \left(\frac{1}{4}\right)\left(\frac{1}{2}\right)^{(n-1)}$, $f(x) = 0.5 \cdot 0.5^x$

4. a. quadratic   **c.** $f(x) = 3x^2$

5. $y = 40.4x + 16,093$; $f(x) \approx 16,093(1.0025)^x$; linear $\approx 18,517$;
   exponential $\approx 18,694$

6. a.

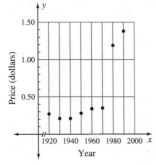

b. Possible answer: exponential; data start off slowly, increase rapidly

c. $f(x) \approx 0.161(1.0263)^x$

7. a. $x = 3$ **b.** $n = -\frac{3}{2}$ **c.** $x = 3$
d. $x = 4$ **e.** $x = -3$
f. $n = \frac{1}{2}$ **g.** $x = \frac{2}{3}$ **h.** $x = \frac{1}{2}$

8. a. $x \approx 0.477$ **b.** $x \approx -0.431$
c. { } or $\varnothing$ **d.** $x = -1$
e. $x = -1$ **f.** $x = \frac{1}{4}$
g. $x = 125$ **h.** $x = 2$
i. $x = 10$ **j.** $x \approx 4.482$

9. a. 0   **b.** 0   **c.** $\log_2 (x^2 + x)$ **d.** $\log_3 (x - 3)$

10. $y = b^2 b^x$; y-intercept $= b^2$

11.

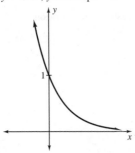

12. $(1 + r) = (1 - 0.07)$; $r = -0.07$

13. $\log (100x) = 2 + \log x$ because $\log 100 = 2$.

14. Possible answer: $(0, 1)$, $(1, 0)$; $(1, 10)$, $(10, 1)$; if $y = 10^x$, then $x = \log y$ and $y = \log x$ have reversed coordinates.

15. a. 8.3   **b.** $\approx 1.26 \times 10^{-12}$ M

16.

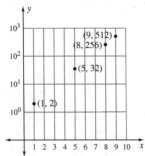

17. $f(x) = 3^x$

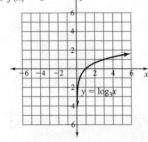

18. $f((x) = \log_2 x$

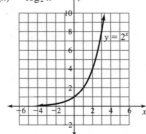

19. ≈ 4.25 yr   **20.** ≈ 8.6 yr

EXERCISES 8.0

1. $w = 2c$, $s = 3c$, $c + s + w = 60$; $c = 10$ hr, $s = 30$ hr, $w = 20$ hr

3. $c = 2p$, $c + f + p = 28$, $4c + 9f + 4p = 222$; $c = 4$ g, $f = 22$ g, $p = 2$ g

5. $f = p$, $c + f + p = 29$, $4c + 9f + 4p = 121$; $c = 27$ g, $f = 1$ g, $p = 1$ g

7. $y = -3x + 4$ **9.** $x = 3y + 7$ **11.** $x = 2y - \frac{5}{2}$

13. $y = 2x + \frac{3}{2}$ **15.** $x^2 = y - 5$ **17.** $y^2 = x^2/2 - x/2 + 3/2$

19. $y = \pm\sqrt{8 - x^2}$ **21.** $x = \pm\sqrt{10 + y^2}$ **23.** $x = 3$, $y = 5$

25. $x = -7.5$, $y = 7.5$ **27.** $x = 5$, $y = 2.2$

29. $x = 3$, $y = -1.2$ **31.** $x = 1.5$, $y = -1.2$

33. $x = 6$, $y = -3$ **35.** $x = -1$, $y = 2$

37. $a = 3$, $b = -2$, $c = \frac{1}{2}$ **39.** $a = 5$, $b = -1$, $c = -2$

41. $x = -9$, $y = 8$, $z = 3$ **43.** $x = -1$, $y = 6$, $z = 5$ **45.** no

47. $x + 0y = 2$; $a = 1$, $b = 0$, $c = 2$

49. $y = \dfrac{ce - af}{bc - ad}$ or $y = \dfrac{af - ce}{ad - bc}$

51. if one variable has a coefficient 1 and is easy to solve for

53. $y = \frac{7}{400,000}x^2 - \frac{1}{25}x + 700$ **55.** $y = \frac{11}{480,000}x^2 - \frac{1}{20}x + 1188$

EXERCISES 8.1

1.

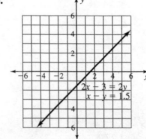

(1, 4)

3.

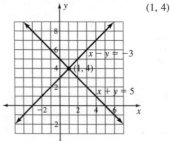

coincident; variables drop out, and result is true.

5.

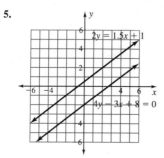

parallel; variables drop out, and result is false.

7. $\left(\frac{10}{3}, \frac{1}{3}\right)$

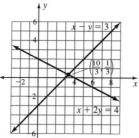

9. $\left(\frac{1}{2}, -3\right)$

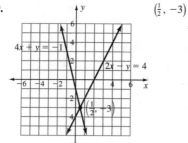

11.

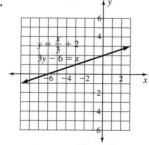

coincident; variables drop out, and result is true.

13. quantity: $1600, $12,000; value: 5% and 8% annual interest

15. quantity: 20 hr, 15 hr; value: $6.50 and $7.25 per hour

17. quantity: 100 mL and 1000 mL; value: 8% and 0% solution

19. What is the total value of the money? Dimes are $0.10; quarters, $0.25.

21. What is the measure of each angle? The sum of the interior angles in a triangle is 180°; an isosceles triangle has two equal interior angles.

23. 3.09 points per credit   **25.** 17 credit hours

27. 400 lb of A, 600 lb of B **29.** $25,000 at 6%, $50,000 at 9%

31. $8000 bank, $2000 credit card

33. 0.45 L water, 0.05 L 50% glucose solution

35. $5000   **37.** none; 4 is always better.   **39.** $7500

41. becomes steeper

43. Write equations for any pair of parallel lines.

EXERCISES 8.2

1. 1×2 **3.** 3×1 **5.** $[4 \quad 7]$ **7.** $\begin{bmatrix} 5 \\ 7 \\ 9 \end{bmatrix}$ **9.** $\begin{bmatrix} 5 & 5 & 5 \\ 5 & 5 & 5 \\ 5 & 5 & 5 \end{bmatrix}$

11. $\begin{bmatrix} x + 3y \\ -x - 2y \end{bmatrix}$ **13.** $\begin{bmatrix} -1 \\ 0 \end{bmatrix}$ **15.** $\begin{bmatrix} 10 \\ 0 \end{bmatrix}$ **17.** $\begin{bmatrix} 14 & 18 \\ -10 & -13 \end{bmatrix}$

19. $\begin{bmatrix} -1 & 0 \\ -1 & 2 \end{bmatrix}$ **21.** $\begin{bmatrix} -10 & -8 \\ 0 & 3 \end{bmatrix}$ **23.** $\begin{bmatrix} -9 & 2 \\ 6 & 2 \end{bmatrix}$ **25.** 1

27. 10   **29.** $\begin{bmatrix} -2 & -3 \\ 1 & 1 \end{bmatrix}$ **31.** $\begin{bmatrix} 0.2 & 0.2 \\ -0.1 & 0.4 \end{bmatrix}$

33. Both are $\begin{bmatrix} 1 & 0 \\ 0 & 1 \end{bmatrix} = I$.   **35.** $x = -2, y = 3.5$

37. det $[A] = 0$; parallel lines   **39.** $x = 2, y = 1.5$

41. $x = 1, y = 1.5$

43. $1x + 2y = 5, ax + 4y = 6$; slope changes; y-intercept

45. $x = 4.2, y = -3.6$ **47.** det $[A] = 0$; parallel lines

49. det $[A] = 0$; lines are coincident.

51. ≈ 156 kg Honduran, ≈ 44 kg Indonesian

53. 800 L of 3% solution, 200 L of 20% solution

55. 850 gal of water, 150 gal of 20% solution

57. dime ≈ 0.1 oz, quarter ≈ 0.2 oz

59. $5600 at 4.5%, $18,000 at 8.5%

61. a. $n = 0$ **b.** $n = 1$ **c.** 1   **d.** $y = x$ or $f(x) = x$
e. When we multiply a matrix by the identity, we get the original matrix.

63. commutative property of multiplication; no

MID–CHAPTER 8 TEST

1. $y = \dfrac{5}{2}x - 2$ **2.** $y = 4x - 10$ **3.** no real-number solution

4. $\left(2, -\frac{2}{3}\right)$ **5.** det $[A] = -0.4$; $(12, -11.4)$

6. det $[A] = 0$; lines are coincident.

7. Answers will vary; det $[A] = 0$.   **8.** Wilt, 31,419; Elgin, 23,149

9. $a = \frac{1}{50,000}, b = -0.05, c = 600$

10. $\frac{1}{3}$ pt of water, $\frac{2}{3}$ pt of 3% hydrogen peroxide solution

EXERCISES 8.3

1. a. plane; dependent   **b.** plane; dependent
c. no common point of intersection; inconsistent
d. no common point of intersection; inconsistent

3. $(2.5, 3, 3.5)$ **5.** inconsistent   **7.** $(-4, -11, 4)$

9. dependent   **11.** inconsistent   **13.** $a = 2, b = 3, c = -1$

15. 12 g protein, 4 g fat, 16 g carbohydrates

17. 17 nickels, 12 dimes, 15 quarters

19. Ayako, $466,034; Curtis, $925,941; Corey, $979,430

21. 50 lb Ethiopian; 200 lb French roast; 250 lb Guatemalan; 500 lb light Colombian

23. $y = 2x^4 + 3x^3 - 2x^2 + x - 5$

25. $\begin{bmatrix} 1 & 1 & 1 \\ 1 & -1 & 0 \\ 0 & 0 & 1 \end{bmatrix} \begin{bmatrix} s \\ m \\ l \end{bmatrix} = \begin{bmatrix} 620 \\ 0 \\ 20 \end{bmatrix}$

27. a. $[A]$ is 1×3, $[B]$ is 3×4, product is 1×4
b. $4800, $5100, $5400, $6000

29. a. $\frac{4.5}{16}$ lb, $\frac{3}{4}$ lb, 3 lb   **b.** 275.6 lb, 299.1 lb, 322.5 lb, 369.4 lb
c. $965, $1047, $1129, $1293
31. $3313, $3546, $3779, $4245

EXERCISES 8.4

1. $(\pm 3, 0)$, $(0, \pm 3)$ **3.** $x^2 + y^2 = 25$ **5.** $\dfrac{x^2}{25} + \dfrac{y^2}{4} = 1$

7. $\dfrac{y^2}{16} - \dfrac{x^2}{9} = 1$ **9.** $x = 3y^2$ **11.** $y = -\frac{1}{9}x^2$

13. $x^2 + y^2 = 16$ **15.** $x^2 + \dfrac{y^2}{25} = 1$ **17.** $\dfrac{y^2}{9} - \dfrac{x^2}{25} = 1$

19. $y = \frac{1}{4}x^2$

21. $r < 0$ has no meaning; circle becomes a point.

23. parabola; $(0, 2)$ **25.** circle; $(\pm 3, 0)$, $(0, \pm 3)$; $y = \pm\sqrt{9 - x^2}$

27. hyperbola; $(\pm 2, 0)$; $y = \pm\sqrt{x^2 - 4}$

29. ellipse; $(\pm 2, 0)$, $(0, \pm 1)$; $y = \pm\sqrt{1 - \dfrac{x^2}{4}}$ or $y = \pm\frac{1}{2}\sqrt{4 - x^2}$

31. ellipse; $(\pm 5, 0)$, $(0, \pm 10)$; $y = \pm\sqrt{100 - 4x^2}$ or $y = \pm 2\sqrt{25 - x^2}$

33. circle; $(\pm 2, 0)$, $(0, \pm 2)$; $y = \pm\sqrt{4 - x^2}$ **35.** straight line; $(0, 0)$

37. hyperbola; $(\pm 1, 0)$; $y = \pm\frac{1}{2}\sqrt{x^2 - 1}$

39. parabola; $(0, 0)$; $y = \pm\frac{1}{2}\sqrt{x}$

41. hyperbola; $(0, \pm 1)$; $y = \pm\sqrt{1 + \dfrac{x^2}{4}}$ or $y = \pm\frac{1}{2}\sqrt{x^2 + 4}$

43. parabola; $(0, 0)$; $y = \pm\sqrt{-x}$

45.

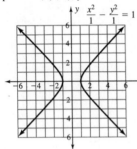
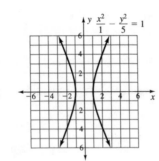
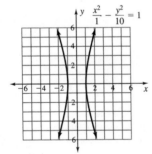

The branches become steeper.

47. a.

x	y
-9	± 3
-4	± 2
-1	± 1
0	0
1	not possible
4	not possible
9	not possible

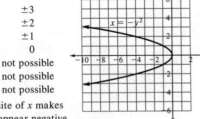

b. Opposite of x makes radicand appear negative.
c. when x is a negative number
49. $D = E = 0$, $F = -169$; $x^2 + y^2 = 169$
51. $D = 2$, $E = -6$, $F = -15$; $x^2 + y^2 + 2x - 6y - 15 = 0$

EXERCISES 8.5

1. parabola, line; $(-1, 6)$, $(3, 2)$

3. hyperbola, parabola; $(\pm 2.5, 2.25)$, $(\pm 2, 0)$

5. parabolas; $\left(-\frac{2}{3}, \frac{8}{9}\right)$, $\left(\frac{1}{2}, \frac{9}{4}\right)$

7. ellipse, parabola; $(\pm 1, 0)$

9. circle, parabola; $(\pm\sqrt{7}, 3)$, $(0, -4)$

11. parabolas; $(\pm 2, -2)$

13. ellipse, parabola; $(0, 1)$, $\approx(\pm 1.323, 0.75)$

15. circle, hyperbola; $(\pm 2, 0)$

17. hyperbola, line; $\approx(3.732, 0.268)$, $\pm(0.268, 3.732)$

19. circle, ellipse; $\left(\pm\dfrac{2\sqrt{15}}{3}, \dfrac{\sqrt{21}}{3}\right)$, $\left(\pm\dfrac{2\sqrt{15}}{3}, -\dfrac{\sqrt{21}}{3}\right)$ or $\approx(\pm 2.581, 1.528)$, $\approx(\pm 2.581, -1.528)$

23. no; coordinates do not satisfy second equation.   **25.** $(0, 1)$

27. $\approx(0.415, 5.333)$ **29.** $(31.6, 1.5)$ **31.** $(4, 1)$ **33.** $\frac{4}{3}$

35. 6   **37.** $\frac{5}{2}$ **39.** $\frac{3}{16}$

41. $x^2 = 2$, $x = \pm\sqrt{2}$; root is the same; $-\sqrt{2}$ must be discarded.

EXERCISES 8.6

1. 2   **3.** 3   **5.** 3 and 4   **7.** 2 and 3

9. $x < 0$ and $y > 0$

11. $(x < 0$ and $y > 0)$ or $(x > 0$ and $y < 0)$

13. $y \leq 2800$, $y \geq 300$, $x \geq 0$ **15.** $y \leq x$, $x \leq 0$

17.

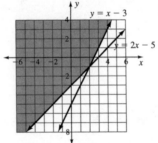

19.

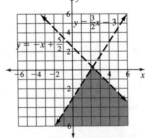

21. y-axis is excluded from solution.

23. 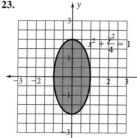 **25.**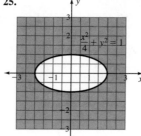

27. 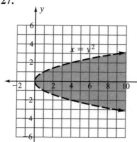 **29.**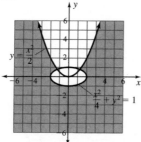

31.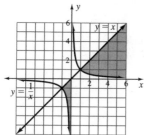

y-axis is excluded from solution.

33.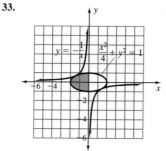

y-axis is excluded from solution.

43. 7 g protein, 16 g fat, 3 g carbohydrates

45. 4 tens, 8 fives, 6 ones **47.** $a = \frac{1}{80,000}$, $b = -\frac{3}{10}$, $c = 286$

49. 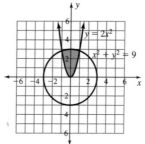 **51.**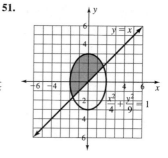

53.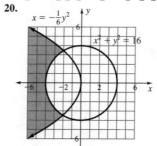

CHAPTER 8 TEST

1. 2 in., 10 in., 12 in. **2.** $x = -3$, $y = -4$

3. infinite number of solutions, coincident lines

4. { } (inconsistent) **5.** $x = 4.5$, $y = -3$, $z = -2.5$

6. parabola, line; $\left(-\frac{1}{3}, -\frac{10}{3}\right)$, $(2, -1)$ **7.** parabola, line; { }

8. circle, hyperbola; $(0, \pm 2)$ **9.** parabola, circle; $(\pm 0.890, 1.791)$

10. $(0.834, 2.5)$ **11.** $(9, 2)$ **12.** Carol, 1956; Peggy, 1966

13. 2 g protein, 19 g fat, 5 g carbohydrate

14. 365 gal of water,, 35 gal of 16% solution

15. 25 pennies, 12 nickels, 22 quarters

16. 6 juice, 3 apples, 2 Oreos

17. $a = \frac{9}{640,000}$, $b = -\frac{3}{100}$, $c = 1500$

18. a. $\begin{bmatrix} 1 & 0 \\ 0 & 1 \end{bmatrix}$ **b.** $\begin{bmatrix} 2 & 5 \\ 1 & 3 \end{bmatrix}$ **c.** 1

19. $\begin{bmatrix} 2 & -3 \\ 4 & 5 \end{bmatrix} \begin{bmatrix} x \\ y \end{bmatrix} = \begin{bmatrix} 6 \\ -32 \end{bmatrix}$; $\begin{bmatrix} x \\ y \end{bmatrix} = \begin{bmatrix} -3 \\ -4 \end{bmatrix}$

20.

$x = -\frac{1}{6}y^2$ $x^2 + y^2 = 16$

CHAPTER 8 REVIEW EXERCISES

1. 5 in., 4 in., 3 in.; $1\frac{2}{3}$ in., $1\frac{1}{3}$ in., 9 in. **3.** 10 in., 6 in., 4 in.

5. $x = 5$, $y = -2$ **7.** $x = 8$, $y = -2$

9. { } or $\varnothing$, parallel lines

11. infinite number of solutions (dependent)

13. $x = 6.2$, $y = -7$, $z = 2.4$ **15.** $\begin{bmatrix} 3 & 6 \\ 5 & 6 \end{bmatrix}$ **17.** $\begin{bmatrix} 18 & 16 \\ 13 & 11 \end{bmatrix}$

19. $\begin{bmatrix} 1 & 0 \\ 0 & 1 \end{bmatrix}$ **21.** $x = -3y^2$ **23.** $\frac{x^2}{9} + \frac{y^2}{4} = 1$

25. parabola, line; $(-1, -2)$ **27.** parabola, line; $(4, -1)$, $(-1, 4)$

29. circle, parabola; $(\pm 1.629, 3.653)$ **31.** parabolas; $\left(\frac{3}{2}, \frac{3}{4}\right)$, $(-1, 2)$

33. parabolas; { } or $\varnothing$ **35.** $x = 1$, $y = 2$ **37.** $x = 10$, $y = 1$

39. Jeanne, 21; Tish, 19

41. 71 mg calcium, 141 mg phosphorus, 3 mg iron

FINAL EXAM REVIEW, PART I

1. -4 **3.** y/x **5.** $x = -4$ **7.** $x \geq -5$

9. coefficient **11.** a constant **13.** $y = -2x + 3$

15. $f(x) \geq 0$ **17.** $y = \frac{9}{5}x + 32$ **19.** $2a^2 + 4ab + 2b^2$

21. parabola **23.** $a^2 + b^2 = c^2$

25. $\left(x + \frac{5}{2}\right)^2 = x^2 + 5x + \frac{25}{4}$

27. Possible answers: π, e, $\sqrt{2}$, $\sqrt{3}$ **29.** $7 - i$ **31.** $4i$

33. $-1 \pm 2i$ **35.** 4   **37.** $x - 1$; other answers are possible.

39. $x \approx 11.9$, $y \approx 20.5$ **41.** $r = \sqrt{P/I}$ **43.** $x + 8$

45. $\dfrac{a(a + 4)}{2}$ **47.** $-\dfrac{2x^2 - 3x + 2}{(x - 2)(x - 1)}$ **49.** $x^2 + x - 4 + \dfrac{8}{x + 1}$

51. all real numbers; $x = 45$ **53.** all real numbers, $x \neq 2$; $x = 7$

55. all real numbers, $x \neq 0$, $x \neq 4$; $\{-4, 7\}$ **57.** $2x^2y^4$

59. 5.67×10^{-7} **61.** $d = \sqrt{(x_2 - x_1)^2 + (y_2 - y_1)^2}$ **63.** 0

65. $9x^2 + 12x + 4$ **67.** $7 + 4\sqrt{3}$

69. a. $(\sqrt{x})^3$ or $\sqrt{x^3}$ **b.** $(\sqrt[m]{x})^n$ or $\sqrt[m]{x^n}$

   c. $\sqrt{x}\,\sqrt{y} = \sqrt{xy}$

71. a. $abc\sqrt[3]{bc^2}$ **b.** $|wxz^3|y^2\sqrt[4]{x}$

73. $x = y$ **75.** $x = \frac{3}{4}$ **77.** $x = -1$ **79.** $x = -\frac{5}{2}$

81. $10^y = x$ **83.** $b^c = a$ **85.** $\log_{10} a / \log_{10} b$ **87.** 2   **89.** 10

91. 10   **93.** $x \approx 3.248$ **95.** $x \approx 1.230$ **97.** $x = 2$

99. $x \approx 0.301$ **101.** $x = 3$ **103.** $x = 1.5$ **105.** $x \approx 7.389$

107. ≈ 9.98 yr or ≈ 10 yr   **109.** $x = 53$, $y = 26$

111. $x = 3.5$, $y = 2.6$, $z = 4.1$ **113.** coincident lines

115. circle, hyperbola; $(\pm 3, \pm\sqrt{2})$

117.

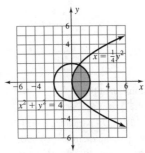

FINAL EXAM REVIEW, PART 2

1. a. 80; 40; 0; -40 (not possible)

   b. $y = a + 0.8(200 - a)$ for $a < 200$; $y = a$

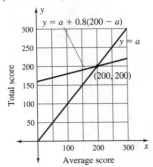

3. all real numbers; $y \leq -4.75$; $x = 0.25$; $(0.25, -4.75)$; highest point determines limit to range; x-coordinate of vertex determines equation of symmetry.

5. $\{0.4, -2\}$

7. $x^2 + 2x - 15 = 0$; $\{-5, 3\}$; two horizontal axis intercepts

9. $y = x^2 + 2x - 5$

11. Replace a with any real number: $a(x - 3)(x + 2) = 0$

13. expression; equation with parabolic graph—infinite number of ordered pairs make it true; equation whose two solutions, $x = -2$ and $x = 5$, are x-intercepts for second equation

15. determines whether the parabolic graph turns up or down, contributes to position of vertex

17. yes; $-3 + 2i$, the conjugate   **19.** $\dfrac{y^2 + 1}{y^2 - 1}$ **21.** $\dfrac{2x}{x^2 + 6}$

23. All are correct; part d contains no fractions in the numerator or denominator and is a single term.

25. $x = 4$ **27.** $x = 25$ **29.** $\approx 7.2\%$, $\approx 9.1\%$, $\approx 13.8\%$

31. geometric: $a_n = 27\left(\frac{1}{3}\right)^{n-1}$; $y = 81\left(\frac{1}{3}\right)^x$; 40.4993; 40.5

33. quadratic: $y = 2x^2 - x$

35. $y \approx 175,200(1.206)^x$, where x is years since 1945; annual growth rate $\approx 20.6\%$

37. f = year of Florence's race, j = year of Jesse's race; $f - j = 52$, $f + j = 3924$; $f = 1988$, $j = 1936$

39. $a = \dfrac{3}{250,000}$, $b = -\dfrac{1}{50}$, $c = 800$

APPENDIX 2 EXERCISES

1. -20 **3.** 1720   **5.** not arithmetic

7. not arithmetic; pairs have different sums.

9. $a_n = 47 + (n - 1)$; 67 seats

11. $a_n = 1896 + (n - 1)4$; 27 games; WWI and WWII

13. $a_n = 1 + (n - 1)3$; 49 tests

17. 265,716   **19.** 127.875; 128   **21.** 88,572   **23.** ≈ 323.9 in.

25. ≈ 404.88 in.   **27.** 2   **29.** 108 ft

APPENDIX 3 EXERCISES

1. $(x + 1)(x^3 + 3x^2 + 3x + 1) = x^4 + 4x^3 + 6x^2 + 4x + 1$

3. $1 - 12a + 54a^2 - 108a^3 + 81a^4$

5. 1, 10, 45, 120, 210, 252, 210, 120, 45, 10, 1

7. $126a^5b^4 + 126a^4b^5 + 84a^3b^6 + 36a^2b^7 + 9ab^8 + b^9$

9. $70a^4b^4 - 56a^3b^5 + 28a^2b^6 - 8ab^7 + b^8$

11. $(a + b)^2 = (a + b)(a + b) = a^2 + 2ab + b^2$; 2nd row; table contains four terms.

13. $x^3 - 3x^2y + 3xy^2 - y^3$

15. $x^6 + 6x^5y + 15x^4y^2 + 20x^3y^3 + 15x^2y^4 + 6xy^5 + y^6$

17. $b^4 - 4b^3 + 6b^2 - 4b + 1$

19. $1 + 5z + 10z^2 + 10z^3 + 5z^4 + z^5$

21. $x^4 + 12x^3 + 54x^2 + 108x + 81$ **23.** $x^3 - 9x^2y + 27xy^2 - 27y^3$

25. 16, 32, 64, 128, 256, 512, 1024; row sums are powers of 2.

27. Digits in 11^0 to 11^4 match Pascal's triangle; digits in 11^5 and 11^6 do not.

Index of Projects

Following are the section numbers, exercise numbers, and titles of the projects throughout this text. Projects marked with an asterisk are particularly suited for small groups, inside or outside class.

Glossary/Index

Absolute value The distance a number is from zero. The absolute value of x is x whenever the input is zero or positive and the opposite of x whenever the input is negative. 201
  and square roots, 199–202, 433
Absolute value function A function that gives the distance of a number from zero on a number line. 136–137
Absolute value symbol The grouping symbol | |, which signifies the absolute value of the quantity within. 10
Adding like terms, 25
Addition
  of polynomials, 153–154
  words that describe, 22
Additive inverses Numbers that are the same distance from zero on a number line and that add to zero; also known as opposites. 4, 49, 355
  denominators containing, 368
Algebraic notation, vocabulary for, 22–24
Altitude of a triangle The perpendicular distance from a vertex (corner) to the opposite side; also known as the height. 174
Antilog, 530
Arithmetic sequence A sequence with a common first difference. 133
Associative property for addition
  $a + (b + c) = (a + b) + c$. 6
Associative property for multiplication
  $a \cdot (b \cdot c) = (a \cdot b) \cdot c$. 6
Asymptote The line that a graph approaches. 382
Axes Two number lines placed at right angles so that they cross at zero. 32
Axis The vertical line through the center of a conical surface. 596
Axis of symmetry A line across which a graph can be folded so that points on one side of the graph match with points on the other side of the graph; also known as a line of symmetry. 183, 184
  of hyperbola, 348

Base The number to which an exponent is applied. 6
Binomial A polynomial with two terms. 152
  squaring, 207–208
Binomial square A two-term expression in the form $(a + b)(b + a)$ or $(a + b)^2$, obtained by squaring a binomial; also known as the square of a binomial. 202, 241

Boundary line The line between two half-planes in a coordinate plane. 64
Braces Grouping symbols that resemble little wires, { }. 10
Brackets Square grouping symbols, [ ]. 10

Calculator, graphing
  absolute value on, 136, 137
  building a table with, 41
  division by zero and, 347
  exponential regression and, 480
  factoring with, 159
  graphing a circle with, 599
  graphing an inequality in two variables with, 65
  grouping symbols and, 136, 166
  linear regression and, 124, 125
  listing an arithmetic sequence with, 133
  logarithms and, 500–501, 508–511, 537
  matrix solution of a system of linear equations and, 578–583, 587–590
  plotting ordered pairs with, 35
  power regression and, 463–464
  quadratic formula and, 216
  quadratic regression and, 258–260
  quartic regression and, 592
  recursion with, 424
  scientific notation and, 413
  solving an equation with, 41, 164, 189, 223
  solving systems of two equations with, 564
  symmetry in hyperbola and, 348
  testing of, 595
  undefined points and, 382
  vertical shifts and, 248
  viewing window distortion and, 348
Change of base formula For any real numbers a, b, and c for which the logarithm is defined,

$$\log_b a = \frac{\log a}{\log b}.$$ 508–509

  proof of, 523
Circle The set of all points $P(x, y)$ equidistant from a point called the center C. 598
Coefficient of correlation A number in the interval $[-1, 1]$ indicating how well data fit a straight line. 124
Coincident lines Two lines described by the same equation. 564–565

689

Symbols

$a + b$	addition of a and b
$\dfrac{a}{b}$	division of a by b
$a \cdot b, a(b), ab,$ $\quad (a)(b)$	multiplication of a and b
$a - b$	subtraction of a and b
-3	negative 3
$-b$	opposite of b
$\pm$	plus or minus (add or subtract)
$+3$	positive three
$\|\ \|$	absolute value
$\{\ \}$	braces
$[\ ]$	brackets
$(\ )$	parentheses
$\circ$	circle on a graph: the point is excluded from the graph
$\bullet$	dot or filled-in circle on a graph: the point is included in the graph
$//$	double slash on a graph: the spacing between the origin and the first number on the axis is different from the spacing between the other numbers
$-\infty$	negative infinity
$+\infty$	positive infinity
$\ldots$	repeats or continues, as in a pattern of numbers
$\mathbb{R}$	set of real numbers
$\varnothing, \{\ \}$	the empty set or null
$a \overset{?}{=} b$	is a equal to b?
$\approx$	is approximately equal to
$=$	is equal to
$>$	is greater than
$\geq$	is greater than or equal to
$<$	is less than

$\leq$	is less than or equal to
$\neq$	is unequal to
(a, b)	the open interval $a < x < b$
$[a, b]$	the closed interval $a \leq x \leq b$
$(a, b]$	the interval $a < x \leq b$
$[a, b)$	the interval $a \leq x < b$
$(-\infty, b]$	the interval $x \leq b$
$(a, +\infty)$	the interval $x > a$
b^n	base b with exponent n
^{2}	square (exponent 2)
^{3}	cube (exponent 3)
$\sqrt{\ }$	principal square root, radical sign
$\sqrt[n]{a}, a^{1/n}$	nth root of a, usually the principal nth root
$i\ (j)$	square root of -1, $\sqrt{-1}$ (j is used in physics)
e	base of the system of natural logarithms, approximately 2.71828
$\ln a$	natural logarithm of a, $\log_e a$
$\log a$	common logarithm of a, $\log_{10} a$
$\log_b a$	logarithm of a with base b
$^\circ$	degree (temperature)
Δ	delta: change
$f(x)$	function of x
$\%$	percent
$\perp$	perpendicular
π	pi, approximately 3.14
$a{:}b$	ratio of a to b
$\llcorner$	square corner at perpendicular lines
x_1	variable x with subscript 1
A^{-1}	inverse of the matrix A
I	an identity matrix

Functions and Sequences

Linear functions, $f(x) = ax + b$, correspond with arithmetic sequences and have common (constant) first differences.

Quadratic functions, $f(x) = ax^2 + bx + c$, correspond with quadratic sequences and have common (constant) second differences.

Exponential functions, $f(x) = ab^x$, correspond with geometric sequences and have both a multiple of the original sequence as a first difference and a common (constant) ratio of consecutive terms.